Preface to the First Edition

In view of the availability of a large number of books on any topic, the authors generally try to explain the rationale for adding one more to the long list. Fortunately in the present instance, we have no qualms of conscience in this direction. In mid-1980's when the present book was planned, there was hardly any single text dealing with the topic as an integrated inter-disciplinary science, which is the status it has attained now.

Till the middle of this century, organometallic chemistry of main group elements was the main focus of attention. In view of their versatility in organic synthesis, organometallic chemistry was generally taught as a branch of organic chemistry. With the discovery of ferrocene in 1951, organotransition metal chemistry has been receiving ever-increasing attention. Simultaneous developments in crystal and ligand field theories of co-ordination compounds had a synergetic effect on a clearer understanding of the naure of bonding in ever-increasing types of excitingly novel organo-transition metal derivatives. Attempts at the elucidation of reaction mechanisms have further revealed their far reaching potential in catalytic industrial processes. Developments in the field of organo-transition metal chemistry continued to be in general included as a part of Inorganic Chemistry texts, particularly in view of their similarities to coordination compounds.

The rapid development in the field of organometallic chemistry as a whole is reflected in the publication during the last decade of a large number of treatises on specalized topics and an excellent nine-volume *Comprehensive Organometallic Chemistry* in 1982. However, the non-availability of any suitable elementary book since 1969 appears to have prompted many authors to fill in the gap simultaneously as evinced by the appearance of an excellent text-book entitled *Basic Organometallic Chemistry* by I. Haiduc and J.J. Zuckerman in 1985, just a little before the submission of the present manuscript to the publishers in 1986, which has been followed by the publication of a couple of other books in the meantime.

The main focus of the authors in this present book has been to reflect the interdisciplinary character of the current status of the subject, which is acting not only as a bridge amongst the conventional branches of chemistry, but is also assuming importance from the points of view of industrial, biological, and environmental aspects of organometallic compounds. Being an introductory and unified text, a historical account of some epoch-making discoveries (including relevant extracts from four more recent Nobel Lectures) has been included to excite the imagination of those who are being exposed to the subject for the first time. With the same objective, considerable, emphasis has been laid on the fundamental and basic aspects of different topics dealt in the book. However, an attempt has been made to illustrate these by a number of examples, chosen carefully to cover the whole subject including the latest developments. This should extend the utility of this book to advanced students and research workers both in the academic as well as applied fields of the subject. A system of cross-referencing has been liberally adopted throughout the text in order to minimize repetition and to emphasize the interconnections amongst different facets of organometallic chemistry—an interdisciplinary science.

At the end, the authors would like to admit that the task of selection, from the existing and fastly exploding literature, to keep the size of this introductory book within reasonable limits has proved to be much more difficult than anticipated originally and necessitated rehashing as well as rewriting of most of the text several times. However, in spite of a sincere effort for utmost objectivity, the choice of the material is bound to be somewhat selective. An attempt has been made to compensate for this lacuna by giving a brief bibliography at the end of each chapter to guide the reader to more comprehensive literature on the topic.

During the preparation of the manuscript, the authors had the benefit of discussing the contents with a number of prominent organometallic chemists like Professor F.G.A. Stone, E.A.V. Ebsworth, E.W. Abel, D.S. Matteson, (Late) J.J. Zuckerman, H. Schumann, M.G. Voronkov, D. Seyferth, J.C. Maire, J.G. Riess, H. Schmidbauer, H.J. Bestmann, H. Nöth, H.W. Roesky, M.M. Taqui Khan, and T.N. Srivastava, who were partcipating in an International Symposium on New Vistas in Organometallic Chemistry organised in our Department of Chemistry (recognised for special assistance in the fields of synthetic inorganic and organometallic chemistry by the University Grants Commission) in 1984 on the occasion of the Golden Jubilee of the Indian National Science Academy. The authors are grateful to them for their valuable advice at the formative stage of the manuscript.

The authors are also grateful to a large number of colleagues like Prof. D.C. Bradley, Drs. G. Srivastava, R. Bohra and B.S. Saraswat who have kindly offered valuable suggestions after going through some portions of the manuscript. The authors would also like to thank the University Grants Commission, New Delhi for a grant towards the preparation of the manuscript and to the National Book Trust for including it in their special programme to make it available to the readers at a subsidized price.

Thanks are also due to Messrs G.M. Tripathi and Megh Singh for typing the drafts of the book repetitively with care and precision and to M/s. Wiley Eastern for publishing the same in this excellent form, in spite of some difficulties with the printing press, necessitating a change after more than half book had been printed.

A careful reperusal of the book for the preparation of an Index reminds the authors of their difficulties in the choice of materials and in many cases, arouses doubts in their minds. We would, therefore, be grateful to the readers (teachers and

ORGANOMETALLIC CHEMISTRY A UNIFIED APPROACH

SECOND REVISED EDITION

R C MEHROTRA
Department of Chemistry
University of Rajasthan
Jaipur, India

A SINGH
Department of Chemistry
University of Rajasthan
Jaipur, India

Committed to Educate the Nation

Publishing Globally
IN INDIA

NEW AGE INTERNATIONAL (P) LIMITED, PUBLISHERS

LONDON • NEW DELHI • NAIROBI

Bangalore • Chennai • Cochin • Guwahati • Hyderabad • Kolkata • Lucknow • Mumbai

Visit us at **www.newagepublishers.com**

Published by New Age International (P) Ltd., Publishers
First Edition: 1981
Second Revised Edition: 2003
Reprint: 2024

GLOBAL OFFICES

- **New Delhi** **NEW AGE INTERNATIONAL (P) LIMITED, PUBLISHERS**
7/30 A, Daryaganj, New Delhi-110002, (INDIA), **Tel.:** (011) 23253472, 23253771, **Mob.:** 9315905300
E-mail: contactus@newagepublishers.com • Visit us at www.newagepublishers.com
- **London** **NEW AGE INTERNATIONAL (UK) LTD.**
27 Old Gloucester Street, London, WC1N 3AX, UK
E-mail: info@newacademicscience.co.uk • Visit us at www.newacademicscience.co.uk

BRANCHES

- **Bangalore** 37/10, 8th Cross (Near Hanuman Temple), Azad Nagar, Chamarajpet, Bangalore- 560 018
Tel.: (080) 26756823, **Mob.:** 9315905288, **E-mail: bangalore@newagepublishers.com**
- **Chennai** 26, Damodaran Street, T. Nagar, Chennai-600 017, **Tel.:** (044) 24353401, **Mob.:** 9315905309
E-mail: chennai@newagepublishers.com
- **Cochin** CC-39/1016, Carrier Station Road, Ernakulam South, Cochin-682 016
Tel.: (0484) 4051304, **Mob.:** 9315905289, **E-mail: cochin@newagepublishers.com**
- **Guwahati** Hemsen Complex, Mohd. Shah Road, Paltan Bazar, Near Starline Hotel
Guwahati-781 008, **Tel.:** (0361) 2513881, **Mob.:** 9315905296
E-mail: guwahati@newagepublishers.com
- **Hyderabad** 105, 1st Floor, Madhiray Kaveri Tower, 3-2-19, Azam Jahi Road, Near Kumar Theater
Nimboliadda Kachiguda, Hyderabad-500 027, **Tel.:** (040) 24652456, **Mob.:** 9315905326
E-mail: hyderabad@newagepublishers.com
- **Kolkata** RDB Chambers (Formerly Lotus Cinema) 106A, 1st Floor, S N Banerjee Road
Kolkata-700 014, **Tel.:** (033) 22273773, **Mob.:** 9315905319
E-mail: kolkata@newagepublishers.com
- **Mumbai** 142C, Victor House, Ground Floor, N.M. Joshi Marg, Lower Parel, Mumbai-400 013
Tel.: (022) 24927869, 24915415, **Mob.:** 9315905282, **E-mail: mumbai@newagepublishers.com**
- **New Delhi** 22, Golden House, Daryaganj, New Delhi-110 002, **Tel.:** (011) 23262368, 23262370
Mob.: 9315905300, **E-mail: sales@newagepublishers.com**

C-23-06-13676

Printed in India at Ajit Printing Press, Delhi.

NEW AGE INTERNATIONAL (P) LIMITED, PUBLISHERS
7/30 A, Daryaganj, New Delhi - 110002
Visit us at **www.newagepublishers.com**
(CIN: U74899DL1966PTC004618)

Preface to the Second Edition

We have no hesitation in admitting that at the time (1982) when we ventured to write a book on Organometallic Chemistry, with the main objective of presenting a brief account of this rapidly developing subject in a unified manner, we had underestimated the dimensions and difficulties of the task. At that time, there was hardly any text-book published on the topic after the excellent text (1979) by G.E. Coates, M.L.H. Green, K. Wade and B.J. Aylett.

Although the authors were in a general way aware of the extra-ordinary pace with which the subject was growing, yet the publication of the *Comprehensive Organometallic Chemistry* in 9 volumes, eds. G. Wilkinson, F.G.A. Stone and E.W. Abel in 1982 gave a clearer picture of the very large volume of knowledge which had been accumulated in the field during the last few years. The first major difficulty, therefore, was the choice of the examples which could reflect the latest developments of the topic, simultaneously keeping the presentation at a level comprehensible to the undergraduate and postgraduate students as well as others who wished to have an initiation in any specialised topic of their current research/development interest.

It took an unbelievably long period of about 10 years when the book could finally came out of the press in 1991. However, the authors have been gratified by the encouraging reactions of a large body of students, colleagues in different institutions and the excellent reviews in a number of journals. The appreciative personal remarks, from prominent active organometallic chemists from within the country such as Professors A. Chakravorty, M.M. Taqui Khan, B.L. Khandelwal, and T.N. Srivastava, as well as from outside such as Professors D.C. Bradley, F.A. Cotton, A.H. Cowley, E.A.V. Ebsworth, M.F. Lappert, P.M. Maitlis, and G. Wilkinson have been a source of considerable encouragement. The general appreciation of the efforts of the authors in trying to depict a unified view of the rapidly growing field of organometallic chemistry is reflected in the remarks (similar to that from many others) of Prof. P.M. Maitlis 'I particularly applaud your aim of taking a view overall, including main group as well as transition metal compounds. That is the only way to approach the subject; it is indeed the way I have taught it for many years here'.

Another source of gratification to the authors has been the excellent demand of the book in international as well as national outlets, necessitating a second edition so soon after the appearance of the first one in mid-1991. In order that the book does not get out of print, the authors have undertaken the revision only on a limited scale in the following directions: (i) obvious corrections; (ii) simplifying the expression at many places; (iii) adding some study questions at the end of each chapter and finally; (iv) adding briefly relevant latest information by changes in the text, as well as by adding or substituting more recent appropriate examples. In the last of these tasks, we have benefitted considerably by the publication in 1995 of the extended 14 volume edition of the *Comprehensive Organometallic Chemistry*.

The authors would appreciate and be highly grateful if their attention is drawn to continuing deficiencies in their presentation. They sincerely solicit suggestions for the improvement of the material by students, teaching fraternity, colleagues in the industry as well as experts in the field, so that these could be incorporated in the next edition which might be necessitated in not too-distant a future.

The authors would like to take this opportunity of thanking reputed publishers New Age International (P) Limited, New Delhi for their patience and unstinted collaboration.

Finally, we would like to thank all those who gave us their comments and valuable suggestions on the first edition.

R.C. MEHROTRA
A. SINGH

students) for their frank and valuable suggestions for improvements of the text, which the authors plan to revise at not too distant a future date.

Last but not the least, the authors are grateful to their wives, Dr. Suman Mehrotra and Sheela Singh for their patient understanding of the ordeal which they had to undergo due to our almost one-sided attention during the completion of this challenging task.

R.C. MEHROTRA
A. SINGH

Abbreviations and Symbols

1. *List of Abbreviations*

The following abbreviations and symbols have generally been used in the text unless indicated otherwise:

Ac	acetyl
acac	acetylacetonate
An	an actinide metal
R, R′	alkyl or aryl (hydrocarbyl radical)
Ar	Aryl
atm	atmosphere
AIBN	azobis (isobutyronitrile)
PhH	benzene
bipy	2, 2′-bipyridine or 2, 2′-bipyridyl
diglyme	bis(2-methoxyethyl) ether
Bu	n-butyl (superscript i or t refer to iso or tertiary-butyl)
9-BBN	9-borabicyclo [3.3.1] nonane cyclohexyl
Cy	cyclohexyl
COD(cod)	1, 5-cyclooctadiene
COT(cot)	Cyclooctatetraene dianion
Cp	*pentahapto* cyclopentadienyl group, η^5-C_5H_5
Cp*	*pentahapto* substituted cyclopentadienyl group
DABCO	1, 4-diazabicyclo [2.2.2] octane
DBU	1, 5-diazabicyclo [5.4.0] undec-5-ene
DME (dme)	1, 2-dimethoxyethane
DMF (dmf)	N, N-dimethylformamide
glyme	1, 2-dimethoxyethane
DMSO (or dmso)	dimethylsulphoxide
dmpe	1, 2-bis(dimethylphosphino) ethane
diphos (dppe)	1, 2-bis(diphenylphosphino) ethane
E^+	electrophile
Et	ethyl
en	ethylenediamine
e.s.r.	electron spin resonance
HMPT	hexamethylphosphoric triamide
i.r. (I.R.)	infrared
Ln	lanthanide metal
L, L′	monodentate ligand (unspecified)
L-L, L″	bidentate ligand
Me	methyl
M, M′	metals of various types
NBD (nbd)	norbornadiene

Nu, Nu^-, Nu:	nucleophile
n.m.r.	nuclear magnetic resonance
n.q.r.	nuclear quadrupole resonance
p-Tol	*para*-tolyl, p-C_6H_4Me
PMDETA, PMDT (pmdeta)	N, N′, N″, N″-pentamethyldiethylenetriamine
Phen	1, 10-phenanthroline
Ph	phenyl
Pr	n-propyl
p.m.r.	proton magnetic resonance
diars	o-phenylenebisdimethylarsine
Py	pyridine
TCNE	tetracyanoethylene
THF(thf)	tetrahydrofuran
TMEDA(tmeda)	N, N, N′, N′-tetramethylethylenediamine
TTFA	thallium(III) trifluoroacetate
PhMe	toluene
TEED	triethylethylenediamine
HTFA	trifluoroacetic acid
U.V.(u.v.)	ultraviolet

2. *Symbols*

δ	delta, chemical shift downfield from TMS, given as ppm
Δ	Delta, symbol for heat supplied to a reaction
$\Delta H^{\ddagger}$	enthalpy of activation
ΔH_f^o	enthalpy of formation from standard states
$\Delta G^{\ddagger}$	Gibb's free energy of activation
ΔG°	standard Gibb's free energy of reaction
ΔH°	standard enthalpy of reaction
$h\nu$	symbol of light
ψ	Greek *psi* wave function or orbital

Contents

1

Introduction, Historical and General

1.1 INTRODUCTION

Organometallic chemistry, an interdisciplinary science, has grown at a phenomenal pace during the last three to four decades. On the academic plane, efforts to elucidate the nature of bonds in the ever-increasing list of exciting novel organometallic compounds have led to a clearer understanding of the nature and variety of chemical bonds. This has significantly, though indirectly, influenced the development of bonding theories in general. For example, the acceptance of the simple sandwich structure of ferrocene raised the status of the molecular orbital theory remarkably, as the alternative valence bond model would have involved consideration of 560(?) resonating structures. From the point of view of applications, beginning with the host of synthetic uses of versatile Grignard reagents at the dawn of this century, a variety (to cite only one example, organoboranes) of organometallic reagents are contributing richly to simpler and more facile routes for the synthesis of multifarious organic and inorganic compounds in the laboratory. On the industrial side, although only a few organometallic compounds like tetraethyllead and silicones have been used in bulk quantities, organometallic catalysts have begun to make unprecedented dents in a number of important industrial processes like those named after Fischer and Tropsch or Ziegler and Natta, and their importance and potential have been increasing at a fast pace.

1.2 DEFINITION

The designation 'organometallic' generally denotes compounds in which organic groups are linked directly to the metal through at least one carbon atom. Compounds like $Ti(OC_4H_9)_4$, $Ca\{N(CH_3)_2\}_2$ and $Fe(SC_5H_{11})_3$ are, therefore, not included in the list of organometallic compounds, although $C_6H_5Ti(OC_4H_9)_3$ and $(C_5H_5)_2Zr(OOCCH_3)_2$, would be organometallics, as both the latter derivatives have organic group(s) linked to metal through carbon atoms, either directly (σ-bond) or through a special (π) type of bonding. The term 'organic derivatives of metals' includes both these types of derivatives and often a new term 'metallo-organic' or 'elemento-organic' is used to signify this broader variety of compounds.

Organic groups can be bound, is one way or the other, to almost all the elements in the periodic table (except probably the noble gases) and potentially, therefore, the number of organometallic compounds is almost unlimited. The suffix 'metallic' in the term 'organometallic' is rather loosely interpreted to include non-metallic elements such as boron, silicon, phosphorus, arsenic, selenium and tellurium. An attempt has been made to justify this usage by interpreting 'metallic' in this context to include elements which are metallic (electropositive) compared to carbon. Based on this interpretation, the term 'organometallic' will exclude derivatives of nitrogen (3.0), oxygen (3.5), sulphur (2.6), fluorine (4.0), chlorine (3.0), bromine (2.8) and iodine (2.7) only, as their electronegativity values (given in parenthesis) are greater than that (2.5) of carbon. Organic derivatives of such elements and conventionally those of phosphorus (2.2) also are dealt within the purview of organic chemistry, and these, therefore, are excluded from this book. On the other hand, a very brief account of the organic chemistry of selenium (2.6) is being included in view of its close similarity with that of tellurium (2.0).

As in many other areas of chemistry, there are some obvious exceptions to this simple definition of organometallic compounds; for example, in spite of species like CaC_2, $Hg(CNO)_2$ and $Fe(CN)_6^{-4}$ having metal-carbon bonds, these are not considered organometallics for traditional or/and some other obvious reasons.

1.3 HISTORICAL ACCOUNT

1.3.1 Main Group Organometallics

'Cacodyl' (tetramethyldiarsine, $As_2(CH_3)_4$) was probably the first organometallic derivative isolated as early as 1760 by a Prussian military apothecary, Cadet de Glaussicourt. It was a chance discovery, while he was working with cobalt solutions on invisible ink. The two common ores of cobalt, i.e. smaltite and cobaltite both contain arsenic and when arsenious oxide (formed as side product in the preparation of cobalt solutions from cobalt ores) was pyrolyzed with potassium acetate, a reddish brown fuming liquid was obtained which was called 'cacodyl' due to its 'bad smell.

After some investigations by other workers, characterization of 'cacodyl' as $As_2(CH_3)_4$ appears to have been made in 1842 by Robert Bunsen, who came to the conclusion that he had discovered an interesting example of an 'organic element' (now called radical) in the form of the methyl group which forms a compound with arsenic.

Following the above, Edward Frankland (a British student of Bunsen) attempted to isolate ethyl radicals in 1849 by the reaction, ($Zn + 2C_2H_5I \rightarrow ZnI_2 + 2C_2H_5^{\cdot}$), which he expected to take place between zinc and ethyl iodide. However, his attempt led to the synthesis of the first organozinc compound, diethylzinc $Zn(C_2H_5)_2$ by the reaction:

$$2C_2H_5I + 2Zn \rightarrow Zn(C_2H_5)_2 + ZnI_2.$$

(It may be interesting to recall at this point that although Frankland was not successful in his objective of isolating ethyl radicals, the pyrolysis of diethylzinc

($Zn(C_2H_5)_2 \rightarrow Zn + 2C_2H_5$) and other metal alkyls did provide many years later the essential source for free ethyl (alkyl) radicals.

Although unsuccessful in his immediate objective, Frankland was soon able to lay the foundations of organometallic chemistry by extending his studies to other elements like mercury, cadmium, tin, lead and silicon. These studies added much not only to the initial development of organometallic chemistry, but incidentally a study of their composition and empirical formulae led to the first reliable assignments of atomic weights of these elements. This was possible because in these alkyls, elements like lead tends to attain their highest valency (often typical of the group to which they belong) and not the lower oxidation states, as depicted in some of their simple inorganic salts.

After a period of low activity for about two decades from 1875, a renewed interest was focussed at the turn of the century on organometallic compounds in view of their uses in organic synthesis and chemotherapy. Prominent workers during this period in the two fields may be mentioned as Grignard and Ehrlich respectively. Kipping's application of Grignard's reagents to the synthesis of silicon derivatives led to a new era of organosilicon chemistry which finally opened up the route to a new class of industrially important organometallic polymers 'silicones'.

The rather unusual anti-knocking properties of tetraethyllead in internal combustion engines led to a new impetus to organometallic chemistry. An important outcome of this development has been to expose organometallic compounds to the public view, which has recently been worried on account of the environmental problems brought about by the use of tetraethyllead in ever-increasing number of automobiles. The effectiveness of dibutyltin dilaurate in preventing discoloration or degradation of polyvinyl chloride polymers is another example of the variety of uses of organometallic compounds. The most notable use of organometallic (mainly organoaluminium) compounds as polymerization catalysts for polyethylene and other hydrocarbon polymers was crowned by the award of Nobel Prize to K. Ziegler (cf. Appendix 1.1) and G. Natta in 1963.

Since the epoch making discovery of the Grignard reagents (organomagnesium halides) in 1900, these compounds are finding ever increasing applications as reactive and versatile intermediates for a wide range of organic and organometallic syntheses. Although lithium alkyls were discovered in 1917 by W. Schlenk and J. Holtz, yet it was only after 1930 (when a convenient synthetic procedure for lithium alkyls was made accessible by Ziegler) that their role in providing even more efficient synthetic routes for organic and organometallic moieties was recognised (as although the chemistry of lithium and Grignard reagents is very similar, yet the former are more reactive). By now organolithium reagents have almost out-classed the Grignard reagents both on synthetic and catalytic (industrial) fronts. For such reactions in which lithium and Grignard reagents are too reactive, the alkyls of aluminium, zinc, boron, and cadmium may be employed because these are less reactive and sometimes, more selective. A signal recognition of the highly useful role of organometallics in synthetic organic chemistry has been in the award of Nobel Prize to Prof. H.C. Brown in 1977 for the discovery of now well-known 'Hydroboration reactions' (cf. Appendix 1.4).

1.3.2 Transition Metal Organometallics

Although the synthesis of the platinum complexes, $[PtCl_2(C_2H_4)]_2$ and $K[PtCl_3(C_2H_4)]$ by W.C. Zeise in 1827 and 1831 respectively even preceded Frankland's discovery of diethylzinc in 1849, the progress in the area of organotransition metal chemistry was rather slow and uneven upto 1950's. For example, although trimethylplatinum iodide was discovered by W.J. Pope, S.J. Peachy and O.S. Gibson as early as 1907, attempts to synthesize stable transition metal alkyls and aryls corresponding to those of the main group elements met with little success and it was assumed for a long time that transition metal-carbon bonds were rather weak in nature.

As in many other fields of science, the above incorrect notion led to ignoring of many useful observations. For example, it appears that although F. Hein had, as early as 1919, succeeded in making some 'polyphenyl' derivatives of chromium and had followed up this finding with the help of a number of coworkers during the next two decades (1919-37), the conclusions of his school were seriously doubted for many years until E.O. Fischer in 1950 prepared independently bis (π-benzene) chromium (0) and proposed that Hein's earlier 'polyphenyl-chromium' compounds were actually arene derivatives containing benzene and/or biphenyl bonded chromium.

Fortunately, another observation made almost simultaneously in 1951-52 by two groups of workers was not lost and in fact, led to the characterization of a novel and exciting derivative, dicyclopentadienyliron, $Fe(C_5H_5)_2$ (which has become famous by the name of 'ferrocene' coined in 1952 by M.C. Whiting) with a sandwich type of structure.

As a compound which revolutionised the progress of organotransition metal chemistry, it may be appropriate to quote a few paragraphs from an account of its authentic history in an article, entitled. 'The Iron Sandwich. A Recollection of the First Four Months' by G. Wilkinson, who later shared the Nobel Prize in 1973 with the German Chemist, E.O. Fischer for their work on such compounds:

> ''Woodward told me that he had seen, which I didn't, in the titles of papers to be published in the Journal of the Chemical Society ... a paper on 'Dicyclopentadiene' (this was the celebrated 'Miller, Tebboth and Tremaine' paper) and he wondered what the hell that could be! If we had got only together, we could have made the compound ourselves weeks before the Kealy and Pauson note in the *Nature* arrived at Harvard.....''

> ''On seeing the structure I, which was also the one Miller, Tebboth and Tremaine had drawn in their paper.., I can remember immediately saying to myself 'Jesus Christ it can't be that!' .... I remember scribbling out on a piece of paper structure II in which both double bonds were coordinated; and almost immediately afterwards III, as the significance

of resonance structures... dawned and the equivalence of the carbons became obvious,.. 'It's a sandwich. ' The thing that really excited me was the thought that if iron did this, the other transition metals must also form sandwich compounds''.

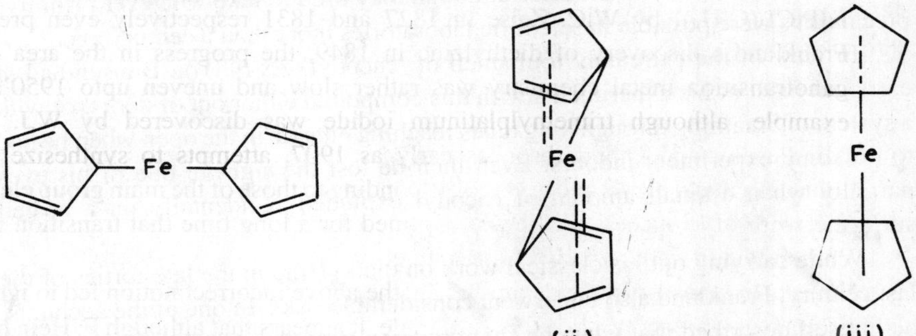

(i) (ii) (iii)

Fig. 1.1 Suggested bonding modes of biscyclopentadienyliron

The extraordinary stability (stable to heat upto 470°C; unreactive towards air; insoluble in water and unaffected by hydrochloric acid as well as boiling alkali hydroxide) and the novel structure of biscyclopentadienyliron excited leading chemists like G. Wilkinson, E.O. Fischer, F.A. Cotton and many others to explore comprehensively the organometallic chemistry of almost all the transition elements including the inner transition metals. Although a large variety of π-bonded (e.g. sandwich, alkene, alkyne, allyl and related) metal complexes of these transition metals were synthesised and studied extensively, yet the simple σ-bonded alkyls and aryls of these metals appeared to be extremely elusive and were considered for a long time to be highly unstable.

This unfounded notion about the inherent thermodynamic instability of σ-oonded transition metal organometallics was corrected by independent suggestions starting in late sixties by a number of groups of workers (cf. Appendix 1.2) who showed that such derivatives can be made kinetically stable by suitable selection of ligands. It was also recognised that out of a number of accessible decomposition pathways, the β-hydrogen pathway is an important route for decomposition of transition metal alkyls, and ligands such as Me_3CCH_2—, Me_3SiCH_2—, or $PhCH_2$—, which are β-hydrogen free often lead to comparatively stable derivatives. Several other ligands (e.g. $PhMe_2CCH_2$—, Ph_3CCH_2—, $(Me_3Si)_2CH$—, Me_3SnCH_2—, o–$MeOC_6H_4SiMe_2CH_2$—, $Me_2PhSiCH_2$—, $MePh_2SiCH_2$—, or $MePh_2CCH_2$—) have since been used to isolate stable neutral or anionic homoleptic d- and f-block metal derivatives.

The synthesis and properties of many of these novel derivatives have not only initiated deep conceptual changes about the nature of chemical bonding and novel structural features (e.g. fluxional molecules, multiple metal-carbon and metal-metal bonds), but have also led to many breakthroughs in synthetic procedures and industrial processes, e.g., homogeneous catalysis of such disparate reactions as alkene hydrogenation and polymerization.

1.3.3 Biological and Environmental Aspects of Organometallic Compounds

Ever since their discovery, organometallic compounds have been known to depict pronounced biological effects. Even the earliest workers, Bunsen and Frankland suffered to some extent from these although they tried to take whatever primitive precautions were possible in scientific laboratories more than a century ago.

In view of the extremely foul smell of 'cacodyl' $As_2(CH_3)_4$, Bunsen had to carry out most of his experiments with this compound either out of doors, or with a system of breathing through long tubes reaching outside in the open when he had to do some experiment indoors. Even then he lost the sight in one of his eyes partially when a small amount of cacodyl exploded accidentally on a heated surface.

While carrying out his classical work on dialkylzinc in the late forties of the last century, Frankland also underwent considerable risks. In one of the accounts, he himself described that when he first prepared methylzinc in July 1849 and in Bunsen's laboratories, he was trying the action of water on the residue in the tube used for preparing the 'methyl' from methyl iodide and zinc and a greenish blue flame several feet long shot out of the tube... Bunsen thought that this was due to cacodyl and, in fact, he warned Frankland that he was 'already irrevocably poisoned'. Things fortunately did not turn out to be as bad and Frankland lived for many years to continue his important research contributions in chemistry!

Systematic investigations on biological effects of organometallics began with the classical work of Paul Ehrlich on organoarsenicals although some organomercurials had been used earlier also in the treatment of syphilis and other infectious diseases. A few well known biologically active compounds are illustrated below:

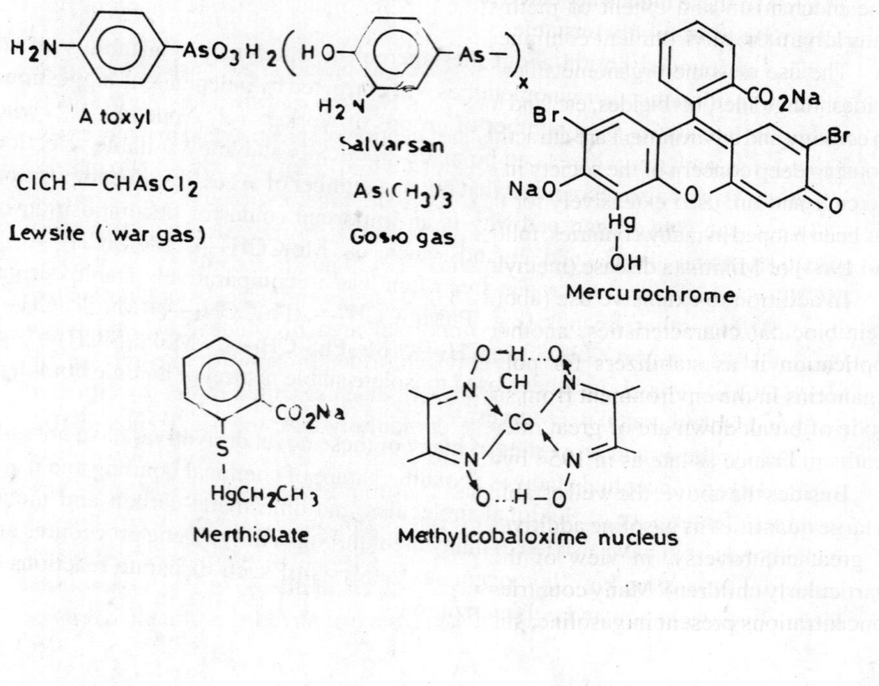

Ehrlich first tried the medicinal properties of 'atoxyl', but in view of its high toxicity (undesirable side-effects), he synthesised a number of other arsenic derivatives, out of which 'salvarsan' proved to be the most effective drug with the lowest toxicity. Mercurochrome and merthiolate are well-known external antiseptics. Lewsite was a notorious gas used in 1914-1918 world war.

Methylcobaloxime is the basic unit which is present in methylcobalamin (Me[CO], a Vitamin B_{12} coenzyme) also. Intense interest has been created is such naturally occurring MeCo species, especially in view of their capacity to transfer the methyl groups to other inorganic metal ions, e.g.:

$$\mathrm{Me[Co]} + \mathrm{Hg}^{2+} \xrightarrow{\mathrm{H_2O}} \mathrm{MeHg}^{+} + \mathrm{H_2O(Co)}^{+}$$

The possibility and mechanism of such methylation reactions in nature have certainly aroused considerable interest from the point of view of the role of organometallics in the environment (see Chapter 7 of this Book).

Historically, an intriguing case of 'arsenic poisoning' was recorded in the last century as a result of the observation that many people died with symptoms characteristic of 'arsenic' poisoning by merely living in certain rooms which came to be nick-named as 'arsenic rooms.' On a closer scrutiny, it was found that the chief characteristics of such rooms were: (i) wall paper using some arsenic compounds for colouring, (ii) moulds growing on them, enhanced by dampness, and (iii) an unpleasant odour. Researches of many chemists including Gossio (1891) and Challenger (1930's) showed that the cause of death under the conditions prevailing in such rooms was a poisonous gas trimethylarsenic $As(CH_3)_3$ produced in minute quantities by a process termed by Challenger as 'biological methylation'. The mechanism and extent of methylation of inorganic moieties which occurs naturally in the environment continues to be studied extensively.

The use of some organometallics employed as biocides, i.e., insecticides, pesticides, fungicides, herbicides, etc. and their possible short as well as long-term effects on ecology and environment are attracting the attention of scientists and in many cases, arousing deep concern of the society in general. Of the organometallics of two metals, mercury and tin, used extensively for their biocidal properties, the use of the former has been banned in many countries, following the first established cases (between 1953 and 1964) of Minimata disease (methylmercury poisoning) in Japan.

In addition to extensive use (about 25,000 tonnes annually) of organotins for their biocidal characteristics, another important area in which they find wide application is as stabilizers for polyvinyl chloride. In view of the release of organotins in the environment from such large-scale uses, their direct toxicity and mode of break down are of great interest, particularly after the occurrence of 102 deaths in France as late as in 1954 by the use of diethyltin diiodide (stalinon).

Besides the above, the well established consumption of organolead derivatives in large quantities as gasoline additives for spark ignition engines has been a matter of great controversy, in view of the possible effects of their use on humans (particularly children). Many countries have reduced, or plan to reduce, organolead concentrations present in gasoline; such measures had already resulted in a reduction of the production of ethyllead anti-knock compounds from the level of 317,000 tonnes in 1974 to about half by 1984.

It is clear from the above account that in spite of an awareness of the problems for a long time, the environmental aspects of organometallics are just beginning to be unravelled but much faster progress is expected in the field in view of the fastly growing interest in this area during the last few years.

1.4 LITERATURE OF ORGANOMETALLIC CHEMISTRY

As in any other area of knowledge, the literature on organometallic chemistry can be divided into three main categories, i.e., (i) original research papers (primary literature), (ii) review literature, (iii) specialized articles, monographs/treatises on individual(s) and surveys as well as dictionaries of compounds, etc., and (iv) text books at different levels, dealing with the subject as a whole.

1.4.1 Primary Literature

An approximate estimate of the number of original papers published in the field of organometallic chemistry in 1950; 1965 and 1995 reflects not only the rapid growth of the subject as a whole, but also the relatively much greater interest being developed in transition metal organometallics during the last 1-2 decades:

Table 1.1 Growth of Literature (Indicated by the Number of Papers Published) **On Organometallic Chemistry**

Class of Elements	1950	1965	1995
Main Group	1,000	1,600	27,000
Transition	50	800	24,000

In order to cope with fastly increasing number of papers appearing in a large number of journals of chemistry published by the various national societies and private publishing houses, a special *Journal of Organometallic Chemistry* (published by Elsevier, Amsterdam under the guidance of an international editorial board) was started in 1964. It is now a leading journal in the field and the editors have commemorated the publication of centenary volumes, (100th, 200th, 300th and 400th) of the journal in 1975, 1980, 1986 and 1990 respectively by inviting each time 20 leading organometallic chemists of the world to write an article on their major contribution(s) in the field; the resulting 79 articles, therefore, represent a salient landmark individually in the growth and development of organometallic chemistry. A second journal *Synthesis and Reactivity in Inorganic and Metal-Organic Chemistry* (published by M. Dekker, New York), started in 1971, deals mainly with the synthetic aspects of organometallic compounds. Another journal *Organometallics in Chemical Synthesis* (also published by Elsevier) and only a short life and was merged in the *Journal of Organometallic Chemistry.* The American Chemical Society has also, since 1982, begun the publication of *Organometallics* which has already become a leading journal in the field in spite of its brief life. The publication of a new journal 'Applied Organometallic Chemistry' by Wiley Publishers since 1986 reflects the increasing applications of organometallic compounds.

In order to remain up to date with the latest researches and findings, the research scholar has to consult the above and other primary journals; this is becoming a formidable task in view of the ever increasing number of original papers. This effort can be assisted by self-consultation of the *Chemical Abstracts*, in which Section 29 deals specially with organometallic chemistry, or through the computerized literature retrieval service of the Chemical Abstracts and other agencies.

1.4.2 Reviews

In view of the enormity of the primary literature on organometallic chemistry sustained efforts have been made particularly since the mid-sixties to present the state-of-art in the fastly growing areas of organometallic chemistry and provide the readers with an up to date overall view of a topic with full bibliography, which enables the searcher to consult the special interest in the primary literature. A few review series of special significance are listed as follows:

(i) *Advances in Organometallic Chemistry* (edited by F.G.A. Stone and R. West) have been published in a book form almost annually (Thirty three volumes have been published up to 1991) since 1964 by Academic Press, New York.

(ii) The journal *Organometallic Chemistry Reviews, Section A. Subject Review,* published 8 volumes between 1964 and 1971 (Elsevier, Amsterdam). In 1972, these were incorporated in the Journal of Organometallic Chemistry, but the series has been reformated since 1976 in book form with the title *'Journal of Organometallic Chemistry Library'*.

(iii) *'Specialist Periodical Reports'* (published by the Royal Society of Chemistry, London) include (a) *Spectroscopic Properties of Inorganic and Organometallic Compounds* (edited by N.N. Greenwood, followed by E.A.V. Ebsworth) and (b) *Organometallic Chemistry* (edited by E.W. Abel and F.G.A. Stone).

In addition to reviews on topical subjects, organometallic chemistry has also been reviewed periodically, element by element, in several publications.

(iv) *'Annual Surveys of Organometallic Chemistry'* (edited by R.B. King and D. Seyferth), started in book form in 1965, but was changed to journal form: *'Organometallic Chemistry Reviews Series B. Annual Surveys'* in 1968. In 1972, it was merged in the *Journal of Organometallic Chemistry* and finally, in its Library series.

1.4.3 Specialised Articles/Monographs on Element(s) Individual and Dictionaries, etc.

Of the various efforts attempting exhaustive coverage of literature on organometallic chemistry element-wise, the best-known is the *Gmelin Handbook of Inorganic Chemistry,* published by Gmelin Institute at Frankfurt, in German and English.

Following the first comprehensive monograph on organometallic chemistry *'Die Chemie der Metall-Organischen Verbindungen'* by A. Krause and A. von Grosse (Borntrager, Berlin, 1937), in which almost all the literature published upto

1935 was systematized, an attempt was initiated by M. Dub to collect all the information between 1937 and 1964 in a volume entitled *'Organometallic Chemistry'* (Springer Verlag, 1964) and has been continued upto 1975 by publication of a few supplementary volumes dealing with different elements (e.g. transition metals in 1966; arsenic, antimony and bismuth in 1968 etc.).

In addition to the above, the following other registers of organometallic compounds may be mentioned:

(i) H.C. Kaufman, *Handbook of Organometallic Compounds,* Van Nostrand, Princeton (1961).

(ii) H. Hagihara, M. Kumada and R. Okawara (eds.), *'Handbook of Organometallic Compounds'* W.A. Benjamin, New York (1968).

(iii) J. Buckingham, *'Dictionary of Organometallic Compounds'* 3 Vols. Chapman and Hall, London, (1984).

1.4.4 Textbooks/Treatises Dealing with the Subject at Various Levels

The first text-book containing any material on organometallics appeared in 1778, '*Elements de Chymie Theorique' et Pratique* by L.B. Guyton de Morreau, H. Maret and J.F. Durande from Dijons, a place which incidentally continues to be active in this field having been the centre of two international conferences on the topic in recent years; it contained a small section on Cadet's fuming liquid discovered in 1760 and later shown to be cacodyl, foul smelling $As_2(CH_3)_4$. Starting with the discovery of Grignard reagents and increasing applications of organometallics in synthetic organic chemistry, most of the texts on 'Organic Chemistry contain an account of organometallic reagents, particularly of the main group elements; special mention may, for example, be made of a chapter on Organometallic Chemistry' by the doyen of organometallic chemists, H. Gilman of the University of Iowa in the 4-volume *'Organic Chemistry'* (edited by H. Gilman), a popular text in 1940's. The third of the 6-volume series *'Comprehensive Organic Chemistry'* (edited by D. Barton and W.D. Ollis; Pergamon Press, Oxford, 1979) presents a detailed account of organometallic chemistry of sulphur, selenium, silicon, boron, and other elements.

As mentioned in the historical account of the development of organometallic chemistry, some facets of transition metal organometallics bear particular relevance to coordination chemistry and hence, these continue to be dealt with in inorganic texts, e.g.,

(i) J.C. Bailar, H.J. Emeleus, R. Nyholm, A.F. Trotman-Dickenson (eds.), *Comprehensive Inorganic Chemistry,* Pergamon Press, Oxford, (1973), vols. 1-5.

(ii) K.F. Purcell and J.C. Kotz, *Inorganic Chemistry,* W.B. Saunders Co., Philadelphia (1977).

(iii) F.A. Cotton and G. Wilkinson, *Advanced Inorganic Chemistry, 5th ed.,* John Wiley & Sons, New York (1988).

(iv) T. Moeller, *Inorganic Chemistry: A Modern Introduction.* John Wiley & Sons, New York (1982).

(v) J.E. Huheey, *Inorganic Chemistry: Principles of Structure and Reactivity,* 3rd ed., Harper & Row, New York (1983).

(vi) W.W. Porterfied, *Inorganic Chemistry: A Unified Approach,* Addison-Wesley, Reading, MA. (1983).

(vii) N.N. Greenwood and A. Earnshaw, *Chemistry of the Elements,* Pergamon Press, Oxford (1984).

The rapid developments in the field of organometallic chemistry since the mid-fifties led to the publication of a series of text-books at various levels on this topic. A few of these are listed below arranged chronologically according to their dates of publication.

(i) G.E. Coates, *Organometallic Compounds,* Methuen, London (1956).

(ii) E.G. Rochow, D.T. Hard and R.N. Lewis, *The Chemistry of Organometallic Compounds,* Wiley, New York (1957).

(iii) H. Zeiss (ed.), *Organometallic Chemistry,* Reinhold, New York (1960).

(iv) J.L. Harwood, *Industrial Applications of Organometallic Compounds* Chapman & Hall, London (1963).

(v) E.G. Rochow, *Organometallic Chemistry,* Reinhold, New York (1964).

(vi) J.J. Eisch, *The Chemistry of Organometallic Compounds,* MacMillan, New York (1967).

(vii) P.L. Pauson, *Organometallic Chemistry,* Arnold, London (1967).

(viii) G.E. Coates and K. Wade, *Organometallic Chemistry, Vol. 1 (The Main Group Elements),* 3rd ed., Methuen, London (1967).

(ix) A.N. Nesmeyanov and Kocheshkov (ed.) *Methods of Elemento-Organic Chemistry Vols. 1-4,* Nauka, Moscow (1967).

(x) G.E. Coates, M.L.H. Green, P. Powell and K. Wade, *Principles of Organometallic Chemistry,* Methuen, London (1968); Chapman & Hall, London (1971).

(xi) M.L.H. Green, *Organometallic Chemistry, Vol. III (The Transition Elements),* Methuen, London (1968).

(xii) R.B. King, *Transition Metal Organometallic Chemistry,* Academic, New York (1970).

(xiii) M. Tsutsui, M.L. Levi, A. Nakamura, M. Ichikawa and K. Mori. *Introduction to Metal-π-Complex Chemistry,* Plenum, New York (1970).

(xiv) R.F. Heck, *Organotransition Metal Chemistry,* Academic, New York (1974).

(xv) I. Bernel (ed.), *Horizons in Organometallic Chemistry,* New York Academy of Sciences, New York (1974).

(xvi) F.R. Hartley, *Elements of Organometallic Chemistry.* The Chemical Society, London (1974).

(xvii) J.W. Swan and D.St.C. *Organometallics in Organic Synthesis,* Chapman and Hall, London (1974).

(xviii) V. Ishii and M. Tsutsui, *Organotransition-Metal Chemistry,* Plenum New York (1974).

(xix) J. Tsuji, *Organic Synthesis by Means of Transition Metal Complexes,* Springer-Verlag, New York (1975).

(xx) H. Alper (ed.), *Transition Metal Organometallics in Organic Synthesis,* Academic, New York (1976).

(xxi) C.E. Carraher, Jr., J.E. Sheats and C.U. Pittaman, Jr., *Organometallic Polymers.* Academic, New York (1978).

(xxii) G.E. Coates, M.L.H. Green, K. Wade and B.J. Aylett, *Organometallic Compounds,* 4th ed., Chapman and Hall, London (1979), Vol. 1, Part 1: Groups I-III (By G.E. Coates, R. Snaith and K. Wade) 1983. Vol. 1 Part 2: Groups IV and V (B.J. Aylett), 1979. Vol. 2: The Transition Metals (by M.L.H. Green) in preparation.

(xxiii) Ei-ichi Nagishi, *Organometallics in Organic Synthesis,* John Wiley. New York (1980).

(xxiv) S.G. Davies, Organotransition Metal Chemistry: *Applications to Organic Synthesis,* Pergamon, Oxford (1982).

(xxv) G. Wilkinson, F.G.A. Stone and E.W. Abel(eds.), *Comprehensive Organometallic Chemistry-I,* Vol. I-IX, Pergamon, Oxford (1982) and *Comprehensive Organometallic Chemistry-II,* Vol. 1-14, Pergamon, Oxford (1995).

(This is a major text covering more than 9,000 pages and 30,000 references; the series has a valuable index of all organometallic structures studied by diffraction techniques and also of review articles in the field of organometallic chemistry).

(xxvi) J.S. Thayer, *Organometallic Compounds and Living Organisms,* Academic, New York (1984).

(xxvii) I. Haiduc and J.J. Zuckerman, *Basic Organometallic Chemistry* Walter de Gruyter, Berlin (1985).

(xxviii) C.M. Lukehart, *Fundamental Transition Metal Organometallic Chemistry,* Brooks/Cole, California (1985).

(xxix) A. Yamamoto, *Organotransition Metal Chemistry,* John Wiley, New York (1986).

(xxx) A.W. Parkins and R.C. Poller, *An Introduction to Organometallic Chemistry,* MacMillan, London (1986).

(xxxi) P.J. Craig, *Organometallic Compounds in the Environment* (Principles and Reactions), Longman, England (1986).

(xxxii) J.P. Collman, L.S. Hegedus, J.R. Norton and R.G. Finke, *Pri nciples and Applications of Organotransition Metal Chemistry,* University Science Books, California (1987).

(xxxiii) J.S. Thayer, *Organometallic Chemistry: An Overview,* VCH Publishers, New York (1988).

(xxxiv) R.H. Crabetree, *The Organometallic Chemistry of the Transition Metals,* John Wiley, New York (1988).

(xxxv) P. Powell, *Principles of Organometallic Chemistry,* Chapman and Hall, London (1988).

(xxxvi) B.J. Wakefield, *Organolithium Methods,* Academic Press, London (1988).

(xxxvii) R.G. Wilkins, *Kinetics and Mechanisms of Reactions of Transition Metal Complexes,* 2nd edn, VCH, Weinheim (1991).

(xxxviii) Ch Elshenbroich and A. Salazer, *Organometallics: A Concise Introduction,* 2nd edition, VCH, Weinheim (1992).

1.5 CURRENT STATUS AND FUTURISTIC ASPECTS OF ORGANOMETALLIC CHEMISTRY

In the preceding pages has been presented a brief account of some of the important landmarks in the development of organometallic chemistry. The choice of a few illustrative instances out of a large number of salient advances in the field is, to some extent, unavoidably personal and subjective, but in doing so a substantial portion of the material has been picked out from similar accounts by well-known authorities in the field (e.g. J.S. Thayer, 'Organometallic Chemistry: A Historical Perspective' in *Advances in Organometallic Chemistry,* Vol. XVII, Eds. F.G.A. Stone and R. West, Academic Press (1975).

By and large, organometallics especially the alkyls and aryls of main group, elements were initially considered as part of Organic Chemistry and some accounts of these, e.g., Grignard Reagents were included in books on Organic Chemistry. By contrast, the metal carbonyls have generally been considered as inorganic compounds. The discovery of ferrocene and related derivatives in 1950's focussed the attention of Inorganic coordination chemists whose contributions to the nature of bonding and mechanism of reactions of transition metal organometallics have added new chapters to what would conventionally have been considered Physical Chemistry. The interest of Organic and Industrial chemists has been accentuated by many novel and effective synthetic routes provided by these new materials functioning as reagents or catalysts. The biological and environmental aspects of these compounds, although recognized quite early, are now beginning to receive ever-increasing attention of specialists in these areas as well as in the allied fields of biological sciences.

In spite of the rapid advances made in understanding of the nature of these compounds on the border-land of inorganic and organic chemistry and revolutionary breakthroughs in their applications, there is an increasing and unprecedented expectation from these compounds that as catalysts, they may be able to stimulate nature in the facile biological processes, such as nitrogen fixation in soil and carbohydrate synthesis in plants. Success in the continuous efforts in these directions appears to be round the corner and would certainly bring about a new era in human progress.

In order to realise these expectations as quickly as possible, much more emphasis is necessary on teaching/learning of Organometallics at the undergraduate as well as postgraduate and research levels. The best way of relieving this would be by treating the subject as an integrated one, rather than as separate portions of the conventional branches of organic and inorganic chemistry. In this connection, the following words of John Thayer are worth quoting:

> "With the sharp, almost hostile demarcation of the two disciplines (organic and inorganic) in earlier years, there always existed a possibility that organometallic chemistry might be partitioned into two (or more interacting areas). Much of the vigour comes from the fact that this has not happened.... On the contrary, such a strong sense of coherence and synergic interaction has emerged that organometallic chemistry has not only retained its (own) identity, but has actually brought the two larger disciplines into closer contact with each other."

Study Questions

1. Identify organometallic compounds amongst the following:
   $NaOOCCH_3$; $Mn(C_5H_5)_2$; $Al(OPr^i)_3$;
   $TiCH_3(OCH_3)_3$; $Mg(Br)CH_3$
2. Write a brief history of the discovery of ferrocene and indicate the importance of this compound in the development of organotransition metal chemistry.
3. Name a few organometallic compounds showing biological activity.

Supplementary Reading

1. *A New Type of Organo-Iron Compound,* by T.J. Kealy and P.J. Pauson, *Nature,* Vol. 168, 1039 (1951).
2. 'Dicyclopentadienyliron', by S.A. Miller, J.A. Tebboth and J.F. Tremaine, *Journal Chemical Society,* 632 (1952).
3. 'The History and Development of Organotransition Metal Chemistry' by M.D. Rausch, *Werner Centennial Advances in Chemistry Series,* No. 62, edited by G.B. Kauffman, p. 486, American Chemical Society, Washington, D.C., (1966).
4. 'My Way in Organometallic Chemistry', by A.N. Nesmeyanov, *Advances in Organometallic Chemistry,* edited by F.G.A. Stone and R. West, Vol. 10, pp. 1-60 Academic Press, London (1972).
5. 'The Iron Sandwich, A Recollection of First Four Months', by G. Wilkinson, *Journal Organometallic Chemistry,* Vol. 100, 273 (1975).
6. *'Organometallic Chemistry: A Historical Perspective'* by J.S. Thayer, *Advances in Organometallic Chemistry,* edited by F.G.A. Stone and R. West, Vol. 13, pp. 1-37, Academic Press, London (1975).

7. 'Perspective and Further Perspectives', *Journal of Organometallic Chemistry,* Vols. 100 (1975) and 200 (1980).
8. 'Trends and Opportunities of Organometallic Chemistry' by G.M. Parshall, *Organometallics,* Vol. 6, 687 (1987).
9. 'Organometallic Chemistry: An Overview' by J.S. Thayer, V.C.H. Publishers, Inc., New York (1988)

2

Classification, Nomenclature and Characteristics of Organometallic Compound

2.1 INTRODUCTION

Almost all the elements (with the exception of noble gases) of the periodic table are now known to form derivatives of one type or another with hydrocarbon groups or other organic moieties; all of these are characterized either by a direct σ-bond or by π-bonds (ionic or covalent; localized or delocalized) in which one or more carbon atoms are within bonding distance of the metal atom. Broadly, organometallic compounds can be classified* in two ways: (i) based on hapticity (from the Greek word 'haptien', meaning to 'fasten'); this can be indicated by the number of carbon atoms of the organic moiety, which are within the bonding distance of the metal such as mono-hapto (h^1 or η^1), dihapto (h^2 or η^2), penta-hapto (h^5 or η^5) organometallics, etc., and (ii) based on the nature (extent of polarity) of the metal-carbon bond, on the basis of which the structure and reactivity of specific organometallic compounds are often explained.

2.2 CLASSIFICATION BASED ON HAPTICITY

In the hapto classification, organic ligands are classified according to the number of carbon atoms within bonding distance of the metal atom.

Organometallic compounds containing mono-hapto (η^1) ligand (linked to the metal atom through one carbon atom only) are predominantly derivatives of main group elements. These are generally covalent compounds, but the polarity of bonds in these increases with decreasing electronegativity of the metal concerned. On the other hand, only a small number of stable transition metal derivatives containing

* It may be relevant to mention another obvious classification of organometallic compounds, into homoleptic and heteroleptic types. The former represent compounds of the type MR_n where all the organic groups are the same and the latter represent compounds of the $MR_aR'_{n-a}$ or MR_aX_{n-a} where R′ is an organic ligand different from R or X is another anionic group such as halide, alkoxide, carboxylate on cyclopentadienide.

mono-hapto (η^1) ligands have been isolated till now, but such compounds are presently coming to the forefront with novel structure and bonding characteristics as well as unusual reactivity patterns. In the lecture entitled *'The long search for stable transition metal alkyls'* delivered by Wilkinson on the occasion (1973) of the award of the Nobel Prize (cf. Appendix 1.3), he has ascribed the instability of such derivatives not to any basic weakness of the transition metal-carbon bonds, but generally to their kinetic lability and thermodynamic instability with regard to possibility of rearrangement to compounds with more than one carbon atom within bonding distance of the metals atom.

Organometallic compounds of transition metals containing poly-hapto ligands have been extensively and intensively studied, ever since the early fifties when the novel bonding mode and structure of 'ferrocene' (dicyclopentadienyliron) were elucidated. However, such compounds of main group metals are extremely rare.

Examples of both mono- as well as poly-hapto organometallics along with such derivatives in which the organic ligands bridge two or more metal atoms are given in Section 2.4 along with their names on the basis of IUPAC nomenclature.

2.3 CLASSIFICATION BASED ON THE POLARITY OF M-C BOND (POSITION OF THE METAL IN THE PERIODIC TABLE)

Following the distinctive tendency of main and transition elements to form mono- and poly-hapto organometallic derivatives respectively a more general relationship between the position of an element in the periodic table and characteristics of its organometallic derivatives is depicted in Table 2.1.

Thus the strongly electropositive s^1 and s^2 metals tend to form essentially ionic derivatives with the exception of lithium and beryllium, the products of which are much less ionic and tend to polymerise.

The main group metals (of the 'p' block excluding elements like N, P, O, S and halogens which are more electronegative than carbon) form essentially volatile covalent compounds, in which the ionic character tends to increase, as expected, on going down any group of the periodic table.

Beryllium and aluminium (circled) as well as lithium and magnesium (to a lesser extent) tend to form electron deficient structures by association, in which there are insufficient number of valence electrons to allow all the atoms to be linked by the conventional two-electron two-centre bonds.

The transition (d) and inner transition (f) metals are notable for the variety of their organic derivatives: ionic, covalent and special 'd' orbital bonded compounds, e.g., the cyclopentadienyl derivatives.

2.4 NOMENCLATURE OF ORGANOMETALLIC COMPOUNDS

Based on the IUPAC (International Union of Pure and Applied Chemistry) rules of naming organometallic compounds (cf. Appendix 2.1; IUPAC Nomenclature of Organic Chemistry, pages 360-372, Pergamon Press, 1979), the names of a few representative example of mono- and poly-hapto types of organometallic com-

Table 2·1

H																	He
Li	Be											B	C	N	O	F	Ne
Na	Mg											Al	Si	P	S	Cl	Ar
K	Ca	Sc	Ti	V	Cr	Mn	Fe	Co	Ni	Cu	Zn	Ga	Ge	As	Se	Br	Kr
Rb	Sr	Y	Zr	Nb	Mo	Tc	Ru	Rh	Pd	Ag	Cd	In	Sn	Sb	Te	I	Xe
Cs	Ba	La *	Hf	Ta	W	Re	Os	Ir	Pt	Au	Hg	Tl	Pb	Bi	Po	At	Rn
Fr	Ra	Ac **															
Lanthanons *		La	Ce	Pr	Nd	Pm	Sm	Eu	Gd	Tb	Dy	Ho	Er	Tm	Yb	Lu	
Actinons **		Ac	Th	Pa	U	Np	Pu	Am	Cm	Bk	Cf	Es	Fm	Md	No	Lw	

..... Principally ionic (s^1 and s^2 metals) derivatives.

_____ Principally covalent (p-block elements), mono-hapto derivatives predominating; however, a few poly-hapto derivatives are also accessible.

- - - Principally poly-hapto (d and f-block) derivatives, but a large number of σ-derivatives are known.

O Metals which tend to form electron-deficient (μ-bonded polymeric) molecules.

pounds are given in Figs. 2.1 and 2.2 respectively. In the first type (mono-hapto, e.g., compounds (a) to (i) in Fig. 2.1), the ligand is bonded to a single metal atom through only one of its carbon atoms and in the second type (poly-hapto, e.g., compounds (j) to (n) in Fig. 2.2), the organic ligand has more than one of its carbon atoms within bonding distance of a single metal atom. In addition to these, a third type of bridged derivatives (e.g., $[Al(CH_3)_3]_2$ and $[Be(CH_3)_2]_a$) are represented in Fig. 2.3 in which the organic ligand functions as a bridging groups between two or more metal atoms.

H_5C_6—Hg—C_6H_5

(a) Diphenylmercury

$(LiC_4H_9)_n$

(b) Butyllithium (polymeric)

H_3C—Zn—C_2H_5

(c) Ethylmethylzinc

$(CH_3)_3SnO_2CCH_3$

(d) Trimethyltin acetate

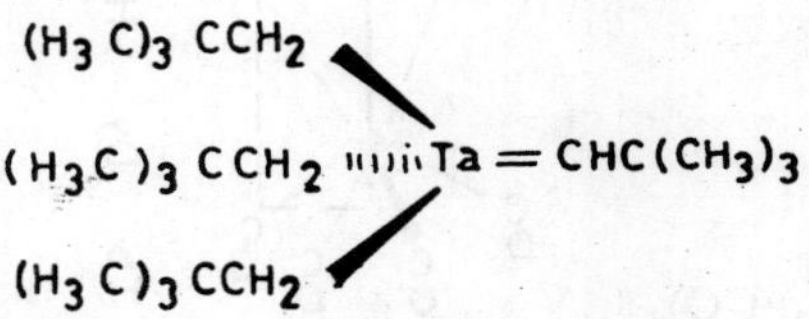

(e) Tris (neopentyl) neopentylidenotantalum

C_2H_5BeH

(f) Ethylberyllium hydride or Ethylhydridoberyllium

CH_3SnH_2Cl

(g) Methyltin chloride dihydride

or

Chlorodihydridomethyltin or Chloro (methyl) stannane

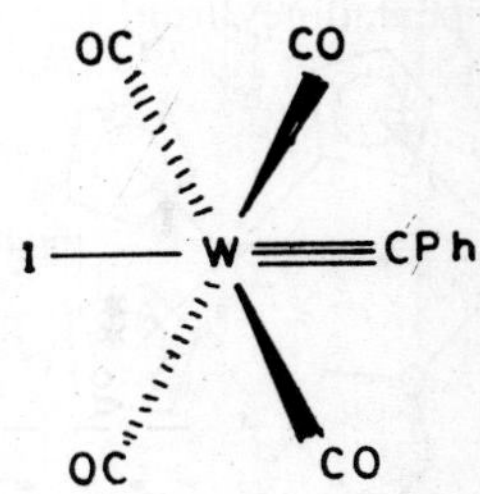

(h) Iodo (phenylmethyne) tungsten tetracarbonyl

$(H_5C_6)_3$ Si—Si$(C_6H_5)_3$

(i) Hexaphenyldisilane

Fig. 2.1 Examples of mono-hapto ligands. Names of simple organometallic compounds like 'diphenylmercury' are written as one word with the name of the organic groups followed by the name of the metal last. Different groups joined to the same metal, are written in alphabetical order 'ethylmethylzine'.

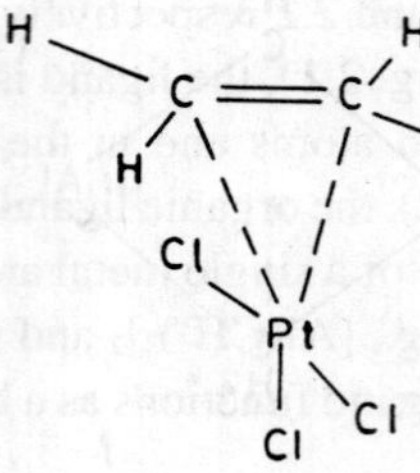

(j) $PtCl_3\,(\eta^2–C_2H_4)^-$
Ethenetrichloroplatinate (II) anion

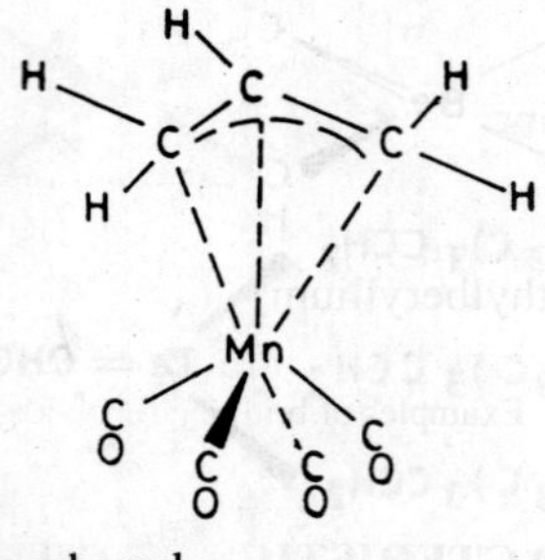

(k) Mn $(\eta^3–C_3H_5)(CO)_4$
Allylmanganese tetracarbonyl

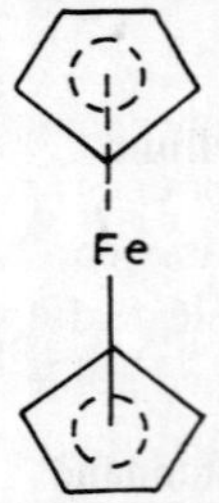

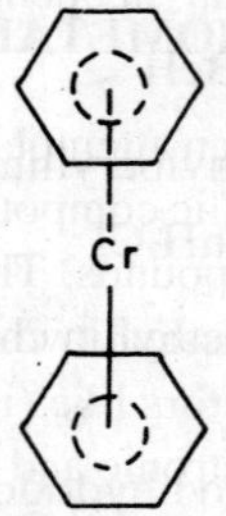

(l) Fe $(\eta^5–C_5H_5)_2$
Ferrocene or Dicyclopentadineyliron

(m) Cr $(\eta^6–C_6H_6)_2$
Dibenzenechromium*

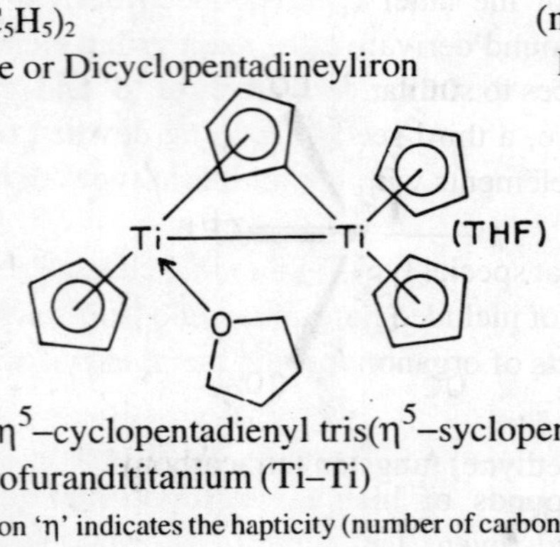

(n) $\eta – (\eta^1 : \eta^5$–cyclopentadienyl tris(η^5–syclopentadienyl)
tetrahydrofuranditītanium (Ti–Ti)

Fig. 2.2 The supercript on 'η' indicates the hapticity (number of carbon atoms within bonding distance of the metal) of the ligand.

---

* **Note**: Alternative name of organometallic derivatives of silicon and also other elements of group IVB (extended sometimes to elements of groups IIIB and VB) are based on their parent hydrides also; e.g., 'chloro (methyl) stannane' and 'hexaphenyl disilane'.

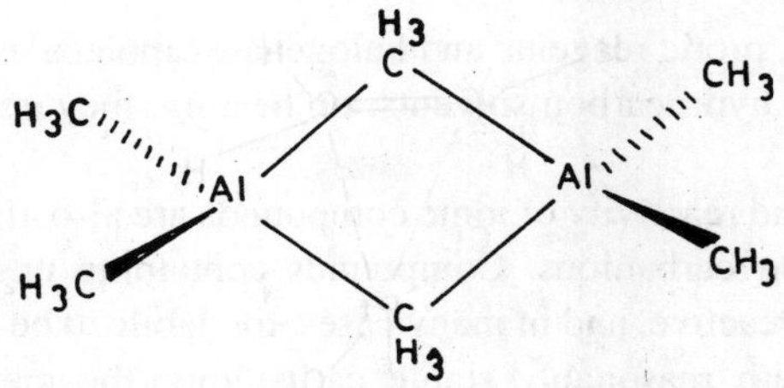

(o) Di-μ-methyl-tetramethyldialuminium
Di-μ-methyl-bi(dimethylaluminium)

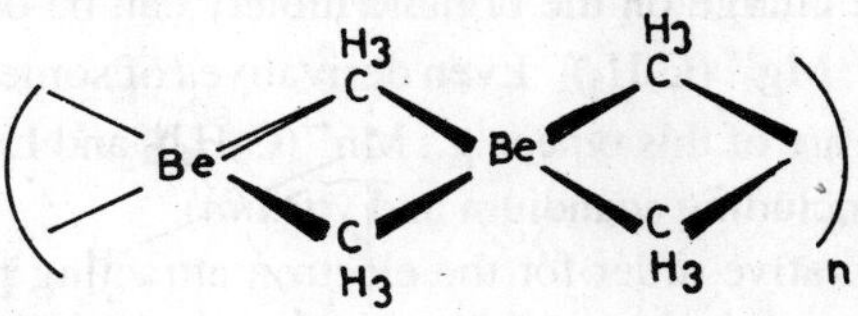

(p) Catena-(di-μ-methylberyllium)

Fig. 2.3 Examples of bridging (μ-bonded) ligands.

2.5 GENERAL CHARACTERISTICS OF DIFFERENT TYPES OF ORGANOMETALLIC COMPOUNDS

In this section, an attempt is being made to present the salient characteristic features of organometallic compounds classified as (i) Ionic Compounds and (ii) σ-Bonded Covalent Compounds. The M—C bond in both these cases can be dealt with in terms of the classical two-electron two-centre bonds. While dealing with the effect of different factors like (i) bond polarity, (ii) electronic and nuclear effects of the metal, (iii) electronic and steric effects of the organic moiety, and (iv) effect of the medium on the reactivity of the latter type σ-bonded derivatives, most of the description would centre around derivatives of main group elements with only a few casual relevant references to similar derivatives of 'd' and 'f' block metals.

In addition to the above, a third section is being devoted to organometallic compounds of main group elements with non-classical types of bonds in electron deficient molecules.

It may by mentioned that special types of bonding characteristics of transition metal organometallics are not included here and would be discussed in Chapter 4, dealing with different aspects of organotransition metal chemistry.

2.5.1 Ionic Compounds

The organometallic compounds of highly electropositive metals (alkali and alkaline earth metals and even lanthanides as well as actinides, with electronegativity values, ~1 or less), are predominantly ionic, with the exception of those of lithium, beryllium and magnesium (which are fairly covalent in most cases). These are generally colourless (except the derivatives derived from an aromatic nucleus or with a system of conjugated double bonds), extremely reactive (towards air, water, protic reagents and halogenated molecules, etc.) non-volatile solids, insoluble in hydrocarbon solvents; on heating, they generally decompose without melting.

The stability and reactivity of ionic compounds are also affected markedly by the stability of their carbanions. Compounds containing unstable anions (e.g., C_nH_{2n+1}) are rather reactive, and in many cases, too labile to be isolated in the pure state. However, with reasonably stable carbanions, the metal derivatives are comparatively more stable and less reactive (e.g., $Na^{+}\,{}^{-}C(C_6H_5)_3$; $Ca^{++}\,(C_5H_5)_2^{-}$).

In some cases, even more polarizing cations (like Mg^{++}) form ionic compounds when the negative charge on the organic moiety can be delocalised over several carbon atoms, e.g., $Mg^{++}\,(C_5H_5)_2^{-}$. Even derivatives of some electropositive 'd' and 'f' block elements are of this type, e.g., $Mn^{++}(C_5H_5^{-})_2$ and $Ln^{3+}(C_5H_5^{-})_3$ (where Ln is lanthanide metal including scandium and yttrium).

A semi-quantitative order for the electron attracting power (and consequent stability) of R in a derivative M–R may be estimated by the acidities of the corresponding hydrocarbons, H–R. This sequence of carbanionic stabilities is given in Table 2.2.

Table 2.2 Relative Carbanion Stability of Organic Groups*

Group	pKa of R-H
Cyclopentyl	44
Isoporpyl	43
n-Propyl	41
Ethyl	40.5
Methyl	39
Cyclopropyl	38
Vinyl($H_2C=CH$)	37
Phenyl (C_6H_5)	37
Allyl ($H_2C=CHCH_2$)	36.5
Diphenylmethyl	33
Triphenylmethyl	31
Pentachlorophenyl (C_6Cl_5)	30.5
Fluorenyl (9)	27
Trifluoromethyl (CF_3)	25.5
Indenyl (3)	23
Pentafluorophenyl (C_6F_5)	23
Phenylethynyl ($C_6H_5C\equiv C$)	18.5
Cyclopentadienyl (C_5H_6)	18

The smaller the value of the pKa of H–R, the more electron attracting 'R' would be and this would be reflected in the comparatively higher stability of its derivatives with metals.

* D.J. Cram, *Fundamentals of Carbanion Chemistry*, Academic Press, New York (1965).

2.5.2 σ-Bonded Compounds

Organometallic compounds, in which a carbon atom of the organic moiety is bonded to the metal by a normal two-electron, two-centred covalent (may be, with considerable ionic character in some cases) bond, are generally formed by most elements with values of electronegativity higher than 1. The chemistry of these compounds is in general similar to that of organic compounds, but there are differences due to factors of the following types:

(a) *Polarity of the Bond*

In all the organometallic compounds, the bond $M^{\delta+}$—$C^{\delta-}$ (unlike the truly covalent C—C bond) is polarized generally in the direction shown. Calculating the percentage ionic character of the M—C bond by the simple Pauling formula: % Ionic Character = $100[1 - \exp\{-1/4\,(x_c - x_m)\}]$ (where x_c and x_m represent the electronegativities of C and M respectively), the Si—C, Al—C, Mg—C and Na—C bonds should have roughly 12, 22, 34 and 45% ionic character respectively. For contrast, it may be pertinent to recall in this connection that the C—Cl bond, which has only 6% ionic character on Pauling's formulation, is considered to be quite polar in organic chemistry and many reactions of chloro-compounds are explained on that basis. Similarly a synchronous four-centre (Fig. 2.4a) or six centre (Fig. 2.4b) transition state of the type:

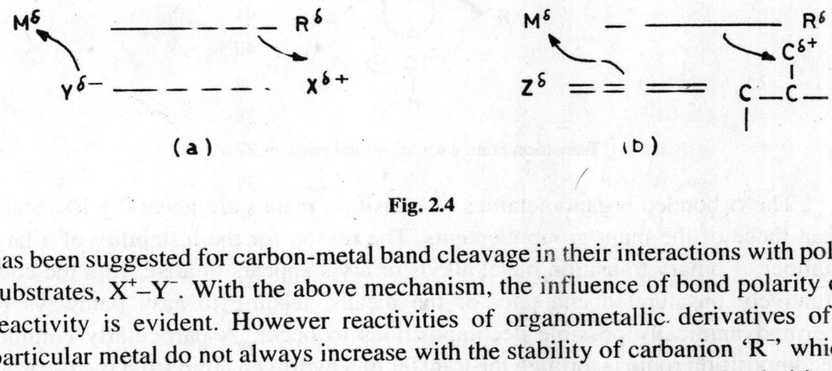

Fig. 2.4

has been suggested for carbon-metal band cleavage in their interactions with polar substrates, X^+–Y^-. With the above mechanism, the influence of bond polarity on reactivity is evident. However reactivities of organometallic derivatives of a particular metal do not always increase with the stability of carbanion 'R^-' which indicates that the reaction mechanism is not solely explained by bond polarities.

(b) *Electronic and Nuclear Effects of the Metal*

Lewis acid character of metal alkyls in general arises from the availability of low energy, vacant electronic orbitals in the metal. For alkyls of main I, II and III group elements, this Lewis acidity would involve *np* orbitals, whereas with metalloids of main group IV, V and VI and some transition metal alkyls, *n* and (*n*–1) *d* orbitals become operative respectively.

Obviously, the Lewis acidity (or the capacity to accept electron pairs from donor molecules like amines, ethers, halide ions, etc.) would increase as one goes from the main group I to III in the same period of elements (i.e., NaR<

$MgR_2 < AlR_3$). In derivatives of main group IV-VI metalloids, the available orbitals are '*nd*'; their simple organometallics like MR_n have too weakly a Lewis acid character to permit isolation of adducts with Lewis bases. The Lewis acid character of 'heteroleptic' derivatives, $MR_{n-a}X_a$, however, increases as the 'R' groups are replaced by electronegative groups (e.g. halogens). This is explained on the basis of contraction of '*d*' orbitals with the increased positive charge on the metal centre so they can participate more effectively in bond formation with donor atoms. An interesting example of this type of difference in reactivity is exhibited by the tetrahedral nature of trialkylsilyl β-diketonates, SiR_3(acac), in which even the strongly chelating β-diketonate moiety is unable to form a chelate by interaction with '*d*' orbitals of silicon, but is able to do so in derivatives of the type, SiR_2X (acac) or $SiRX_2$ (acac).

Although alkyls of main IV to VI group metalloids do not generally form stable isolable adducts with Lewis bases, yet a transition state with '*d*' orbital participation is generally postulated in their cases also as the initiating step in the cleavage of carbon-metalloid bonds by nucleophiles (e.g., OH^-, R^- or H^-):

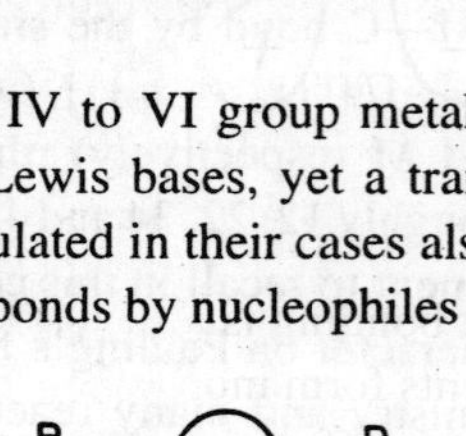

Transition State with 'd' orbital participation

The σ-bonded organometallics of transition metals are generally less stable than those of the main group elements. The reason for the instability of a large number of binary transition metal alkyls or aryls appears to arise from the coordinatively unsaturated character of the metals, leading to easy pathways for thermodynamically possible decompositions to occur. A particularly common decomposition route is through the transfer of a hydrogen atom from the β-carbon atom of the alkyl chain (β-elimination, discussed later in the Section 4.1 dealing with σ-bonded transition metal organometallics).

The coordinatively unsaturated nature of the transition metal centres in their alkyls can be overcome by forming complexes with Lewis bases (e.g., R_2O, R_2S, R_3P, R_3As, RO^- and X^-) in which $(n-1)d$ orbitals, hybridised with *ns* and *np* orbitals, play an important role. Along with electronic effects, steric hindrance to the approach of the reagent to the metal-carbon bond can also contribute to the increased chemical (kinetic) stability.

Alternatively, the filled transition metal $(n-1)d$ orbitals with proper symmetry tend to overlap with the available antibonding π-orbitals of attached carbonyl, vinyl or phenyl group. Such back-bonding tends to disperse the negative charge on the metal and to enhance the electron-density between the metal and carbon centres. The structure of pentacarbonylpentafluorophenylmanganese (Fig. 2.5) illustrates this type of situation. In this molecule, the strong inductive withdrawal of electrons in the carbon-manganese σ-bond by C_6F_5 is also important in stabilising the compound.

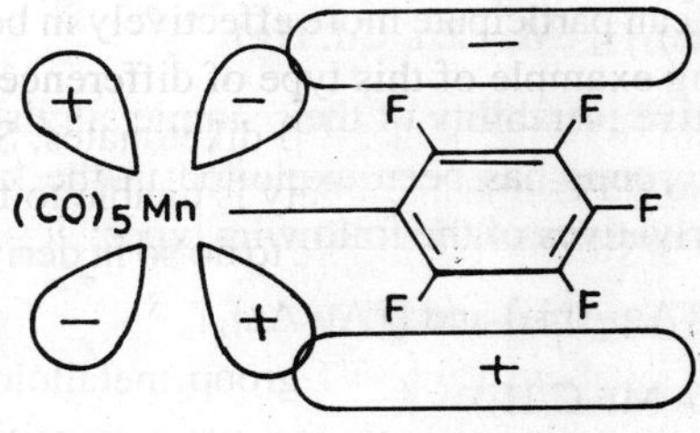

Fig. 2.5

The above type of *dπ-pπ* bonding is able to explain why unsaturated groups with electronegative substituents form more stable M–C bonds, e.g., in $C_6F_5Ti(OPr)_3$.

The converse electronic effect, i.e., the donation of π-electrons by unsaturated groups attached to the metal (metalloid), explains the weaker Lewis acidity and inertness towards oxygen of trivinylborane:

$$(H_2C=CH)_2\,B—HC=CH_2 \rightarrow (H_2C=CH)_2B^-=CH—CH^+$$

Another important metal parameter is its covalent radius. In any group of elements, the members with lower atomic numbers have smaller radii and hence steric factors may impede many chemical reactions at the carbon-metal bond. Consequently, carbon bonds to mercury, thallium and lead will be chemically more labile than other members of their individual groups respectively.

(c) *Electronic and Steric Effects of the Organic Moiety*

The degree of association of aluminium alkyls, AlR_3, in inert solvents is in the sequence, R = Me > Et > Pr > Bu^t, (the last two being essentially monomeric) could be explained on the basis of the higher + I inductive effect of the branched alkyl groups (tending to enhance the electron density at aluminium atoms) as well as on steric factors (impeding the tendency of association).

In addition to the above simple effects of ramification of the organic moiety, the unsaturation and its location in the organic groups also influences the stability of M—C bond in MR_n appreciably. For α, β unsaturation, the stability of the potential carbanion in MR_n increases with the enhancement of the 's' character of the α-carbon atom:

$$\underset{sp^3}{M-CH_2-R} < \underset{sp^3}{M-CH=CH-R};\ \underset{sp^2}{M-C_6H_5} < \underset{sp}{M-C\equiv C-R}$$

This results in the increased salt-like character of metal acetylides over metal alkenyls and of metal aryls over metal alkyls.

The effect of increased steric demands of some groups, resulting in the stability of some rare coordination states in main group organometallics, can be illustrated by the following examples:

$[Li(thf)_4][Li\{C(SiMe_3)\}_2]$

$[M\{CH(SiMe_3)_2\}_3]$ (M = Ga, In, Bi)

$[M\{C(SiMe_3)_3\}_2]$ (M = Zn, Cd, Hg)

In view of the comparative instability of their simple alkyls, the advantage of the steric factors of bulky groups has been exploited in the kinetic stability of organotransition metal derivatives of the following types:

$[Cu\ Ar_2]^-$; $[\{Ag\ Ar\}_4]$ and $[\{Au\ Ar\}_4]$

(Ar = 2, 4, 6–$Me_3C_6H_2$)

$[MR_3]$; $[M'R_4Cl]$ and $[M''R_3\ Cl]^-$

(M = Ti(III), V(III), Cr(III), Sc(III);

M′ = Zr, Hf; M″ = Yb; R = $(Me_3Si)_2CH$—)

(d) *Solvent Effects*

Donor solvents are well known for their ability to reduce molecular association of metal alkyls and even to change the molecular geometry of monomeric metal alkyls. The increased electron density around the metal atom on coordination ($R_nM \leftarrow L$) with the donor (Lewis base; L = OR_2, NR'_3, X^- and R^-) solvent molecules should tend to enhance the polarization of the metal-carbon bond in the direction, $M^{\delta+} - C^{\delta-}$ providing more carbanionic character to the organic groups (R).

The greater reactivity of organolithium compounds in donor solvents (e.g., ethers or amines) compared to that in hydrocarbon solvents (cf. Section 2.4.4) is an example supporting the above view-point.

Alternatively, this type of coordination should lead to lessening of the tendency of back bonding ($M^{\delta+} \leftleftarrows C^{\delta-}$) from the organic moiety, and thus weaken the metal-carbon bond.

In addition, this type of coordination would be more pronounced in electron deficient metal alkyls with Lewis acidic character (e.g., BR_3, AlR_3, ZnR_2, etc.) and should lead to lowering of the reactivity of the concerned metal alkyl. For example, in ether solvents, the reactivity of triphenylaluminium to alkynes is reduced due to lessening of the electrophilicity of the central metal atom.

Further, the presence of donor solvents sometimes tends to generate novel and fascinating structural types (*cf.* Chapter 4) in organotransition metal chemistry.

2.5.3 Some Electron Deficient Bridged Molecules

A comparison of the physical properties of Main group III alkyls reveals that aluminium trialkyls do not fit in the normal and expected gradation in their characteristics; this is illustrated, for example, by the m.p.'s and b.p's of the methyls:

	BMe_3	$AlMe_3$	$GaMe_3$	$InMe_3$	$TlMe_3$
m.p. °C	–153	15	–16	89.5	38.5
b.p. °C	–20	126	56	136	76/85 mm

It may be mentioned that whereas all the other trimethyls are essentially monometric in the vapour state, trimethylaluminium is known to be dimeric (except at higher temperatures). In the solid state trimethylaluminium is dimeric whereas trimethylindium (and also trimethylthallium) is tetrameric. The oligomerization tendency in these alkyls is dependent of the steric bulk of the alkyl groups (R); for example, the degree of association of aluminium alkyls, AlR_3, in hydrocarbon solvents decreases in the sequence: $R = Me > Et > Pr^i > Bu^t \approx Me_3CCH_2$ derivatives.

In its dimerization tendency, trimethylaluminium shows a similarity to aluminium trichloride and aluminium tertiary butoxide. However, in both the latter compounds lone pair electrons are available on the halogen or the alkoxy oxygen atoms giving rise to donor-acceptor linkages of the type:

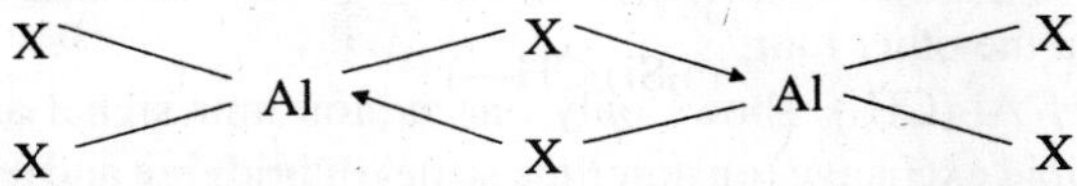
in the dimer.

However, in $\{Al(CH_3)_3\}_2$, the methyl moieties have no lone pair electrons; thus, there are insufficient number of electrons for formation of conventional donor-acceptor bridges in $\{Al(CH_3)_3\}_2$. This structural problem which is associated with deficiency of electrons (to provide two-electron two-centred bonds between the atoms in this molecule) has been a source of curiosity and investigations for many years.

The main interpretations of bonding in $Al_2(CH_3)_6$ closely follows those of diborane, and suggestions had been put forward that linkages probably involve hydrogen bonding, or resonance (between forms such as I to IV):

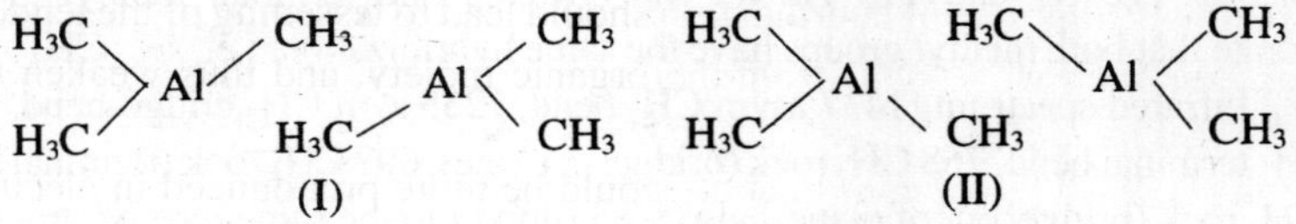
(I) (II)

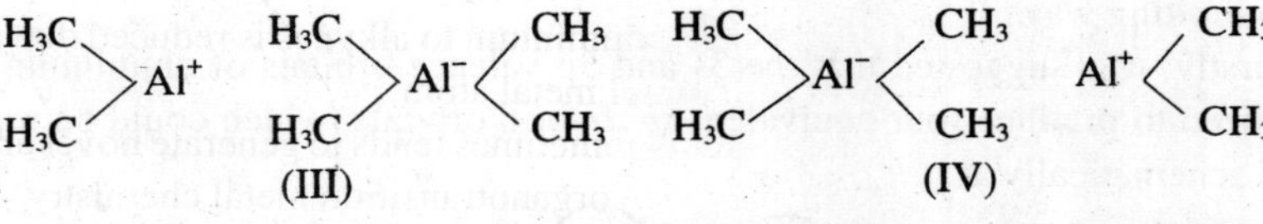
(III) (IV)

An alternative structure which may be referred to as a "methylated double bond", was suggested by analogy with the protonated double bond description of diborane:

The C^+H_3 groups are regarded as being embedded in the π-electron charge clouds of an Al—Al double bond.

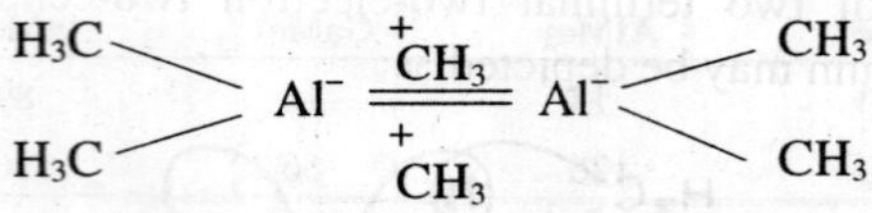

Probably the most convincing explanation for the bridge bond in the trimethylaluminium uses two-electron, three-centre bond description (*cf.* B_2H_6) (Fig. 2.6).

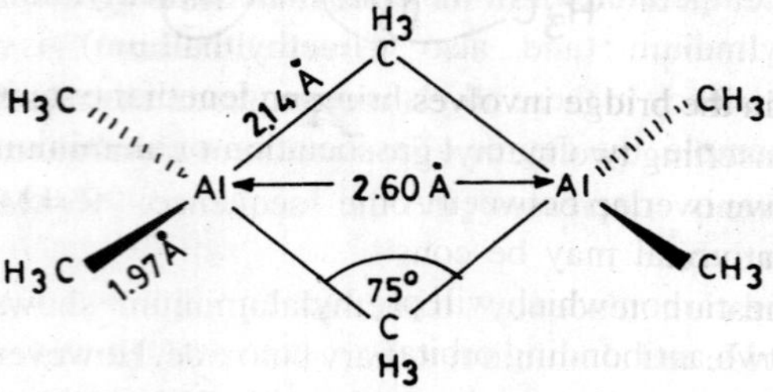

Fig. 2.6 Structure of $Al_2(CH_3)_6$

This bridge model suggests that two of the six methyl groups in $Al_2(CH_3)_6$ are attached differently from the other four.

An ^1H nmr study of $Al_2(CH_3)_6$ shows only one proton nmr signal at room temperature due to the rapid exchange (on nmr time scale) of bridging and terminal methyl groups. However, at –75°C (when interchange process is sufficiently slowed down) two separate resonances in the intensity ratio 1 : 2 for bridging and terminal methyl groups respectively have been observed.

The above proposed symmetrical structure for trimethylaluminium dimer has been elucidated by X-ray diffraction and has been further confirmed by more precise information about the position of the hydrogen atoms of the bridging methyl groups with the help of low temperature X-ray crystallographic studies and by ^{27}Al n.q.r. measurements.

Further n.m.r. studies, i.e., by ^{13}C n.m.r. spectrum at –75°C, the $^{13}C—^1H$ coupling constants as 112 Hz and 114 Hz for the bridging and terminal methyl groups, corroborate that both methyl groups have the same hybridization, i.e., sp^3 (CH_4 : J_{CH} = 125 Hz). Infrared spectrum [1437 asym CH_3 bend, 1255 sym CH_3 bridge-bend, 1201 sym CH_3 terminal bend, 768 CH_3 rock (bridge, in plane), 697 CH_3 rock (terminal), 608 cm^{-1} CH_3 rock (bridge, out of plane)] also lend support to the same type of structure.

The bonding situation in dimeric aluminium alkyls can, therefore, be described in two limiting ways.

Firstly, it is suggested that the 3s and 3p valence orbitals of aluminium are hybridized to produce four equivalent sp^3 hybrid orbitals, which could be represented schematically as:

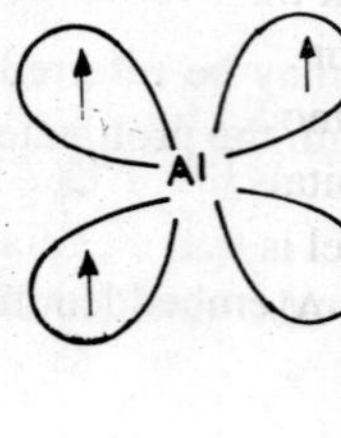

The formation of two terminal two-electron two-centre Al—CH_3 bonds in trimethylaluminium may be depicted as:

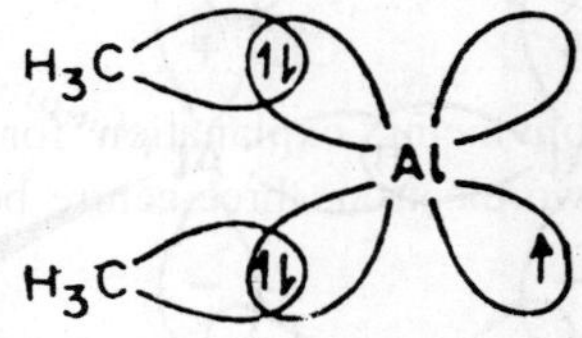

The bonding in the bridge involves bringing together two units of the above type, followed by inserting two methyl groups in between the units in such a manner as to give effective overlap between Al(sp^3), C(sp^3), and Al(sp^3) hybrid orbitals, i.e., the molecular orbital may be constructed from the sp^3 hybrid orbitals of the aluminium and carbon which will produce two bonding orbitals, two nonbonding orbitals, and two antibonding orbitals; each bonding molecular orbital encompasses two aluminium and one carbon atom, and is completely filled:

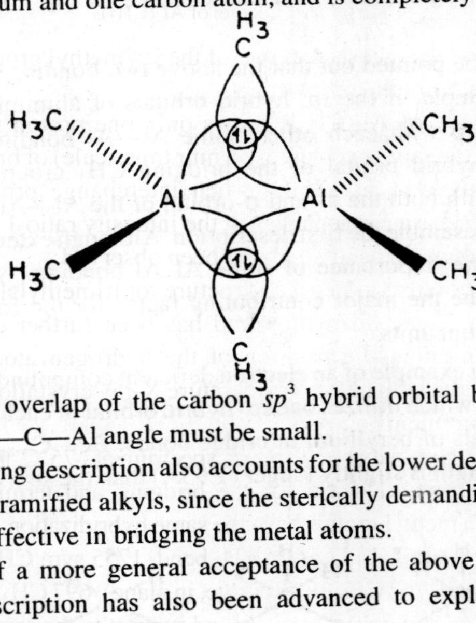

For maximum overlap of the carbon sp^3 hybrid orbital by two Al(sp^3) hybrid orbitals, the Al—C—Al angle must be small.

This bonding description also accounts for the lower degree of polymerization found with the ramified alkyls, since the sterically demanding alkyl groups would be much less effective in bridging the metal atoms.

In spite of a more general acceptance of the above mode of bonding, an alternative description has also been advanced to explain the short (2.60Å) aluminium-aluminium distance.

In this, the bridge frame-work is constructed from two aluminium trigonal (sp^2) hybrid orbitals, two aluminium p atomic orbitals, and two carbon sp^3 hybrid orbitals, generating two molecular orbitals each of bonding, non-bonding, and anti-bonding character. The terminal Al—CH_3 bonds are normal two-electron two-centre bonds resulting from the overlap of a sp^2 hybrid orbital of aluminium and a sp^3 hybrid orbital of carbon.

One of the features of this model for trimethylaluminium dimerization is that one the occupied molecular orbitals involves Al—Al bonding.

The advantage of this model is that the relatively short Al—Al distance (2.60 – 2.62Å) is interpretable as Al—Al σ-bond, and the very acute Al—C (bridge)—Al

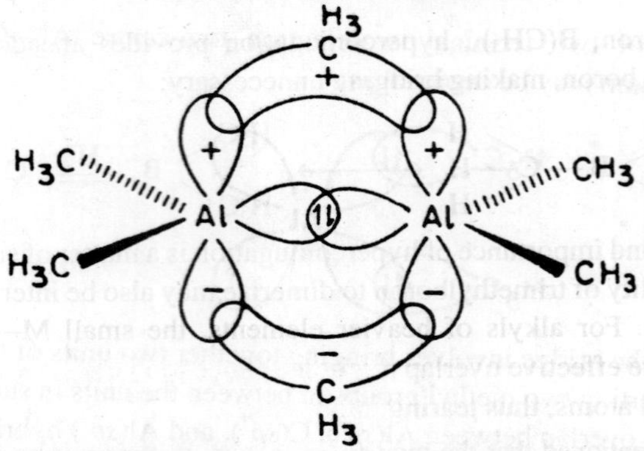

angle (75°) may also imply the existence of an Al–Al bond. The acute bridging angle is, incidentally, a characteristic feature of organobridged electron deficient compounds of aluminium, beryllium, and magnesium, and the relatively wide external C—Al—C angle (123°) is a consequence of sp^2 hybridization of aluminium.

It may, however, be pointed out that the above two bonding patterns are not incompatible. For example, if the sp^3 hybrid orbitals of aluminium in the first description also overlap with each other, some Al—Al bonding would arise. Similarly, if the sp^3 hybrid orbital of the bridging CH_3 group in the second description overlaps with both the π- and σ-orbital of the Al—Al bond, then the bridge bonding would resemble the first description. Although extensive molecular calculations point to the importance of some Al–Al bonding., yet Al—C—Al interactions appear to be the major contributing factor for the formation of the dimeric trialkylaluminium units.

Another interesting example of an electron deficient compound is $[Be(CH_3)_2]_n$ (a long chain polymer), which utilizes one sp^3 hybrid orbital of each methyl groups and all of the sp^3 orbitals of beryllium in bridge bonds. Be—C_{bridge}—Be angle is 60° and Be—C bond length is slightly longer (1.93Å) than the sum of the covalent radii (1.68Å):

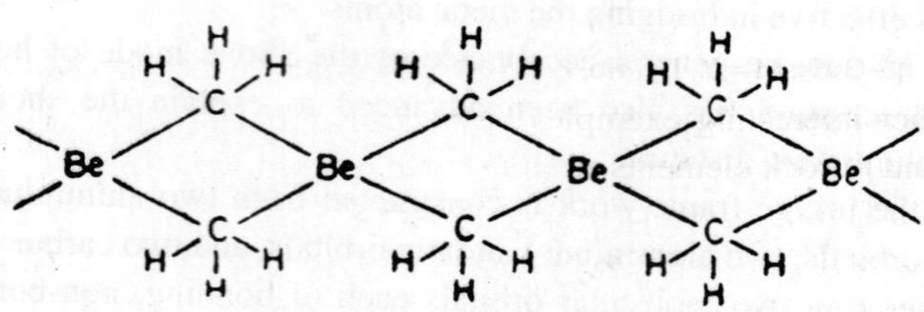

The dimethylmagnesium has also been found to assume a polymeric structure similar to that of dimethylberyllium. Unlike trimethylaluminium and dimethylberyllium, however, polymeric units of dimethylmagnesium appear to be less susceptible to dissociation on being complexed with a weak Lewis base such as Et_2O.

As indicated above none of the other members of main group III form associated organometallics (particularly in gas phase) except those of aluminium.

For trimethylboron, $B(CH_3)_3$, hyperconjugation provides an additional pair of electrons for the boron, making bridging unnecessary:

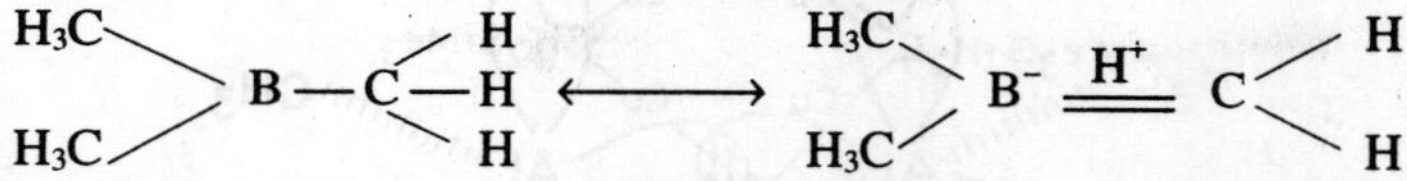

The extent and importance of hyperconjugation is a matter of some controversy, and the inability of trimethylboron to dimerize may also be interpreted in terms of steric factors. For alkyls of heavier elements, the small M—C—M angles required to secure effective overlap might introduce too large a repulsion between the bulkier metal atoms, thus tearing them, apart.

It may be mentioned that the monomeric nature of dimethylberyllium in the gas phase coupled with quite high Be—C stretching frequency (1081 cm^{-1}) can be explained in terms of hyperconjugation (similar to that in the case of R_3) in the gas phase:

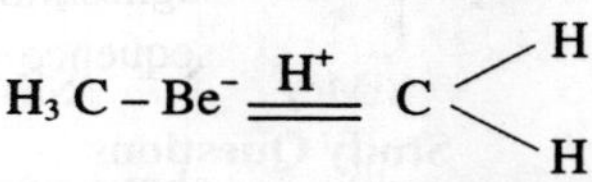

The formation of electron-deficient alkyl bridges is not an exclusive property of light *s*-block (Li, Be, or Mg). aluminium and (in the solid state) the heavier man group III element, as demonstrated by the synthesis and structural characterization of a series of new classes of electron-deficient alkyl bridged transition and inner transition metal organometallics (see Section 4.2).

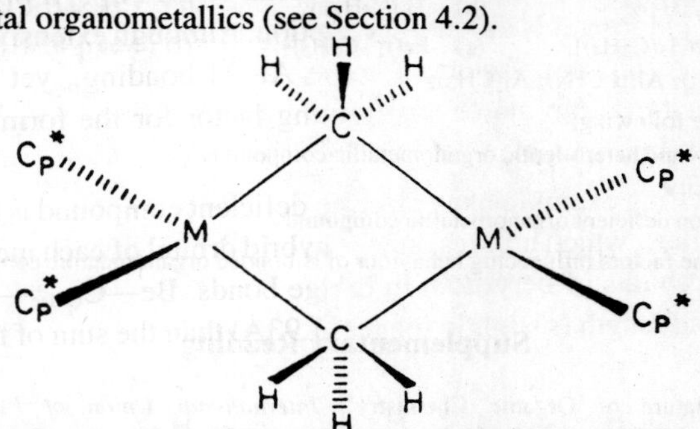

($C_p^* = \eta^5$–C_5H_5; M = Y, Dy, Ho, Er, Tm, Yb; $Cp^* = (Me_3Si)_2C_5H_3$, M = Nd).

A few other interesting examples of electron-deficient alkyl bridged compounds of *d*- and *f*-block elements are illustrated below:

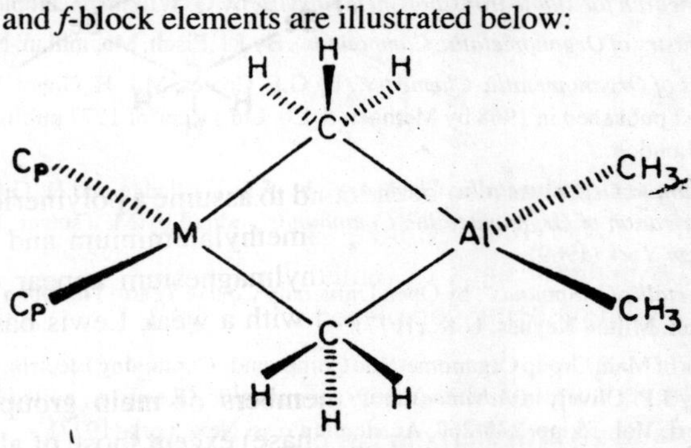

(Cp = η^5–C_5H_5; M = Sc, Y, Gd, Dy, Ho, Er, Tm, or Yb)

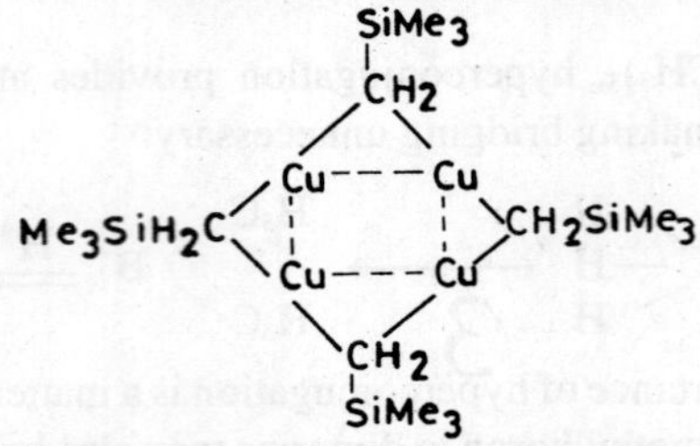

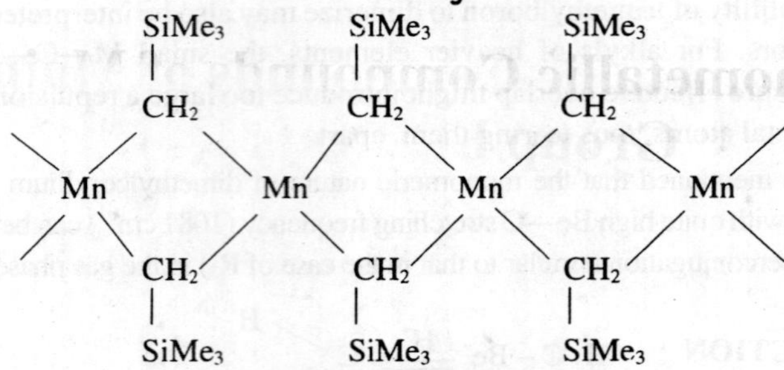

Study Questions

1. Discuss the classification of organometallic compounds on the basis of the position of the metal in the periodic table.
2. Describe briefly the general characteristics of organometallic compounds.
3. Based on the IUPAC rules, name the following organometallic compounds:
   (a) $Zn(C_2H_5)_2$; (b) $(LiC_4H_9)n$; (c) CH_3SnH_2Cl;
   (d) $K[PtCl_3(C_2H_4)]$; (e) $Fe(\eta^5\text{-}C_5H_5)_2$; (f) $Mn(\eta^3\text{-}C_3H_5)(CO)_4$;
   (g) $(CH_3)_2Al(\mu\text{-}CH_3)_2Al(CH_3)_2$
4. Define the following:
   (a) Homo- and hetero-leptic organometallic compounds
   (b) Hapticity
   (c) Electron deficient organometallic compounds
5. Discuss the factors influencing behaviour of σ-bonded organometallic compounds.

Supplementary Reading

1. 'Nomenclature of Organic Chemistry', *International Union of Pure and Applied Chemistry* (IUPC), pp. 360-466, Pergamon Press (1979).
2. *'Organometallic Chemistry'* by E.G. Rochow, Reinhold, New York (1964).
3. *'The long search for stable transition metal alkyls'*, by G. Wilkinson, Noble Lecture (1973).
4. *'The Chemistry of Organometallic Compounds'*, By J.J. Eisch, Macmillan, New York (1967).
5. *'Principles of Organometallic Chemistry'*, by G.E. Coates, M.L.H. Green, P. Powell, and K. Wade (First published in 1968 by Methuen & Co. Ltd.) Reprint 1977 published by Chapman and Hall, London.
6. *'Introduction to Organometallic Chemistry'*, by F.K. Cartledge and H. Gilman, pp. 1-33 in *'Characterization of Organometallic Compounds'*, edited by M. Tsutsui, Interscience Publishers, New York (1969).
7. *'Organometallic Compounds'*, by Open University Course Team, The Open University Press, Walton Hall, Milton Keynes, U.K. (1977).
8. 'Structures of Main Group Organometallic Compounds Containing Electron-Deficient Bridge Bonds', by J.P. Oliver, in *Advances in Organometallic Chemistry*, edited by F.G.A. Stone, and R. West, Vol. 15, pp. 235-267, Academic Press, New York (1977).

3

Organometallic Compounds of Main Group Elements

3.1 INTRODUCTION

Although the first organoarsenic derivative was synthesised in 1769 by L.C. Cadet de Glassicourt, yet it was only in mid-19th century that the nature of the compound was established as $(H_3C)_2As{\bullet}As(CH_3)_2$. The real beginnings of the non-transition (main group) metal organometallic chemistry can, however, be traced to the synthesis of zinc alkyls by Frankland in 1848 which was followed a few decades later by Grignard's synthesis and exploitation of organomagnesium halides (Grignard received Nobel Prize for this work in 1912) in organic synthesis.

Besides structural and other features, a wide and in fact fastly growing interest in organometallic compounds of main group elements is sustained by their established and potential applications as reagents (e.g. organometallics of lithium, magnesium, zinc, cadmium, mercury, boron, aluminium, thallium, silicon, and tin) in organic and organometallic synthesis (see Chapter 6).

This chapter provides with the aid of a few illustrative examples, a general account of the salient (synthetic, structural and bonding) characteristics of this class (main group organometallics) of compounds, beginning with a few selected chronological developments:

1760 L.C Cadet de Glassicourt's accidental preparation methylarsenical liquid (''cacodyl'').

1840 R.W. Bunsen characterizes cacodyl as $As_2(CH_3)_4$.

1849 E. Frankland (generally regarded as founder of organometallic chemistry) discovers alkylzinc compounds.

1852 E. Frankland describes the preparation of methylmercury iodide from mercury and iodomethane in sunlight.

1863 C. Friedel and J.M. Crafts prepare tetramethylsilane (the first organosilane).

1898 O. Dimroth studies mercuration of aromatic compounds

1900 V. Grignard describes direct preparation of organomagnesium halides now known as Grignard reagents.

1901 F.S. Kipping lays the foundation of organosilicon chemistry and prepares $[(C_6H_5)_2SiO]_x$ (diphenylsilicone).

1909 P. Ehrlich (Nobel Prize, 1908) introduced Salvarsan for treatment of syphilis and laid the foundation of chemotherapy.

1912 V. Grignard and P. Sabatier share the Nobel Prize in chemistry for discovery of the Grignard reagent, and for developing method of hydrogenating organic compounds in the presence of finely disintegrated metals respectively.

1914 W. Schlenk and coworkers investigate organoalkali compounds.

1922 T. Midgley and T.A. Boyd describe the anti-knocking behaviour of $Pb(C_2H_5)_4$.

1929 T.A. Paneth generates alkyl radicals through pyrolysis of PbR_4.

1930 K. Ziegler and H. Colonius describe direct preparation of lithium alkyls.

1934 H. Gilman and coworkers describe metallation of aromatic hydrocarbons with organolithium reagents.

1938 H. Gilman as well as G. Wittig prepare aryllithium reagents independently from aryl halides and butyllithium (halogen-lithium exchange reaction).

1943 E. G. Rochow describes the 'direct synthesis' of chlorosilanes, extensively used for preparation of silicones (which have become industrially very important).

1955 K. Ziegler and coworkers prepare aluminium alkyls by the direct combination of aluminium hydrogen, and alkenes.

1956 H.C. Brown and B.C. Subba Rao discover hydroboration.

1956 G. Wittig describes methylenation of carbonyl compounds with methylenetriphenylphosphoranes.

1960 M.F. Hawthrone and coworkers discover carboranes.

1976 M.F. Lappert and coworkers laid the foundation of subvalent Group 14 metal alkyls and stable organometal (III) free radicals MR_3.

1979 Nobel Prize to H.C. Brown and G. Wittig for their contributions on the applications of organoboranes and methylenephosphoranes, respectively, in organic synthesis.

1981 R. West and coworkers isolate the first stable compound $Si_2(C_6H_2Me_3\text{-}1,3,5)_4$ with Si = Si double bond.

1981 A.G. Brook, et al, report a stable compound containing Si = C bond, $(Me_3Si)_2Si{=}C(OSiMe_3)C_{10}H_{15}$.

1981-90 The synthesis of Ar*P = PAr* (Ar* = $C_6H_2Bu^t_3$-2, 4, 6) in M. Yoshifuji's group, followed by isolation of —P = As—, —P = Sb—, and —As = As— derivatives by A.H. Cowley, et al. and further extension to material chemistry.

3.2 GENERAL CHARACTERISTICS

Most organometallic compounds (simple alkyls and aryls) of main group elements (where the organic group is bonded to the element by a normal 2-electron 2-centre bond, with few exceptions such as alkyls of lithium, beryllium, magnesium, aluminium and zinc, which in view of their electron deficient character undergo association through two-electron three-centre bonds) depict properties characteristic of organic rather than inorganic compounds. The majority of these have discrete molecular structure (Fig. 3.1).

These, therefore, exist at ambient temperatures either as low-melting crystalline solids or as volatile liquids, which are soluble (miscible) in weakly polar (e.g. OEt_2 THF, etc.) or non-polar (e.g. PhH, PhMe, etc.) organic solvents.

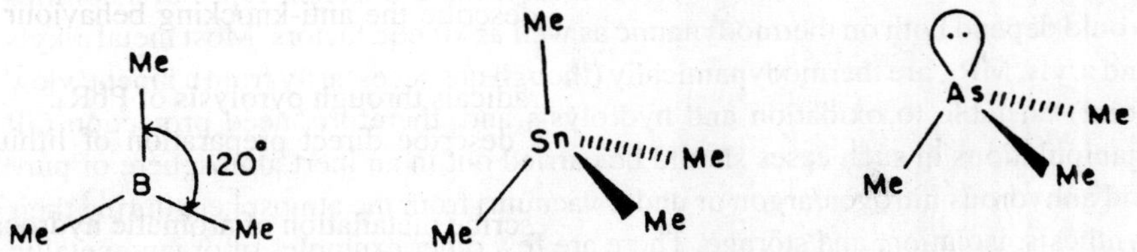

Fig. 3.1 Some simple metal alkyls.

The lower alkyls, particularly the methyls tend to be amongst the most volatile (Table 3.1) derivatives of the metal concerned.

Table 3.1 Boiling Points of Methyl Derivatives $M(CH_3)_n$ of Some Main Group Elements

GROUP II		GROUP III		GROUP IV		GROUP V		GROUP VI	
$M(CH_3)_2$	b.p. (°C)	$M(CH_3)_3$	b.p. (°C)	$M(CH_3)_4$	b.p (°C)	$M(CH_3)_3$	b.p. (°C)	$M(CH_3)_2$	b.p. (°C)
Be	217*[a]	B	−22	C	10	N	3	O	−25
Mg	—	Al	226**	Si	27	P	40	S	37
Zn	44	Ga	56	Ge	43	As	52	Se	58
Cd	106	In	136	Sn	77	Sb	79	Te	82
Hg	93	Tl	147*	Pb	110	Bi	110	Po	—

* *Extrapolated boiling point.* [a]Unlike their group congeners, which have discrete molecular structures, $Be(CH_3)_2$ and $Mg(CH_3)_2$ have associated polymeric structures and hence, t hese are comparatively non-volatile.

**Trimethylaluminium vaporises as dimeric $Al_2(CH_3)_6$.

A perusal of the data in Table 3.1 shows that with the exception of highly associated methyls of beryllium and magnesium and to a lesser extent of dimeric trimethylaluminium, the values of boiling points of the metal methyls show the expected decrease in volatility with increasing relative molecular mass for the members of a group of elements in the Periodic Table.

The chemical properties of these organometallic compounds vary widely and can be correlated with the difference in electronegativity (i.e. % ionic character) between the metal and carbon, and by the donor characteristics of the central atom. At the one extreme, alkyls of more electropositive metals (e.g. Li, Na, Mg) including aluminium, zinc and even boron are spontaneously inflammable in air and react violently with water (except that of boron, which requires a higher temperature); at the other end, the tetra-alkyls of group IVB elements are relatively unreactive species.

3.3 STABILITY OF ORGANOMETALLIC COMPOUNDS

In dealing with the stability of a compound, one should be clear about the type of stability under consideration. In general terms, the stability of an organometallic compound may refer to either its thermal stability, or resistance to chemical attack (particularly by air and moisture). Obviously, these different types of stabilities would depend both on thermodynamic as well as kinetic factors. Most metal alkyls and aryls, MR_n, are thermodynamically (though not necessarily from a kinetic view point) unstable to oxidation and hydrolysis and, therefore, need protection (all manipulations in such cases should be carried out in an inert atmosphere of pure and anhydrous nitrogen/argon or under vacuum) from the atmosphere during their synthesis, isolation, and storage. There are few other examples of organometallic compounds like tetramethyllead which are thermodynamically unstable with respect to oxidation and hydrolysis, yet they survive exposure to air or/and moisture possibly because of kinetic factors.

3.3.1 Thermal Stability

The thermodynamic stability of a compound relative to its constituent elements (or for practical purposes, relative to the products of decomposition) would obviously depend upon the standard free energies of formation, ΔG_f° of the different species concerned. In view of the non-availability of accurate data on ΔG_f° values for organometallic compounds, one has to rely on the known standard enthalpies of formation, ΔH_f°, making due allowance for plausible estimates for the contributions of the entropy terms $T\Delta S$, in the well-known relationship:

$$\Delta G = \Delta H - T\Delta S$$

In Table 3.2 are listed the standard values of molar enthalpies of formation (ΔH_f° / kJ mol^{-1} as well as of metal-carbon average bond enthalpies E_m(M—C) of some volatile metal methyls, $M(CH_3)_n$.

The decomposition of the metal methyls into the component elements, i.e., metal, carbon and hydrogen could be expected to be an exothermic process in the case of heavier metal alkyls, e.g., $MgMe_2$, $CdMe_2$, $InMe_3$, $PbMe_4$ and $BiMe_3$, for which ΔH_f°, i.e., the enthalpy of formation is positive. Conversely for methyls of lighter elements such as B, Al, Si, Ge and P, it should be an endothermic process.

In an actual decomposition process, however, the metal alkyls yield the metal, hydrogen and mixtures of hydrocarbons rather than simple elementary carbon. In such situations, therefore, we would have to take into account the enthalpies of formation of the hydrocarbons involved. For example, let us for the purpose of simplicity consider that tetramethyllead decomposes into lead, methane and ethene only, i.e.,

$$\begin{array}{lccccc} & Pb(CH_3)_4(g) & \longrightarrow & Pb(s) + & 2CH_4(g) + & C_2H_4(g) \\ \Delta H_f^\circ \text{ (kJ mol}^{-1}) & +137^\circ & & 0 & 2 \times (-73) & +52 \end{array}$$

The standard enthalpy change ΔH_m° for this reaction would be given by the difference between the standard enthalpies of formation, ΔH_f, of the products and the reactants, i.e.,

$$\Delta H_m^\circ = -150 + 52 - 137 = -235 \text{ kJ mol}^{-1}$$

Table 3.2 Standard Molar Enthalpies of Formation (ΔH_f in kJ mol^{-1}) and Metal Carbon Average Molar Bond Enthalpies E_m(M—C) in kJ mol^{-1} of some Metal Methyls, $M(CH_3)_n$.

Compound	ΔH_f°	E_m
Group IIB (12)		
$ZnMe_2$	+55	176
$CdMe_2$	+110	138
$HgMe_2$	– 93	121
Group IIIB (13)		
BMe_3	–122	364
$AlMe_3$	–88	276
$GaMe_3$	–39	247
$InMe_3$	+172	163
$TlMe_3$	...	...
Group IVB (14)		
CMe_4	–167	347
$SiMe_4$	–238	293
$GeMe_4$	–108	247
$SnMe_4$	– 19	218
$PbMe_4$	+137	155
Group VB (15)		
NMe_3	– 24	314
PMe_3	– 96	276
$AsMe_3$	+15	230
$SbMe_3$	+31	218
$BiMe_3$	+192	114

The reaction, therefore, is highly exothermic. Moreover, the reaction should involve an increase of entropy as three gaseous product molecules are formed for every reagent molecule decomposed. The term, $T\Delta S$ in the relationship, $\Delta G = \Delta H - T\Delta S$ would, therefore, be negative and enhance the overall negative value of ΔG°. This clearly indicates a higher stability for the products than the reactant molecules.

Similarly for the reaction:

$$\begin{array}{lccccc} & Si(CH_3)_4(g) & \longrightarrow & Si(s) + & 2CH_4(g) + & C_2H_4(g) \\ \Delta H_f^\circ \text{ (kJ mol}^{-1}) & -238 & & 0 & 2 \times (-75) & 52 \end{array}$$

From the above data, ΔH_m for $Si(CH_3)_4$ works out to be equal to $-150 + 52 + 238 = +140$ kJ mol^{-1}. The value of ΔG° for $Si(CH_3)_4$ also can be estimated from the relationship, $\Delta G = \Delta H - T\Delta S$. For reason stated in the case of $Pb(CH_3)_4$, the value of $-T\Delta S$ would be negative in this case also. The value of ΔG° for $Si(CH_3)_4$, therefore, would be lower than +140 kJ mol^{-1}, but it would become negative (indicating the thermodynamic instability of tetramethylsilicon also on these simple considerations), only if the absolute value of $T\Delta S$ is greater than 140 (which is rather improbable).

The question naturally arises as to how species like tetramethyllead with such a high negative value of ΔG° (indicating thermodynamic instability) can be actually isolated.

For resolving such anomalies, one would have to take into consideration whether a suitable reaction path is available for the decomposition process to occur. The kinetic stability factor, therefore, appears to play a more predominant role. Although thermodynamically unstable, there appears to be no route in such cases with sufficiently low activation energy by which decomposition could occur at ordinary (and sometimes even at elevated) temperatures with a noticeable rate. The point is illustrated schematically in Figure 3.2, which represents the decomposition of a metal alkyl by homolytic dissociation of its metal carbon bonds.

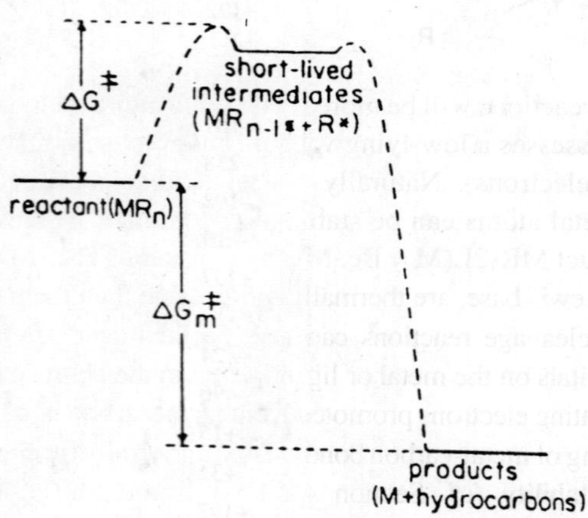

Fig. 3.2 Schematic representation of the thermal decomposition of a metal alkyl MR_n by unimolecular homolytic dissociation ($\Delta G^\ddagger$, the free energy of activation of the first stage of the reaction is a measure of the kinetic stability of MR_n, whereas $\Delta G_m^\ddagger$, the free energy of the overall decomposition indicates the thermodynamic instability of MR_n

The value of activation energy, $\Delta G^\ddagger$ will depend largely on the metal-carbon bond strength, i.e., on E_m (M—C). As the data in Table 3.2 shows, the value of E_m(M—C) decreases, making the thermal decomposition of the metal alkyls more facile as the atomic weight increases for members of the same group of the periodic table. In homolytic dissociation processes of this type, radicals like $MR^\bullet_{n-1}$ and $R^\bullet$, once formed, are highly reactive and undergo rapid irreversible changes further with the formation of stable products, e.g., formation of R—R by association of R° or of alkane RH by abstraction of hydrogen. The thermodynamic stability of these stable final products provides the driving force for the reaction, once it is initiated. By contrast, dissociation of MX_n, say $M(CO)_n$ into $M(CO)_{n-1}$ and CO could be reversible, as the CO molecules, even after being dissociated from the parent metal carbonyl, remain available for recombination with $M(CO)_{n-1}$.

Homolytic dissociation of metal-carbon bond is not the only way by which organometallic compounds may decompose thermally. Another important route by which such decompositions often occur is called 'β' elimination, in which the hydrogen atom on a β-carbon atom can migrate to the metal atom, allowing elimination of alkene and the formation of a metal hydride:

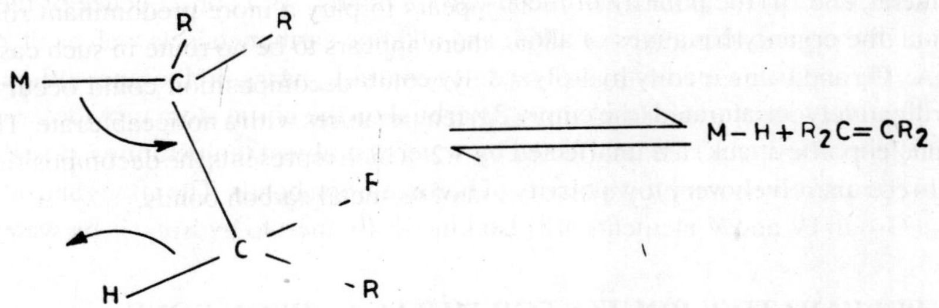

Such β-elimination reactions will be more facile if the metal atom is coordinatively unsaturated (and possesses a low-lying vacant orbital to accommodate the metal-hydrogen bonding electrons). Naturally one would expect that derivatives of Group II and III metal atoms can be stabilised by adduct formation with Lewis bases. Thus the adduct $MR_2 2L$(M = Be, Mg, or Zn) and $M'R_3L$(M′ = B, Al, Ga, In or Tl) where L is a Lewis base, are thermally more stable than the parent alkyls. The metal carbon bond cleavage reactions can also be facilitated by the presence of low-lying vacant orbitals on the metal or ligand, even in the absence of a β-hydrogen atom, by accommodating electrons promoted from metal carbon bonding orbitals with consequent weakening of metal-carbon bonds. This is generally the predominant factor for the thermal instability (cf. Section 4.2.1) of some transition metal alkyls; e.g. tetramethyltitanium, $TiMe_4$, decomposes above –78°C.

3.3.2 Stability to Oxidation

The thermodynamic instability of organometallic compounds may be ascribed to the large value of free energies of formation of the metal oxides, carbon dioxide and water. Actually the combustion of organometallic compounds is a highly exothermic process, e.g.,

$$Zn(CH_3)_2(g) + 4O_2(s) \longrightarrow ZnO(s) + 2CO_2(g) + 3H_2O(g)$$

$$\Delta H_m^* = -1918 \text{ kJ mol}^{-1}$$

$$Sn(CH_3)_4(g) + 8O_2(g) \longrightarrow SnO_2(s) + 4CO_2(g) + 6H_2O(g)$$

$$\Delta H_m^* = -3591 \text{ kJ mol}^{-1}$$

It is worth noting that dimethylzine is a spontaneously inflammable compound whereas tetramethyltin is air stable. The greater reactivity of dimethylzinc can be easily understood on the basis of its strong Lewis acid character. In addition to $ZnMe_2$, the lighter alkyls of Li, Na, Be, Mg, Zn, B, Al, Ga, In, Tl, As, Sb and Bi are also spontaneously inflammable. Thus the presence of empty low-lying orbitals (e.g. 4*p* orbital of $GaMe_3$) or a lone pair of electrons (as in $SbMe_3$) appears to be a source for kinetic reactivity to oxidation by air. The Group IVB tetraalkyls such as $SnMe_4$ possess neither of these characteristics and, therefore, they resist oxidation at ambient temperatures.

3.3.3 Stability to Hydrolysis

The organometallic compounds are generally hydrolysed via uncleophilic attack by water, which is facilitated by: (i) the presence of empty low-lying orbitals on the metal, and (ii) the polarity of metal-carbon bonds. As a consequence of these factors, the organoderivatives of alkali and alkaline earth metals as well as of Zn, Cd, Al, Ga and In are readily hydrolyzed. By contrast, the trialkylboranes although coordinatively unsaturated (the empty 2*p* orbital on the boron atom provides a site for uncleophilic attack) are unaffected by water at ambient temperatures, possibly due to comparatively very low polarity of boron-carbon bonds. The alkyls and aryls of the Group IV and V elements are also kinetically inert to hydrolysis by water.

3.4 PREPARATIVE ROUTES FOR METAL-CARBON BOND FORMATION

The types of reactions generally employed for the synthesis of non-transition metal organometallic compounds have been summarized in Table 3.3.

Out of a very large number of preparative methods available now for the synthesis of organometallic compounds (homoleptic or heteroleptic) of non-transition elements, only a few methods of more general applicability are being presented in Sections 3.4.1 to 3.4.8. Under each section, a few illustrative examples of the application of the specific method have been presented, the method of carbanion-halide exchange reaction (Section 3.4.3) is of such wide applicability to organometallics of groups 2 to 16 element that these have been presented in different sub-sections.

3.4.1 Oxidative-Addition Reactions

These reactions constitute one of the principal routes for the formation of metal-carbon bonds; these can be formally termed as 'oxidative addition' types, as they involve the oxidation of a metal 'M' by the addition of a group 'RX'. Naturally, such reactions are more prone to occur with metals of high electropositivity (metals with low ionization energy). The following equations represent one- and two-electron oxidative addition reactions respectively:

$$2M + RX \rightarrow MR + MX$$

$$M + RX \rightarrow \underset{}{M(R)} - X$$

Table 3.3 Preparation of Main Group Organometallics

Types of Reaction (Sub-section)	Preparation
Oxidative-Additive (3.4.1)	$2M + RX \rightarrow MR + MX$ or $M + RX \rightarrow M(R)X$
Metal-Metal Exchange (Transmetallation) (3.4.2)	$M + M'R_n \rightarrow MR_n + M'$ ($M' = Mg, Zn, Hg, Al, Sn$)
Carbanion-Halide Exchange (3.4.3)	$nM'R_m + mRX_n \rightarrow mMR_n + nM'X_m$ ($M' = Li, Mg, Zn, Cd, Hg, Al, Sn$)
Metal-Hydrogen Exchange (3.4.4)	$MR' + RH \rightarrow MR + R'H$
Metal-Hydride-Alkene (Addition) (3.4.5)	$M-H + {>}C=C{<} \rightarrow M-\overset{\mid}{\underset{\mid}{C}}-\overset{\mid}{\underset{\mid}{C}}-H$
Methylene-Insertion (3.4.6)	$MCl_4 + CH_2N_2 \rightarrow MCl_3(CH_2Cl) + N_2$
Miscellaneous (3.4.7 & 3.4.8)	$As(OH)_3 + PhN_2X \rightarrow PhAsO(OH)_2 + N_2 + HX$ $M(O_2CMe)_2 \xrightarrow{\text{heat}} MMe_2 + 2CO_2$ $M(O_2CMe)_n + C_6H_6 \rightarrow M(O_2CMe)_{n-1}(C_6H_5) + MeCO_2H$

These reactions involve changes in the formal oxidation states. Some specific examples of these two types of reactions are described in the following pages:

$$2M + RX \xrightarrow{\text{ether* or hydrocarbon}} MR + MX^*$$

(R = alkyl or aryl; X = Cl, Br (in most cases), or I (in some specific cases); M = Li).

With more electropositive metals, e.g., sodium to caesium, the initially formed metal alkyl has a strong tendency to undergo Wurtz-Fittig reaction (e.g. $NaR + RX \rightarrow R—R + NaX$), which can be avoided to some extent by modifications, such as the use of finely divided metals and efficient stirring, which enhance the formation of metal alkyls and also by slow addition of alkyl halides to minimize the side-reaction. The alkyls of less electropositive metals have generally been prepared, by using the metal in the form of an alloy:

$$2EtI + 2Hg/Na \xrightarrow{\text{in EtOAc}} HgEt_2 + 2NaI$$

* Ethers are generally prone to attack by organolithium derivatives, but there are wide variations in the reaction rates. For example, more basic tetrahydrofuran reacts faster with an organolithium reagent than diethyl ether. It is, therefore, preferred to keep the temperature of the reaction as low as possible, but sufficiently high to keep it going.

$$4RX + 4Sn/Na \longrightarrow SnR_4 + 3Sn + 4NaX$$

(R = alkyl)

$$4EtX + 4Pb/Na \longrightarrow PbEt_4 + 4NaX + 3Pb$$

(X = Cl, I)

The main driving force for these reactions appears to be the reaction between a halogen and an electropositive (alkali) metal.

The well known example of two-electron oxidative-addition reaction is the formation of Grignard reagents and related compounds:

$$RX + Mg \xrightarrow{\text{ether/THF}} Mg(R)X$$

(R = alkyl, aryl or fluoroalkyl; X = halogen) (N.B. the reactivity of the alkyl halide increases from chloride to iodide; bromides are generally chosen as the reactions with chlorides may be too slow and the use of iodides is entailed by some side-reactions).

This type of oxidative-addition reactions has been extended to metals other than magnesium.

Reactions with Ca*, Sr and Ba are generally not as facile as for Mg and special attention to experimental conditions is necessary to obtain satisfactory yields. A slight change in the reaction conditions as illustrated by the following equations gives rise to some potentially useful homoleptic- and heteroleptic-organometallics of metals arranged groupwise.

$$Me_3ECH_2Cl + Mg \xrightarrow{\text{ether}} MgCl(CH_2EMe_3)$$

(E = C, Si)

$$CF_3I + Mg \xrightarrow[-70^\circ C]{\text{ether}} MgI(CF_3)$$

$$RI + Zn\ ZnI(R) \xrightarrow{\text{heat}} 1/2ZnR_2 + 1/2ZnI_2$$

(R = alkyl)

$$C_6H_5Br + Zn^* \xrightarrow{\text{THF}} ZnBr(C_6H_5)$$

(Zn* = finally divided zinc prepared from Zn Br_2 and potassium in THF).

$$2RX + 2Hg\ (Na) \rightarrow Hg\ R_2 + Hg + 2\ Na\ X$$

(X = Cl, Br; R = Me, Et, CH_2Ph; usually Hg R_2, but unsymmetrical, Hg(R)X may also be formed).

$$CF_3I + Hg \xrightarrow{260^\circ C} HgI(CF_3)$$

* For example, successful preparations of calcium compounds require either very pure metal with low sodium content or slightly less pure calcium in the presence of mercury.

$$RI + Hg \xrightarrow{h\nu} HgI(R)$$
(R = alkyl)

$$3RX + 2Al \rightarrow AlRX_2 + AlR_2X \xrightarrow{Na} AlR_3 + 3NaX + Al$$
(R = alkyl; X = halogen)

$$MeCl(g) + Si(Cu) \xrightarrow{300\text{–}400^\circ C} SiCl_{4-n}\,Me_n$$

The above reaction produces a mixture of products ($SiCl_4$, $SiCl_3$ (Me), $SiCl_2$ Me_2, and SiCl Me_3, with $SiCl_2$ Me_2 predominating).

$$2EtCl + Ge \xrightarrow[\text{over copper}]{350^\circ C} GeCl_2Et_2$$

$$2MeCl + Sn \xrightarrow{\text{under pressure}} SnCl_2Me_2$$

$$3MeBr + 2As \xrightarrow{\text{heat}} AsBrMe_2(Me) + As\,Br_2\,(Me)$$

$$Te + 2RX \xrightarrow{Cu} TeR_2X_2$$
($R = CH_3$, Et)

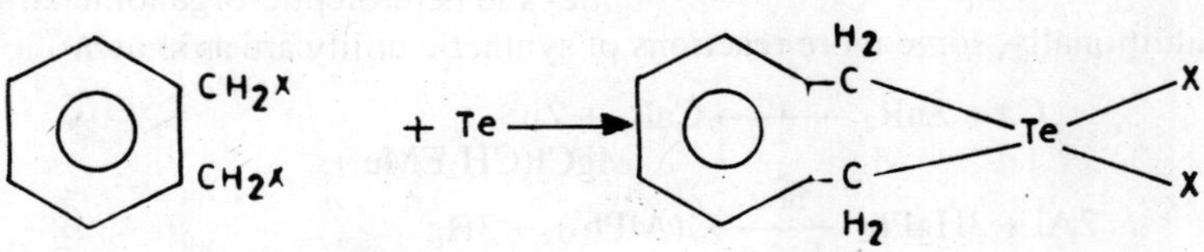

(x = halogen)

$$(CH_2)_2I_2 + Te \longrightarrow (CH_2)_nTeI_2$$
(n = 4 or 5)

3.4.2 Metal-Metal Exchange Reactions (Transmetallation)

These have been generally represented by the following equation:

$$M + M'R \rightarrow MR + M'$$

The facility of this route depends upon the difference in the free energies of the formation of the two species MR and M′ R. In view of the unavailability of reliable values for ΔG° for organometallic compounds in general, $\Delta \overset{\circ}{H}_f$ (heats of formation) may be used to estimate the thermodynamic feasibility of such reactions. A perusal of thermodynamic data in Table 3.2 indicates that heavier metal alkyls (and aryls) should be more prone to lose organic groups.

There are two ways of estimating the feasibility of the above reactions. Firstly, it may be considered formally as an oxidation-reduction reaction in which the more electropositive metals (main groups I, II and III, except boron) should be more prone to participate in such transformations. Secondly, the reaction of any metal whose metal-carbon (M—C) bond energy is higher than that of M′—C bond energy in M′R used, would tend to occur as:

$$2Li + Mg\,R_2 \rightarrow 2\,LiR + Mg$$

$$2Na + HgR_2 \rightarrow 2\,NaR + Hg$$

$$M + Hg\,R_2 \rightarrow MR_2 + Hg\ (M = Ca, Sr, Ba)$$

The reaction of sodium with mercury alkyls has often been used when pure organoalkali metal compounds (including compounds of heavier alkali metals) are required; although free from Wurtz coupling, yet the reaction is slow and inconvenient. Recently (1980) a more versatile method has been developed in which an unfavourable equilibrium is displaced by the solubility of lithium alkoxides and insolubility of organoalkali metal compounds in the solvents employed:

$$LiR + MOR' \xrightarrow{\text{hydrocarbon}} MR\downarrow + LiOR'$$

(R′ = Bu, menthyl; M = Na, K, Cs)

More recently (1990), an opposite type of solubility behaviour of the products has been exploited for the synthesis of some soluble magnesium alkyls:

$$2Li\,CH\,(Si\,Me_3)_2 + Mg\,(OC_6\,H_2\,Bu^t_2\text{-}2,\,6\,Me\text{-}4\,)_2$$
$$\rightarrow Mg\,\{CH\,(Si\,Me_3)_2\}_2 + 2Li\,(OC_6\,H_2\,Bu^t_2\text{ - }2,\,6\,Me\text{-}4)\downarrow$$

Additionally, some more reactions of synthetic utility are as shown:

$$Ca + ZnR_2 \longrightarrow CaR_2 + Zn$$

$$2Al + 3HgPh_2 \xrightarrow{\text{heat}} (AlPh_3)_2 + 3Hg$$

$$2M + 3HgR_2 \xrightarrow{\text{heat}} (MR_3)_2 + 3Hg$$

(M = Ga or In; especially for R = alkyl)

$$\underset{\text{(excess)}}{Zn} + Hg\,R_2 \rightarrow Zn\,R_2 + Hg\,(Zn)$$

The last reaction is not applicable for cadmium, because only a mixture of products is obtained:

$$Cd + HgR_2 \rightarrow CdR_2 + Hg$$

(R = alkyl, 75%; R = phenyl, 25%)

As the Hg—C bonds are exceptionally weak, such transformation should be feasible for many metals using HgR_2 derivatives. For example, organometallics of main group I, II or III metals, Sn, Bi (using metal-amalgam), and Te may be prepared by this route. However, experimentally there are a large number of exceptions. One of the advantages of the above general reaction is its feasibility in non-ethereal solvents resulting thereby in ether-free compounds (that are potentially strong Lewis acids) of Be, Al or Ga.

3.4.3 Carbanion-Halide Exchange Reactions

This type of reaction will obviously involve the interchange of the carbanion (alkyl/aryl) and halide groups present on two different metals and can be represented in most general terms by the following type of equation:

$$MX_n + M'R_m \rightarrow MX_{n-a}\,R_a + M'R_{m-a}\,X_a$$

(X = halogen and R = alkyl or aryl)

This is a general procedure of considerable synthetic value and is probably the most versatile and widely used for all laboratory methods. A perusal of the thermochemical information given in Figure 3.3 leads to the conclusion that such reactions proceed effectively in the direction leading to the bonding of the halogen with more electropositive and of the carbanion with the less electropositive metal.

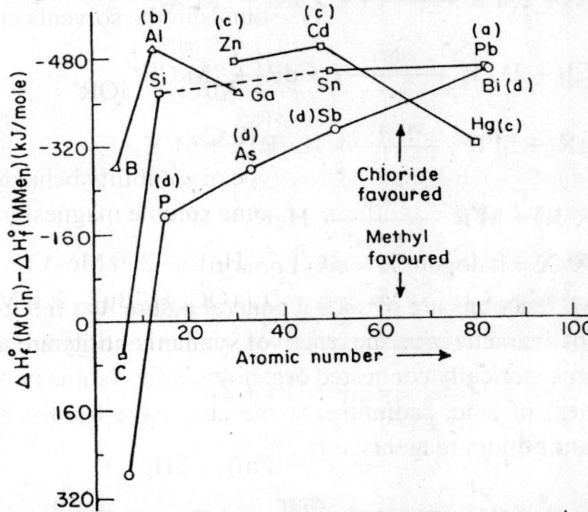

Fig. 3.3 Plot of $\Delta H^\circ_f\,[MCl_n(g)] - \Delta H^\circ_f\,M(CH_3)_n(g)]$ versus atomic number.

Amongst the halides (MX_n) employed for such reactions, chlorides have been commonly used. In some cases, metal derivatives like hydrides, alkoxides, etc. have also been employed in place of halides. For the reactants represented by $M'R_m$, the Grignard reagents MgRX or the organolithium reagents, LiR are the most common ones, but alkyls (aryls) of other metals like zinc, cadmium, aluminium, tin, and lead have also been used. The versatility of this synthetic route is being indicated in the following sub-sections by its applications to the preparation of organometallics of a few metals arranged groupwise:

3.4.3.1 Synthesis of Organometallics of Beryllium, Magnesium, Zinc, Cadmium and Mercury

A large number of reactions of anionic alkylating (arylating) agents with halides of the above metals are known, a few of which are indicated by the following equations:

$$BeX_2 + 2M'R \rightarrow BeR_2 + 2M'X_2$$

(M′= Li or MgX, X = usually the chloride, R = alkyl or aryl).

(In the above reactions, the products are generally complexed species with ether in view of the strongly Lewis acid character of BeR_2 and a hydrocarbon medium will have to be employed for obtaining ether-free BeR_2 species).

$$MgX_2 + LiR \xrightarrow{\text{ether}} Mg(R)X + LiX$$

(X = Br, I; R = alkyl or aryl).

(This reaction may be a route of potential value for Grignard reagents, which are difficult to be prepared by the direct reaction: $Mg + RX \rightarrow Mg\,RX$)

$$ZnX_2 + M'R \xrightarrow{\text{ether}} Zn(R)X + M'X$$

$$ZnX_2 + 2M'R \xrightarrow{\text{ether}} ZnR_2 + 2M'X$$

$$CdCl_2 + 2M'R \xrightarrow{\text{ether}} CdR_2 + 2M'Cl$$

$$HgX_2 + M'R \xrightarrow{\text{ether}} Hg(R)X + M'X$$

$$HgX(R) + M'R \xrightarrow{\text{ether}} HgR_2 + M'X$$

(M′ = Li or MgX, X = halogen; R = alkyl or aryl)

(The last two reactions are of such a general nature that these are sometimes used as a means of characterizing the reactive Grignard or organolithium reagents).

Recently some sterically congested organometallics (some of them exhibiting unusual properties) of zinc, cadmium, or mercury have been synthesized using Grignard or organolithium reagents, e.g.,

$$MX_2 + 2M'R \xrightarrow{\text{ether}} MR_2 + 2M'X$$

(M = Zn, Cd, or Hg; M′ = Li or MgX; X = a halogen, but usually the chloride; R = a highly ramified moiety like $(Me_3Si)_2\,CH$ or $(Me_3Si)_3C$).

$$HgCl_2 + LiAr' \xrightarrow{\text{ether}} HgCl(Ar') + LiCl$$

$$2HgCl(Ar') + Cu + 2Py \longrightarrow HgAr'_2 + CuCl_2(Py)_2 + Hg$$

(Ar′ = 2, 4, 6 – tri–tert–butylphenyl)

3.4.3.2 Organometallics of Main Group III Metals (B, Al Ga, In and Tl)

The organometallic compounds (alkyls, aryls, alkenyls or alkynyls) of Li, Na, K, Al, Si, Pb, Sn, Cd, or Hg have proved to be more popular reagents to introduce an organic group in boron compounds, BX_3(X = halogens, hydrogen, alkoxy, thioloxy, or amido group). Some typical reactions are represented by the following equations:

$$BF_3.OEt_2 + 3LiMe \rightarrow BMe_3 + 4LiF + Et_2O$$

$$BCl_3 + 3LiCH_2SiMe_3 \rightarrow B(CH_2SiMe_3)_3 + 3LiCl$$

$$BCl_3 + 2LiCH(SiMe_3)_2 \rightarrow BCl[CH(SiMe_3)_2]_2$$
$$BF_3 + 3MgBr(R) \rightarrow BR_3 + 3MgBrF$$
$$B(OR)_3 + Mg(C \equiv CH)Br \rightarrow B(C \equiv CH)(OR)_2 + Mg(OR)Br$$
$$2BF_3 + Zn(CH = CH_2)_2 \rightarrow 2B(CH = CH_2)F_2 + 2ZnF_2$$
$$BBr_3 + HgBr(Ph) \rightarrow BBr_2(Ph) + HgBr_2$$
$$2BCl_3 + Al_2Et_6 \rightarrow 2BEt_3 + Al_2Cl_6$$
$$4B_2H_6 + Al_2Me_6 \rightarrow 2BMe_3 + 2Al(BH_4)_3$$
$$4BCl_3 + Sn(CH = CH_2)_4 \rightarrow 4B(CH = CH_2)Cl_2 + SnCl_4$$
$$4BCl_3 + 3PbMe_4 \rightarrow 4BMe_3 + PbClMe_3$$

Synthetic routes to other group III organometallics can be represented by the following typical reactions:

$$AlX_3 + 3Mg(R)X \xrightarrow[-3MgX_2]{Et_2O} AlR_3 \cdot Et_2O \xrightarrow[-Et_2O]{heat} AlR_3$$

$$AlCl_3 + 3LiR \xrightarrow{heptane} AlR_3 + 3LiCl$$

$$MX_3 + 3Mg(R)X \xrightarrow{Et_2O} MR_3 + 3MgX_2$$

(M = Ga or In; specially applicable for R = phenyl, substituted phenyl or other organic group)

$$MX_3 + 3LiR \rightarrow MR_3 + 3LiX$$

or NaR

(R = alkyl; M = Ga or In; X = a halogen)

$$TlCl_3 + 2Mg(R)X \xrightarrow{Et_2O \text{ or } THF} Tl(R)_2Cl + MgX_2 + MgCl_2$$
$$\downarrow + Mg(R)X \text{ (or LiR)}$$
$$TlR_3 + MgXCl \text{ (or LiCl)}$$

The role of the steric demands of the ligands on the course of the reaction of group III metal halides can be demonstrated by the reactions described below. Whereas the reactions of $LiCH_2SiMe_3$ proceed to completion in the cases of chlorides of all the group III metals, i.e., B, Al, Ga and In, the same is able to proceed to the formation of only a bis-product in the case of a more hindered ligand, $Li\{CH(SiMe_3)_2\}$ with the smaller elements of the group, i.e., B and Al:

$$MCl_3 + 3LiCH_2SiMe_3 \xrightarrow{Et_2O} M(CH_2SiMe_3)_3 + 3LiCl$$

(M = B, Al, Ga, In)

$$MCl_3 + 2LiCH(SiMe_3)_2 \xrightarrow{Et_2O} MCl\{CH(SiMe_3)_2\}_2 + 2LiCl$$

(Excess) (M = B or Al)

$$MCl_3 + LiCH(SiMe_3)_2 \xrightarrow{Et_2O \text{ or } Ph\,Me} M\{CH(SiMe_3)_2\}_3 + 3LiCl$$

(M = Ga or In)

It would, however, be of interest to note that on using even a more ramified ligand, i.e., tris (trimethylsilyl) methyllithium, only one group can be introduced around the boron atom:

$$BCl_3 + LiC(SiMe_3)_3 \rightarrow BCl_2\{C(Si\ Me_3)_3\} + LiCl$$

3.4.3.3 Organometallics of Main Group IV and V Elements (Si, Ge, Sn, Pb, As, Sb, and Bi)

The synthetic reactions employed for group IV organometallics are also of the same pattern as those described in the preceding sub-section for elements of group II and III and can be exemplified by

$$SiCl_4 \xrightarrow[-\ MgCl_2]{+\ MgCl(Me)} SiCl_3(Me) \xrightarrow[-\ MgCl_2]{+\ MgCl(Me)} SiCl_2Me_2 \xrightarrow[-\ MgCl_2]{+\ MgCl(Me)} SiClMe_3$$

$$SiMe_4 \xleftarrow[-\ MgCl_2]{+\ MgCl(Me)} SiClMe_3$$

$$SiCl_4 + 4LiCH_2\ SiR_3 \longrightarrow Si(CH_2SiR_3)_4 + 4LiCl$$

$$SiCl_4 + 4RCl + 8Na \longrightarrow SiR_4 + 8\ NaCl$$

Note: Preparations by Wurtz reactions do not require isolation of the intermediate organometallic reagents.

$$GeR_3X + MR'_n \xrightarrow{Et_2O} GeR_3R' + MR'_{n-1}X$$

(unsymmetrical derivatives)

(M = Li, Na, K, Mg, Zn, Hg, or Al; n = valency of the metal).

$$GeCl_4 + 4MR \xrightarrow{Et_2O/PhH} GeR_4 + 4MCl$$

(M = Li or MgX; X = Cl or Br; R = alkyl, C_5H_5 or C_5H_4Me)

$$GeCl_4 + 4MgBr(Ph) \xrightarrow{PhMe} GePh_4 + 4MgBrCl$$

$$GeCl_4 + 4RX + 8Na \longrightarrow GeR_4 + 4NaCl + 4NaX$$

$$SnCl_4 + 4MR \longrightarrow SnR_4 + 4MCl$$

(M = Li, Na, K, MgX, Zn, or Al, R = alkyl, aryl, C_5H_5 or C_5H_4Me)

$$3SnCl_4 + 4AlR_3 + 4NaCl \longrightarrow 3SnR_4 + 4AlCl_3 \cdot NaCl$$

or $4R_3N$ / or $4AlCl_3 \cdot R_3N$

$$SnR_2Cl_2 + 2Mg(R')X \longrightarrow SnR_2R'_2 + 2MgClX$$

(Unsymmetrical derivatives)

Because of the industrial interest in tetraethyl- and tetramethyllead as antiknock additives for gasoline, the synthesis of these tetralkyls has been intensively investigated, using lead (II) salts (usually $PbCl_2$) as starting materials:

$$2PbCl_2 + 4LiR \rightarrow PbR_4 + Pb + 4LiCl$$

$$2PbCl_2 + 4Mg(R)X \rightarrow PbR_4 + Pl + 4MgClX$$

$$6PbX_2 + 4AlEt_3 \rightarrow 3PbE_4 + 3Pb + 4AlX_3$$

The deposition of the lead metal may be avoided by carrying out the reactions in the presence of an alkyl iodide:

$$2PbC_2 + 6MR + 2RI \rightarrow 2PbR_4 + 4MCl + 2MI$$

(M = Li or MgX)

$$PbCl_2 = AlEt_3 + EtI \rightarrow PbEt_4 + AlCl_2I$$

Although the use of lead (IV) salts (such as $Pb(OCOCH_3)_4$ or $K_2[PbCl_6]$) in place of lead (II) chloride, does not appear to offer any special advantage, yet these continue to be sometimes used in the synthesis of a few derivatives:

$$K_2[PbCl_6] + 4LiC \equiv CR \rightarrow Pb\,(C \equiv CR)_4 + 4LiCl + 2KCl$$

$$Pb(O_2CCH_3)_4 + 4Mg(R)Cl \rightarrow PbR_4 + 4Mg(O_2CCH_3)Cl$$

Some preparative routes for the synthesis of important organometallic halide precursors have already been mentioned earlier (Section 3.4) and other methods which utilize either metal-carbon bond cleavage and oxidative addition reactions (*cf.* Section 3.5.7) or redistribution reactions (Section 3.5.11) will be discussed later in this chapter.

The applications of carbanion-halide exchange method for the synthesis of organometallic derivatives of main group V elements can be exemplified by the following reactions:

$$MCl_3 + 3MgI(Me) \rightarrow MMe_3 + 3MgClI$$

$$MCl_3 + 2LiPh \rightarrow MPh_3 + 3LiCl$$

$$4MX_3 + 3LiAlEt_4 \rightarrow 4MEt_3 + 3LiAlX_4$$

(M = As, Sb, or Bi)

$$AsCl_3 + AlR_3 \rightarrow AsR_3 + AlCl_3$$

$$2AsCl_3 + 3ZnMe_2 \rightarrow 2AsMe_3 + 3ZnCl_2$$

$$AsX_3 + 3RX + 6Na \rightarrow AsR_3 + 6NaX$$

$$SbF_3 + 3AlEt_3 \rightarrow SbEt_3 + 3AlEt_2F$$

$$BiCl_3 + 3LiCH(SiMe_3)_2 \xrightarrow{\text{ether}} Bi\{CH(SiMe_3)_2\}_3 + 3LiCl$$

Under controlled conditions and/or with sterically crowded Grignard or lithium reagents, the reaction of MX_3 is generally stopped at partial substitution e.g.,

$$MCl_3 + 2LiCH(SiMe_3)_2 \xrightarrow{\text{either}} MCl\{CH(SiMe_3)_2\}_2 + LiCl$$

(M = As or Sb)

$$SbBr_3 + MgCl(Bu^t) \xrightarrow{-50^\circ C} SbBr_2(Bu^t) + MgBrCl$$

$$SbCl_3 + 3MgCl(Bu^t) \xrightarrow{0^\circ C} SbCl(Bu^t)_2 + 2MgCl_2$$

The organo-arsenic (V), -antimony (V), and-bismuth (V) derivatives can be prepared by the following reactions:

$$MMe_3 + X_2 \rightarrow MX_2Me_3$$

$$MX_2Me_3 + 2LiMe \rightarrow MMe_5 + 2LiX$$

(M = As, Sb, Bi; X = Cl or Br)

(It may be mentioned that amongst these, $SbMe_5$ is the most stable and the bismuth analogue is the least stable).

$$MPh_3 + Cl_2 \rightarrow MCl_2Ph_3$$

$$MCl_2Ph_3 + Mg(Cl)Ph \rightarrow [MPh_4]^+Cl^- + MgCl_2$$

$$[MPh_4]^+Cl^- + LiPh \rightarrow MPh_5 + LiCl$$

(M = As, Sb, or Bi)

Although pentahalides are not suitable precursors for penta-alkyls or-aryls of main Group V elements, yet the vinyl derivative of antimony can be prepared from $SbCl_5$:

$$SbCl_5 + 5LiCH=CHMe \xrightarrow[-60^\circ C]{Et_2O} Sb(CH=CHMe)_5 + 5LiCl$$

3.4.3.4 Organo-Derivatives of Main Group VI Elements

The applications of carbanion-halide exchange reactions for the synthesis of the simpler organo-derivatives of tellurium can be illustrated by the following equations:

$$Te\,X_2 + 2RMgX \xrightarrow{\text{ether}} TeR_2 + 2MgX_2$$

$$TeCl_2 + 2C_6H_5MgBr \xrightarrow{\text{ether}} Te(C_6H_5)_2 + MgCl_2 + MgBr_2$$

$$Te(C_6H_5)_2 \xrightarrow{X_2} (C_6H_5)_2TeX_2 \xrightarrow{+2LiC_6H_5} Te(C_6H_5)_4 + 2LiX$$

$$Te(CH_3)_4 + XeF_2 \xrightarrow[(-Xe)]{} TeF_2(CH_3)_4 \xrightarrow[(-ZnF_2)]{Zn(CH_3)_2} Te(CH_3)_6$$

Diphenylselenium can also be synthesized by the following reaction which might involve a carbanion (in situ):

$$\underset{\text{excess}}{SeCl_2 + C_6H_6} + AlCl_3 \rightarrow Se(C_6H_5)_3^+ Cl^- + H_2 + H[AlCl_4]$$
$$\downarrow 230^\circ$$
$$Se(C_6H_5)_2 + C_6H_5Cl$$

Carbanion-halide reactions accompanied by reduction of selenium/tellurium derivatives from tetra- to di-valent states can be represented by the following equations:

$$SeBr_4 + 3HgR_2 \rightarrow SeR_2 + RBr + 3RHgBr$$
$$TeX_4 + 3HgRX \rightarrow TeR_2 + RX + 3HgX_2$$

Reactions of tellurium tetrachloride with third and fourth group organometallics appear to proceed without reduction of tellurium (IV):

$$TeCl_4 + MR_2Cl \rightarrow TeRCl_3 + MRCl_2 (M = Ga, In, Tl)$$
$$TeCl_4 + MR_4 \rightarrow TeRCl_3 + MR_3Cl (M = Ge, Sn)$$

In view of the electronegative nature of selenium and tellurium, the reverse type of reactions of alkali selenides and tellurides with alkyl (aryl) halides can also be used for the synthesis of organo-selenium and-tellurium derivatives. For example, dialkyls (aryls) or selenium (tellurium) can be prepared by the interaction of sodium (potassium) selenide (telluride) with alkyl halides (or sulphonates):

$$2RX + Na_2Se \rightarrow SeR_2 + 2NaX$$
$$2NaOS(O_2)OCH_3 + Na_2Se \rightarrow Se(CH_3)_2 + 2Na_2SO_4$$
$$Ba(OS \cdot O_2 \cdot OCH_3)_2 + K_2Te \rightarrow Te(CH_3)_2 + K_2SO_4 + BaSO_4$$

Aryl derivatives of tellurium can also be readily prepared by the following types of reactions:

$$C_6H_5X + TeCl_4 \rightarrow TeCl_3(C_6H_4X) + HCl$$
$$2C_6H_5X + TeCl_4 \rightarrow TeCl_2(C_6H_4X)_2 + 2HCl$$
$$(X = OH, OCH_3, NO_2, NH_2)$$

3.4.4 Metal-Hydrogen Exchange (Metallation) Reactions

These metallation reactions represented by the following general equation:

$$RH + MR' \underset{}{\overset{\text{ether}}{\rightleftharpoons}} MR + R'H$$

are reversible but in many cases the equilibrium lies far to one side or the reaction can be pushed to one side to make the metallation effectively complete.

The metallation reactions are believed to occur through nucleophilic attack of the hydrocarbon moiety of the organometal reagent on a hydrogen atom of the organic substrate. The natural expectation would, therefore, be that the hydrogen atom undergoing metallation must be quite acidic, and the organic moiety of the metal-containing reagent should have pronounced carbanionic character. Some examples of metallation reactions are represented as follows:

$$(C_6H_5)_3CH + LiC_2H_5 \rightarrow LiC(C_6H_5)_3 + C_2H_6$$

$$C_6H_6 + NaBu^i \rightarrow NaC_6H_5 + Bu^iH$$

$$C_6H_5CH_3 + KC_6H_5 \rightarrow KCH_2C_6H_5 + C_6H_6$$

It has been observed that alkali amides are potentially more effective metallating reagents than the thermodynamically more basic alkyl- and aryl-alkali metals:

$$RH + MNR'_2 \rightleftharpoons MR + HNR'_2$$

The increased kinetic basicity of the amido group is a consequence of the use of nitrogen lone pair in the metallation reaction involving a four-centred transition state:

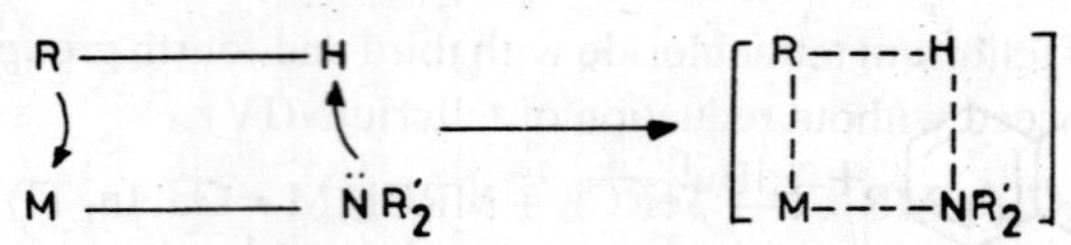

In view of the strong polarization ($M^{\delta+} - C^{\delta-}$) of the alkali metal carbon bond, the carbon atom bound to an alkali metal should be rather a strong base, capable of abstracting protons from more acidic hydrocarbons; consequently many alkali metal organometallics may be used as powerful metallating agents. However, in practice only a few commercially available reagents such as $LiBu^n$, $LiBu^t$, and LiMe find common usage for their synthetic utility. As the organic groups like Bu^n, Bu^t, and Me have pKa values of nearly 40, these common alkylithium reagents should be capable of metallating hydrocarbons with pKa < 35 (*cf.* Table 2.2). For metallating organic substrates with less acidic proton, some modification in the lithium reagent is essential, which could enhance the effective basicity of the alkyllithium reagents. Such enhancement of carbanionic character can be achieved by the presence of electron-donating (cation-solvating) solvents or ligands such as THF, TMEDA, PMDT, HMPT, DABCO, etc., leading to metallation of less acidic organic substrates, which could not be metallated by the use of simple alkyllithium reagents:

$$(Me_3Si)_2CH_2 + LiBu^n + PMDT \rightarrow LiCH(SiMe_2)_2 \cdot PMDT + Bu^nH$$

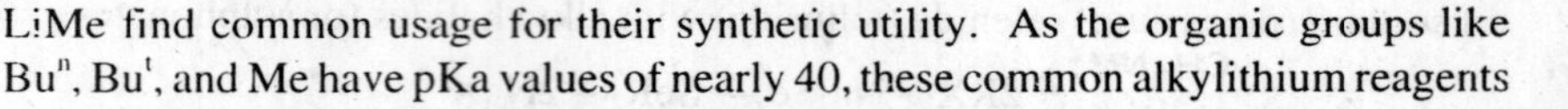

$$2(Me_3Si)_3\,CH + 2LiMe \xrightarrow[\text{reflux}]{\text{THF}} [Li(THF)_4]^+ [Li\{C(SiMe_3)_2\}]^- + 2MeH$$

$$(C_6H_5)_3CH + LiBu^n + TMEDA \xrightarrow{n\text{–}C_6H_{14}} Li(TMEDA)^+C^-(C_6H_5)_3 + n\text{–}BuH$$

The polarization $M^{\delta+}$ — $C^{\delta-}$ should increase from lithium to sodium to potassium and hence, the organoalkali reagents of the latter alkali metals should be more effective metallating reagents. As the alkyls of heavier alkali metals are not generally available commercially, they can be produced in situ by the addition of $NaOBu^t/KOBu^t$ along with alkyllithium reagent, e.g.,

$$LiR + KOBu^t \text{ (or } NaOBu^t) \rightarrow LiOBu^t + KR \text{ (or NaR)}$$

In spite of such a varied list of metallating agents, lithium organometallics are still finding a wide application. These reactions are generally inductively controlled. For example, lithiation of phenoxathin with $LiBu^n$ occurs ortho to the oxygen rather than to the sulphur:

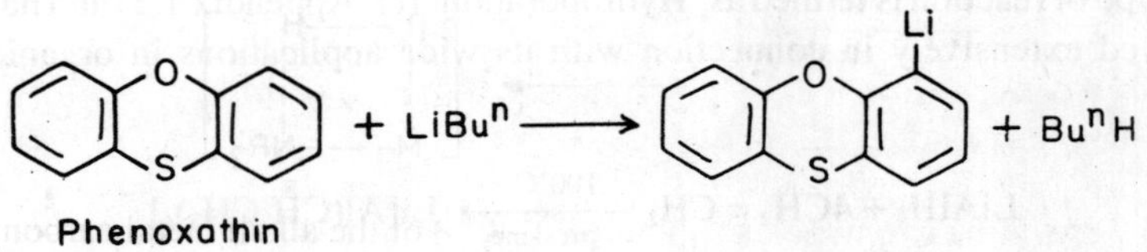

When in a disubstituted benzene ring both the substituents are ortho directing, the metallation is controlled by the group which has stronger directing properties:

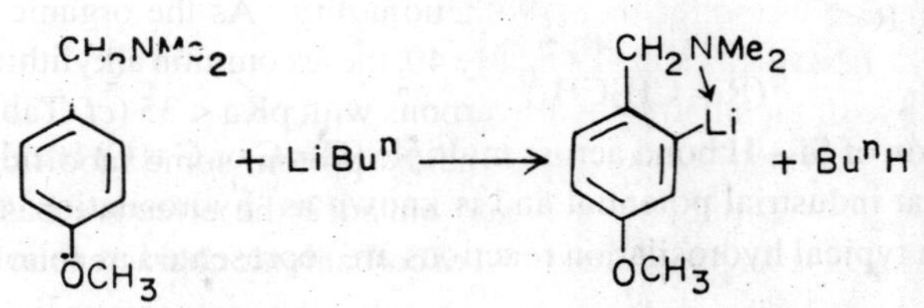

The effect of coordinating solvents and ligands on enhancing the effectiveness of organolithium reagents has already been discussed above.

3.4.5 Metal-Hydride Additions to Alkenes

This synthetic route for organometallic derivatives involves insertion of an alkene in a metal-hydrogen bond. Generally speaking the concept of insertion reactions is of wide applicability in chemistry and can be defined as a reaction in which any atom or group of atoms (often unsaturated) is inserted between two atoms bound together:

$$L_{n-1}MX + YZ \rightarrow L_{n-1}M(YZ) - X$$

$$\text{(e.g., } R_3SnNR_2 + CO_2 \rightarrow R_3SnOC(O)NR_2)$$

$$L_{n-1}ML + YZ \rightarrow L_{n-1}M(YZ) - L \xrightarrow{\text{so on}} M\{(YZ) - L\}_n$$

$$\text{(e.g., } Ti(NR_2)_4 + 4CS_2 \rightarrow Ti\{SC(S)NR_2\}_4)$$

The insertion of an alkene in a metal-hydrogen bond results in the formation of a metal-carbon bond. This is represented as follows:

$$\gt M-H + \gt C=C\lt \rightarrow \gt M-\overset{|}{\underset{|}{C}}-\overset{|}{\underset{|}{C}}-H$$

Such reactions are of interest in connection with the initiating reaction in the polymerization of alkenes by Ziegler catalysts. A few typical insertion reactions of synthetic importance in organometallic chemistry are represented by the following equations:

$$B_2H_6 + 6RCH = CH_2 \rightarrow 2B(CH_2CH_2R)_3$$

(This type of reaction is termed as 'Hydroboration' (*cf.* Appendix 1.3) and has been employed extensively in connection with its wide applications in organic synthesis):

$$LiAlH_4 + 4CH_2 = CH_2 \xrightarrow[\text{pressure}]{100^\circ C} Li[Al(CH_2CH_3)_4]$$

$$(AlH_3)_n + 3n(CH_2 = CHR) \longrightarrow nAl(CH_2CH_2R)_3$$

$$R_2Al-H + H_2C = CH_2 \longrightarrow Al(CH_2CH_3)_3$$

$$(R = CH_2CH_3)$$

The addition of Si—H bond across multiple ($C = C$ or $C \equiv C$) bonds is another reaction of great industrial potential and is known as 'hydrosilation' (*cf.* Section 6.4.2.2). Some typical hydrosilation reactions are represented as follows:

$$Ph_3\,Si-H + CH_2 = CHR \xrightarrow{Bz_2O_2} Si(CH_2CH_2R)Ph_3$$

$$X_3Si{-}H + RC \equiv CH \rightarrow X_3SiCH = CHR \rightarrow X_3SiCHRCH_2SiX_3$$

$$(X = Cl, \text{Alkyl or aryl})$$

$$R_3Si{-}H \xrightarrow[Co_2(CO)_8,\ 0^\circ C,\ 1\ atm]{CH_2 = CH_2} SiR_3(CH_2CH_3)$$

$$R_3Si{-}H \xrightarrow[CH_2 = CHCH_2NH_2]{H_2PtCl_6} SiR_3(CH_2)_3NH_2$$

Insertion reactions of $C = C$ double bonds into Sn—H and Ge—H bonds are illustrated by the following equations respectively:

$$Ph_3Sn{-}H + C_6H_5CH = CH_2 \rightarrow Sn(CH_2CH_2C_6H_5)Ph_3$$

$$Bu^n_3\,Ge{-}H + CH_2 = CHCN \rightarrow GeBu^n_3\,(CH_2CH_2CN)$$

3.4.6 Methylene-Insertion Reactions (Methylenations)

Methylenation by diazomethane is yet another route to produce metal-carbon bonds, e.g.,

$$HgCl_2 + CH_2N_2 \rightarrow HgCl(CH_2Cl) + N_2 \xrightarrow{CH_2N_2} Hg(CH_2Cl)_2 + N_2$$

$$SiCl_4 + CH_2N_2 \xrightarrow[-50^\circ C]{\text{ether}} SiCl_3(CH_2Cl) + N_2$$

$$GeCl_4 + CH_2N_2 \xrightarrow{\text{Cu (catalyst)}} GeCl_3(CH_2Cl) + N_2$$

$$SnCl_2Me_2 + CH_2N_2 \xrightarrow{\text{ether}} SnCl(CH_2Cl)Me_2 + N_2$$

Replacement of all the metal-chlorine bonds is often difficult in the above reactions, which, therefore, generally result finally in mixed derivatives.

3.4.7 Reactions of Aryl Diazonium Salts with Metal Halides or Oxides in Aqueous Medium

This route is of great synthetic utility for a wide range of organometallics of heavier metals and metalloids. The accessibility of the reaction depends on two conditions: (a) the use of aqueous media possibly due to the solubility of the reactants requires the products to be hydrolytically stable and (b) the concerned central element must be prone to electrophilic attack by the diazonium ion, ArN_2^+. Probably the most successful applications are in arsenic (Bart reaction) and antimony (Schmidt reaction) chemistry:

$$As(OH)_3 + PhN_2X \xrightarrow[\text{aq. acetone}]{\text{alkaline}} AsO(OH)_2(Ph) + N_2 + HX$$

$$Sb(OH)_3 + PhN_2X \xrightarrow[\text{aq. acetone}]{\text{alkaline}} SbO(OH)_2(Ph) + N_2 + HX$$

In these reactions, along with the formation of phenyl derivatives, arsenic and antimony, are oxidized from trivalent to pentavalent state.

3.4.8 Some Other Methods for Metal-Carbon Bond Formation

3.4.8.1 Elimination Reactions

These reactions in most cases are the reverse of insertion reactions. They proceed with the elimination of CO_2, SO_2, SO_3, etc., from their corresponding carboxylate, sulphinate, sulphonate derivatives respectively. For example, thermal (or photolytic or radical initiated) decomposition of mercury carboxylates, sulphinates and sulphonates can lead to mercury-carbon bonded species:

$$—HgO_2CR \xrightarrow[\text{peroxide}]{\text{heat, U.V. or}} —Hg - R + CO_2$$

$$—HgO_2SR \xrightarrow{\text{heat}} —Hg - R + SO_2$$

$$—HgO_3SR \xrightarrow{\text{heat}} —Hg - R + SO_3$$

Some typical examples of radical initiated reactions are represented as follows:

$$Hg(O_2C_{11}H_{21})_2 - \text{cyclo} \xrightarrow[\text{(ii) KBr}]{\text{(i) (PhCO}_2)_2\text{, PhH, 80°C}} Hg(Br)C_{11}H_{21} - \text{cyclo}$$

A noteworthy feature of such reactions occurring *via* thermal decarboxylation with reasonable facility, is the requirement of the presence of electron withdrawing substituents, e.g.:

$$M(O_2CC_6F_5)_2 \xrightarrow{-200°C} M(C_6F_5)_2 + 2CO_2$$

(M = Zn or Cd)

$$Hg(O_2CMe_2COCMe)_2 \xrightarrow[\text{(ii) KCl}]{\text{(i) boiling water}} MeCOCMe_2HgCl$$

$$Hg(O_2C_6Cl_5)_2 \xrightarrow{\text{boiling pyridine}} Hg(C_6Cl_5)_2 + 2CO_2$$

$$Hg(O_2CCF_3)_2 \xrightarrow{300°C} Hg(O_2CCF_3)\ CF_3 + CO_2$$

$$Hg(O_2CCF_3)_2 \xrightarrow[\text{K}_2\text{CO}_3]{200°C} Hg(O_2CCF_3)\ CF_3$$

The last reaction occurs at much lower temperatures if carried out photolytically:

$$Hg(O_2CCF_3)_2 \xrightarrow[-196°C]{h\nu} Hg\ (CF_3)_2 + 2CO_2$$

$$Hg(O_3SHC_6Cl_4)_2 \xrightarrow{200°C} Hg\ (C_6Cl_4H)_2 + 2SO_3$$

(The presence of pyridine in this reaction facilities SO_3 elimination at lower temperatures and also results in higher yields).

The thermal decomposition reactions in the synthesis of tin, lead, arsenic and antimony organometallics are of limited utility as compared to those for mercury and are essentially confined to the types of reactions described as follows:

$$MR_3\ (O_2CR') \rightarrow MR_3\,R' + CO_2$$

$$(M = Sn \text{ or } Pb,\ R' = CH_2CN)$$

$$SnBu_3(O_2C-C \equiv CPh) \xrightarrow{60°C} SnBu_3\ (C \equiv CPh) + CO_2$$

$$PbPh_3\ (O_2C-C \equiv CPh) \xrightarrow{153°C} PbPh_3\ (C \equiv CPh) + CO_2$$

$$AsMe_2\ (O_2CCF_3) \xrightarrow{\text{heat}} AsMe_2\ (CF_3) + CO_2$$

$$SbR_4\ (O_2SAr) \xrightarrow{\text{heat}} SbR_4\ (Ar) + SO_2$$

(R = alkyl, Ar = aryl)

3.4.8.2 *Metallation of Unsaturated Hydrocarbons by Heavy Metal Salts (Especially Mercuration and Thallation of Aromatic Compounds)*

Benzene and toluene are easily mercurated in ethanol (methanol) to afford mono-substituted products:

$$C_6H_6 + Hg\,(O_2CCH_3)_2 \xrightarrow{\text{ethanol}} Hg\,(O_2CCH_3)\,C_6H_5 + CH_3COOH$$

The mercuration reactions are more facile with aromatic amines, phenols, and ethers than those with unsubstituted benzene, while halogeno- and nitro-substituted benzenes react very sluggishly. The facility of mercuration of thiophene has been utilized to remove thiophene from benzene, since the dimercurated product is insoluble:

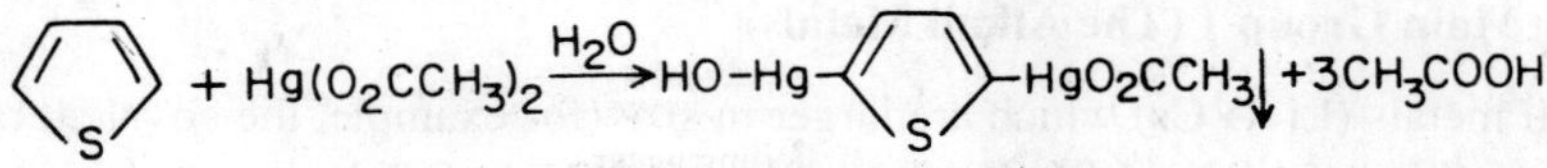

Similarly benzene (and thiophene) can be thallated (but at higher temperatures) in the absence of solvent:

$$C_6H_6 + Tl\,(O_2CCH_3)_3 \xrightarrow{100°C} Tl\,(O_2CCH_3)_2C_6H_5 + CH_3COOH$$

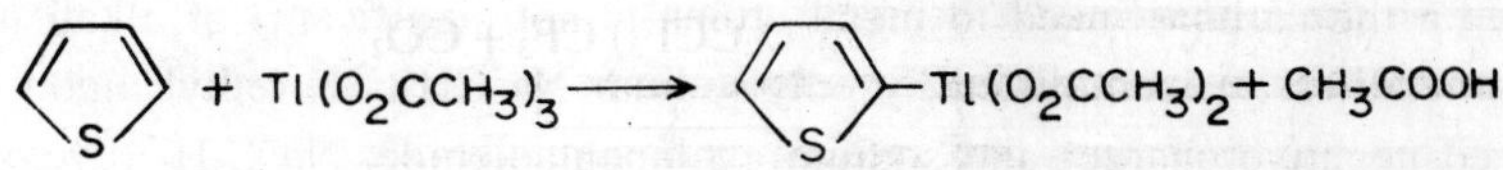

The thallation reactions essentially lead to regiospecific synthesis, e.g.:

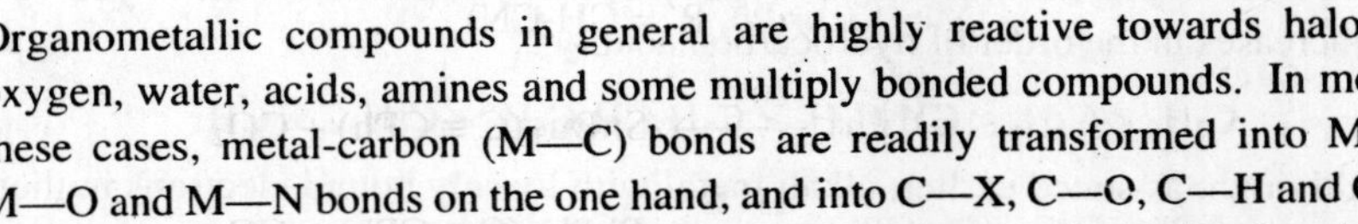

3.5 GROUP TRENDS (INCLUDING CHEMISTRY OF A FEW SELECTED METALS) AND SOME TYPICAL REACTIONS

Organometallic compounds in general are highly reactive towards halogens, oxygen, water, acids, amines and some multiply bonded compounds. In most of these cases, metal-carbon (M—C) bonds are readily transformed into M—X, M—O and M—N bonds on the one hand, and into C—X, C—O, C—H and C—C bonds on the other, which reflects the relative weakness of metal-carbon bonds. Generally metal alkyls and aryls, MR_n, are thermodynamically (but may not be kinetically) unstable. However, their reactivity varies to a great extent depending upon the nature of the metal; e.g., organometallics of the alkali and alkaline-earth metals are extremely reactive, whereas main group IV organometallics (e.g. $SiMe_4$)

are relatively inert to attach by air or water, and some group III (e.g. $InMe_3$) and group V (e.g. $SbMe_3$) inflame in air.

The factors which can obviously affect the reactivity of organometallics may be associated with (i) nature of the metal, (ii) the steric aspects, and (iii) the presence or absence of a vacant orbital or a lone pair of electrons on the metal. The comparative effect of these factors on the general properties of organometal compounds is presented in the following sub-sections groupwise along with slightly more detailed chemistry in a schematic form of at least one element from each group. These discussions are followed by a few examples of typical reactions on main group organometallic compounds.

3.5.1 Main Group 1 (The Alkali Metals)

Alkali metals (Li → Cs) which are larger in size (for example, the covalent radii are 1.35 (Li), 1.54 (Na), 1.96 (K), 2.11 (Rb) and 2.24 A (Cs) depict low ionization energies (ranging from 5.4 to 3.9 eV) and highly electropositive character (e.g. the electronegativity values of these metals are in the range 1.0–0.8). The resulting univalent cations, M^+, are spherical with low polarizability: These, therefore, generate essentially ionic chemistry (especially when delocalization of negative charge on the carbon linked to metal atoms is not significant) of alkali metal organometallics as exemplified by ethylsodium $Na^+CH_2CH_3^-$ (ethyl anion with localized negative charge) and sodium cyclopentadienide, $Na^+C_5H_5^-$ (cyclopentadienide anion with delocalization of negative charge).

The degree of ionic character increases with increasing atomic number of alkali metal. For example, percentage ionic character (given in parenthesis) of metal-carbon bond of metal alkyls increases in the order:

$$\text{Li—C}(\sim 43) < \text{Na—C}(\sim 47) < \text{K—C}(\sim 52) < \text{Rb—C}\ (\sim 52) < \text{Cs—C}(\sim 57).$$

As a consequence of this increase in ionic character, their reactivity (as nucleophile or carbanionic reagent) and solubility in organic solvents follows the same pattern: LiR > NaR > KR ≈ RbR > CsR. In fact, due to high reactivity of organometallics of sodium and its heavier congeners (K, Rb or Cs), their organometallic chemistry has been explored to a lesser degree compared to the current extensive organometallic chemistry of lithium. However, their pronounce carbanionic (R^-) character of the attached alkyl group confers upon them strong metallating reactivity. For example, the reactivity of organosodium compound towards hydrocarbons increases in the order of hydrocarbon acidity:

$$C_2H_6 < C_6H_6 < CH_3C_6H_5 < C_6H_5CH_2C_6H_5 < C_6H_5C \equiv CH$$

Although, as stated earlier, alkali metal with loosely bound electrons in their valence shell form organometallic compounds with essentially an ionic structure, M^+R^-, yet the small size (0.60 Å) of the lithium cation results in its greater polarizing power and frequently confers special properties on its compounds. For this reason, the element is sometimes considered "anomalous" and shows similarities with magnesium (an element diagonally situated in the periodic table). For example, organolithium derivatives are appreciably covalent and soluble in hydrocarbons or non-polar solvents, and are often volatile. By contrast, organometallic compounds of heavier alkali metals are predominantly ionic, non-volatile, insoluble in non-polar solvents, and are extremely reactive. The electronic configuration of the valence shell or σ-bonded alkali metal (M) alkyl (or aryls) can be represented as follows:

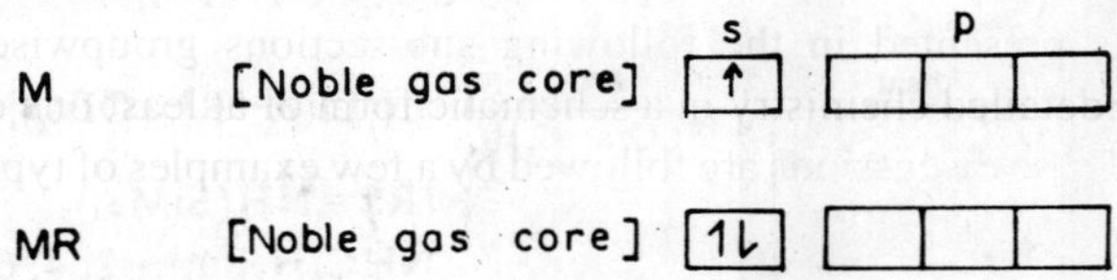

The oxygen and moisture sensitivity, the Lewis acidic character and associated nature of organoalkali compounds (especially organolithium compounds) may be explained to a large extent on the atomic size and by the possibility of the acceptance of electron pairs into their vacant p-shells from donor species. For example, in dilute diethyl ether solutions two molecules of ethyllithium are associated with two molecules of the solvent. A three-centre molecular orbital containing two electrons extends over the Li—C_2H_5—Li bridge and each lithium now has a share in two electrons from the coordinated solvent molecule:

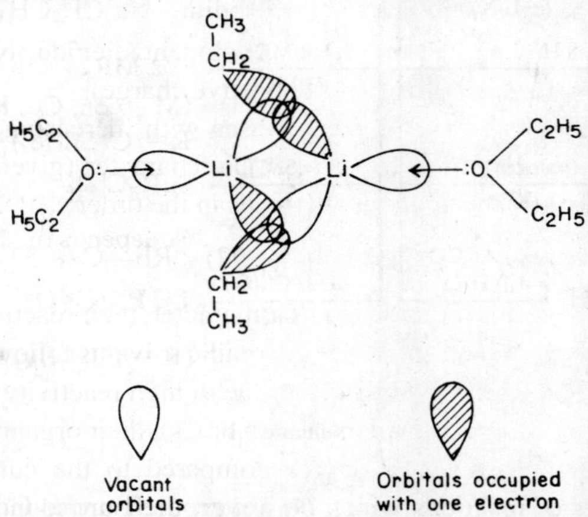

In addition to the above, the valence orbitals participate in *sp*- and *sp*^{3}- hybridizations, giving rise to linear and tetrahedral coordination geometries (Table 3.4).

Table 3.4 Some Typical Coordination Geometries of Lithium Organometallics

Type of hybridization	Coordination Number	Geometry	Examples
sp	2	Linear	$[Li\{C(SiMe_3)_3\}_2]^-$ $[Li\{C(SiMe_2Ph)_3\}(thf)]$
sp^2	3	Trigonal planar	$[\{LiC_2H_5(OEt_2)\}_2]$, $[LiCPh_3{\cdot}TMEDA)]$,
sp^3	4	Tetrahedral	$[\{LiPh(TMEDA)\}_2]$,
sp^3	7-coordinate environment of each lithium atom	Li_4C_4 cluster	$[LiCH(SiMe_3)_2\ (pmdeta)]$ Li_4Me_4

The higher ionicity of Li—C bond (~43%) compared to that of Mg—C (~34% in Grignard reagents) is responsible for greater reactivity and versatility of lithium reagents than Grignard reagents by a factor of about 100. A summary of some typical reactions of organolithium compounds, LiR, is given in Scheme 3.1.

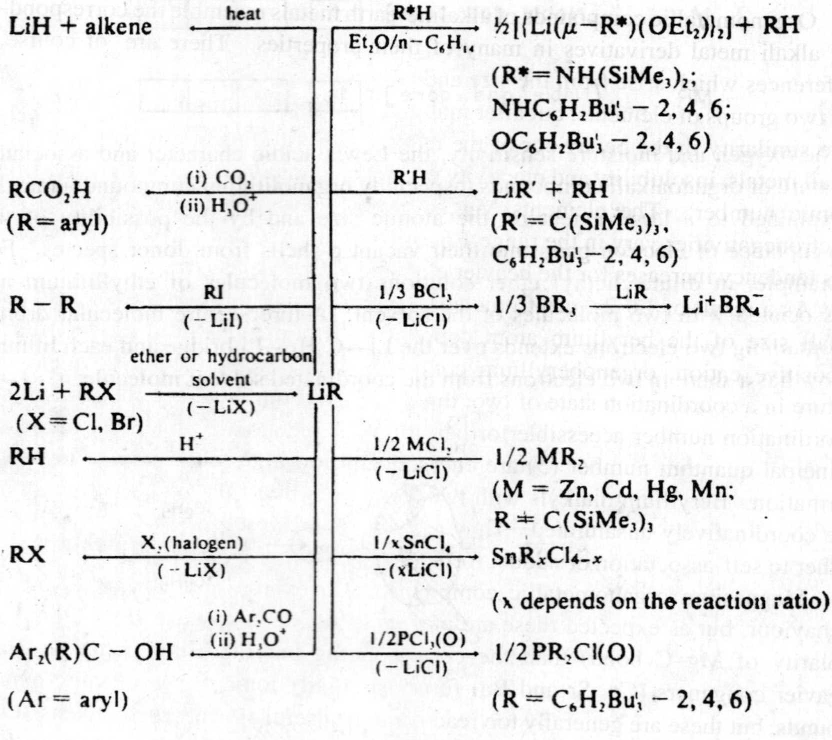

Scheme 3.1

Indeed, a number of reactions which do not proceed under Grignard conditions may be effected with the corresponding organolithium compounds, which exhibit varied types of reactivity depending upon the nature of the solvents or/and coordinating ligands (*cf.* Section 3.4.4).

The chemistry of organosodium compounds and its heavier congeners parallels that of the organolithium compounds to a very large extent. The differences arise from the insolubility of these compounds, the greater metal-carbon bond polarity, an increasing cationic character in these aggregates or ion pairs, and lower values of M—C (M = Na, K, Rb, or Cs) bond dissociation energy. The last three properties together enhance the tendency of metal-hydrogen (or halogen) exchange and other nucleophilic reactions. Sometimes such higher reactivity can lead to complications such as metallation instead of addition across and alkene and Wurtz coupling instead of metal alkyl bond formation.

3.5.2 Group 2 Metals

Be — Mg < Ca — Sr — Ba ; Zn — Cd — Hg

Compounds containing M—C bonds are well established for Be, Mg, Zn, Cd and Hg. By contrast, only a few organometallic compounds of Ca, Sr, or Ba have been isolated.

Organometallic compounds of alkaline-earth metals resemble the corresponding alkali metal derivatives in many of their properties. There are, of course, differences which arise from the size and varying number of valence electrons in the two groups of elements, but after making allowance for these, there is remarkable similarity in the behaviour of the two families of compounds. As with the alkali metals, insolubility and reactivity within the group increases with increasing atomic number. The elements Mg → Ra are strongly electropositive (i.e., electronegativities vary in the range 1.6–0.9) and form ionic metal-carbon bonds; this tendency increases for the heavier elements of the group.

As a result of $1s^2\ 2s^2$ electronic configuration, high ionization energies, and small size of the beryllium atom coupled with high polarizing power of its dipositive cation, organoberyllium compounds are predominantly covalent in nature in a coordination state of two, three or four. Four would be the maximum coordination number accessible for beryllium, as the atomic orbitals of the next principal quantum number (3) are energetically too high to participate in bond formation. Beryllium dialkyls with two vacant *p* orbitals on the beryllium atom are coordinatively unsaturated. They are, therefore, strong Lewis acids, leading either to self-association or adduct formation with donor ligands.

Magnesium organometallic compounds also exhibit appreciable covalent, behaviour, but as expected these are more reactive (as a consequence of greater polarity of Mg–C bond) than the corresponding beryllium compounds. The heavier congeners (Ca, Sr and Ba) form essentially ionic organometallic compounds, but these are generally too reactive to be useful as synthetic reagents. On the other hand, organomagnesium compounds (e.g., Grignard reagents) with moderate reactivity find wide applications in organic and organometallic transformations, which are represented schematically in Scheme 3.2.

The dialkyls of Zn, Cd, and Hg are all highly toxic, volatile liquids and are extremely air- and moisture-sensitive, except for mercury alkyls (HgR_2), which are stable towards air and water. Their non-reactivity with oxygen (of the air) and water arises from the comparative weakness of Hg—O bond compared to the strength of the Hg—C bond. In fact, the latter is weak, being generally of the order of only 121 kJ mol^{-1} (compared to 176 and 138 kJ mol^{-1} respectively for Zn—C and Cd—C bonds). This weakness of the Hg–C bonds is reflected in their thermal and photochemical instability, requiring mercury organometallics to be stored (in some cases) at lower temperature in the dark. Indeed, the weakness of Hg—C bonds renders them to be useful as synthetic reagents, as the mercury in these, can be replaced with facility by other metals, the bonding energy of which with carbon is comparatively larger (*cf.* Section 3.3.1).

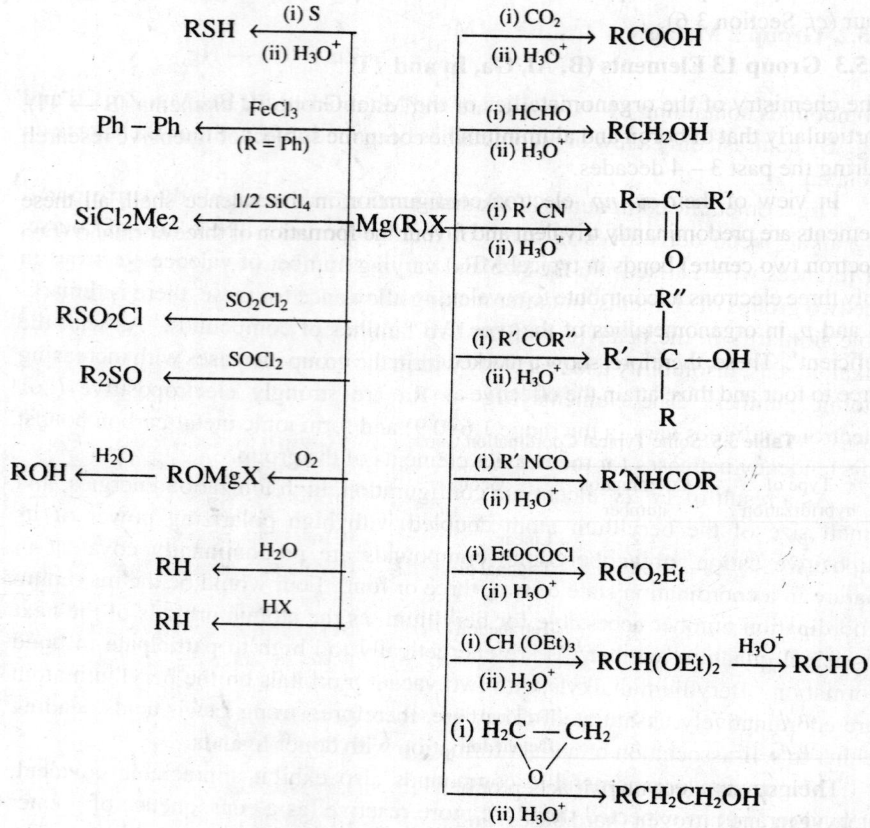

Scheme 3.2 Some Typical Reactions of Grignard Reagents

In general chemical reactivity of Group II organometallics, including both 'the sub-groups (A and B)' varies according to the variation in the electronegativity of the elements:

Ba	Sr	Ca	Mg	Be	Zn	Cd	Hg
(most electropositive)					(least electropositive)		

← Electropositive character increases

← reactivity of M–C bond increases

The elements of group II use *ns* and *np* orbitals only in bonding. The $(n-1)d$ orbitals in Zn, Cd, and Hg are completely filled and do not participate in chemical bonding. The group II elements with $ns^2\,np^0$ electronic configuration in the valence shell can adopt varied types of hybridizations (such as sp, sp^2 and sp^3), giving rise to linear, trigonal-planar, and tetrahedral coordination geometries (see Table 3.5). In contrast to Be, the accessibility of suitable orbitals for Mg, Zn, Cd and Hg in their dialkyl compounds permits them to exhibit coordination numbers higher than four (*cf.* Section 3.6).

3.5.3 Group 13 Elements (B, Al, Ga, In and Tl)

The chemistry of the organometallics of the Main Group III elements (B → Tl), particularly that of boron and aluminium has been the subject of intensive research during the past 3 – 4 decades.

In view of their $ns^2 np^1$ electron configuration in the valence shell, all these elements are predominantly trivalent and favour the formation of three normal σ (two electron two centre) bonds in typical MR_3 compounds. However, the availability of only three electrons to contribute to covalent bonding involving the four orbitals' s, p_x, p_y and p_z in organometallics of the type $(MR_3)_2$ renders their compounds 'electron deficient'. These, therefore, show a marked tendency to increase their covalency from three to four and thus, attain the effective atomic number of the next noble gas.

Table 3.5 Some Typical Coordination Geometries of Group II Organometallics

Type of hybridization	Coordination number	Geometry	Examples
sp	2	Linear	$BeBu^t$, $ZnMe_2$, $HgMe_2$
sp^2	3	Trigonal planar	$[\{Pr^iBe(\mu — Pr^i)\}_2]$, $BeMe_2 \cdot NMe_3$, $BePh_3^-$, $[\{MeZn(\mu — NPh_2)\}_2]$
sp^3	4	Tetrahedral	$BeMe_2{\cdot}2NMe_3$, $BeMe_2{\cdot}TMEDA$; $BeMe_4^{2-}$, $[\{EtMg(\mu — Br)(OEt_2)\}_2]$ $ZnMe_2{\cdot}DME$
sp^3d	5	Trigonal bipyramidal	$Mg(Br)Et{\cdot}3THF$
sp^3d^2	6	Octahedral	$[\{Mg_2Cl_3(Et)(THF)_3\}_2]$

Their strong electron pair acceptor behaviour (Lewis acidity) and high affinity for oxygen and nitrogen (hard bases) and even for carbanions lead to formation of a large number of compounds (including adducts) in different geometries, as illustrated in Table 3.6 :

Table 3.6 Some Typical Coordination Geometries of Main Group III Organometallics

Type of hybridization	Coordination number	Geometry	Examples
sp^2	3	Trigonal planar	BMe_3; $Al(CH_2SiMe_3)_3$; $[Ga\{CH(SiMe_3)_2\}_3]$; $[In\{CH(SiMe_3)_2\}_3]$
sp^3	4	Tetrahedral (neutral derivatives)	$BMe_3 \cdot OEt$; $AlMe_3 \cdot OEt_2$; $Me_2Al(\mu - Me)_2AlMe_2$; $Ga(CH_2SiMe_3)_3 \cdot NMe_3$; $In(CH_2SiMe_3)_3 \cdot THF$; $TlMe_3 \cdot PMe_3$
sp^3	4	Tetrahedral (anionic) (derivatives)	$[BPh_4]^-$; $[AlMe_4]^-$; $[GaMe_4]^-$; $[InMe_4]^-$; $[TlMe_4]^-$
sp^3	4	Tetrahedral (cationic derivatives)	$[BPh_2(bipy)]+$; $[AlEt_2\ (o\text{–}phen)]^+$
sp	2	Linear	$[TlMe_2]^+X^-$ (X = Cl, I)
sp^2d	5	Trigonal bibyramid	$AlMe_3(bipy)$

The general reactivity pattern and Lewis acid strength of main group III trialkyls vary in the sequence; $BR_3 < AlR_3 < GaR_3 > InR_3 > TlR_3$ and this variation can be easily understood, in terms of the electronegativity of the elements (except in the pair $BR_3 < AlR_3$, in which steric factors play a predominant role). As expected triorganoaluminiums achieve four-coordination in the presence of donor species such as an amine, ether, phosphine, sulphide, selenide, telluride, and a carbanion ($:R^-$), with the donor ability towards aluminium decreasing in the order :

$$NMe_3 > PMe_3 > OMe_2 > SMe_2 > SeMe_2 > TeMe_2$$

The formation of $[MR_2]^+$ cations, isoelectronic with main group II, can be expected for main group III elements. Such cations are particularly stable in the case of Tl (R_2Tl^+ is isoelectronic with HgR_2). The R_2B^+ cations, can be stabilized by bidentate ligands (such as bipy), generating a tetrahedral species, $[R_2B\ (bipy)]^+$. Similarly ortho-phenanthroline can stabilize R_2Al^+ cations, otherwise incapable of free existence. The tendency to form R_2M^+ cations and their stability increase from B → Tl. Although aluminium in purely inorganic compounds can utilize 3d-orbital in achieving higher coordination numbers (e.g. AlH_3, biby, $[AlF_6]^3$, $Al(acac)_3$), yet organometallic compounds of aluminium in coordination numbers higher than four are extremely rare (two such examples of pentacoordinated aluminium are known in compounds of the following types):

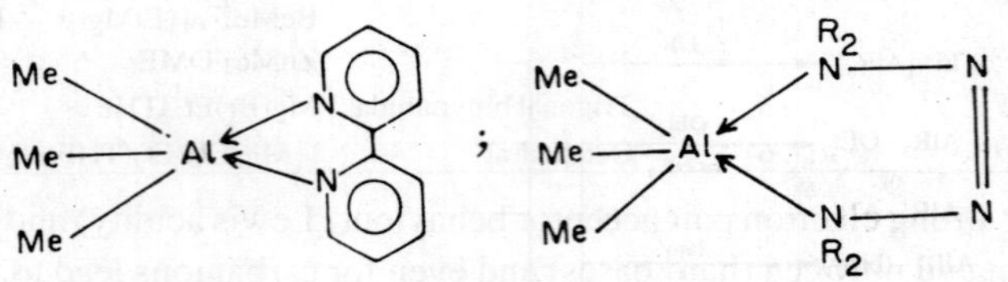

The element-carbon bond distance increases from boron to thallium and consequently element-carbon bond energies in these trialkyls fall with increasing atomic numbers (*cf.* Section 3.3.1). The same sequence, B > Al > Ga > In > Tl is followed in the thermal stability of the trialkyls.

The electron deficient character of these trialkyls could be expected to lead to their dimerization, but the reasons for this behaviour being observed only in the case of some aluminium trialkyls have been discussed earlier (*cf.* Section 2.5.3).

The electron deficient characteristics of these alkyls (particularly those of boron and aluminium) and their sensitivity to oxygen (most are pyrophoric in air) and oxidizing agents as well as to compounds with active hydrogen (such as water, alcohols, amines, and hydrogen halides) have been exploited extensively in 'hydroboration' and other synthetic reactions, as illustrated in schemes 3.3 and 3.4.

Organoaluminium compounds find unique applications in organic as well as organometallic synthesis and coordination catalysts for alkene polymerization and other reactions. The properties of organoaluminium compounds that make them a

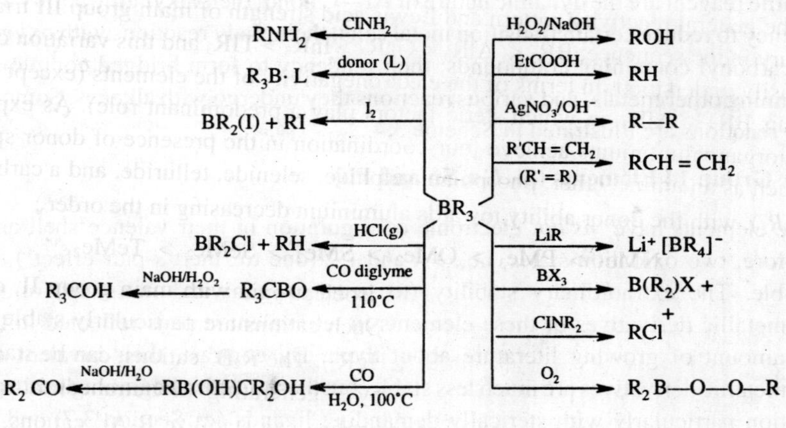

Scheme 3.3 Some Reactions of Triorganoboranes

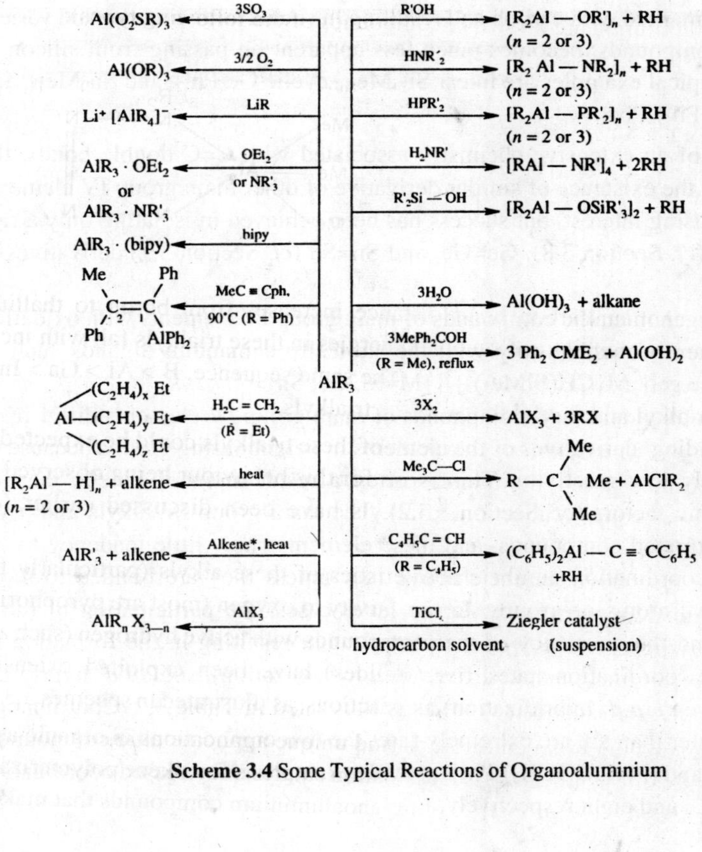

Scheme 3.4 Some Typical Reactions of Organoaluminium

versatile reagent are the dynamic nature of Al—C bond; their alkylating ability and tendency to reduce certain transition metal salts; their ready reactions with oxygen and carbonyl containing compounds; their tendency to form bridged complexes containing other metals; and various reactions they undergo with alkenes. Some of these reactions are illustrated in Scheme 3.4.

3.5.4 Group 14 Elements (Si, Ge, Sn and Pb)

These elements have ns^2np^2 electronic configuration in their valence shell and, therefore, two oxidation states, i.e., + 4 and + 2 (due to 'inert *s*-pair effect') are possible. The extraordinary stability (to heat, oxygen, and moisture) of organometallic derivatives of these elements in tetravalent state is reflected in the vast amount of growing literature about them. By contrast, their bivalent organometallic derivatives are much less stable, but these are also beginning to attract attention particularly with sterically demanding ligands (*cf.* Section 3.7) and π-bonding ligands (*cf.* Section 3.8).

As in the other groups of the periodic table, the metal-carbon bonds (which have pronounced covalent character) become weaker from silicon to lead (*cf.* Section 3.3.1). In contrast to elements of other groups, members of main group IV show a special characteristic of element-element bond formation. This 'catenation' tendency, so marked with carbon and resulting in such a large number and variety of organic compounds, becomes much less apparent on passing from silicon to lead; some typical examples are linear $Si_{24}Me_{50}$, cyclic Ge_6Ph_{12}, and Sn_9Me_{18}, and branched $(PbPh_3)_4Pb$.

In view of an extensive chemistry associated with C=C double bonds, the possibility of the existence of similar derivative of other main group IV elements has been arousing interest, but success has been achieved in isolating only a few Si=Si, Si=C (*cf.* Section 3.8), Ge=Ge, and Sn=Sn (*cf.* Section 3.7) derivatives in recent past.

Stable organometallic compounds of main group IV elements with oxidation number + 3 are accessible only with the sterically demanding ligands such as $CH(SiMe_3)_2$; e.g., $[^{\bullet}M\{CH(SiMe_3)_2\}_3]$ [M=Si, Ge, Sn].

The tetra-alkyl and -aryl compounds of main group IV elements differ from the corresponding derivatives of the elements in neighbouring groups because of their relatively low reactivity. This is often a result more of kinetic than of thermodynamic factors (*cf.* Section 3.3.2). The main group IV alkyls and aryls behave as saturated compounds, and these elements show little tendency to increase their coordination numbers above four unless they are bonded to more electronegative atoms or groups. In the latter cases, the participation of the *d* orbitals expand the covalency of the central atom resulting in the formation of derivatives in coordination states, five (sp^3d hydridization), six (sp^3d^2 hybridization), and seven (sp^3d^3 hybridization) as is illustrated in Table 3.7. Coordination numbers greater than six are extremely rare; the two organometallic compounds, $SnMe(NO_3)_3$ and $[Pb(OCOCH_3)_3Ph_2]^-$ provide examples of increased coordination numbers seven and eight respectively.

Table 3.7 Some Typical Coordination Geometries of Main Group IV Organometallics

Type of hybridization	Coordination number	Geometry	Examples
sp^3	4	Tetrahedral	$SiMe_4$; $GeMe_4$; $SnMe_4$; $PbMe_4$
sp^3d	5	Trigonal bipyramidal	$SiF_4(Me)^-$ 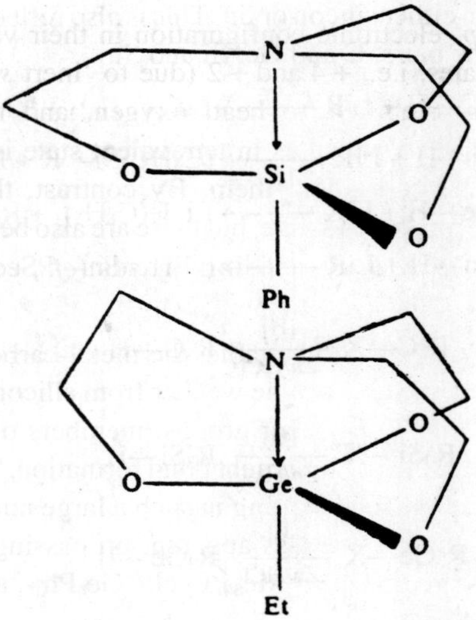 $SnClMe_3(C_6H_5N)$; $[PbClPh_3]_n$
sp^3d^2	6	Octahedral	$[SiF_4Ph_2]^{2-}$; $Sn(acac)_2Me_2$; $[PbCl_2Me_2]_n$, $PbCl_2Ph_2(bipy)$
sp^2	3	Pyramidal	$[^{\bullet}M\{CH(SiMe_3)_2\}_3]$ (M = Si, Ge or Sn)
	3	Environment around the central atom intermediate between pyramidal and planar	$Si_2(C_6H_3Me_2\text{-}2, 6)$, M_2R_4 (M = Ge, Sn, or Pb); (R = CH $(Si Me_3)_2$)

Within main group IV, the reactivity of the M—C bond in tetra-alkyls or -aryls (MR_4) increases progressively from silicon to lead as (i) the bond energy decreases in the same sequence (*cf.* Section 3.3.1):

$$Si(C_2H_5)_4 \quad Ge(C_2H_5)_4 \quad Sn(C_2H_5)_4 \quad Pb(C_2H_5)_4$$

decreasing M—C bond energy $\longrightarrow$

decreasing thermal stability $\longrightarrow$

and (ii) as expansion of the coordination number of the metal (M) becomes easier with increasing atomic size and decreasing difference between *np* and *nd* orbitals.

The electronegativity values of elements in the same group generally decrease with atomic number. However, the values of electronegativity of main group IV elements (C = 2.5; Si = 1.8; Ge = 1.8; Sn = 1.8 and Pb = 1.8) on Pauling scale showed hardly any change from silicon to lead. However, the differences in the stability of organometallic and other derivatives led Alfred and Rochow in early sixties to suggest a modified scale of values (C = 2.50; Si = 1.74; Ge = 2.02; Sn = 1.72; and Pb = 1.55) for these elements, according to which germanium is more electronegative than either silicon or tin. This is also reflected in the similarity between derivatives of the pairs: C and Ge; Si and Sn:

$$(C_6H_5)_3C\text{—}H + LiR \longrightarrow LiC(C_6H_5)_3 + RH$$

$$(C_6H_5)_3Si\text{—}H + LiR \longrightarrow (C_6H_5)_3Si\text{—}R + LiH$$

$$(C_6H_5)_3Ge\text{—}H + LiR \longrightarrow LiGe(C_6H_5)_3 + RH$$

$$(C_6H_5)_3Sn\text{—}H + LiR \longrightarrow (C_6H_5)_3Sn\text{—}R + LiH$$

$$R_3C\text{—}X \xrightarrow[Zn/HCl]{[H]} R_3C\text{—}H$$

$$R_3Si\text{—}X \xrightarrow[Zn/HCl]{[H]} R_3Si\text{—}H$$

$$R_3Ge\text{—}X \xrightarrow[Zn/HCl]{[H]} R_3Ge\text{—}H$$

$$R_3Sn\text{—}X \xrightarrow[Zn/HCl]{[H]} R_3Sn\text{—}H$$

Of the four elements (Si, Ge, Sn and Pb), silicon and tin have been receiving much greater attention. This may be due to a much larger abundance of silicon and growing applications of organometallics of both these elements. On the other hand, germanium is comparatively scarce and its derivatives have not shown any special features or possibilities of application distinguishing them from those of silicon. In spite of their earlier wide applications as gasoline additives, the organolead compounds tend to be avoided recently due to concomitant environmental problems. This position is unlikely to change for the reasons already stated. In the following two Schemes 3.5 and 3.6, therefore, are illustrated some typical reactions of organo-silicon and -tin compounds, leading to the synthesis of useful precursors (or reagents) for a plethora of important organic and organometallic transformations (*cf.* Section 6.3):

$$6Me_3Si\text{—}Cl + 8NaNH_2 \xrightarrow[heat]{PhH} 3(Me_3Si)_2NNa + 4NH_3 + 5NaCl + NH_4Cl$$

$$2Me_3Si\text{—}Cl + 3NH_3 \longrightarrow (Me_3Si)_2NH + 2NH_4Cl$$

$$Me_3Si\text{—}Cl + 2NHMe_2 \longrightarrow Si(NMe_2)Me_3 + Me_2NH_2^+Cl^-$$

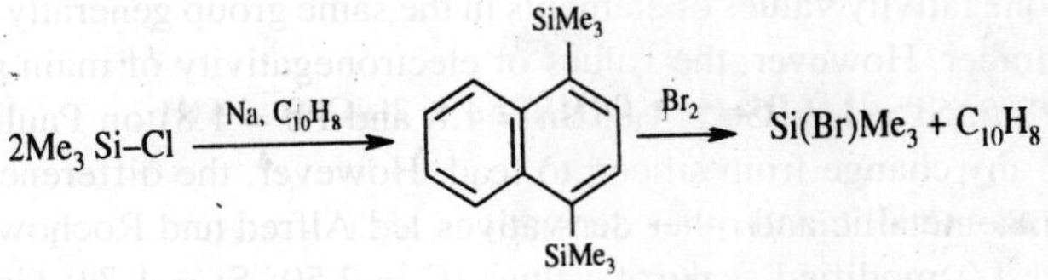

$$2Me_3Si{-}Cl + 2\,Na \longrightarrow Me_3Si{-}SiMe_3 + NaCl$$

$$Me_3Si{-}SiMe_3 \xrightarrow{I_2,\ 60\text{–}70^\circ C} 2Si(I)Me_3$$

$$Me_3Si{-}SiMe_3 \xrightarrow{LiMe,\ HMPT} LiSiMe_3 + SiMe_4$$

$$3Me_3Si - Cl + Li_3N \xrightarrow[\text{reflux}]{THF} (Me_3Si)_3N + 3LiCl$$

$$Me_3Si - Cl + H_2O \xrightarrow{PhNMe_2} \tfrac{1}{2}(Me_3Si)_2O \xrightarrow{I_2,\ Al,\ 60^\circ C} Si(I)Me_3$$

$$Me_3SiCH_2 - Cl + 2Na + Cl - CH_2SiMe_3 \longrightarrow (SiMe_3)_2CH_2 + 2NaCl$$

$$3Me_3Si - Cl + CHCl_3 + 6Li \xrightarrow{THF} (SiMe_3)_3\,CH + 6LiCl$$

$$2Me_3Si - Cl + CH_2\,Cl_2 + 3LiBu \xrightarrow[\text{(ii) } H_2O]{\text{(i) }-100 \text{ to } -60^\circ C} (SiMe_3)_2\,CHCl + 3LiCl + BuCl + 2BuH$$

$$Me_3Si - Cl + LiAlH_4 \longrightarrow Si(H)Me_3 + LiCl + AlH_3$$

$$Me_3Si - Cl + AgCN \longrightarrow Si(CN)Me_3 + AgCl$$

$$Me_3Si - Cl \xrightarrow[-LiCl]{LiCn,\ Et_2O} Si(CN)Me_3$$

$$Me_3Si\ Cl + LiSH \longrightarrow SiMe_3(SH) + LiCl$$

$$Me_3Si - Cl + NaPPh_2 \longrightarrow SiMe_3(PPh_2) + NaCl$$

$$Me_3Si - Cl + ROH + Et_3N \xrightarrow{Et_2O} Si(OR)Me_3 + Et_2N \cdot HCl$$

(a)

$$SiMe_4 \xrightarrow{Cl_2,\ h\nu} Me_3Si - CH_2Cl$$

$$Me_3Si - CH_2Cl \xrightarrow{NaI} Si(CH_2I)Me_3 + NaCl$$

$$Me_3Si - CH_2Cl \xrightarrow{Mg,\ Et_2O} MgCl(CH_2SiMe_3)$$

$$Me_3Si - CH_2Cl \xrightarrow{NaCN} Si(CH_2CN)Me_3 + NaCl$$

$$Me_3Si - CH_2Cl \xrightarrow[\text{(ii) } HNO_2]{\text{(i) } NaNHAc} \xrightarrow{\text{base}} Si(CHN_2)Me_3$$

(b)

$$Ph_2Si(Cl)_2 + 2LiAlH_4 \longrightarrow Si(H)_2Ph_2 + 2LiCl + 2AlH_3$$

$$Ph_3Si{-}Cl + LiAlH_4 \longrightarrow Si(H)Ph_3 + LiCl + AlH_3$$

$$Ph_3Si{-}Cl + 2Li \xrightarrow{THF} LiSiPh_3 + LiCl$$

$$Ph_3Si{-}H \xrightarrow{Br_2} Si(Br)Ph_3 + HBr$$

(c)

Scheme 3.5 Some Typical Reactions of (a) Chlorotrimethylsilane, (b) Chloromethyltrimethylsilane, and (c) Chloro- and phenylsilanes

3.5.5 Group 15 Elements (As, Sb, and Bi)

All the elements with ns^2np^3 valence electronic configuration form basically covalent organometallic compounds in both the +3 and +5 oxidation state. The general types of group V organometallics and their stereochemical possibilities are depicted in Table 3.8·

Table 3.8 Some Typical Coordination Geometries of Main Group V Organometallics

Formal valence	Types of hybridization	Coordination number	Geometry	Examples
1	sp^2	3	Trigonal planar	$As\{Cr(Co)_5\}_2Ph$
3	sp^2	3	Trigonal pyramidal	MR_3 (M = As, Sb, or Bi; R = Me, Ph, or p-ClC_6H_4), As $R_2(X)$(R = Me or Ph; X = Cl, Br, or I); $AsI_2(Me)$
	sp^3	4	Tetrahedral	$[AsMe_4]^-$
5	sp^3d	5	Trigonal bipyramidal	$AsMe_5$, $AsPh_5$, $AsBr_2(CH_3)_3$ $Sb(p\text{-}MeC_6H_4)_5$, $SbCl_2(CH_3)_3$ $BiPh_5$, $BiCl_2Ph_3$
	sp^3d	5	Square pyramidal	$SbPh_5$
	sp^3d^2	6	Octahedral	$[SbPh_6]^-$, $[SbFMe_4]_n$

As a consequence of the well known 'inert s-pair effect', the stability of organometallics in trivalent state generally increases from arsenic to bismuth. Although the reverse trend in stability should be expected for pentavalent derivatives, yet $SbMe_5$ is more stable than $AsMe_5$ as well as $BiMe_5$. In addition to these two oxidation states, organometallics of these elements are also known in other oxidation states such as +1, +2, and +4, which have been stabilized in the form of a complex or radical, e.g., $\{(OC)_5Cr\}_2AsPh$, $^{\bullet}As\{CH(SiMe_3)_2\}_2$, $^{\bullet}AsPh_4$, respectively. In a particular oxidation state, the polarity of metal-carbon bond increases on descending (As $\rightarrow$ Bi) the group.

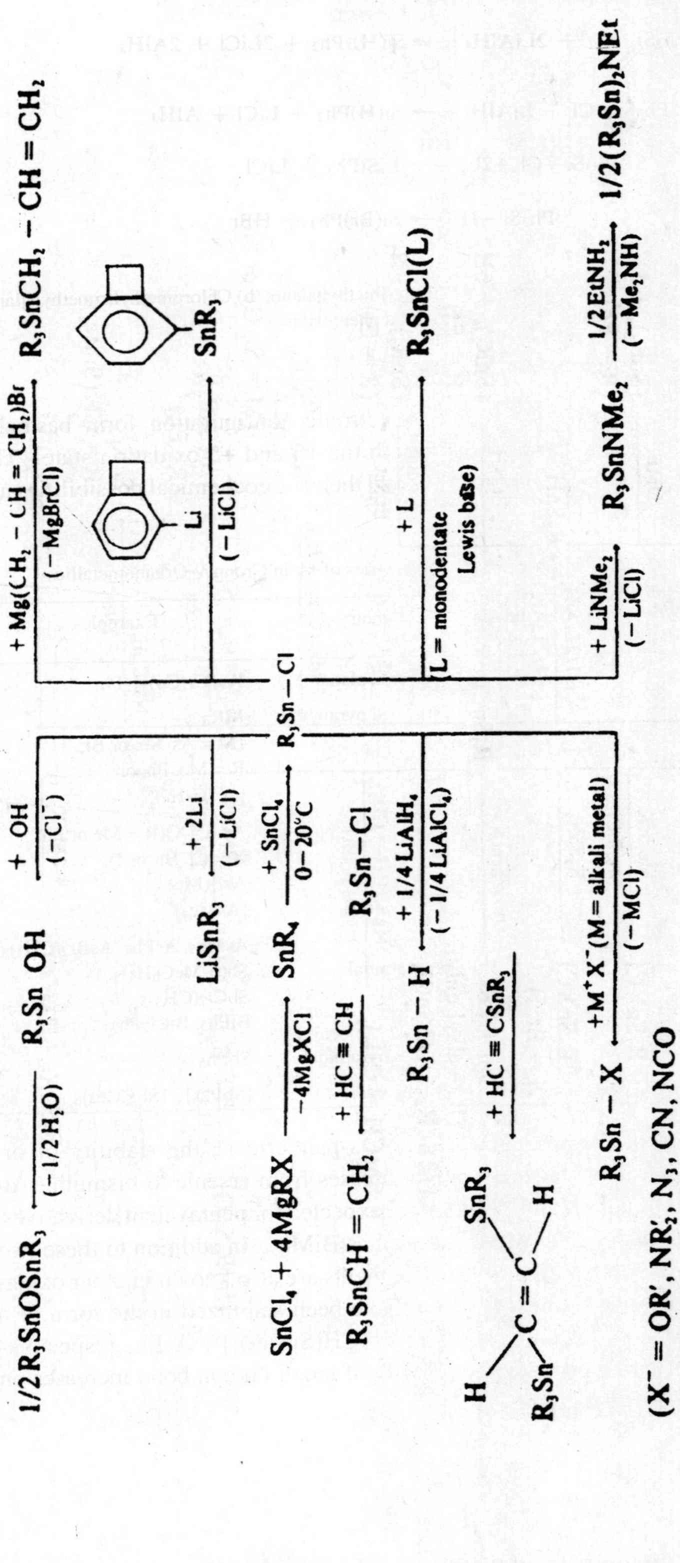

Scheme 3.6a Some Typical Reactions of Chlorotriorganotin.

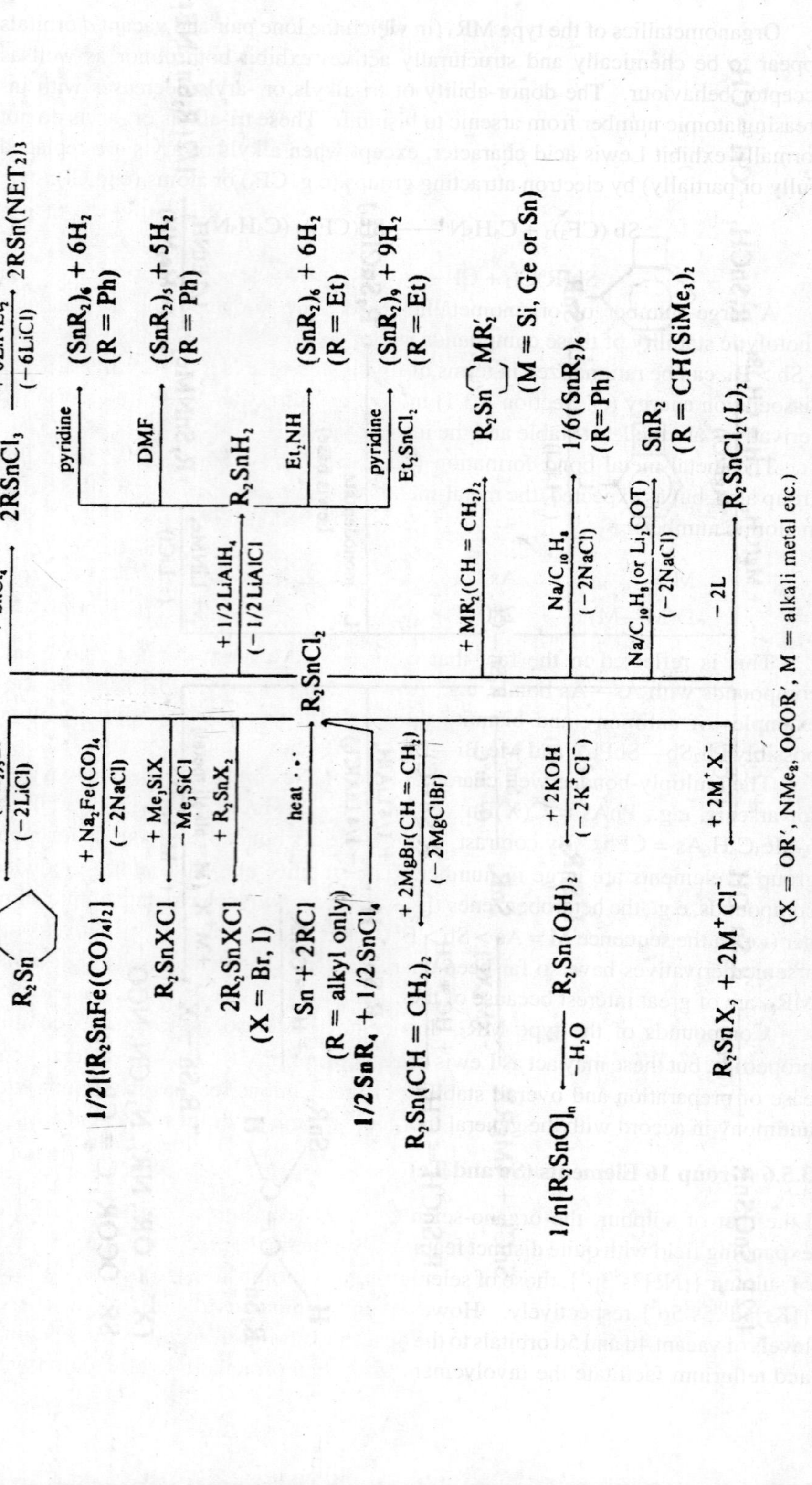

Scheme 3.6b Some Representative Reactions of Diorganotin Compounds (especially of Dichloride).

Organometallics of the type MR_3 (in which the lone pair and vacant *d* orbitals appear to be chemically and structurally active) exhibit both donor as well as acceptor behaviour. The donor-ability of tri-alkyls or -aryls decreases with increasing atomic number from arsenic to bismuth. These tri-alkyls or -aryls do not normally exhibit Lewis acid character, except when alkyls or aryls are replaced (fully or partially) by electron attracting groups (e.g. CF_3) or atoms (e.g. Cl):

$$Sb(CF_3)_3 + C_5H_5N \longrightarrow Sb(CF_3)_3(C_5H_5N)$$

$$Sb(R)Cl_2 + Cl^- \longrightarrow Sb(R)Cl_3^-$$

A large number of organometallics, MR_3, are known. The thermal and photolytic stability of these compounds which decreases in the sequence, M = As > Sb > Bi, can be rationalized in terms of steady decrease in the mean M—C bond dissociation energy (*cf.* Section 3.3.1) in the same direction. The organobismuth derivatives are the least stable and the most reactive.

The metal-metal bond formation (i.e. catenation) ability does exist in this group also, but as expected, the metal-metal bond strength decreases with increase in atomic number:

M =	As	Sb	Bi
D(M—M)	280	267	200 kJ mol^{-1}

This is reflected in the fact that compared to a number of organoarsenic compounds with As—As bonds, e.g., As_2Me_4, $(AsPh)_6$, or $(AsPh)_5$, well-defined examples of antimony and bismuth derivatives are only $Me_2Sb—SbMe_2$ (and possibly $Ph_2Sb—SbPh_2$) and $Me_2Bi—BiMe_2$ (the least stable respectively).

The multiply-bonded well characterized acyclic compounds are known only for arsenic, e.g., PhAs = C(X)Bu^t (when X = $OSiMe_3$ or N(R)$SiMe_3$) or 2, 4, 6-$Me_3C_6H_6As = CPh_2$. By contrast, cyclic multiply bonded compounds for main group V elements are large in number. The stability of cyclic multiply bonded compounds, e.g., the heterobenzenes  decreases with increasing atomic number (i.e. in the sequence M = As > Sb > Bi). In the case of acyclic derivatives only arsenic derivatives have so far been characterized. The penta-alkyls and -aryls, MR_5, are of great interest because of their structural (*cf.* Section 3.6.5) features.

Compounds of the type MR_5, do not obviously exhibit electron donating properties, but these may act as Lewis acids in forming MR_6^- anions. Generally the ease of preparation and overall stability of these anions reaches maximum with antimony in accord with the general inorganic chemistry of these elements.

3.5.6 Group 16 Elements (Se and Te)

Like that of sulphur, the organo-selenium and -tellurium chemistry is a fastly expanding field with quite distinct features. Similar to the electronic configuration of sulphur {[Ne]$3s^2 3p^4$}, those of selenium and tellurium are {[Ar]$3d^{10} 4s^2 4p^4$] and {[Kr]$4d^{10} 5s^2 5p^4$] respectively. However, the comparative similarity in energy levels of vacant 4d and 5d orbitals to the 4p and 5p orbitals respectively in selenium and tellurium facilitate the involvement of these d orbitals in hybridization with the corresponding s and p orbitals. This results in an increasing tendency of attainment of higher (5 or 6) covalencies (coordination states) from sulphur to tellurium in their covalent organo-derivatives. In Table 3.9 are listed the coordination geometries of a few typical organo-derivatives of these elements.

Table 3.9 Typical Coordination Environments of Organo-selenium and -tellurium Compounds

Compound	Hybridiztion	Shape	
ER_2	sp^3	Bent	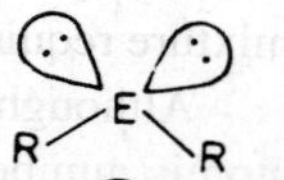
$[EMe_3]^+[BPh_4]^-$	sp^3	Pyramidal cation	
$[EMe_3]^+I^-$	sp^3d^2	Pseudo-trigonal bipyramidal configuration due to close association of the cation with iodide	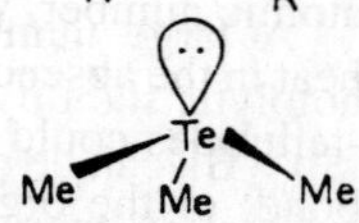
Me_2ECl_2 $[EMe_2]^+ [TeI_4Me]^-$	sp^3d sp^2d^3	Pseudo-trigonal bipyramidal cation and anion both six-coordinate pseudooctahedral (*cf.* Section 3.6.6)	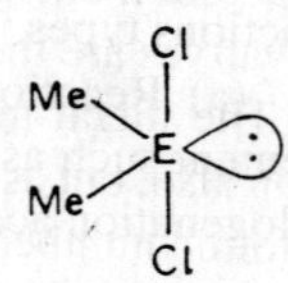
$PhTeCl_3$	sp^3d	Two independent tellurium atoms in square based pyramidal geometry which are bridged by chlorine atoms to give a chain structure (*cf.* Section 3.6.6).	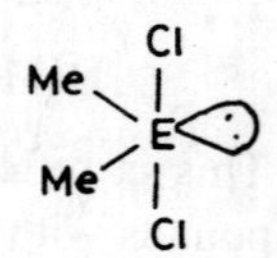

In the first two pairs of species (i.e., SeR_2, TeR_2; $[SeMe_3]^+$ and $[TeMe_3]^+$), the central elements, selenium and tellurium attain an octet (noble gas) configuration. In addition to the above, elements of this group (apparently even polonium which is most metallic) exhibit a decreasing tendency to attain noble gas (ns^2np^6) configuration by accepting two electrons from metals and forming dinegative ions, E^{2-}.

As in the case of main group elements N, P, As, Sb and Bi, the Lewis base character of compounds ER_2 of main group VI elements decreases with the rise of atomic number from oxygen to tellurium, with the exception of selenium compounds which in some isolated cases appear to be slightly stronger nucleophiles than their sulphur analogues.

The presence of lone pair of electrons on ER_2(E = Se or Te) permits the formation of trivalent derivatives such as $[SeR_3]^+X^-$ and $[TeR_3]^+X^-$.

$$ER_2 + RX \longrightarrow [ER_3]^+X^-$$

This type of reaction is more facile with selenium and tellurium derivatives compared to that with sulphides because of the greater nucleophilicity (due to the presence of lone pair electrons on $4p$ and $5p$ orbitals) of the easily polarizable selenium and tellurium atoms respectively. Predictably alkyl (aryl) selenides are alkylated to give selenonium salts with greater difficulty than the alkylation of dialkylselenides, e.g., interaction of $Se(CH_3)C_6H_5$ with CH_3I gave dimethylpheny selenonium salts in only a poor yield.

The order of nucleophilicity appears to follow the sequence: $CH_3SeR > CH_3SR > C_6H_5SeR > C_6H_5SR > CH_3TeR > C_6H_5TeR$.

The reactions of diorganotellurides are generally more facile than those of selenium analogues and follow the same pattern:

$$TeRR' + R''X \longrightarrow [TeRR'R'']^+X^-$$

These combinations proceed quite vigorously with lower aliphatic tellurides and methyl iodide; in the case of aromatic and higher aliphatic tellurides, the mixture requires heating to facilitate the reaction.

Although the stability of E—C bond in the group decreases with increasing atomic number, yet Se—C and Te—C bonds are surprisingly stable to bases and heat in the absence of oxidizing agents. The reactive site in diorgano-selenides and -tellurides could be the chalcogen element (Se or Te) itself; the element-carbon bond; or the organic group attached to the element, resulting in the following reactions types.

(a) Reactions involving increase in the oxidation state of the chalcogen element such as oxidative addition of organic and inorganic halides, halogens or halogenation accompanied by oxidation:

$$ER_2 + R'X \longrightarrow [ER_2R']^+X^-$$

$$ER_2 + SnI_4 \longrightarrow [ER_2(I)\,(SnI_3)]$$

$$ER_2 + X_2 \longrightarrow ER_2X_2$$

$$ER_2 + SO_2X_2 \longrightarrow ER_2X_2 + SO_2$$

$$ER_2 + SOCl_2 \longrightarrow ER_2Cl_2$$

(E = Se or Te; R = alkyl or aryl)

Halogenolysis works well for both aliphatic and aromatic-diselenides or -tellurides.

(b) *Cleavage of E—C bond:* The lower strength of tellurium-carbon bond is depicted by the following reactions in which dialkyl-selenide undergoes merely oxidation (with retention of selenium-carbon bond) whereas in the tellurium analogues, the tellurium-carbon bond undergoes cleavage:

$$2SeR_2 + O_2 \longrightarrow 2R_2Se = O;$$

$$Te(CH_2R)_2 + 2O_2 \longrightarrow RCH_2Te{-}O{-}O{-}H + RCOOH$$

or

alkaline H_2O_2

With sulphur, however, both selenium and tellurium derivatives undergo cleavage reactions:

$$ER_2 + S \longrightarrow R_2S + E$$

(c) *Formation of addition complexes*

$$ER_2 + HgX_2 \longrightarrow HgX_2{\cdot}ER_2$$

Other metal halides which have been found to form complexes are CdI_2, CuCl, AgI, $AuCl_3$, $PdCl_2$, $PtCl_2$, etc. By contrast, organotellurium halides exhibit Lewis acid character:

$$TeR_2X_2 + nL \longrightarrow TeR_2X_2{\cdot}nL$$

$$TeRX_3 + nL \longrightarrow TeRX_3{\cdot}nL$$

$$(n = 1 \text{ or } 2;\ L = \text{Lewis base})$$

In addition to these, a few types of reactions are more pronounced with organotellurium halides:

(d) *Nucleophilic substitution*

$$TeR_2X_2 + AgX' \longrightarrow TeR_2(X')_2 + 2AgX$$

$$(X' = NCO, NCS, NC, F, \text{etc.})$$

(e) *Reduction elimination*

$$TeR_2X_2 + N_2H_4 \longrightarrow TeR_2$$

$$\text{or}$$

$$Na_2S{\cdot}9H_2O$$

(f) *Anionic complex formation:*

$$TeR_2X_2 + [NR_4]^+ X^- \rightarrow [NR_4]^+ [TeR_2X_3]^-$$

In addition to nitrogen reagents, those of P, As, and Sb could also be used. Various permutations and combinations have yielded a large number of water soluble biologically active derivatives.

3.5.7 Reactions With Oxygen and Halogens

Most organometallic compounds of the first three main group elements react exothermally with oxygen. Although many of these (especially the lower alkyls of alkali and alkaline-earth metals, Be, Mg, Zn, Cd, B, Al, Ga, In and Tl) are spontaneously inflammable in air, yet under carefully controlled milder conditions, alkyls of some of these metals may be oxidized to the alkoxide stage only. Tetra-alkyls of main group IV elements, on the other hand, generally show stability towards molecular oxygen. By contrast, triorgano-derivatives, MR_3 of main group V elements (M = P, As, Sb and Bi) and diorgano-derivatives, ER_2 of main group VI elements (E = Se, or Te) are oxidized in limited supply of oxygen to $MR_3(O)$, $ER_2(O)$ respectively with the retention of element-carbon bonds:

$$MR_n + \text{excess } O_2 \xrightarrow[\text{combustion}]{} M_2O_n + CO_2 + H_2O$$

$$AlEt_3 + \text{limited } O_2 \rightarrow Al(OEt)_3$$

$$Mg(R)X + \text{limited } O_2 \rightarrow MgX(O_2R)$$

$$2MR_3 + \text{limited } O_2 \rightarrow 2MR_3(O)$$

$$(M = As, Sb)$$

$$2ER_2 + \text{limited } O_2 \rightarrow 2ER_2(O)$$

$$(E = Se, Te)$$

Halogens generally cleave (the ease and completion of cleavage reactions are largely dependent on the electropositive character of the concerned element, i.e., more electropositive the metal, more reaction will go to completion) the metal-carbon bonds:

$$LiR + X_2 \longrightarrow LiX + RX$$

$$Mg(R)X + X_2 \longrightarrow MgX_2 + RX$$

$$BR_3 + I_2 \xrightarrow{160°C} BIR_2 + RI$$

$$AlR_3 + 3X_2 \longrightarrow AlX_3 + 3RX$$

$$GaR_3 + X_2 \longrightarrow GaR_2X + RX$$

(R = alkyl or aryl; X = halogen)

The action of chlorine on tetra-alkyls of main group IV elements provides an interesting illustration of gradation in chemical reactivity with increasing atomic weight:

$$SiMe_4 + Cl_2 \xrightarrow{h\nu} Si(CH_2Cl)Me_3 + MeCl$$

$$GeMe_3 + Cl_3 \xrightarrow[\text{temperature}]{\text{room}} GeClMe_3 + MeCl$$

$$GeMe_4 + Br_2 \xrightarrow[\text{25 h}]{\text{reflux}} GeBrMe_3 + MeBr$$

$$GeMe_4 + 2Br_2 \xrightarrow[\text{catalyst, reflux}]{AlBr_3} GeBr_2Me_2 + 2MeBr$$

$$SnMe_4 + Cl_2 \xrightarrow[\text{conditions}]{\text{under controlled}} SnClMe_3 + MeCl$$

(The reaction with tetramethyltin is carried out under controlled conditions to get monochlorotrimethyltin, otherwise formation of bis-product is more favourable).

$$PbMe_4 + 2Cl_2 \xrightarrow{\text{room temperature}} PbCl_4 + 4MeCl$$

When the above reaction is carried out at low temperature (–60°C) a mixture both of $PbMe_3X$ and $PbMe_2X_2$ is obtained:

$$2PbMe_4 + 3X_2 \xrightarrow{\text{low temperature}} PbMe_3X + 3MeX + PbMe_2X_2$$

An interesting feature to note is that when both alkyl and phenyl groups are present in the same molecule, the latter is preferentially cleaved, e.g.:

$$GeEt_2Ph_2 + Br_2 \rightarrow GeBr_2Et_2 + 2PhBr$$

By contrast, the trialkyls and the triaryls of P, As and the triaryls of Bi as well as diorgano derivatives of Se and Te undergo oxidative addition with halogens. A few illustrative examples are given below:

$$MR_3 + X_2 \rightarrow MR_3X_2$$

(M = P, As, Sb, R = alkyl; M = P, As, Sb, or Bi; R = aryl)

$$ER_2 + X_2 \rightarrow ER_2X_2$$

(E = Se or Te; R = alkyl or aryl; X = halogen)

For trialkylbismuth, however, cleavage of a bismuth-alkyl bond by halogens occurs in preference to oxidative-addition:

$$BiR_3 + X_2 \rightarrow BiR_2X + RX$$

(R = alkyl, X = halogen)

Phenyl groups are readily cleaved from pentaphenyls of arsenic, antimony, and bismuth on reactions with halogens, giving rise to tetraphenyl-arsonium, stibonium, and -bismuthonium trihalides respectively:

$$AsPh_5 + 2X_2 \rightarrow [AsPh_4]^+ X_3^- + PhX$$

(X = Cl, Br, or I)

Reaction of bromine or iodine with pentaphenylantimony takes place in two stages, e.g.:

$$SbPh_5 + I_2 \rightarrow [SbPh_4]^+I^- + PhI$$

$$[SbPh_4]^+I^- + I_2 \rightarrow [SbPh_4]^+ I_3^-$$

By contrast, interaction of pentaphenylbismuth with bromine at room temperature gives $BiPh_3$:

$$BiPh_5 + Br_2 \rightarrow BiPh_3 + 2PhBr$$

However, at –70°C bismuthonium tribromide is formed:

$$BiPh_4 + 2Br_2 \xrightarrow{-70^\circ C} [BiPh_4]^+Br_3^- + PhBr$$

Tetraphenylbismuthonium tribromide decomposes in two different ways on raising the temperature:

$$[BiPh_4]^+ Br_3^- \rightarrow \begin{cases} \xrightarrow{-30^\circ C} [BiBr_2Ph_3] + PhBr \\ \xrightarrow[25^\circ C]{} [BiBrPh_2] + 2PhBr \end{cases}$$

3.5.8 Alkylation (or Arylation) Reactions

As discussed earlier (Section 3.4.3), the reactions of some alkylating or arylating reagents, which are typical organometallic compounds, e.g., LiR, Mg(R)X, ZnR_2, AlR_3, SnR_4, PbR_4, etc., with the halide (or sometimes alkoxide/amide) of another metal can under suitable conditions lead to the synthesis of new organometallic derivatives:

$$MX_n + LiR \xrightarrow[-\,LiX]{} MRX_{n-1} \xrightarrow[-\,(n-1)\,LiX]{+\,(n-1)\,LiR} MR_n \xrightarrow{+\,aLiR} Li_aMR_{n+a}$$

The synthesis of organometallic derivatives by the above types of reactions can be termed as alkylation (or arylation). The reactions of these alkylating agents with main group metal halides, etc., have already been discussed in the earlier Section 3.4.3.2, and we shall, therefore, concentrate in the following pages on the reactions of these alkylating agents with transition metal halides or other derivatives.

In contrast to the reactions of a bulky reagent like $LiCH(SiMe_3)_2$ leading to the formation of only disubstituted products, $[MCl]\{CH(SiMe_3)_2\}_2]$ in cases of B and Al and trisubstituted products $[M\{CH(SiMe_3)_2\}_3]$ in case of Ga, In and Bi (*cf.* Section 3.4.3.2), the reaction proceeds further with the formation of 'ate' products in the case of larger atoms like those of lanthanoids:

$$MCl_3 + 3LiCH(SiMe_3)_2 \xrightarrow{THF} [Li(THF)_4]^+ [MCl\{CH(SiMe_3)_2\}_3]^-$$

It is noteworthy to mention that trichlorides of lanthanoid metals react with a less sterically demanding ligand, like LiMe in the presence of chelating diamines like TMEDA to afford 'ate' complexes with octahedral environment around the metal atom:

$$MCl_3 + 6LiMe + 3TMEDA \rightarrow [Li(TMEDA)]_3^+ [M(CH_3)_6]^{3-} + 3LiCl$$

Alkylation of tetrachlorides of titanium, zirconium and hafnium with suitable lithium alkyls (e.g., neopentyls; *cf.* Section 4.2.2.1) generates isolable (at ambient temperatures) homoleptic derivatives:

$$MCl_4 + 4LiCH_2CMe_3 \xrightarrow{Et_2O} M(CH_2CMe_3)_4 + 4LiCl$$

Some other alkylation reactions by lithium alkyls can be represented by the following reactions:

$$CrCl_3(THF)_3 \xrightarrow{LiCH_2SiMe_3} Li\{Cr(CH_2SiMe_3)_4\} \xrightarrow{-e} Cr(CH_2SiMe_3)_4$$

$$MnI_2 + 2LiCH_2C_6H_4NMe_2\text{–}o \xrightarrow{THF} Mn(CH_2C_6H_4NMe_2\text{–}o)_2 + 2LiI$$

$$4CuI + 4LiCH_2SiMe_3 \xrightarrow[-\,10^\circ C]{hexane\text{–}ether} (CuCH_2SiMe_3)_4 + 4LiI$$

Complementary to organolithium reagents are organomagnesium compounds, with which also a few typical reactions are represented by the following equations:

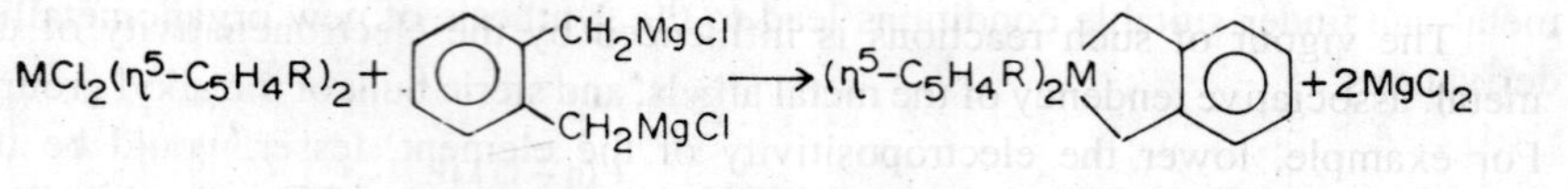

(M = Ti, Zr, Hf or Nb)

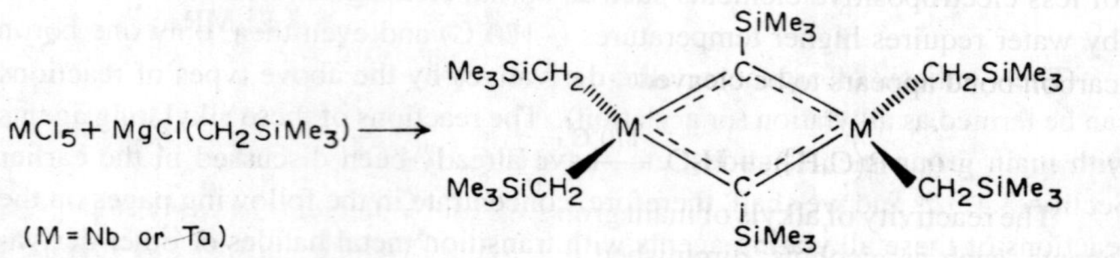

(M = Nb or Ta)

$$2MoCl_4(THF)_2 + 3Mg(CH_2SiMe_3)_2 + 2PMe_3 \rightarrow 2MoCl(CH_3SiMe_3)_3(PMe_3) + 3MgCl_2$$

$$NiCl_2(Py)_2 + 2MgCl(CH_2SiMe_3) \rightarrow Ni(CH_2SiMe_3)_2\,(py)_2 + 2MgCl_2$$

Although organolithium or organomagnesium reagents are much more versatile and widely used alkylating or arylating agents, the use of other alkylating (arylating) agents (such as ZnR_2, HgR_2, BR_3, AlR_3, PbR_4, SnR_4) has been found advantageous where comparatively a milder non-basic reagent is required due to the following reasons: (a) the need of partial alkylation or arylation, and (b) to avoid reduction of the transition metal ions. A few examples of such reactions are represented below:

$$NbCl_5 + ZnMe_2 \xrightarrow[\text{12h; r.t.}]{\text{pentane}} NbCl_3Me_2 + ZnCl_2$$

$$WCl_6 + SnR_4 \xrightarrow{\text{hexane}} WCl_5(R) + SnClR$$

(R = Me, Et, Bu, or Ph)

$$TiCl_4 + PbEt_4 \rightarrow TiCl_3(Et) + PbEt_3Cl$$

$$3WCl_6 + BR_3 \xrightarrow{\text{ether}} 3WCl_5(R) + BCl_3$$

(R = ME, Et, Bu)

$$TiCl_3(Me) + AlMe_3 \rightarrow TiCl_2Me_2 + AlMe_2Cl$$

$$4PtCl_4 + 6HgMe_2 \rightarrow (PtClMe_3)_4 + 6MgCl_2$$

$$2WCl_6 + HgMe_2 \rightarrow 2WCl_5Me + HgCl_2$$

$$6RuCl_2 + 6(diphos)_2 + (AlMe_3)_2 \rightarrow 5RuClMe(diphos) + 2AlCl_3$$

3.5.9 Reactions with Protic Reagents

Alkyls (aryls) of the first three main group elements are highly reactive towards water and in most cases lead to violent reactions. These reactions can be represented by the following general equation:

$$MR_n + nH_2O \rightarrow M(OH)_n + nRH$$

(M = elements of group IA, IIA, and IIIB where n = 1, 2, or 3 respectively).

The vigour of such reactions is influenced by the electronegativity of the metal, associative tendency of the metal alkyls, and steric bulk of the alkyl groups. For example, lower the electropositivity of the element, lesser would be the reactivity of its metal alkyls with protic reagents like water. Thus, for the alkyls of less electropositive elements such as boron, cleavage of element-carbon bond by water requires higher temperatures (~170°C) and even then, only one boron-carbon bond appears to be cleaved:

$$B(C_4H_9^n)_3 + H_2O \xrightarrow{170^\circ C} B(C_4H_9^n)_2(OH) + C_4H_{10}$$

The reactivity of alkyls of main groups IV and V elements towards other protic agents, such as alcohols, diminished remarkably (which may be due to kinetic factors) and these do not appear to react with water under normal conditions.

The reactivity of metal alkyls towards other protic agent, such as alcohols, amines, cyclopentadiene and other reagents having acidic hydrogen atoms is generally milder and some of these can be illustrated by the following equations:

$$BeMe_2 + MeOH \rightarrow Be(OMe)Me + MeH$$

$$AlMe_3 + MeOH \rightarrow Al(OMe)Me_2 + MeH$$

$$4AlPh_3 + 4PhNH_2 \rightarrow 4\ Al(NPh)\cdot Ph + 8PhH$$

$$LiPh + C_5H_6 \rightarrow LiC_5H_5 + PhH$$

$$LiPh + RC \equiv CH \rightarrow LiC \equiv CR + PhH$$

With stronger protic agents like hydrogen halides, the alkyls of first three group metals are generally converted into corresponding metal halides, but alkyls of even fourth main group metals tend to react in a limited manner with hydrogen halides, e.g.,

$$SnPh_4 + HX \rightarrow SnPh_3X + PhH$$

$$PbMe_4 + HX \rightarrow PbMe_3X + MeH$$

The organo-derivatives of main group V elements show an interesting variation in their reactivity towards hydrogen halides. For example, the reaction of $AsPh_3$ in liquid hydrogen chloride has been represented by the following equation on the basis of conductance measurements:

$$AsPh_3 + 3HCl \rightleftharpoons [Ph_3AsH]^+[HCl_2]^-$$

Antimony trialkyls, however, are such strong reducing agents that these are oxidised even by hydrogen halides to pentavalent antimony derivatives:

$$SbMe_3 + 2HCl \rightarrow SbMe_3Cl_2 + H_2$$

By contrast, interaction of triphenylbismuth with hydrogen chloride yields bismuth chloride:

$$BiPh_3 + 3HCl \rightarrow BiCl_3 + 3PhH$$

This can be rationalized in terms of (i) the greater stability of bismuth in its trivalent state, and (ii) the weakest nature of Bi—C bond amongst main group V elements.

3.5.10 Reactions with Hydridic Hydrogen

A few organometallic reagents such as organolithium or organomagnesium compound can substitute an orgnaic group for hydrogen in hydrides of some metalloids, e.g.:

$$B_2H_6 + 6Mg(Cl)Et \rightarrow 2BEt_3 + 6Mg(H)Cl$$

$$Si(H)Ph_3 + LiMe \rightarrow Si(Me)Ph_3 + LiH$$

$$Sn(H)Ph_3 + LiMe \rightarrow Sn(Me)Ph_3 + LiH$$

By contrast, the reaction of $Ge(H)Ph_3$ with methyllithium can be represented as:

$$Ge(H)Ph_3 + LiMe \rightarrow LiGePh_3 + MeH$$

This reaction has been cited as an evidence of germanium showing greater electronegativity compared to that of tin as well as silicon (*cf.* Section 3.5.4).

3.5.11 Redistribution Reactions

The redistribution reaction (exchange of groups between two derivatives of the same elements) is a general feature of the main group organometallics and the course of the reaction can generally be monitored by n.m.r. techniques. A few examples of reactions which have been investigated in detail are represented by the following equations:

$$ZnR_2 + ZnX_2 \rightarrow 2Zn(R)X$$

(R = Me or Et, X = Cl, Br, or I)

$$HgR_2 + HgX_2 \rightarrow 2Hg(R)X$$

(R = alkyl, aryl, etc.; X = Cl, Br, I)

$$HgMe_2 + HgPh_2 \xrightarrow{150^\circ C} 2Hg(Me)Ph$$

$$BR_3 + BX_3 \rightarrow BR_2(X) + BR(X)_2$$

(R = alkyl or aryl; X = F, Cl, OR′, SR′, NR′$_2$, etc.)

$$BR_3 + \underset{\text{(excess)}}{BX_3} \rightarrow BR(X)_2$$

(R = alkyl or aryl; X = H, Cl, SR′, etc.)

$$AlR_3 + 2AlR'_3 \rightarrow 3AlR'_2(R)$$

(R = iso-octyl, R′ = iso-butyl)

$$2GaR_3 + GaCl_3 \rightarrow 3GaCl(R)_2$$

(R = Ph or Bu^i)

$$SiEt_4 + SiPr^n_4 \xrightarrow[\text{heat}]{AlCl_3\text{ (catalyst)}} SiEt_xPr^n_{4-x}$$

$$GeR_4 + GeCl_4 \rightarrow GeClR_3 + GeCl_3R$$

$$Sn\,R_4 + SnCl_4 \xrightarrow[\text{rapid}]{0-20^\circ C} SnClR_3 + SnCl_3R$$

$$SnClR_3 + SnCl_3R \xrightarrow[\text{slow}]{180^\circ C} 2SnCl_2R_2$$

$$\underset{\text{(excess)}}{3SnR_4 + SnCl_4} \rightarrow 4SnClR_3$$

$$SnR_4 + SnCl_4 \xrightarrow{200^\circ C} 2SnCl_2R_2$$

$$SnR_4 + \underset{\text{(excess)}}{3SnCl_4} \rightarrow 4SnCl_3R$$

Redistribution reactions have also been proved useful synthetic route for making mixed alkyl (aryl) halide derivatives of arsenic and antimony:

$$AsR_3 + AsX_3 \rightleftharpoons AsR_2X + AsRX_2$$

(R = larger alkyl groups, trimethylarsine does not behave in this way; X = Cl, Br, I)

$$SbPh_3 + 2SbCl_3 \xrightarrow[\text{reflux}]{CH_2Cl_2} 3SbCl_2Ph$$

$$2SbPh_3 + SbCl_3 \xrightarrow[\text{reflux}]{CH_2Cl_2} 3SbClPh_2$$

3.5.12 Pyrolysis

The thermally unstable alkyls of some metals like lead (or bismuth) on heating lead to the generation of radicals, which play a key-role in the antiknock properties of $PbEt_4$. Pyrolysis of more reactive alkyls generally leads to production of a metal hydride and the elimination of an alkene (a type of reaction which is more prevalent in transition metal organometallics; *cf.* Section 4.1), e.g.,

$$LiCH_2CH_2CH_2CH_3 \xrightarrow{140^\circ C} LiH + CH_3CH_2CH = CH_2$$

$$Be(CMe_3)_2 \xrightarrow{100^\circ C} BeH_2 + 2H_2C = CMe_3$$

$$Al(CH_2CHMe_2)_3 \xrightarrow{150-180^\circ C} Al(H)(CH_2CHMe_2)_2 + Me_2C = CH_2$$

Reactions of the above type generally lead to the formation of very complex mixtures of products, due to the fact that alkenes so produced undergo rearrangement and polymerization reactions.

By contrast, dimethyls of Be, Mg and Hg, decompose on heating in quite different manner:

$$BeMe_2 \xrightarrow{200^\circ C} (BeCH_2)_n \xrightarrow{225^\circ C} CH_4 + Be_4C$$

$$MgMe_2 \xrightarrow{250^\circ C} CH_4 + (MgCH_2)_n$$

$$HgMe_2 \xrightarrow{300°C} Hg + 2Me^{\cdot} \rightarrow \text{Products}$$

3.5.13 Complex Formation

The organometallic compounds of main groups II and III elements are coordinatively unsaturated. Hence they exhibit a tendency for coordination with nucleophilic ligands as well as for self-association. These, therefore, generally form complexes with amines, ethers, halide ions, and carbanions, e.g., $BeR_2{\cdot}NMe_3$; $BeR_2.2NMe_3$; $BR_3.NR_3'$; $AlR_3.NR_3$; $GaR_3.NR_3'$; $BeR_2.Et_2O$; $BeR_2{\cdot}2Et_2O$; MgBrEt. $2Et_2O$; $MgBrPh{\cdot}2Et_2O$; $AlR_3{\cdot}Et_2O$; $(Bu^tMgCl{\cdot}Et_2O)_2$; AlR3X; Li2[$BeMe_4$]; Na[BPh_4]; Li[B(H)Ph_3]; Na[B(OR)Ph_3]; K[$B(CN)_4$] and Li[$AlMe_4$].

Heats of dissociation have been measured for a variety of the above complexes e.g., $R_3M.\ NR_3'$ and from a comparison of such data, the following order of acceptor strength has been postulated: B < Al > Ga > In > Tl. Steric congestion generally considered to be responsible for the boron-aluminium relation (B < Al). The Lewis acidity sequence for group II metal alkyls appears to be similar, i.e.,

$$Be < Mg > Zn > Cd > Hg$$

It is noteworthy that although diorganozinc compounds are electron deficient, yet they are generally incapable of association involving bridging organic groups, and also stable isolable complexes are formed only with cyclic ethers (which are sterically less demanding), or with comparatively donating stronger ligands such as 1, 4-dioxane, DME, DMP, TMEDA. As expected cadmium alkyls are in general weak acceptors, but $CdEt_2$ forms strong complexes with cyclic sulphides (which can be ascribed to the soft acid character of cadmium); and curiously, trimethylsilyl-derivative, $Cd(CH_2SiMe_3)_2$ forms a stable volatile complex with bipyridyl (bipy) of an unusual stoichiometry, $R_2Cd.1.5.$ bipy.

The methyls, MMe_3(M = Al and Ga), of aluminium and gallium belong to the class 'a' type acceptor (N > P > As; O > S). Dimethylmercury, trimethylthallium, and more surprising trimethylborons also exhibit some class 'b' character (S > O) also.

In contrast to the organometallic compounds of main group II and III elements in which coordinative unsaturation arises from unfilled 'p' orbitals in their valence shells, group IV elements have s^2p^6 electronic configuration in their tetravalent covalent organometallic derivatives. These elements, however, have a tendency of achieving higher coordination numbers by utilizing their 'd' orbitals, which are, however, rather diffused in derivatives like tetra-alkyls to participate effectively in adduct formation. This diffuseness of 'd' orbitals can be considerably reduced by induction of positive charge on the central metal atoms as the alkyl groups are gradually replaced by electronegative moieties. This tendency is reflected in the octahedral nature of $Cl_2Si(acac)_2$ in contrast to the tetrahedral configuration of $Me_2Si(acac)_2$ (*cf.* Section 2.5.2). Similarly, in the derivatives of the type, R_nSnX_{4-n} the tendency of increasing the coordination numbers through complex formation increases as the value of 'n' is reduced, e.g., $SnClR_3.C_5H_5N$; $SnCl_2R_2.2C_5H_5N$;

$SnCl_3R \cdot 2C_5H_5N$. The higher adduct forming tendency of tin compared to that of silicon can be illustrated by the octahedral 6-coordinate structure of $Me_2Sn(acac)_2$ in contrast to the tetrahedral form of $Me_2Si(acac)_2$.

The presence of a lone pair of electrons in bivalent compounds of tin enables them to function as donor molecules to suitable acceptor species, e.g., $R_2Sn.Cr(CO)_5$ (*cf.* Section 3.7.1.2).

In view of a lone pair being available in triorganometal (III) compounds of main group V elements (E), these also function as Lewis bases to a variety of acceptor molecules forming complexes of the type:

$$EMe_3.BF_3;\ EMe_3.BMe_3;\ EMe_3.GaMe_3 \text{ and } EMe_3 \cdot BH_3$$

In all these complexes the base strength decrease in the order $NMe_3 > PMe_3 > AsMe_3 > SbMe_3$. However, in the last complex $EMe_3.BH_3$, the order of base strength is $PMe_3 > NMe_3 > AsMe_3 > SbMe_3$.

In addition to formation of adducts with Lewis acids (generally hard acid type), these triorgano derivatives of main group V elements form adducts with transition metals also, e.g.,

$$(EMe_3)_2 \cdot PdX_2;\ EMe_3 \cdot Ni(CO)_3;\ (EMe_3)_2 \cdot Mo(CO)_4$$

The stability of these transition metal adducts is controlled by ($E \rightleftarrows M$) (d – d) π interactions (involving filled transition metal d-orbitals and vacant d-orbitals on the group V element), in addition to the decreasing basic strength of ER_3 from nitrogen to bismuth. The stability order, therefore, generally follows the pattern $PR_3 > NR_3 > AsR_3 > SbR_3 > BiR_3$. In the $PR_3 > NR_3$ pair, the availability of 'd' orbitals in phosphorus resulting in the possibility of back-donation from the 'd' orbitals of transition metals overdominates the weakening of the donor ability of phosphorus compared to nitrogen. However, the tendency of this back-donation is not able to recompensate the sharp decrease in the basic strength from P to Bi.

For the same central metal atom, the donating capacity of an alkyl derivative will decrease with increasing electronegative character of the alkyl group. For example, tristrifluoromethylstibine, $Sb(CF_3)_3$ has hardly any donating capacity and it does not form an adduct with $AlCl_3$ or $PdCl_2$. In fact, by contrast it acts as an acceptor molecule and forms the adduct, $Sb(CF_3)_3C_5H_5N$ with pyridine.

In the absence of lone pair of electrons, organoderivatives of main group elements in their pentavalent state do not exhibit any Lewis base character. On the other hand, some of these behave as Lewis acids as exemplified by the formation of adducts like $PhSbCl_4 \cdot Me_2SO$ and $SbCl_5 \cdot Ph^-$ as well as derivative in higher coordination state like $SbCl_{4-n}Ph(acac)_n$. This type of tendency would be enhanced by the electron withdrawing effect of the electronegative group inducing a positive charge on the central metal atom.

3.5.14 Insertion Reactions

Insertion reactions have proved a powerful tool in synthetic chemistry leading to new classes of compounds; these show unique importance in catalytic system involving transition metal-carbon bonds (*cf.* Chapter 6). A number of reaction involving insertion of alkene, SO_2, SO_3 or CO_2 into a main group metal-carbon bond have also been known. Synthetic reactions involving insertion of multiple bonded organic or inorganic substrates into metal-carbon bonds have shown phenomenal potential in organic synthesis involving lithium- and magnesium-carbon bonds (*cf.* Sections 3.5.1, 3.5.2 and 6.3).

Alkene insertion reactions into Al–C bonds, which have been exploited immensely in industry are of two main types:

(a) growth reactions to synthesize unbranched primary alcohols and alkene e.g.:

$$AlEt_3 \xrightarrow{+\,nC_2H_4} Al(CH_2CH_2R)_3 \xrightarrow[\text{(ii) } H_3O^+]{\text{(i) } O_2} 3RCH_2CH_2OH$$

$$Al(CH_2CH_2R)_3 \xrightarrow{\Delta} RCH=CH_2$$

and (b) low-pressure stereoregular polymerization of ethene and propene in the presence of mixed catalysts such as a mixture of $TiCl_4$ and Al_2Et_6 in heptane (*cf.* Section 6.4.2.6). Some illustrative examples of SO_2 and SO_3 insertion reaction into metal-carbon bonds are given below:

$$AlR_3 + 3SO_2 \rightarrow Al(OSOR)_3$$

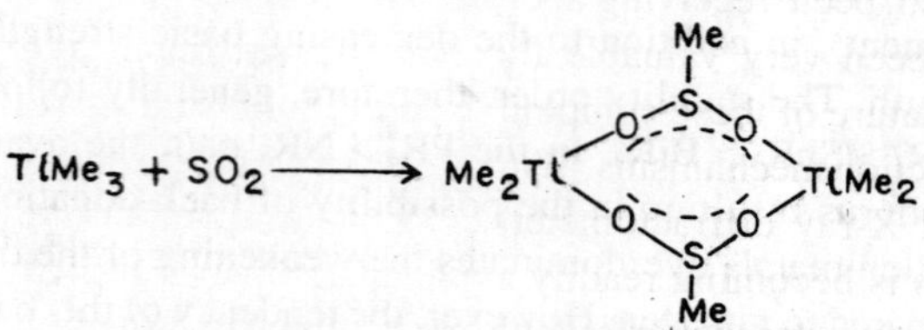

$$SnR_4 + SO_2 \xrightarrow{65^\circ C} (R_3Sn)_2SO_4 + R_2Sn(OSOR)_2$$

$$SnR_4 + SO_2(\text{liq.}) \longrightarrow R_3Sn(OSOR)$$

$$GeR_4 + SO_3 \longrightarrow R_2Ge(OSO_2R)$$

$$\underset{\text{(excess)}}{PbR_4 + 2SO_2} \xrightarrow{\text{r.t.}} R_2Pb(OSOR)_2$$

$$PbR_4 + SO_2 \xrightarrow{0^\circ C} R_3PbOSOR$$

$$PbR_4 + SO_3 \xrightarrow{\text{ether}} R_3Pb(OSO_2R) \xrightarrow{SO_3} R_2Pb(OSO_2R)_2$$

Insertion of CO_2 into metal-carbon bonds has shown remarkable variation in reactivity depending upon the nature of the metal. For example, alkali and alkaline earth metal alkyls (or aryls) react extremely vigorously to give salts of carboxylic acids, RCO_2Li, $Be(OCOR)_2$, etc. Zinc dialkyls are much less reactive towards CO_2, e.g.,

$$ZnEt_2 + 2CO_2 \xrightarrow{150^\circ C} Zn(O_2CEt)_2$$

and organo-cadmium and -mercury are even less reactive.

With AlR_3, one CO_2 inserts at the room temperature and a second one at 220°C under pressure, e.g.,

$$AlR_3 + CO_2 \xrightarrow{r.t.} AlR_2(O_2CR)$$

$$AlR_3 + 2CO_2 \xrightarrow{220^\circ C} AlR(O_2CR)_2$$

The B—C, Si—C, Ge—C and Sn—C bonds, are, however, rather inert to CO_2 insertion.

These trends in the reactivity of insertion reactions are consistent with (i) the extent of bond polarities $M^{\delta+}$ — $C^{\delta-}$ and $O^{\delta-}$=$X^{\delta+}$=$O^{\delta-}$ (where X = C or S), and (ii) the relative metal-carbon bond strength. Thus weaker and more polar is the M—C bond, the greater would be its reactivity towards insertion reactions.

3.6 STRUCTURAL ASPECTS

With the fastly growing facilities for structural elucidation, organometallic compounds have also been receiving increasing attention in these directions; this knowledge has been very valuable for a clearer and more quantitative understanding of the nature of these compounds in the solid state as well as in solution and also their reaction mechanisms. With a fast development and easier availability of computerized X-ray diffractometers, increasing quantitative information on crystal structures is becoming readily available. However, other physiochemical (e.g. i.r. and n.m.r. spectroscopic) techniques continue to provide confirmatory and even elucidatory structural information particularly for liquid and amorphous compounds which can not be crystallized easily. In addition reactions of a large number of compounds are studied in solution, wherein these undergo a variety of structural transformations, a study of which by spectroscopic techniques provides useful information not only on the structural features of individual species, but also on the mechanism of the reactions involved (*cf.* n.m.r. studies of group III methyls at the end of Section 3.6.3).

Detailed information available upto 1980 has been included in the form of an 'Index of structures determined by Diffraction Methods' on pages 1209-1520 of volume IX of the 'Comprehensive Organometallic Chemistry', edited by G. Wilkinson, *et al.* (Pergamon, 1982). Similarly the structural information obtained from other physico-chemical techniques has been included in a number of reference books and review articles (*cf.* list of Supplementary Reading at the end of this chapter). Due to limitation of size in view of the introductory nature of this text and to maintain coherency, relevant structural information has been briefly mentioned in the succeeding sections on only a few typical organometallic derivatives of metals, arranged groupwise; for sake of convenience, however, a little more background information on some spectroscopic techniques has been provided in

Section 3.6.4.1 while dealing with silicon and tin organometallics for which these have been most extensively employed (and it might be in some cases advisable to read this Section 3.6.4.1 before following the group-wise account).

3.6.1 Organolithium Compounds

For sake of simplicity, organolithium compounds have been represented by the simple monomeric formula 'LiR' in the preceding pages, but in actuality, their structures are generally more complex; e.g., n-alkylithium compounds are often hexameric in hydrocarbon solvents. This associative tendency can be lowered in some cases by (i) coordination with electron donors, (ii) steric demands of the organic group, and (iii) the capacity of the organic group to delocalise negative charge. Thus, $LiBu^n$ is hexameric in cyclohexane and tetrameric in diethyl ether (or THF); $LiBu^t$ is tetrameric in hydrocarbon solvents; $LiCH = CH_2$ is trimeric in tetrahydrofuran (THF); LiC_6H_5 is dimeric in diethyl ether or THF; $LiCH_2C_6H_5$ is dimeric in benzene and monomeric in THF; and $(Me_3Si)_2CHLi(PMDETA)$ and $Li\{C(SiMe_2Ph)_3\}(THF)$ are monomeric even in the solid state.

Additionally the first lithium 'ate' complex, $[Li(THF)_4]^+[Li\{C(SiMe_3)_3\}_2]^-$ in which the lithium is two coordinate has been crystallographically authenticated.

The simple Li—R bond represents lithium atom (having only two electrons in its valence shell) as electron deficient, which explains its tendency for polymerization like other electron deficient molecules.

Infrared studies show that the strong tendency of methylithium to associate persists in the vapour phase also. The low C—H stretching vibrational frequencies of tetrameric methyllithium, $(LiMe)_4$ (2840 and 2780 cm^{-1}) have been reported to be characteristic of methyl groups involved in electron-deficient bonding. The metal-carbon symmetric stretching mode is observed at 558 and 530 cm^{-1} due to $^6LiCH_3$ and $^7LiCH_3$ respectively.

Raman bands (in diethyl ether solutions) of methyllithium at 486, 299, 209 and 171 cm^{-1} are characteristic of cluster modes of the tetrameric units.

Nuclear magnetic resonance studies (in ether and THF as solvents) indicate that $LiCH_3$ retains the tetrameric solid state structure in solution. The ^{1}Hn.m.r. studies of diamine-chelated lithium alkyl compounds (primarily butyllithium) in paraffinic solvents showed marked upfield chemical shifts of the α–CH_2 protons of LiBu. The magnitude of this shift may be correlated with the stability of the chelated LiBu and presumably reflects a polarization and subsequent loosening of the Li—C bond. There is also a smaller downfield chemical shift of the N–alkyl group protons of the chelating agent caused by a general transfer of electron density from the chelating nitrogen atom to lithium; this type of downfield shift in n.m.r. spectrum of the chelating agent protons is a useful indication for chelation of lithium by the diamine ligand.

The structures (determined by X-ray techniques), or both methyl- and ethylithium can be depicted by tetrameric. $(LiR)_4$ units. These comprise of four lithium atoms at the corners of a tetrahedron; the alkyl groups are located at each face of the tetrahedron (Figs. 3.4 and 3.5). Each carbon atom is associated through

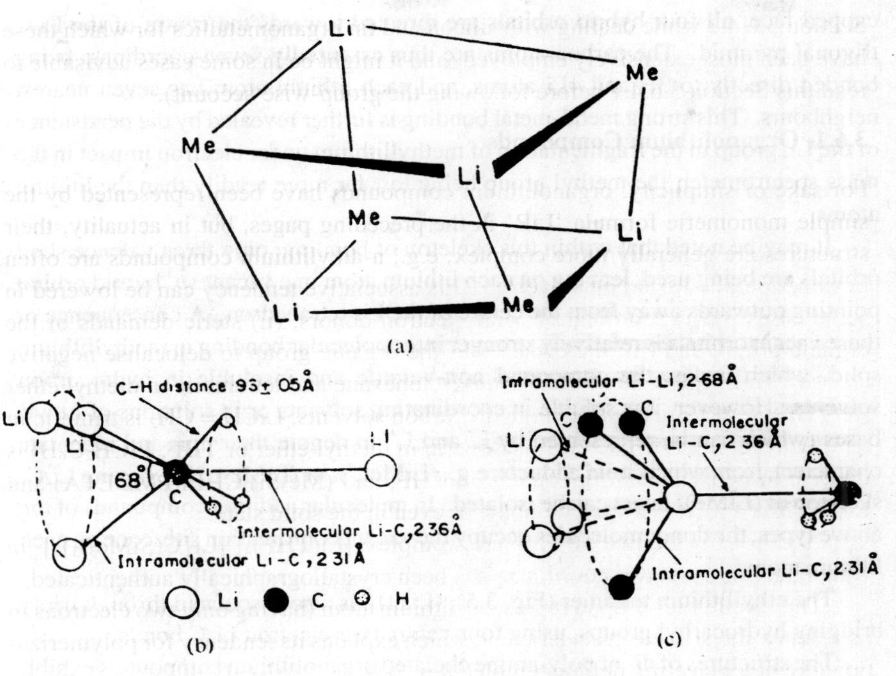

Fig. 3.4 (a) The structure of $(LiCH_3)_4$; (b) The seven-coordinate environment of each C atom; and (c) The seven-coordinate environment of each lithium atom.

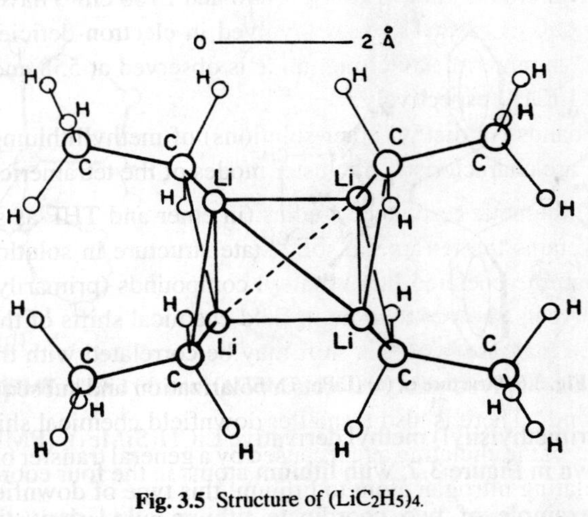

Fig. 3.5 Structure of $(LiC_2H_5)_4$.

three equidistant lithium atoms, thus acting as triply bridging hydrocarbyl groups, using four-centre Li_3C bonds. This means that there are only four electron pairs to form a total of twelve lithium-carbon bonds which is characteristic of electron deficient compounds. Each carbon atom utilizes one sp^3 hybrid orbital to overlap with the three sp^3 hybrid atomic orbitals provided by the three lithium atoms in the capped face; all four hybrid orbitals are directed towards the centre of the Li_3C trigonal pyramid. The carbon atoms are thus essentially seven-coordinate being bonded directly to 3H and 4Li atoms, and each lithium atom has seven nearest neighbours. This strong metal-metal bonding is further revealed by the persistence of the Li_4 group in the fragmentation of methyllithium under electron impact in the mass spectrometer, the methyl group being lost far more readily than the lithium atoms.

It may be noted that within this skeleton of bonding, only three valence shell orbitals are being used, leaving on each lithium atom one vacant sp^3 hybrid orbital pointing outwards away from the centre of the Li_4 tetrahedron. A consequence of these vacant orbitals is relatively stronger intermolecular bonding in methyllithium solid, which makes the compound non-volatile and insoluble in hydrocarbon solvents. However, it is soluble in coordinating solvents or in solutions of Lewis bases (which can be represented by L′ and L″ to denote their uni- and bidentate character), from which solid adducts, e.g., $(LiMeL')_4$ or $(LiMe)_4.2L''$, having Li_4C_4 skeleton of $(LiMe)_4$ intact can be isolated. In molecular addition compounds of the above types, the donor molecules occupy the vacant coordination sites, one on each lithium atom.

The ethyllithium tetramer (Fig. 3.5), $(LiEt)_4$, is also associated through triply bridging hydrocarbyl groups, using four-centre two-electron Li_3C bonds.

The structures of di- or poly-amine chelated organolithium compounds exhibit quite different patterns. For example, tetramethylethylenediamine (TMEDA) complex of phenyllithium is a dimer (Fig. 3.6a) with phenyl bridges, whereas TMEDA complex of $LiCPh_3$ is a monomer (Fig. 3.6b) in the solid state.

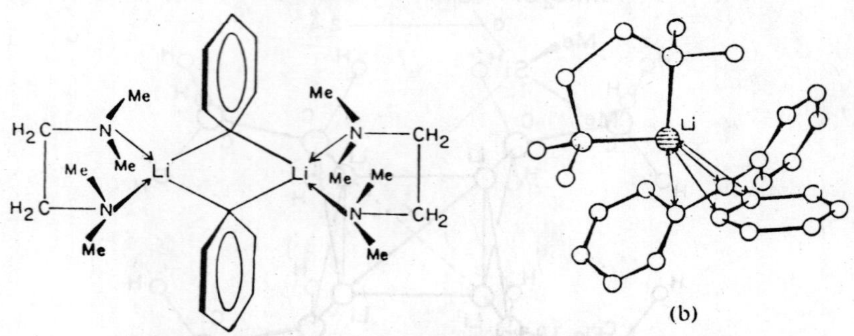

Fig. 3.6 Structure of (a) $(LiPh.\ TMEDA)_2$ and (b) $LiCPh_3$ TMEDA.

The bis(trimethylsilyl) methyl derivative $LiCH(SiMe_3)_2$ (PMDETA) adopts a structure shown in Figure 3.7, with lithium atoms in the four coordination state.

A rare example of two coordinate lithium alkyl derivative $[Li(THF)_4]^+$ $[Li\{C(SiMe_3)_3\}_2]^-$ possesses (shown by single X-ray crystallography) linear discrete type of anionic structure shown in Figure 3.8.

Crystal structure determination on (tetrahydrofuran) tris (dimethylphenylsilyl) methyl-lithium has shown it to be a monomeric complex. The structure is shown in Figure 3.9.

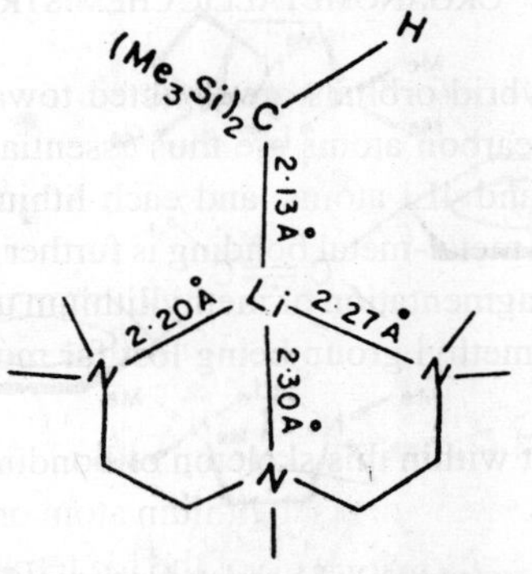

Fig. 3.7 Structure of $LiCH(SiMe_3)_2$ (PMDETA).

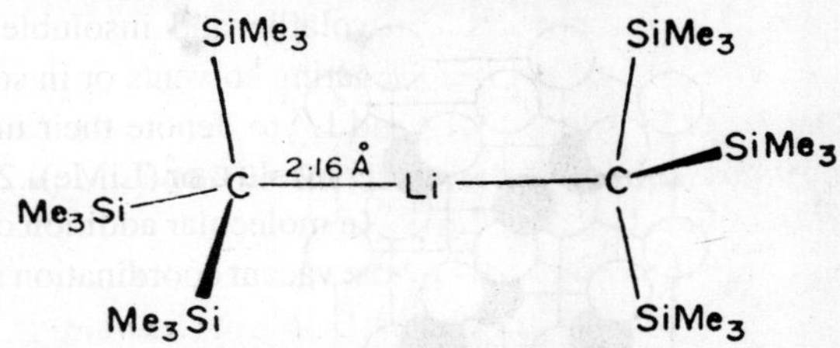

Fig. 3.8 Structure of $[Li\{C(SiMe_3)_3\}_2]^-$.

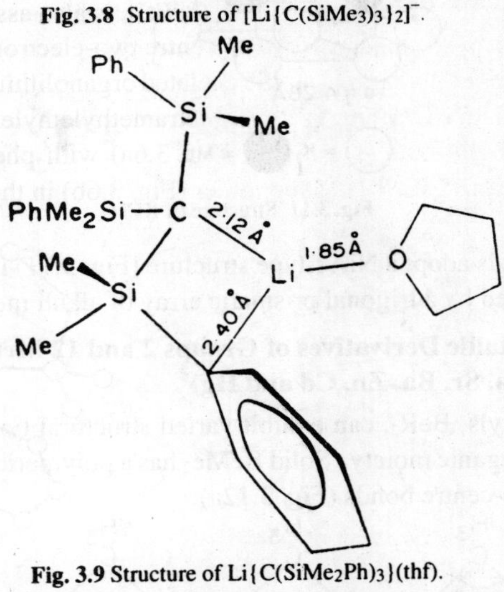

Fig. 3.9 Structure of $Li\{C(SiMe_2Ph)_3\}$(thf).

Polyamine chelated stilbene derivatives (Fig. 3.10) which are ionic in the solid state exhibit interesting structural features with two chelated lithium atoms located above and below the olefinic bond of the planar *trans*-stilbene molecule.

Due to significant decrease in ionisation potentials of heavier alkali metals, the structures of their organometallic derivatives tend to be typical of ionic compounds. Thus, although $NaCH_3$ is tetrameric like $LiCH_3$, NaC_2H_5 depicts a layer structure with the CH_2 groups in a trigonal pyramidal array of Na neighbours. In view of even more ionic character of K—C (as well as Rb—C and Cs—C)

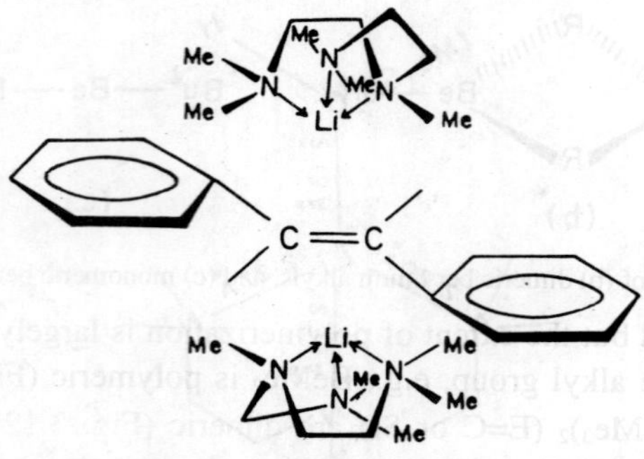

Fig. 3.10 Structure of PMDETA chelated lithium stilbene derivative.

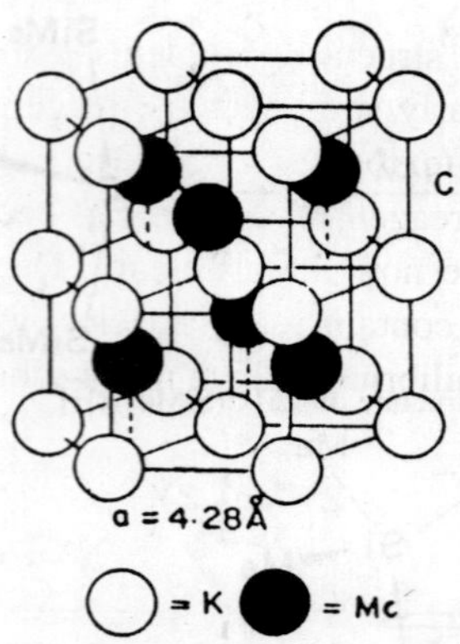

Fig. 3.11 Structure of KCH_3.

bonds, their methyls adopt a NiAs-type structure (Fig. 3.11) in which each methyl group is surrounded by a trigonal prismatic array of alkali metals.

3.6.2 Organometallic Derivatives of Groups 2 and 12 Metals (Be, Mg, Ca, Sr, Ba, Zn, Cd and Hg)

The beryllium alkyls, BeR_2, can exhibit varied structural types, depending upon the nature of the organic moiety. Solid $BeMe_2$ has a polymeric structure, involving two-electron three-centre bonds (Fig. 3.12a):

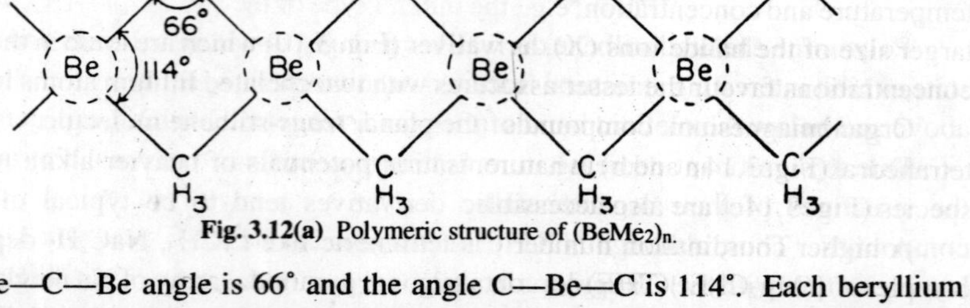

Fig. 3.12(a) Polymeric structure of $(BeMe_2)_n$.

The Be—C—Be angle is 66° and the angle C—Be—C is 114°. Each beryllium atom has a tetrahedral environment of four methyl groups. Other beryllium alkyls

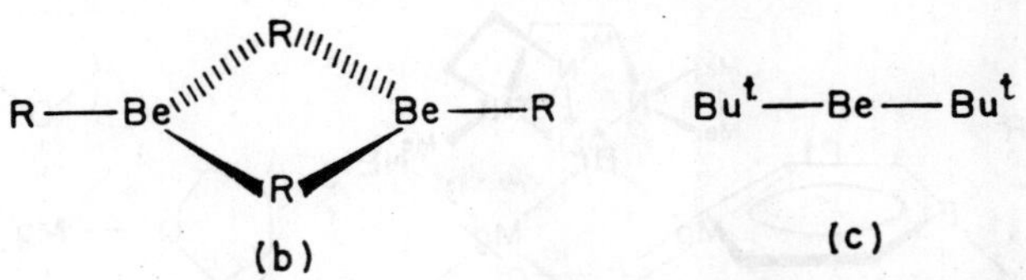

Fig. 3.12 Structure of (b) dimeric beryllium alkyls, and (c) monomeric beryllium alkyl(s).

are also polymerized but the extent of polymerization is largely governed by the steric demand of the alkyl group, e.g., $BeMe_2$ is polymeric (Fig. 3.12a), $BeEt_2$, $BePr_2$ and $Be(CH_2EMe_3)_2$ (E=C or Si) are dimeric (Fig. 3.12b), and $BeBu^t_2$ is monomeric (Fig. 3.12c). The monomeric alkyls are linear molecules (Fig. 3.12c) and Be—C bonds are formed by normal two-electron two-centre bonds, utilizing sp hybrid orbitals:

In an attempt to resolve the structural problems associated with Grignard reagents, it was realized at an early stage that the reagent is not (as generally represented in elementary organic textbooks) a simple monomeric species in ether. The constitution of the Grignard reagent, therefore, has been a subject of intense scrutiny, but its main features are now almost clear. The solution of Grignard reagents, MgRX, is believed to contain many species, interconvertible via an extended form of the Schlenk equilibrium shown in the scheme:

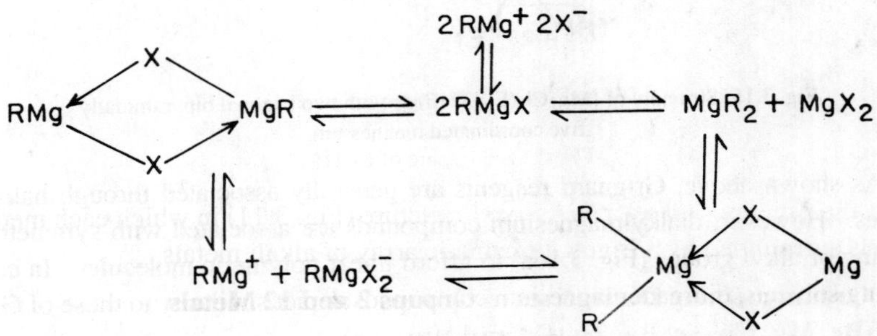

Fig. 3.13 Schlenk equilibrium in MgRX system.
(*Note*: All species are normally solvated; solvent molecules are not shown for convenience.)

The concentration of the various species shown in the scheme is largely dependent on a number of factors such as the nature of R and X, the solvent, temperature and concentration; e.g., the bulkier type of the alkyl group (R), and the larger size of the halide ions (X), as well as strongly donating solvents and lower concentrations favour the lesser associated or monomeric species.

Organomagnesium compounds (Grignard reagents, MgRX) are generally tetrahedral (Fig. 3.14 a and b) in nature, but penta-coordinated trigonal bipyramidal species (Fig. 3.14c) are also accessible.

A higher coordination number 6 for magnesium has also been found in the complex $[(Mg_2Cl_3Et)(THF)_3]_2$ obtained by crystallizing MgClEt from tetrahydrofuran (THF).

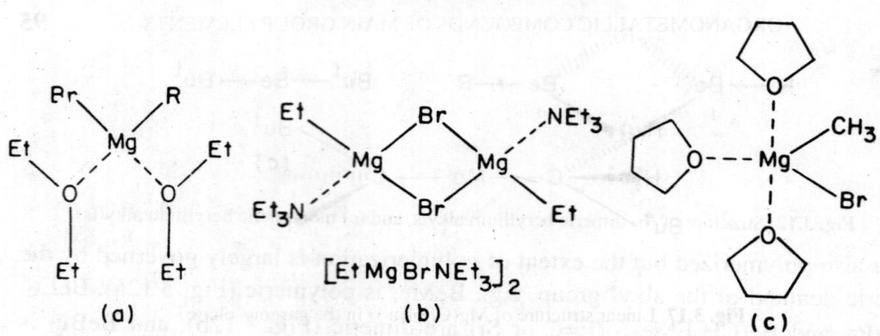

Fig. 3.14 Structure of (a) $Mg(Br)R(Et_2O)_2$, (b) $[Mg(Br)Et(NEt_3)]_2$, and (c) $MgBr(Me)(THF)_3$.

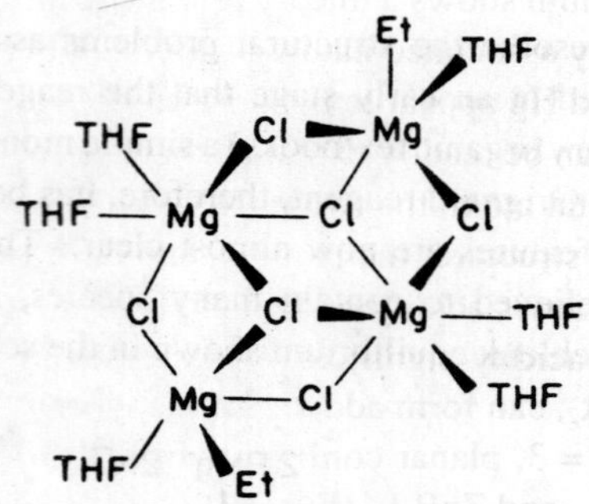

Fig. 3.15 Structure of $[Mg_2Cl_3(Et)(THF)_3]_2$ with two trigonal bipyramidally five coordinated magnesium.

As shown above, Grignard reagents are generally associated through halogen bridges. However, dialkylmagnesium compounds are associated with symmetrical bridging of alkyl groups (Fig. 3.16a) to afford linear polymeric molecules. In coordinating solvents, diorganomagnesium compounds depict similarity to those of Grig-

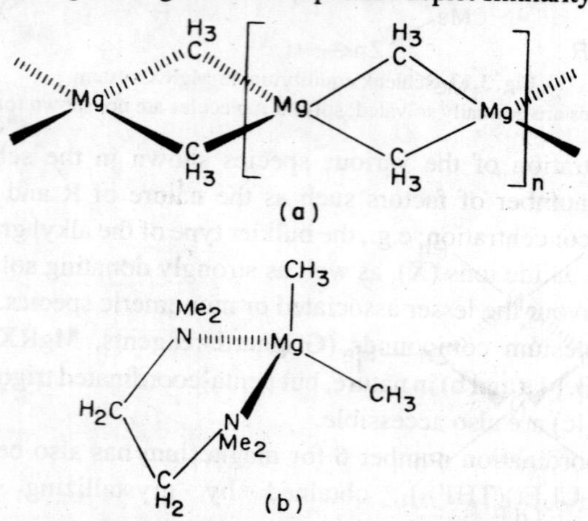

Fig. 3.16 Structure of (a) $[Mg(CH_3)_2]_n$, and (b) $Mg(CH_3)_2(TMEDA)$.

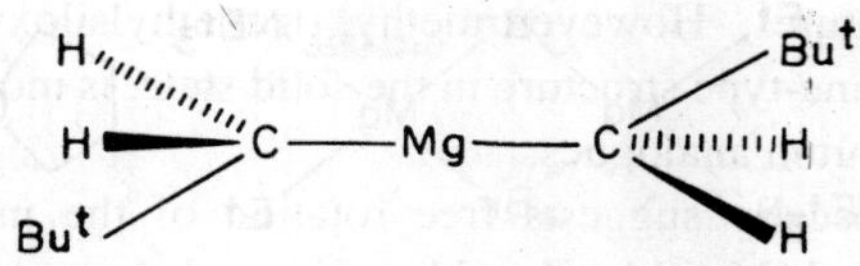

Fig. 3.17 Linear structure of $Mg(CH_2Bu^t)_2$ in the gaseous phase.

nard reagents in forming monomeric structures with tetrahedral environment (Fig. 3.16b) around the magnesium. Contrary to the lower alkyls of magnesium, bis (neopentyl) magnesium shows a linear monomeric structure in the gas phase.

In contrast to highly associated nature of alkyls of beryllium and magnesium the alkyls of Zn, Cd, and Hg appear to be monomeric, volatile, and covalent. The shapes of these alkyls can be readily explained in terms of VSEPR theory, with *sp* hybridization of the metal atom, the two bond pairs being as far apart as possible (Fig. 3.18a). In such a structure, the metal atoms have two vacant valence shell orbitals, which confer Lewis acidity on the alkyls of these metals particularly on the zinc alkyls (Lewis acidity significantly decreases in going from Zn to Hg). Thus zinc dialkyls, ZnR_2, can form adducts of the types:

ZnR_2L (with C.N. = 3, planar configuration Fig. 3.18b; L = a monodentate Lewis base), and ZnR_2L_2 and ZnR_2L' (Fig. 3.18c) (C.N. = 4, tetrahedral geometry; L′ = chelating diethers or amines). Heteroleptic derivatives of the type ZnXR (X = halogen, OR, SR, NR_2 or PR_2) generally adopt dimeric structure. $(ZnRX)_2$, (Fig. 3.18d) or cubane-type tetramers $(ZnRX)_4$ (Fig. 3.18e). These points are illustrated by a few typical examples in Figure 3.18.

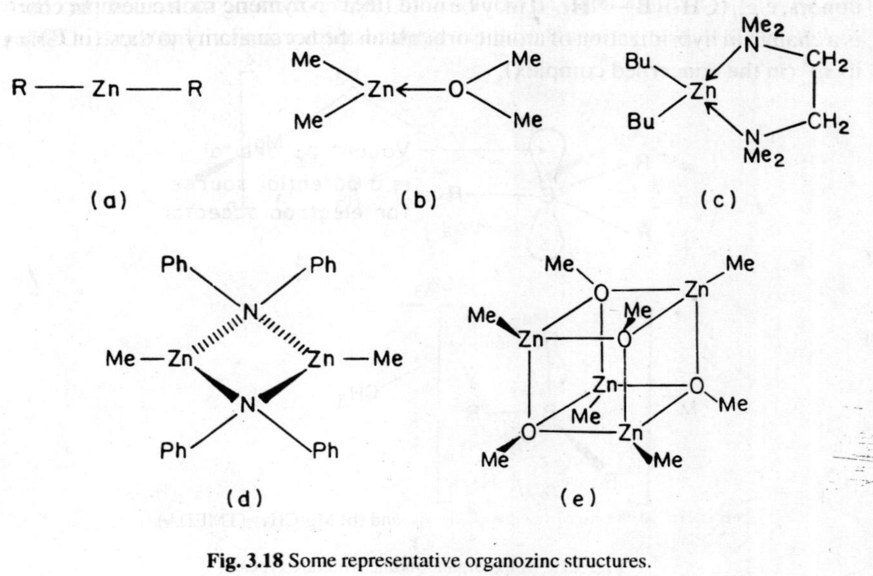

Fig. 3.18 Some representative organozinc structures.

It should be noted that organometallics of cadmium and mercury exhibit similar structural features. However, methyl (trimethylsiloxy) mercury which is tetrameric, with cubane-type structure in the solid state, is monomeric in solution, unlike zinc and cadmium analogues.

Infrared spectroscopy suggests free rotation of the methyl groups about metal-carbon bonds in MMe_2($M = Zn$, Cd, or Hg) and some structurally significant i.r. absorption bands are depicted in Table 3.10:

Table 3.10

	ν_a (M—C)	ν_s (M—C)	δ(CMC)
$ZnMe_2$	613	511	157
$CdMe_2$	536	473	140
$HgMe_2$	540	518	160

3.6.3 Organometallics of Group 13 Elements (B, Al, Ga, In and Tl)

One of the most interesting features of alkyls of main group III elements (similar to that of Li, Be, and Mg alkyls) is their electron deficient nature. These have a formal valency of three. Organometallics of these elements, in their monomeric forms, have only three pairs of electrons in their valency shell, i.e., two electrons short of an octet. These (e.g. monomeric trialkylboranes) exhibit trigonal planar configuration (Fig. 3.19), utilizing sp^2 hybrid orbitals of the central element. Such molecules with only six electrons in the valence shell are, therefore, coordinatively unsaturated.

Accordingly, the alkyls of these elements are Lewis acids and tend to form molecular-addition complexes (unless precluded on steric grounds) with elec-donors, e.g., $(CH_3)_3B{\leftarrow}NH_3$. It may be noted that on forming such a complex there is a change in hybridization of atomic orbitals on the boron atom from sp^2 (in BMe_3) to sp^3 (in the concerned complex).

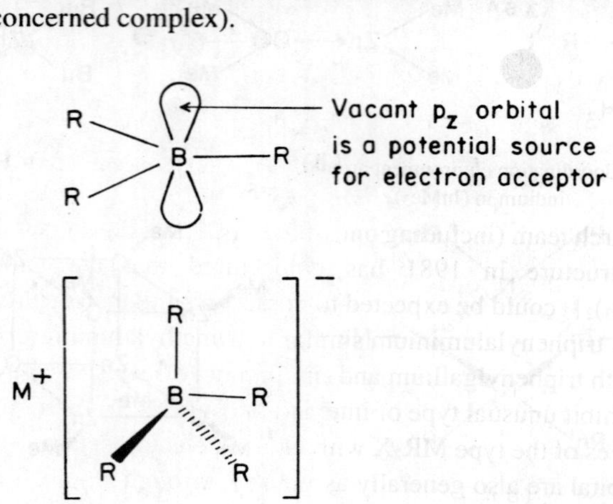

Fig. 3.19 Some organoborane structure types.

The main features of the bonding in trimethylaluminium dimer, Al_2Me_6, already been dealt with in Chapter 2 (Section 2.5.3). The structure of trime aluminium as determined by X-ray diffraction is illustrated in Figure 2.6, lengthening of the bridging Al–C bonds, the acute angle at the bridging carbon atom, and the orientation of the bridging methyl groups, etc., all support bonding characteristics discussed earlier in Section 2.5.3.

The heavier congeners of aluminium have markedly lesser tendency to electron deficient bridges involving carbon atoms. Even their trimethyl derivatives MMe_3(M = Ga, In or Tl) are monomeric in the vapour phase and in solution although their trigonal planar molecules (M = In or Tl) are packed in the solid state in a net work of tetrameric units, held together with very weak M—C bond.

The crystal structure of $GaMe_3$ is yet to be determined, but that of $InMe_3$ in the solid state consists of tetrameric In_4Me_{12} species, consisting of planar trigonal $InMe_3$ units, in which each indium atom is apparently bonded to two carbon atoms of two neighbouring $InMe_3$ molecules, thus each indium atom becomes surrounded by five methyl groups (i.e., an effective coordination number of five in a highly distorted trigonal-bipyramidal geometry (Fig. 3.20).

The extent of association in such molecules is markedly influenced by the ligands. For example, with a sterically more demanding ligand, $In\{CH(SiMe_3)_2\}_3$ exists even in the solid state as a monomer (Fig. 3.21).

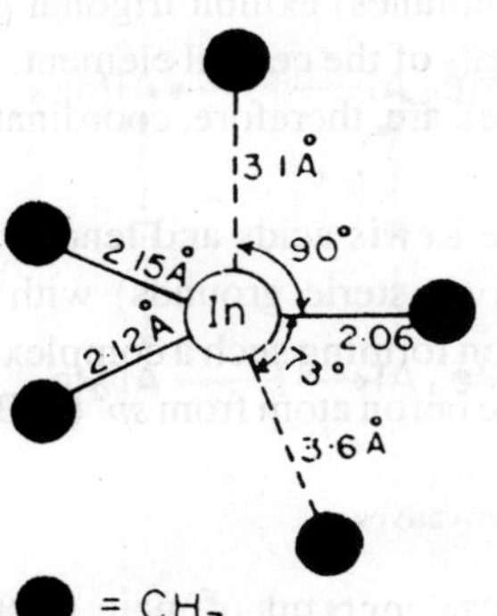

Fig. 3.20 Coordination environment around indium in $(InMe_3)_4$.

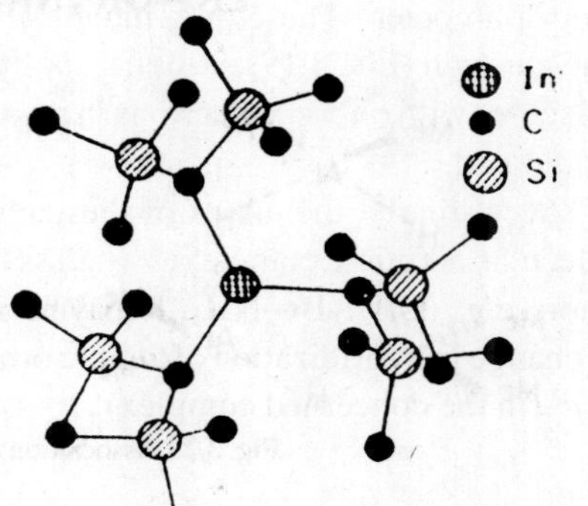

Fig. 3.21 Structure of $In\{CH(SiMe_3)_2\}_3$.

The research team (including one of the authors of this book) who determined the above structure in 1981 has conjectured that the gallium analogue, $Ga\{CH(SiMe_3)_2\}_3$ could be expected to possess a similar structure.

Although triphenylaluminium similar to trimethylaluminium is a dimer (Fig. 3.22a), yet both triphenylgallium and -indium have been shown by X-ray crystallogrphy to exhibit unusual type of interaction (Fig. 3.22b) in the solid phase.

Derivatives of the type MR_2X with only six electrons in the valence shell and a vacant p orbital are also generally associated, through anionic (X) bridges. The formation of dimers is the most common mode of association, although trimers or tetramers can be formed with relatively smaller bridging groups (H, F, OMe or NHMe).

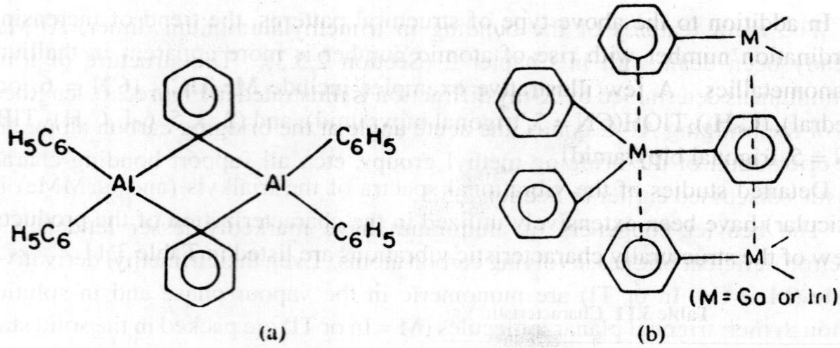

Fig. 3.22 Different modes of association in triphenyls of aluminium, gallium and indium.

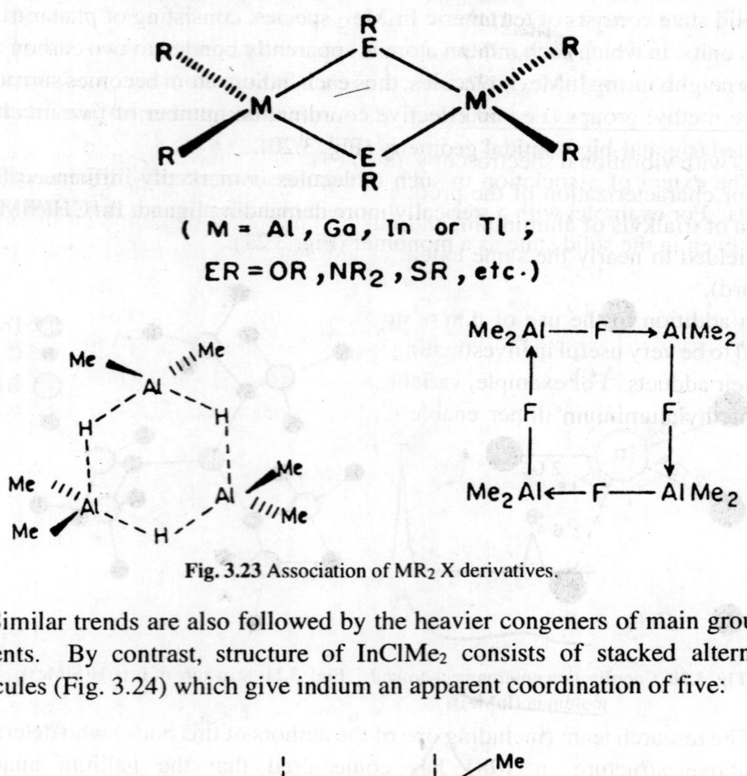

Fig. 3.23 Association of MR_2 X derivatives.

Similar trends are also followed by the heavier congeners of main group III elements. By contrast, structure of $InClMe_2$ consists of stacked alternative molecules (Fig. 3.24) which give indium an apparent coordination of five:

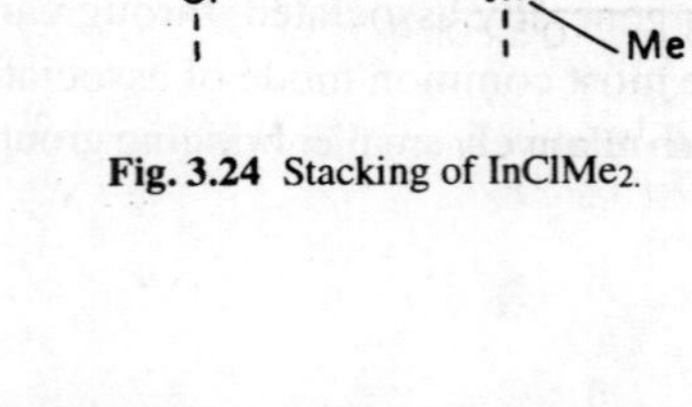

Fig. 3.24 Stacking of $InClMe_2$.

In addition to the above type of structural patterns, the trend of increasing coordination number with rise of atomic number is more apparent in thallium organometallics. A few illustrative examples include Me_2TlCN (CN = 6, octahedral), $(C_6H_5)_2TlOH$(CN = 5, trigonal bipyramid), and (2, 3, 5, 6-$F_4C_6H)_2$ TlBr (CN = 5, trigonal bipyramid).

Detailed studies of the vibrational spectra of the trialkyls (and of MMe_3 in particular) have been extensively utilized in the characterization of the products. A few of the structurally characteristic vibrations are listed in Table 3.11.

Table 3.11 Characteristic Variations of Group III Metal Alkyls

Compound	ν_a(MC)	ν_s(MC)	δ (CMC)	
$^{10}BMe_3$	1177	680	341	321
$^{11}BMe_3$	1149	680	341	321
$AlMe_3$	760	530	...	170
$GaMe_3$	577	522	...	163
$InMe_3$	500	467	...	132

As with vibrational spectroscopy, the n.m.r. techniques have also been widely used for characterization of the products. One significant point about ^1H n.m.r. spectra of trialkyls of aluminium, gallium, and indium is that alkyl group protons are shielded to nearly the same extent as are those in tetramethylsilane (used as standard).

In addition to the use of n.m.r. studies for diagnostic purposes, these have proved to be very useful in investigating the solution behaviour of MR_3 compounds and their adducts. For example, variable temperature ^1H n.m.r. studies (Fig. 3.25) of trimethylaluminium dimer enable us to demonstrate (i) the equivalence (as

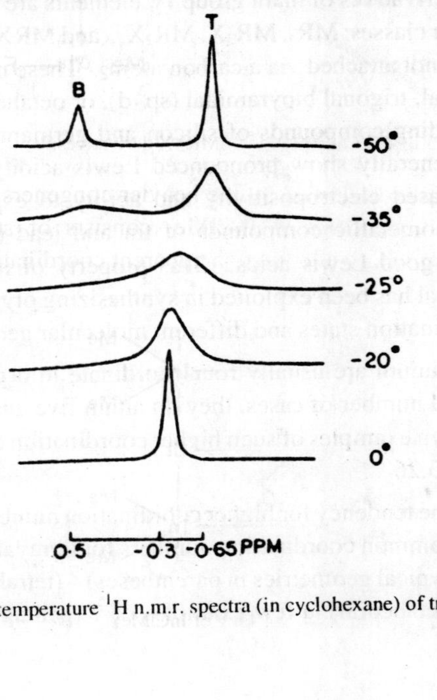

Fig. 3.25 Variable temperature ^1H n.m.r. spectra (in cyclohexane) of trimethylaluminium dimer.

shown by the appearance of a sharp singlet only) of all six methyl protons at room temperature, due to the rapid exchange (on n.m.r. time-scale) of terminal and bridging methyl groups; (ii) that on slowly cooling toluene or cyclohexane solution of trimerhylaluminium upto –50°C (i.e. on slowly freezing the dynamic process), the methyl resonance broadened and separated (at –50°C) into two distinct singlets having a 2 : 1 intensity ratio, arising respectively from the terminal and bridging methyl groups, and (iii) that the activation energy for exchange process calculated from the line shapes is 65 kJ mol^{-1} (in cyclohexane) or 71 kJ mol^{-1} (in toluene).

The n.m.r. spectra of monomeric coordination compounds, $MR_3.L$, have been used to assess the relative strength of donors (L) and electronic influence of the metal (M) centre. N.m.r. studies have also proved of great assistance in the studies of the exchange processes of the following types:

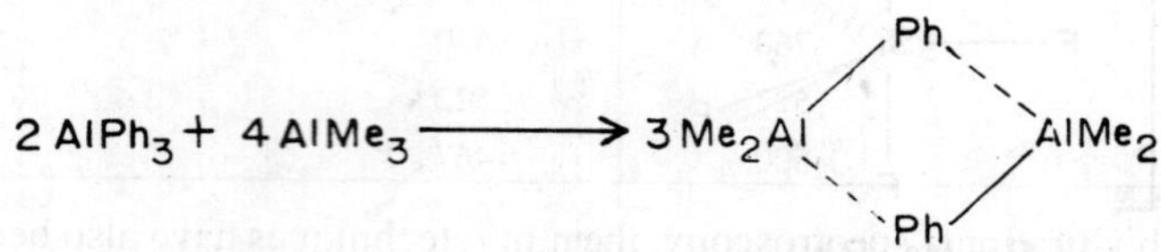

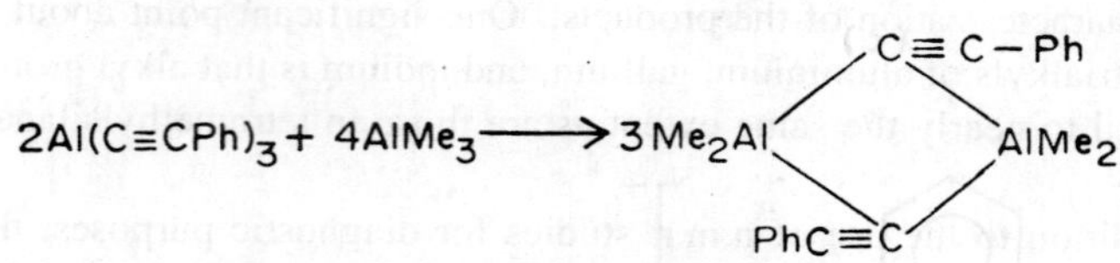

3.6.4 Organometallics of Group 14 Elements (Si, Ge, Sn, and Pb)

The organometallic derivatives of main group IV elements are often conveniently divided into four major classes: MR_4, MR_3X, MR_2X_2, and MRX_3(R = alkyl or aryl; X = an atom or group not attached via a carbon atom). These may generally have regular (sp^3) tetrahedral, trigonal bipyramidal (sp^3 d), or octahedral (sp^3 d^2) structures. The corresponding compounds of silicon and germanium, except MRX_3 derivatives, do not generally show pronounced Lewis acidity. However, as a consequence of increased electropositivity and ability to expand coordination numbers, many organometallic compounds of tin and lead (except tetraalkyls (aryls)) are generally good Lewis acids. This property of main group IV organometallics in general has been exploited in synthesizing organometallics complexes in higher coordination states and different molecular geometries.

Silicon and germanium are usually four-coordinate in organometallic compounds but in a limited number of cases, they do attain five and six-coordination states. The better known examples of such higher coordination derivatives may be cited as shown in Fig. 3.26.

As stated earlier, the tendency for higher coordination numbers increases from silicon to lead. Most common coordination numbers for tetravalent compounds of tin, and lead are (with typical geometries in parentheses) 4 (tetrahedral), 5 (trigonal bipyramidal), and 6 (octahedral), a few examples of 7 (pentagonal bipyramidal)

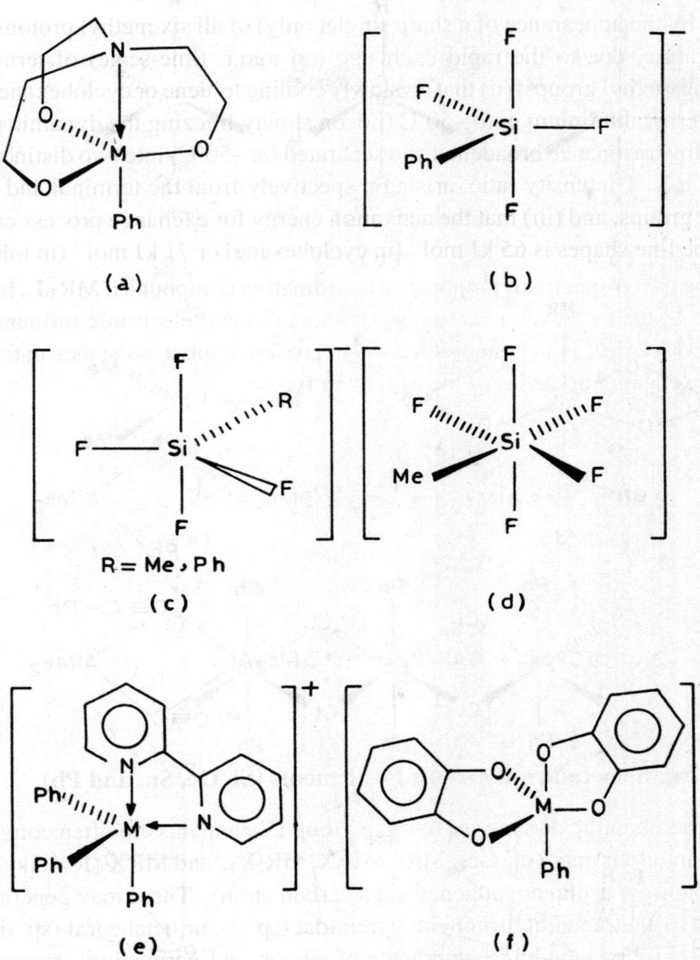

Fig. 3.26 Organo-silicon and -germanium derivatives in 5 or 6 coordination states.
Note: Only three structures, i.e., 3.26a (for Si & Ge both), and 3.26f have been established by X-ray diffraction.

and 8 (hexagonal bipyramidal) coordination are also known; a few typical examples are shown in Fig. 3.27.

In illustrating the structural features of a few group IV organometallics in different coordination states, the unique contribution of X-ray diffraction technique has been re-emphasized. As indicated above, no six-coordinated derivative of organo silicon or -germanium has been crystallographically established.

In case of organotin derivatives, the most important structural feature of the solid compounds of a general formula, SnR_3X or SnR_2X_2 (where X = a halogen or pseudohalogen) is their pronounced tendency to raise the coordination number of tin from four to five (for the SnR_3X compounds) or six (for the SnR_2X_2 and $SnRX_3$ compounds) via bridging X groups (i.e. by intermolecular association). As ex-

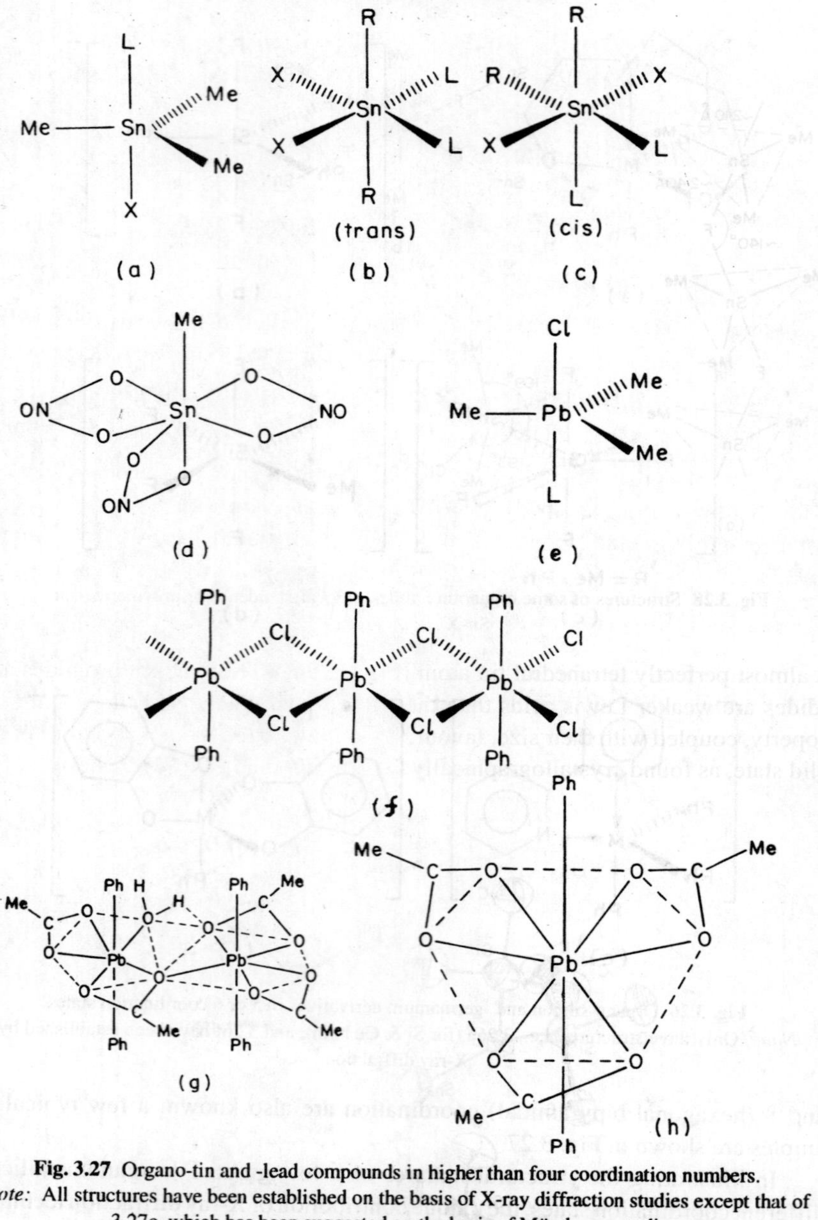

Fig. 3.27 Organo-tin and -lead compounds in higher than four coordination numbers.
Note: All structures have been established on the basis of X-ray diffraction studies except that of 3.27c, which has been suggested on the basis of Mössbauer studies.

pected, fluorine is more effective in this role than other halogens. For example, $Sn(X)Me_3$ (X = F or Cl) has a chain structure with trigonal bipyramidal coordination about tin (Figure 3.28a) and $Sn(F)_2Me_2$ has a layer structure with octahedral tin atom and *trans*-methyl groups above and below the F-bridged layers (Fig. 3.28b). The weaker Cl-bridging in $SnCl_2Me_2$ leads to the more distorted structure (Fig. 3.28c). The sterically hindered trialkyltin chloride $SnCl[CH(SiMe_3)_2]_3$ shows

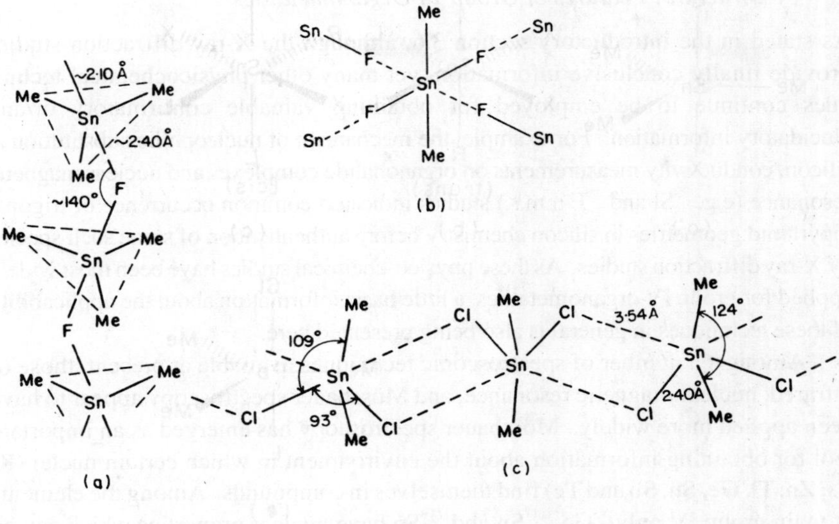

Fig. 3.28 Structures of some organotin halides, showing tendency to polymerize *via* Sn-X . . . Sn bands.

an almost perfectly tetrahedral tin atom (Fig. 3.29a). Triorganotin bromides and iodides are weaker Lewis acids than their fluoro and chloro analogues, and this property, coupled with their size, favours a monomeric tetrahedral structure in the solid state, as found crystallographically for $SnBrPh_3$ and $SnI(CH_2SiMe_3)_3$.

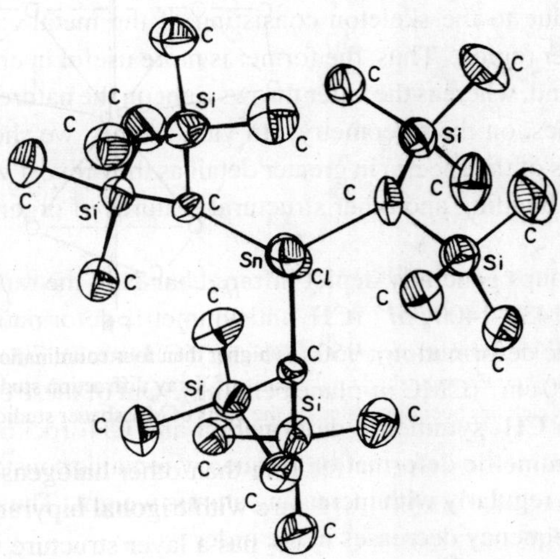

Fig. 3.29 Structure of $SnCl[CH(SiMe_3)_2]_3$.

3.6.4.1. Applications of Spectroscopic Techniques in the Elucidation of Structural Features of Group 14 Organometallics

As stated in the introductory section 3.6, although the X-ray diffraction studies provide finally conclusive information, yet many other physicochemical techniques continue to be employed for obtaining valuable confirmatory or/and elucidatory information. For example, the mechanism of nucleophilic substitution at silicon, conductivity measurements on organohalide complexes and nuclear magnetic resonance (e.g. ^{29}Si and ^{19}F n.m.r.) studies indicated common occurrence of trigonal bipyramid geometries in silicon chemistry before authentication of some such species by X-ray diffraction studies. As these physico-chemical studies have been most widely applied for group IV organometallics, a little basic information about the applicability of these techniques in general is also being presented here.

Amongst a number of spectroscopic techniques available at present, those of infrared, nuclear magnetic resonance, and Mössbauer spectroscopy appear to have been applied more widely. Mössbauer spectroscopy has emerged as an important tool for obtaining information about the environment in which certain nuclei (K, Cs, Zn, Tl, Ge, Sn, Sb and Te) find themselves in compounds. Among the elements of main group IV only ^{73}Ge, ^{117}Sn and ^{119}Sn have nuclear properties which can be investigated by Mössbauer effect, and out of these, ^{119}Sn derivatives have received maximum attention.

(*I*) *Infrared Spectroscopy*

Amongst the common spectroscopic techniques, infrared spectroscopy has been employed for the longest period. The infrared spectra of organometallic compounds show characteristic absorptions mainly in two regions: The high frequency or NaCl region (4000-650 cm^{-1}), and the low frequency or far-infrared or CsI region (650-50 cm^{-1}). The bands due to the organic moiety appear generally in the former region, and those due to the skeleton consisting of the metal-carbon bonds are observed in the latter region. Thus, the former is more useful in characterizing the organic moiety/ligand, whereas the latter throws light on the nature of metal-ligand bonds and sometimes, on their geometry. In view of this, we should discuss the far-infrared portions of the spectra in greater detail as they reveal vital information about the nature of bonding and other structural features of organometallic compounds.

The M-CH_3 groups generally depict infrared bands in the ranges: 3000-2900 cm^{-1} (C-H stretch); 1430-1400 cm^{-1} (CH_3 antisymmetric deformation); 1300-1100 cm^{-1} (CH_3 symmetric deformation); 950-700 cm^{-1} (CH_3 rock); 700-400 cm^{-1} (M-C stretch); and 300-100 cm^{-1} (CMC in-plane bending). Out of these characteristic i.r. bands, those due to CH_3 symmetric deformation and CH_3 rocking are the most useful. The CH_3 symmetric deformation frequency remains constant for the same metal, but decreases regularly with increasing atomic weight. Thus for main group IV elements, this frequency decreases in the order:

Si	>	Ge	>	Sn	>	Pb
1260-50		1240-30		1200-1180		1170-50 cm^{-1}

The metal attached CH_3 group rocking is sensitive to the nature of the metal, but it also varies with the structural types (even if the metal is the same.)

Metal carbon stretching frequencies (Table 3.12) are sensitive to the metal and their values as expected decrease regularly with increasing atomic weights. For group IV organometallics the general range for metal-carbon stretching frequency is

νSi-C	>	νGe-C	>	νSn-C	>	νPb-C
800-600		630-550		600-500		500-460 cm^{-1}

For tetramethyls metal-carbon skeleton frequencies (Table 3.12) follow the same trend:

Table 3.12 Metal-Carbon Skeleton Frequencies (cm^{-1}) for Group 14 Tetramethyls.

Compound	Structure	ν_a (MC)	ν_s (MC)	δ (CMC)
$Si(CH_3)_4$	Td	696	598	239, 202
$Ge(CH_3)_4$	Td	595	558	195, 175
$Sn(CH_3)_4$	Td	529	508	157
$Pb(CH_3)_4$	Td	476	459	120

In phenyl-metal derivatives a characteristic band appears in the region 1200-1050 cm^{-1}; this is a benzene ring vibration interacting with M-C stretching vibration and, therefore, it is sensitive to the nature of the metal. For example, in tetraphenyls of group IV metals, this stretching vibration decreases with the rise of atomic weights:

$SiPh_4$	>	$GePh_4$	>	$SnPh_4$	>	$PbPh_4$
1100		1080		1065		1052 cm^{-1}

In a series of compounds of the type, $SnCl_{4-x}(C_6H_5)_x$, as the number of the phenyl groups increases, in-plane and out-of-plane bending modes shift to lower frequencies and CH wagging vibrations to higher frequencies.

In vinyl-metal compounds, CH_2 wagging (which appears in the region 960-940 cm^{-1}) is much less sensitive to the nature of the metal compared to methyls. For example, in a series of compounds, $M(C_6H_5)_3(CH = CH_2)$, CH_2 wagging is reported to decrease in the order

M = Si	>	Ge	>	Sn	>	Pb
960		952		950		942 cm^{-1}

The applications of the above general principles to a few illustrative examples of organometallics of silicon, germanium, and tin are now described in the following pages.

(a) *Organosilicon compounds:* A large amount of infrared spectral data has been accumulated on a variety of organosilicon compounds, and these results give the following frequency regions for various structurally characteristic bonds containing silicon: ν(Si-H), 2236-2150 cm^{-1}; δ_s (Si-CH_3), 1280-1255 cm^{-1} (a very strong sharp band); ν (Si-F), 970-850 cm^{-1}; ν(Si-Cl), 600-450 cm^{-1}; ρ(Si-CH_3)

860-760 cm^{-1} (one methyl on a silicon usually absorbs near 765 cm^{-1}; two methyls, near 855 and 800 cm^{-1}; and three methyls, near 840 and 765 cm^{-1}); and ν(Si-C), 800-600 cm^{-1}. It is interesting to note that the Si-C stretching frequencies vary markedly as the other group(s) attached to silicon change (Table 3.13).

Table 3.13 Si-C Stretching Frequencies (cm^{-1})

Compound	ν_s (Si-C)	ν_a (Si-C)
$SiH(CH_3)_3$	623	718
$SiF(CH_3)_3$	620	695
$SiCl(CH_3)_3$	636	700
$SiH_2(CH_3)_2$	658	732
$SiF_2(CH_3)_2$	670	786
$SiCl_2(CH_3)_2$	688	805

The absorption in the range, 1125-1100 cm^{-1} in the i.r. spectrum of Si-C_6H_5 derivatives is ascribed to a planar ring vibration having some contribution of Si-C stretching also. Absorption bands due to phenyl groups themselves are found near 1430, 730, and 695 cm^{-1}.

The characteristic absorptions due to vinyl group attached to silicon appear at 1615-1590 cm^{-1} (C = C stretch), 1410-1390 cm^{-1} (CH_2 deformation), 1020-1000 cm^{-1} (*trans* CH wag), and 980-950 cm^{-1} (CH_2 wag).

(b) *Organogermanium compounds:* A number of germanium organometallics containing the Ge-CH_3 group show the following absorption: δ_{as} (CH_3), 1437-1405 cm^{-1}; δ_s(CH_3), 1259-1229 cm^{-1}; ρ_r(CH_3), 850-787 cm^{-1}; and ν(Ge-C), 641-535 cm^{-1}. Infrared evidence has also been obtained for both *trans* and gauche forms in n-butyl and n-propylgermanes: ν(Ge-C) *trans* at 645, and ν(Ge-C) gauche near 550 cm^{-1}. The phenylgermanium compounds exhibit vibrations characteristics of Ge-Ph stretching mode at about 300 cm^{-1}: ν_{as} (Ge-Ph), ~ 325 cm^{-1}; ν_s (Ge-Ph), ~ 300 cm^{-1}. Other characteristic infrared absorptions of some germanium-ligand stretching vibrations are: ν(Ge-F), 744-662 cm^{-1}; ν(Ge-Cl), 424-362 cm^{-1}; ν(Ge-Br), 327-228 cm^{-1}; ν(Ge-I), 283-200 cm^{-1}; ν(Ge-H), 2100-2000 cm^{-1}; and ν(Ge-O), 926-789 cm^{-1}.

(c) *Organotin compounds:* A large number of tin oranometallics containing Sn-CH_3 group exhibit the following group vibrations: δ_a(CH_3), 1430-1400 cm^{-1}; δ_s(CH_3), 1230-1180 cm^{-1}; ρ_r(CH_3), 800-700 cm^{-1}. Characteristic infrared absorptions of some tin-ligand stretching vibrations are: ν(Sn-C), 600-470 cm^{-1}; ν(Sn-Cl), 390-310 cm^{-1}: ν(Sn-Br), 270-220 cm^{-1}; and ν(Sn-I), 210-170 cm^{-1}.

Infrared-spectroscopy has provided valuable information on the molecular geometry of organotin compounds. Particularly important is the criterion established for determining the configuration of SnC_3 and SnC_2 moieties in chlorotrimethyltin and dichlorodimethyltin derivatives. For example, if the spectrum reveals two Sn-C stretching vibrations (both the symmetric and the asymmetric modes), the configuration of SnC_3 group is nonplanar as is shown by $SnX(CH_3)_3$ type compounds (with X = Cl, Br, I, NO_3, etc) in Table 3.14 or that of SnC_2 is nonlinear.

Table 3.14 Infrared Absorption due to Sn-C Stretching Frequencies (cm^{-1}) in $SnX(CH_3)_3$ Type Derivatives

X	Structure	ν(Sn-C)
Cl	Tetrahedral	545, 513
Br	Tetrahedral	543, 512
I	Tetrahedral	540, 509
NO_3	Tetrahedral	557, 520
F	Planar	552
OH	Planar	540
ClO_4	Planar	552
CH_3COO	Planar	552

However, the appearance of only one band assignable to the Sn-C stretching vibration, the configuration of SnC_3 group is planar or that of SnC_2 group is linear.

Thus, the infrared spectra of $Sn(OH)(CH_3)_3$, $Sn(ClO_4)(CH_3)_3$ and $Sn(CH_3COO)(CH_3)_3$ show only one strong absorption in the Sn-C stretching region around 550 cm^{-1}. This indicates a planar arrangement of the SnC_3 moiety (Fig. 3.28a). For the other halides two absorptions assigned to the symmetric and asymmetric Sn-C stretching vibrations are observed, in accord with the tetrahedral configuration about the tin atom.

In dihalogenodimethyltin compounds, a band characteristic of ν_s(Sn-C) mode is observed at 515, 514 and 511 cm^{-1} respectively in the chloride, bromide and iodide derivatives. The corresponding ν_a(Sn-C) mode is observed at 567, 563, 547 cm^{-1} respectively. It is noteworthy that for tetrahedral configuration in compounds of the type $SnX_2(CH_3)_2$, two Sn-C stretching modes are infrared active. However, if it possesses the linear $Sn(CH_3)_2^{++}$ cation and two X anions, only one stretching mode is allowed in the infrared spectrum. For example, compounds $Sn(O_2CH)_2(CH_3)_2$ and $Sn(OH)(NO_3)(CH_3)_2$ exhibit ν(Sn-C) at 592 and 585 cm^{-1} respectively.

In view of organotin compounds showing higher coordination states of 5 and 6 quite frequently, a brief account of the infrared characteristics of these is now being presented.

(i) *Pentacoordinated tin compounds:* Triorganotin derivatives SnR_3X (X = anionic ligand) achieve pentacoordination either by intermolecular tin-ligand bridge bonds (e.g. in $SnFMe_3$) or by intramolecular dative tin ligand interaction (e.g. in $SnClMe_3{\cdot}HMPA$). The pentacoordinated complexes of the type $SnXMe_3{\cdot}L$ (L = unidentate Lewis base) in the trigonal bipyramidal arrangement can exist in the following two forms (Fig. 3.30a and 3.30b):

These two different structures can be distinguished on the basis of their infrared spectra. For example, the planar arrangement of Sn-Me_3 moiety (Fig. 3.30a) should exhibit only one band in the Sn-C stretching region. Thus for the complex $SnXMe_3(Py)$ (X = Cl or NCS), one band is observed at ~ 560 cm^{-1} due to Sn-C stretching frequency. The Sn-Cl stretching vibration in chlorotrimethyltin observed around 330 cm^{-1}, is shifted to below 250 cm^{-1} in the adducts. This

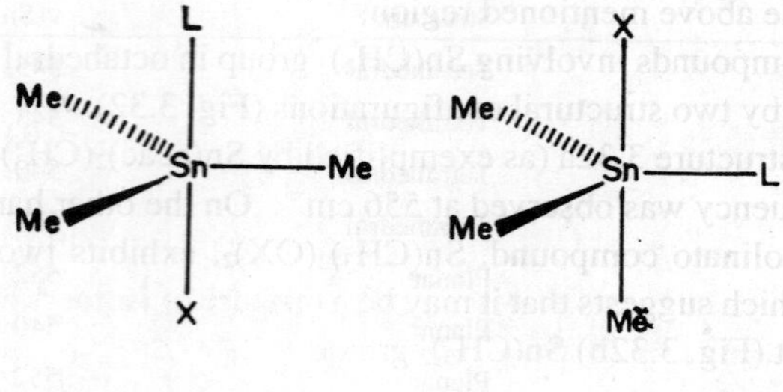

(a) (b)

Fig. 3.30 Trimethyltin compounds with trigonal bipyramidal geometry.

lowering in Sn-Cl stretching frequency is understandable in terms of the increase in electron density on the tin atom, which causes enhancement in the polarity of Sn-Cl bond. The derivatives in nonplanar arrangement (Fig. 3.30b) are not accessible because of the preferential occupation of axial sites by electronegative atoms or groups.

(ii) *Hexa- and penta-coordinated compounds,* $R_2SnX_2{\cdot}2L$ *and* $R_2SnX_2{\cdot}L$: Infra-red spectroscopy has been particularly effective in elucidating the structural features of several octahedral complexes with the general formula $Me_2SnX_2{\cdot}2L$ (where X = anionic ligand, and L = monodetate Lewis base). For example, the absence of a symmetric Sn-C stretching vibration (in the 580-500 cm^{-1} region) in the i.r. spectrum of $SnCl_2(CH_3)_2{\cdot}2NC_5H_5$ strongly suggests that the $Sn(CH_3)_2$ moiety is linear in this derivative with a trans alignment of the methyl groups. The molecular geometry of this and many other similar complexes can be represented in the following two forms (Figures 3.31a and 3.31b).

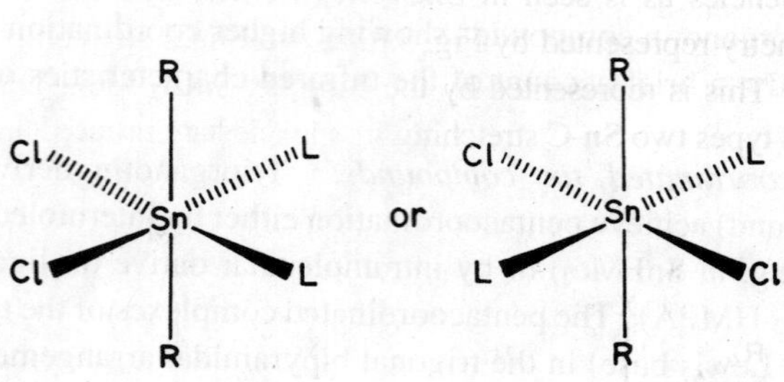

(a) (b)

Fig. 3.31 ***Cis*** **and** ***trans*** **forms of octahedral** $SnCl_2R_2{\cdot}2L$**.**

Distinction between the above two possible configurations can also be made from the Sn-Cl absorptions in the i.r. spectra. For example, compounds having the structure 3.31(a) should have two infrared active Sn-Cl stretching frequencies, as have been observed in $SnCl_2(CH_3)_2(2NC_5H_5)$ in the 230 cm^{-1} region. However, compounds in configuration, Fig. 3.31(b), should exhibit only one Sn-Cl stretching frequency in the above mentioned region.

Chelate compounds involving $Sn(CH_3)_2$ group in octahedral environment can be represented by two structural configurations (Fig. 3.32).

For *trans* structure 3.32a (as exemplified by $Sn(acac)_2(CH_3)_2$), only one Sn-C stretching frequency was observed at 556 cm^{-1}. On the other hand, the analogous 8-hydroxyquinolinato compound, $Sn(CH_3)_2(OX)_2$, exhibits two Sn-C stretching frequencies, which suggests that it may be a mixture of isomers having linear (Fig. 3.32a) and bent (Fig. 3.32b) $Sn(CH_3)_2$ groups.

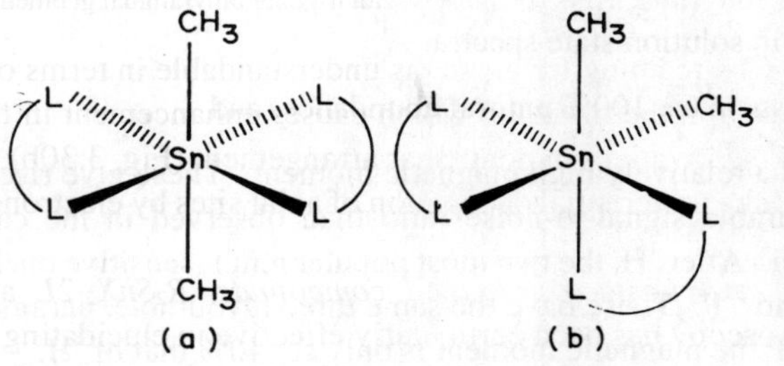

Fig. 3.32 Chelated octahedral $Sn(CH_3)_2(L\text{-}L')_2$ *cis* and *trans* forms.

Pentacoordinated geometry for $R_2SnCl_2{\cdot}L$ dichlorodiorganotin derivatives, is not very common, but some complexes in trigonal bipyramidal arrangement (such as $SnCl_2Me_2{\cdot}Me_2SO$ and $SnCl_2Me_2{\cdot}DMF$) are known. These types of complexes can be represented in two ways (Figs. 3.33a and 3.33b):

The compounds represented by Fig. 3.33a should show one infrared active Sn-Cl stretching frequencies as is seen in $SnCl_2Me_2{\cdot}Me_2SO$, whereas for compounds in molecular geometry represented by Fig. 3.33b, two Sn-Cl stretching frequencies are infrared active. This is represented by the complex $SnCl_2Me_2 \cdot DMF$. In both the above structural types two Sn-C stretching frequencies are infrared active.

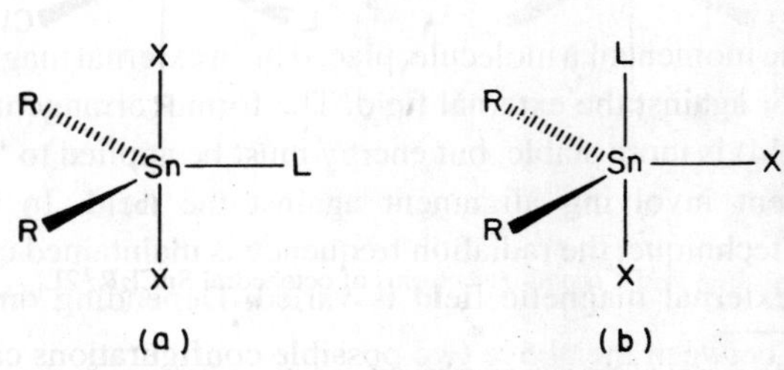

Fig. 3.33 Isomeric forms of penta-coordinated $SnCl_2Me_2{\cdot}L$ derivatives (R = Me).

(II) *Nuclear Magnetic Resonance Spectroscopy*

Since the discovery (in 1945) of n.m.r. phenomenon by two groups of scientists (Bloch, *et al.* at Stanford University, and Purcell, *et al.* at Harvard University) for which Block and Purcell were jointly awarded the Nobel Prize for Physcis in 1952, the applications of n.m.r. spectroscopy have been increasing at a fast pace in chemistry. With the introduction of the first commercial ^{1}H n.m.r. spectrometer in 1953, and gradual improvements in its resolving power, proton spectrometers have become an integral part of every research and teaching laboratory for many years. This is because the ^{1}H nucleus is most suitable for n.m.r. studies due to the following factors:

(i) it has spin quantum number $I = \frac{1}{2}$, and therefore normally gives sharp lines in solution state spectra.

(ii) it exists in ca. 100% natural abundance, and

(iii) it has a relatively high magnetic moment. These give rise to in a more favourable signal-to-noise ratio than observed in the cases of other nuclei. After ^{1}H, the two most popular n.m.r. sensitive nuclei have been ^{19}F and ^{31}P. These have the same three favourable, parameters, though for ^{31}P the magnetic moment is only ca. 40% that of ^{1}H.

The advent (in 1966) and development (in 1976) of Fourier Transform (FT) n.m.r. spectroscopy have paved the way for a faster progress in the applications of n.m.r. spectroscopy with other nuclei such as ^{13}C; ^{29}Si, ^{119}Sn, ^{207}Pb and ^{15}N, all of which have nuclear spin $I = \frac{1}{2}$.

In spite of less sharp resonances with nuclei like ^{11}B and ^{27}Al with spin values 3/2 and 5/2 respectively, many useful studies have been made employing n.m.r. spectroscopy with nuclei having $I > 1/2$ also, taking the assistance of more sophisticated instrumentation facilities.

In the following pages, the applications of ^{1}H n.m.r. spectroscopy to main group organometallics are discussed first, followed by a brief account* of the salient features of ^{13}C and ^{119}Sn n.m.r. studies relevant to the structural elucidation of organometallic compounds of nontransition metals in general and that of tin in particular.

The magnetic moment of a molecule, placed in an external magnetic field, may be aligned with or against the external field. The former arrangement (i.e., alignment with the field) is more stable, but energy must be applied to 'flip' to the less stable arrangement involving alignment against the field. In the commonly employed n.m.r. technique, the radiation frequency is maintained constant and the strength of the external magnetic field is varied. Depending on the electronic

---

* For a more detailed account of the n.m.r. technique and its applications to elucidation of structural problems in chemistry, the reader is advised to consult a large number of special books available on the topic (see references in supplementary reading).

environment of the proton (or some other n.m.r. sensitive nuclei such as ^{13}C, ^{19}F and ^{119}Sn), at some specified value of the field strength, an absorption occurs indicated by the appearance of a signal (resonance) when the energy required to flip the spin state coincides with the energy of radiation. The effective field seen by the nucleus is affected by its electronic environment. The chemical shift of the resonance (difference in the shielding constants of the sample and a reference compound) is, therefore, indicative of the chemical environment and hence, the n.m.r. spectroscopy is a very useful tool in the elucidation of the molecular geometry of the absorbing species.

Nuclear magnetic resonance studies provide information on the following parameters: (i) chemical shifts, (ii) coupling constants, (iii) relaxation times, (iv) exchange life times, and (v) intensities.

The chemical shift provides valuable information on the kinds of protons (or other nuclei) present in a molecule, but it does not provide us with any quantitative information concerning the ionicity of bonds.

A second structurally useful parameter derivable from n.m.r. spectroscopy is the spin-spin coupling constant. It is a measure of the interaction of two nuclei through the electrons between them. As such it is sensitive to the electron density at the nucleus and might be a useful probe of ionicity effects.

(a) *^1H n.m.r. (p.m.r.) spectra.* The solution of a typical structural analysis problem by n.m.r. techniques in general and ^1H n.m.r. in particular utilizes at least four kinds of information obtained from n.m.r. spectrum: (i) Chemical shifts (δ), which indicate the electronic environment of each kind of proton nucleus; (ii) intensities, which provide information about the number of nuclei (^1H) of each kind; (iii) coupling constants and splitting of a signal into several peaks (these parameters tell us about the environment of a nucleus (^1H) with respect to other, nearby nuclei (^1H)), and (iv) number of signals, which is indicative of the different environments of the same nuclei (^1H) in a molecule.

The applications of ^1H n.m.r. spectroscopy for organometallic species are being illustrated by the simplest members of the series, i.e., metal methyls.

Chemical Shifts: The proton resonances (δ) $M–CH_3$ groups generally lie at the highly screened end of the chemical shift scale in the region + 2 → – 1.5. In organic compounds there is a relationship between the proton chemical shift and the electronegativity of the substituents (i.e., the polarity of the C-X bond):

CH_3–X	CH_3–H	CH_3–I	CH_3–Br	CH_3–Cl	CH_3–F
δ(CH_3), ppm	0.13	1.98	2.45	2.84	4.13
Electronegativity (Pauling) value of X	2.1	2.5	2.8	3.0	4.0

A similar correlation between the methyl proton chemical shifts and the polarity of the M–C bond has been observed for methyl derivatives of metals, $M(CH_3)_n$, Table 3.15.

Table 3.15 Chemical Shifts of Metal-Attached Methyl Protons in $M(CH_3)_n$

Metal	Range of chemical shift (δ)
Li	– 1.3 to – 1.5
Be, Mg	– 1.0 to – 1.5
Zn, Cd	0 to – 1
Al, Ga, In, Tl	0.6 to – 0.5
Hg	0 to 1
Si, Ge, Sn, Pb	0 to 1
As, Sb. Bi	0.3 to 0.6
Se, Te	2.0

As is evident from Table 3.15, the increased shielding of the protons in the series M = Hg (1.9), Sn (1.8), Cd (1.7), Zn (1.6), Al (1.5), Mg (1.2) and Li (1.0) parallels the increasing ionic character of the corresponding metal-carbon bonds, as indicated by the decreasing electronegativity values (given in parenthesis) of the different metals on the Pauling scale.

The electronegativity effect on the methyl chemical shifts can also be observed in a series of compounds of the type $MX_{4-n}(CH_3)_n$(M = Si, Ge, Sn; X = Cl; n = 1, 2, 3 or 4) shown in Table 3.16.

Table 3.16 Chemical Shifts (δ) of Metal-Attached Methyl Protons in $MX_{4-n}(CH_3)_n$

Compound	δ	Compound	δ	Compound	δ
$SiMe_4$	0.0	$GeMe_4$	– 0.14	$SnMe_4$	0.07
$SiClMe_3$	0.42	$GeClMe_3$	0.64	$SnClMe_3$	0.63
$SiCl_2Me_2$	0.80	$GeCl_2Me_2$	1.30	$SnCl_2Me_2$	1.17
$SiCl_3Me$	1.14	$GeCl_3Me$	1.83	$SnCl_3Me$	1.65

The [1]H n.m.r. studies have provided valuable information for other systems also. For example, the saturated carbanions exhibit pronounced shielding effects only for the α-protons with resonances at high field. On the other hand, extensive delocalization of charge is indicated by the δ values of the aromatic protons in benzyllithium where the *orth-* and *para*-protons are the most affected (Table 3.17):

Table 3.17

$LiCH_3$ ↑ δ(ppm) – 1.3	$LiCH_2CH_3$ ↑ ↑ – 0.99 1.33	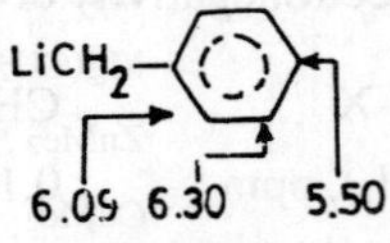
$Mg(CH_3)_2$ ↓ δ(ppm) –1.3	$Mg(CH_2CH_3)_2$ ↑ ↑ 0.64 1.26	$Mg(CH_2\ CH_2\ CH_3)_2$ ↑ ↑ ↑ – 0.57 1.50 0.90

Coupling Constants: The measurement of spin-spin coupling constants in metal alkyls has proved to be of considerable significance in providing information about the molecular geometries. For example, $^2J(^{119}Sn–C–^1H)$ coupling constants provide valuable information on structures of organotin compounds on the whole and on structures of methyltin halides in particular, with values varying between ~ 37–120 Hz. Simple theoretical considerations lead to the inference that the coupling constant would be directly proportional to the degree of 's' character of the metal orbital bonded to the methyl group. Thus, in $SnMe_4$, J(^{119}Sn–(^{1}H)) is 54 Hz, while in $SnClMe_3$, J is 58.5 Hz reflecting a redistribution of s-character while retaining approximately tetrahedral coordination for tin. In the adduct of $SnClMe_3$ with pyridine, $SnClMe_3Py$, which is known (by X-ray crystallographically) to contain tin in a trigonal bipyramidal environment of three methyl groups in equatorial sites and two ligands chloride and pyridine in axial positions, J value increases to 67 Hz.

(b) ^{13}*C n.m.r. Spectroscopy*: Carbon occurs in 3 isotopic forms ^{12}C, ^{13}C and ^{14}C with natural abundances of 98.9, 1.1, and 10^{-10} % respectively. Carbon–12 (and so is the case incidentally with the predominant isotopic form of oxygen) has fortunately zero spin, which makes the ^{1}H n.m.r. spectroscopy of organic (and organometallic) compounds rather simple and convenient.

In spite of the low abundance ^{13}C and unfavourable sensitivity (compared to ^{1}H), n.m.r. studies with these nuclei with spin 1/2 value have in some cases added useful information to that derived by other physico-chemical studies including p.m.r. spectroscopy. For example, ^{13}C chemical shifts generally provide information about the hybridization state of the carbon atom. The general trend in δ (^{13}C) values, i.e., σ (sp^3) > σ (sp) > σ (sp^2), parallels the order found in ^{1}H n.m.r. The normal range of ^{13}C chemical shifts is 0–200 ppm, so that the spread of chemical shifts is about twenty times that of the protons.

The small upfield ^{13}C chemical shift of methyllithium, relative to methane, can be cited as an evidence for a predominantly sp^3 hybridized carbon (in metal methyls), with some excess electronic charge density on the metal. The experimental observations depicted in Table 3.18 serve to illustrate this point.

Table 3.18 ^{13}C Chemical Shifts for Some Metal Methyls

Compound	(δ, ppm)
LiMe	– 9.2 to – 16.2 (solvent dependent)
$LiCH_2Ph$	18.3 to 29.8
$Mg(CH_2Ph)_2$	21.9
$ZnMe_2$	– 4.2
$CdMe_2$	1.0
$HgMe_2$	23.5
$SiMe_4$	0
$GeMe_4$	– 0.8 to – 1.4
$SnMe_4$	– 8.6 to – 9.3
$PbMe_4$	– 2.6 to – 3.4

Further, the measurement of ^{13}C coupling constants to other n.m.r. sensitive nuclei (such as ^{1}H) shows the structure dependence of the $^{1}J\,(^{13}C\text{–}^{1}H)$ data, as shown in Table 3.19.

Table 3.19 $^{1}J\,(^{13}C\text{–}^{1}H)$ Coupling Constants of Some Metal Methyls

Compound	$^{1}J\,(^{13}C\text{–}H)$ in Hz
$(LiMe)_4$	98
$BeMe_2$	106
$CdMe_2$	127
$HgMe_2$	130
Al_2Me_6	113
$GaMe_3$	122
$InMe_3$	120
$SeMe_2$	140
$TeMe_2$	141
$SiMe_4$	118
$GeMe_4$	125
$SnMe_4$	128
$PbMe_4$	133

To account for the $^{1}J\,(^{13}C\text{–}^{1}H)$ values (ca. 98 Hz), which are less than normally observed (123–205 Hz) for sp^3 hybridized carbon, an increased s-character of the M–C bond (40%) was advocated.

It may be worthwhile to mention that splitting pattern due to ^{13}C coupling with other n.m.r. active nuclei has often proved to be of particular importance in elucidating structures of main group organometallics. For example, the number of lines observed in the ^{13}C n.m.r. spectrum of $(LiMe)_4$ in THF shows coupling of ^{13}C with three equivalent ^{7}Li atoms, which is in conformity with its solid state structure (*cf.* Section 3.6.1).

(c) *^{119}Sn n.m.r. spectroscopy*: Out of ten natural isotopes of tin, only three: ^{115}Sn, ^{117}Sn and ^{119}Sn with abundances of 0.34, 7.07 and 8.58% respectively are amenable to n.m.r. spectroscopy as all the three have nuclear spin values of 1/2. However, ^{115}Sn is of too low an abundance to be of significance. Metal chemical shifts have generally been observed for ^{119}Sn and data are reported relative to $SnMe_4 = 0$, with downfield shifts from the reference compound having a positive sign.

^{119}Sn chemical shifts, $\delta\,(^{119}Sn)$ of organotin derivatives are spread over a range of ~ 600 ppm. In any series of organotin compounds, factors leading to increase in electron density (shielding) of the tin atom should shift the $\delta\,(^{119}Sn)$ to higher field; this is exemplified by the values of $\delta\,(^{119}Sn)$ given in parentheses for a series of SnR_3Cl derivatives; $SnClMe_3$ (+ 164), $SnClEt_3$ (+ 155), $SnBu_3Cl$ (+ 141), $SnClBu^t_3$ (+ 50), $SnClPh_3$ (– 48), which follow the relative order of the electron-releasing ability (i.e., + I indicative effect) of the alkyl (aryl) groups.

Structurally, a more informative property of ^{119}Sn chemical shifts is the growing upfield shift (for similar reasons) of δ (^{119}Sn) with increase in coordination number of the tin atom from four to five, six or seven. For example, δ (^{119}Sn) for tetrahedral $SnClMe_3$ at + 159 ppm moves to – 9 ppm on addition to excess pyridine; this upfield shift is consistent with the formation of trigonal bipyramidal adduct (which is confirmed by X-ray studies Fig. 3.30a). The observed ^{119}Sn chemical shifts for some other tetracoordinated organotin compounds compared with those of the corresponding derivatives in octahedral geometries given in parentheses confirm the same trend: $SnMe_2 (SMe)_2$, + 114 ($SnMe_2 (SCSNEt_2)_2$, – 336); $SnBu_2 Cl_2$, + 122($Sn(CH_2CH_2OCOBu)_2Cl_2$, – 68.5); $SnBuCl_3$, + 6 ($SnBu (Cl)(Ox)_2$ – 395).

Further evidence for tetrahedral, trigonal bipyramidal, octahedral, and other higher coordination geometries has been obtained from the *coupling constants* involving tin and hydrogen, ^2J (^{119}Sn–C–H) (especially for methyltin compounds) as well as tin and carbon, ^1J (^{119}Sn–^{13}C). For example, ^2J (^{119}Sn–C–H) values for pentacoordinate (trigonal bipyramidal) trimethyltin derivatives (59–71 Hz) are higher than those for analogous tetracoordinate (tetrahedral) derivatives (e.g., $SnClMe_3$, 58.5).

A similar trend is indicated by the ^1J (^{119}Sn–^{13}C) coupling constant values of methyltin complexes. For example, tetrahedral chlorotrimethyltin shows ^1J (^{119}Sn–^{13}C) coupling value of 372 Hz, but in the presence of excess of pyridine which forms the pentacoordinate complex ($SnClMe_3 \cdot py$), it increases to 472 Hz.

A further increase in coordination number raises the ^1J (^{119}Sn–^{13}C) coupling constant values more significantly, as is shown by the value of 900–770 Hz in octahedral SnR_2X_2 derivatives.

(III) *Mössbauer Spectroscopy*

Mössbauer spectroscopy is a physico-chemical technique to estimate the electron density at the nuclei of some atoms by measurements of transitions between nuclear energy levels with the help of emission and absorption of γ-rays. Discovered in 1957, by R.L. Mössbauer, the technique began to be used for elucidation of chemical structures with the nucleus, ^{57}Fe. Although work has been concentrated mainly around ^{57}Fe and ^{119}Sn, there are now over thirty other isotopes which are now known to exhibit the effect and chemically useful information has been collected on the compounds of Ni, Ru, Sb, Te, I, Xe, W, Ir, Au, Np, etc.

In view of the complications caused by the concomitant recoil of the nuclei emitting high energy γ-rays, the Mössbauer spectroscopy is generally performed on solid samples (often at low, say liquid nitrogen temperatures) so that the crystal forces render the relevant nuclei almost recoil-less. This limits the utility of the Mössbauer spectroscopy generally to solid compounds.

The hyperfine interactions between the nucleus and its orbital electrons affect the energy of the x-ray necessary to affect the nuclear transition in the sample from the ground to excite state. The energy of γ-ray from the source is, therefore, varied employing the well-known Doppler effect over the range of the above energy differences by moving the source relative to the sample. The energy of the γ-ray (E_γ) is thus varied by ΔE according to the Doppler formula,

$$\Delta E = (v/c)\, E_\gamma$$

where v and c are velocities of the nucleus and light respectively.

The amount of absorption is generally plotted against the energy of the source photon (expressed usually in velocity units for Mössbauer measurements). The same nucleus gives different spectra due to the following types of hyperfine interactions between the nuclear charge distribution and the extra-nuclear electric and magnetic fields :

(i) The Is omer Shift (I.S.) results from the charge distribution of the nucleus and those electrons which have a finite probability of being found in the region of the nucleus. The isomer shift δ can be computed by considering the effect of the overlap of '*s*' electron density with the nuclear charge density,

$$\delta = \frac{4\pi}{5} Ze^2 R^2 \frac{dR}{R} \{[\psi_{0\,\text{absorber}}]^2 - [\psi_{0\,\text{source}}]^2\}$$

where *dR is the* difference between the radii of the excited and ground states of the nucleus, *Z* is the nuclear charge, *e* is the electronic charge and $[\psi_0]^2$ is the total electron density at the nucleus.

The value of isomer shift, therefore, depends on a nuclear characteristic *dR/R* and an extranuclear parameter—the *s* electron density. For some isotopes such ^{57}Fe, *dR* has negative value and hence the value of δ would be positive when $[\psi_{0\,\text{source}}]^2 > [\psi_{\text{absorber}}]^2$.

Although the value of I.S. is generally determined by the *s* electron density at the nucleus, yet the removal or addition of *p* or *d* electrons also indirectly affect the '*s*' electron density and hence, the value of I.S. For example, the removal of a '*d*' electrons in the change of d^6 ion (Fe^{2+}) to d^5 ion (Fe^{3+}) would result in the *s* electrons to expand, with a consequential reduction in the '*s*' electron density at the nucleus of the absorber ion nucleus and an increase in the observed value of isomer shift (I.S.).

For ^{119}Sn, the sign of (*dR/R*) is now recognized as positive and hence contrary to the case ^{57}Fe, the isomer shift should increase with an increase in '*s*' electron density but decrease with an increase in *p* electron density. On the same arguments, tetrahedral covalent tin ($5s5p^3$) should be expected to depict a more positive shift than 'ionic' tin (IV) compounds; the isomer shift in a tin compound would, therefore, throw light on the covalent character of its bonds. The measurements of values of isomer shift along with those of quardrupole splitting have been extensively employed in structural elucidation of organotin species as described briefly later in the section.

(ii) The Quadrupole Splitting (Q.S.) arises from the splitting of nuclear energy levels due to the interaction of noncubic extranuclear electric fields for nuclei which are not spherical and for which $I > 1/2$. For example, for ^{57}Fe and ^{119}Sn nuclei, the level, $I = 3/2$ splits into two, while the $I = 1/2$ level remains degenerate. A characteristic two line spectrum is obtained (Fig. 3.34) from the two allowed transition; the separation of the peaks is quadrupole splitting (Q.S.) and the centroid of two peaks relative to the source is for practical purposes almost equivalent to the isomer shifts (I.S).

The application of the above two parameters (I.S. and Q.S.) to structural elucidation of organotin moieties is now being presented in the following pages.

(a) *Organotin compounds:* Tin nucleus occurs in ten isotopic forms, out of which the five more abundant ones are ^{116}Sn (14.2%); ^{117}Sn (7.6%); ^{118}Sn (24.0%); ^{119}Sn (8.6%) and ^{120}Sn (33.0%). Of the above isotopes, only ^{117}Sn and ^{119}Sn have nuclear spin values of 1/2 making them useful for Mössbauer spectroscopy, ^{119}Sn being generally preferred (due to slightly greater abundance and higher value of nuclear magnetic moment). ^{119}Sn isomer shifts are usually expressed relative to SnO_2 or $BaSnO_3$ standard absorbers.

Isomer shifts (δ) which are generally a measure of the '*s*' electron density on ^{119}Sn nucleus have been found to be in range of ± 5 mm s^{-1} for tin compounds, the values of δ, increasing with the *s* electron density. The decrease in the '*s*' electron density on the tin nucleus with the increasing electron affinity of the ligands X and Y in derivatives of the formula $M_2^+ [BuSnX_{5-n}Y_n]^{2-}$ is reflected in the gradual lowering in the values of isomer shifts in the following series of compounds: $(Et_4N)_2^+ [BuSnBr_5]^{2-}$ 138; $(Et_4N)_2^+ [BuSnBr_2Cl]^{2-}$, 1.20: $(Et_4N)_2^+ [BuSnCl_5]^{2-}$, 1.03; $K_2^+ [BuSnClF_4]^{2-}$, 0.47 and $K_2^+ [BuSnF_5]^{2-}$, 0.27 mm s^{-1}.

Fig. 3.34 *A typical Mössbauer Spectrum.* **The magnitude of the quadrupole splitting (Q.S) is proportional to the interaction of electric field gradient with the quadrupole moment of the nucleus. While the isomer shift measures the *s* electron density at the nucleus, the quadrupole splitting gives an idea of the distortion from cubic symmetry of the electron distribution and ligands about the Mössbauer nucleus.**

Similarly in organotin derivatives of the general formulae SnR_4 and $SnR_3(OH)$, the value of δ should increase with electron donating tendency of the group R: $SnPh_4$, 1.15; $SnMe_4$, 1.20; $SnEt_4$, 1.30; $SnBu_4$, 1.35 and $SnPh_3(OH)$, 1.16; $SnMe_3(OH)$, 1.20; $SnEt_3(OH)$, 1.30; $SnBu_3(OH)$, 1.37 mm s^{-1}.

As stated earlier also, the '*s*' electron density should also be influenced by a change in stereochemistry at the tin atom. For example, a change of sp^3 hybridization in SnI_2Ph_2 to sp^3d^2 hybridization in SnI_2Ph_2(bipy) lowers the value of δ from 1.51 to 1.41 mm s^{-1}. In addition to the effect of coordination number, the value of δ appears to be considerably dependent on the *cis-trans* isomeric forms in octahedral tin compounds. The cis complexes like $SnMe_2(ox)_2$ and $Sn(NCS)_2Ph_2$(bipy) have much lower δ values (around 0.8 mm s^{-1}) whereas the *trans* compounds like $Sn(acac)_2Me_2$ and $SnCl_2Ph_2$(bipy) have higher δ values (around 1.2 mm s^{-1}).

Quadrupole splitting (represented by Q.S. or ΔE_Q in mm s^{-1}) parameters have proved to be of greater value in structural elucidation of organotin compounds.

Perfectly symmetric molecules like $SnMe_4$ show only a single line in the Mössbauer spectrum, indicating as expected a value 0 for quadrupole splitting. The range of the Q.S. for organotin compounds in different stereochemistries is illustrated in Table 3.20.

Table 3.20 Ranges of Quadrupole Splittings (Q.S.) for Organotin Compounds (in mm s^{-1} at liquid nitrogen temperature)

Structural Type		Value of Q.S.
Tetrahedral R_nSnX_{4-n} (n = 1-3)		0.0-2.3
Trigonal Bipyramidal R_3SnX_2		
	cis	1.7-2.4
	trans	3.0-4.0
	mer	3.5-4.1
Octahedral R_2SnX_4		
	cis	1.6-2.3
	trans	3.4-4.3
Pentagonal bipyramidal R_2SnX_5 (R *trans*)		~ 4

As reflected from the above table, measurement of quadrupole splitting is helpful in distinguishing not only the coordination number of tin, but also amongst the various possible stereochemical forms in the same coordination state of the central tin atom.

It would be valuable to record that the value of quadrupole splitting has been successfully correlated with C-Sn-C bond angle in a number of octahedral organotin derivatives (Table 3.21).

Table 3.21 Changes in the Value of Q.S. with C-Sn-C Angle in Dimethyltin Octahedral Complexes

Complex	C-Sn-C Angle	Q.S. mm. s^{-1}
(i) $Me_2Sn(ox)_2$	110°	2.0
(ii) $Me_2Sn(SCSNMe_2)_2$	136°	3.1
(iii) $Me_2Sn(ONMeCOMe)_2$	146°	3.3
(iv) Me_2Sn (salen)	160°	3.5
(v) $Me_2Sn(acac)_2$	180°	4.0

The data in Table 3.21 is corroborative of the ranges indicated for *cis* (~2.0) and *trans* (~ 4.0) for complexes (i) and (v) respectively; the remainder three are obviously in distorted geometries.

Mössbauer spectroscopy of organotin (II) compounds has also received some attention. For example, Cp_2Sn shows a large isomer shift value of 3.74 mm s^{-1} indicative of a high '*s*' electron density in bivalent tin organometallics.

3.6.5 Organometallics of Group 15 Elements (As, Sb and Bi)

(i) *Infra-red spectroscopy:* As expected, the infra-red spectra of AsR_3, SbR_3 and BiR_3 depict νM—C bands in the regions of 599-478 and 460-400 cm^{-1} respectively.

Out of the organometallics of the above elements in pentavalent state the i.r. spectra of antimony derivatives have been intensively studied, a brief account of which is presented in Table 3.22.

Table 3.22 Some Characteristic Infra-red Group Frequencies (cm^{-1}) of Organoantimony Compounds

Compound	Band (cm^{-1})	Assignment
SbS_2Me_3	590-570	ν_a(Sb-C)
SbX_2Me_3	546-527	ν_s(Sb-C)
$SbMe_5$	516	ν(Sb-C_{eq})
	456	ν(Sb-C_{ax})
SbF_2R_3	509-480	ν_a(Sb-F)
$SbCl_2R_3$	275-272	ν_s(Sb-Cl)

The infrared spectrum of $Sb(CH_3)_5$ shows two Sb-C stretching frequencies at 456 cm^{-1} for axial Sb-C stretching and 516 cm^{-1} for equatorial Sb-C stretching; the results favour the trigonal bipyramidal structure. This is also the case for AsR_5 (R = Me or Ph) (from vibrational spectrum). The compounds of the type $SbX(CH_3)_4$, have the ionic structure, $Sb(CH_3)_4^+ X^-$, and because of the tetrahedral symmetry of the cation, these exhibit only one Sb-C stretching band near 570 cm^{-1} $SbX_2(CH_3)_3$ (X = O_2CH or NO_3) also assumes trigonal bipyramidal structure with the methyl groups in the equatorial plane and X's at the axial positions. Furthermore, these X groups are covalently bonded to the antimony atom. The Sb-C stretching frequencies are 581 cm^{-1} for X = OOCH, and 573 cm^{-1} for X = NO_3 in CCl_4 solutions.

(ii) *Proton nuclear magnetic resonance spectroscopy:* In solution, the methyl groups in MMe_5 and in $(p\text{-}MeC_6H_4)_5$ M(M = As or Sb) have been shown to be equivalent by ^{1}H n.m.r. spectroscopy even at low temperatures, e.g., at −100°C for $SbMe_5$, although two peaks in the ratio of 3 : 2 could be expected on the basis of trigonal bipyramidal structure of these derivatives. These equivalences have been explained to be due to a rapid exchange of the methyl and tolyl groups between axial and equatorial positions, involving square pyramidal intermediates or other transition states.

(iii) *X-Ray diffraction:* In the past two decades structures with pentacoordinated central atoms have been the subject of increasing interest. This development was stimulated by the progress of dynamic n.m.r. spectroscopy. Theoretical considerations point to the possibility of the existence of only three of the five possible pentacoordinated structures having at least one symmetry element, i.e., the trigonal bipyramid, the rectangular pyramid and the monocapped tetrahedron (Fig. 3.35).

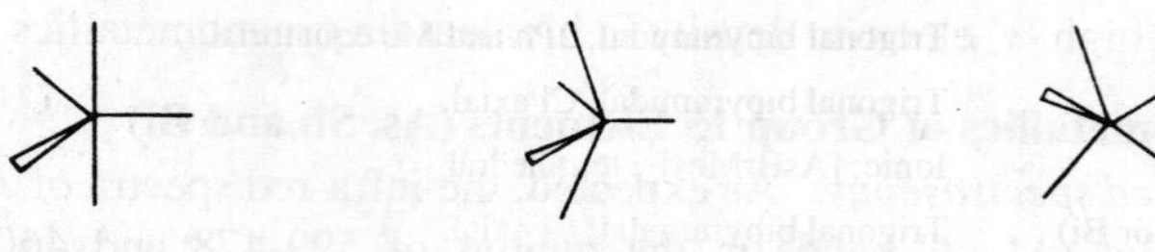

D_{3h} C_{4v} C_s

Fig. 3.35 Pentavalent structures having at least one symmetry element.

All these basic structures have been proposed for different compounds, but crystallographically authenticated compounds belong so far to the first two categories (Fig. 3.36) only.

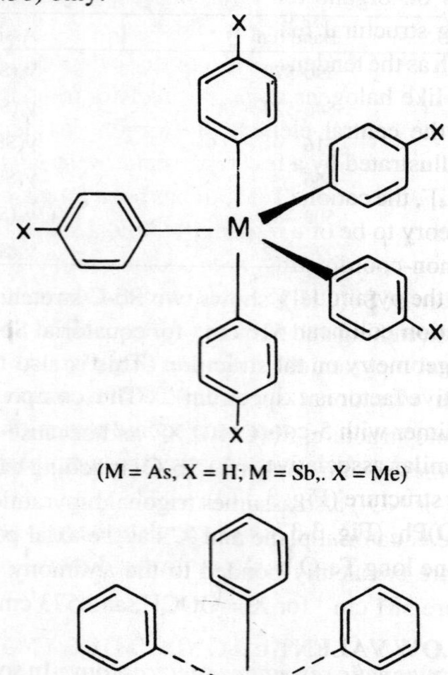

(M = As, X = H; M = Sb,. X = Me)

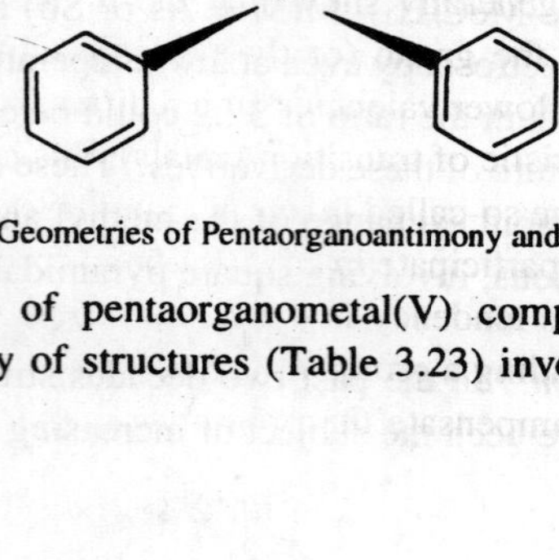

Fig. 3.36 Different Geometries of Pentaorganoantimony and -arsenic Compounds.

Compared to that of pentaorganometal(V) compounds, organometal(V) halides exhibit a variety of structures (Table 3.23) involving four, five and six coordinate metal.

Table 3.23 Some Coordination Environments of As, Sb, and Bi in their Organometallic (V) Halides

Compound	Geometry
$AsBrMe_4$	Ionic; tetrahedral $(AsMe_4)^+$
$SbFMe_4$	Structure involving polymeric chains with bridging fluorines; six-coordinate Sb.
$SbFMePh_3$	Trigonal bipyramidal; 2Ph and Me equatorial.
$AsCl_2Me_3$	Trigonal bipyramidal; Cl axial.
$AsBr_2Me_3$	Ionic; $[AsBrMe_3]^+$, tetrahedral
MCl_2Ph_3 (M = Sb or Bi)	Trigonal bipyramidal; Cl axial.
$(SbCl_3Me_2)_2$	Two forms: (1) Covalent with bridging chlorines, octahedral (2) Ionic; $[SbMe_4]^+ [SbCl_6]^-$, tetrahedral cation
$(SbCl_3Ph_2)_2$	Dimer; bridging Cl, Octahedral

3.6.6 Organotellurium (selenium) Derivatives

Physico-chemical studies on organo-tellurium and -selenium compounds have revealed many interesting structural features which appear to be governed by a number of parameters such as the tendency of weak dissociation of their bonds with electronegative elements like halogens and conversely, a trend of increasing the coordination number of the central element by forming associated molecules. Some of these would be illustrated by a few representative structures:

(i) In $[TeMe_3]^+ [BPh_4]^-$, the cation $[TeMe_3]^+$ has been found, as expected from electron pair repulsion theory to be of a regular pyramidal shape (Fig. 3.37a).

(ii) However, if the non-coordinating large counter-anion $[BPh_4]^-$ is replaced by an iodide [I] ion, then the pyramidal $[SeMe_3]^-$ in $SeMe_3I$ tends to be associated with the iodide in a direction collinear with one of the Se-Me bonds confering a distored pseudo-trigonal geometry on the molecule (Fig. 3.37b).

The effect of associative factor is exhibited in $TeClPh_3$ which has been shown to be a chlorine bridged dimer with 5-coordinate square pyramidal tellurium (Fig. 3.37c). In $TeCl_2Ph_2$, a similar associative tendency through chlorine atoms leads to a bridged 5-coordinate structure (Fig. 3.37d).

The structure of $Te(O)Ph_2$ (Fig. 3.37e) also depicts the monomeric units to be linked by one short and one long Te-O bonds to form a dimeric unit.

3.7 LESS FAMILIAR LOW VALENT σ-BONDED ORGANOMETALLICS

The main group elements generally show that the highest valency which corresponds to the number of the group (of the periodic table) in which they are situated, along with a set of lower valencies which differ from the above by units of 2(rather than 1, characteristic of transition metals); this is explained in general terms to be mainly due to the so-called innert '*s*' pair effect, i.e., the lesser ability of the pair of *s*-electrons to participate in covalent bond formation.

The prime cause of this tendency is understood to be due to the decreasing strength of bonds on descending a group, as the energy gained from forming two additional bonds fails to compensate the promotional energy necessary for say,

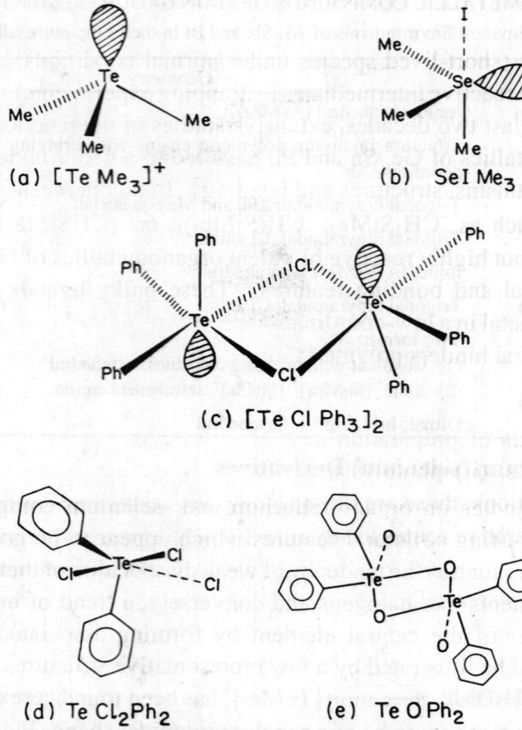

Fig. 3.37 Some representative structures of organo-selenium and -tellurium halides.

M(II)→M(IV) transformation. In elements of group V, the higher valency of '5' as well as the lower valency of '3' are of comparative stability. By contrast, for the elements of group III(B, Al, Ga, In and Tl), the lower valency of '1' is much less stable although as in other groups the stability of the lower valency increases with the increase of atomic size of the element. In group IV also, the stability of +2 oxidation state increases from C to Pb in ionic compounds, but in covalent organometallic derivatives the tetravalent compounds (e.g., PbR_4) appear to be much more stable than divalent derivatives (e.g., PbR_2).

As the organometallics of As(III), Sb(III) and Bi(III) are fairly stable, these have already been described in earlier sections dealing with comparatively more stable derivatives. This section will, therefore, primarily deal with the chemistry of the divalent organometallic compounds of Si, Ge, Sn, and Pb. However, a brief account would also be presented of (i) Organometallic (III) radicals (i.e., seven electron organometal species) of Ge, Sn and Pb, and (ii) monovalent organometallics of gallium and indium.

3.7.1 Bi-Valent Organometallics of Si, Ge, Sn and Pb

The significant role of bi-valent carbon species (i.e., carbenes) in organic reactions, has paved the way for an enormous interest in the bi-valent analogues of the heavier main group IV elements. Although the chemistry of bi-valent silicon compounds (silylenes) has been notably developed during the last two decades, yet these have a very low stability (short-lived species under normal conditions; non-isolable; recognised as highly reactive intermediates by trapping experiments) similar to that of carbenes. In the last two decades, extensive studies of the reactions leading to bi-valent organometallics of Ge, Sn and Pb have led to a better understanding of their reaction mechanisms, structures and bondings. In fact in recent years the use of bulky ligands such as $^{-}CH_2SiMe_3$, $^{-}CH(SiMe_3)_2$, or $^{-}(CHSiMe_3)_2C_6H_4^{-}$-o has generated isolable, but highly reactive bi-valent organometallics of Ge, Sn and Pb with novel structural and bonding features. These bulky ligands stabilise the compounds of the metal in a low-coordination geometry, as the congested environment around the metal hinders polymerization due to steric factors.

3.7.1.1 Preparation

The general methods of preparation of bivalent Si, Ge, Sn and Pb derivatives involve (i) pyrolysis, (ii) photolysis, (iii) reductive chemical elimination, and (iv) metathesis reactions; these are illustrated by the following examples.

Available preparative routes to silylenes generally involve either pyrolysis of silanes,

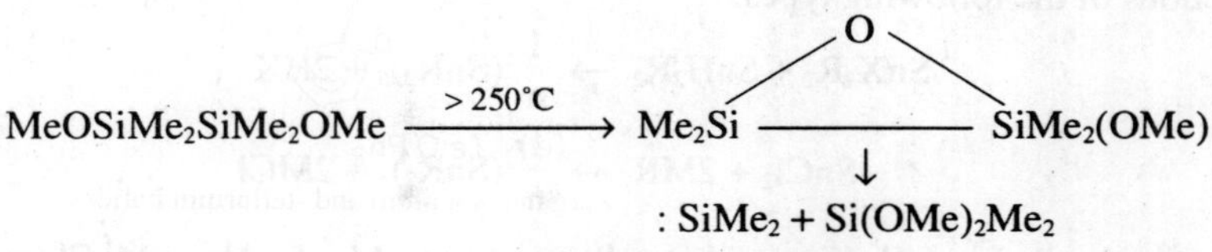

$$Si_2R_6 \rightleftharpoons :SiR_2 + SiR_4$$

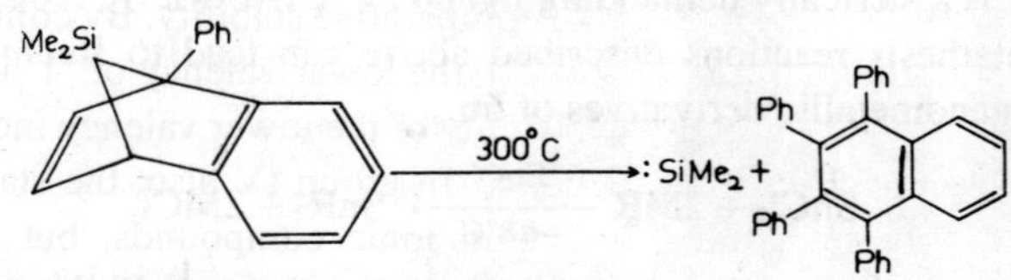

or ring contraction reaction of cyclic polysilanes:

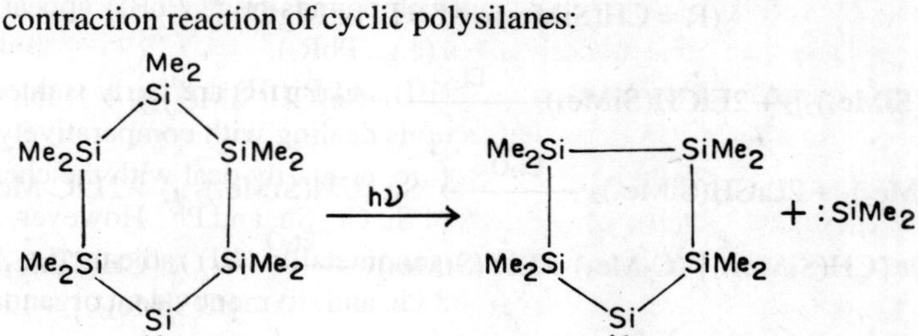

The pyrolysis reaction of silicon compounds (e.g. disilanes) is generally not as simple as represented by the equation ($Si_2R_6 \rightleftharpoons :SiR_2 + SiR_4$), because it may also produce silyl radicals:

$$Si_2R_6 \rightleftharpoons 2\ ^{\bullet}SiR_3$$

The reverse of the above reaction is a radical coupling reaction, whereas the reaction of $:SiR_2$ with SiR_4 may be considered as an insertion reaction.

Dehalogenation (or reductive elimination) reactions provide another route of generating $:SiMe_2$, e.g.,

$$SiCl_2Me_2 + 2Na/K \xrightarrow[\text{gas phase}]{260^\circ C} :SiMe_2 + 2NaCl/KCl$$

It may be noted that $:SiMe_2$ readily undergoes insertion or addition reactions with a wide variety of compounds, e.g.,

$$:Si(CH_3)_2 \xrightarrow{SiH(CH_3)_3} (CH_3)_3\,SiSiH(CH_3)_2$$

$$:Si(CH_3)_2 \xrightarrow{H_2C=CH_2} (CH_3)_2Si\begin{matrix} CH_2-CH_2 \\ CH_2-CH_2 \end{matrix}Si(CH_3)_2$$

$$:Si(CH_3)_2 \xrightarrow{HCl} SiClH(CH_3)_2$$

In the absence of such 'trapping agents', $:SiMe_2$ may be obtained in the form of a polymeric solid $(SiMe_2)_n$.

The simple alkyl- or aryl-stannylenes, originally formulated as SnR_2 are actually linear, branched, or polystannanes, and can generally be prepared by reactions of the following types:

$$SnX_2R_2 + SnH_2R_2 \rightarrow \frac{1}{n}(SnR_2)_n + 2HX$$

$$SnCl_2 + 2MR \rightarrow \frac{1}{n}(SnR_2)_n + 2MCl$$

(R = simple alkyl or aryl; X = amido or alkoxy groups; M = Li, Na or MgCl (or Br)

When R is a sterically demanding ligand lie $^-CH(SiMe_3)_2$, reductive elimination and metathesis reactions described above can lead to stable and isolable subvalent organometallic derivatives of tin:

$$SnCl_2 + 2MR \xrightarrow[-18^\circ C]{Et_2O} SnR_2 + 2MCl$$

$$(R = CH(SiMe_3)_2,\ M = Li \text{ or } MgCl)$$

$$Sn[N(SiMe_3)_2]_2 + 2LiCH(SiMe_3)_2 \xrightarrow{Et_2O} Sn[CH(SiMe_3)_2]_2 + 2LiN(SiMe_3)_2$$

$$Sn\,(C_5Me_5)_2 + 2LiCH(SiMe_3)_2 \xrightarrow{Et_2O} Sn\,\{CH(SiMe_3)_2\}_2 + 2LiC_5Me_5$$

$$Ge\{CH(SiMe_3)_2\}\,(C_5Me_5) + LiC(SiMe_3)_3 \xrightarrow{Et_2O} Ge\{CH(SiMe_3)_2\}\{C(SiMe_3)_3\} + LiC_5Me_5$$

$$SnCl_2[CH(SiMe_3)_2]_2 + Li_2(COT) \xrightarrow{Et_2O} Sn[CH(SiMe_3)_2]_2 + 2LiCl + (COT)$$

(COT = cyclooctatetraene anion)

Syntheses of dialkylgermanium derivatives from $GeCl_2$ and its complexes by metathesis reactions are accompanied with extensive decomposition. Synthesis of even a sterically hindered derivative like $Ge[CH(SiMe_3)_2]_2$ was also not promising by using $GeCl_2$, but the reactions of (i) $Ge[N(SiMe_3)_2]_2$ with $LiCH(SiMe_3)_2$, and (ii) $GeCl_2{\cdot}$dioxane with $MgCl\{CH(SiMe_3)_2\}$ yielded the desired product:

$$Ge[N(SiMe_3)_2]_2 + 2LiCH(SiMe_3)_2 \xrightarrow{Et_2O} Ge[CH(SiMe_3)_2]_2 + 2LiN(SiMe_3)_2$$

$$GeCl_2 \cdot dioxane + 2MgCl\{CH(SiMe_3)_2\} \xrightarrow{Et_2O} Ge[CH(SiMe_3)_2]_2 + 2MgCl_2$$

The first tin(II) metallocycle was obtained in 1982 by the reaction of $SnCl_4$ with $[\{Li(tmeda)\}_2\{o\text{–}C_6H_4(CHSiMe_3)_2\}]$ (represented by A in scheme 3.7) at –78°C in Et_2O. These synthetic procedures are illustrated in Scheme 3.7.

The dialkyllead derivative, $Pb[CH(SiMe_3)_2]_2$, can be prepared by the following reactions:

$$PbX_2 + 2LiCH(SiMe_3)_2 \xrightarrow{Et_2O} Pb[CH(SiMe_3)_2]_2 + 2LiX$$
$$(X = Cl \text{ or } N(SiMe_3)_2$$

3.7.1.2 Properties, Structure and Bonding

The physical properties of the compounds $M[CH(SiMe_3)_2]_2$ (M = Ge, Sn or Pb) are summarised in Table 3.24.

Table 3.24 Physical Data for $M[CH(SiMe_3)_2]_2$

Compound	M = Ge	M = Sn	M = Pb
Colour	Yellow[a]	Red[a]	Purple
M.P. (°C)	179–181[b]	135-137[b]	43-45
^{1}H n.m.r. (δ)[c]	Si-CH3 9.51	9.68	9.78
	M – CH 6.40	8.10	8.85

[a]Thermochromic; colours deepen on heating
[b]Sublimes under reduced pressure,
[c]C_6H_6 solution

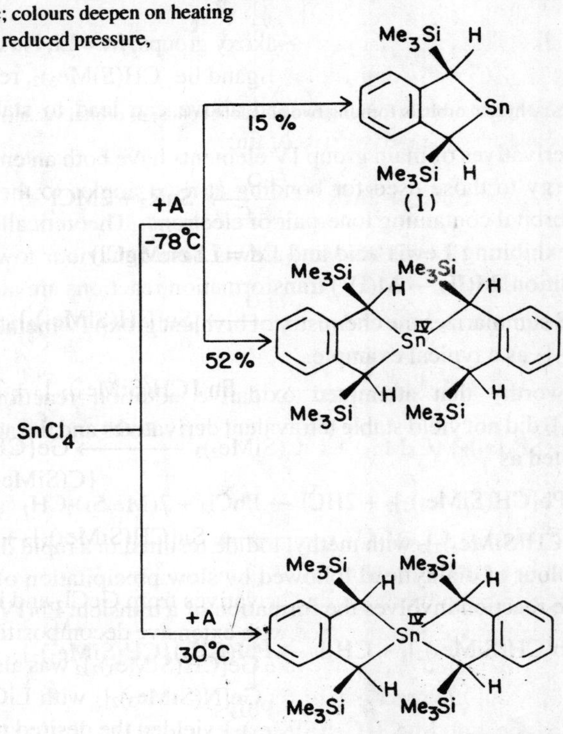

Scheme 3.7 *Synthetic Routes to Tin(II) Metallocycles.*

Tin and lead derivatives of the above types are moderately soluble in hexane, benzene, or diethyl ether whereas the germanium analogues are less soluble. The lead derivative is pyrophoric, whereas derivatives of germanium and tin may be handled briefly in air without significant decomposition. However, in the latter two cases also, the derivatives are extremely air-sensitive in solution.

The ground state valence shell electronic configurations of germanium, tin, and lead are $4s^2 4p^2$, $5s^2 5p^2$ and $6s^2 6p^2$ respectively,

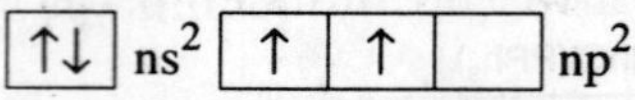

and these elements could form bivalent organometallics simply by using the two unpaired electrons for bonding. The stereochemistry of the molecule so formed can be explained in terms of sp^2 hybridization of the valence shell electrons (see figure below) and the strong colours of the compounds, $M[CH(SiMe_3)_2]_2$ (M = Ge, Sn or Pb) are presumably associated with the open-shell electronic configuration of the metal(s).

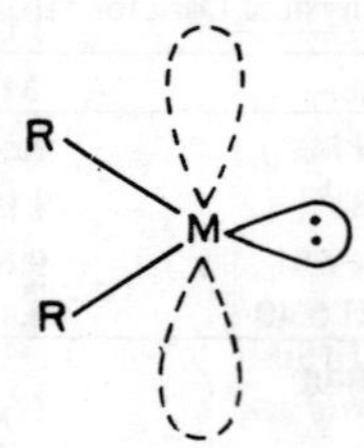

(sp_xp_y hybrid orbitals forming two covalent bonds; p_z orbital vacant).

Bivalent derivatives of main group IV elements have both an empty p_z orbital (of similar energy to those used for bonding at right angles to the plane of the molecule) and orbital containing lone-pair of electrons. Theoretically they should be capable of exhibiting Lewis acid and Lewis base behaviour towards suitable ligands. In addition, M(II) → M(IV) transformation reactions are also possible.

Scheme 3.8 summarizes the chemistry of bivalent group IV metal alkyls taking $Sn\{CH(SiMe_3)_2\}_2$ as a typical example.

It is noteworthy that attempted oxidative addition reactions involving $Pb[CH(SiMe_3)_2]_2$ did not yield stable tetravalent derivatives and the actual reaction can be represented as

$$Pb[CH(SiMe_3)_2]_2 + 2HCl \rightarrow PbCl_2 + 2(Me_3Si)_2CH_2$$

Reaction of $Pb[CH(SiMe_3)_2]_2$ with methyl iodide resulted in a rapid disappearance of the purple colour of dialkyllead followed by slow precipitation of PbI_2; which suggests that the reaction involves the formation of a transient Pb(IV) adduct:

$$Pb[CH(SiMe_3)_2]_2 + CH_3I \rightarrow PbI(CH_3)\{CH(SiMe_3)_2\}_2 \downarrow PbI_2 + \ldots$$

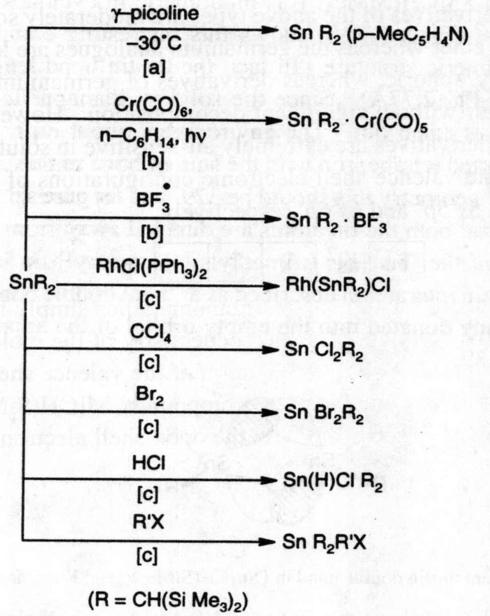

[a] = Lewis acidic character;
[b] = Lewis base behaviour;
[c] = oxidative addition

***** **Unlike stannocene (cf. Section 3.6), Sn{CH(SiMe$_2$)$_2$}$_2$ does not form adducts with main group Lewis acids such as BF$_2$ but it does act as a Lewis base towards several transition metal complexes.**

Scheme 3.8 Some Reactions of Sn{CH(SiMe$_2$)$_2$}$_2$.

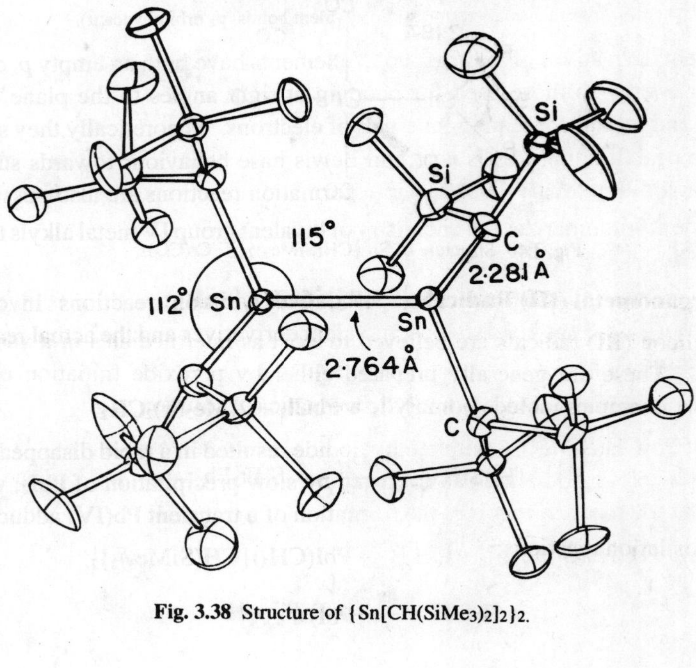

Fig. 3.38 Structure of {Sn[CH(SiMe$_3$)$_2$]$_2$}$_2$.

The compound $Sn[CH(SiMe_3)_2]_2$ is monomeric in cyclohexane solution but the crystal structure (Figure 3.38) shows this interesting compound to have a centrosymmetric dimeric structure. In fact, the tin-tin bond length of 2.764Å is close to that in Sn_2Ph_6(2.77Å). Since the solid is diamagnetic, a conventional tin-tin single bond is ruled out. The environment about each tin atom can be described as a distorted tetrahedron with the sum of bond angles at tin as 341° (for ideal tetrahedral sp^3 geometry this should be 329° and for pure sp^2, 360°). The two sets of alkyl groups at both the tin atoms are directed away from each other. The unusual geometry of the [bis {bis (trimethylsily)} methyl] tin led to suggest an unusual kind of tin-tin interaction described as a "bent double bond" with the lone pair on each tin being donated into the empty orbital of the adjacent tin atom, as shown in Figure. 3.39.

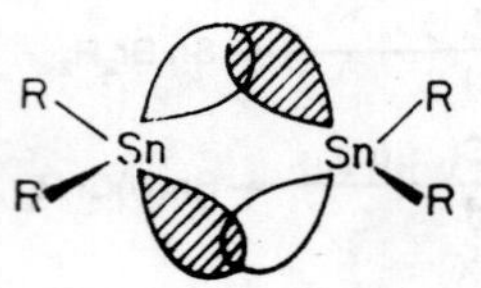

Fig. 3.39 Bent tin-tin double bond in $\{Sn[CH(SiMe_3)_2]_2\}_2$; R = $(Me_3Si)_2CH$.

The strong donor characteristic of $Sn[CH(SiMe_3)_2]_2$ results in complexes such as $(CO)_5Cr \cdot Sn[CH(SiMe_3)_2]_2$ with tin in an unusual trigonal planar environment (Figure 3.40); both Sn—C bond length and the C—Sn—C angle are less than those in the derivative $\{Sn[CH(SiMe_3)_2]_2\}_2$ itself.

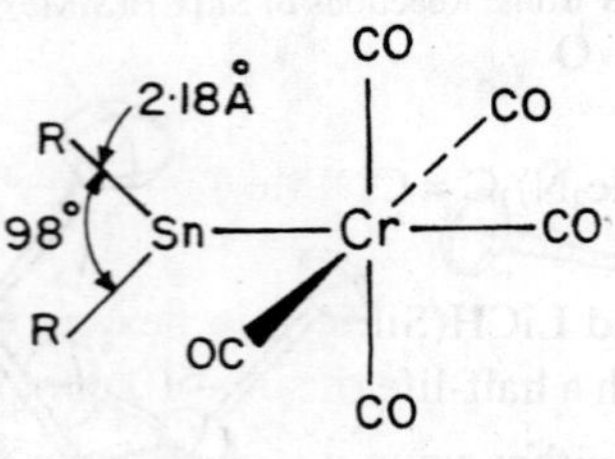

Fig. 3.40 Structure of Sn $\{CH(SiMe_3)_2\}_2 \cdot Cr(CO)_5$.

3.7.2 Organometal (III) Radicals ($^{\bullet}SiR_3$, $^{\bullet}GeR_3$, $^{\bullet}SnR_3$)

Organosilicon (III) radicals are believed to exist as intermediates in a variety of reactions. These are generally prepared either by peroxide initiation or U.V. irradiation; examples include homolytic aromatic substitution, e.g.:

$$Et_3SiH \xrightarrow[C_6H_6;\ 135^\circ C]{Bu^t_2O_2} SiEt_3Ph$$

and hydrosilation reactions:

$$R_3SiH \xrightarrow{h\nu} R_3Si^{\bullet} + H^{\bullet}$$

$$R_3Si^{\bullet} + R'CH=CH_2 \xrightarrow{h\nu} R_3SiCH_2CHR' \xrightarrow{R_3SiH} R_3SiCH_2CH_2R' + R_3Si^{\bullet}$$

$^{\bullet}SiMe_3$ radicals may also be prepared by photolysis of $Hg(SiMe_3)_2$ which can react with a number of substrates, e.g.,

$$Hg(SiMe_3)_2 + PhMe \xrightarrow{h\nu} MeC_6H_4SiMe_3 +$$ 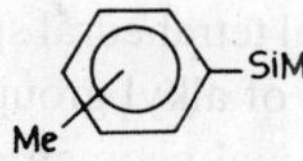

$$Hg(SiMe_3)_2 + CF_3C \equiv CCF_3 \xrightarrow{h\nu} trans-F_3C(SiMe_3)C = C(SiMe_3)CF_3$$

Reactions of the above type with other Group IV radicals $^{\bullet}MR_3$ have shown the following order of reactivity:

$$^{\bullet}SiR_3 > {}^{\bullet}GeR_3 \approx {}^{\bullet}SnR_3 > {}^{\bullet}PbR_3$$

The organosilicon radicals are very short-lived when less sterically demanding groups are involved. However, radicals of longer life can be generated either by photolysis of the corresponding hydride in the presence of a peroxide or by photolysis of $SiBr(SiMe_3)_3$ with an electron-rich olefin:

$$(Me_3Si)_3SiH + PhC(=O)-O-O-C(=O)Ph \xrightarrow{h\nu} {}^{\bullet}Si(Me_3)_3$$

$$(Me_3Si)_3SiBr + (Me_2N)_2C = C(NMe_2)_2 \xrightarrow[-70^{\circ}C]{h\nu} {}^{\bullet}Si(SiMe_3)_3$$

When a mixture of Si_2Cl_6 and $LiCH(SiMe_3)_2$ in hexane is photolysed, a bulkier radical $^{\bullet}Si[CH(SiMe_3)_2]_3$, with a half-life of some 10 min. at 30°C is produced.

These radicals can adopt either a pyramidal or planar geometry, which can be distinguished by the chiral character of the former type of derivatives.

Organogermanium (III) radicals $^{\bullet}GeR_3$, have been prepared by various routes.

(1) Radio- or photo-lysis of germanium-hydrogen bonds using high energy radiations (γ-rays, X-rays, even U.V. light), e.g.,

$$GeHR_3 \xrightarrow{h\nu} {}^{\bullet}GeR_3 + {}^{\bullet}H$$

$$Bu^t_2O_2 \xrightarrow{h\nu} 2Bu_tO^{\bullet} \xrightarrow[-70^{\circ}C]{GeHMe_2} {}^{\bullet}GeMe_3 + Bu^tOH$$

(2) Hydrogen abstraction process, e.g.,

$$GeHR_3 + R' \rightarrow R' - H + {}^{\bullet}GeR_3$$

(3) Chemical reduction method, e.g.,

$$GeClR_3 + (Me_2N)_2C = C(NMe_2)_2 \xrightarrow{h\nu} \bullet GeR_3 + [C_2(NMe_2)_4]^{2+} + (Cl^-)$$

(4) Electrochemical reduction process, e.g.,

$$Ph_3GeX \xrightarrow{e} Ph_3Ge^{\bullet} + X^- \quad (X = Cl, Br)$$

$$Ph_3Ge^{\bullet} \xrightarrow{\text{dimerisation}} (GePh_3)_2 \qquad Ph_3Ge^{\bullet} \xrightarrow{MeOCH_2CH_2OMe} GeHPh_3$$

(V) Photolysis of germanium-mercury bonded derivatives, e.g.,

$$Hg(GeR_3)_2 \xrightarrow{h\nu} Hg + 2\dot{G}eR_3$$

Steric hindrance is generally a key factor in stabilisation of organogermanium (III) radicals. However, factors such as the non-availability of disproportionation pathways and the possibility of delocalisation into the d orbitals of main-group IV elements, may also contribute to the stability. For example, trimethylgermanium (III) radical is observed at –70°C (e.s.r. spectrum shows that it is pyramidal in shape); the trimesitylgermanium (III) radical at 20°C, and the tris {bis (trimethylsilyl) methyl}germanium (III) radicals are much longer lived, with a half-life of more than 4 months at room temperature.

The general characteristics of organotin (III) radicals are very similar to those of their lighter analogues (Si or Ge). When R is a less sterically demanding group, the radicals are highly reactive and can only be studied at low temperatures. When R is a more ramified group sufficient to restrict dimerisation, persistent species with greater life-time may be produced. The most stable of such species is $^{\bullet}Sn[CH(SiMe_3)_2]_3$.

Some preparative routes to short-lived $^{\bullet}SnR_3$ species are illustrated in Scheme 3.9.

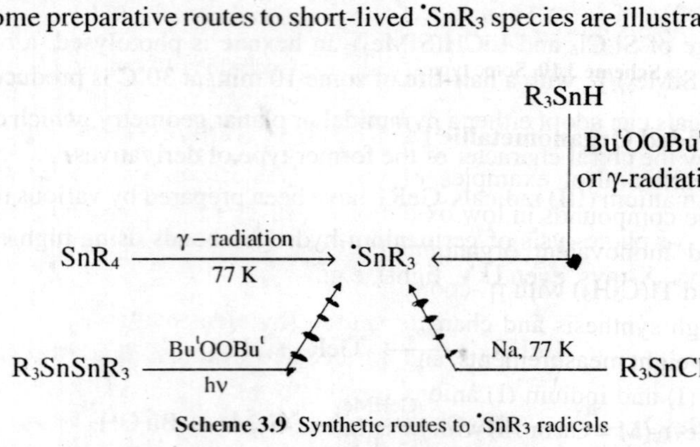

Scheme 3.9 Synthetic routes to $^{\bullet}SnR_3$ radicals

When R is a bulky group, other routes are possible, e.g.,

$$R_3SnSnR_3 \xrightarrow{h\nu} 2R_3\dot{S}n$$

$$(R_3Sn)_2Hg \xrightarrow{h\nu} 2R_3\dot{S}n + Hg$$

$$(R = PhCMe_2CH_2—)$$

$$R_2Sn \xrightarrow{h\nu} R_3Sn.+ \ldots$$

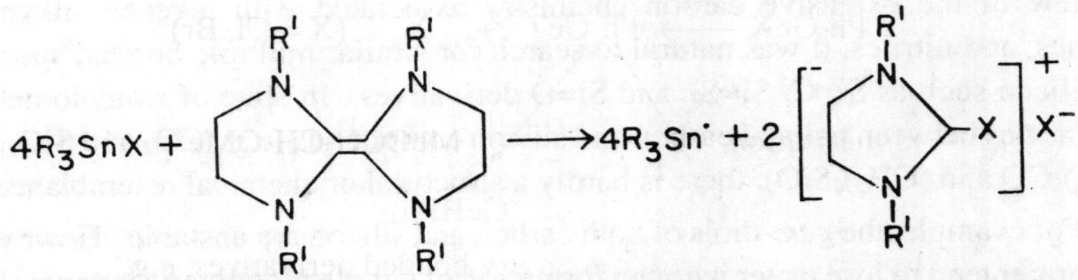

The resulting $^{\bullet}Sn[CH(SiMe_3)_2]_3$ radical has a half-life of about 1 year at 20°C.

These $R_3Sn^{\bullet}$ radicals are pyramidal (by e.s.r. studies) in shape. Some of the principal reactions of the organotin (III) radicals are illustrated in the following scheme:

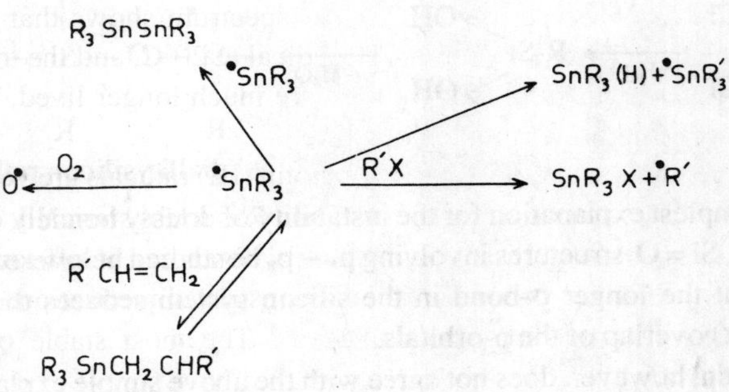

Scheme 3.10 Some typical reactions of trialkyltin (III) radicals.

3.7.3 σ-Bonded Organometallic Derivatives of Gallium (I) and Indium (I)

As stated earlier also, examples of well-characterized, kinetically stable, organometallic compounds in low oxidation states are rare. For example, only two well-defined monovalent organometallics of group III were known such as $In(C_5H_5)$ and $Tl(C_5H_5)$ with η^5-coordination.

Although synthesis and characterization (by elemental analysis, cryoscopic molecular weight measurements, and infrared and ^{1}H n.m.r. data) of organometallic-gallium (I) and indium (I) anions were reported employing the interaction of $M(CH_2SiMe_3)_3$ (M = Ga or In) with M′H(M′ = Na or K) in benzene or ether solution by O.T. Beachley, Jr., and co-workers, in 1980-81, yet the same workers in 1984 have contradicted their original findings and have now identified the products from the above reactions in hydrocarbon solvents as $KGa(CH_2SiMe_3)_3H$ and

$KIn(CH_2SiMe_3)_3H$. Consequently, the synthesis of organogallium (I), and organoindium (I) compounds remains an unresolved challenge for chemists.

3.8 MULTIPLE BONDED SILICON AND ARSENIC ORGANOMETALLICS

In view of the extensive carbon chemistry associated with alkenes, alkynes, ketones, and nitriles, it was natural to search for similar multiple bonded species for silicon such as Si=C, Si=Si, and Si=O derivatives. In spite of stoichiometric similarities between pairs of carbon and silicon compounds (e.g., CO_2 and SiO_2, or $(CH_3)_2CO$ and $(CH_3)_2SiO$), there is hardly a structural or chemical resemblance.

For example, the *gem*-diols of both carbon and silicon are unstable. However, the former tend to lose water with the formation of simple monomeric ketones, but the latter tend to form dialkylsilicone linear polymers;

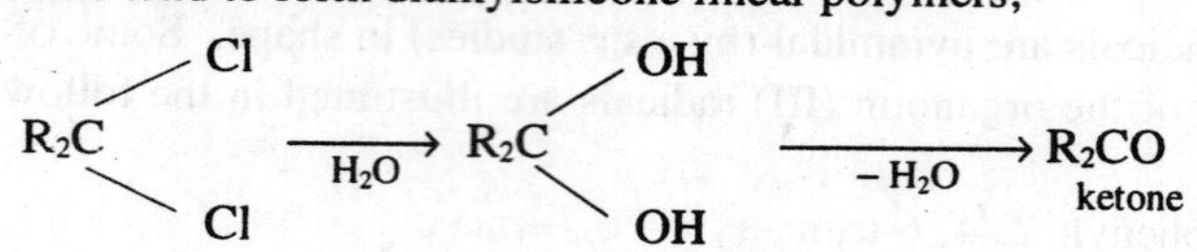

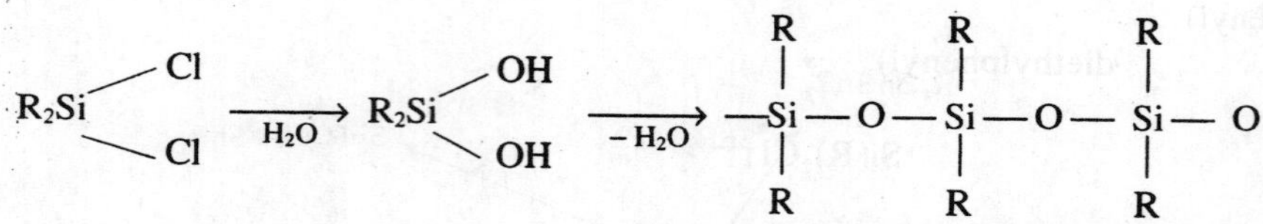

diallkysilicon polymer

The simplest explanation for the instability of doubly bonded, $>Si=C<$, $>Si=Si<$, $Si=O$ structures involving $p_\pi - p_\pi$ bonds can be offered from the point of view that the longer σ-bond in the silicon system reduces the possibility of sideways (π) overlap of the p-orbitals.

Mulliken, however, does not agree with the above simple explanation. He has suggested that the availability of 'd' orbitals of comparative energy in case of silicon may favour two single bonds over one double bond. This view-point further emphasizes that all Si-X single bonds (where X is N, O, or halogens) also have some double bond character arising from the back-bonding from the electronegative element to the 'd' orbitals of silicon. This type of back-bonding should make the formal single bonds in the case of silicon actually stronger than a simple (without back-bonding as in the case of carbon) σ-bond and more than half as strong as the hypothetical $p_\pi - p_\pi$ double bonds.

Similarly poor p_π–p_π bonding was advocated for the non-isolability of compounds containing P=P, As=As, or Sb=Sb bonds. However, it has recently been demonstrated that multiple-bonded compounds of heavier group IV and V elements can be kinetically stabilized by attaching bulkier groups on these atoms.

3.8.1 Preparation

Multiple bonded (Si=C or Si=Si) silicon derivatives have been prepared by the reactions illustrated by the following equations:

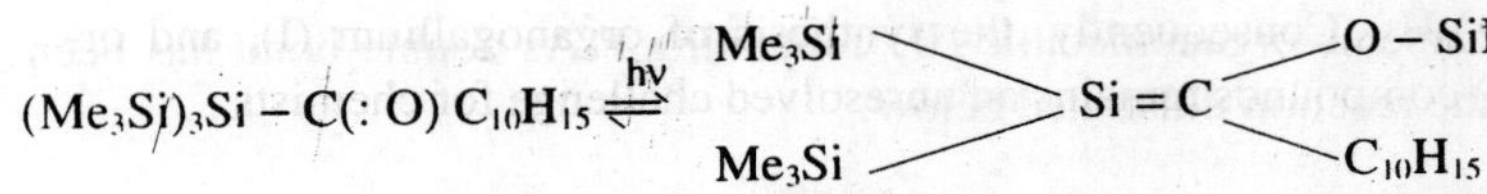

($C_{10}H_{15}$ represents admantyl)

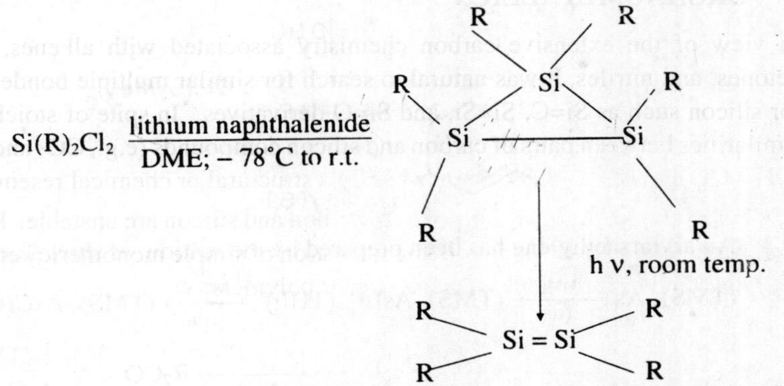

(**R** = 2, 6-dimethylphenyl; 2, 4, 6-trimethylphenyl (mesityl); and 2, 6-diethylphenyl)

diethylphenyl)

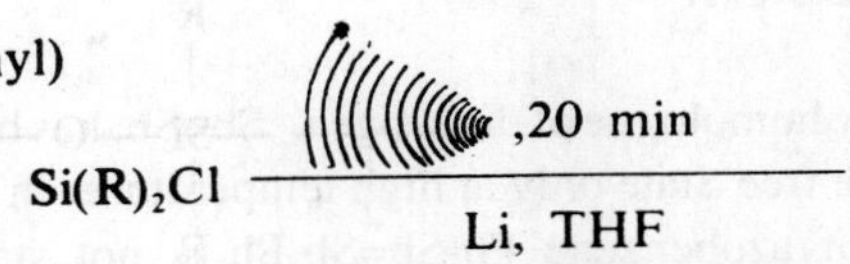

(* ultrasonic waves; R = 2, 4, 6-trimethylphenyl)

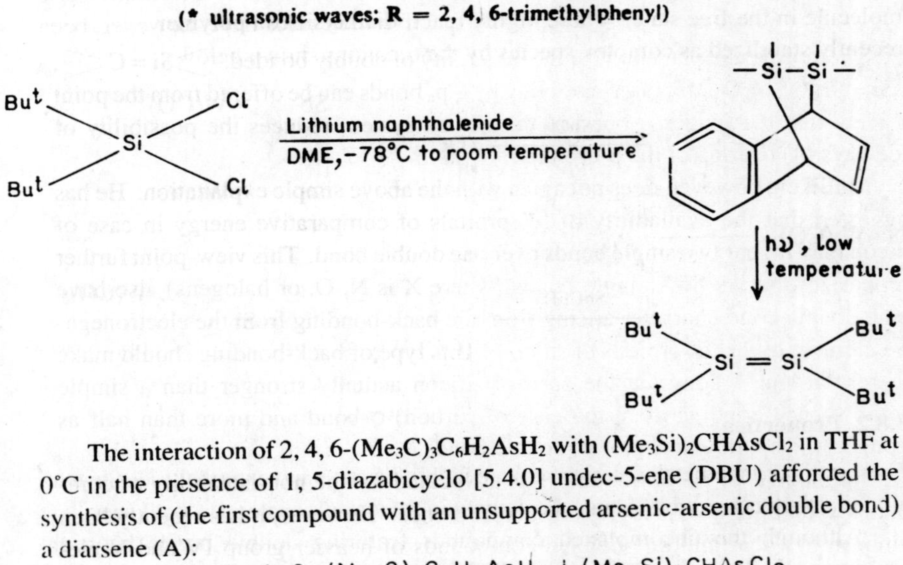

The interaction of 2, 4, 6-$(Me_3C)_3C_6H_2AsH_2$ with $(Me_3Si)_2CHAsCl_2$ in THF at 0°C in the presence of 1, 5-diazabicyclo [5.4.0] undec-5-ene (DBU) afforded the synthesis of (the first compound with an unsupported arsenic-arsenic double bond) a diarsene (A):

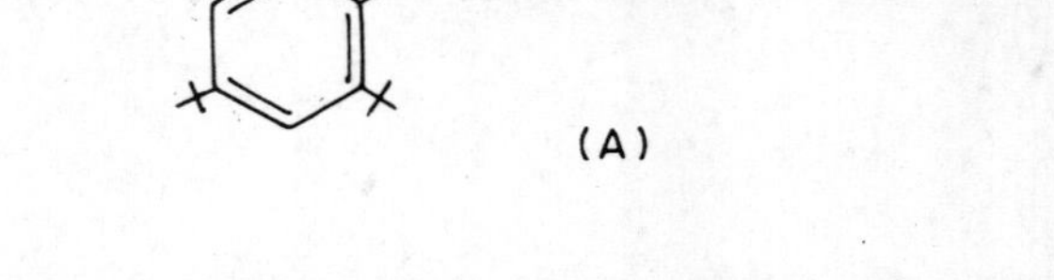

(A)

Another arsenic organometallic (B) containing P-As double bond has been prepared by the reaction illustrated below:

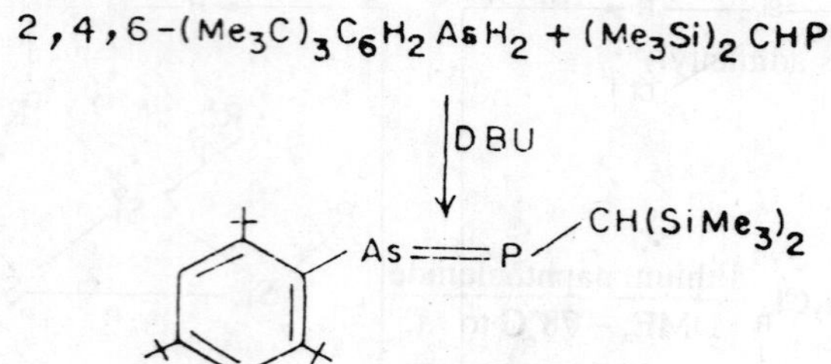

(B)

An arylarsaethylene has been prepared by the reation, as shown below:

$$(TMS)_3\,As \xrightarrow[THF]{LiMe} (TMS)_2\,AsLi_2\ (THF) \xrightarrow[THF]{ArCOCl} (TMS)_2\,AsC(O)\,Ar$$

$$\downarrow (TMS)_2O$$

$$ArC{\equiv}As$$

TMS = Me_3Si
Ar = 2, 4, 6-$Bu^t{}_3C_6H_2$

The antimony homologue of dinitrogen, :Sb≡Sb: has been shown to be capable of existence in the free State only at high temperatures in the gas phase; similarly the homologue of azobenzene Ph-$\ddot{S}$b=$\ddot{S}$b-Ph is not stable as the monomeric molecule in the free state. These highly reactive molecules have, however, been recently stabilized as complex species by the reactions shown below:

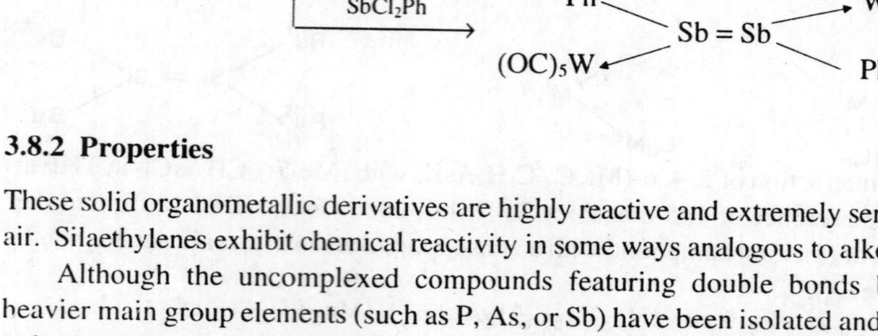

3.8.2 Properties

These solid organometallic derivatives are highly reactive and extremely sensitive to air. Silaethylenes exhibit chemical reactivity in some ways analogous to alkenes.

Although the uncomplexed compounds featuring double bonds between heavier main group elements (such as P, As, or Sb) have been isolated and studied only since 1981, yet their complexed forms, e.g., $Mo(\eta^2{-}P_2H_2)\,(\eta^5{-}C_5H_5)$, $Pd(\eta^2{-}Ph_2P_2)(Ph_2PCH_2PPh_2)$, $Ni\{\eta^2{-}(Me_3Si)_2P_2\}(Et_3P)_2$, $Pt\{\eta^2{-}(C_6F_5)_2P_2\}(Ph_3P)_2$, $Fe\{\eta^2{-}(C_6F_5)_2As_2\}(CO)_4$, $Pt\{\eta^2{-}(C_6F_5)_2As_2\}(Ph_3P)_2$, and (PhM=MPh) $\{M'(CO)_5\}_3$ (M = As, M′ = Cr) have been known since 1971. Since 1981, other

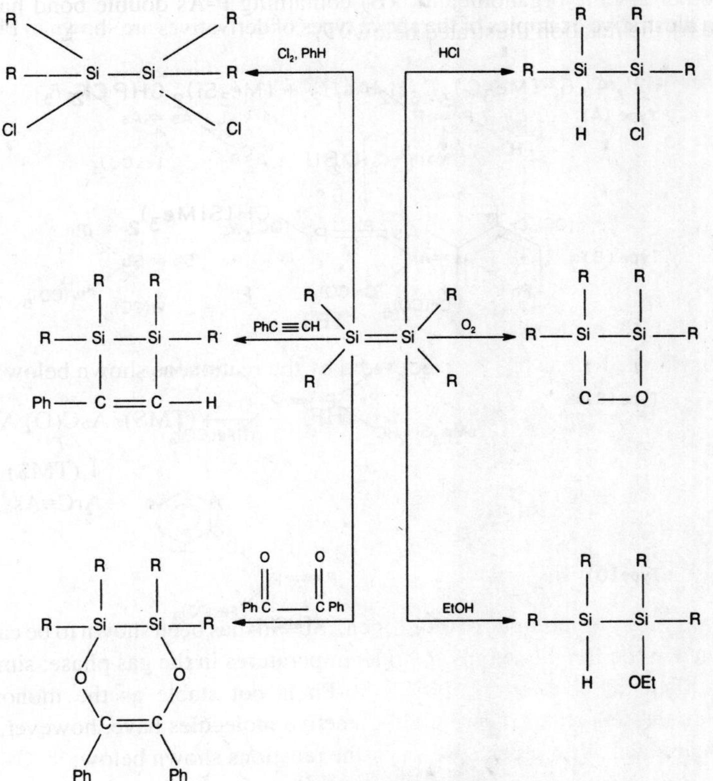

Fig. 3.41 Some reactions of silaethylenes.

complexed species with novel bonding characteristics have been isolated and characterized. These compounds can exhibit six bonding modes (A, B, C, D, E and F) in their coordination complexes (Fig. 3.42).

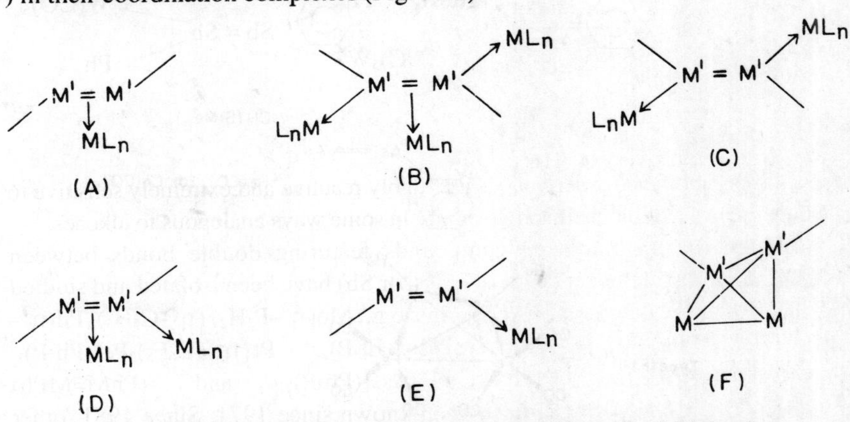

Fig. 3.42 Bonding modes in coordination complexes with metal multiple bonds.

Some illustrative examples of the above types of derivatives are shown in Fig. 3.43.

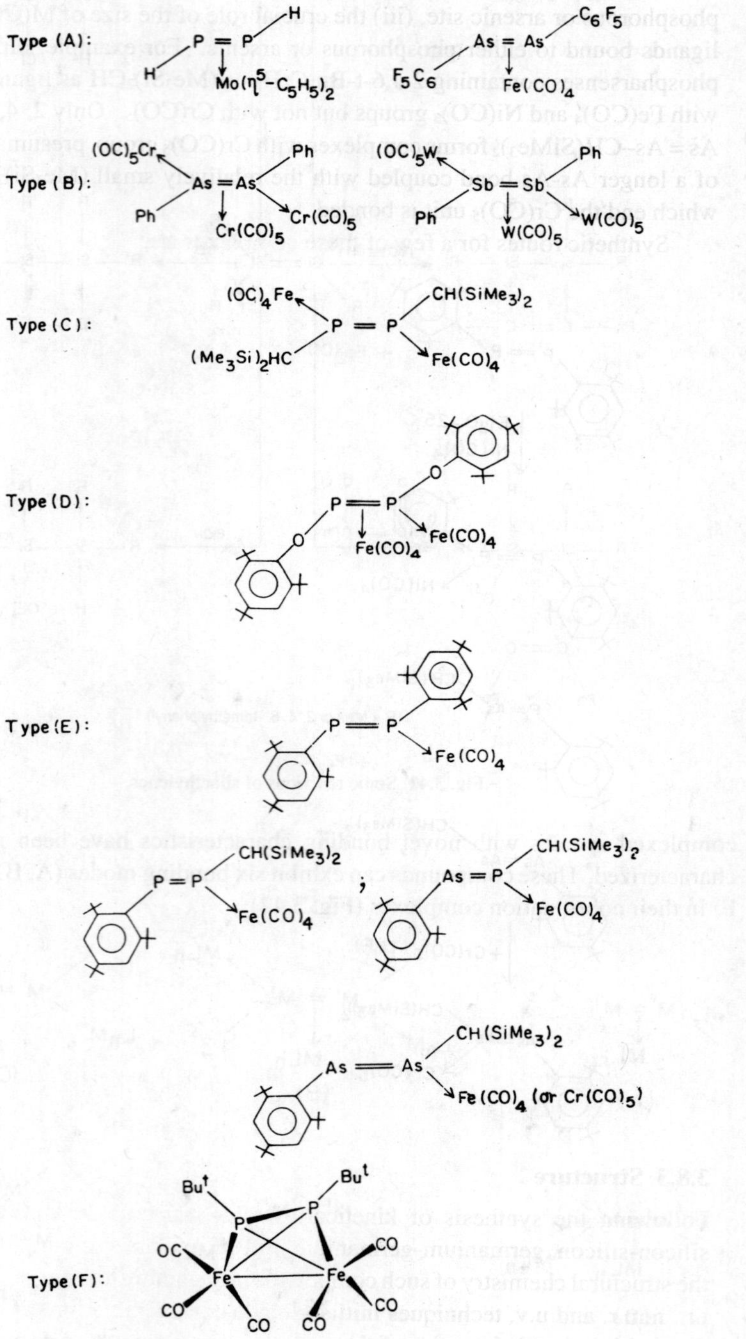

Fig. 3.43 Examples of multiple bonded organometallics.

Some characteristic features of such reactions are (i) attachment of only one $M(CO)_x$ group, (ii) bonding of the group $M(CO)_x$ at the least sterically hindered phosphorous or arsenic site, (iii) the crucial role of the size of $M(CO)_x$ and of the ligands bound to either phosphorous or arsenic. For example, diphosphenes or phospharsenses containing 2,4,6-t-$Bu_3C_6H_2$ or $(Me_3Si)_2CH$ as ligands complexes with $Fe(CO)_4$ and $Ni(CO)_3$ groups but not with $Cr(CO)_5$. Only 2, 4, 6-t-$Bu_3C_6H_2$-As = As–$CH(SiMe_3)_2$ forms complexes with $Cr(CO)_5$ group, presumably as a result of a longer As-As bond coupled with the relatively small $(Me_3Si)_2CH$ group, to which end the $Cr(CO)_5$ unit is bonded.

Synthetic routes for a few of these complexes are:

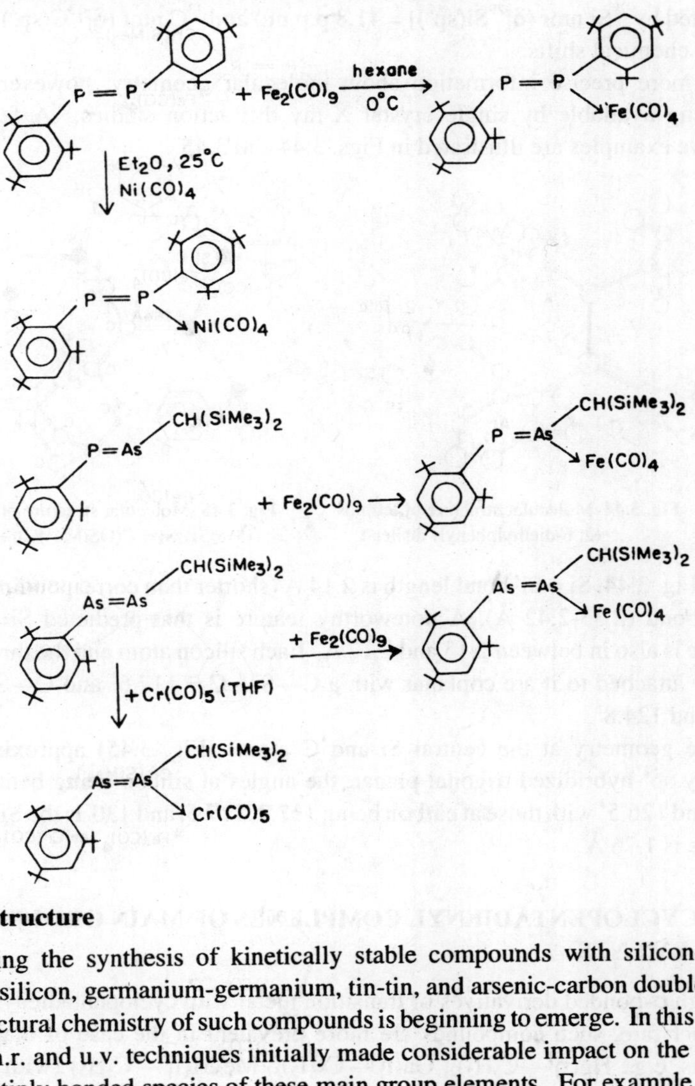

3.8.3 Structure

Following the synthesis of kinetically stable compounds with silicon-carbon, silicon-silicon, germanium-germanium, tin-tin, and arsenic-carbon double bonds, the structural chemistry of such compounds is beginning to emerge. In this context, i.r., n.m.r. and u.v. techniques initially made considerable impact on the study of the multiply bonded species of these main group elements. For example, the σ-π nature of the silicon-silicon double bond is manifested by a $\pi \rightarrow \pi^*$ transition in the visible region (near 400 nm) as well as by the ^{29}Si nmr chemical shift (15 ppm. downfield from Me_4Si). It may be mentioned that in ethylene ^{13}C n.m.r. chemical shift occurs at 24 p.p.m. (downfield from Me_4Si) which is indicative of a close similarity between the electronic configuration of the Si=Si to that of the C=C double bond.

Further infrared and nuclear magnetic resonance studies have shown considerable utility in distinguishing species having Si=C bonds, e.g., the compound $(Me_3Si)_2Si{=}C(OSiMe_3)C_{10}H_{15}$ exhibits in infrared spectrum an absorption at 1135 cm^{-1} attributable to ν(Si = C) and the presence of Si—C double bond is further supported by ^{29}Si nmr ($\delta[^{29}Si(sp^2)] = 41.8$ p.p.m.) and ^{13}C nmr ($\delta[^{13}C(sp^2)] = 214.2$ p.p.m.) chemical shifts.

A more precise information about molecular geometry, however, is now becoming available by single crystal X-ray diffraction studies. A few representative examples are illustrated in Figs. 3.44 and 3.45.

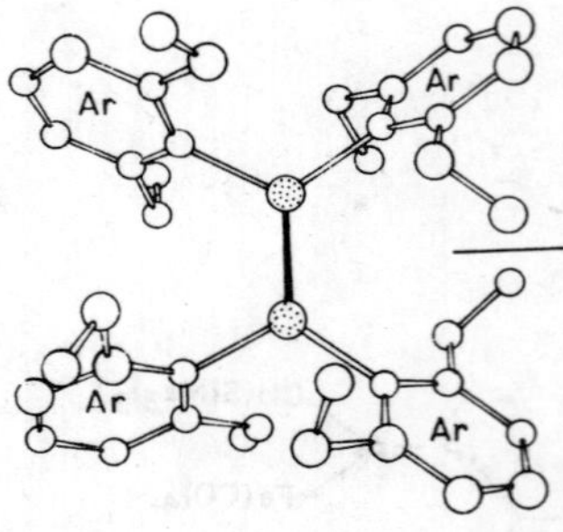

Fig. 3.44 Molecular structure of tetrakis (2, 6-diethylphenyl) disilene.

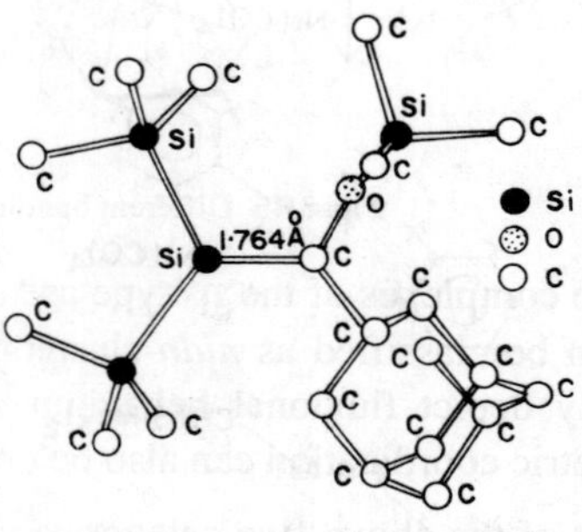

Fig. 3.45 Molecular structure of $(Me_3Si)_2Si = C(OSiMe_3)C_{10}H_{15}$

In Fig. 3.44, Si = Si′ bond length is 2.14 Å (shorter than corresponding Si—Si single bond (2.33-2.42 Å). A noteworthy feature is that predicted Si=Si bond distance is also in between 2.13 and 2.17 Å. Each silicon atom and the three atoms directly attached to it are coplanar with a C—Si—C = 117.6° and C—Si—Si′ = 117.7 and 124.8°.

The geometry at the central Si and C atoms (Fig. 3.45) approximates to formally sp^2-hybridized trigonal planar, the angles at silicon being being 114.5, 118.9 and 126.5° with those at carbon being 112.2, 117.5, and 130.1; the Si=C bond distance is 1.76 Å.

3.9 π-CYCLOPENTADIENYL COMPLEXES OF MAIN GROUP ELEMENTS

Although σ-bonded derivatives of transition metal with cyclopentadienyl ligands are rather rare, such compounds are more prevalent in the case of main group elements, e.g., $Hg(\eta^1{-}C_5H_5)_2$, $Ga(\eta^1{-}C_5H_5)_3$, $Me_3M(\eta^1{-}C_5H_5)$ (with M = Si, Ge, Sn or Pb), $Sn(\eta^1{-}C_5H_4Me)_4$, etc. By contrast, the π-bonded cyclopentadienyl complexes of transition metals are of common occurrence (cf. Section 4.3.6) whereas the chemistry of π-cyclopentadienyl complexes of main group elements is still in its infancy.

In addition to the simple σ-bonding involving only one carbon atom (η^1) and π-bonding of the η^5 type bringing all the five carbon atoms of the cyclopentadiene ring within bonding distance of the metal, intermediate types (η^2, η^3 and η^4) of bonding are also possible as shown in Fig. 3.46.

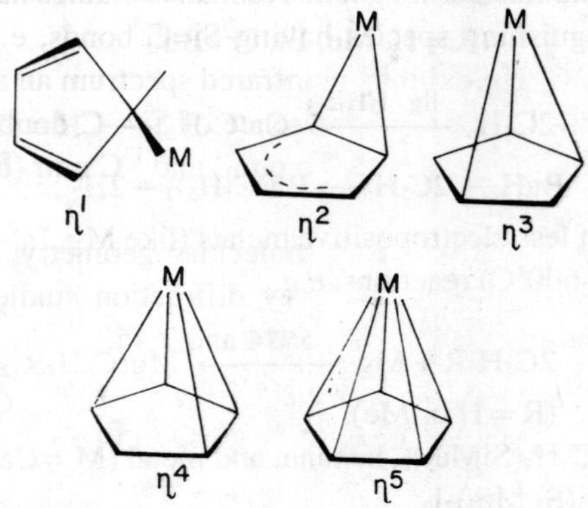

Fig. 3.46 Different bonding modes of cyclopentadiene.

The complexes of the η^1 type are σ-bonded, whereas π-complexes of the η^5 type can be classified as *nido*-clusters. Compounds of the types η^2, η^3 and η^4 generally depict fluxional behaviour in solution. In some isolated cases, an asymmetric coordination can also be observed, e.g., in $(\eta^5—C_5H_5)Be(\eta^1—C_5H_5)$.

Out of the above five categories of bonding modes, types η^1, η^2, η^3 and η^5 appear to be known with the main group elements. Some illustrative examples are:

η^1 type: $MR_3(\eta^1—C_5H_5)$ (M = Si, Ge, Sn, Pb; R = alkyl), $Ga(\eta^1—C_5H_5)_3$, $Sn(\eta^1—C_5H_5)_4$

η^2 type: $[Sn(C_5H_5N)(\eta^2—C_5Me_5)]^+ CF_3SO_3^-$

η^3 type: $Al(\eta^3—C_5H_5)Me_2$, $GeCl(\eta^3—C_5H_5)$, $SnCl(\eta^3—C_5H_5)$.

η^5 type: $Be(\eta^5—C_5H_5)Y$ (Y = H, Me, Cl, Br, C ≡ CH and BH_4),
$Zn(\eta^5—C_5H_5)Me$.
$Tl(\eta^5—C_5H_5)$ (Vapour),
$In(\eta^5—C_5H_5)$ (Vapour)
$M(\eta^5—C_5H_5)_2$ (M = Ge, Sn, Pb)
$[Sn(\eta^5—C_5Me_5)]^+ BF_4^-$
$[Li\{\eta^5—C_5H_4SiMe_3\}(TMEDA)]$
$M(\eta^5—C_5Me_5)_2$ and
$M(\eta^5—C_5HPr^i_4)_2$ (M = Ca, Ba)

3.9.1 Preparation

The cyclopentadienide/cyclopentadienyl complexes of main group elements can be prepared by a variety of methods; the most useful procedures include:

(i) Interaction of the cyclopentadiene with highly electropositive metals (usually Na) or their derivatives in donor solvents. Although a wide range of solvents have been used, yet the most versatile solvents are THF or DME, e.g.,

$$C_5H_5R + Na \xrightarrow{THF} Na^{+}\,{}^{-}C_5H_4R + 1/2H_2$$

$$(R = H, Me, Et, Pr^i, Bu^t)$$

$$CaC_2 + 2C_5H_6 \xrightarrow{lig \cdot NH_3} Ca(C_5H_5)_2 + CH \equiv CH$$

$$BaH_2 + 2C_5H_6 \rightarrow Ba(C_5H_5)_2 + 2H_2$$

The derivatives of less electropositive metals (like Mg, In, or Tl) are prepared by vapour phase (400-600°C) reactions, e.g.,

$$2C_5H_5R + Mg \xrightarrow{550°C} Mg(C_5H_4R)_2 + H_2$$

$$(R = H \text{ or } Me).$$

Condensation of $C_5H_4(SiMe_3)_2$, hexane, and metal (M = Ca, Sr, Ba) vapour at −196°C yields $M[C_5H_3(Si\,Me_3)_2]_2$:

$$M_{(g)} + 2C_5H_4(Si\,Me_3)_2 \longrightarrow M[C_5H_3(Si\,Me_3)_2]_2 + H_2$$

(ii) Reaction of cyclopentadiene with either lithium or Grignard reagents, e.g.,

$$C_5H_5R + LiBu^n \xrightarrow{THF} Li^{+}\,{}^{-}C_5H_4R + Bu^nH$$

$$(R = H, Me, Et, Pr^i, Bu^t, SiMe_3)$$

$$2C_5H_6 + 2MgRBr \rightarrow Mg(C_5H_5)_2 + Mg\,Br_2 + 2RH.$$

(iii) Reaction of cyclopentadiene with thallous hydroxide, e.g.,

$$2C_5H_5R + Tl_2SO_4 + 2KOH \rightarrow 2Tl^{+}\,{}^{-}C_5H_4R + K_2SO_4 + H_2O$$

$$(R = H, Me, Pr^i, Bu^t, SiMe_3), \text{ and}$$

(iv) Reaction of anhydrous metal halides with solutions of alkali metal cyclopentadienides, e.g.,

$$BeCl_2 + 2Na^{+}\,{}^{-}C_5H_5 \xrightarrow{THF} Be(C_5H_5)_2 + 2NaCl$$

$$MI_2 + 2K^{+}\,{}^{-}C_5Me_5 \longrightarrow M\,(C_5Me_5)_2 + 2KI$$

$$MI_2 + 2K^{+}\,{}^{-}C_5H(Pr^i)_4 \xrightarrow[THF]{Et_2O \text{ or}} M(C_5H(Pr^i)_4)_2 + 2KI$$

$$(M = Ca \text{ or } Ba)$$

$$InCl_3 + 3Na^{+}\,C_5H_5^{-} \rightarrow In(C_5H_5)_3 + 3NaCl$$

$$In(C_5H_5)_3 + C_5H_5 \rightarrow InC_5H_5 + 3/2(C_5H_5)_2$$

InC_5H_5 can be best prepared by the reaction of LiC_5H_5 with a slurry of InCl in Et_2O:

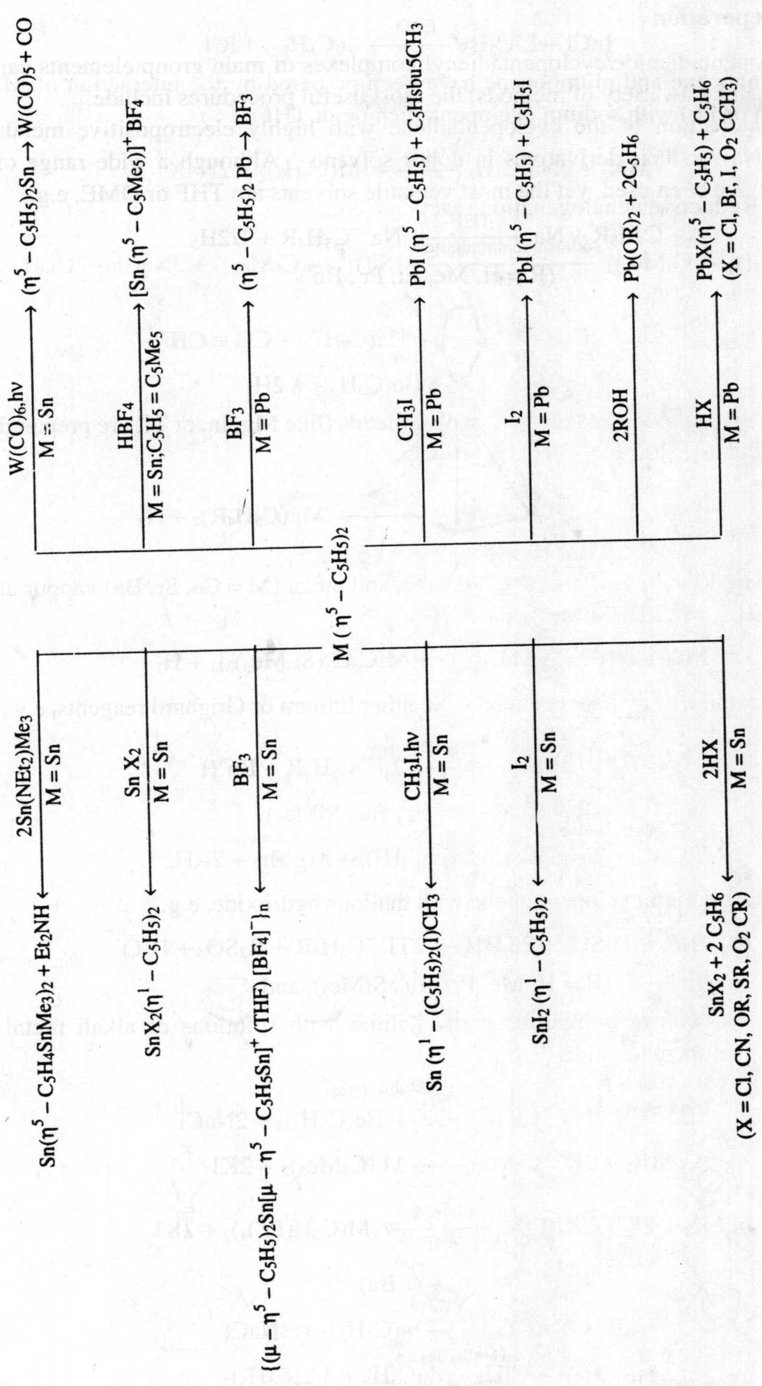

Scheme 3.10 Some reactions of stannocene and plumbocene.

$$InCl + LiC_5H_5 \xrightarrow{Et_2O} InC_5H_5 + LiCl$$

Stannocene and plumbocene have been prepared by the interaction of MX_2 (M = Sn or Pb) with sodium cyclopentadienide in THF:

$$MX_2 + 2Na^+C_5H_5^- \xrightarrow{THF} M(C_5H_5)_2 + 2NaX$$

(v) Reductive dehalogenation reactions, e.g.,

$$SiCl_2(\eta^1 - C_5Me_5)_2 \xrightarrow[THF]{\text{sodium nephthalenide}} Si(\eta^5 - C_5Me_5)_2 + 2NaCl + 2C_{10}H_8$$

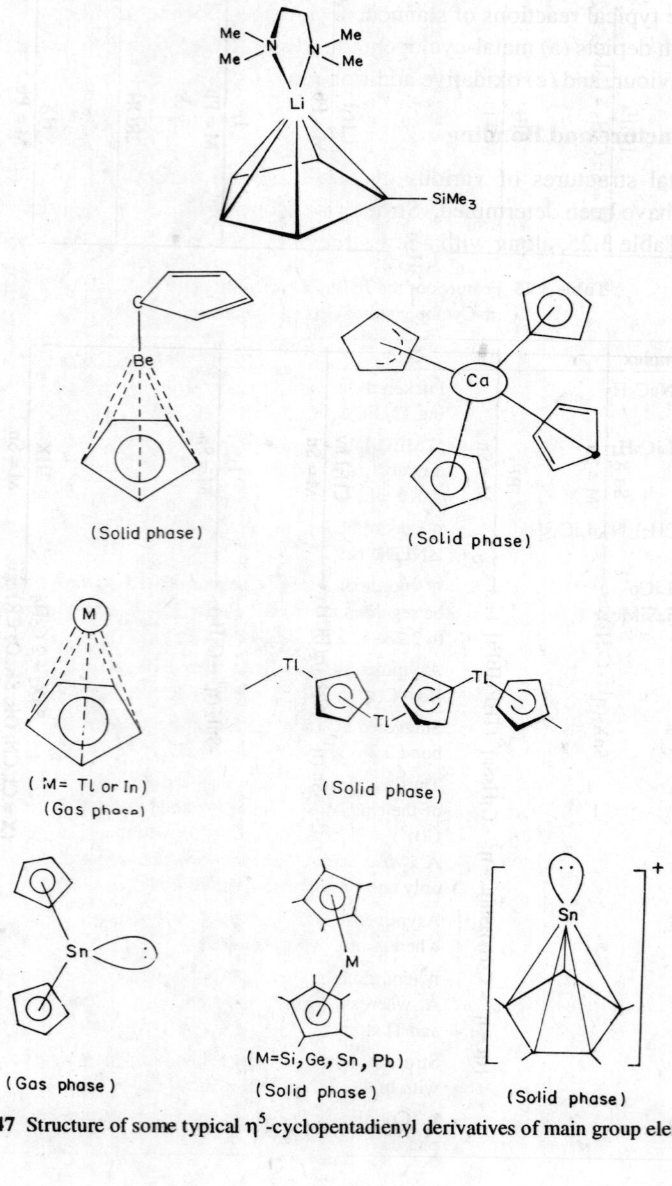

Fig. 3.47 Structure of some typical η^5-cyclopentadienyl derivatives of main group elements.

3.9.2 Properties

All these π-cyclopentadienyl metal compounds are extremely sensitive towards air and moisture, except TlC_5H_5 which is air-stable. The alkali metal, magnesium or thallium cyclopentadienides are useful sources for cyclopentadienide anions; thus these have been widely used in the synthesis of new and varied classes of cyclopentadienyl metal derivatives (cf. Section 4.3.6). Although $M(C_5H_5)_2$ and $M(C_5Me_5)_2$ (M = Ca, Ba) are as usual air and moisture sensitive, the corresponding $M(C_5HPr^t_4)_2$ derivatives have been shown to exhibit marked stability in moist air.

Some typical reactions of stannocene and plumbocene are shown in Scheme 3.10 which depicts (a) metal-cyclopentadienyl bond cleavage reactions, (b) Lewis base behaviour, and (c) oxidative addition reactions.

3.9.3 Structure and Bonding

The crystal structures of various cyclopentadienyl derivatives of main group elements have been determined. Structural features of a few typical examples are given in Table 3.25, along with a few structures shown in Fig. 3.47.

Table 3.25 Features of the X-Ray Crystal Structure of Typical π-Cyclopentadienylmetal Complexes

Complex	Features
$(TMEDA)NaC_5H_5$	Puckered chains; Cp rings bridging Na(TMEDA) units; chelating TMEDA; Na—C, 2.856-2.909Å; ∠NNaN, 70.5°
$(TMEDA)LiC_9H_{17}$ (indenyl)	(TMEDA)Li above five-membered ring; approximate η^3 coordination of Li to the indenyl (C_5 ring); 3Li—C, 2.298 ± 0.019, 2Li—C, 2.377 ± 0.002 Å, ∠NLiN, 86.40°
$\{HC(CH_2CH_2)_3N\}_2LiC_{13}H_9$ (fluorenyl)	η^3 coordination of Li; Li—C, 2.333, 2.471 and 2.794Å; ∠NLiN, 123.7°
(TMEDA)LiCp′ (Cp′ = $C_5H_4SiMe_3$)	η^5 coordination of Li, chelating TMEDA; the lithium atom may be regarded as trigonal planar; Li—C bonds range from 2.257 to 2.286Å; ∠NLiN, 84.9°.
$Be(C_5H_5)_2$	Asymmetrical structure with one ring η^5 and other η^1-bonded, Be—C(η^1) = 1.81 Å and Be—C(η^5) = 1.94 Å
$Mg(C_5H_5)_2$	Staggered (D_{5d}) sandwich structure with Mg—C and C—C bond 2.3 Å and 1.39 Å respectively.
$Ca(C_5H_5)_2$	The Ca is surrounded by four planar cyclopentadienyl rings; two of the rings are η^5, third η^3, and fourth η^1-coordinated, Ca—C(η^5) = 2.75 Å; Ca—C(η^3) = with one Ca—C distance, 2.7 Å, two at 2.79 Å, and other two are longer 2.95 Å; Ca (η^1) with only one Ca—C within the bonding distance (3.1 Å).
$Zn(C_5H_5)_2$	Asymmetric sandwich structure; one ring is pentahapto bonded whereas other is η^1-bonded.
TlC_5H_5	η^5-coordination of Tl(C_{5v} symmetry), with Tl—C 2.70 Å, whereas Zigzag chains equidistant alternating C_5H_5 rings and Tl atoms.
InC_5H_5	Structure similar to that of thallium compound in each phase with In—C 2.62 Å (gas phase).
$Sn(C_5H_5)_2$	η^5-Coordination of tin; the two rings are non-parallel in the gas phase.

(Contd.)

Complex	Features
$[Me_5C_5Sn)^+BF_4^-$	η^5-Coordination of tin cation, which exists in the form of an axially symmetrical pentagonal pyramidal *nido*-cluster.
$Pb(C_5H_5)_2$	Non-parallel η^5-Coordination of the cyclopentadienyl rings in the gas phase; two crystalline forms are known; the orthorhombic polymorph has the unusual chain-like structure.
$Sn(C_5Me_5)_2$	η^5-Coordination of the cyclopentadienyl rings, the two rings are non-parallel in the solid phase.

Study Questions

1. Define the following with suitable example(s):
   (a) Organometallic compounds; (b) Metallation;
   (c) Transmetallation; (d) Hydrometallation;
   (e) Hydroboration; (f) Exchange (or metathetic) reactions
2. Suggest a reason for wide variation in the thermal stability of organometallic compounds, e.g., tetramethylsilane is unchanged after many days at 500°C, whereas tetramethyltitanium decomposes rapidly at room temperature.
3. Explain why $Be(CH_3)_2$ or $LiCH_3$ has a tendency for oligomerization and the resulting structure is electron-deficient in nature? Sketch the appropriate orbitals that are suitable to form both terminal and bridging types of bonding in the example chosen.
4. Using suitable starting materials, write equations for the preparation of the following, indicating the reaction conditions required:
   (a) $LiBu^n$; (b) $MgI(CH_3)$; (c) R_3SnOH; (d) $Pb(C_2H_5)_4$; (e) $Hg(CH_3)_2$.
5. Provide suitable explanations for the following:
   (a) The reaction: $M + M'R \longrightarrow MR + M'$ proceeds spontaneously if M′ is less electropositive than M.
   (b) $Ga(CH_3)_3$ and $Sb(CH_3)_3$ undergo aerial oxidation rapidly, whereas $Sn(CH_3)_4$ resists such effects at ambient temperatures.
   (c) The stability of $Si(CH_3)_4$ and CCl_4 compared to that of $SiCl_4$ to hydrolytic reaction.
   (d) Thermodynamic instability of multiply bonded silicon compounds.
6. What product(s) would you expect from the following reactions? Indicate necessary reaction conditions, where appropriate:
   (a) $4C_2H_5Cl + 4\ Na/Pb \longrightarrow$
   (b) $CH_3Cl + Si\ (Cu) \longrightarrow$
   (c) $2CH_3Cl + Sn \longrightarrow$
   (d) $3LiCH(SiMe_3)_2 + AlCl_3 \longrightarrow$
   (e) $6PbCl_2 + 4Al(C_2H_5)_3 \longrightarrow$
   (f) $[Bi(C_6H_5)_4]^+ Cl^- + LiC_6H_5 \longrightarrow$
   (g) $Si(CH_3)_4 + Cl_2 \xrightarrow{h\nu}$
   (h) $Sn(CH_3)_4 + Cl_2 \longrightarrow$
   (i) $WCl_6 + Sn(C_6H_5)_4 \longrightarrow$
   (j) $LaCl_3 + 6LiCH_3 + 3TMEDA \longrightarrow$
7. Describe suitable reactions for the preparation of organometallic compounds of Li, Mg, Zn, Hg, B, Al, Tl, Si, Sn or Pb.
8. Describe a few salient applications (cf. Chapter Six) of main group organometallic compounds.
9. Sketch with brief comments the structures of the following:
   (a) $[LiC_2H_5]_4$; (b) $[Be(CH_3)_2]_n$; (c) $Al_2(CH_3)_6$;
   (d) $Zn(CH_3)_2$; (e) $Sn(CH_3)_4$.

Supplementary Reading

1. *Organometallic Compounds*, by G.E. Coates and K. Wade, 4th ed., Volume 1, Part 1: *Groups I-III*, Chapman and Hall, London (1981).
2. *Polyamine Chelated Alkali Metal Compounds*, edited by A.W. Langer, Advances in Chemistry Series, No. 130, American Chemical Society, Washington, DC (1974).
3. *The Chemistry of Organolithium Compounds*, by B.J. Wakefield, Pergamon Press, London (1974).
4. *Organoaluminium Compounds*, by T. Mole and E.A. Jeffery, Elsevier Publishing Company (1972).
5. *Organometallic Compounds*, by B.J. Aylett, 4th ed., Volume I, *The Main-Group Elements*, Part 2 (Groups IV and V), Chapman and Hall, London (1979).
6. *Organometallic Compounds of the Group IV Elements*, edited by A.G. MacDiarmid, Volume 1 (Parts 1 and 2) (1968); Volume 2 (Parts 1 and 2) (1972), M. Dekker, Interscience, New York.
7. *Organotin Compounds*, edited by A.K. Sawyer, Volume I and II, M. Dekker, New York (1971).
8. *Basic Organometallic Chemistry*, by I. Haidue and J.J. Zuckerman, Walter de Gruyter, Berlin/New York (1985).
9. *Comprehensive Organometallic Chemistry*, edited by G. Wilkinson, F.G.A. Stone and E.W. Abel, Volume 1 (Chapters 1-8) and 2 (Chapters 9-17), Pergamon Press, Oxford (1982).
10. The following chapters in the *Comprehensive Organic Chemistry*, edited by D.H.R. Barton and W.D. Ollis, Volume 3, Pergamon Press, Oxford (1979):
    (i) *Organic Compounds of Alkali Metals*, by B.J. Wakefield, pp. 943-968.
    (ii) *Organic Compounds of Group II Metals*, by B.J. Wakefield, pp. 969-1012.
    (iii) *Organic Boron Compounds*, by A. Pelter and K. Smith, pp. 943-968.
    (iv) *Organic Compounds of Group III Metals*, by G. Zweifel, pp. 1013-1060.
    (v) *Organic Silicon Compounds*, by I. Fleming, pp. 539-633.
    (vi) *Organic Compounds of Group IV Metals*, by R.C. Poller, pp. 1061-1110.
    (vii) *Organic Compounds of Antimon and Bismuth*, by R.C. Poller, pp. 1111-1125.
    (viii) *Organic Selenium and Tellurium Compounds*, by P.D. Magnus, pp. 491-538.
11. *Spectroscopic Properties of Inorganic and Organometallic Compounds*, 1968-1985 (Specialist Periodical Reports), Royal Society of Chemistry, London.
12. *Nuclear Magnetic Resonance Spectroscopy of Nuclei Other than the Proton*, edited by T. Exenrod and G.A. Webb, Wiley-Interscience, London (1974).
13. *NMR and the Periodic Table*, edited by R.K. Harris and B.E. Mann, Academic Press, London (1978).
14. ^{13}*C NMR Data for Organometallic Compounds*, edited by B.E. Mann and B.F. Taylor, Academic Press, London (1981).
15. *Characterization of Organometallic Compounds*, edited by M. Tsutsui, Part I (1969) and Part II (1970), John Wiley, New York.
16. *Vibrational Spectra of Organometallic Compounds*, by E. Maslowsky, Jr., John Wiley, New York (1977).
17. *An Introauction to Organometallic Chemistry*, by A.W. Parkins and R.C. Poller, Macmillan, London (1986)..
18. *Principles of Organometallic Chemistry*, by P. Powell, Chapman and Hall, London (1988).
19. *Organolithium Methods*, by B.J. Wakefield, Academic Press, London (1988).
20. *Organometallics*, by Ch. Elschenbroich and A. Salzer, VCH Publishers, Weinheim, 2nd Edition (1992).

4

Organometallic Compounds of Transition Metals

4.1 INTRODUCTION AND NATURE OF BONDING

Although some organotransition metal derivatives (e.g. $K[PtCl_3(C_2H_4)]$: Zeise, 1827 and $[PtI(CH_3)_3]$: Pope, 1909) were known long ago, yet a real momentum in this area of research was attained only after the chance discovery (cf. Section 1.3.2) of ferrocene independently by Kealy and Pauson, and by Miller, Tebboth and Tremaine in 1951. Organo-transition metal derivatives have in general been of greater interest not only due to their fascinating structural and bonding variations, but also due to their potential applicability as catalytic and stoichiometric reagents along with their role in biological processes. There is hardly any other field of chemistry in which such a large variety of derivatives with novel types of bonding, structure and reactivity have been added in a span of about four decades only, as illustrated by the following few salient developments arranged chronologically:

1827 W.C. Zeise prepares a platinum complex containing ethylene, $K^+[PtCl_3(C_2H_4)]^-$.

1890 L. Mond discovers $Ni(CO)_4$.

1891 M.C.R. Berthelot and L. Mond prepare $Fe(CO)_5$.

1909 W.J. Pope reports the formation of the first transition metal alkyl, $[PtI(CH_3)_3]_4$.

1919 F. Hein prepares η^6-arene chrominum (not well-characterized) complexes.

1925 F. Fischer and H. Trospsch use metals to prepare hydrocarbon fuels from CO and H_2 (The Fischer-Tropsch Process).

1930 H. Reihlen and A. Grühl prepare 1,3-butadieneiron-tricarbonyl.

1938 O. Roelen develops the Oxo-process [terminal alkenes + CO + H_2 $\xrightarrow[\text{Catalyst}]{\text{Metal complex}}$ (C + 1) aldehyde].

1939-48 W. Reppe's group develops metal-catalyzed acetylene chemistry and describes the catalytic cyclotrimerization of acetylene to cyclooctatetraene.

1951-53 Development of M.J.S. Dewar, J. Chatt, and L.A. Duncanson concepts for metal-alkene bonding.

1951 The research groups of P.L. Pauson and S.A. Miller discover ferrocene, $Fe(\eta^5\text{–}C_5H_5)_2$ independently.

1951 L.E. Orgel, L. Pauling, and H. Zeiss describe the back-bonding in metal carbonyls.

1952 G. Wilkinson, M. Rosenblum, M.C. Whiting, and R.B. Woodward propose a sandwich structure for ferrocene.

1953-54 K. Ziegler and G. Natta independently discover metal catalyzed polymerization of ethylene (in 1953) and propylene (in 1954) respectively.

1955 F.A. Cotton and G. Wilkinson characterize fluxional behaviour in organometallic compounds.

1955 J. Halpern initiates a study on the mechanism of homogeneous catalytic hydrogenation.

1955 E.O. Fischer's group characterizes the first authentic η^6-arene complexes.

1956 H.C. Longuet-Higgins and L.E. Orgel predict stable cyclobutadiene complex formation.

1958 The structure of $[(\eta^5\text{–}C_5H_5)Mo(CO)_3]_2$ reveals a covalent metal-metal bond without any bridging ligand.

1958 R. Criegee and W. Hübel describe an oxidative-addition reaction and also prepare stable cyclobutadiene complex.

1959 J. Smidt and W. Hafner prepare the first π-allyl transition metal complex, $[Pd(\eta^3\text{–}C_3H_5)Cl]_2$.

1961 X-ray structure determination of vitamin B_{12} coenzyme (by P.G. Lonhort and D.C. Hodgkin) reveals a Co-C bond.

1962 X-ray structural studies confirm existence of metal-carbonyl clusters.

1962 L. Vasaka and J.W. Diluzio discover the "Vasaka Complex", $IrCl(CO)(PPh_3)_3$.

1963 Nobel prize to K. Ziegler and G. Natta for their contributions to alkene polymerization.

1963-66 Metal-metal bonds of the orders: 1, 2, 3 and 4 are well established.

1964 E.O. Fischer and A. Maasböl isolate the first transition metal carbene complex.

1964 R.L. Banks and G.C. Bailey report alkene metathesis.

1965 G. Wilkinson and R.S. Coffey independently discover the homogeneous catalyst, $RhCl(PPh_3)_3$, for hydrogenation of alkenes.

1968 A Streitwieser isolates ''uranocene'' $U(\eta^8-C_8H_8)_2$.

1969 P.L. Timms reports the synthesis of organotransition metal complexes by means of metal-atom ligand-vapour condensation.

1970 G. Wilkinson's group reports kinetically inert transition metal alkyls through blockage of β-elimination.

1971 Monsanto develops rnodium-catalyzed acetic acid process.

1972 Industrial L-dopa synthesis by using a soluble rhodium catalyst.

1972 H. Werner reports the first triple-decker sandwich complex, $[Ni_2(\eta^5-C_5H_5)_3]^-$.

1973 E.O. Fischer's group reports the first transition metal carbyne complex, $I(CO)_4Cr(CR)$.

1973 G. Wilkinson's group reports isolation of $W(CH_3)_6$.

1973 E.O. Fischer and G. Wilkinson receive Nobel Prize for their contributions to organotransition metal chemistry.

1974 Industrial rhodium-catalyzed carbonylation of CH_3OH to acetic acid.

1974 R.R. Schrock's group discovers high-valent alkylidene and alkylidyne complexes.

1976 Industrial rhodium-catalyzed, low pressure Oxo-process.

1976 Stoichiometric reduction of a CO ligand with H_2.

1980 Industrial use of coal-derived CO and H_2 gases announced.

1980 J.M. Brown and J.J. Halpern independently elucidate the mechanism of asymmetric catalytic alkene hydrogenation.

1981 R.R. Schrock's group discovers a homogeneous catalyst for acetylene metathesis.

1981 Nobel Prize to R. Hoffmann and K. Fukul for ''Semiempirical MO concepts in a unified discussion of structure and reactivity of inorganic, organic and organometallic molecules, isolobal analogies''.

1982 R.G. Bergman and A.H. Janowicz describe intermolecular oxidative-addition of saturated hydrocarbons.

1983 P. Watson and W.A.G. Graham report independently the activation of methane by oxidative-addition reaction.

1983 The concept of ''agostic'' structures is formulated by M. Brookhart and M.L.H. Green.

1983 P.G. Garrou reports degradation of phosphines during catalytic hydroformylation.

1984 G.J. Kubas and coworkers isolate a molecular hydrogen complex, $W(CO)_3(PPr^i_3)_2(\eta^2\text{–}H_2)$.

1987 F.G.N. Cloke and coworkers describe the synthesis of bis(η^6-1, 3, 5-tri-t-butylbenzene) sandwich complexes of yttrium(o) and gadolinium(o).

1988 W.J. Evans group reports the first example of a dinitrogen complex of a f-group metal, $[(C_5Me_5)_2Sm]_2(\mu\text{-}\eta^2\text{-}\eta^2\text{-}N_2)$.

The organotransition metal chemistry is dominated by the presence of partially filled 'd' orbitals of suitable energy in the transition metal atoms. The σ-bonded transition metal alkyls tend to be rather unstable compared to their main group analogues. In fact, the transition metal-carbon σ-bond is now recognized to be similar in thermodynamic stability to its main group metal-carbon analogue. However, the tendency of transition metals to increase their coordination number leads to facile decomposition pathways, e.g., through the transfer of a β-hydrogen atom from the alkyl moiety (cf. Section 4.2.1). On the other hand, the possibility of back donation of these very 'd' electrons to suitable anti-bonding orbitals of the ligand moieties, results in π-bond formation, conferring a special stability on such derivatives.

In this chapter, we shall discuss various aspects (e.g., synthetic, reactivity, nature of the metal-carbon bond and structural characteristics) of organo-transition metal compounds. For a systematic understanding of the nature of these fastly growing and often, novel derivatives, our emphasis would be on elucidation of general principles, illustrating them by classical as well as by other selected examples of some special significance. For convenience, transition metal organometallics can be divided into two broad classes:

(*i*) *σ-Bonded organometallics* (σ-hydrocarbyls), in which a single carbon atom of the ligand is attached directly to the metal through a σ-(two electron two centre) bond (Section 4.2), and

(*ii*) *π-Bonded organometallics*, which are formed between a metal atom (ion or complex species) and ligands such as CO, RNC, :CRR′, ⋮CR, $RC\equiv CR$, $RR'C=CRR'$, $C_6H_{6-n}R_n$, and other unsaturated molecules. These ligands in addition to the presence of a lone pair of electrons have in common the presence of low-lying vacant orbitals of correct symmetry to form π-bonds by accepting electrons from transition metal d-orbitals, a phenomenon known as back-bonding. These can be further sub-divided into the following two types:

(a) In a large number of π-bonded transition metal organometallics, generally the entire unsaturated hydrocarbon molecule is involved in the electron donation, with no preferential bonding between the metal atom and any single carbon atom of the organic moieties:

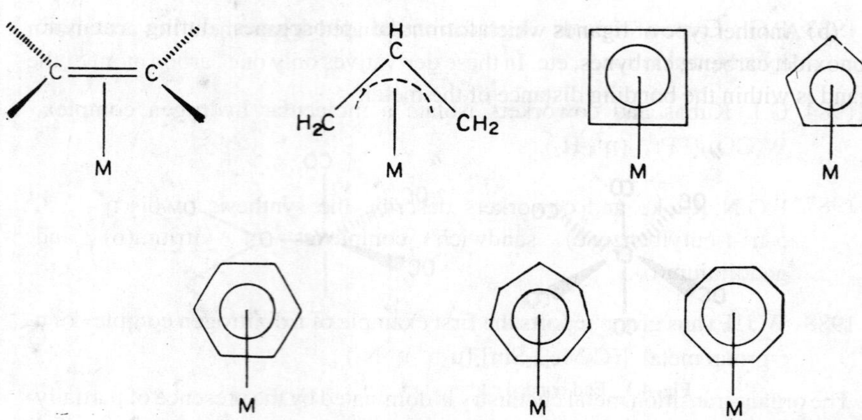

Fig. 4.1 π-Bonded transition metal organometallies in which all the carbon atoms of the ligand are within bonding distance of the metal atom.

However, a significant number of examples are now becoming available in which (as demonstrated by X-ray diffraction studies) preferential bonding sites (i.e., out of a number of carbon atoms available for bonding in a ligand only a few of them are within the bonding distance of the metal) have been established; for example, Figure 4.2 illustrates the additional bonding possibilities of some unsaturated hydrocarbon ligands to metals (in addition to the π-electron cloud of the ligand as a whole bonding the metal).

A characteristic feature of the above modes of bonding is their transverse (side-bonded) nature. In addition these ligands can be end-bonded also through simple σ-bonding. e.g., M– 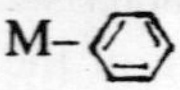

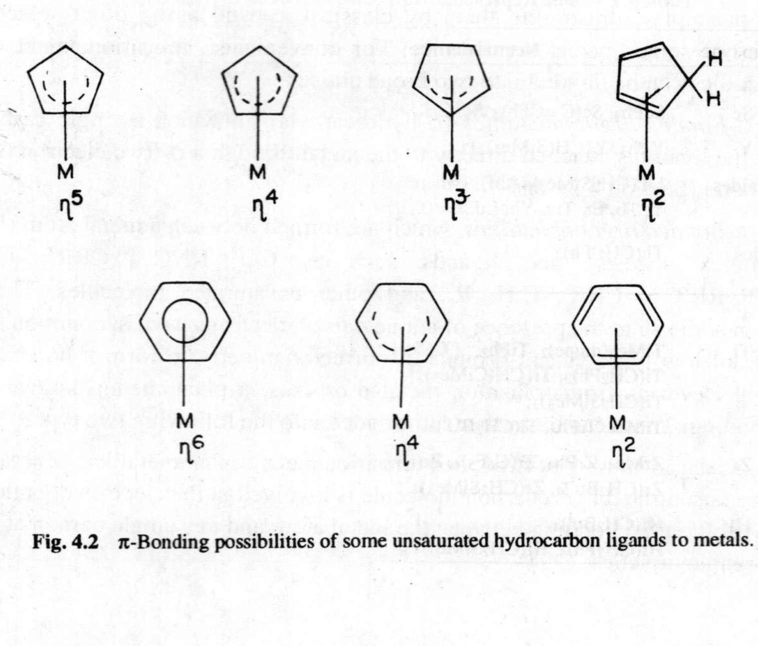

Fig. 4.2 π-Bonding possibilities of some unsaturated hydrocarbon ligands to metals.

(b) Another type of ligands which form π-bonded organometallics are carbon monoxide, carbene, carbynes, etc. In these derivatives only one carbon atom of the ligand is within the bonding distance of the metal:

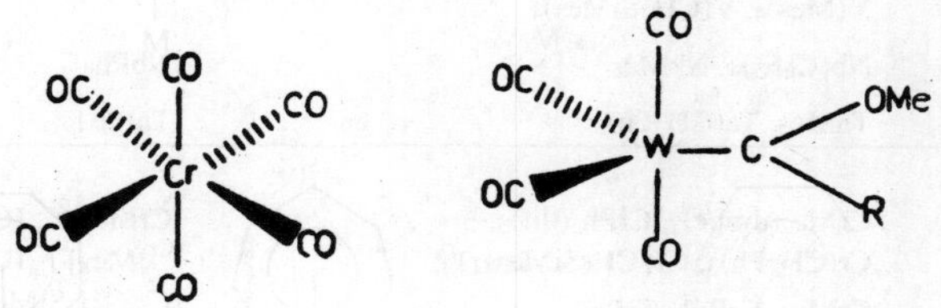

Fig. 4.3 End bonded carbonyl and carbene complexes.

4.2 σ-BONDED ORGANOMETALLICS

4.2.1 Introduction

The early literature is full of reports indicating failures to prepare simple derivatives such as tetraethyltitanium or diethylnickel. The title "The long search for stable transition metal alkyls" chosen by Wilkinson for his Nobel Lecture in 1973 itself indicates the extended and ardous efforts by chemists for a clearer understanding of the methods of synthesis and stability of the transition-metal carbon σ-bond. Before proceeding further it may be appropriate to mention that the progress in the synthesis of even simple transition metal alkyls has been phenomenal during the last 2 decades as depicted by typical examples of σ-bonded homoleptic hydrocarbyls of a large number of transition and inner transition metals listed in Table 4.1.

Table 4.1 Some Representative Transition Metal σ-Hydrocarbyls

Group type	Neutral derivatives	Anionic derivatives
3		
Sc	$ScPh_3$, $Sc(C\equiv CPh)_2$, $Sc\{CH(SiMe_3)_2\}_3$	
Y	YPh_3, $Y\{CH(SiMe_3)_2\}_3$	
Lanthanides	Ln $(CH_2SiMe_3)_3$ $(thf)_x$ (where Ln = Sc, Y, Tb, Er, Tm, Yb, Lu, x = 0, 2, or 3)	$[LnMe_6]^{3-}$, $Ln(CH_2SiMe_3)_4^-$ $[LaPh_4]^-$
Actinides	$Th(CH_2Ph)_4$	$[U(CH_2SiMe_3)_6]^{2-}$ $[U(CH_2SiMe_3)_8]^{3-}$
4		
Ti	$TiMe_2(dmpe)_2$, $TiPh_2$, $[TiMe_5]^-$, $Ti(CH_2Ph)_2$; $Ti(CH_2Ph)_3$, $Ti\{CH(CiMe_3)_2\}_3$, $Ti(CH_2SiMe_3)_3$; $TiMe_4$, $TiPh_4$, $Ti(CH_2BU^t)_4$, $Ti(CH_2SiMe_3)_4$	
Zr	$ZrMe_4$, $ZrPh_4$, $Zr(C_6F_5)_4$, $Zr(CH_2Ph)_4$ $Zr(CH_2Bu^t)_4$, $Zr(CH_2SiMe_3)_4$	$[ZrMe_6]^{2-}$
Hf	$Hf(CH_2Bu^t)_4$, $Hf(CH_2Ph)_4$, $Hf(CH_2SiMe_3)_4$	

Group type	Neutral derivatives	Anionic derivatives
5		
V	$VMe_2(dmpe)_2$, $V\{CH(SiMe_3)_2\}_3$, $V(CH_2SiMe_3)_3$; VPh_4, $V(CH_2Ph)_4$, $V(Mes)_4$, $V(CH_2SiMe_3)_4$	$[V Ph_6]^{4-}$
Nb	$Nb(C_6F_5)_4$; $NbMe_5$	$[NbPh_6]^-$
Ta	$TaMe_5$, $Ta(CH_2Ph)_5$	$[TaPh_6]^-$
6		
Cr	$CrMe_2(dmpe)_2$, $CrPh_3(thf)_3$; $Cr(CH_2Ph)_3$, $Cr\{CH(SiMe_3)_2\}_3$; $CrMe_4$, $CrBu^t_4$, $CrPr^i_4$, $Cr(CH_2SiMe_3)_4$	$[CrMe_4)^{2-}$, $[CrMe_6]^{2-}$ $[Cr_2Me_8]^{4-}$, $[CrPh_4]^-$ $[Cr(CH_2SiMe_3)_4]^-$
Mo	$Mo(CH_2Ph)_4$, $Mo(Mes)_4$ (Mes=mesityl); $Mo_2(CH_2SiMe_3)_6$	
W	WMe_6, $W(CH_2C_6H_4CH_2\text{–}o)_3$; $W(C_6F_5)_5$, $W(CH_2Ph)_4$; $W_2(CH_2SiMe_3)_2$	$[WMe_8]^{2-}$, $[W_2Me_8]^{2-}$ $[WPh_6]^{2-}$
7		
Mn	$MnMe_2(dmpe)_2$, $\{Mn(CH_2Ph)_2\}_n$ $\{Mn(CH_2Bu^t)_2\}_4$, $\{Mn(CH_2SiMe_3)_2\}_n$, $Mn\{C(SiMe_3)_3\}_2$, $\{Mn(CH_2CMe_2Ph)_3\}_2$	$[MnMe_4]^{2-}$, $[Mn(CH_2SiMe_3)_4]^{2-}$
Re	Re_8Me_3, $Re_3(CH_2SiMe_3)_{12}$	$[Re_2Me_8]^{2-}$
8		
Fe	$FeMe_2(dmpe)_2$, $Fe(Mes)_2$	$[FeMe_4]^{2-}$, $[FePh_4]^{2-}$
9		
Co	$Co(C_6F_5)_2$ (bipy), CoR_4 (R = norbornyl)	$[CoMe_4]^{2-}$, $[Co(C{\equiv}CR)_6]^{3-}$
10		
Ni	$NiBu^t_2$, $Ni(CH_2SiMe_3)_2$ (bipy) $NiEt_2$(bipy)	$[NiMe_4]^{2-}$, $[Ni(C_6F_5)_4]^{2-}$, $[NiPh_4]^{2-}$
11		
Cu	$\{CuC_6F_5\}_4$, $\{Cu(CH_2\ SiMe_3)\}_4$ CuR_2 ($R{=}CH_2\ CH_2\ PMe_2$)	$[Cu(Mes)_2]$, $[Cu\{C(SiMe_3)_3\}_2]$ $[Cu(C\equiv CR)_2]^-$ $[CuMe_3]^-$,
Ag	$\{AgMes\}_5$, $\{AuMes\}_5$,,	$[Ag\{C(SiMe_3)_3\}_2]^-$ $[Ag(C\equiv CR)_2]^-$
Au	$Au\{CH(SiMe_3)_2\}$ (L), $Au(C\equiv CR)$, $Au(C_6F_5)_3$	$AuMe_2$, $AuMe_4$ $[Au(C\equiv CR)_2]^-$

Although the thermodynamic data on the metal-carbon bond energies is still quite scanty, yet the picture which emerges out is that the transition metal-carbon bonds are not significantly weaker than main group metal-carbon bonds. With the realization that the actual bonding energy of transition metal-carbon σ-bonds is sufficiently high, the observed instability and tendency to decomposition of transition metal alkyls has to be explained on the basis of kinetic factors. One of the distinguishing characteristics of transition (from main group) metal alkyls is the coordinative unsaturation due to the presence of vacant 'd' orbitals facilitating their decomposition via routes in which the transition metal increases its coordination number. Out of the several pathways (cf. Section 4.2.3.2) by which transition metal alkyls can decompose, one of the most facile routes preferred by them is through the so-called hydride transfer accompanied with alkene-elimination. This type of reaction involves the transfer of a hydrogen from the second or β-carbon atom of the alkyl chain to the metal, increasing its coordination number. The resulting intermediate hydrido-alkene complex tends to lose alkene with the formation of metal hydride which in some cases may decompose to metal and hydrogen:

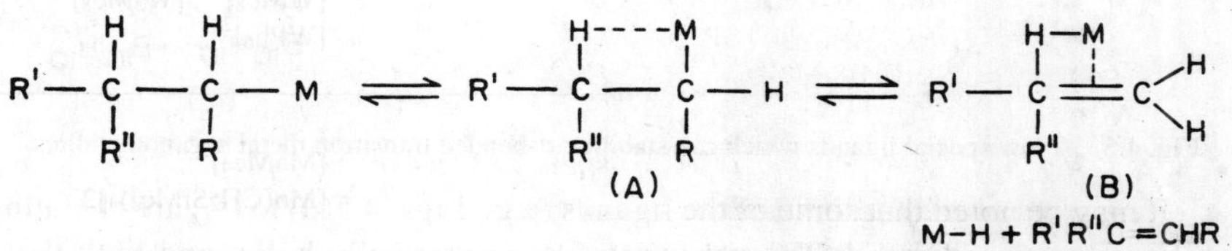

Fig. 4.4 Decomposition of a transition metal alkyl into metal hydride and olefin via β-hydrogen transfer.

The reverse of the above reaction, i.e., the formation of alkyls by addition (insertion) of alkenes to M – H bonds is also of great significance in a number of catalytic processes discussed in Chapter 6.

If the possibility of β-hydrogen transfer could be eliminated or hindered, the stability of the resulting alkyls should increase. The simplest examples of this type of stability are the simple metal methyls e.g., $Ti(CH_3)_4$ in which the β-hydrogen transfer pathway is obviously ruled out. The comparative stability of metal benzyles and neopentyls, i.e., $M(CH_2C_6H_5)_n$ and $M(CH_2C(CH_3)_3)_n$ can also be explained on the same lines, i.e., the absence of any hydrogen atom at the β-position. The influence of steric hindrance in inhibiting the approach of β-hydrogen in crowded tert.butyl/isopropyl groups to the central metal atom explains the comparative stability of derivatives like $Cr\{C(CH_3)_3\}_4$.

Another additional factor can be cited to contribute to the stability of derivative like $M(CH_2SiMe_3)_n$, which are distinctly more stable than the analogous neopentyls $M(CH_2CMe_3)_n$. This can be understood as the substitution of β-carbon atom in the ligand by silicon precludes the possibility of β-hydrogen elimination as the formation of C = Si analogue of the alkene (cf. Fig. 4.4) is very unlikely. Some other examples of ligands of this type can be cited as $–CH_2GeR_3$ and $–CH_2SnR_3$ as well as 1-norbornyl (Fig. 4.5(a)) in which case also the resulting bridge head alkene is highly unstable). A few more ligands of this type are given in Fig. 4.5.

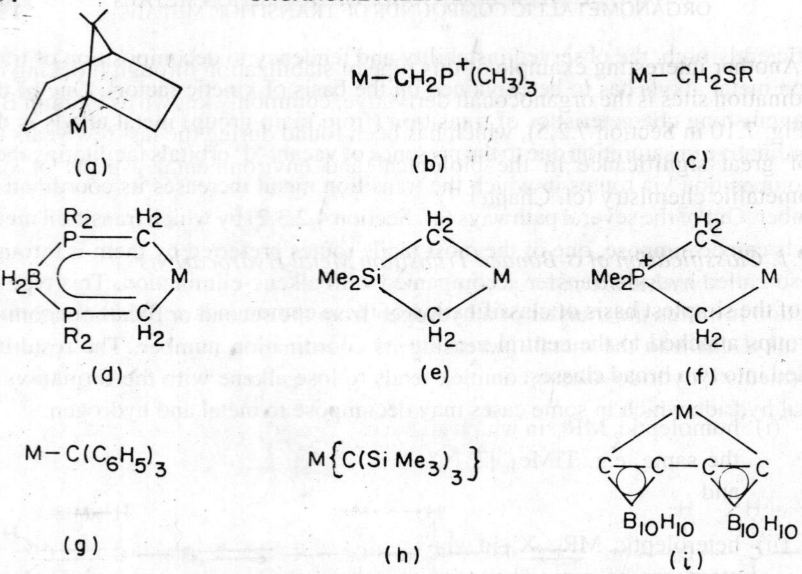

Fig. 4.5 A few special ligands which can stabilize σ-bonded transition metal organometallics.

It may be noted that some of the ligands (e.g., Figs. 4.5(d) to (f)) are chelating while others (e.g. Figs. 4.5(a) and (g) to (i)) are sterically bulky and both these factors would have significant effects on the stability of the corresponding alkyl derivatives.

It may be pointed out that the decomposition of metal alkyls through β-hydrogen pathway (Fig. 4.4) involves the formation of intermediate hydridoalkene complex(B) in which the coordination number of the metal has been increased. It, therefore, follows that if the coordination sites of the metal were already saturated by the presence of other ligands on the metal, then the formation of the complex (B) would be precluded. A few examples of derivatives of this type stabilized by the presence of the other ligands are: $Mo(CH_2CH_3)(CO)_3(\eta^5\text{-}C_5H_5)$; $Mn(CH_2CH_3)(CO)_5$; and $Fe(\eta^1\text{-}C_5H_5)(CO)_2(\eta^5\text{-}C_5H_5)$. Such stabilizing ligands are generally π-bonding or π-acid type, but could also be other Lewis bases like ammonia in $[Rh(C_2H_5)(NH_3)_5](ClO_4)_2$ or bissalicylaldehyde-ethylenediimine (salen) in $(Co(C_2H_5)(salen))$. The earlier notion that the presence of these ligands (π-acid or other type) produces electronic changes in stabilizing the metal-carbon bond is no longer considered valid and the prime role of these ligands appears to be merely to saturate the coordination sites, precluding the formation of an intermediate (B in Fig. 4.4) in which the metal has increased its coordination state further. Some additional examples illustrating stabilization of metal-carbon σ-bond by blocking the coordination sites are as follows:

$TiMe_4(dipy)$, $Ti(CH_2Ph)_4$, WMe_6 (neutral complexes)

$Mn(CH_2SiMe_3)_4^{2-}$, $ReMe_8^{2-}$, MMe_6^{3-} (anionic complexes)

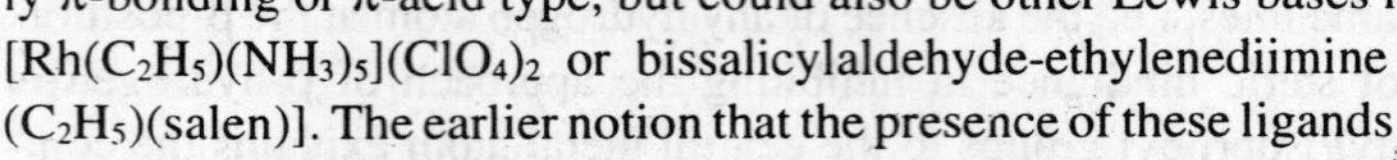

(Chelated complexes)

Another interesting example of this type of stabilization through blocking of coordination sites is the organocobalt derivative, commonly known as vitamin B_{12} (cf. Fig. 7.10 in Section 7.2.5), which has been found during the last few years to be of great significance in the biological and environmental aspects of organometallic chemistry (cf. Chapter 7).

4.2.1.1. *Classification of σ-Bonded Transition Metal Hydrocarbyls*

One of the simplest basis of classification of these compounds could be the nature of groups attached to the central metal atom. These hydrocarbyls could thus be divided into two broad classes:

(i) homoleptic, MR_n, in which all the organic groups on the metal atom are the same, e.g. $TiMe_4$, $[Ti\{CH(SiMe_3)_2\}_3]$, and $\{Cr\{CH_2C(Me)_2Ph\}_4]$; and

(ii) heteroleptic, $MR_{n-a}X_a$, in which the groups attached to the central metal atom are of more than one type, e.g. $TiCl_nMe_{4-n}$, $ZrMe_n(OPh)_{4-n}$, $ZrMe_2(\eta^5-C_5H_5)_2$, and $ZrCl\{CH(SiMe_3)_2\}(\eta^5-C_5H_5)_2$.

Another basis of classification could be the nature of ligand as well as on the number of metal atoms present in a molecular species. A few examples of ligands for mononuclear species are given in Table 4.2.

Table 4.2 Types of Mononuclear σ-Bonded Derivatives of Transition Metals

Ligand	Hybridization of ligating carbon atom	Examples
Alkyl	sp^3	$[Cr(CH_2SiMe_3)_4]$, $[Ni(CH_2SiMe_3)_2(Py)_2]$
Ylide	sp^3	$[Cr\{CH_2PMe_2CH_2\}_3]$
Alkenyl	sp^2	$[Cr(PhC{=}CMe_2)_4]$
Aryl	sp^2	$[Cu(C_6H_2Me_3-2,4,6)_2]^-$, $\{Au(C_6H_2Me_3-2,4,6)\}_5$
Carbane	sp^2	$[W\{C(OMe)Ph\}(CO)_5]$,
Acyl	sp^2	$[Re(COMe)(CO)_5]$,
Iminoacyl	sp^2	$[Zr\{C(N\ Tol-p)Me\}(\eta^5-C_5H_5)_2]$
Alkynyl	sp	*trans*-$[Pt(C\equiv CPh)Cl(PMe_2Ph)_2]$
Ketenide	sp	$[Ag_2C_2O(Py)]$
Vinylcarbene	sp	$[MoCl(\eta^5-C_5H_5)\{C{=}C(CN)_2\}]$
Carbyne	sp	*trans*-$[W(CPh)I(CO)_4]$,
Carbonyl	sp	$Fe(CO)_5$, $Ni(CO)_4$
Isonitrile	sp	$[Fe(CNBu^t)_5]$

In addition to the mononuclear species (Table 4.1), polynuclear σ-bonded organometallic can also be formed either through bridging alkyl(aryl) groups or through metal-metal interactions. A few examples of the former (bridging) type are given in Fig. 4.6.

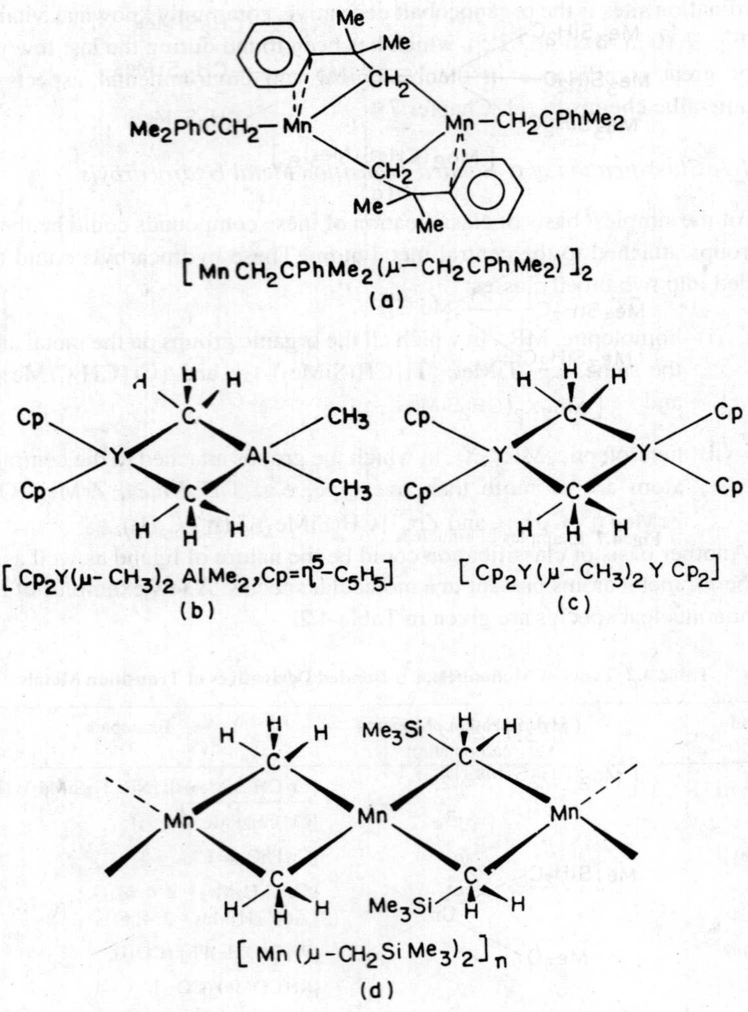

Fig. 4.6 Examples of poly- or binuclear bridged hydrocarbyls.

In some cases (specially for 4*d* and 5*d* metals), the formation of metal-metal bond appears to be preferred for achieving binuclearity. For example, instead of bridging trimethylsilylmethyl ($-CH_2SiMe_3$) groups as shown in the case of the manganese derivative (Fig. 4.6 d) the multiple metal-metal bonding appears to bind the two $Mo(CH_2SiMe_3)_3$ moieties in the binuclear derivative (Fig. 4.7 a). A few other examples of this type are shown in Fig. 4.7.

In addition, there are quite a few examples (Fig. 4.8) in which both metal-metal bonding and bridging alkyl (aryl or some other ligand) groups contribute to the binuclearity of the species.

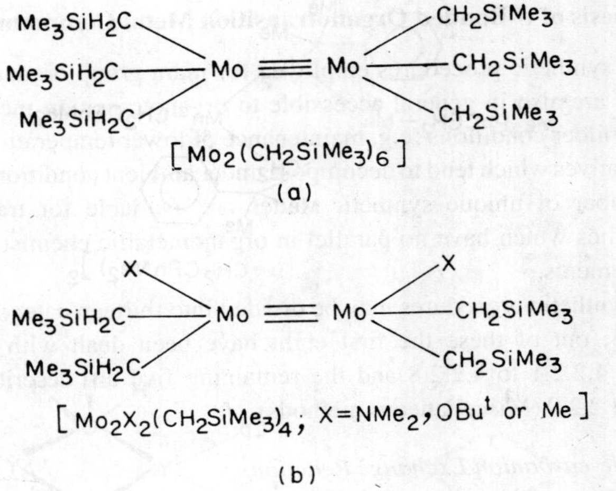

Fig. 4.7 Examples of binuclear hydrocarbyls with metal-metal bonds.

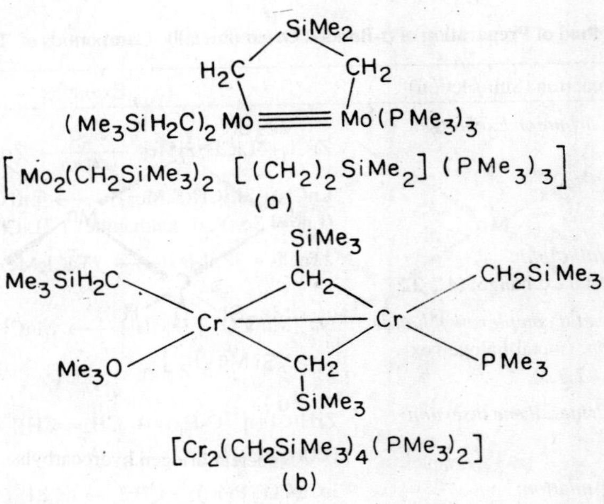

Fig. 4.8 Examples of binuclear hydrocarbyls with both bridging groups and metal-metal bonds.

In the following pages are described the general methods of preparation, properties and structures of σ-bonded organometallics of transition metals, followed by a description of some special classes of compounds.

As the organometallic derivatives of inner transition metals have been less extensively explored, and have been recently emerging out as one of the frontier areas of research, a brief description of these along with those of their congeners (Sc & Y) comes after that of the transition metal, d^2 to d^9.

4.2.2 Synthesis of σ-Bonded Organotransition Metal Compounds

Most of the synthetic procedures employed for main group organometallics (cf. Section 3.4) are also in general accessible to organotransition metal complexes except that milder conditions (e.g. maintenance of lower temperatures) are essential for derivatives which tend to decompose under ambient conditions. In addition, a large number of unique synthetic routes are available for transition metal organometallics which have no parallel in organometallic chemistry of the non-transition elements.

These synthetic procedures may be divided into thirteen categories as shown in Table 4.3; out of these the first eight have been dealt with separately in sub-sections 4.2.2.1 to 4.2.2.8 and the remaining five are described briefly in sub-section 4.2.2.9 (Miscellaneous methods).

4.2.2.1 Halide-carbanion Exchange Reactions

The most intensively studied and widely used synthetic procedure for transition metal alkyls (aryls) is the interaction of alkylating (arylating) reagents with transition metal halides (alkoxides, aryloxides or amides have also been sometimes used).

Table 4.3 Method of Preparation of σ-Bonded Organometallic Compounds of Transition Metals

Type of reaction (Sub-section)	Example
(1) *Halide-Carbanion Exchange:* (4.2.2.1)	$ZrCl_4 + 4LiCH_2SiMe_3 \xrightarrow[-4LiCl]{Et_2O} Zr(CH_2SiMe_3)_4$
	$LnCl_3 + 3LiCH(SiMe_3)_2 \longrightarrow Ln\{CH(SiMe_3)_2\}_3$ (Ln = Sc, Y, or lanthanide) + 3LiCl
(2) *Transmetallation:* (metal-metal exchange) (4.2.2.2)	$2TaCl_5 + 3ZnMe_2 \longrightarrow 2TaCl_2Me_3 + 3\,ZnCl_2$
(3) *Anionic metal complexalkyl halide interaction:* (metal-halogen exchange) (4.2.3.3)	$Na^+[Mn(CO)_5]^- + CH_3I \longrightarrow MnCH_3(CO)_5 + NaI$
(4) *Metal hydride-alkene insertion:* (4.2.2.4)	$ZrH(Cl)(\eta^5\text{–}C_5H_5)_2 + CH_2{=}CH_2 \longrightarrow Zr(H)(Cl)(Et)(\eta^5\text{–}C_6H_5)_2$
(5) *Oxidative-addtion:* (4.2.2.5)	$IrCl(CO)(PPh_3)_2 + CH_3I \rightarrow IrCl(I)(CH_3)(CO)(PPh_3)_2$
(6) *Elimination reaction:* (4.2.2.6)	$Ni(O_2CPh)(bipy) \longrightarrow NiPh_2(bipy) + 2CO_2$
(7) *Cyclometallation:* (orthometallation) (4.2.2.7)	$RhCH_3(PPh_3)_2 \xrightarrow{heat} Rh(C_6H_4PPh_2)(PPh_3)_2 + CH_4$
(8) *Metal vapour synthesis:*	$Pd + CF_3I \rightarrow (PdICF_3)_n$
(9a) *Metal alkene-nucleophile interaction:* (4.2.2.9)	$[(Fe(CO)_2(C_2H_4)(\eta^5\text{–}C_5H_5)]^+ \xrightarrow{MeOH,\ Na_2CO_3} Cp(CO)_2Fe\text{–}CH_2CH_2OMe$

Type of reaction (Sub-section)	Example
(9b) *Ylide synthesis:* (4.2.2.9)	$Ag^+ + Me_3P{\equiv}CH_2 \longrightarrow$ 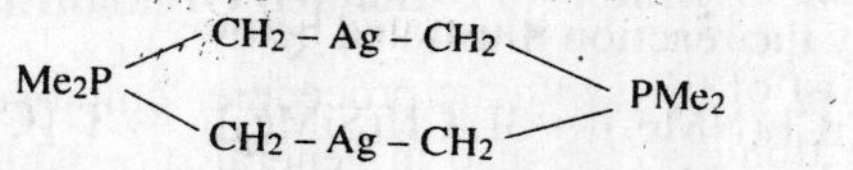
(9c) *Paramagnetic metal ion-free radical reaction:* (4.2.2.9)	$Cr(H_2O)_6^{2+} + Bu^tOOH \longrightarrow [CrBu^t(H_2O)_5]^{2+}$
(9d) *Oxidative deamination:* (4.2.2.9)	$2AuCl_4 + 2PhNHNH_2HCl \xrightarrow[-2N_2]{} [AuCl_2Ph_2]^- + AuCl_2 + 6HCl$
(9e) *Cocondensation:* (reaction of compounds with active methylene group) (4.2.2.9)	*trans*-$PtCH_3(OH)(PPh_3)_2 + CH_3NO_2 \xrightarrow{-H_2O} PtCH_3(CH_2NO_2)(PPh_3)_2$

The organometallic reagents employed are generally organolithium or magnesium compounds; the reactivity generally follows the order LiR>Mg (R)X>Mg R_2. The reductive ability of these reagents also follows the same order. Although reductive side reactions have been observed particularly in the case of some earlier 3*d* metals (e.g. Ti and V), the reducing nature of these reagents does not generally pose a problem in the cases of 4*d* and 5*d* metal derivatives.

In the following pages are given a few illustrative examples of this synthetic procedure for alkyls (aryls) of transition metals arranged in the sequence d^2 to d^p metals, followed by those of lanthanides (including scandium and yttrium) and actinides:

Ti, Zr and Hf derivatives

$$TiCl_4 + 4LiMe \xrightarrow[-80^\circ C]{ether} TiMe_4 + 4LiCl$$

$$TiCl_4 + 4Mg(CH_2Ph)X \xrightarrow[-60^\circ C]{ether} Ti(CH_2Ph)_4 + 4MgClX$$

$$TiCl_3 + MgBr(R) \xrightarrow[-30^\circ C]{pyridine} TiCl_2R(Py)_2 + MgBrCl$$

$$(R = Me, Ph)$$

$$2TiCl(\eta^5{-}C_5H_5)_2 + 2Mg(R)X \xrightarrow[0\ to\ -20^\circ C]{pentane} 2TiR(\eta^5{-}C_5H_5)_2 + 2MgClX$$

$$(R = Ph;\ 2{,}6\text{-}Me_2C_6H_3;\ 2{,}4{,}6\text{-}Me_3C_6H_2,\ etc.)$$

$$TiCl_3(\eta^5{-}C_5H_5) + 3LiMe \xrightarrow[-30^\circ C]{ether} TiMe_3(\eta^5{-}C_5H_5) + 3LiCl$$

$$\begin{array}{l} MCl_4 + 4LiCH_2EMe_3 \\ \quad or \\ \quad 4MgCl(CH_2EMe_3) \end{array} \xrightarrow{Et_2O\ or\ n-C_6H_{14}} \begin{array}{l} M(CH_2EMe_3)_4 + 4LiCl \\ \quad or \\ \quad 4MgCl_2 \end{array}$$

$(M \equiv Ti, Zr, or Hf; E = C or Si)$

The use of sterically demanding alkyl group, $(Me_3Si)_2CH$, leads to the isolation of thermally stable three-coordinate titanium (III) alkyl, $Ti\{CH(SiMe_3)_2\}_3$ by the reaction illustrated below:

$$TiCl_3(NMe_3)_3 + 3LiCH(SiMe_3)_2 \rightarrow Ti\{CH(SiMe_3)_2\}_3 + 3LiCl + 3NMe_3$$

$$TiCl_4 + 4LiCH(SiMe_3)_2 \rightarrow Ti\{CH(SiMe_3)_2\}_3 + 4LiCl + 1/2(Me_3Si)_2CH$$

The analogous reaction with $ZrCl_4$ or $HfCl_4$ gives corresponding heteroleptic alkyl $MCl\{CH(SiMe_3)_2\}_3$:

$$MCl_4 + 4LiCH(SiMe_3)_2 \longrightarrow MCl\{CH(SiMe_3)_2\}_3 + 3LiCl + LiCH(SiMe_3)_2$$
$$(M = Zr \text{ or } Hf)$$

Alkylation of $TiCl_2(Me_2PCH_2CH_2PMe_2)_2$ with LiMe(2 mols) in ether leads to the formation of dimethyl complex of titanium(II):

$$TiCl_2(dmpe)_2 + 2LiMe \xrightarrow{Et_2O} trans\text{–}TiMe_2(dmpe)_2 + 2LiCl$$

V, Nb, and Ta derivatives

Interest in the isolation of stable alkyls and aryls of vanadium has been aroused only recently. Authenticated examples of derivatives containing V-C σ-bonds were not known upto 1955 and even in 1976 less than ten such complexes were known. However, sterically demanding alkyl or aryl ligands or those lacking β-hydrogen atoms have yielded stable compounds:

$$VCl_3{\cdot}(NMe_3)_2 + 3LiCH(SiMe_3)_2 \longrightarrow V\{CH(SiMe_3)_2\}_3 + 3LiCl + 2NMe_3$$

$$VCl_3{\cdot}(THF)_3 + 3MgX(2,4,6\text{–}Me_3C_6H_2) \xrightarrow[-30^\circ C]{THF} V(2,4,6\text{–}Me_3C_6H_2)_3 + 3MgXCl$$

$$VCl_3(THF)_3 + 3LiPh \longrightarrow VPh_3(THF)_3 + 3LiCl$$

$$VCl_3(dmpe)_2 + 2LiMe \xrightarrow{Et_2O} trans\text{–}VMe_2(dmpe)_2 + 2LiCl$$

Although reactions of $NbCl_5$ and $TaCl_5$ with Grignard reagents were carried out as early as 1940, yet an isolable alkyl was not reported until 1964. Some preparations of niobium and tantalum alkyls and aryls are illustrated below:

$$NbCl_3Me_2 + 3LiMe + dme \xrightarrow[-78^\circ C]{ether} NbMe_5\,(dme) + 3LiCl$$

(cf. Sect. 4.2.2.2)

$$TaCl_2Me_3 + 2LiMe \xrightarrow[-78^\circ C]{Et_2O} TaMe_5 \xrightarrow{dme} TaMe_5(dme)$$

(cf. Sect. 4.2.2.2)

The interaction of $NbCl_5$ or $TaCl_5$ with $LiCH_2EMe_3$ (E = C or Si) gave unusual results which seems to be possibly due to the fact that accommodation of more than four (such) bulky ligands around a single atom is sterically inhibited. Thus the reaction of $MCl_2(CH_2CMe_3)_3$ (M = Nb or Ta) with $LiCH_2CMe_3$(2 mols) in pentane proceed according to the equation:

$$MCl_2(CH_2CMe_3)_3 + 2LiCH_2CMe_3 \xrightarrow{\text{pentane}} (Me_3CCH_2)_3M{=}C\begin{smallmatrix} H \\ CMe_3 \end{smallmatrix} + 2LiCl + Me_3CCH_3 \quad (M = Nb, Ta)$$

Cr, Mo, and W derivatives

Simple methyls of chromium and molydenum and tungsten are unstable; attempted preparations of these using a reactive alkylating agent (such as lithium methyl) lead to the formation of alkylate anion, e.g.:

$$CrCl_3 + 6LiMe \xrightarrow[-18^\circ C]{\text{dioxane}} Li_3[CrMe_6]\cdot 3C_4H_8O_2 + 3LiCl$$

$$CrCl_3(THF)_3 + \underset{\text{(excess)}}{4LiCH_2SiMe_3} \longrightarrow Li[Cr(CH_2SiMe_3)_4] + 3LiCl \xrightarrow{-e} Cr(CH_2SiMe_3)_4$$

$$MoCl_3(THF)_3 + 5LiMe \xrightarrow{\text{THF}} Li_4[Mo_2Me_8](THF)_4$$

$$WCl_4 + \underset{\text{(excess)}}{8LiMe} \xrightarrow{Et_2O} Li_4[W_2Me_8]\cdot 4Et_2O + 4LiCl$$

Although hexamethyl of tungsten could not be obtained by reacting WCl_6 with LiMe, possibly due to alkylate anion formation, yet the reaction with a milder reagent, aluminium methyl can lead to the formation of WMe_6:

$$WCl_6 + 6AlMe_3 \rightarrow WMe_6 + 6AlClMe_2$$

The use of lithium reagents derived from sterically demanding ligands (such as $-CH_2SiMe_3$, $-CH(SiMe_3)_2$, or Bu^t) has led to the formation of some elusive type of compounds of Cr, Mo, or W. e.g.,

$$CrCl_3(THF)_3 + 4LiR \xrightarrow[Et_2O]{\text{pentane or}} Li^+[CrR_4]^- \xrightarrow{-e} CrR_4$$

$$R = CH_2SiMe_3,\ CH_2CMe_3,\ CH_2C(Ph)Me_2,\ CH_2CPh_3,\ \text{or } Bu^t)$$

$$CrCl_3 + 3LiCH(SiMe_3)_2 \rightarrow Cr\{CH(SiMe_3)_2\}_3 + 3LiCl$$

$$MoCl_5 + MgCl(CH_2SiMe_3) \rightarrow Mo_2(CH_2SiMe_3)_6 + \cdots$$

$$MoCl_5 + LiCH_2CMe_3 \rightarrow Mo_2(CH_2CMe_3)_6 + \cdots$$

$$WCl_6 + MgCl(CH_2SiMe_3) \rightarrow W_2(CH_2SiMe_3)_6 + \cdots$$
$$\text{or}$$
$$LiCH_2SiMe_3$$

$$WCl_4(O) + o\text{-}C_6H_4(CH_2MgCl)_2 \text{ (excess)} \xrightarrow{THF} W(CH_2C_6H_4CH_2-o)_3 + 3MgCl_2$$

Interaction of $CrCl_2(dmpe)_2$ with LiMe (2 mols) in ether results in the formation of *trans*-$CrMe_2(dmpe)_2$:

$$CrCl_2(dmpe)_2 + 2LiMe \xrightarrow{Et_2O} trans\text{-}CrMe_2(dmpe)_2 + 2LiCl$$

Mn and Re derivatives

Simple alkyls of manganese MnR_2(R = Me or Ph) were described in the earlier literature, but their existence is seriously doubted. However, a substantial number of binary anionic and neutral alkyls of manganese are now accessible by the reactions illustrated below:

$$MnCl_2 + 4LiMe \xrightarrow{Et_2O} Li_2[MnMe_4]\cdot 0.5Et_2O + 2LiCl$$

$$MnCl_2 + MgMe_2 + \underset{\text{(excess)}}{dmpe} \rightarrow trans\text{-}MnMe_2(dmpe)_2 + 2LiCl$$

$$MnCl_2 + 4LiCH_2SiMe_3 \rightarrow Li_2[Mn(CH_2SiMe_3)_4] + 2LiCl$$

$$MnCl_2 + 2MgCl(CH_2C(Ph)Me) \xrightarrow{Et_2O} Mn\{CH_2C(Ph)Me_2\}_2 \text{ (dimer)} + 2MgCl_2$$
$$\text{or}$$
$$Mg(CH_2C(Ph)Me_2)_2$$

$$MnCl_2 + 2MgCl(CH_2EMe_3) \xrightarrow{Et_2O} Mn(CH_2EMe_3)_2 + 2MgCl_2$$
$$\text{or}$$
$$Mg(CH_2EMe_3)_2$$

(when E = C, tetramer; E = Si, polymer)

Principal precursors for rhenium σ-bonded alkyls are $ReCl_5$, $ReOCl_4$ or $ReCl_4(THF)_2$:

$$\underset{\text{or } 4MgRX}{ReOCl_4 + 4LiR} \xrightarrow[-4LiCl \text{ or } -4MgClX]{} ReOMe_4 \xrightarrow{+AlMe_3} ReMe_6 \xrightarrow{+TMEDA,\ +2LiMe} Li_2TMEDA][Re_2Me_8]$$

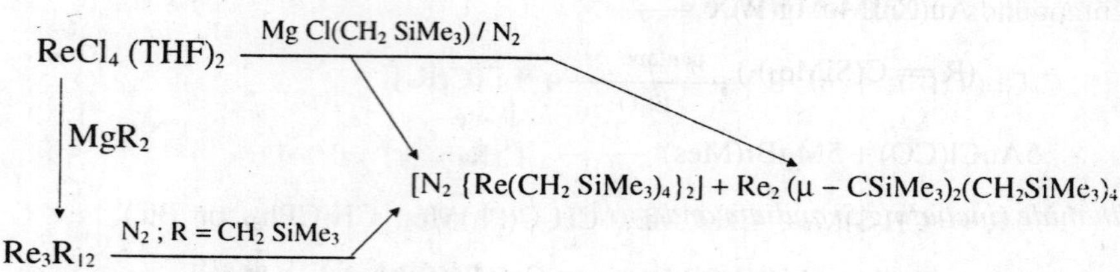

Fe, Rh, Co, Ni, Pd and Pt derivatives

$$FeCl_2(dmpe)_2 + 2LiMe \xrightarrow{Et_2O} cis\text{-}FeMe_2(dmpe)_2 + 2LiCl$$

$$FeCl_3(OEt_2) + 4LiMe \xrightarrow[-20°C]{Et_2O} Li_2[FeMe_4] + 3LiCl$$

$$RhCl_3\,(SC_4H_6) + 3MgBr(Mes) \xrightarrow[-SC_4H_6]{} Rh(Mes)_3 + 3MgBrCl$$

$$NiBr(\eta^3\text{–}C_3H_5)_2 + 4MgBr(Mes) \longrightarrow 2Ni(Mes)_2 + 2Mg(C_3H_5)Br + 2MgBr_2$$

$$NiCl_2(PEt_3)_2 + 2MgBr(Mes) \longrightarrow Ni(Mes)_2(PEt_3)_2 + 2MgBrCl$$

$$\downarrow \text{vacuum}$$

$$Ni(Mes)_2 + 2PEt_3$$

(Mes = 2, 4, 6-trimethylphenyl)

$$NiCl_2(Py)_2 + 2MgCl(CH_2SiMe_3) \rightarrow Ni(CH_2SiMe_3)_2(Py)_2 + 2MgCl$$

$$PdCl_2(L_2) + 2LiC_6F_5 \rightarrow Pd(C_6F_5)_2(L_2) + 2LiCl$$

(L_2 = N–, As–, or S-bidentate ligand)

$$PdCl_2(C_6H_5CN)_2 + 2LiCH_2SC_6H_5 \rightarrow [\{Pd(CH_2SC_6H_5)_2\}_4] + 2LiCl + 2C_6H_5CN$$

$$PtCl_2(PEt_3)_2 + 2LiMe \rightarrow PtMe_2(PEt_3)_2 + 2LiCl$$

$$PtCl_2(PR_3)_2 + R'C \equiv CH + Et_3N + CuI \rightarrow trans\text{–}Pt(C \equiv CR')Cl(PR_3)_2 + HI + \tfrac{1}{2}\,Cu_2Cl_2$$

Cu, Ag, and Au derivatives

$$2CuX + ZnR_2 \xrightarrow[-10°C]{Et_2O} 2\{CuR\}_n + ZnX_2$$

(X = Cl or Br; R = Ph, 2–MeC_6H_4, 2,6–$Me_2C_6H_4$)

CuPh is more polymerised while others are tetrameric in benzene

$$CuCl + 2LiMe \xrightarrow{Et_2O} Li[CuMe_2] + LiCl$$

$$4CuI + 4LiCH_2SiMe_3 \longrightarrow [\{Cu(CH_2SiMe_3)\}_4] + 4LiI$$

$$CuI + [Li(thf)_4]^+[LiR_2]^- \longrightarrow [Li(thf)_4]^+\,[CuR_2]^- + LiI$$

$$AgI + [Li(thf)_4]^+[LiR_2]^- \longrightarrow [Li(thf)_4]^+[AgR_2]^- + LiI$$

$$Au(X)L + Mg(R)X \longrightarrow Au(R)(L) + MgX_2$$

(R = $C(SiMe_3)_3$)

$$5AuCl(CO) + 5MgBr(Mes) \xrightarrow{THF} [\{AuMes\}_5] + 5MgBrCl$$

Lanthanide (including Scandium and Yttrium) and Actinide derivatives

$$LnCl_3 + 6LiMe + 3TMEDA \xrightarrow{Et_2O} [Li(TMEDA)]_3^+\,[LnMe_6]^{3-} + 3LiCl$$

(Ln = a lanthanide metal)

$$LnCl_3 + 3LiCH_2SiMe_3 \xrightarrow{THF} [Ln(CH_2SiMe_3)_3(thf)_x + 3LiCl$$

$$Ln = Sc, Y, Tb, Er, Tm, Yb, Lu; x = 0, 2, 3)$$

$$LnCl_3 + 3LiCH(SiMe_3)_2 \longrightarrow [Ln\{CH(SiMe_3)_2\}_3] + 3LiCl$$

$(Ln = Sc, Y)$

$$LnCl_3 + 3LiCH(SiMe_3)_2 \longrightarrow [Li(thf)_4]^+ [LnCl\{CH(SiMe_3)_2\}_3]^- + 2LiCl$$

$(Ln = Tb, Er, Tm, or Yb)$

$$LnCl_3 + 4LiBu^t \xrightarrow{THF} [Li(thf)_4]^+ [LnBu^t_4]^- + 3LiCl$$

$(Ln = Y, Sm, Er, Yb \text{ or } Lu)$

$$LuCl_3 + 4LiC_6H_3Me_2 - 2,6 \longrightarrow [Li(thf)_4]^+[Lu(C_6H_3Me_2)_4]^- + 3LiCl$$

$$UCl_4 + \underset{(excess)}{6LiR} \xrightarrow{S} Li_2[UR_6]{\cdot}8S + 4LiCl$$

$(R = Me, Ph, \text{ or } CH_2SiMe_3;\ S = Et_2O \text{ or } THF)$

$$ThCl_4 + 4LiCH_2Ph \xrightarrow[-20°C]{THF} Th(CH_2Ph)_4 + 4LiCl$$

4.2.2.2 Transmetallation Reactions

These reactions are really of the same type as halide-carbanion exchange described in the preceding sub-section. However, in view of the milder (arising possibly from lesser ionic) nature of alkylating (arylating) agents like those of Zn, Cd, Hg, B, Al, Sn, or Pb (which commonly bring about only partial substitution), these have been included by some authors under a separate heading which is being followed here also.

In view of their milder activity, these reagents generally give rise to heteroleptic derivatives as exemplified by the following reactions:

$$TiCl_4 + AlMe_3 \rightleftharpoons TiCl_3Me + AlClMe_2$$

$$TiCl_4 + AlClMe_2 \xrightleftharpoons{pentane} TiCl_3Me + AlCl_2Me$$

The above equilibria can be pushed towards right by addition of excess NaCl which effectively removes $AlCl_2Me$ from the system as insoluble $Na^+[AlCl_3Me]^-$. A noteworthy point is that $TiCl_3Me$ can also be prepared from reaction of $TiCl_4$ with MgClMe in hexane, but the preparation is more smooth with the use of $ZnMe_2$ or PbR_4. e.g.,

$$2TiCl_4 + ZnMe_2 \rightarrow 2TiCl_3Me + ZnCl_2$$

$$TiCl_4 + PbR_4 \rightarrow TiCl_3R + PbClR_3$$

$(R = Me \text{ or } Et)$

The mixed (alkyl/halide) complexes of niobium and tantalum are obtained pure in high yields by the use of appropriate quantity of ZnR_2:

$$NbCl_5 + ZnR_2 \xrightarrow{\text{pentane}} NbCl_3R_2 + ZnCl_2$$

$$2TaCl_5 + 3ZnR_2 \xrightarrow{\text{pentane}} 2TaCl_2R_3 + 3ZnCl_2$$

($R = Me$, CH_2SiMe_3).

Use of organoaluminium reagents in THF results in the formation of a series of chromium complexes:

$$CrCl_3(THF)_3 + AlR_3 \text{ (or } AlR_2(OEt)) \xrightarrow{\text{THF}} CrRCl_2(THF)_3 + \text{traces of } CrCl_2(THF)$$

($R = Me$, Et, Pr or Bu^i)

Alkylation of WCl_6 with a variety of reagents gives $WRCl_5$:

$$2WCl_6 + MMe_2 \xrightarrow{-35^\circ C} 2WCl_5Me + MCl_2$$

($M = Zn$ (in ether) or Hg (in CH_2Cl_2))

$$2WCl_6 + ZnPh_2 \xrightarrow{\text{ether}} WCl_5Ph \text{ (incomplete)}$$

$$WCl_6 + SnR_4 \xrightarrow{\text{hexane}} WRCl_5 + SnR_3Cl$$

($R = Me$, Et, Bu, or Ph)

$$3WCl_6 + BR_3 \xrightarrow{\text{ether}} 3WRCl_5 + BCl_3$$

($R = Me$, Et, Bu)

$$WCl_6 + 2SnPh_4 \xrightarrow[\text{reflux}]{\text{pentane}} WCl_5Ph + 2SnClPh_3 + PhCl$$

The formation WMe_6 by reaction of WCl_6 with $AlMe_3$ has already been mentioned earlier (4.2.2.1).

Extent of alkylation (or arylation) of platinum-chlorine bond by reagents like $HgPh_2$ or $SnMe_4$ is generally dictated both by the nature of the solvent and the ligands in the precursor platinum complexes. Few typical examples are illustrated by the following reactions:

$$PtCl_2(PR_3)_2 + HgPh_2 \xrightarrow[\text{reflux}]{\text{PhH or EtOH}} trans\text{–}PtCl(Ph)(PR_3)_2 + HgClPh$$

$$PtCl_2(1,5\text{–cod}) + SnMe_4 \xrightarrow[100^\circ C]{Cl_2CHCHCl_2} PtClMe_2(1,5\text{–cod}) + SnClMe_3$$

$$PtCl_2(1,5\text{–}Cod) + SnMe_4 \xrightarrow{DMSO} PtMe_2\,(1,5\text{–}cod) + SnCl_2Me_2$$

$$cis\text{–}Pt(O_2CCF_3)_2(PMe_2Ph)_2 + SnMe_4 \rightarrow trans\text{–}PtMe(O_2CCF_3)(PMe_2Ph)_2 + SnMe_3(O_2CCF_3)$$

$$2\ cis\text{–}Pt(O_2CCF_3)_2(PPh_3)_2 + 2SnMe_4 \rightarrow cis\text{–}PtMe_2(PPh_3)_2 + trans\text{–}PtMe(O_2CCF_3)(PPh_3)_2 + SnMe_2(O_2CCF_3)_2 + SnMe_3(O_2CCF_3)$$

4.2.2.3 Reactions of Anionic Metal Complexes with Organic Halides

This is a synthetic route of considerable promise when precursors such as sodium salts of carbonylate or other anions are readily accessible. A few illustrative examples are given below:

$$Na^+[Mo(CO)_3(\eta^5\text{–}C_5H_5)]^- + PhCH_2Cl \rightarrow Mo(CH_2Ph)(CO)_3(\eta^5\text{–}C_5H_5) + NaCl$$

$$Na^+[Mn(CO)_5]^- + CH_3I \rightarrow MnCH_3(CO)_5 + NaI$$

$$Na^+[Mn(CO)_5]^- + CF_3COCl \rightarrow Mn(COCF_3)(CO)_5 + NaCl \xrightarrow{heat} Mn(CF_3)(CO)_5 + CO$$

$$Na^+[Fe(CO)_2(\eta^5\text{–}C_5H_5)]^- + C_6F_6 \rightarrow Fe(C_6F_5)(CO)_2(\eta^5\text{–}C_5H_5) + NaF$$

$$Na^+[Fe(CO)_2(\eta^5\text{–}C_5H_5)]^- + RCl \xrightarrow{hexane} Fe(R)(CO)_2(\eta^5\text{–}C_5H_5) + NaCl$$

$$Na_3^+[Co(CN)_5]^{3-} + PhCH_2Br \rightarrow Na_2^+[Co(CH_2Ph)(CN)_5]^{2-} + NaBr$$

$$Na^+\,[Co(CO)_4]^- + CH_2{=}CH\text{–}CH_2Br \rightarrow \underset{intermediate}{Co(CH_2\text{–}CH{=}CH_2)(CO)_4} + NaBr$$

$$\downarrow$$

$$Co(CO)_3(\eta^3\text{–}C_3H_5) + CO$$

4.2.2.4 Insertion Reactions

These reactions are of great significance in catalytic processes and are discussed in detail in Section 6.4, but a few examples of metal-carbon σ-bond syntheses involving such reactions are cited below:

(a) *Insertion of multiply bonded substrates into M–H bonds*:

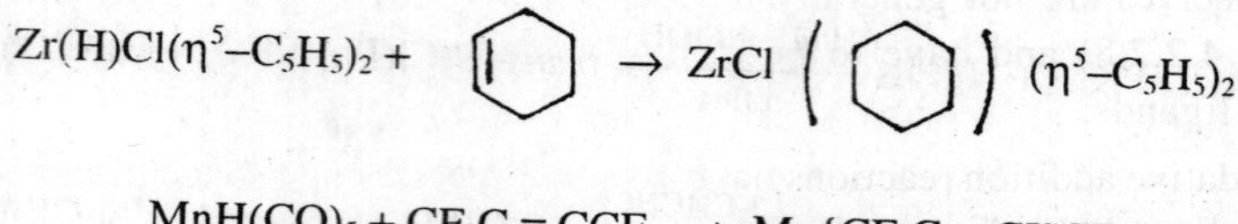

$$MnH(CO)_5 + CF_3C\equiv CCF_3 \rightarrow Mn\{CF_3C{=}CHCF_3\}(CO)_5$$

$$FeH(CO)_2(\eta^5\text{–}C_5H_5) + CH_2{=}CHCN \rightarrow Fe\{CH(CH_3)CN\}(CO)_2(\eta^5\text{–}C_5H_5)$$

$$CoH(CO)_4 + C_2F_4 \rightarrow Co(CF_2CF_2H)(CO)_4$$

$$[RhH(NH_3)_5]^{2+} + C_2H_4 \rightarrow [Rh(C_2H_5)(NH_3)_5]^{2+}$$

$$trans\text{–}PtH(Cl)(PEt_3)_2 + C_2H_4 \rightarrow trans\text{–}PtCl(C_2H_5)(PEt_3)_2$$

(b) *Insertion of unsaturated system into metal-metal bond*:

$$Co_2(CO)_8 + C_2F_4 \rightarrow (OC)_4Co\text{–}CF_2CF_2\text{–}Co(CO)_4$$

$$Me_3Sn\text{–}Mn(CO)_5 + C_2F_4 \rightarrow Mn(CF_2CF_2SnMe_3)(CO)_5$$

(c) *Insertion of unsaturated compounds into M–C* bonds:

$$MnCH_3(CO)_5 + CO \rightarrow Mn\{C(:O)CH_3\}(CO)_5$$

$$MnPh(CO)_5 + PPh_3 \rightarrow Mn\{C:O)Ph\}(CO)_4(PPh_3)$$

$$trans\text{–}PtCl(CH_3)(PEt_3)_2 + CO \rightarrow trans\text{–}Pt(C(:O)CH_3)Cl(PEt_3)_2$$

4.2.2.5 *Oxidative Addition Reactions*

Oxidative addition (or *oxad*) reaction (cf. Section 3.4.1) which are useful for establishment of new transition metal-carbon σ-bonds involve the oxidation (i.e., increase in the formal oxidation number), accompanied by an increase in the coordination number of the central metal atom. Complexes of metals in low spin d^7, d^8, or d^{10} configurations generally undergo this type of reaction and in the process of enhancing their coordination number, the central metal tends to attain the stable closed-shell configuration corresponding to 18 valence electrons.

Coordinative unsaturated state of the central transition metal appears to facilitate an 'oxad' type of reaction. For example, in the following reaction, iridium(I) changes from 4-coordinate 16-electron to 18-electron iridium(III):

$$IrCl(CO)(PPh_3)_2 + CH_3I \rightarrow Ir(CH_3)(Cl)(I)(CO)(PPh_3)_2$$

Similarly, in the following reaction, 5-coordinate 18-electron ruthenium (0) is converted into 6-coordinate 18-electron ruthenium(II), accompanied by the loss of a CO molecule:

$$Ru(CO)_3(PPh_3)_2 + CH_3I \rightarrow Ru(CH_3)(I)(CO)_2(PPh_3)_2 + CO$$

To facilitate oxidative addition type of reactions, the metals should preferably be in low oxidation state. Unlike the main group metals (e.g. the formation of Grignard reagents from metallic magnesium and alkyl halides), the transition metals themselves are not generally reactive enough under ambient conditions (cf. Section 4.2.2.8) and have to be in low oxidation state in the presence of coordinated ligands.

The oxidative addition reactions have assumed special importance in catalytic processes and would be discussed in detail in Section 6.4.1.1.

4.2.2.6 Elimination Reactions

As the name signifies, this class of reactions involves elimination of a molecular species from appropriate complexes with or without change of oxidation state of the central metal atom. The former type is accompanied with a lowering of the oxidation state of generally an intermediates species involved in catalytic processes (cf. Section 6.4.2). The latter type, involving elimination of molecules like N_2, CO, CO_2, SO_2 and H_2 from appropriate species such as azoderivatives, acyls, carboxylates, sulphinates and hydrides respectively, leads to the formation of a new transition metal-carbon σ-bond as illustrated by the following typical examples:

Azo-derivatives

$$trans\text{–}PtCl(N{=}N\text{—}C_6H_4NO_2\text{—}p)(PEt_3)_2 \xrightarrow{120^\circ C} trans\text{–}PtCl(C_6H_4NO_2\text{—}p)(PEt_3)_2 + N_2$$

Acyls

$$Mn(COC_6H_4Me\text{—}p)(CO)_5 \xrightarrow{h\nu} Mn(C_6H_4Me\text{—}p)(CO)_5 + CO$$

$$(OC)_5Mn\text{—}CH_2\text{—}CH{=}CH_2 \xrightarrow[\text{or } h\nu]{80^\circ C} HC\langle{}^{CH_2}_{CH_2}\rangle\text{—}Mn(CO)_4 + CO$$

$$Fe(COPh)(CO)_2(\eta^5\text{—}C_5H_5) \xrightarrow{h\nu} Mn(Ph)(CO)_2(\eta^5\text{—}C_5H_5) + CO$$

$$Li^+[CO(CO)_4]^- + F_3CCOCl \rightarrow Co(COCF_3)(CO)_4 + LiCl$$
$$\downarrow < 25^\circ C$$
$$Co(CF_3)(CO)_4 + CO$$

Carboxylates

$$Ni(CO_2Ph)_2(bipy) \xrightarrow{heat} NiPh_2(bipy) + CO_2$$

$$K_2PtCl_4 + 2TlO_2CC_6F_5 + 2py \xrightarrow{heat} trans\text{–}Pt(C_6F_5)_2(Py)_2 + 2KCl + 2TlCl + 2CO_2$$

Sulphinates

$$Fe(SO_2R)(CO)_2(\eta^5\text{—}C_5H_5) \xrightarrow{heat} FeR(CO)_2(\eta^5\text{—}C_5H_5) + SO_2$$

$$IrCl_2(SO_2C_6H_4Me\text{—}p)(CO)(PPh_3)_2 \xrightarrow{\text{heat}} IrCl_2(C_6H_4Me\text{—}p)(CO)(PPh_3)_2 + SO_2$$

$$Pt(SO_2Ph)(Cl(PR_3)_2 \xrightarrow{\text{heat}} PtCl(Ph)(PR_3)_2 + SO_2$$

Hydrides

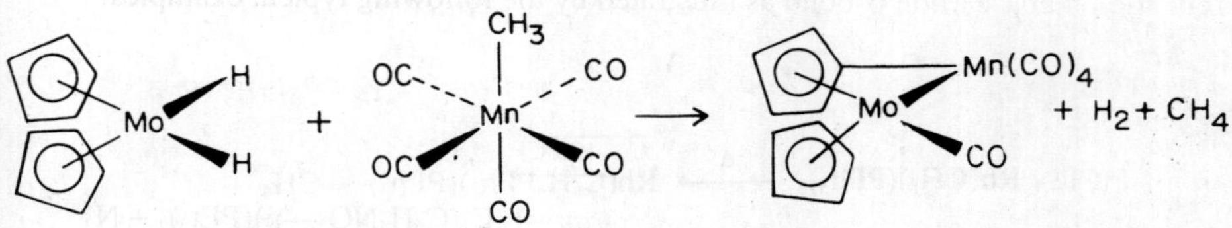

4.2.2.7 Cyclometallation Reactions

A general class of problem that has continued to attract the attention of inorganic/organometallic chemists over many years lies in the coordination and activation of small molecules such as H_2, CO, CO_2, $SiHR_3$, RX, RCH CHR′, X_2, HX, etc., by transition metal complexes. During 1960's, interest in the alkane activation problem grew as a consequence of several related developments. The "oxidative addition" reaction, eg.,

$$M + X-Y \longrightarrow M\begin{matrix} \diagup X \\ \diagdown Y \end{matrix}$$

$$(X\text{–}Y = H_2, HCl, MeI, Cl_2, R_3Si\text{–}H)$$

was recognized, in which low-valent metal complexes were shown to insert into a variety of X-Y bonds. However, the oxidative addition of C-H bonds remained elusive until the discovery of *cyclometallation* (by M.I. Bruce in 1977) initiated considerable activity in these directions.

The cyclometallaltion reaction can be viewed to proceed initially through an intramolecular oxidative addition of a ligand C—H bond to the metal:

$$L_nM \leftarrow :EC-H \longrightarrow L_nM\begin{matrix} \diagup EC \\ \diagdown H \end{matrix}$$

Alkyl/aryl groups attached to a coordinated nitrogen/phosphorus(E) ligand species often undergo such oxidative additions involving C—H bond fission and M—C bond formation. As indicated above, such reactions result in the formation of a heterocyclic product, which is generally called a cyclometallated derivative. The hydorgen from the C—H species is in the first stage generally bonded to the metal. In a few cases, this hydrogen remains unchanged but generally, it is eliminated (cf. some illustrative examples below) as HCl, CH_4, H_2 etc., by combining with Cl, CH_3 or another H bound to the metal.

One of the more common examples of intramolecular C—H oxidative addition is the orthometallation reaction of aryl groups:

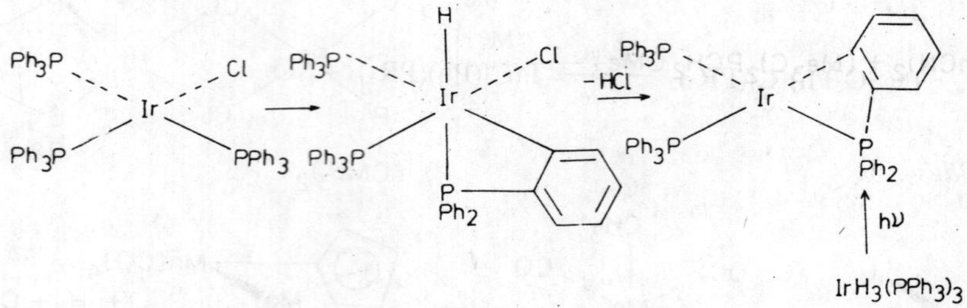

$$Rh(CH_3)(PPh_3)_3 \xrightarrow{\Delta} Rh(C_6H_4PPh_2)(PPh_3)_2 + CH_4$$

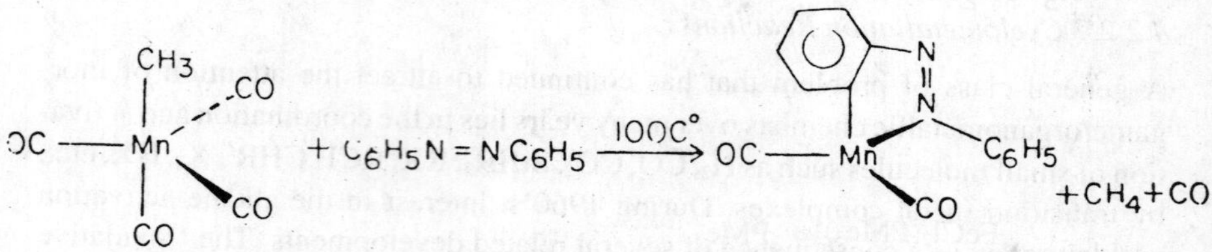

The C—H bonds of the coordinated cyclopentadienyl rings are also susceptible to undergo cyclometallation reactions, e.g.:

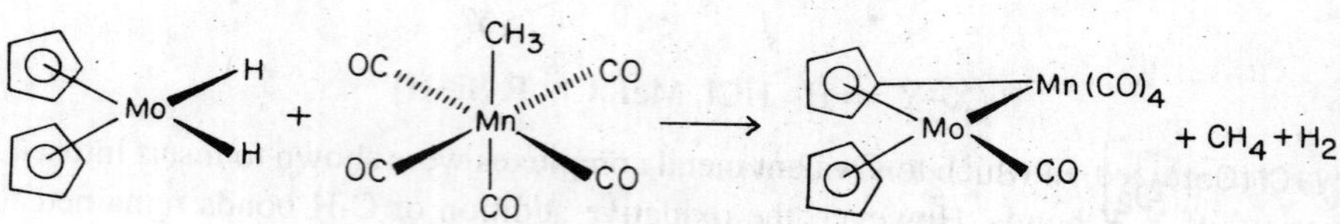

As can be seen from the above reactions that in a large number of cases, the presence of an adjacent leaving group (e.g., an alkyl, hydride, or halide), provides a driving force for such reactions.

The first example of the cyclometallation of sp^3C—H bonds came from Chatt and Davidson's study (1965) of Ru (dmpe)$_2$:

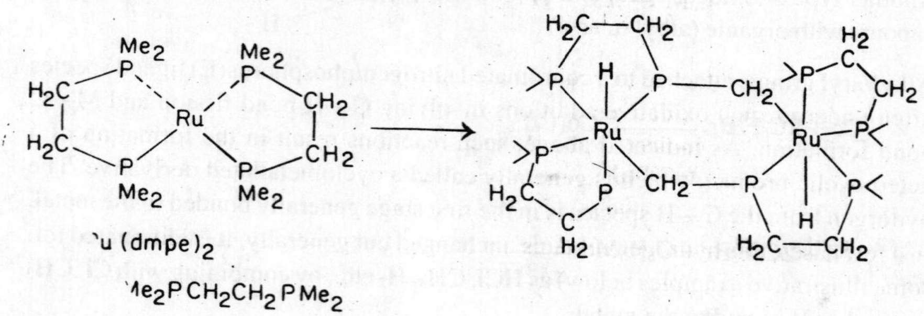

A few other examples of oxidative addition of sp^3 C—H bonds may be cited as:

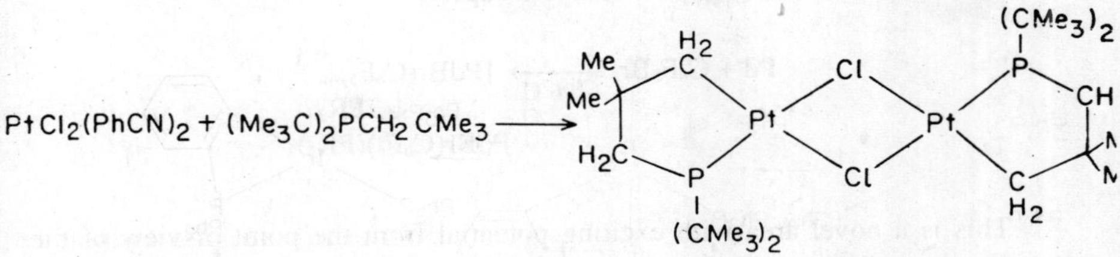

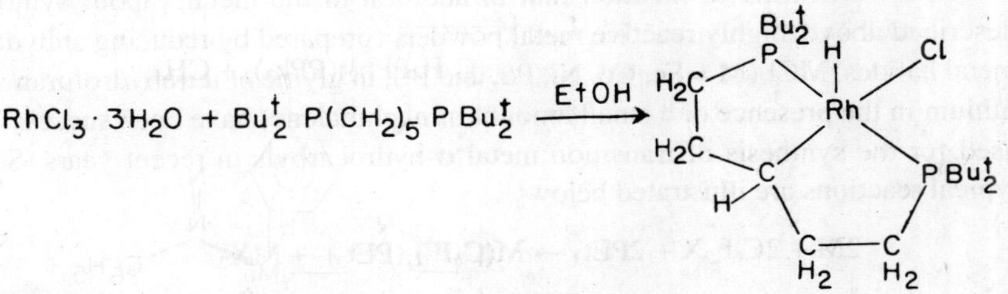

$$FeCl_2(PMe_3)_3 + \underset{(\text{excess})}{PMe_3} \xrightarrow{\text{Na/Hg}} [FeH\{CH_2PMe_2\}(PMe_3)_4] + 2NaCl$$

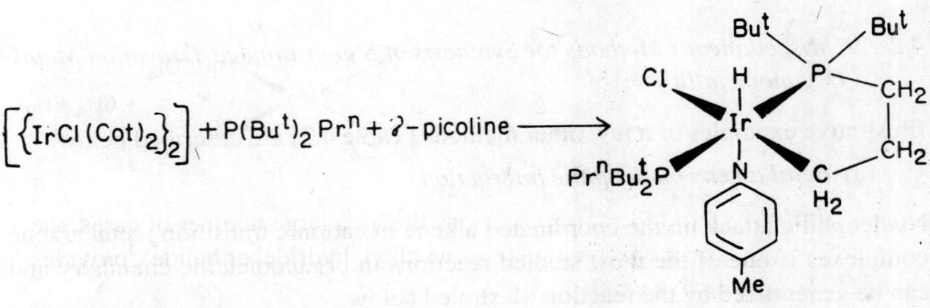

4.2.2.8 Metal Atom Reactions

Another type of oxidative reaction (cf. Section 4.2.2.5) is cocondensation of metal vapours with organic (alkyl or aryl) halides. A few examples are illustrated below:

$$Co + 2C_6F_6Br \xrightarrow[(-CoBr_2)]{-196^\circ C} Co(C_6F_6)_2 \xrightarrow{\text{PhMe}} Co(C_6F_5)_2(\eta^6\text{—}C_6H_4Me)$$

$$Ni + 2C_6F_5Br + C_6H_5Me \xrightarrow{-196^\circ C} Ni(C_6F_5)_2(\eta^6\text{—}C_6H_5Me) + NiBr_2$$

$$Ni + C_6F_5Br \xrightarrow[-196^\circ C]{} NiBr(C_6F_5) \xrightarrow{2PEt_3} NiBr(C_6F_5)(PEt_3)_2$$

$$Pd + C_6F_5Br \xrightarrow[-196^\circ C]{} [PdBr(C_6F_5)]_n \xrightarrow{2PR_3} PdBr(C_6F_5)(PR_3)_2$$

This is a novel area with exciting potential from the point of view of the development of more efficient techniques for exploitation in general chemical syntheses including those of organometallic species.

It is worthwhile to mention that in addition to the metal vapour synthesis described above, highly reactive metal powders (prepared by reducing anhydrous metal halides, MCl_2(M = Fe, Co, Ni, Pd, and Pt), in glyme or tetrahydrofuran with lithium in the presence of a small amount of naphthalene) have been successfully used for the synthesis of transition metal σ-hydrocarbyls in recent years. Some typical reactions are illustrated below·

$$2M + 2C_6F_5X + 2PEt_3 \rightarrow M(C_6F_5)_2(PEt_3)_2 + MX_2$$

(M = Fe, Co; X = Br, I)

$$2Ni + 2C_6F_5I + 2C_5H_5N \rightarrow Ni(C_6F_5)_2(C_6H_5N)_2 + NiI_2$$

$$Pd + C_6F_5I + L—L \rightarrow PdI(C_6F_5)(L—L)$$

(L—L = dme or bipy)

4.2.2.9 Miscellaneous Methods for Synthesis of Sigma-Bonded Transition Metal Organometallics

Illustrative examples of a few other methods (Table 4.2) are described below:

(a) *Metal alkene nucleophile interaction*

Nucleophilic attack on the coordinated alkene in cationic transition metal alkene complexes is one of the most studied reactions in organometallic chemistry and can be generalized by the reaction illustrated below:

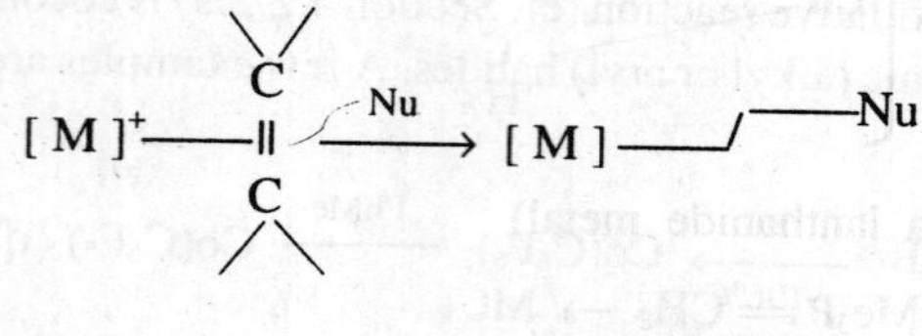

Many different types of cationic transition metal alkene complexes have been reacted with a wide range of nucleophiles (e.g. H^-, R^-, CH_3^-, MeO^-, etc.) to generate σ-bonded organometallic derivatives. Some typical reactions of these types are illustrated by the following reactions:

$$[(Fe(CO)_2(\eta^2{-}C_2H_4)(\eta^5{-}C_5H_5)]^+ + BH_4^- \rightarrow [FeEt(CO)_2(\eta^5{-}C_5H_5)] + \tfrac{1}{2}B_2H_6$$

$$[Fe(CO)_2(\eta^5{-}C_5H_5)(\eta^2{-}C_2H_4)]^+ + PPh_3 \rightarrow [Fe(CH_2CH_2PPh_3)(CO)_2(\eta^5{-}C_5H_5)]$$

$$[RhClEt(\eta^5{-}C_5H_5)(\eta^2{-}C_2H_4)] \xrightarrow[10^\circ C]{L} [Rh(Bu)Cl(L)(\eta^5{-}C_5H_5)]$$

$$[Pd(PPh_3)(\eta^5{-}C_5H_5)(\eta^2{-}C_2H_4)]^+ \xrightarrow[10^\circ C]{NaOMe/MeOH} [Pd(CH_2CH_2OMe)(PPh_3)(\eta^5{-}C_5H_5)]$$

$$cis\text{-}[PtCl_2(Py)(\eta^2{-}C_2H_4)] + Py \rightarrow cis\text{-}[PtCl_2(CH_2CH_2Py)(Py)]$$

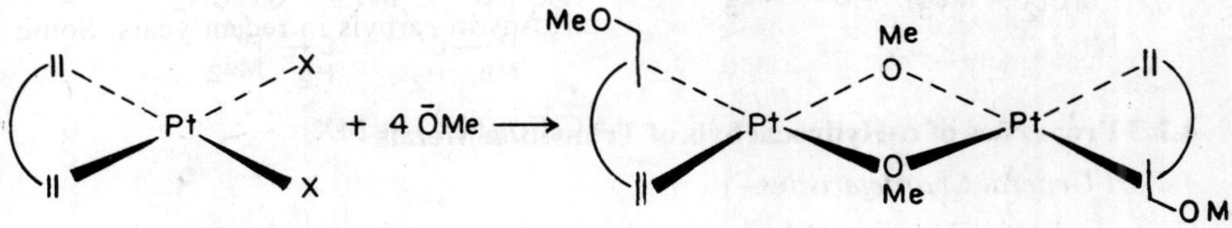

(b) *Ylide synthesis*

It has already been mentioned in the earlier Section 4.2 that chelate ring formation has proven to be an effective means of stabilizing the transition metal-alkyl bond. The use of phosphorus ylide ligands has, therefore, proved to be a versatile strategy of enhancing the stability of transition metal-alkyls, e.g.:

$$LnCl_3 + 3Me_3P = CH_2 \longrightarrow Ln(CH_2\overset{+}{P}Me_3)_3\ 3Cl^-$$

$\downarrow$ 3LiBu

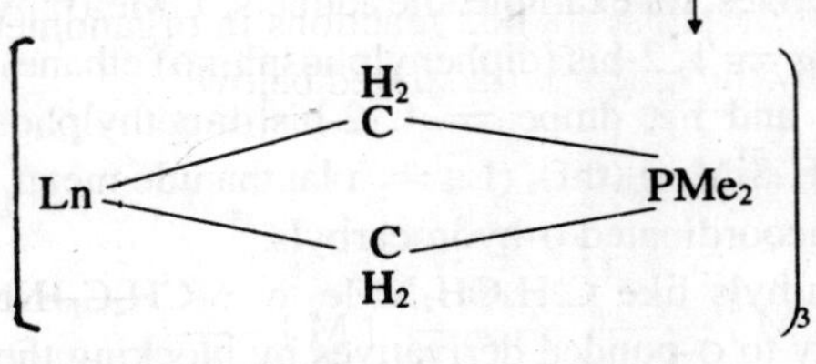

$+ 3C_4H_{10} + 3LiCl$

(Ln = a lanthanide metal)

$$MCl_4 + nMe_3P{=}CH_2 \rightarrow MCl_4(CH_2PMe_3)_n$$

(M = Ti, n = 2 and 3; M = Zr, n = 2, 3, and 4)

$$MCl_2(\eta^5{-}C_5H_5)_2 + 2Me_3P{=}CH_2 \rightarrow [M(CH_2PMe_3)_2(\eta^5{-}C_5H_5)_2]^{2+}\ 2Cl^-$$

(M = Ti, Zr)

$$Cr(CO)_5(PPh_3) + Ph_3P{=}CH_2 \rightarrow Cr(CH_2PPh_3)(CO)_5 + PPh_3$$

$$Me_4PCl + 2LiMe \rightarrow Li(CH_2)_2PMe_2 + LiCl + 2CH_4$$

$$CrCl_3(THF)_3 + 3Li(CH_2)_2PMe_2$$

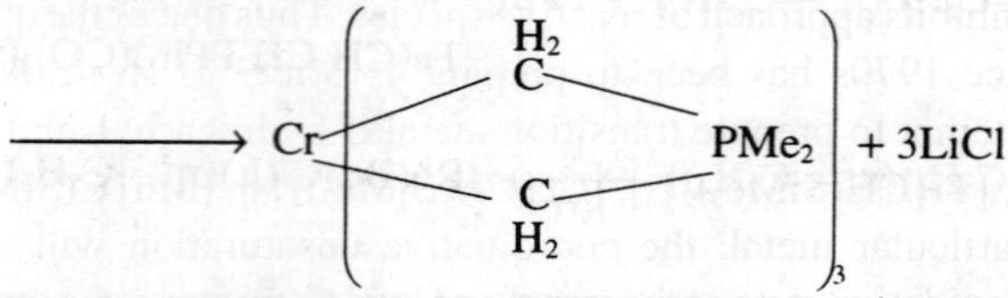

$$Li_4[Mo_2Me_8] + 4Me_4PCl \rightarrow [Mo_2\{(CH_2)PMe_2\}_4] + 4CH_4 + 4LiCl$$

In addition to the above, double ylides of phosphorus react with metal chlorides to form spirocyclic derivatives:

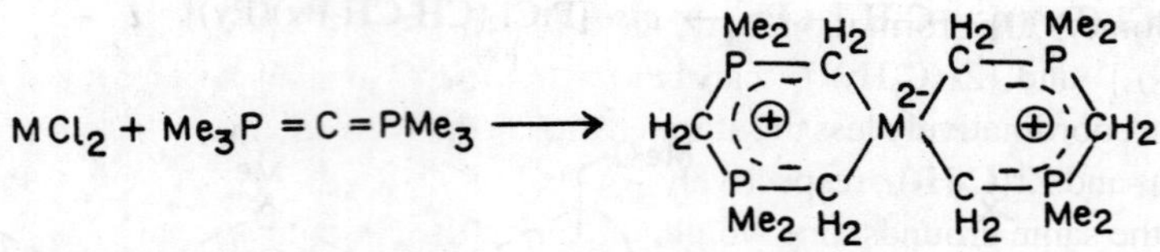

4.2.3 Properties of σ-Hydrocarbyls of Transition Metals

4.2.3.1 General Characteristics

Compared to main group σ-hydrocarbyls (cf. Section 3.5), the transition metal analogues are generally much less stable, although the instability is mainly due to kinetic rather than thermodynamic factors. In view of much greater diversity in their nature, it is rather difficult to attempt a simple generalization of their properties, but an attempt would be made to correlate some of these on the basis of their coordinative unsaturation and steric congestion.

One of the distinguishing characteristics of transition metal σ-hydrocarbyls is their coordinative unsaturation, which leads to their instability through facile decomposition paths (cf. Section 4.2.3.2). On the other hand, this unsaturation also explains the tendency of these compounds to act as Lewis acids with a variety of donor ligands; the presence of these ancillary ligands confers a comparative stability on these derivatives, for example: the adducts, $TiMe_4$(bipy); $TaMe_5$ (dme); CoMe $(diphos)_2$ (diphos = 1, 2-bis (diphenylphosphino) ethane); MMe_2 $(dmpe)_2$ (M = Ti, V, Cr, Mn, and Fe, dmpe = 1, 2-bis(dimethylphosphino) ethane), $NiEt_2$(bipy); and $Ln(CH_3SiMe_3)_3(thf)_n$ (Ln = a lanthanide metal, n = 2 or 3) are more stable than the uncoordinated σ-hydrocarbyls.

Chelating hydrocarbyls like $C_6H_4CH_2NMe_2$-o, o-$CH_2C_6H_4NMe_2$; etc., also tend to provide stability to σ-bonded derivatives by blocking the available coordination sites, e.g.,

$$[Cr\{CH_2C_6H_4NMe_2{-}o\}_3] \text{ and } [Co\{C_6H_4CH_2NMe_2\}_3].$$

Another mode by which these derivatives tend to satisfy their coordination environment is by forming polynuclear species through alkyl or other bridging ligands or through multiple metal-metal bonding (cf. Section 4.2.1). The association through bridging would obviously be hindered by more ramified ligands, thus resulting in mononuclear species. Formally these would be coordinatively unsaturated yet they attain thermal stability probably due to steric congestion of the ligands which inhibits approach of reactive species. Thus one of the attractive areas of research since 1970s has been to prepare a variety of sterically demanding ligands in an attempt to prepare transition metal σ-hydrocarbyls in low coordination states, e.g., $[Ti\{CH(SiMe_3)_2\}_3]$, $[Cr\{CH(SiMe_3)_2\}_3]$, $[Mn\{C(SiMe_3)_3\}_2]$, etc.

For any particular metal, the coordinative unsaturation will increase with lowering of the oxidation state of the metal and lead to greater associative tendency. For example, compared to the monomeric character of $Ti(C_6H_5)_4$, the polymeric nature of $[Ti(C_6H_5)_2]_n$ can be cited.

Transition metal σ-hydrocarbyls may be of three types, i.e., neutral, anionic, and cationic. The formation and stability of anionic species of the types. $[Ti(C_6H_5)_5]^-$ and $[Zr(C_6H_5)_6]^{2-}$ can be understood as in these, the central metal atoms are coordinatively less unsaturated than in the corresponding neutral species, $Ti(C_6H_5)_4$ and $Zr(C_6H_5)_4$, respectively.

On the same grounds, one would not expect the cationic species to be formed easily; in fact, these often require a chelating hydrocarbyl of the type $Me_3P{=}CH_2$ which for example, confers stability on cationic species of the types:

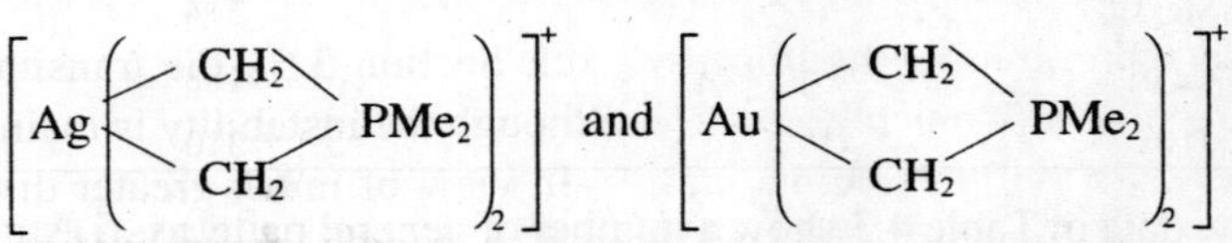

Another characteristic feature of transition metal σ-hydrocarbyls has been observed in their magnetic resonance spectra. Both ^{1}H and ^{13}C chemical shifts at σ-carbon atoms (attached to the transition metals) occur at higher fields than in the parent hydrocarbons and the direct ^{13}C-metal coupling constants tend to be unusually large (~50 Hz).

A wide variation in chemical shifts (δ) is also observed, depending on the metal involved and the ancillary ligands. A few examples of such variations are given below:

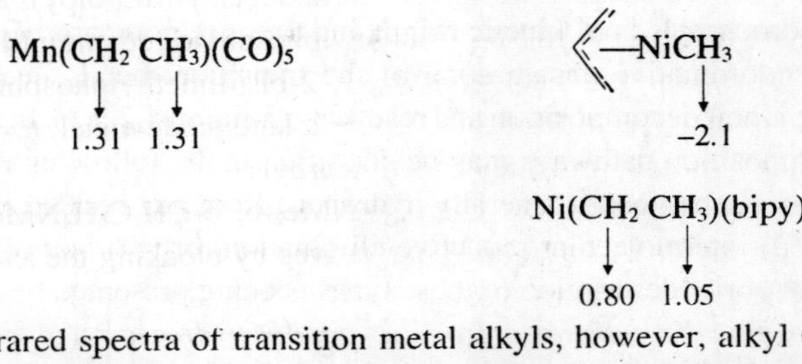

In the infrared spectra of transition metal alkyls, however, alkyl C—H frequencies show little variation from those of the parent organic compounds.

4.2.3.2 Thermal Stability and Decomposition Pathways

Homoleptic σ-hydrocarbyls of transition metals, MR_n are generally unstable under ambient conditions and are often extremely sensitive to moisture and protic as well as oxidizing reagents including oxygen. The earlier notion that this might be essentially due to comparative thermodynamic instability of metal-carbon σ-bond was proved wrong as the actual thermochemical bond energies 'E' were actually found (Table 4.3) to be of the same order for transition and main group elements.

Table 4.3

Reference Compounds	Metal-Carbon (M-C) Bond Energy ('E' kJ mol^{-1})
$GeMe_4$	247
$GeEt_4$	239
$SnMe_4$	218
$SnEt_4$	193
$SnPh_4$	257
$TiMe_4$	...
$Ti(CH_2SiMe_3)_4$	268
$Ti(CH_2CMe_3)_4$	184
$Ti(CH_2C_6H_5)_4$	264
$ZrMe_4$	...
$Zr(CH_2SiMe_3)_4$	314
$Zr(CH_2CMe_3)_4$	188
$Zr(CH_2C_6H_5)_4$	310

The few data in Table 4.3 show a number of general patterns: (i) similar order of 'E' values for transition and main group metals; (ii) with increasing mass of the metals, 'E' values decreases in the main group, but increase in the case of transition metals; (iii) for the same metal, 'E' values follow the order Ph > Me > Et. The interesting observation of the higher value of 'E' for M-C(CH_2SiMe_3) compared to that for M-C(CH_2CMe_3) bond could probably be ascribed to the greater steric effect of the latter (neopentyl) group, which could possibly lead to lengthening of the M-C bond.

As discussed earlier (Section 4.2.3.1), the comparative instability of transition metal σ-hydrocarbyls is of kinetic origin and they are comparatively more labile due to the coordinative unsaturation of the transition metals, leading to facile pathways for their decomposition and reactivity (Section 4.2.3.3). In simpler terms, these decomposition pathways may be classified in the following three types: (i) migration of a substituent (generally hydrogen) from a σ-bonded hydrocarbyl to the metal; (ii) intermolecular reductive elimination; or (iii) homolytic fission of M–C bond. A brief description of these types is being presented below:

(i) *Migration of a substituent from a σ-bonded hydrocarbyl to the metal* occurs more commonly from the β-carbon atom (β-elimination) and less frequently from α-carbon atom (α–elimination). A third type, i.e. γ-elimination, which has been observed for some main group hydrocarbyls, does not have any established example in transition metal chemistry.

(a) β-*Hydrogen elimination* has already been discussed in Section 4.2.1.

(b) α-*Elimination reactions*: These are of considerable synthetic utility in main group σ-hydrocarbyls and involve the transfer of a substituent 'R' from the α-carbon atom to the metal with elimination of a carbene which may remain coordinated to the metal:

$$\mathrm{LMRR'R'' \rightarrow LMR + {:}CR'R''}$$

Although the main focus of our attention in this chapter is transition metal organometallic chemistry, yet for sake of explaining the α-elimination process, two examples for reactions for main group derivatives are given below:

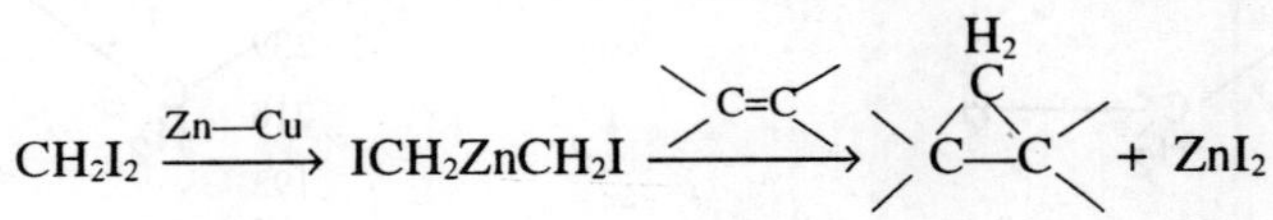

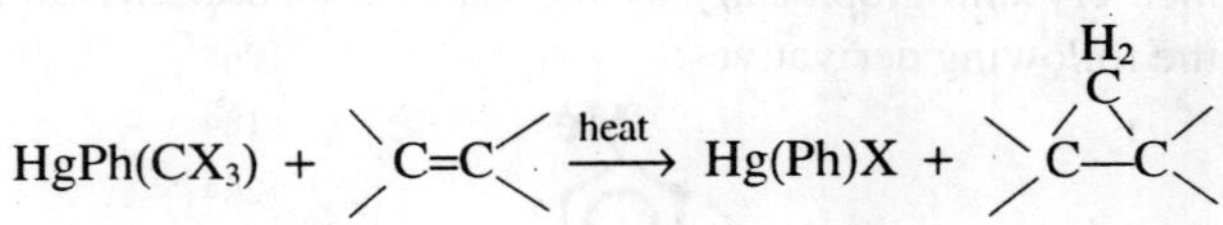

α-*Elimination pathway* has been much less observed in transition metal hydrocarbyls and the main examples involve metal methyls. The binary permethyls such as $TaMe_5$, WMe_6, and $ReMe_6$ are not very stable and tend to decompose explosively with the formation of methane, e.g.,

$$\mathrm{WMe_6 \rightarrow 3CH_4 + ``W(CH_2)_3"}$$

In view of the metal being in its highest oxidation state, an internal oxidative addition of the type:

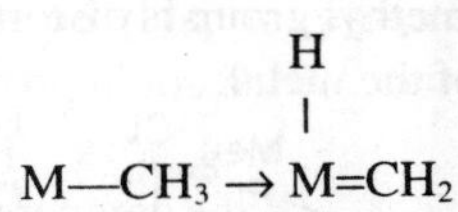

is not feasible for the above decompositions which could occur either *via* homolytic fission (free radical mechanism) or α-elimination through a four-centre transition state:

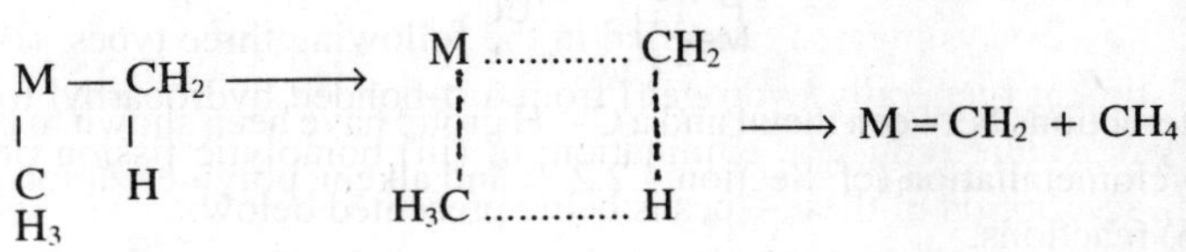

This pathway plays a key-role in the synthesis of alkylidene complexes (cf. Section 4.4). A number of metal benzyls and neopentyls such as those of titanium and tantalum have been shown to decompose by this pathway. Another mechanism of α-elimination, which is recently (1982) recognized, involves agostic hydrogens (C–H ⇀ M).

The term "agostic" is used particularly in those situations where a hydrogen atom is covalently bonded both to a carbon atom and a transition metal atom. Evidence for such interaction (C–H $\rightarrow$ M) has been studied spectroscopically and is being illustrated below:

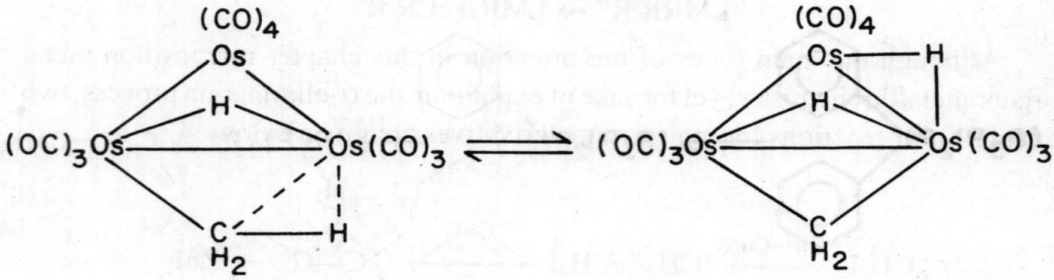

The role of spectroscopically recognized 'agostic' C-H interactions has now been confirmed crystallographically by the near octahedral environment around rhodium in the following derivatives:

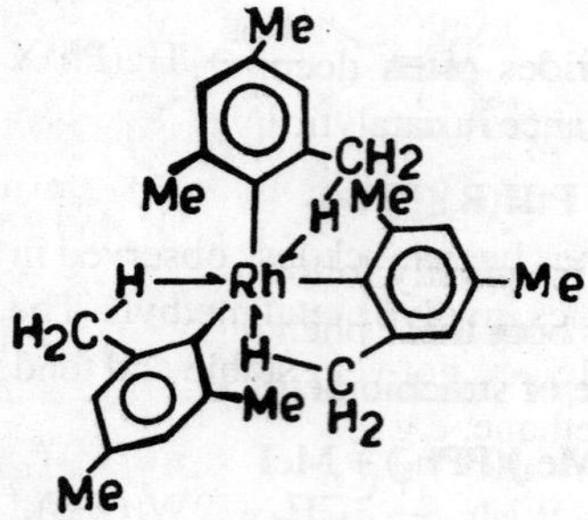

It has also been shown by crystallographic analysis that when a methyl group is attached to a d° metal centre which has a less than 18-electron environment, then one of the C—H bond of the methyl group is distorted in such a manner that it lies within the bonding distance of the metal:

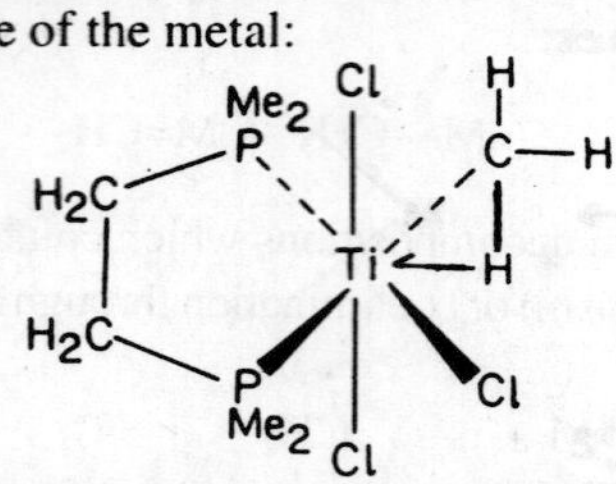

Such strong interactions between metal and a C—H group have been shown to play a key role in cyclometallation (cf. Section 4.2.2.7) and alkene polymerization (cf. Section 6.4.2.6) reactions.

(ii) *Intermolecular reductive elimination pathway* is the reverse of oxidative addition reactions (cf. Section 4.2.2.5) and can be represented as:

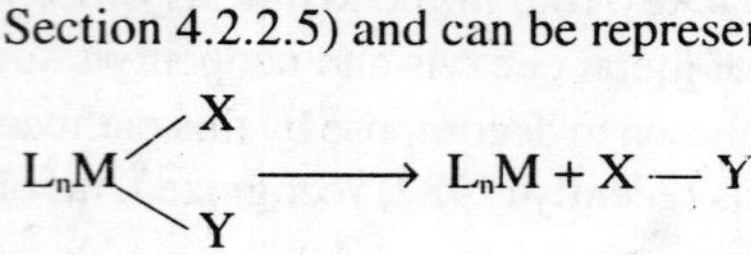

Both the oxidation state as well as the coordination number of the central metal 'M' are reduced by two units and hence, this type of pathway would be confined to metals with stable oxidation states differing by two units, which is more common with later transition metals.

A few examples of this type of pathway are given below:

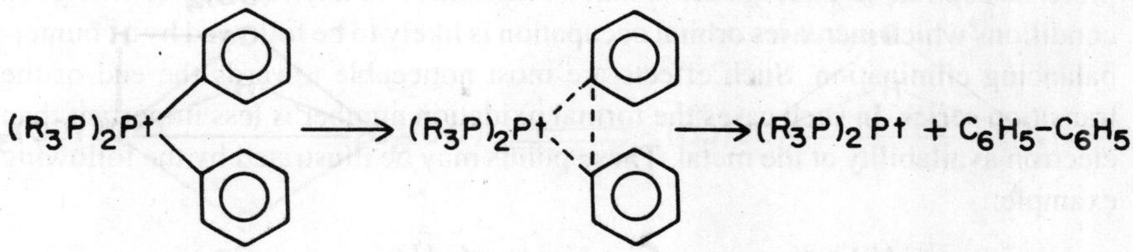

$$AuMe_3(PPh_3) \rightarrow AuMe(PPh_3) + C_2H_6$$

It may be important to mention that the latter reaction occurs by intramolecular elimination pathway in Me_2SO solvent, but changes to inter-molecular route in benzene or decalin.

Organometal hybrides often decompose by this route and such reactions assume special significance in catalytic hydrogenation (Section 6.4.2.1) processes:

$$PtH(R)(PPh_3)_2 \rightarrow Pt(PPh_3)_2 + RH$$

Reductive elimination frequently follows oxidative addition, and in many important catalytic processes both these phenomena play a crucial role (cf. Section 6.4). An illustrative example of stoichiometric reaction is depicted below:

$$Au(CH_2SiMe_3)(PPh_3) + MeI \rightarrow AuIMe(CH_2SiMe_3)(PPh_3) \downarrow AuI(PPh_3) + SiEtMe_3$$

A formal case of reductive elimination can also be exemplified by the loss, dimerization, polymerization, and ring closure of an alkene involving transition metal η^2-alkene complexes:

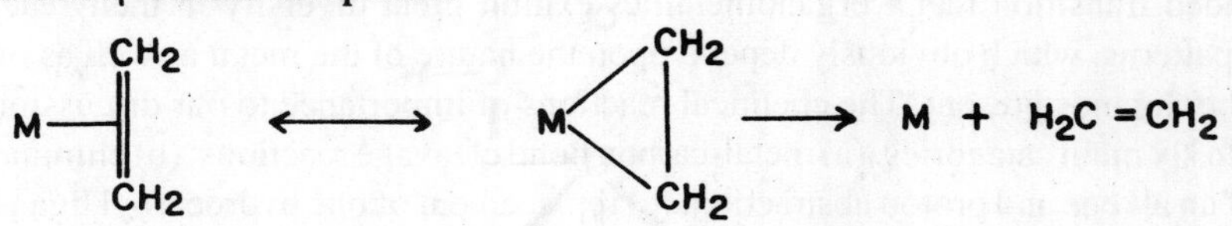

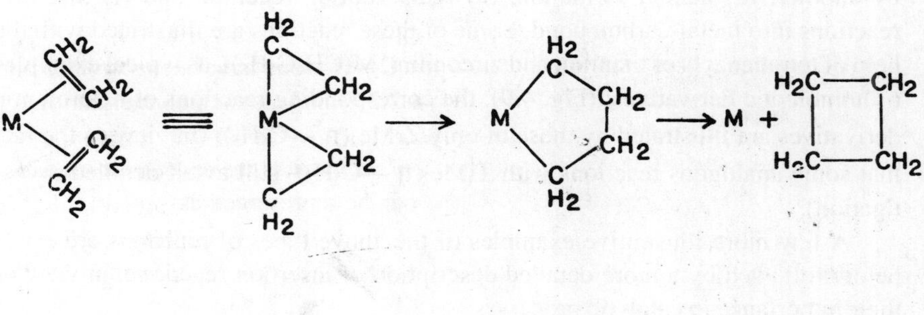

In these two examples, the number of electrons formally donated to the metal is reduced by two, wherein 18-electron complexes are transformed into a 16-electron species. These types of reductive elimination processes are more favourable for coordinatively saturated derivatives. Alternatively complexes in which the orbital occupation is higher than usual for the concerned metal will be more susceptible to undergo elimination reactions, and any reaction or change in conditions which increases orbital occupation is likely to be followed by a counter-balancing elimination. Such effects are most noticeable towards the end of the transition series. In such cases the formal oxidation number is less important than electron availability at the metal. These points may be illustrated by the following example:

$$NiEt_2\,(bipy) \rightarrow Ni + bipy + C_4H_{10}$$

16e $\qquad$ Ea = 274 kJ mol^{-1}

$\downarrow H_2C=CHX$

$$NiEt_2\,(bipy)(\eta^2\text{—}H_2C{=}CHX) \rightarrow Ni\,(bipy)(\eta^2\text{—}H_2C{=}CHX) + C_4H_{10}$$

$$Ea = 65\ kJ\ mol^{-1}$$

(iii) *Homolytic fission of M—C bond* appears to be a common pathway for decomposition of main group alkyls, e.g.:

$$HgR_2 \rightarrow 2R^{\cdot} + Hg$$

In view of more facile alternative pathways with lower activation energies being available for the decomposition of transition metal hydrocarbyls, the homolysis route should be much less preferred in their cases. In fact, evidence for the homolytic fission in a few cases, e.g., $Zr(CH_2Ph)_4$ has not been so far substantiated by the observation of $PhCH_2^{\cdot}$ radicals using physico-chemical techniques like ESR.

4.2.3.3 Chemical Reactions

σ-Bonded transition metal organometallics exhibit great diversity in their reactivity patterns, which obviously depend upon the nature of the metal as well as on the other ligands present. The chemical reactions of importance to our discussion fall into six main categories: (a) metal-carbon bond cleavage reactions, (b) elimination of an alkene and proton abstraction, (c) replacement of one hydrocarbyl ligand by another, (d) adduct formation, (e) redistribution reaction, and (f) insertion reactions into metal-carbon bond. Some of these reactions are illustrated with the help of tetrabenzyls of titanium and zirconium, $M(CH_2C_6H_5)_4$, as typical examples of homoleptic derivatives; (Fig. 4.9), the corresponding reactions of heteroleptic derivatives are illustrated by those of only $ZrMe_2(\eta^5\text{—}C_5H_5)_2$ (in view of the fact that some analogous reactions with $TiMe_2(\eta^5\text{—}C_5H_5)_2$ still await detailed investigation).

A few more illustrative examples of the above types of reactions are given below followed by a more detailed description of insertion reactions, in view of their importance in catalytic processes.

(a) *Metal-carbon bond cleavage reactions.* Transition metal alkyls are only weakly nucleophilic; for example, these do not react readily with aldehydes or ketones, thus differing from main group metal alkyls. Protons cleave most transition metal alkyl bonds, presumably by direct attack on the metalcarbon bond.

$$M{-}C\lessgtr + H^+ \rightarrow \gtrless C{-}H + M^+$$

Some specific examples of metal-carbon bond cleavage reactions by protonic reagents are:

$$TiMe_4 + \underset{(excess)}{4MeCO_2H} \xrightarrow{ether} Ti(O_2CMe)_4 + 4MeH$$

$$Zr(CH_2CMe_3)_4 + 3Me_3PCH_2 \rightarrow 1/2Zr_2[(CH_2)_2PMe_2](\mu{-}CPMe_3)_2 + 4CMe_4$$

$$Zr(CH_2SiMe_3)_4 + 4HOAr \rightarrow Zr(OAr)_4 + 4SiMe_4$$

$$OAr = 2,6{-}Pr^i_2C_6H_3O)$$

$$CrPh_3{\cdot}(THF)_n + 3AcacH \rightarrow Cr(acac)_3 + 3PhH + nTHF$$

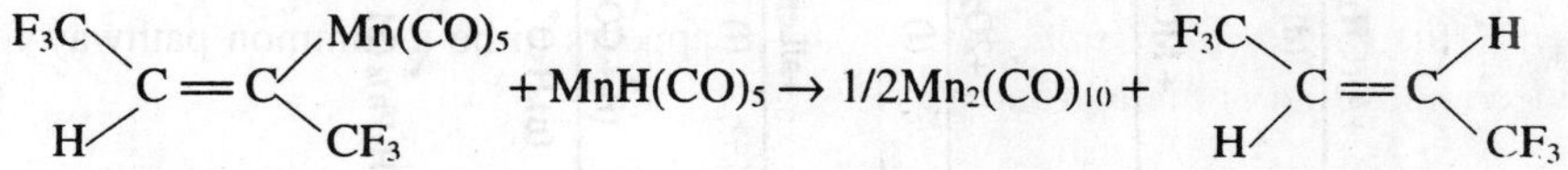

$$CoMe_2(PMe_3)_3 + 2ROH \rightarrow 1/n[Co(OR)_2]_n + 3MeH + 2PMe_3$$

$$(ROH = MeOH, PhOH)$$

$$cis{-}PtMe_2(PEt_3)_2 + HCl \rightarrow cis[PtCl(Me)(PEt_3)_2] + CH_4$$

$$cis{-}[PtMe_2(PEt_3)_2] + 2HCl \rightarrow cis{-}[PtCl_2(PEt_3)_2] + 2CH_4$$

The susceptibility of metal alkyls to nucleophilic attack under oxidizing conditions can be represented by the following reactions:

$$TiMe_4 + 4I_2 \rightarrow 4MeI + TiI_4$$

$$MnMe(CO)_5 + Br_2 \rightarrow MeBr + MnBr(CO)_5$$

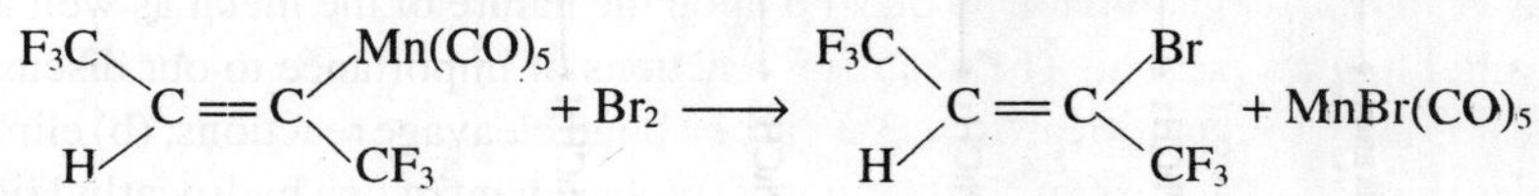

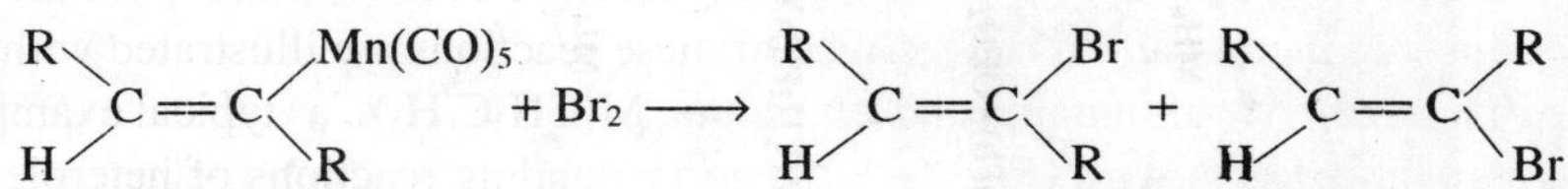

(R = CO$_2$Me) (Product ratio, 3:7) + MnBr(CO)$_5$

$$CoMe_2(PMe_3)_3 + X_2 \rightarrow CoX(Me)(PMe_3)_3 + MeX$$

$$(X_2 = Cl_2, Br_2)$$

$$AuBrEt_2 + Br_2 \rightarrow AuBr_2Et + EtBr$$

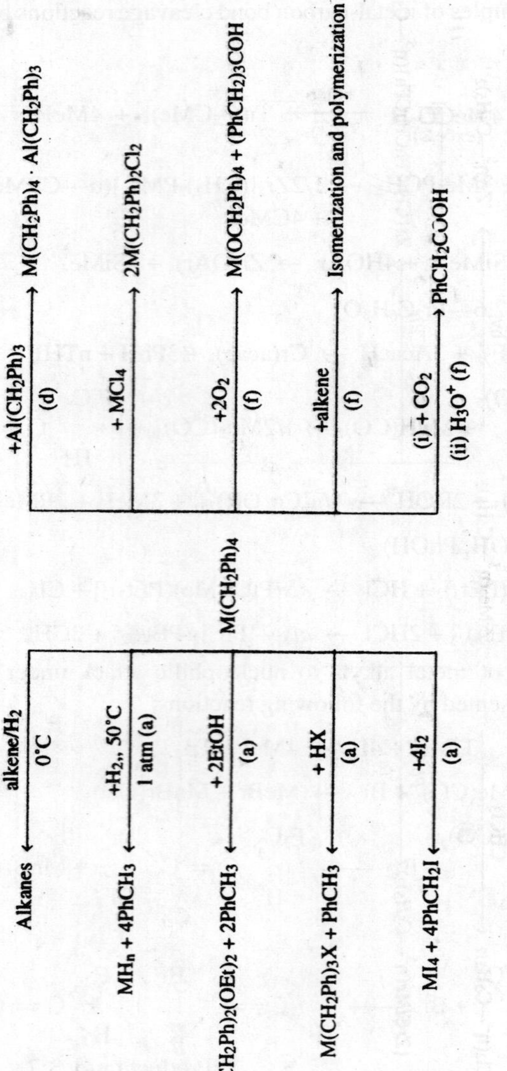

Fig. 4.9 Some reactions of $M(CH_2Ph)_4$ (M = Ti or Zr)

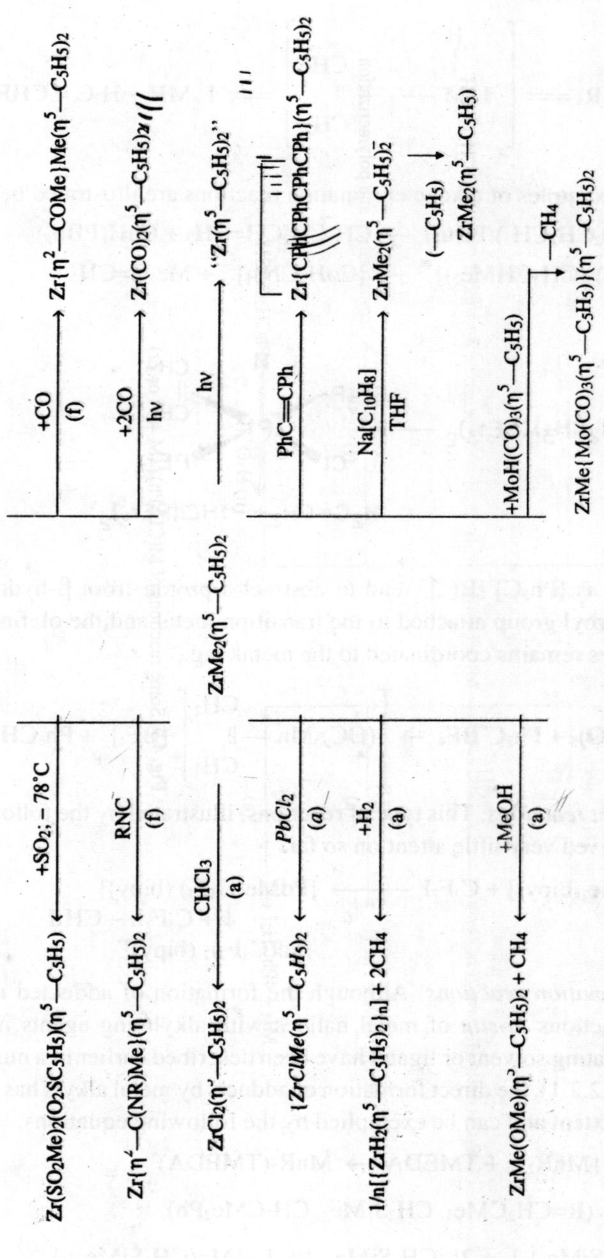

Fig. 4.10 Some typical reactions of $ZrMe_2(\eta^5-C_5H_5)_2$

(b) *Elimination of an alkene and proton abstraction*. One such reaction pathway is the so-called β-elimination of alkene with concomitant formation of the metal hydride (cf. Section):

$$L_nM-\overset{\alpha}{C}H_2\overset{\beta}{C}H_2R \rightleftharpoons \left[L_n\overset{\displaystyle H \atop |}{M} ---- \begin{matrix} CHR \\ \| \\ CH_2 \end{matrix} \right] \rightleftharpoons L_nMH + H_2C=CHR$$

Some typical examples of alkene elimination reactions are illustrated below:

$$Cu(CH_2CH_2CH_2CH_3)(PBu^n_3) \rightarrow CH_3CH_2CH{=}CH_2 + CuH(PBu^n_3)$$

$$[Co(CN)_5(CH_2CHMe_2)]^{3-} \rightarrow [CoH(CN)_5]^{3-} + Me_2C{=}CH_2$$

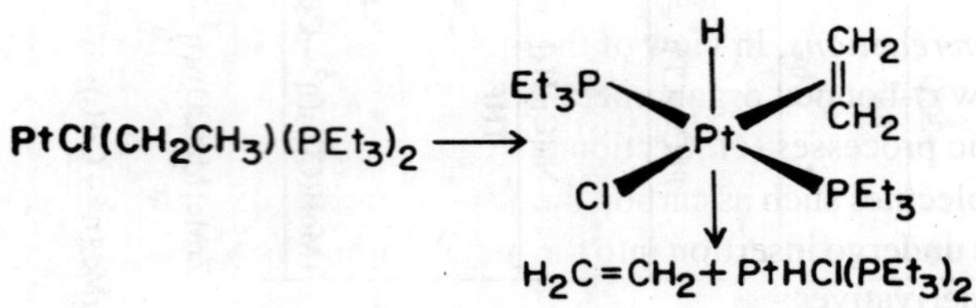

Reagents such as $[Ph_3C]^+[BF_4]^-$ tend to abstract a proton from β-hydrogen containing hydrocarbyl group attached to the transition metal and the olefin thus generated sometimes remains coordinated to the metal, e.g.,

$$MnCH_2CH_3(CO)_5 + Ph_3C^+BF_4^- \rightarrow \left[(OC)_5Mn - \begin{matrix} CH_2 \\ \| \\ CH_2 \end{matrix} \right]^+ [BF_4]^- + Ph_3CH$$

(c) *Replacement reactions*. This type of reactions, illustrated by the following example, have received very little attention so far.

$$[PdMe_2(bipy)] + C_3F_7I \xrightarrow[-CH_3I]{} [PdMe(C_3F_7)\,(bipy)]$$

$$\downarrow + C_3F_7I, - CH_3I$$

$$[Pd(C_3F_7)_2\,(bipy)]$$

(d) *Adduct formation reactions*. Although the formation of adducted metal alkyls from the reactions ***in-situ*** of metal halides with alkylating agents in the presence of coordinating solvent or ligand have been described earlier in a number of cases (e.g. Sec. 4.2.2.1), the direct formation of adducts by metal alkyls has been studied to a lesser extent and can be exemplied by the following equations:

$$\{MnR_2\}_n + TMEDA \rightarrow MnR_2(TMEDA)$$

$$(R{=}CH_2CMe_3,\ CH_2SiMe_3,\ CH_2CMe_2Ph)$$

$$\{Mn(CH_2SiMe_3)_2\}_n + 2LiCH_2SiMe_3 \rightarrow Li_2[Mn(CH_2SiMe_3)_4]$$

(e) ***Redistribution reactions.*** A redistribution reaction can be defined in general terms as a reaction in which an exchange takes place of the type illustrated below:

$$MX_n + M'Y_m \rightleftharpoons MX_{n-a}\,Y_a + M'X_aY_{m-a}$$

Alternatively, the metals M and M′ and the ligands X and Y may be the same or different. Such reactions have also been described as disproportionation, symmeterization, metathesis, scrambling or transfer reactions.

Alkyl transfers from main group to transition metals are well studied (cf. Sections 4.2.2.1 and 4.2.2.2). Such transfer reactions in transition metal organometallics are much less known and a few recent examples are given below:

cis–$PtMe_2(PMe_2Ph)_2$ + *cis*–$PdCl_2(PMe_2Ph)_2$

→ *trans*–$PtCl(Me)(PMe_2Ph)_2$ + *trans*–$PdCl(Me)(Me_2Ph)_2$

cis–$PtMe_2(PEt_3)_2$ + *cis*–$PtCl_2(PEt_3)_2$ → 2 *cis*–$PtCl(Me)(PEt_3)_2)$

$$2Ni(R)X(PEt_3)_2 \rightarrow NiR_2(PEt_3)_2 + NiX_2(PEt_3)_2$$

$$\downarrow$$

$$R\text{—}R + Ni(PFt_3)_2$$

(f) *Insertion reactions*. In view of the applicability of insertion reactions in the synthesis of new σ-bonded organometallics (cf. Section 4.2.2.4) and their importance in catalytic processes (cf. Section 6.4), these are dealt in a little more detail. Unsaturated molecules such as carbon monoxide, carbon dioxide, isocyanides and sulphur dioxide undergo insertion into the metal-carbon bonds, producing new and often, exciting derivatives.

(i) *Insertion of carbon monoxide*. The CO insertion products are especially important, as they appear to be key intermediates (see Chapter 6) in a variety of catalytic and industrial processes, such as hydroformylation reaction:

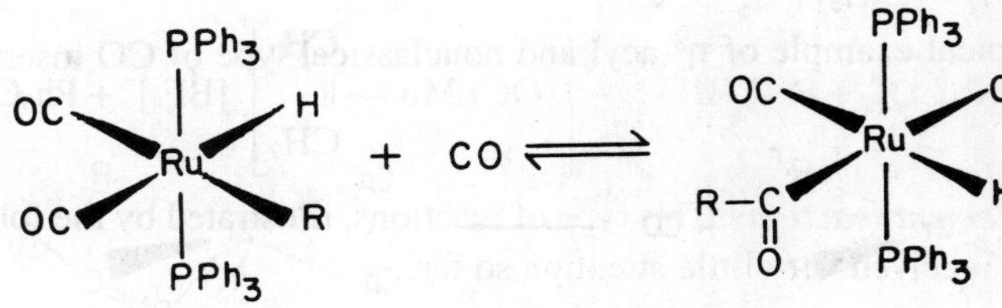

In view of the high oxgen affinity of early transition metals, lanthanides as well as actinides, their alklys gives η^2-acyls and other nonclassical types of CO insertion products; this oxophilicity results in dihapto oxgen bound migratory insertion products *A* and *B*:

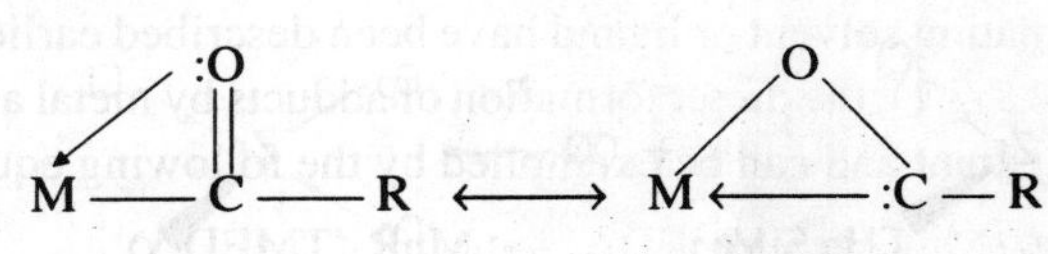

(A) (B)

which exhibit carbene-like chemical, structural, and spectroscopic properties. This η^2-coordination mode has an interesting aspect, namely, the formation of two possible isomers in the carbonylation of bis (cyclopentadienyl) dialkyls/diaryls as well as haloalkyls of group 4 transition metals and of the actinides, $MR_2(\eta^5—C_5H_5)_2$ and $MRX(\eta^5—C_5H_5)_2$; one of these isomers has C—O vector pointing away from the additional ligand X(*C*) and the other one depicts the opposite C—O orientation (*D*). Acyl derivatives of group 4 metals, such as $Ti(COCH_3)Cl(\eta^5—C_5H_5)_2$ and $Zr(COCH_3)(CH_3)(\eta^5—C_5H_5)_2$, show structures of types (*D*). By contract, the actinide analogues favour structure (*C*) or at least (*C*) can be as stable as (*D*):

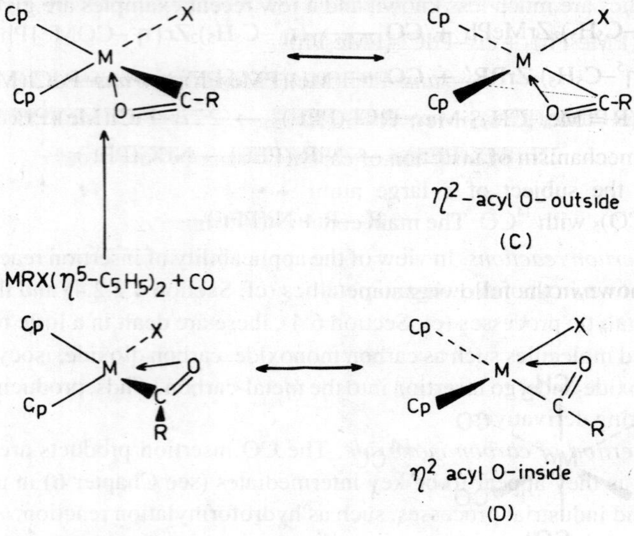

(where $Cp = \eta^5—C_5H_5$)

A few typical example of η^2-acyl and nonclassical type of CO insertion are shown below:

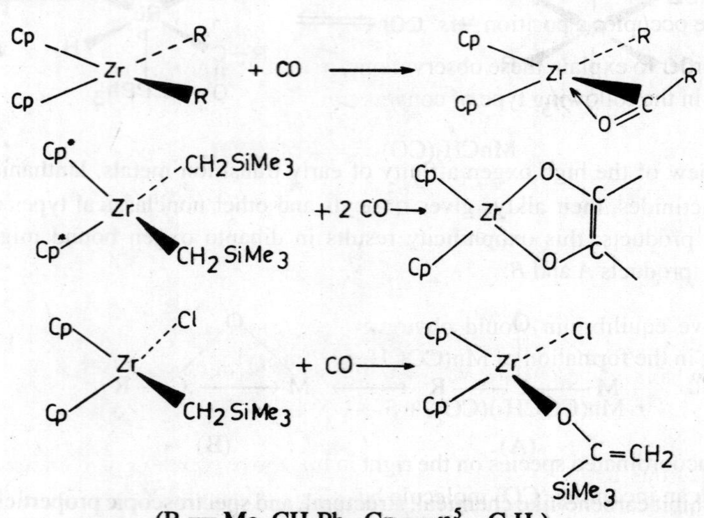

($R = Me, CH_2Ph,\ Cp = \eta^5—C_5H_5$)

($Cp^* = \eta^5–C_5Me_5$)

Another interesting feature of carbon monoxide insertion is the regio-selectivity which is observed when two different hydrocarbyls are present, as for example, in $ZrRR'(\eta^5{-}C_5H_5)_2$ (where $R/R' = CH(SiMe_3)_2/Me$, Ph/Me, $CH_2SiMe_3/CHPh_2$):

$$(\eta^5{-}C_5H_5)_2ZrMe\{CH(SiMe_3)\} + CO \longrightarrow (\eta^5{-}C_5H_5)_2ZrMe\{\eta^2{-}COCH(SiMe_3)_2\}$$

$$(\eta^5{-}C_5H_5)_2ZrMePh + CO \longrightarrow (\eta^5{-}C_5H_5)_2Zr\{\eta^2{-}COMe\}Ph$$

$$\eta^5{-}C_5H_5)_2ZrRR' + CO \longrightarrow (\eta^5{-}C_5H_5)_2Zr(\eta^2{-}COR)R'$$

$(R = Me,\ CH_2SiMe_3;\ R' = CHPh_2)$

The mechanism of insertion of carbon monoxide into the metal-carbon bond has been the subject of a large number of penetrating studies, e.g., that of $MnCH_3(CO)_5$ with ^{14}CO. The main conclusions from these may be summarized as follow:

As shown in the following step,

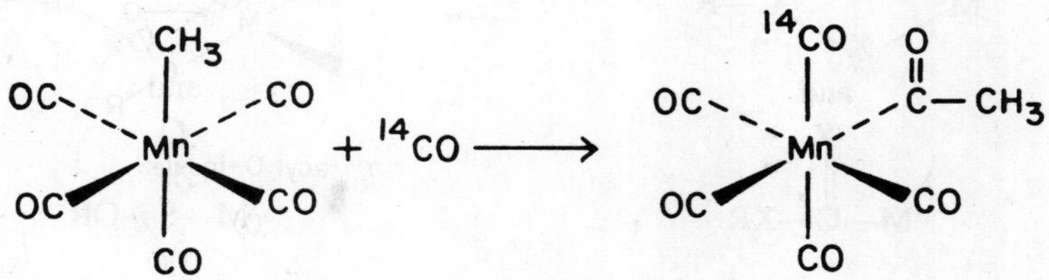

the CO molecule which assumes acyl character is not the external one, but it is one of the CO groups already coordinated to the metal. Interestingly, the incoming CO molecule occupies a position '*cis*' to the acyl group.

In order to explain these observations, it can be assumed that the primary step consists in the following type of conversion:

$$MnCH_3(CO)_5 \rightleftharpoons \begin{array}{c} Mn(CO)_4 \\ | \\ CO \\ | \\ CH_3 \end{array}$$

The above equilibrium would obviously be facilitated by a donor solvent 'S' resulting in the formation of $Mn(COCH_3)(CO)_4(S)$:

$$Mn(COCH_3)(CO)_4 + S \rightarrow Mn(COCH_3)(CO)_4(S)$$

The hexacoordinated species on the right in the above equation would interact not only with an incoming 'CO' molecule but also with another donor ligand (L) like phosphines:

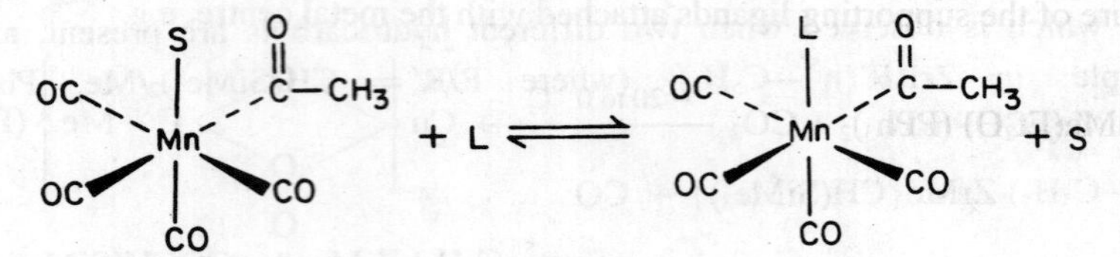

(ii) *Insertion of carbon dioxide, carbon disulphide and sulphur dioxide*

Possible modes of insertion of CX_2 (i.e. CO_2 or CS_2) and SO_2 in transition metal-carbon bonds are represented below:

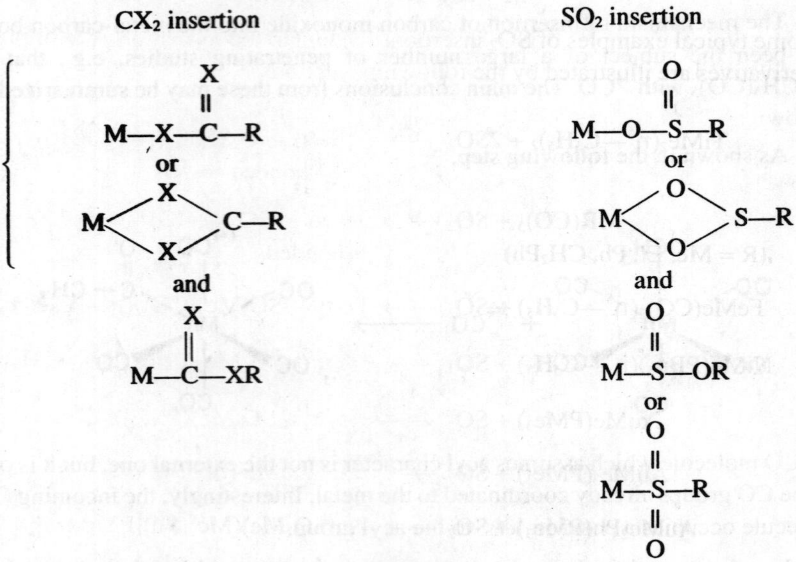

A few typical examples of such insertion reactions are given below:

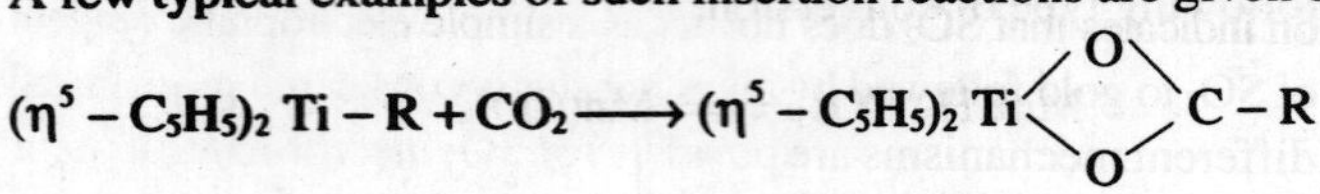

$$2VR_3 \cdot nTHF + CO_2 \rightarrow V(OCOR)_2 + VR_4 \cdot nTHF$$

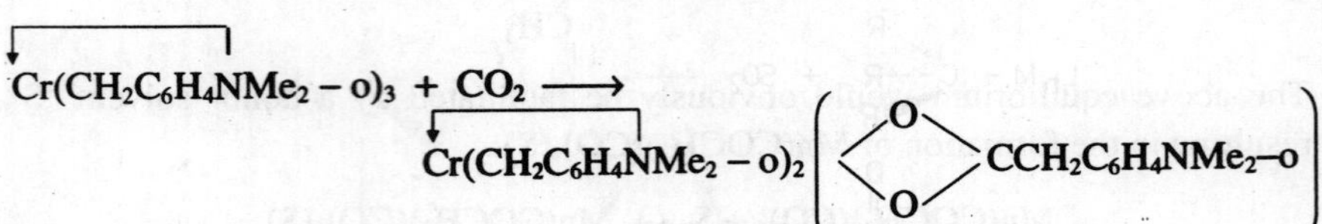

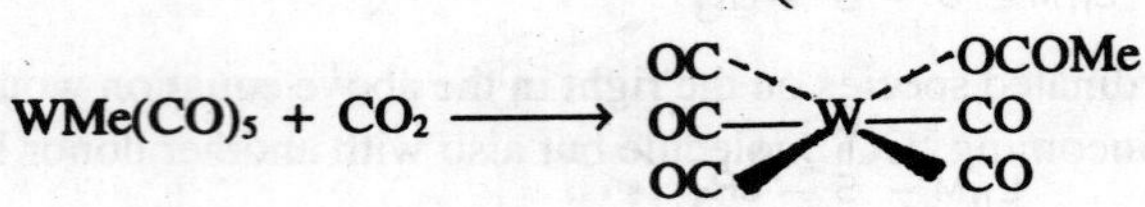

$$Mn(CH_2Ph)_2 + 2CO_2 \rightarrow Mn(OCOCH_2Ph)_2$$

Insertion modes of CO_2 into Cu—R bonds are generally dependent on the nature of the supporting ligands attached with the metal centre, e.g.:

$$CuMe(Et_2O)(PPh_3)_2 + CO_2 \xrightarrow{-20 \text{ to } 0\,^\circ C} Cu\left\{ \begin{matrix} O \\ O \end{matrix} \!>\! C - Me \right\}(PPh_3)_2$$

$$CuMe(PPh_3)_3 + CO_2 \longrightarrow Cu(O\overset{\overset{O}{\|}}{C}Me)(PPh_3)_3$$

Although products from SO_2 insertion into transition metal-carbon bonds can also vary, yet the usual product is S-sulphinate:

$$M - R + SO_2 \rightarrow M(SO_2R)$$

Some typical examples of SO_2 insertions leading to the formation of S-sulphinate derivatives are illustrated by the following reactions:

$$TiMe_2(\eta^5\text{—}C_5H_5)_2 + 2SO_2 \xrightarrow{-78\,^\circ C} \underset{\text{O-bonded}}{Ti(\eta^1\text{—}O_2SMe)_2(\eta^5\text{—}C_5H_5)_2}$$

$$\underset{(R = Me,\ Et,\ Ph,\ CH_2Ph)}{MR(CO)_5} + SO_2 \longrightarrow \underset{\text{S-bonded}}{Mn(\eta^1\text{—}SO_2R)(CO)_5}$$

$$FeMe(CO)_2(\eta^5\text{—}C_5H_5) + SO_2 \longrightarrow Fe(\eta^1\text{—}SO_2Me)(CO)_2(\eta^5\text{—}C_5H_5)$$

$$NiMe(PBu_3)(\eta^5\text{—}C_5H_5) + SO_2 \longrightarrow Ni(\eta^1\text{—}SO_2Me)(PBu_3)(\eta^5\text{—}C_5H_5)$$

$$AuMe(PMe_3) + SO_2 \longrightarrow Au(\eta^1\text{—}SO_2Me)(PMe_3)$$

$$AuMe_3(PMe_3) + SO_2 \longrightarrow cis\text{—}AuMe_2(SO_2Me)(PMe_3)$$

$$AuMe_2Ph(PMe_3) + SO_2 \longrightarrow Au(SO_2Me)(Me)(Ph)(PMe_3)$$

The selective insertion of SO_2 into a gold methyl rather than a gold phenyl bond in the last reaction indicates that SO_2 does not act as a simple electrophilic reagent. Rather coordination of SO_2 to gold followed by alkyl transfer is considered more plausible.

Several different mechanisms are possible for SO_2 insertion into a metalcarbon bond; one of these involving *exo* attack, is illustrated below:

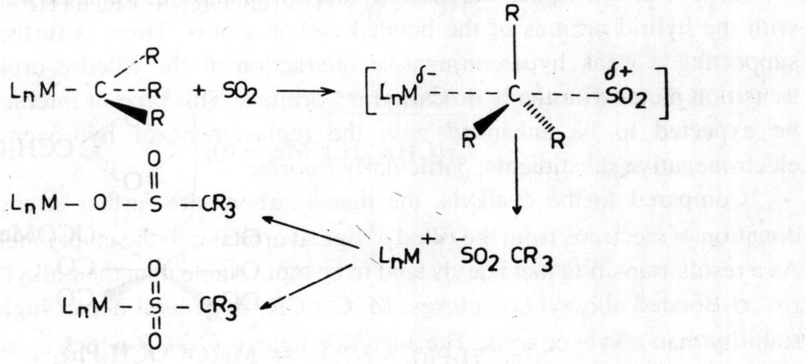

By contrast, SO_2 insertion into the Co—R groups of cobaloximes, has been explained, on the basis of a radical chain mechanism:

$$\text{Co—R} \rightarrow \text{Co}^0 + \text{R}^0$$

$$\text{Co}^0 + SO_2 \rightarrow \text{CoSO}_2^0$$

$$\text{CoSO}_2^0 + \text{Co—R} \rightarrow \text{CoSO}_2\text{R} + \text{Co}^0$$

Some illustrative examples of CS_2 insertion into transition metal-carbon bonds are the following:

$$\text{MnR(CO)}_5 + CS_2 \rightarrow \text{Mn}(\eta^2\text{—S}_2\text{CR})(\text{CO})_4 + \text{CO}$$

(R = Me, Ph, p-Tol)

(iii) *Insertion of isocyanides (RNC)*. Although isocyanides are isoelectronic with CO, yet examples of their insertions into the transition metal-carbon bond are relatively fewer in number. Few examples of isocyanide insertion reactions are:

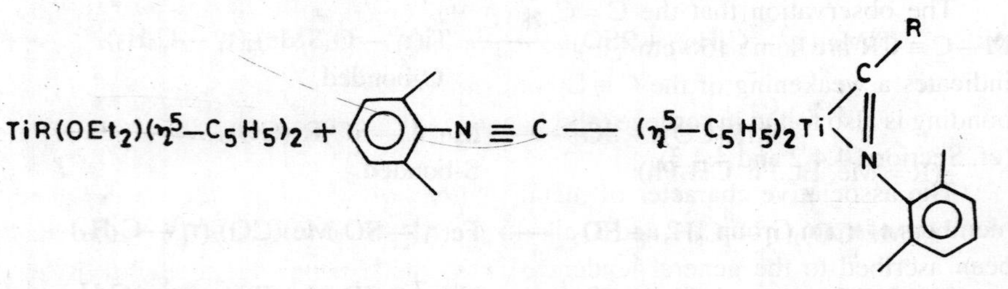

(R = Me, Ph, o–C_6H_4Me)

$$(\eta^5\text{—C}_5\text{H}_5)_2\text{ZrRR}' + \text{R}''\text{NC} \rightarrow (\eta^5\text{—C}_5\text{H}_5)_2\text{ZrR}\{\eta^2\text{—C(NR}'')\text{R}'\}(\eta^5\text{—C}_5\text{H}_5)_2$$

In contrast to CO insertion in the above system, the insertion of an isocyanide R″NC into the more hindered (R′) of the two R and R′ hydrocarbyl groups under ambient conditions appears to be more general.

4.2.3.4 *Bonding and Structure*

In general, the bonding in transition metal σ-hydrocarbyls can be described by an appropriate overlap of a hybrid orbital on the transition metal with the sp^3 orbitals of the carbon atoms. Photoelectron spectral studies on homoleptic species like WMe_6 have indicated that the metal *d*- and *s*-orbitals contribute the major overlap with the hybrid orbitals of the bonded carbon atoms. There is further evidence supporting a weak hyperconjugative interaction of the filled *d*-orbitals of the transition metal with the hydrocarbyl σ*-orbitals. This type of interaction would be expected to be enhanced with the replacement of hydrogen atoms by electronegative substituents, particularly fluorine.

Compared to the σ-alkyls, the metal-carbons are further strengthened by donation of electrons from the filled p^π ligand orbital into the empty metal orbitals. As a result, transition metal aryls tend to be more stable than their alkyl analogues.

σ-Bonded alkynyl complexes, M–C ≡ CR, in general depict higher thermal stability than alkyls or aryls. The acetylide ligand $^-$C ≡ CR which is isoelectronic with CN^- and CO may be expected to have both σ-donor π-acceptor character (Fig. 4.11):

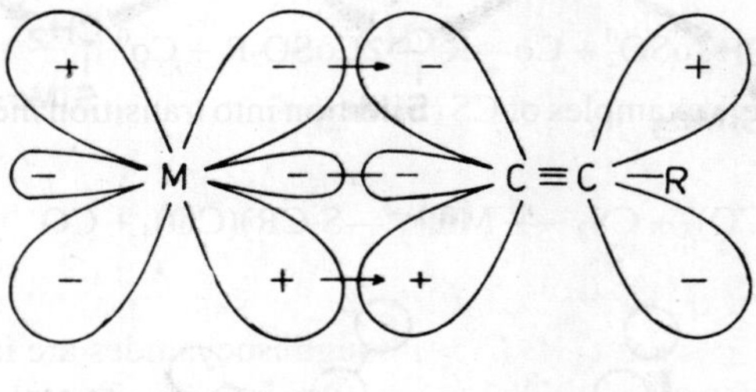

Fig. 4.11 Bonding in σ-alkynyl complexes of transition metals

The observation that the C = C stretching frequencies of metal acetylides M—C ≡ CR are some 150 cm^{-1} lower than those of related acetylenes RC ≡ CR′ indicates a weakening of the C ≡ C bonds in the former. Metal-carbon multiple bonding is also found in carbene (alkylidene) and carbyne (alkylidyne) complexes (cf. Sections 4.4.2 and 4.4.3).

The associative character of metal (particularly those of electron deficient members of main Group 1, 2 and 3) alkyl, alkenyl, alkynyl and aryl derivatives has been ascribed to the general tendency of the metal centre for attaining higher coordination states.

Its occurrence is being recognized more and more in the transition metal organometallic chemistry. A few illustrative examples of associated transition metal hydrocarbyls are shown in Fig. 4.12.

Such association, however, can be broken by adding electron donating ligands to yield monomolecular derivatives; for example, the dimeric $[Mn\{CH_2(CH_3)_2Ph\}_2]_2$ (Figure 4.12a) is converted into monomeric species (Figure 4.12b) in the presence of TMEDA:

$$1/2[Mn\{CH_2(CH_3)_2Ph\}_2]_2 + TMEDA \quad (a)$$
$$\downarrow$$
$$[Mn\{CH_2(CH_3)_2Ph\}_2(TMEDA)] \quad (b)$$

Fig. 4.12 Conversion of $[Mn\{CH_2(CH_3)_2Ph\}_2]_2$ into a mono-nuclear species with an ancillary ligand

Even in the absence of donor ligands, the associative tendency of simpler metal alkyls is sterically hindered by bulkier organic moieties. This is reflected in the case of manganese(II) alkyls of the polymeric $[Mn(CH_2SiMe_3)_2]_n$ (Figure 4.13a) and monomeric $[Mn\{C(SiMe_3)_3\}_2]$ (Fig. 4.13b).

Although the X-ray structure of $[Mn\{CH(SiMe_3)_2\}_2]_x$ with a ligand of steric bulk between the above two extremes is yet not known, yet it is expected to be least associated.

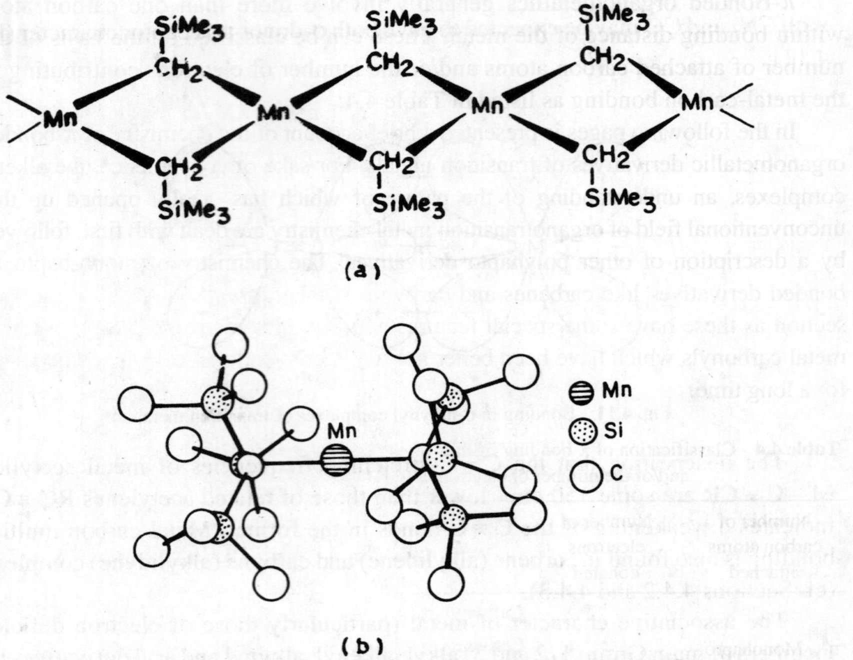

Fig. 4.13 Structure of (a) $[Mn(CH_2SiMe_3)_2]_n$, and (b) $[Mn\{C(SiMe_3)_3\}_2]$

4.3 π-BONDED ORGANOMETALLICS

4.3.1 Introduction

As stated earlier also, compared to σ-hydrocarbyls, π-bonded organometallics of transition metals are much more common and occur in a variety of fascinating structural patterns. In fact the interactions of the orbitals of unsaturated hydrocarbons with these metals often bring about such novel and non-classical types of bonding, which do not have a counterpart elsewhere in chemistry. In view of the lability of many such species, they play an important role in the catalytic reactions of alkenes, alkynes, and their derivatives. Many of these catalytic reactions involves transition metal-carbon monoxide bonding, the nature of which is in some ways similar to that in π-bonded organometallics; both of these generally follow the Sidgwick effective atomic number (EAN) or 18-electron rule.

The nature of bonding in different types of π-bonded organometallics would be dealt with in appropriate sections; and of these are based on the basic Dewar-Chatt-Duncanson approach, followed by applications of molecular orbital techniques (cf. Sec. 4.3.2.3). The readers are advised to note, for example, the elegant manner in which R. Hoffmann (cf. Nobel Lecture, 1982: Appendix 1.4) has used molecular orbital theory in the development of the simple concept of isolobal fragements, with the help of which the nature of bonding and properties of a variety of organic and organometallic/inorganic moieties can be explained in an analogous manner.

π-Bonded organometallics generally involve more than one carbon atom within bonding distance of the metal. These can be classified on the basis of the number of attached carbon atoms and/or the number of electrons contributing to the metal-carbon bonding as listed in Table 4.4:

In the following pages is presented a brief account of the chemistry of π-bonded organometallic derivatives of transition metals. For sake of convenience, the alkene complexes, an understanding of the nature of which has really opened up this unconventional field of organotransition metal chemistry are dealt with first, followed by a description of other polyhapto derivatives. The chemistry of mono-hapto π-bonded derivatives like carbenes and carbynes is taken up at the end in a separate section as these have some special features and also in some senses, they resemble metal carbonyls which have been better known in conventional inorganic chemistry for a long time.

Table 4.4 Classification of π-bonding ligands on the basis of the number of attached carbon atoms and/or the number of electrons denoted to metal-carbon bonding.

Number of carbon atoms attached	Number of electrons donated	Examples: Ligand	Examples: Complex
		Carbonyl (:C ≡ O)	$Cr(CO)_6$, $Fe(CO)_5$, $Ni(CO)_4$
η^1, Monohapto	2	Carbene {:CR_2, :C(OR)R′, C(OR)(NHAr), :Cyclo}	$TaCH_3$ (:CH2) Cp_2
	3	Carbyne (:CR)	Cr (: CMe)$(CO)_4$ Cr (: $CNEt_2$)$(CO)_4$
			$Fe(CO)_4(\eta^2—C_2H_4)$ $Ni(PPh_3)_2(\eta^2—C_2H_4)$
η^2, Dihapto	2	Alkene (>C = C<)	$[PtCl_3(\eta^2—C_2H_4)]^-$
		Alkyne (—C≡C—)	$Pt(PPh_3)_2(\eta^2—C_2Ph_2)$
η^3, Trihapto	3	π–Allyl (>C ∴ C ∴ C<)	$Ni(\eta^3—C_3H_5)_2$
η^4, Tetrahapto	4	Conjugated diene (e.g. butadiene)	$Fe(CO)_4$ 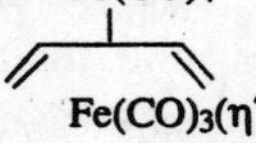
		Cyclobutadiene	$Fe(CO)_3(\eta^4—C_4H_4)$
η^5, Pentahapto	5	Dienyl (e.g. cyclopentadienyl and cycloheptadienyl derivatives)	$Fe(\eta^5—C_5H_5)_2$ $[Fe(CO)_3(\eta^5—C_6H_7)]^+$
η^6, Hexahapto	6	Arene (e.g. benzene and substituted benzenes)	$V(\eta^6—C_6H_6)_2$ $Cr(\eta^6—C_6H_6)_2$
		Cycloheptatriene	$Mo(CO)_3(\eta^6—C_7H_8)$ $Mn(\eta^5—C_5H_5)(\eta^6—C_7H_8)$
		Cycloocta—1, 3, 5 triene	$Mo(CO)_3(\eta^6—C_8H_8)$ $Mn(\eta^5—C_5H_5)(\eta^6—C_8H_8)$

4.3.2 η^2-Alkene Complexes

The complexation of alkenes with transition metals (in which C = C double bond contributes two electrons) has aroused considerable interest (theoretically and practically) since early fifties. The alkene-metal bond involves both σ-donor and π-acceptor liknages of these ligands, which appear to play a synergetic role on the strength of these bonds. The bonding of alkenes to metals has been shown to be similar to that in metal carbonyls; both the ligands are two electron donors and in fact, the replacement of carbon monoxide in metal carbonyls by olefins provides a versatile route for the synthesis of the latter type of compounds. Again, derivatives of the alkenes are also generally more stable with the metal in low oxidation state and are governed by the simple E.A.N. rule of Sidgwick.

The transition metal-alkene complexes may be of two general types representing the two extremes (a and c) of bonding in such complexes (Figure 4.14):

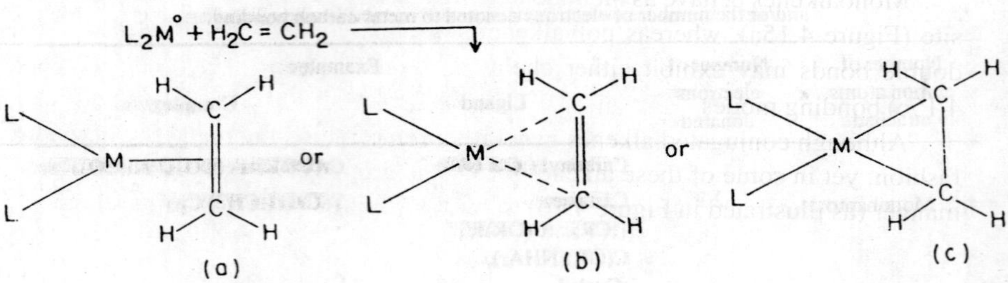

Fig. 4.14 Diagramatic representation showing the change in hybridisation from sp^2 to sp^3 at the olefinic carbon atom.

As shown in the following figure only the later transition metals (especially in their lower oxdation states) are known to form complexes with mono alkenes:

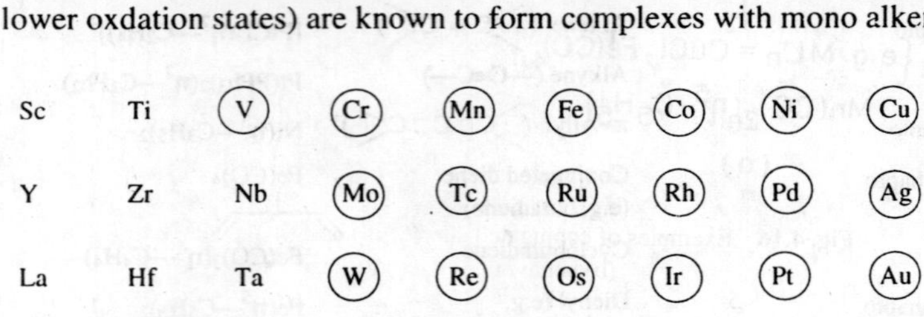

Fig. 4.14A Transition metals within circles are known to form alkene complexes

Alkene complexes of transition metals are generally heteroleptic in nature, e.g., $Re(CO)_4$ (alkene)$^+$, $MoX(CO)_2$(alkene)$_3$, and $Fe(CO)_2$ (alkene) (η^5—C_5H_5), although a few homoleptic derivatives also are known, e.g., Co(η^2—C_2H_4)$_n$ (where n = 1 or 2); Ni(η^2—C_2H_4)$_n$, Ni(η^2—C_2F_4)$_n$ and Pd(η^2—C_2H_4)$_4$ (where n = 1, 2 or 3) have been isolated in a low-temperature matrix using metal-vapour synthetic procedures.

The metal-alkene complexes formed by the transition metal shown in Figure 4.14A encompass a variety of structural types depending upon the ligating behaviour of alkenes, as shown in Figure 4.15.

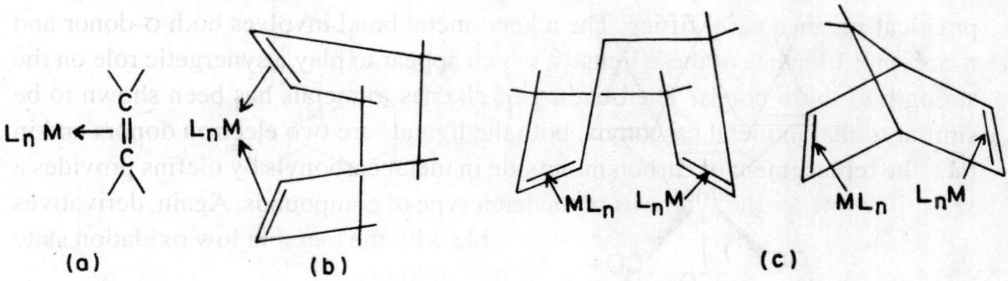

Fig. 4.15 Classification of non-conjugated (isolated) alkene complexes based on their ligating behaviour.

Monoalkenes behave as monodentate ligands, by occupying one coordination site (Figure 4.15a), whereas polyalkenes with non-conjugated (isolated) C = C double bonds may exhibit either chelating (Figure 4.15b) or bridging (Figure 4.15c) bonding modes.

Although conjugated alkenes generally prefer to bind the metal in polyhapto fashion, yet in some of these alkenes the C = C groups can behave in an isolated manner (as illustrated in Figure 4.16) to form bridging or nonbridging systems:

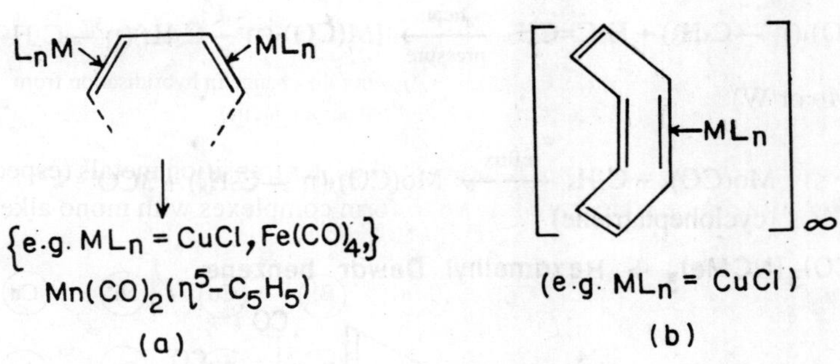

Fig. 4.16 Examples of conjugated alkenes (a) bridging metal atoms, and (b) behaving as a monoalkene

Cyclic alkene with several C = C bonds, as in cyclooctatetraene, may bind a metal in double bidentate bridging fashion as shown in Fig. 4.17:

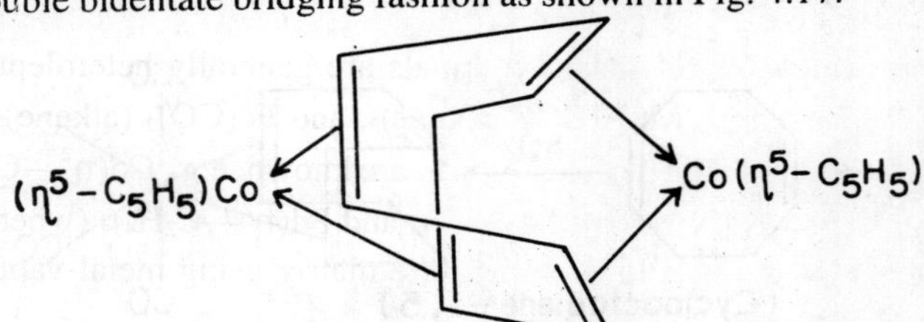

Fig. 4.17 Cyclooctatetraene chelating 2 metal atoms

Examples of complexes are known exhibiting tri- and tetra-dentate ligating behaviour of alkenes (i.e. with 3 or 4 isolated double bonds attached to the same metal atom):

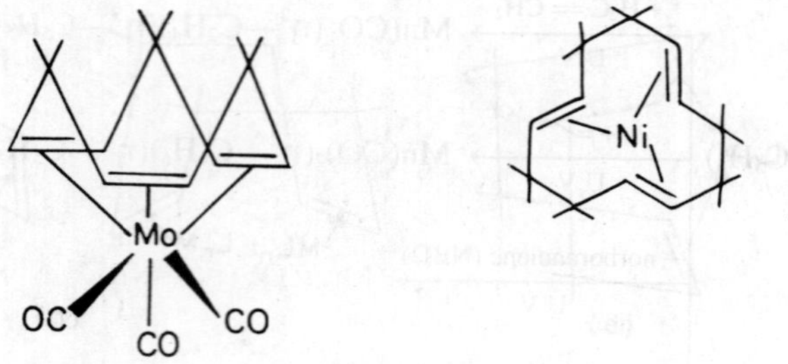

Fig. 4.18

4.3.2.1 Preparative Methods

The general methods for synthesis of transition metal-alkene (both mono- and poly-dentate) complexes can be classified as follows:

(*a*) *Interaction of a metal compound (usually a halide or carbonyl) with an alkene (is the most common preparative route)*:

$$M(CO)_6 + (NC)_2C = C(CN)_2 \rightarrow M(CO)_5\{\eta^2—C_2(CN)_4\} + CO$$
(M = Cr, Mo or W)

$$MX(CO)_3(\eta^5—C_5H_5) + H_2C{=}CH_2 \xrightarrow[\text{pressure}]{\text{heat}} [M(CO)_3(\eta^2—C_2H_4)(\eta^5—C_5H_5)]^+X^-$$

(M = Mo or W)

$$Mo(CO)_6 + C_7H_8 \xrightarrow{\text{reflux}} Mo(CO)_3(\eta^6—C_7H_8) + 3CO$$
(cycloheptatriene)

$M(CO)_3(NCMe)_3$ + Hexamethyl Dewar benzene

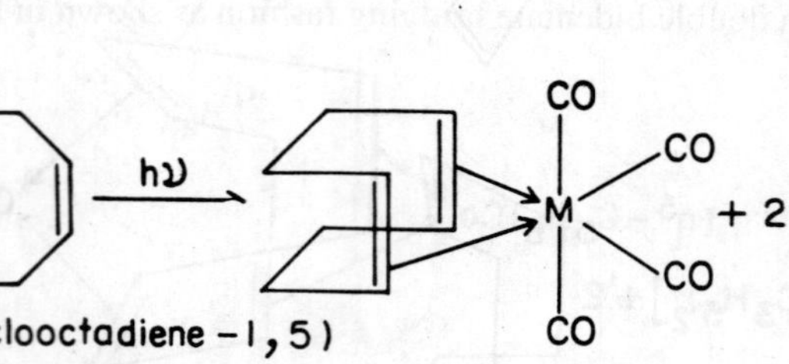

(M = Cr, Mo, W)

(M = Cr, Mo, W)

$$Mn(CO)_3X + H_2C = CH_2 \xrightarrow{140\ atm} [Mn(CO)_3(\eta^2\text{—}C_2H_4)]^+X^-$$

$$Mn(CO)_3(\eta^5\text{—}C_5H_5) \xrightarrow[U.V.]{+H_2C=CH_2} Mn(CO)_2(\eta^2\text{—}C_2H_4)(\eta^5\text{—}C_5H_5) + CO$$

$$Mn(CO)_3(\eta^5\text{—}C_5H_5) \xrightarrow[U.V.]{+ =\!\!/\!\!=} Mn(CO)_2(\eta^2\text{—}C_4H_6)(\eta^5\text{—}C_5H_5) + CO$$

$$Mn(CO)_3(\eta^5\text{—}C_5H_5) \xrightarrow[U.V.]{+\ norbornadiene\ (NBD)} (Mn(CO)_2(\eta^2\text{—}NBD)(\eta^5\text{—}C_5H_5) + CO$$

$$Mn(CO)_3(\eta^5\text{—}C_5H_5) \xrightarrow[U.V.]{+\ cyclohexadiene-1,3} Mn(CO)_2(\eta^2\text{—}C_6H_8)(\eta^5\text{—}C_5H_5) + CO$$

$$Re(CO)_5Cl + 2\ H_2C = CH_2 \xrightarrow{250\ atm} [Re(CO)_4(\eta^2\text{—}C_2H_4)_2]^+Cl^-$$

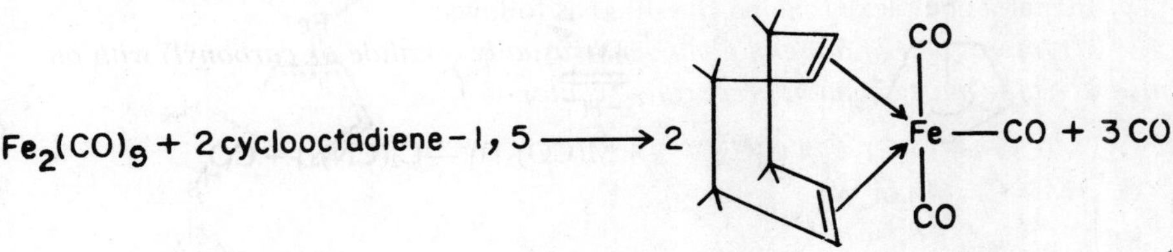

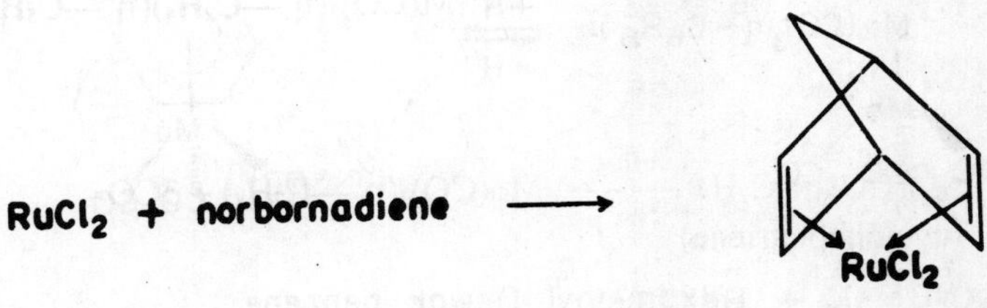

$$Fe(CO)_5 + \underset{(butadiene)}{C_4H_6} \xrightarrow[in\ bomb]{80°C} Fe(CO)_3(\eta^4\text{—}C_4H_6) + 2\ CO$$

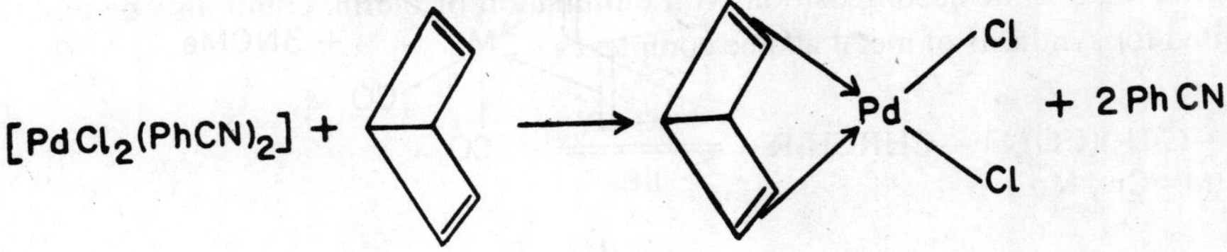

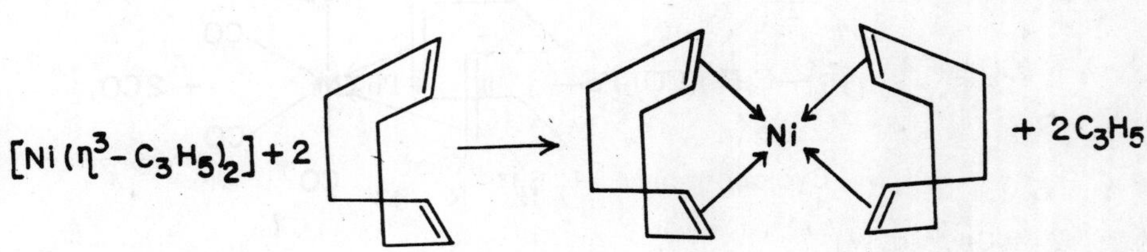

$$(PdCl_4)^{2-} + C_2H_4 \rightarrow [PdCl_3(C_2H_4)]^- + Cl$$

$$FeCl(\eta^5{-}C_5H_5)(CO)_2 + C_2H_4 \rightarrow [Fe(\eta^5{-}C_5H_5)(CO)_2(C_2H_4)]^+ AlCl_4^-$$

It may be mentioned that the displacement of ligands such as CO, RCN and PR_3 is often quite facile and proceeds smoothly. However, in cases where the π-bonding between the metal and carbon atoms is rather strong, displacement could be achieved either thermally or more conveniently by UV photolysis.

(b) *Reaction of hydride ion with metallocenes or half-sandwich compounds (nucleophilic addition)*:

$$(\eta^5{-}C_5H_5)_2Co^+ + H^- \xrightarrow[THF]{NaBH_4} Co(\eta^5{-}C_5H_5)C_5H_6$$

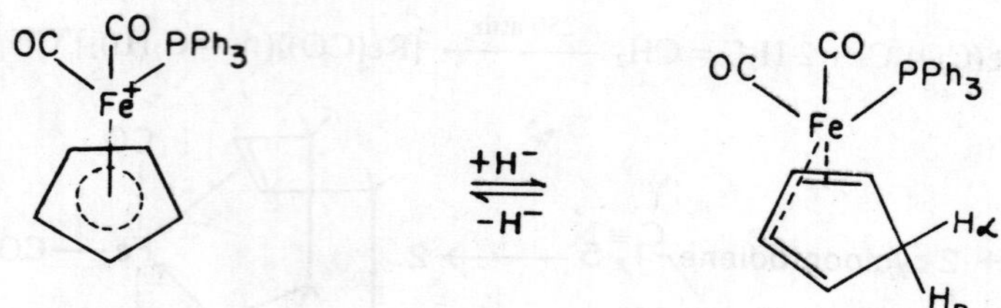

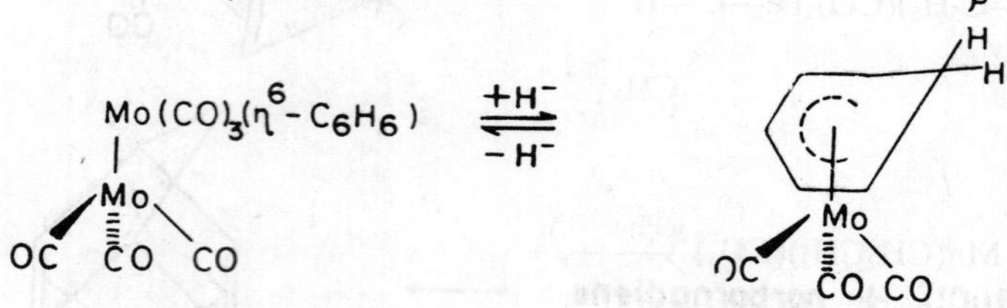

(c) *Hyaride ion abstraction or protonation reactions of σ-bonded organometallics,* β-Hydrogen elimination, which provides a common route (cf. Section 4.2.3.2) of decomposition with elimination of olefin, could also be exploited for synthesis of metal alkene complexes:

$$(\eta^5{-}C_5H_5)(CO)_2Fe{-}CHRCH_2R' \underset{BH_4^-}{\overset{Ph_3C^+BF_4^-}{\rightleftharpoons}}$$

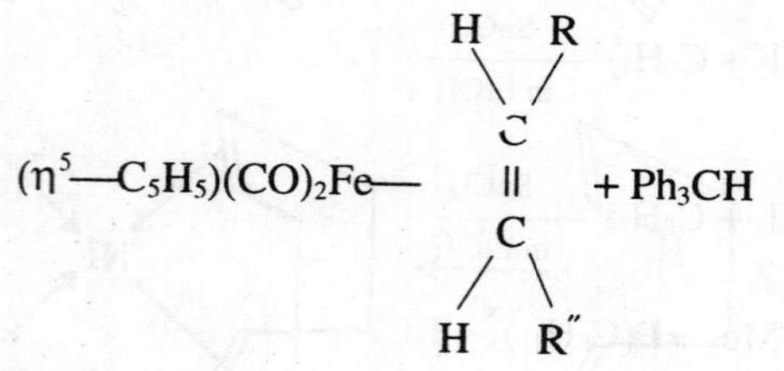

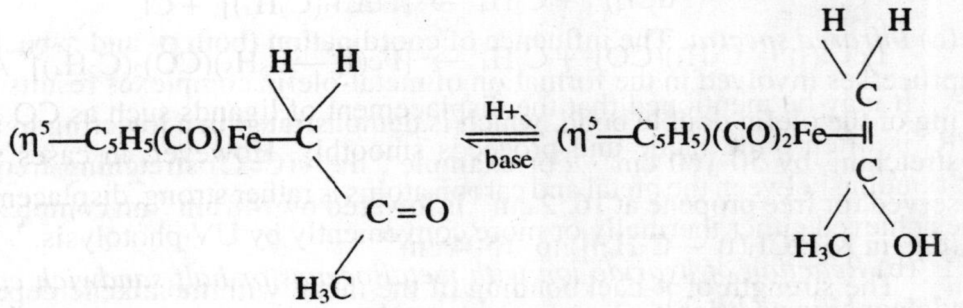

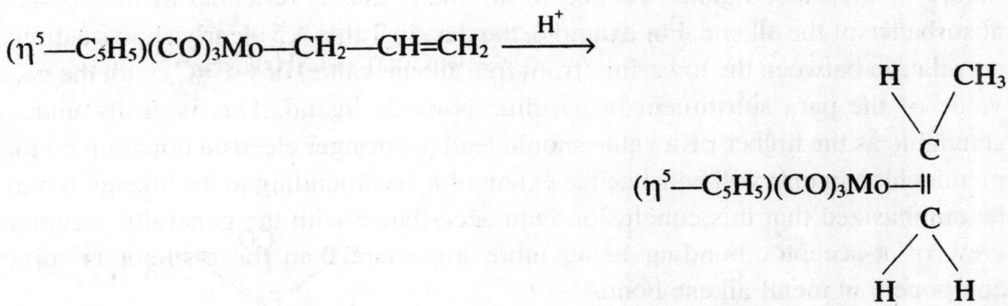

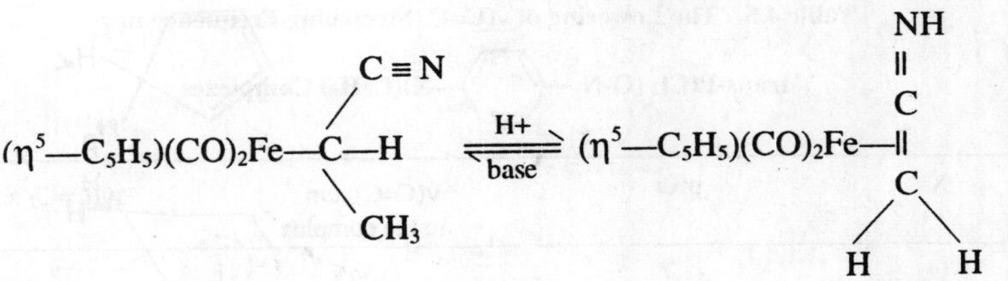

$$Mn(CH_2CH_3)(CO_5) \xrightarrow{Ph_3C^+BF_4^-} [Mn(CO)_5(CH_2{=}CH_2)]^+ BF_4^-$$

Salts of the organic cations Ph_3C^+ and R_3O^+ are particularly useful in these preparative procedures, because by-products (Ph_3CH or $R_2O + RH$) do not interfere in the isolation of products.

(d) *Reductive alkenation.* In these, the transition metal is reduced simultaneously to a lower oxidation state either by the olefin or some reducing agent added to the system:

$$RhCl_3 + C_2H_4 \xrightarrow{EtOH} [\{RhCl(\eta^2{-}C_2H_4)\}_2]$$

$$RhCl_3 + C_8H_{12} \xrightarrow[\text{in EtOH}]{SnCl_2} [\{RhCl(\eta^2{-}C_8H_{12})_2\}_2]$$

$$IrCl_3 + C_8H_{12} \xrightarrow[\text{in EtOH}]{SnCl_2} [IrSnCl_3(\eta^2{-}C_8H_{12})]$$

$$Ni(acac)_2 + AlMe_3 + P(C_6H_{11})_3 + RCH = CH_2 \rightarrow [\{Ni(\eta^2{-}CH_2 = CHR)P(C_6H_{11})_3\}_2]$$

4.3.2.2 Physical Properties

(a) *Infrared spectra.* The influence of coordination (both σ- and π-back donation) processes involved in the formation of metal-olefin complexes results in weakening of the olefin double bond, which is demonstrated in a lowering in the ν(C=C) stretching by 50-160 cm^{-1}. For example , the ν(C=C) stretching frequency, observed for free propene at 1652 cm^{-1} is lowered by 148 cm^{-1} on complex formation (e.g. in $K[PtCl_3(\eta^2—C_3H_6)]$) to 1504 cm^{-1}.

The strength of π-backbonding of the metal with the alkene depends on the nature of the other ligands bonded to the metal and is reflected in the ν(C=C) absorbance of the alkene. For example, the data in Table 4.5 clearly shows a direct correlation between the lowering (from free alkene value) of ν(C=C) with the pKa value of the para substituent in pyridine N-oxide ligand. This is easily understandable as the higher pKa value should lead to stronger electron donation on the metal which in turn will enhance the extent of π-backbonding to the alkene. It may be emphasized that this conclusion is in accordance with the generally accepted view of π-acceptor bonding being more important than the σ-donor bonding component in metal-alkene bonds.

Table 4.5 The Lowering of ν(C=C) Stretching Frequency in

trans-$PtCl_2$ (O-N — 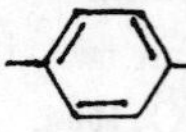 **—X)(C_2H_4) Complexes**

X	pKa	ν(C=C) cm^{-1} in the complex	Δν(C=C) cm^{-1}
NO_2	−1.7	1545	78
Cl	−0.4	1515	108
H	0.8	1510	113
Me	1.3	1500	123

Note: *ν(C=C) in free C_2H_4 is observed at 1623 cm^{-1} (Raman). Although it is inactive in free ethylene, yet ν(C=C) stretching mode is observed as a weak absorption in the infrared spectra of the metal complexes.*

Additionally, the data in Table 4.6 reveal some observed trends of transition metal-alkene complexes. For example, the magnitude of shift could be in g[illegible] correlated with the strength of metal-alkene bond although care must be exercised to make allowance for factors such as coupling (cf. CH_3 scissoring mode) with other modes of vibration.

Table 4.6 The ν(C=C) stretching Vibrations of some Transition Metal-Alkene Complexes (cm^{-1})

	Complex	ν(C=C), cm^{-1}	Δν(C=C), cm^{-1}
1.	$PdCl_2(C_2H_4)_2$	1527	96
2.	$PtCl_2(C_2H_4)_2$	1516	107
3.	$K[PdCl_3(C_2H_4)]$	1525	98
4.	$K[PtCl_3(C_2H_4)]$	1516	107

	Complex	ν(C=C), cm^{-1}	$\Delta\nu$(C=C), cm^{-1}
5.	Mo (η^5—C_5H_5)$(CO)_3(C_2H_4)$	1511	112
6.	W(η^5—C_5H_5)$(CO)_2(C_2H_4)$	1510	113
7.	$Cr(CO)_2(C_2H_4)$ (mesitylene)	1497	126
8.	$Mo(CO)_2(C_2H_4)$ (mesitylene)	1490	133
9.	$Ag(C_2H_4)$	1550	73

The observed shift (Table 4.6) indicate the Pt-olefin bond is stronger than Pd-olefin bond. This in fact appears to be a general trend that the π-bonding ability of transition metals (in the same oxidation state) increases as one transcends from 3*d* to 4*d* to 5*d* species. The same trend is observed for Mo-W pair (items 5, 6) and even more so in Cr-Mo pair (items 7, 8). The comparatively larger value of shifts in the last pair also reflects a general tendency that strength of M → L π-bonding increases with the lowering in the oxidation state of the metal.

The magnitude of shift for silver-ethylene complex shows that it is the weakest (in agreement with other data) amongst the complexes listed in the Table 4.6. In fact, this weakness in silver-olefin bonding has been found to be quite useful for the purification of olefins. For example, white crystals of the stoichiometry $[Ag(olefin)_n]^+ClO_4^-$ (n = 1 or 2) can be readily obtained by shaking a reaction mixture containing aqueous $AgClO_4$ and appropriate olefin, from which the regeneration of olefin is quite easy.

The use of infrared spectra in structural elucidation can be appreciated by examining ν(C=C) stretching frequencies (listed in Table 4.7) of a series of compounds $M(CO)_4L$ (where M = Cr, Mo or W, L = 2-allylphenylphosphine ($C_{21}H_{19}P$)).

The significant lowering in ν(C=C) values in the complexes (Table 4.7) does confirm the presence of olefin-metal bonds also in these complexes. The values of Δν(C=C) clearly demonstrate that the strength of metal-olefin bonds varies in the sequence W > Cr > Mo, which is in accordance with the trend observed for hexacarbonyls of chromium group metals (cf. Section 4.5). This trend is slightly at variance with the general trend that the strength of metal ligand-bonding increases from 3*d* to 4*d* and 5*d* series and points out that the generalisations are in many cases still only approximately true and require much more detailed and refined investigations to throw light on the impact of different parameters on such systems.

Table 4.7 The ν(C=C) Stretching Frequencies of some of Chromium Molybdenum and Tungsten Complexes

Compound	ν(C=C), cm^{-1}	$\Delta\nu$(C=C), cm^{-1}
$C_{21}H_{19}P$	1637 w	—
$Cr(CO)_4(C_{21}H_{19}P)$	1520 vw	117
$Mo(CO)_4(C_{21}H_{19}P)$	1524 vw	113
$W(CO)_4C_{21}H_{19}P)$	1502 vw	135

(b) *Nuclear magnetic resonance studies.* Alkenes are in general rigidly bonded to the transition metals and these complexes are, therefore, often fluxional (cf. Chapter 5) in nature. As discussed in the next chapter, nmr (particularly ^1H) studies provide an insight into such behaviour of complexes in solution. It has been found that the solid state structures do not often persist in solution and these studies, therefore, assume special importance in such cases where complexes are used in solution state in their industrial applications. Non-rigid attachment of alkenes to transition metals generally permits rotation either of a barrel-type (i.e. the rotation of metal-alkene axis as shown in *B* of Figure 4.19).

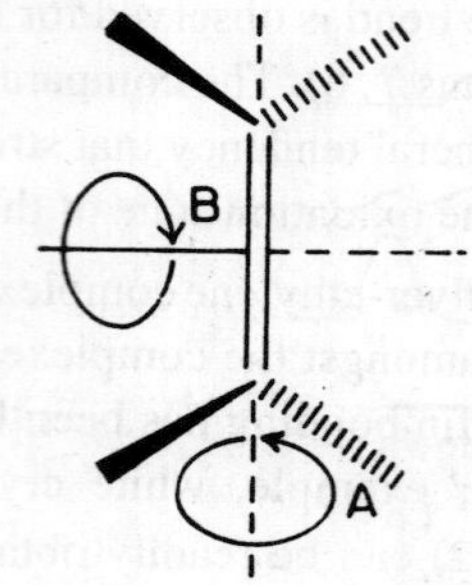

Fig. 4.19 Diagrammatic representation of rotational processes in metal-alkene complexes.

The above phenomenon of the non-rigid attachment of alkenes to transition metals has been investigated for the complex, (η^5—C_5H_5)Rh(C_2H_4)$_2$ (Fig. 4.20a) by the application of ^1H n.m.r. spectroscopy. The spectrum of the complex was found to vary with temperature (Fig. 4.20b). The spectrum recorded at – 19°C depicts two distinct types of environments about the protons. The protons labelled as Ho (outer) appear to give the downfield multiplets (d) and the upfield multiplets (u) could be ascribed to the inner protons, Hi. These discrete signals were explained on the basis that at the low temperature, the rotation of the olefin becomes so slow that it does not affect the observation in the n.m.r. spectrum of the two distinct environments for the protons in the molecule. However, as the temperature is raised, the rotation of ethene molecule about the metal-alkene bond tends to make all the ethene protons magnetically equivalent and the separate lines collapse into a broad signal. An analysis of the line shape (cf. Section 5.4.3) indicates that the activation energy of the rotation process is ~63 kJ.

In addition to the explanation of the nature of the variable temperature n.m.r. spectrum observed for the molecule on the basis of rotations (A and B in Fig. 4.19), another possible interpretation could be on the basis of intermolecular exchange of ethene molecule. The latter possibility is, however, ruled out by the high rhodium-ethene dissociation bond energy which has been separately measured as ~130 kJ mol^{-1}.

A study of the variable temperature n.m.r. study of the molecule, (Fig. 4.20) helps to distinguish between the two types of rotation shown in Fig. 4.19. The ^1H

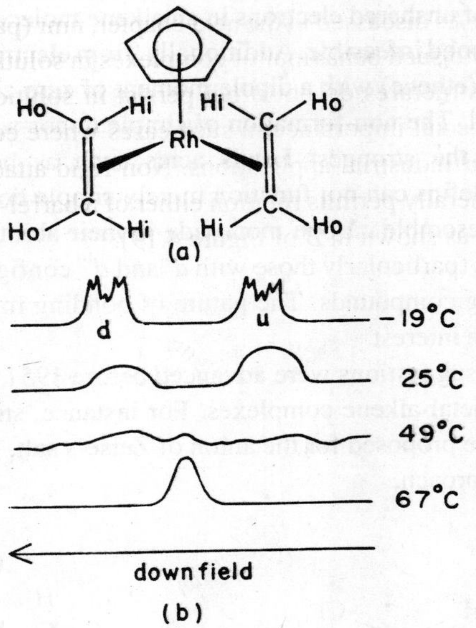

Fig. 4.20 Variable temperature ^{1}H n.m.r. of ($\eta^5 - C_5H_5$) $Rh(C_2H_4)_2$.

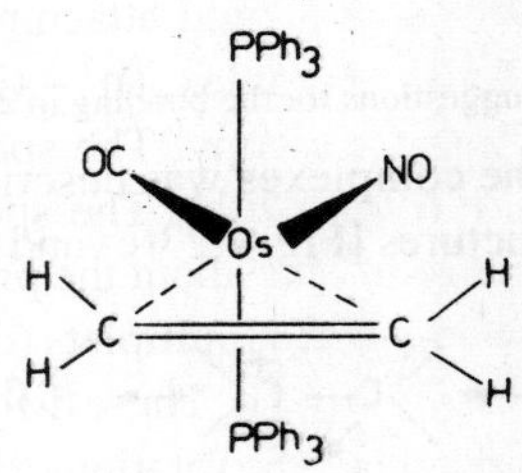

Fig. 4.21

n.m.r. spectrum of the molecule at – 90°C does show to sets of resonance lines of equal intensity (one for two ethene protons closer to CO and the other for the two protons in the proximity of NO), but these coalesce to a broad signal at – 65°C. However, compared to the simple complex, (η^5—C_5H_5)$Rh(C_2H_4)_2$ this osmium complex (4.21) incorporates two PPh_3 ligands also, the ^{31}P atoms of which begin to show spin-spin coupling with the four equivalent ethene protons at still higher temperatures to give a triplet with intensities in 1 : 2 : 1 ratio. This coupling between protons and phosphorus atoms is observed in higher temperature spectrum. This observations lends support to B rather than A type rotation (Fig. 4.19). These conclusions are further corroborated by the ^{13}C spectrum, which shows for the ethene ligand at 20°C only one ^{13}C line, which splits into two equivalent lines when the temperature is lowered.

4.3.2.3 Bonding and Structure

The unavailability of unshared electrons in an alkene molecule, makes the formation of usual dative bond infeasible. Additionally, from electrostatic considerations the simplest alkene (ethene) with a dipole moment of zero can not be expected to be a potential ligand. The non-formation of simple donor-acceptor complexes of alkenes even with the strongest Lewis acids such as boron halides clearly demonstrates that olefins can not function merely simple donors or Lewis bases. However, alkenes resemble carbon monoxide in their ability to complexes with transition metal ions (particularly those with d^8 and d^{10} configurations) affording a variety of interesting compounds. The nature of bonding in such complexes has aroused considerable interest.

Many tentative suggestions were advanced before 1951 to explain the nature of the bonding in metal-alkene complexes. For instance, structures of the types 4.22a and 4.22b were proposed for the anion of Zeise's salt, $[PtCl_3(C_2H_4)]^-$. Later on, in a modified approach,

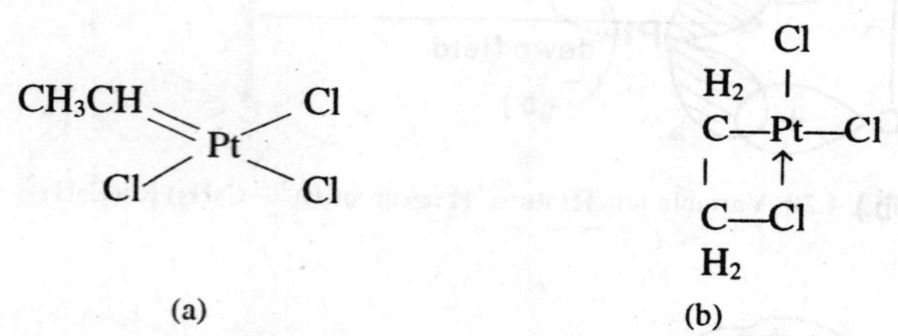

Fig. 4.22 Suggestions for the bonding in Zeise's salt.

the bonding in platinum ethene complexes was described in terms of a resonance hybrid of the contributing structures [Fig. 4.23(c) and (d).]

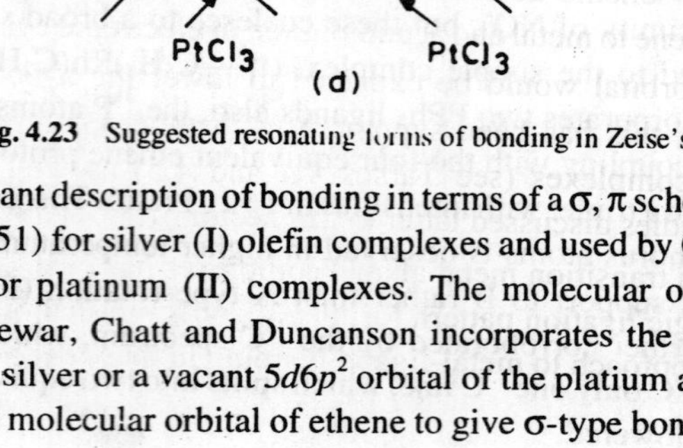

Fig. 4.23 Suggested resonating forms of bonding in Zeise's salt.

A more elegant description of bonding in terms of a σ, π scheme was suggested by Dewar (in 1951) for silver (I) olefin complexes and used by Chatt and Duncanson (in 1953) for platinum (II) complexes. The molecular orbital approach as developed by Dewar, Chatt and Duncanson incorporates the ideas: (i) a vacant 5*d*-orbital of the silver or a vacant $5d6p^2$ orbital of the platium atom is overlapped by the (bonding) molecular orbital of ethene to give σ-type bond (Fig. 4.24a), and

(ii) a π-type bond (Fig. 4.24b) is formed by overlapping a filled $4d$-orbital of silver atom or a $5d6p$-hybrid, orbital of the platinum atom with antibonding $\pi_z^* 2p$ molecular orbital of the alkene. In Fig. 4.24(c), an attempt has been made to depict the synergic strengthening of the metal-alkene bonds in such complexes.

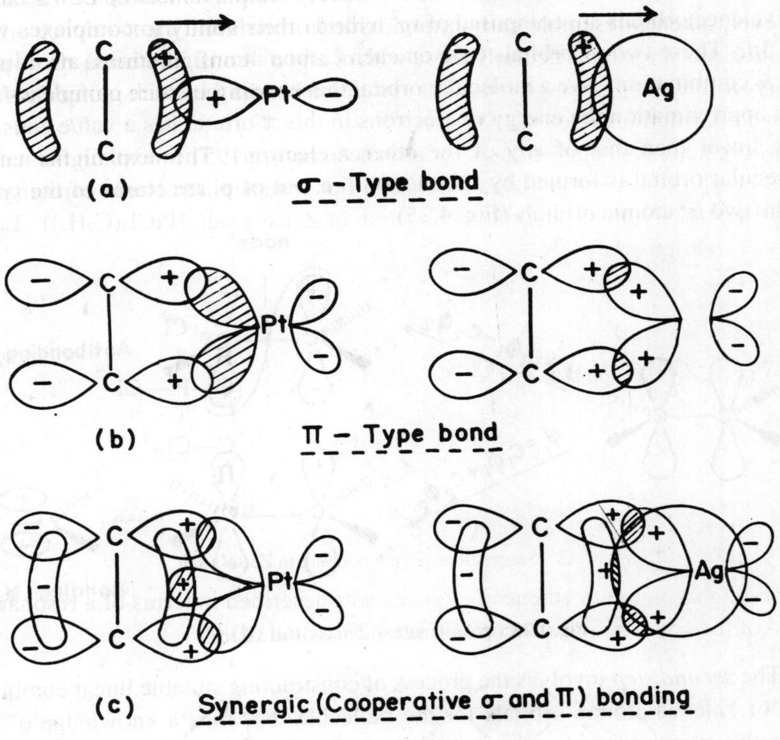

Fig. 4.24 The bonding in $[PtCl_3(C_2H_4)]^-$ or $[Ag(C_2H_4)]^+$.

In the bonding scheme described in Fig. 4.24, both donation of π-electron density from the alkene to metal and accumulation of electronic charge in the olefin antibonding $\pi_z^*\, 2p$ orbital would be expected to lower the π-bond order of the coordinated alkene; this has been supported by bathochromic shifts in the I.r. spectra of alkene complexes (see Tables 4.4 and 4.5) as well as by X-ray a crystallographic studies discussed latter in this section.

The bonding in transition metal alkene complexes has been explained earlier in terms of a synergic ligation pattern (cf. Fig. 4.24c). We shall now consider the molecular orbital approach to metal-ligand bonding in these derivatives.

In general the application of LCAO-MO method to complexes of transition metals with unsaturated molecules follows the following three steps, as in the case of complexes of simpler molecules.

Firstly, one has to pick out the valency atomic orbitals of the atoms in the molecule under consideration. For example, in a transition metal alkene complex such as Zeise's salt. $[PtCl_3(H_2C{=}CH_2)]^-$, the metal valence atomic orbitals are *5d*, *6s* and *5p*. The ligands may have σ and π valence orbitals; for example, in an alkene the valence orbitals are composed of sp^2 hybrid orbitals and the π-valence orbitals are *3pz*. These two *pz* orbitals (one on each carbon atom), in ethene, on in-phase linear combination, give a molecular orbital that covers the entire molecule. In the first approximation the energy or electrons in this π-orbital has a value considerably lower than that of any of the other π-electrons. The next higher energy molecular orbital is formed by subtraction (i.e. out-of-phase linear combination) of the two *pz* atomic orbitals (Fig. 4.25):

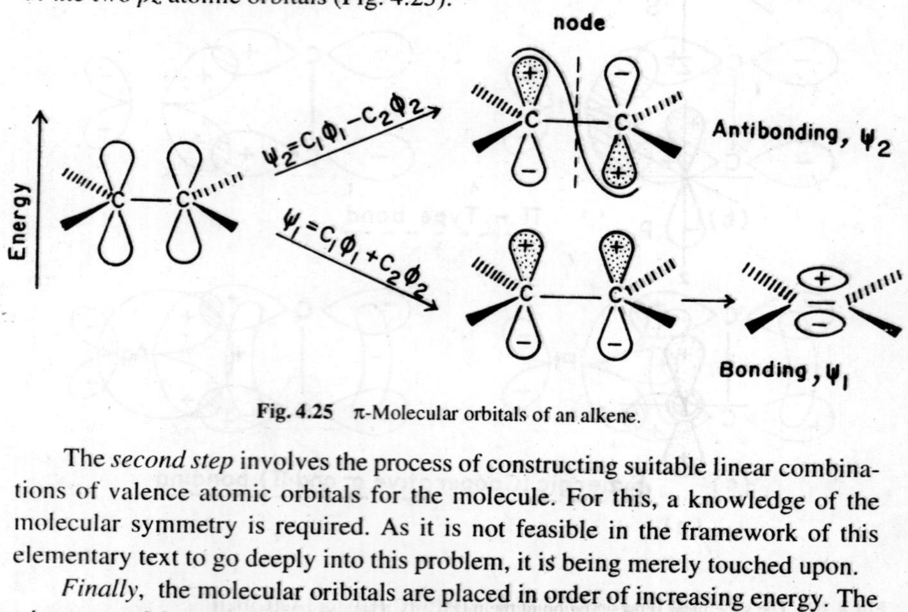

Fig. 4.25 π-Molecular orbitals of an alkene.

The *second step* involves the process of constructing suitable linear combinations of valence atomic orbitals for the molecule. For this, a knowledge of the molecular symmetry is required. As it is not feasible in the framework of this elementary text to go deeply into this problem, it is being merely touched upon.

Finally, the molecular oribitals are placed in order of increasing energy. The advantage of the molecular orbital approach to metal-ligand bonding is that we can picture the filling of molecular orbitals with electrons in the order of their energy, irrespective of the origin of the electrons.

Metal-ligand bonding occurs when orbitals on the metal and an alkene match in energies and symmertry. This means that depending upon the molecular geometry, only certain metal and alkene orbitals can be involved in bonding, as illustrated in Fig. 4.26.

The predictions from the above theoretical considerations have received ample support by the actual structures of metal-alkene complexes. For example, the crystal structures of $K[PtCl_3(C_2H_4)]$ (Zeise's salt) and $Pt\{(NC)_2C{=}C(CN)_2\}(PPh_3)_2$ are depicted in Figs. 4.27 and 4.28, respectively.

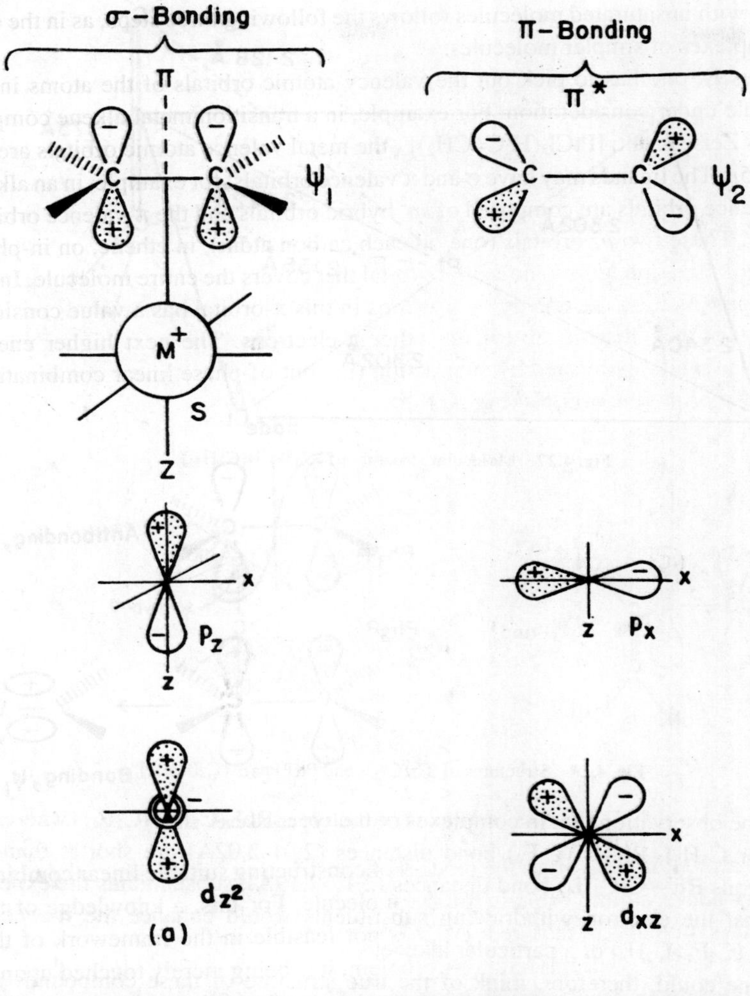

Fig. 4.26 The schematic representation of the metal orbitals having appropriate π-molecular orbital symmetry for bonding with the alkene.

The C—C bond distance (1.37Å) in the complex (Fig. 4.27) is slightly longer than in the free ethene (1.34Å) which can be explained to arise from the population of electrons in the π^* orbital of ethene, which brings about a change in the nature of the hybrid orbitals at carbon atoms. For example, a perusal of Fig. 4.28 shows that the C-C bond length and (NC) C(CN) angle in coordinated tetracyanoethylene are increased considerably with a concomitant change in the sp^2 hybridization in the parent alkene to sp^3 type in the complex.

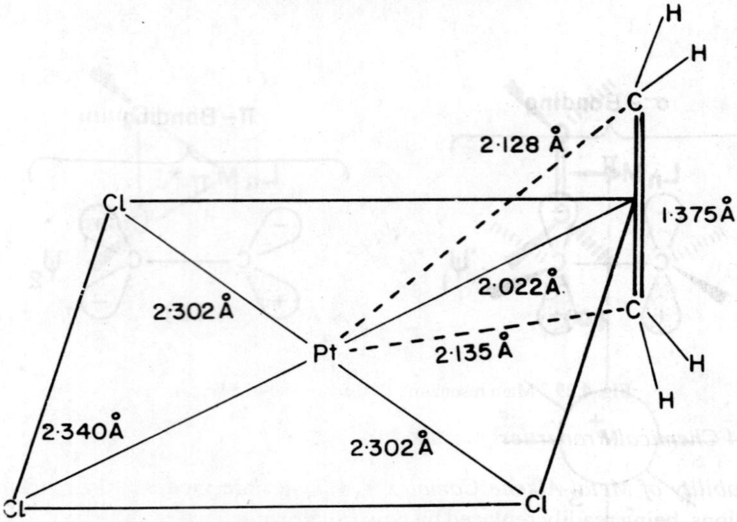

Fig. 4.27 Molecular structure of $K^{+}[PtCl_3(C_2H_4)]^{-}$

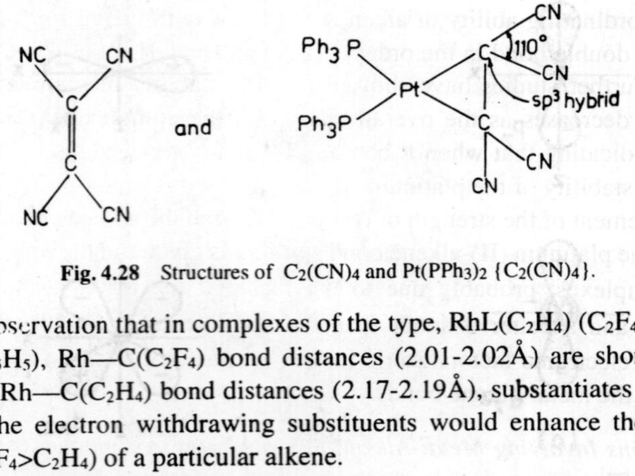

Fig. 4.28 Structures of $C_2(CN)_4$ and $Pt(PPh_3)_2\{C_2(CN)_4\}$.

The observation that in complexes of the type, $RhL(C_2H_4)(C_2F_4)$ (where L= acac or C_5H_5), Rh—$C(C_2F_4)$ bond distances (2.01-2.02Å) are shorter than the analogous Rh—$C(C_2H_4)$ bond distances (2.17-2.19Å), substantiates the expectation that the electron withdrawing substituents would enhance the π-acceptor ability (C_2F_4>C_2H_4) of a particular alkene.

One could, therefore, think of the true structure of these compounds as a resonance hybrid of two extreme representations (Fig. 4.29) of bonding.

A considerable amount of X-ray structural data, which has become available in more recent years, supports the following general conclusions:

(a) the C—C bond length in the alkene is increased on complexation.

(b) the planar alkene molecule becomes non-planar on complexation with the substituents bonding away from the metal as well as other ligands,

(c) the C—C bond is perpendicular to the molecular plane in 16 electron molecules of the type ML_3 (alkene), e.g., $[PtCl_3(C_2H_4)]^{-}$, and.

(d) the carbon atoms of the ligated alkene molecules are symmetrically situated (i.e., equidistant) with respect to the central metal atom.

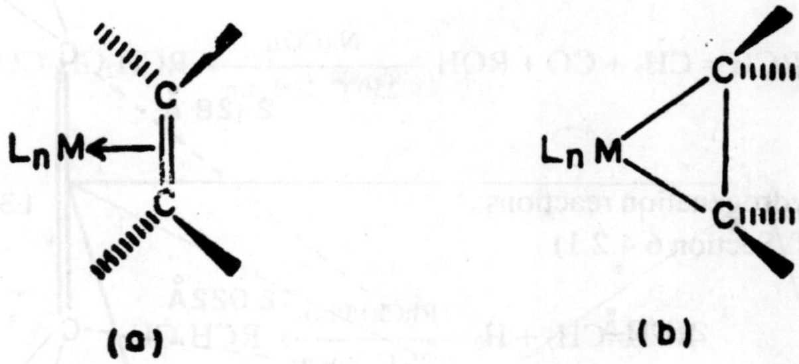

Fig. 4.29 Main resonating forms of metal-alkene complexes.

4.3.2.4 Chemical Properties

(a) *Stability of Metal-Alkene Complexes.* Alkenes are modest donor ligands for metal ions, being readily replaced by powerful ligands such as cyanide and tertiary phosphines as well as halide anions, in high concentration. The qualitative information on the stability of, for example, platinum (II)-alkene complexes, indicates that the coordinating ability of alkenes decreases with increasing substitution of the alkenic double bond in the order: C_2H_4>PhCH=CH_2> Ph_2 C=CH_2>Ph (Me) C=CH_2. Further studies have showed that the stability of platinum (II)-alkene complexes decreases as the overall charge on the complex ion becomes more positive, indicating that when π-bonding is greatly reduced, as in cationic complexes, the stability of the platinum (II)-alkene bond is greatly decreased, despite the enhancement of the strength of σ–bond between the alkene and the metal. For example, the platinum (II)-alkene bond strength is greater in the bromo than in the chloro complexes, probably due to the fact that the $PtBr_3^-$ group is a poorer π-acceptor and a better σ-donor than the $PtCl_3^-$ group. It is also found that in addition to electronic factors, steric factors are also important in determining the strength of the metal-alkene bonds.

(b) *Reactions Involving Metal-Alkene Intermediates in Homogeneous Catalyzed Reactions.* The consequence of alkene-metal interaction (Fig. 4.24) include a lowering of the electron density of the π-bonding orbital and an increase in the electron density of the π^* orbital. The overall effect of alkene coordination a metal centre brings about excitation of alkene π-bonding electrons to the π^* level resulting in activation of the alkene molecule, which tends to undergo a wide variety of reactions, many of which are of great industrial importance. Some typical examples involving metal-alkene intermediates in the homogeneous catalyzed organic syntheses are illustrated as follows:

(1) Hydroformylation or *oxo* reaction:
(cf. Section 6.4.2.3)

$$RCH=CH_2+CO+H_2 \xrightarrow[150^\circ C,\ 100\text{–}400\ atm]{Co_2(CO)_8} RCH_2CH_2CHO$$

(2) **Carboxylation reactions:**

$$RCH=CH_2 + CO + ROH \xrightarrow[250^\circ C,\ 200\ atm]{Ni(CO)_4} RCH_2CH_2COOR$$

(3) **Hydrogenation reactions:**
(cf. Section 6.4.2.1)

$$RCH{=}CH_2 + H_2 \xrightarrow[25^\circ C,\ 1\ atm]{RhCl(PPh_3)_3} RCH_2CH_2$$

(4) **Hydrosilylation reactions:**
(cf. Section 6.4.2.2)

$$R_3SiH + CH_2=CH_2 \xrightarrow[0^\circ C,\ 1\ atm]{Co_2(CO)_8} R_3SiCH_2CH_3$$

(5) **Oligomerization of alkenes:**
(cf. Section 6.4.2.6)

$$2CH_2{=}CH_2 \xrightarrow[55^\circ C,\ 500\ atm]{RhCl_3/HCl} CH_3CH{=}CHCH_3$$

(6) **Dismutation (or metathesis) of alkenes:**
(cf. Section 6.4.2.7)

$$2CH_3CH=CH_2 \xrightarrow{Mo(CO)_6\ \text{on alumina or}\ WCl_6 + LiBu} CH_2=CH_2 + CH_3CH{=}CHCH_3$$

(7) **Isomerization of alkenes:**
(cf. Section 6.4.2.5)

$$CH_3CH_2CH=CH_2 \xrightarrow{[RhCl_2(CO)_2]/HCl} CH_3CH{=}CHCH_3$$

Reactions of this type play a major role in the industrial uses of organotransition metal compounds and will be discussed in detail in Chapter 6.

(c) *Stoichiometric reactions of transition metal alkene complexes*: (i) Nucleophilic attack on η^2-alkene complexes. The nucleophilic attack can occur in three different ways: (a) the nucleophile displaces the alkene, (b) the nucleophile attacks the metal but displaces ligands other than the alkene, and (c) the nucleophile attacks at an alkenic carbon of the alkene. These general types of reactions are summarized in Fig. 4.30.

$$K^+[PtCl_3(C_2H_4)]^- \xrightarrow{CN^-} K_2[Pt(CN)_4]$$
$$K^+[PtCl_3(C_2H_4)]^- \xrightarrow{CO} K[PtCl_3(CO)]$$
$$K^+[PtCl_3(C_2H_4)]^- \xrightarrow{RC\equiv CR} K[PtCl_3(RC\equiv CR)]$$
$$K^+[PtCl_3(C_2H_4)]^- \xrightarrow{C_2D_4} K[PtCl_3(C_2D_4)]$$
$$K^+[PtCl_3(C_2H_4)]^- \xrightarrow{RNH_2} \textit{trans}\text{-}[PtCl_2(RNH_2)(C_2H_4)]$$
$$K^+[PtCl_3(C_2H_4)]^- \xrightarrow{H_2O} K[PtCl_2(H_2O)(C_2H_4)] \xrightarrow{OH^-} Pt + KCl + 2HCl + CH_3CHO$$
$$K^+[PtCl_3(C_2H_4)]^- \xrightarrow{PR_3} \textit{trans}\text{-}[PtCl_2(PR_3)_2]$$
$$K^+[PtCl_3(C_2H_4)]^- \xrightarrow{PhCH=CH_2} [PtCl_3(PhCH=CH_2)]^-$$

Fig. 4.30 Some reactions of Zeise salt, $K^+[PtCl_3(\eta^2—C_2H_4)]^-$

Further examples of displacement of coordinated alkenes by a variety of nucleophiles are illustrated as follows:

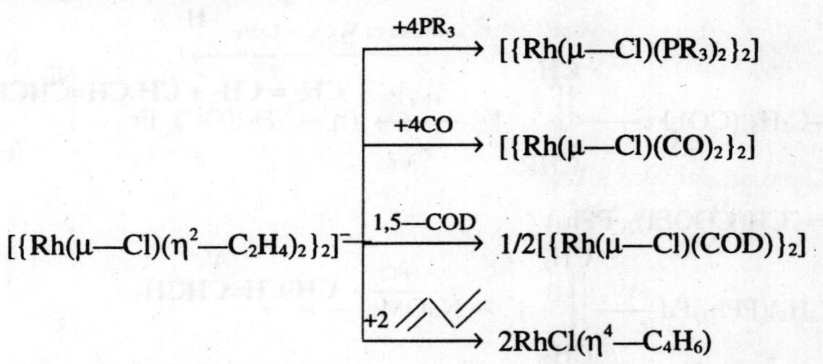

$$Rh(\eta^2—C_2H_4)_2(\eta^5—C_5H_5) \xrightarrow{CO,\ (—C_2H_4),\ h\nu} Rh(CO)(\eta^2—C_2H_4)(\eta^5—C_5H_5) \xrightarrow[h\nu]{+CO} Rh(CO)_2(\eta^5—C_5H_5) + C_2H_4$$

$$Rh(\eta^2—C_2H_4)_2(\eta^5—C_5H_5) + SO_2 \xrightarrow[-80^\circ C]{} Rh(SO_2)(\eta^2—C_2H_4)(\eta^5—C_5H_5) + C_2H_4$$

$$\mathrm{Ni(PPh_3)_2(\eta^2\text{—}C_2H_4)} \xrightarrow{\mathrm{PhCH{=}CH_2}} \mathrm{Ni(PPh_3)_2(\eta^2\text{—}PhCH{=}CH_2)}$$

$$\mathrm{Ni(PPh_3)_2(\eta^2\text{—}C_2H_4)} \xrightarrow{\mathrm{CH_3C{\equiv}CCH_3}} \mathrm{Ni(PPh_3)_2(\eta^2\text{—}CH_3C{\equiv}CCH_3) + C_2H_4}$$

Coordination of alkenes to metal centres make them highly susceptible to nucleophilic attack. The nucleophilic attack on alkene complexes can either by direct attack of the nucleophile on the alkene or by an initial attack of the nucleophile at the metal centre followed by a transfer on the coordinated alkene. Evidence for the first path has been derived from the following types of reactions:

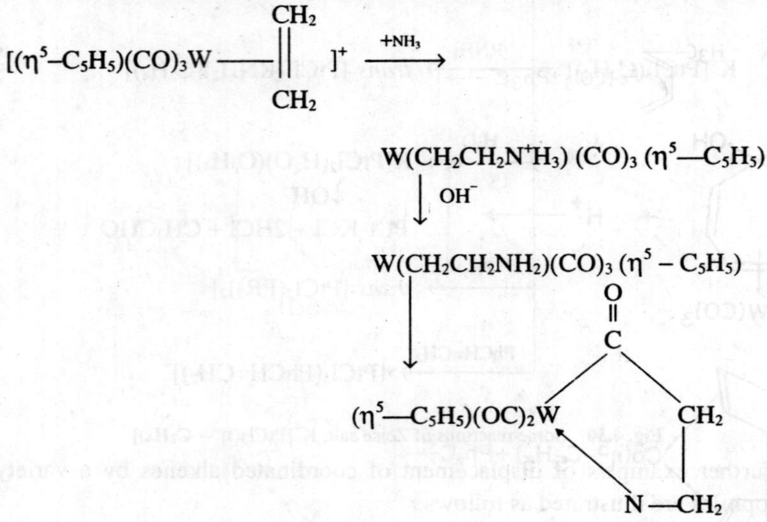

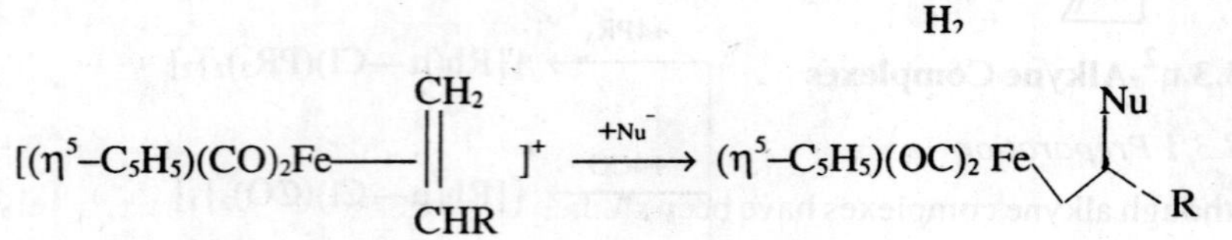

($Nu^- = {}^-CH(COOEt)_2$, PPh_3)

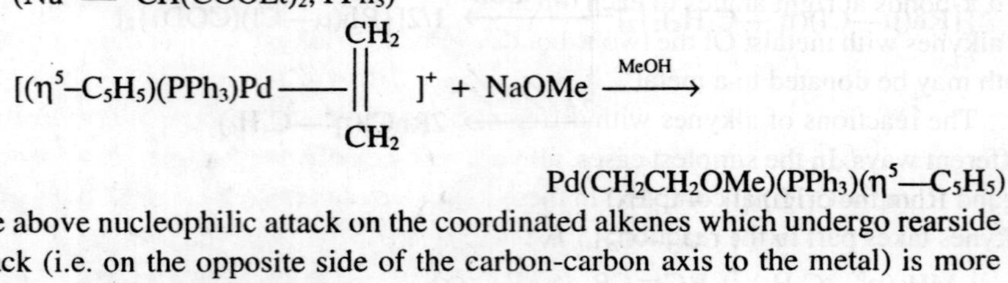

The above nucleophilic attack on the coordinated alkenes which undergo rearside attack (i.e. on the opposite side of the carbon-carbon axis to the metal) is more pronounced for cationic species.

The evidence for the second path, i.e., the initial attack of the nucleophile followed by a transfer on the coordinated alkene has been deduced from kinetic studies and has been discussed in Section 6.4.2.4, dealing with Pd(II) catalyzed oxidation of C_2H_4 to CH_3CHO.

(ii) *Reactions with electrophiles.* In contrast with the ready susceptibility of coordinated alkenes to undergo attack by nucleophiles, attack by electrophiles is relatively uncommon. Indeed, such properties of coordinated alkenes contrast markedly with the reactions of free alkenes, which normally proceed by electrophilic addition. Electrophilic reactions, however, may be illustrated by direct protonation and hydrogen abstraction reactions, which result in a change in the hydrocarbon ligand and its attachment to the metal. A few illustrative examples are:

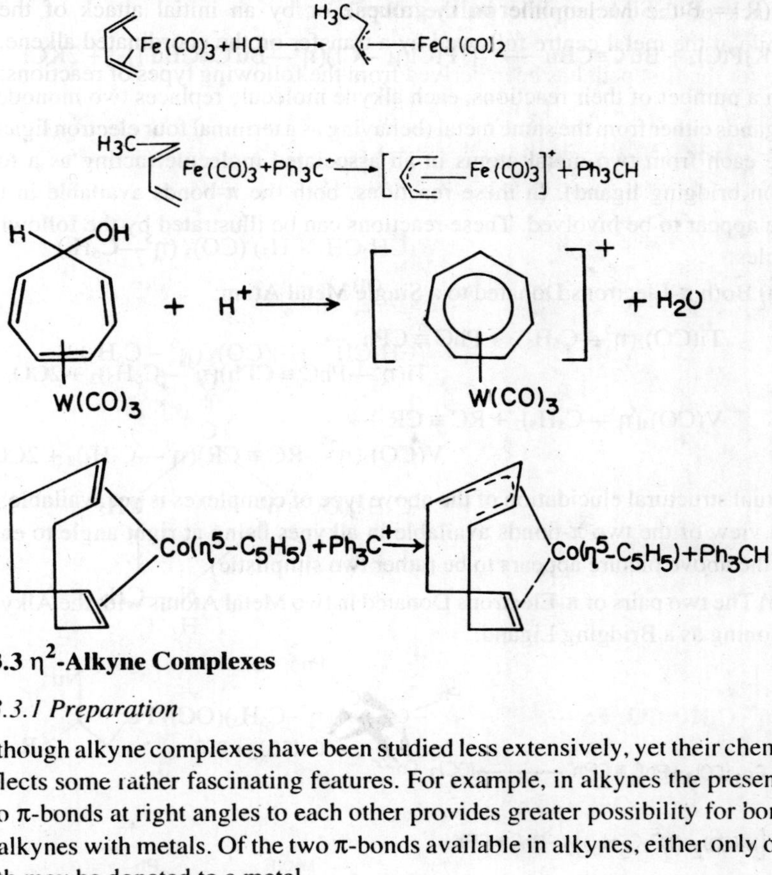

4.3.3 η^2-Alkyne Complexes

4.3.3.1 Preparation

Although alkyne complexes have been studied less extensively, yet their chemistry reflects some rather fascinating features. For example, in alkynes the presence of two π-bonds at right angles to each other provides greater possibility for bonding of alkynes with metals. Of the two π-bonds available in alkynes, either only one or both may be donated to a metal.

The reactions of alkynes with different transition metal species can occur in different ways. In the simplest cases, alkynes replace only one two-electron donor ligand from the original complex. In these, only one of the two π-electron pairs of alkynes takes part in the reaction:

$$MH_3(\eta^5{-}C_5H_5)_2 + RC \equiv CR \rightarrow MH(\eta^2{-}RC \equiv CR)(\eta^5{-}C_5H_5)_2 + H_2$$

(M = Nb, Ta)

$$M(CO)_6 + RC = CR' \xrightarrow{\text{U.V.}} M(CO)_5(RC \equiv CR') + CO$$

(M = Mo, W)

$$Mo(CO)(\eta^5\text{–}C_5H_5)_2 + RC \equiv CR \rightarrow Mo(\eta^2\text{–}RC \equiv CR)(\eta^5\text{–}C_5H_5)_2 + CO$$

$$Mn(CO)_3(\eta^5\text{–}C_5H_5) + PhC \equiv CPh \xrightarrow{\text{U.V.}} Mn(CO)_2(\eta^2\text{–}PhC \equiv CPh) + CO$$

$$Fe(CO)_5 + RC{\equiv}CR \rightarrow Fe(CO)_4(\eta^2\text{—}RC{\equiv}CR) + CO$$

($R = Bu^t$, $SiMe_3$ or other bulky groups)

$$K_2PtCl_4 + Bu^tC{\equiv}CBu^t \rightarrow \tfrac{1}{2}[\{PtCl(\mu\text{—}Cl)(\eta^2\text{—}Bu^tC{\equiv}CBu^t)\}_2] + 2KCl$$

In a number of their reactions, each alkyne molecule replaces two monodentate ligands either from the same metal (behaving as a terminal four electron ligand) or one each from two metal atoms in an associated molecule (acting as a four electron bridging ligand). In these reactions, both the π-bonds available in the alkyne appear to be involved. These reactions can be illustrated by the following examples:

(a) Both π-Electrons Donated to a Single Metal Atom:

$$Ti(CO)_2(\eta^5\text{—}C_5H_5)_2 + PhC \equiv CPh \rightarrow Ti(\eta^2\text{—}PhC \equiv CPh)(\eta^5\text{—}C_5H_5)_2 + 2CO$$

$$V(CO)_4(\eta^5\text{—}C_5H_5)_2 + RC \equiv CR \rightarrow V(CO)_2(\eta^2\text{—}RC \equiv CR)(\eta^5\text{—}C_5H_5)_2 + 2CO$$

(no actual structural elucidation of the above type of complexes is yet available; in fact in view of the two π-bonds available in alkynes being at right angle to each other, the above picture appears to be rather two simplistic).

(b) The two pairs of π-Electrons Donated in two Metal Atoms with the Alkyne Functioning as a Bridging Ligand:

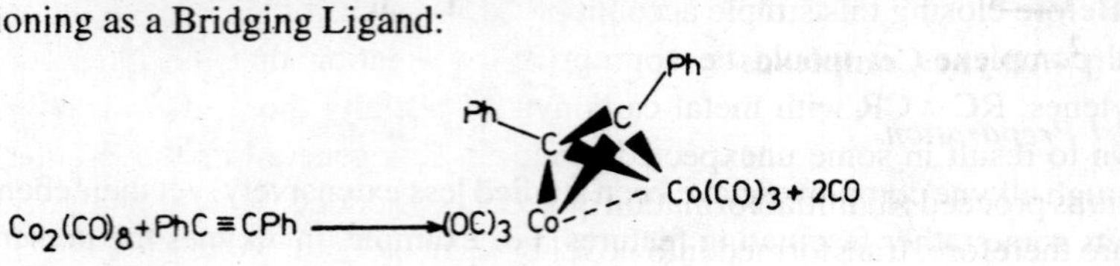

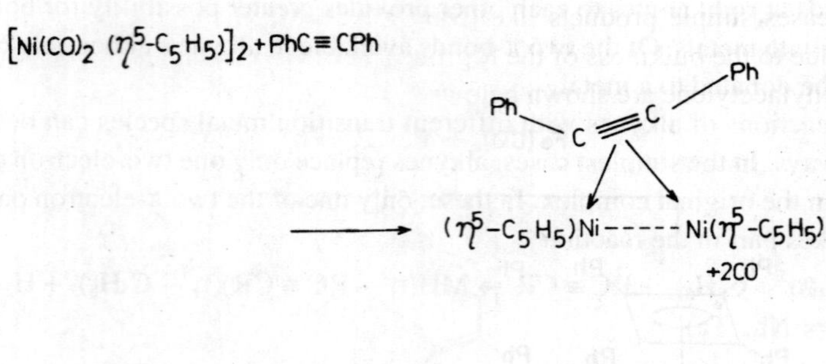

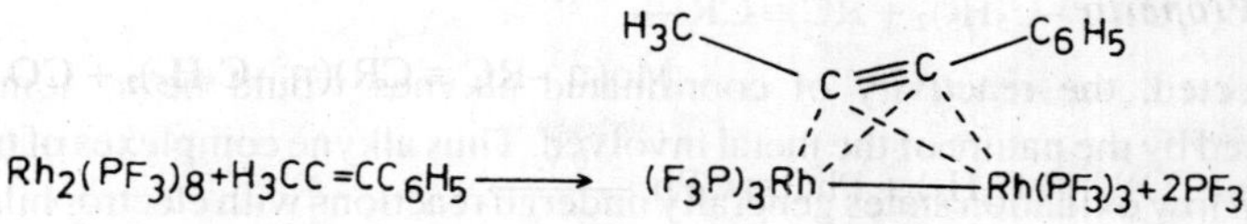

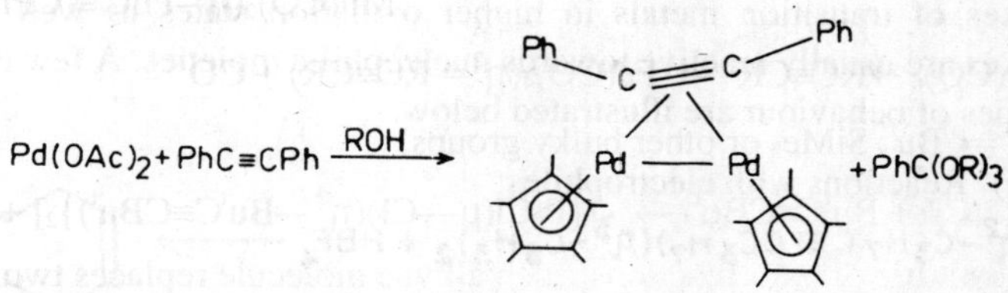

Reactions of metal carbonyls with alkynes often form polynuclear metal clusters:

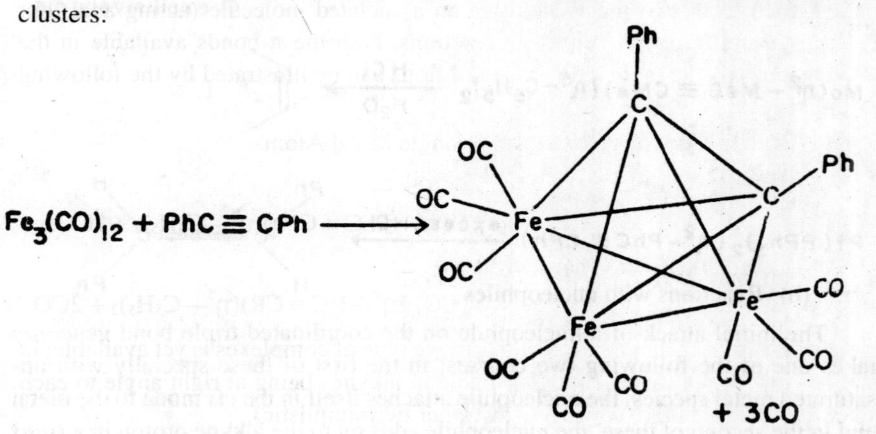

Before closing this simple account of the reactions of alkynes with transition metal complexes, it would be appropriate to mention that the interactions of acetylenes, RC ≡ CR with metal carbonyls (especially those of iron) have been shown to result in some unexpected products. It is generally believed that these reactions proceed *via* initial formation of acetylene adducts, which are highly labile and are therefore, transformed into novel products, except when R = Bu^t, $SiMe_3$, in which cases simple products like $(Me_3SiC{\equiv}CSiMe_3)Fe(CO)_4$ can be isolated probably due to the bulkiness of the R groups. A few typical reactions of $Fe(CO)_5$ with diphenylacetylene are shown below:

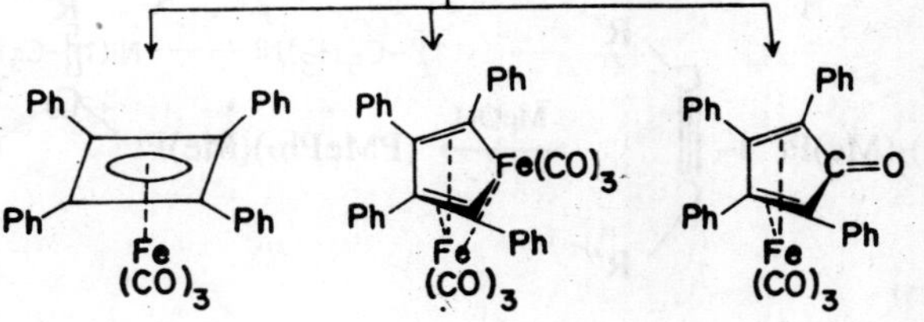

4.3.3.2 Properties

As expected, the reactivity of coordinated alkynes would be to some extent influenced by the nature of the metal involved. Thus alkyne complexes of transition metals in low oxidation states generally undergo reactions with electrophiles, while complexes of transition metals in higher oxidation states as well as cationic complexes are usually reactive towards nucleophilic moieties. A few examples of these types of behaviour are illustrated below:

(a) Reactions with electrophiles:

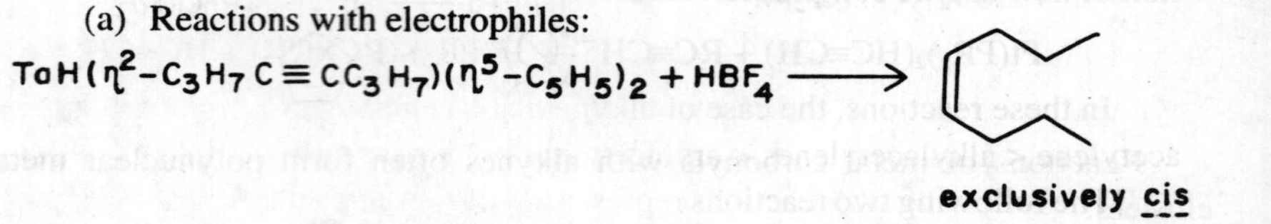

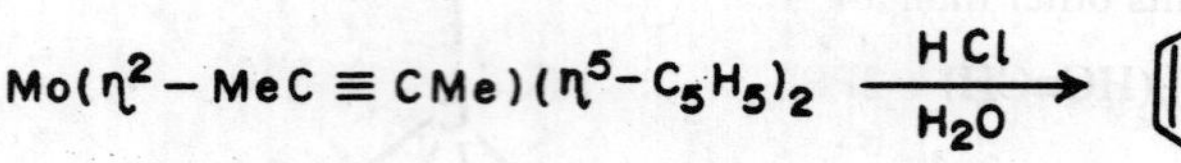

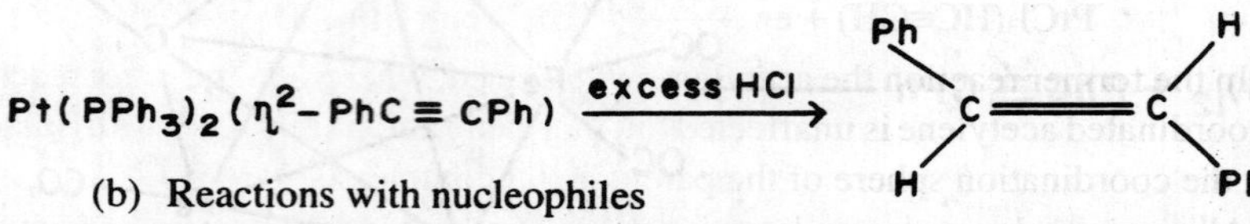

(b) Reactions with nucleophiles

The initial attack of a nucleophile on the coordinated triple bond generally takes one of the following two courses; in the first of these specially with unsaturated metal species, the nucleophile attaches itself in the *cis* mode to the metal and in the second of these, the nucleophile adds on to the alkyne proton in a *trans* mode. Some typical examples are illustrated below:

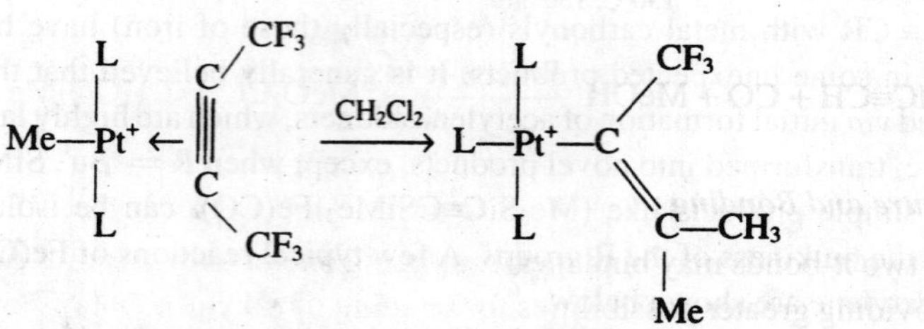

L = PMe₂Ph, AsMe₃

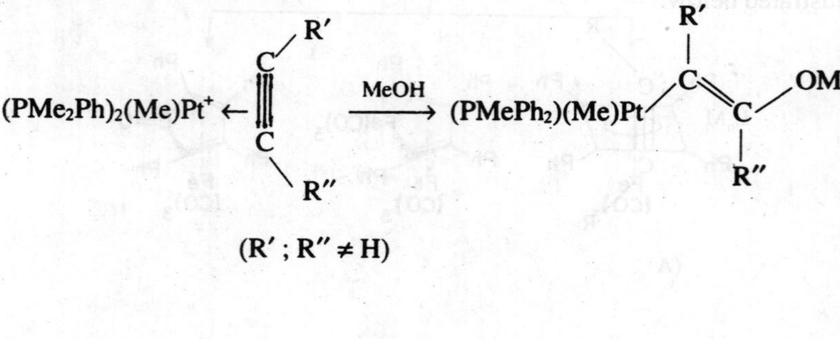

(R′ ; R″ ≠ H)

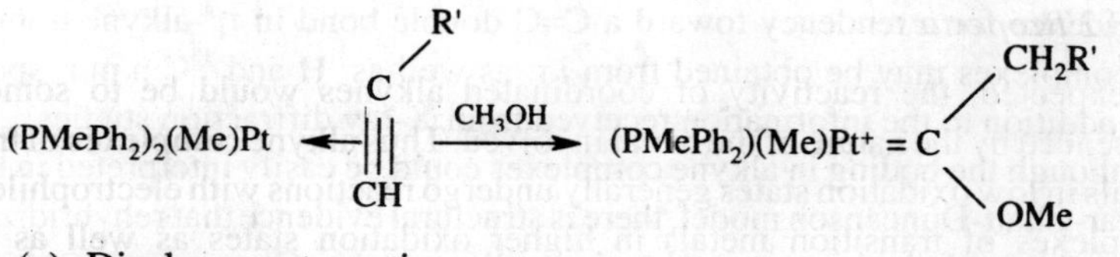

(c) Displacement reactions

The displacement of one alkyne by another has been exploited for the preparation of new alkyne complexes:

$$Pt(PR_3)_2(HC{\equiv}CH) + RC{\equiv}CH \rightarrow Pt(PR_3)_2(RC{\equiv}CH) + HC{\equiv}CH$$

In these reactions, the case of alkyne displacement can be represented as: acetylene < alkylacetylenes < arylacetylenes < nitroacetylenes.

The following two reactions represent two types of displacements from alkyne complexes with reagents other than alkynes:

$$Pt(PPh_3)_2(HC{\equiv}CH) + \underset{(excess)}{2PPh_3} \rightarrow Pt(PPh_3)_4 + HC{\equiv}CH$$

$$PtCl_3(HC{\equiv}CH) + en \xrightarrow{THF} [PtCl(en)(HC{\equiv}CH)]^+Cl^-$$

In the former reaction the acetylene is displaced, whereas in the latter reaction the coordinated acetylene is unaffected but inorganic anions (e.g. Cl^-) are displaced from the coordination sphere of the parent metal complex.

Alkyne complexes also undergo a number of other interesting reactions, involving, for example, oligomerization (cf. Section 6.4.2.6) and hydrometallation, e.g.:

$$RC{\equiv}CH \xrightarrow{Pd(II)} [RC{\equiv}C{-}Pd^{II}] \xrightarrow[R'OH]{CO} RC{\equiv}CCO_2R'$$

$$HC{\equiv}CH + H_2O \xrightarrow[150^\circ C,\ 130\ atm]{Ni(CO)_4} H_2C{=}CHCO_2H$$

$$HC{\equiv}CH + CO + MeOH \xrightarrow[O_2]{PdCl_2/thiourea} MeO_2CCH{=}CHCO_2Me$$

4.3.3.3 Structure and Bonding

Alkynes with two π-bonds may bind metals utilizing one or both sets of π-electron pairs, thus providing greater possibilities for bonding by alkynes. Out of a number of structural types the two extreme cases (A & B) of η^2-alkyne bonding are illustrated below:

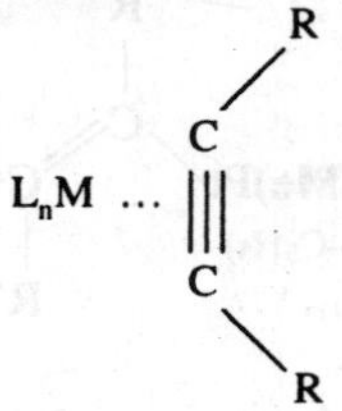

(A)

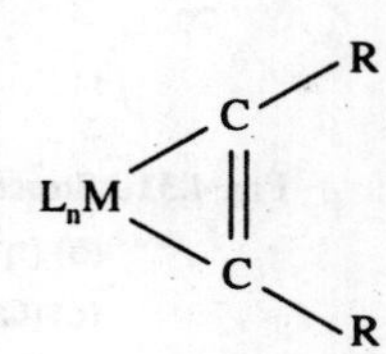

(B)

Evidence for a tendency toward a C=C double bond in η^2-alkyne transition metal complexes may be obtained from i.r. as well as ^1H and ^{13}C n.m.r. spectral data in addition to the information received from X-ray diffraction studies.

Although the boding in alkyne complexes could be easily interpreted in terms of Dewar-Chatt-Duncanson model, there is structural evidence that rehybridization occurs to form relatively strong σ-bonds, as illustrated in C:

(C)

In transition metal derivatives, alkynes may function as (i) a *dihapto*-terminal ligand (Fig. 4.31a) or (ii) a bridging ligand where the alkyne ligand forms a bridge between two (Fig. 4.31b & c) or more atoms (Fig. 4.31d). Another form of bridge is shown in Figure 4.31e.

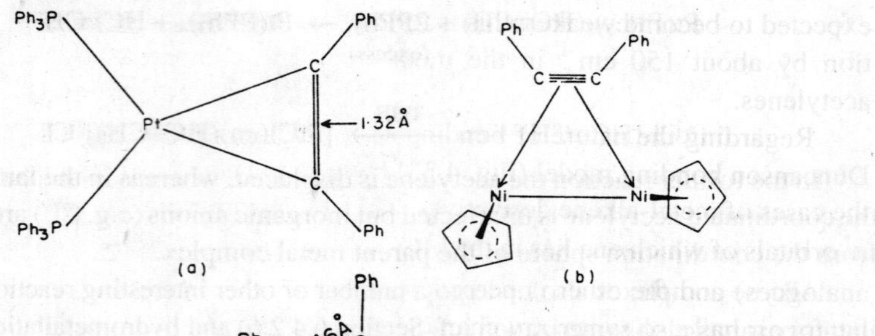

Fig. 4.31 Structures of (a) $Pt(PPh_3)_2(\eta^2—PhC \equiv CPh)$,
(b) $(\eta^5—C_5H_5)Ni(\mu—PhC\equiv CPh)Ni(\eta^5—C_5H_5)$
(c) $(OC)_3Co(\mu—\eta^2—PhC \equiv CPh)Co(CO)_3$
(d) $Fe_3(CO)_9(\eta^2—PhC \equiv CPh)_2$ and
(e) $(\eta^5— C_5H_5)(OC)Rh(\mu— \eta^1— CF_3C \equiv CCF_3)Rh(CO)(\eta^5— C_5H_5)$

It would not be out of place to mention that in addition to the η^2-complexes, alkynes can form σ-bonded alkynylide derivatives (cf. Section 4.2.1). As the alkynylide $RC{\equiv}C^-$ is isoelectronic with NC^- and CO ligands, η^1-alkynyl transition metal complexes containing M—C≡CR groups, may also be expected to have both σ-donor and α-acceptor character (Fig. 4.32).

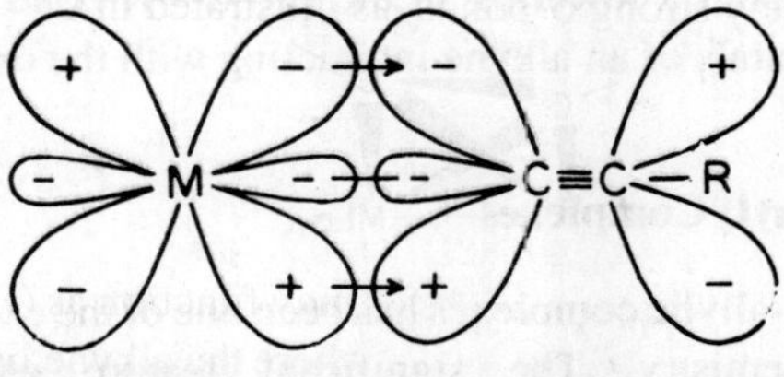

Fig. 4.32 Bonding in transition metal-alkynyl complexes.

In the formation of a σ-bond of the above type, the C≡C bond should be expected to become weaker; this is reflected in the lowering of the νC≡C absorption by about 150 cm^{-1} in the metal acetylide compared to that in the free acetylenes.

Regarding the nature of bonding of η^2-alkyne complexes, the Dewar-Chatt-Duncanson bonding model (Fig. 4.33) appears to be applicable in these also as in the cases of metal-alkene derivatives. The alkyne molecule has two pairs of π and π^* orbitals of which one lies in the MCC plane ($\pi_{\parallel}$ and $\pi^*_{\parallel}$, similar to that in alkene analogues) and the other is perpendicular to the MCC plane ($\pi_{\perp}$ and $\pi^*_{\perp}$). This latter pair has also symmetry characteristics appropriate for bond formation with transition metal *d* orbitals.

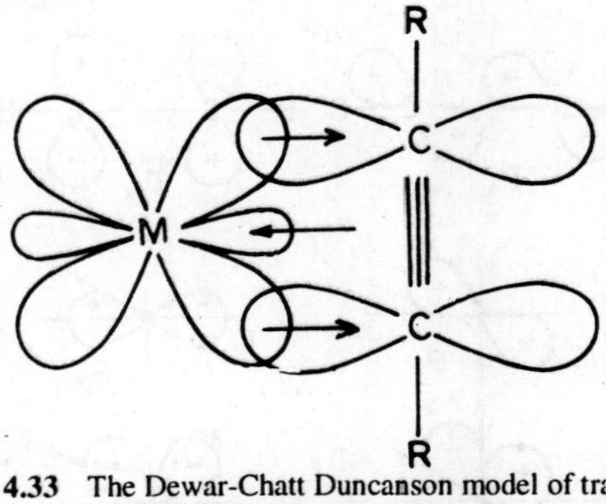

Fig. 4.33 The Dewar-Chatt Duncanson model of transition metal-alkyne bonding.

The strength of $\pi^*_{\perp}$ interaction involving a δ type overlap is further enhanced by the overlap involving the filled $\pi_{\perp}$ molecular orbitals. Thus the alkyne molecules are able to function both as π-acceptor as well as π-donor ligands. For the simple electron counting rules, care has to obviously taken to see if the alkyne is functioning as a two or four electron donor ligand. For example, molecules like $Mo(S_2CNR_2)_2(CO)(HC{\equiv}CH)$, $MoO(S_2CNR_2)_2(HC{\equiv}CH)$, $MoMe(CO)(HC{\equiv}CH)(\eta^5{-}C_5H_5)$, and $Pt(HC{\equiv}CH)_2$ appears to conform to 18-electron rule only if the alkyne is postulated to function as a four-electron ligand.

As in the case of alkene complexes (cf. Section 4.3.2.4), coordination of an alkyne to a metal will result in a depletion of electron density in the π-bonding orbitals, and in an increase of density in the π^*-antibonding orbital. This should bring about the lowering of the bond-order in the alkyne, resulting in an increase in the carbon-carbon bond length. Further, the substituents are pushed away from the metal bringing about a lowering in the C—Ĉ—R bond angles. Figure 4.34 shows the atomic orbitals of an alkyne interacting with the orbitals of a transition metal.

4.3.4 η^3-Allyl (or Enyl) Complexes

The preparation of η^3-allylic complexes has been one of the exciting developments in coordination chemistry. The significant feature of the allyl group, $H_2C{=}CH{-}CH_2{-}$, is its dual (mono- as well as *tri-hapto*) character of ligation. A large number of η^3-allyl complexes of almost all $d^1 - d^8$ metals in the three (3*d*, 4*d* and 5*d*) transition series, with the possible exception of technetium, have been described in recent years.

The bonding of the allylic group to the metal is believed to be similar to that of metal-alkene as explained on Dewar-Chatt-Duncanson model. For example, the allylic group donates its π-electrons to the metal atom forming a σ-type bond and accepts electrons into its antibonding π^*-molecular orbitals in a π-type bond. These allylic complexes, therefore, involve a three electron delocalized allylic system. $H_2C \doteq CH \doteq CH_2$, as illustrated in Figure 4.35. The η^3-allyl group is in general symmetrically displaced around the transition metal.

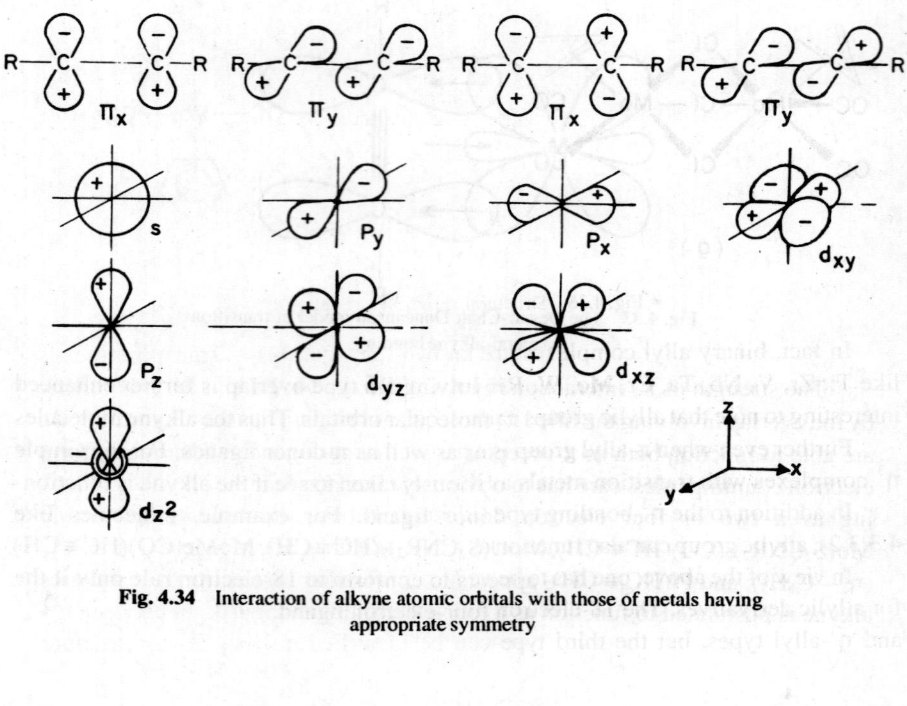

Fig. 4.34 Interaction of alkyne atomic orbitals with those of metals having appropriate symmetry

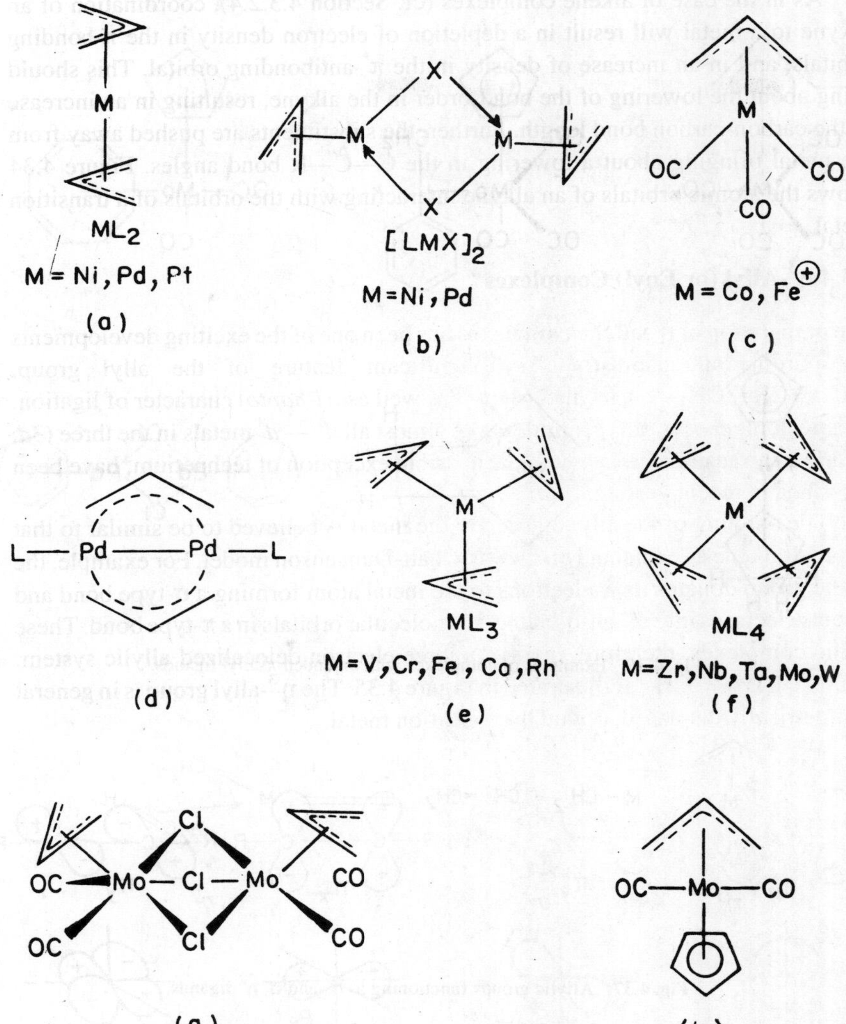

Fig. 4.35 Structural varieties in η^3–allyl

In fact, binary allyl complexes are known for a number of transition metals like Ti, Zr, V, Nb, Ta, Cr, Mo, W, Re, Fe, Co, Rh, Ir, Ni, Pd and Pt. It is further interesting to note that allylic groups can also act as bridging ligands (Figure 4.35d).

Further even when π-allyl group is a part of a larger ligand, it is known to form η^3-complexes with transition metals as illustrated in Fig. 4.36.

In addition to the η^3-bonding type (illustrated above and discussed in Section 4.3.3.2), allylic group can also function as a σ-ligand or as a σ, η^2 ligand (Fig. 4.37).

In view of the above, one has to be very careful while suggesting the structures for allylic derivatives. The 18-electron rule is useful in distinguishing between η^3- and η^1-allyl types, but the third type can be identified only by X-ray structure

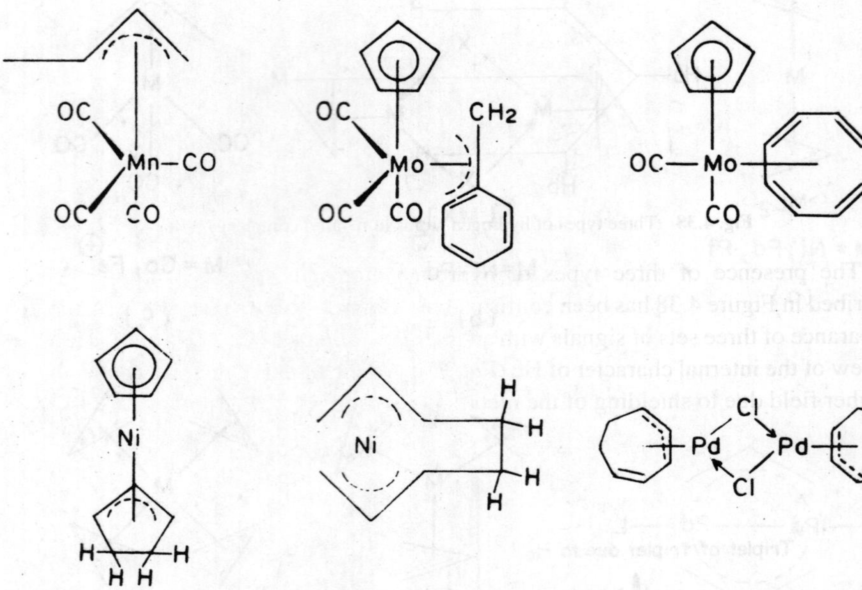

Fig. 4.36 Examples of complexes with extended π-allyl ligands.

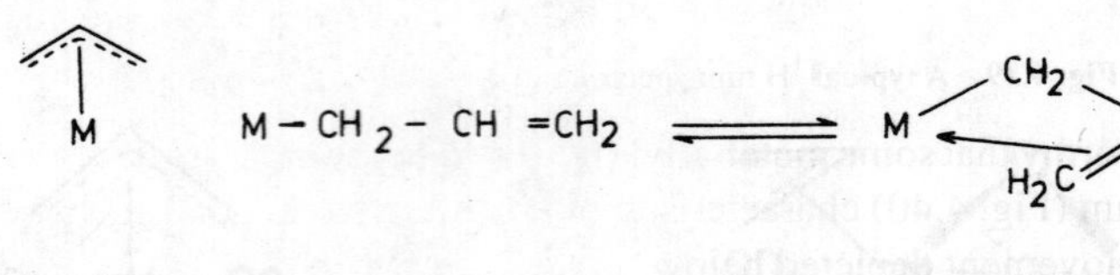

Fig. 4.37 Allylic groups functioning as σ- and σ, η^2-ligands.

determination. The σ, η^2-allyl derivatives are fluxional (cf. Chapter 5) with the metal changing positions from one side to another of the plane comprising the three allylic carbon atoms:

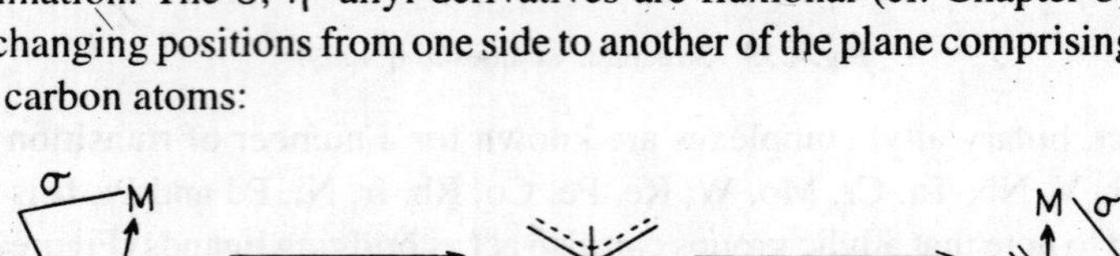

Just as alkene complexes show similarity to carbonyls, the delocalized allylic system $CH_2 \cdots CH \cdots CH_2$ acting as a formal three-electron donor can be likened to NO ligands in metal nitrosyl derivatives. These allylic derivatives have also been visualized as a sort of half sandwich, i.e., intermediate between alkene and the sandwich compounds.

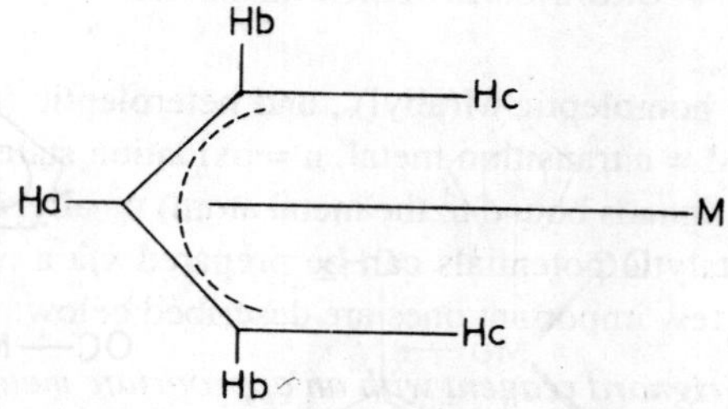

Fig. 4.38 Three types of hydrogen atoms in η^3-allyl complexes.

The presence of three types of hydrogen atoms in η^3-allylic complexes described in Figure 4.38 has been confirmed by ^{1}H n.m.r. studies (Fig. 4.39) by the appearance of three sets of signals with an intensity ratio of 1 : 2 : 2 (Ha : Hb : Hc). In view of the internal character of Hc (Fig. 4.38), their signals generally appear at a higher field due to shielding of the metal:

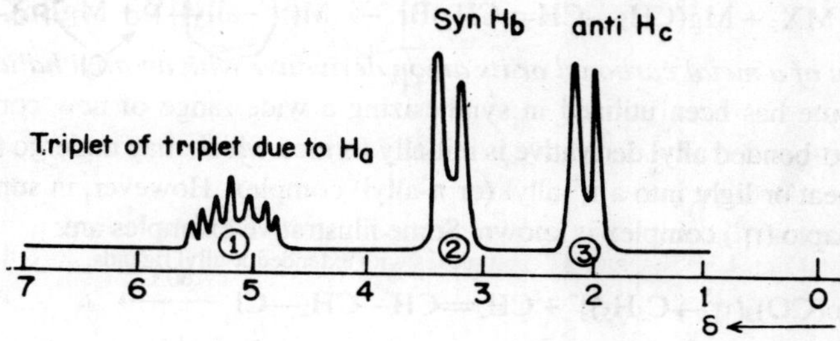

Fig. 4.39 A typical ^{1}H nmr spectrum of a π-allyl A_2M_2X group.

It is noteworthy that some metal-allyl (η^1 or σ-type) complexes give a different type of spectrum (Fig. 4.40) characteristic of an A_4X system, arising as a result of a head to tail movement depicted below:

$$\begin{array}{c} M—CH_2—CH=CH_2 \\ \updownarrow \\ M—CH_2—CH=CH_2 \end{array} \quad \text{(Head to tail movement)}$$

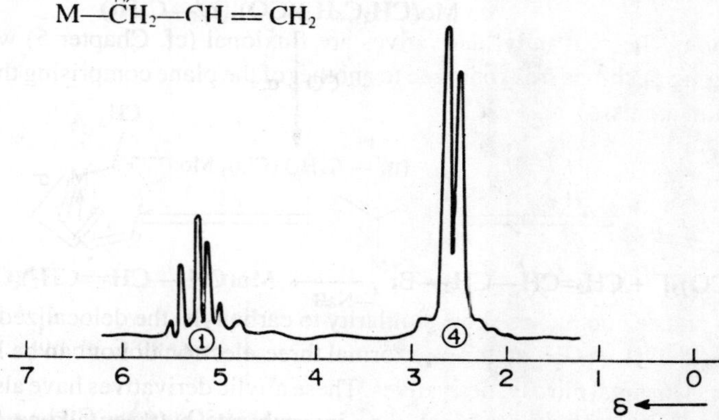

Fig. 4.40 ^{1}H nmr spectrum of some allyl-metal complexes of an A_4X system.

4.3.4.1 Preparation

A wide range of both homoleptic M(allyl)$_n$, and heteroleptic (or mixed ligand), M(allyl)$_{n-a}$X$_a$ (where M = a transition metal, n = oxidation state of the metal, a = number of the anionic ligands bound to the metal atom) η^3-allyl complexes having both synthetic and catalytic potentials can be prepared via a number of general routes, out of which a few important ones are described below:

(a) *Reaction of allyl Grignard reagent with an appropriate metal halide, e.g.:*

$$MX_n + nMg(CH_2{-}CH{=}CH_2)Br \xrightarrow{\text{ether}} M(\eta^3\text{-allyl})_n + nMgBrX$$

(M = Ti, Zr, Hf, Th, V, Nb, Tb, Cr, Mo, W, Fe, Co and Pt;
n = valency of the metal).

In some cases particularly for divalent metals (like Ni and Pd), the above reactions have been carried out in equimolar ratios and compounds of the type M(η^3–allyl)X have also been isolated:

$$MX_2 + Mg(CH_2{-}CH{=}CH_2)Br \rightarrow M(\eta^3\text{-allyl})X + MgBrX$$

(b) *Reaction of a metal carbonyl or its anion derivative with an allyl halide*

This route has been utilized in synthesizing a wide range of new complexes. Usually the σ-bonded allyl derivative is initially formed which may undergo transformation by heat or light into a η^3-allyl (or π-allyl) complex. However, in some cases only the trihapto (η^3) complex is known. Some illustrative examples are:

$$Na^+[Mo(CO)_3(\eta^5{-}C_5H_5)]^- + CH_2{=}CH{-}CH_2{-}Cl \xrightarrow{80^\circ C} Mo(CH_2{-}CH{=}CH_2)(CO)_3(\eta^5{-}C_5H_5) + NaCl$$

$$\downarrow h\nu, -CO$$

$$Mo(CO)_2(\eta^3{-}C_3H_5)(\eta^5{-}C_5H_5)$$

$$Na^+[Mo(CO)_5(\eta^5{-}C_5H_5)]^- + C_6H_5CH_2Cl \xrightarrow[-NaCl]{} Mo(CH_2C_6H_5)(CO)_3(\eta^5{-}C_5H_5)$$

$$\downarrow -CO,\ u.v.$$

$(\eta_5 - C_5H_5)(CO)_2$ Mo

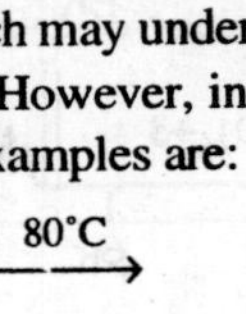

$$Na^+[Mn(CO)_5]^- + CH_2{=}CH{-}CH_2{-}Br \xrightarrow[-NaBr]{} Mn(CH_2{-}CH{=}CH_2)(CO)_5$$

$$\downarrow \text{heat, } 80^\circ C \text{ or } h\nu$$

$$Mn(CO)_4(\eta^3{-}C_3H_5) + CO$$

$$Na^+[Fe(CO)_2(\eta^5-C_5H_5)]^- + H_2C{=}CH_2-CH_2Cl \xrightarrow[-NaCl]{} Fe(CH_2-CH{=}CH_2)(CO)_2(\eta^5-C_5H_5)$$

$$\downarrow -CO \text{ u.v.}$$

$$Fe(CO)(\eta^3-C_3H_5)(\eta^5-C_5H_5)$$

$$Na^+[Co(CO)_4]^- + H_2C{=}CH-CH_2Br \xrightarrow[-NaBr]{} Co(CH_2-CH{=}CH_2)(CO)_4$$

$$\downarrow -CO$$

$$Co(CO)_3(\eta^3-C_3H_5)$$

$$Ni(CO)_4 + H_2C{=}CH-CH_2Cl \longrightarrow NiCl(\eta^3-C_3H_5) + 4CO$$

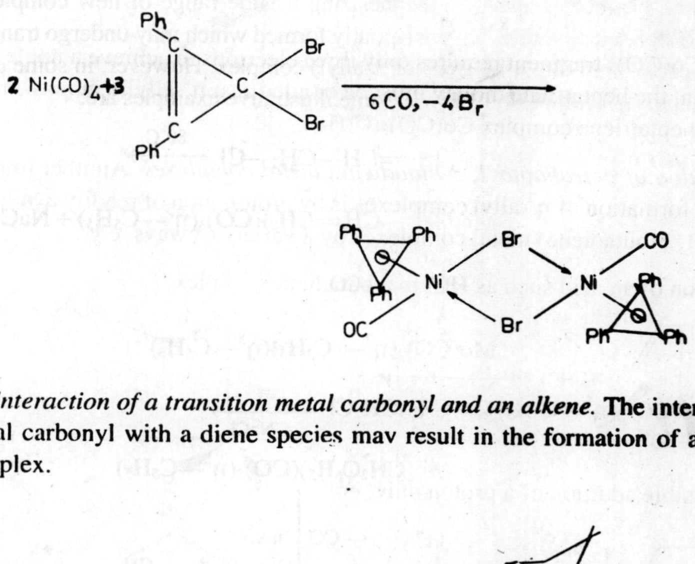

(c) *Interaction of a transition metal carbonyl and an alkene.* **The interaction of a metal carbonyl with a diene species mav result in the formation of an η^3-metal complex.**

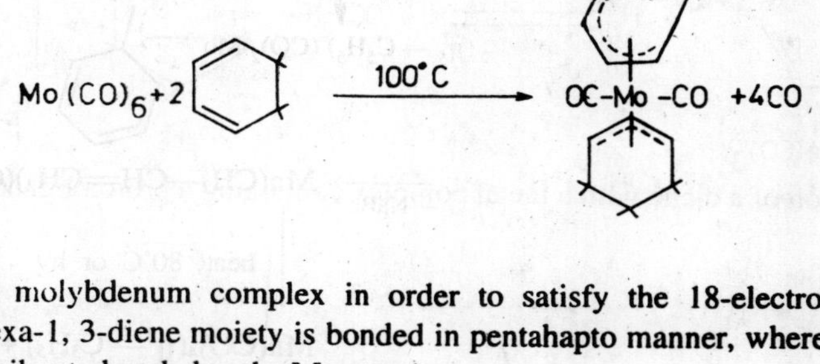

(In the molybdenum complex in order to satisfy the 18-electron rule, one cyclohexa-1, 3-diene moiety is bonded in pentahapto manner, whereas the other ring is trihapto bonded to the Mo atom).

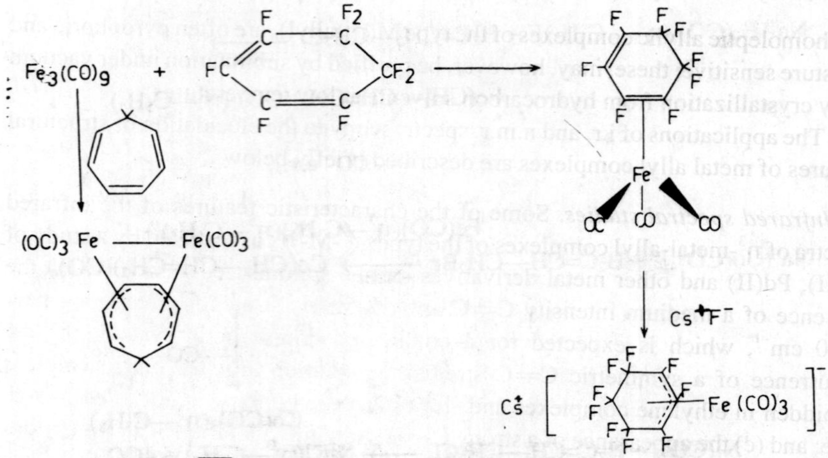

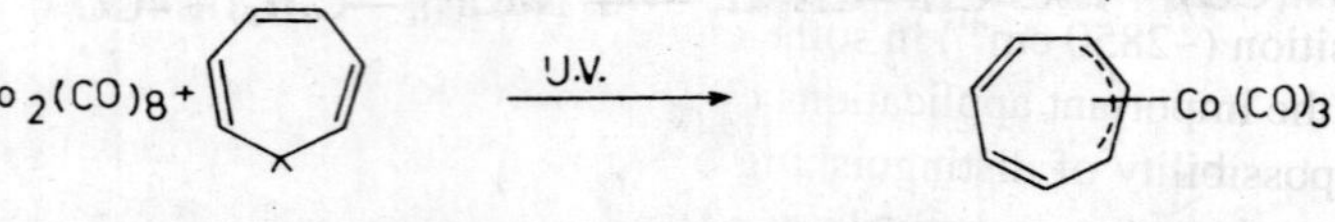

(Since $Co(CO)_3$ fragment requires only three electrons to achieve a noble gas configuration, the heptatriene moiety must be bonded as a η^2-allylic moiety in the above cycloheptatriene complex $Co(CO)_3(C_7H_6)$).

(d) *Protonation of (tetrahapto-1, 3-butadiene) metal complexes.* Another important route to formation of η^3-allyl complexes is by protonation of readily available (tetrahapto-1, 3-butadiene) metal complexes by a variety of ways, e.g.:

(i) by addition of an acid such as HCl to a butadiene complex,

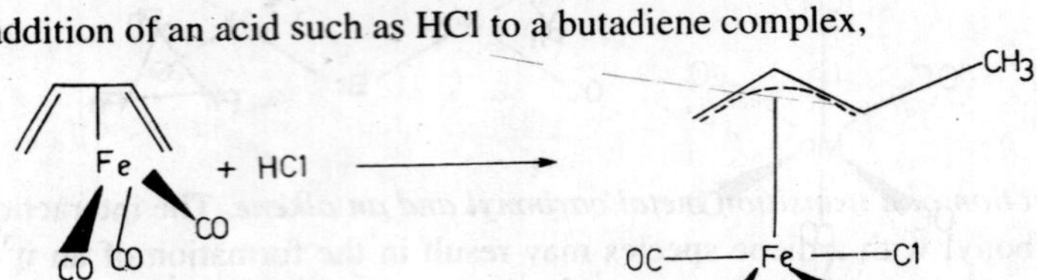

(ii) by reversible addition of a proton only, or

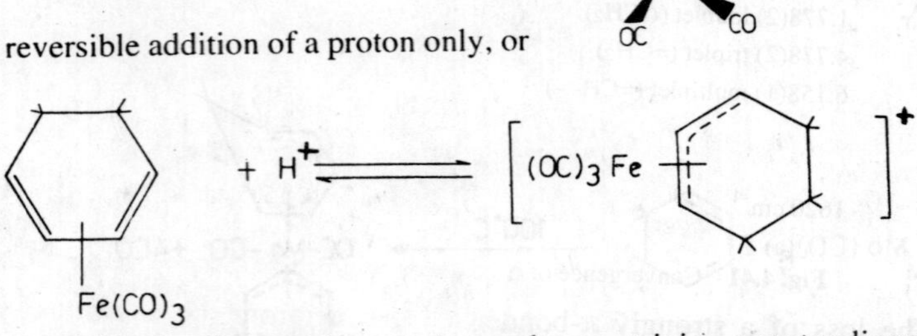

(iii) by reaction of a diene with a metal complex containing a proton ligand,

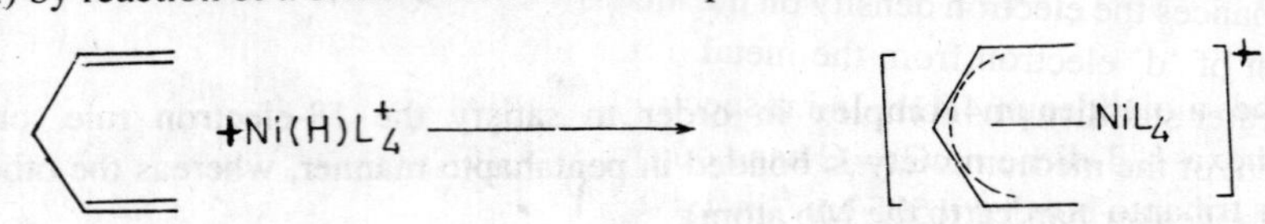

4.3.4.2 Physical Properties

The homoleptic allylic complexes of the type M(η^3-allyl)$_n$ are often pyrophoric and moisture sensitive; these, may, however, be purified by sublimation under vacuum or by crystallization from hydrocarbon solvents at low temperatures.

The applications of i.r. and n.m.r. spectroscopy to the elucidation of structural features of metal allyl complexes are described briefly below:

(a) *Infrared spectral studies.* Some of the characteristic features of the infrared spectra of η^3-metal-allyl complexes of the type, ⟨-M-⟩, as revealed by a study of Ni(II), Pd(II) and other metal derivatives can be summarized as below: (a) the presence of a medium intensity C=C antisymmetric stretching frequency near 1460 cm^{-1}, which is expected for a conjugated double bond system, (b) the occurrence of a symmetric C=C stretching vibration at 1020 cm^{-1}, which is forbidden in ethylene complexes and shows that the bonding is not of σ or σ-η^2 type; and (c) the appearance of a strong C—H stretching vibration at an exceptionally low position (~2850 cm^{-1}) in some allylic complexes.

One of the important applications of i.r. spectroscopy to metal allylic derivatives is the possibility of distinguishing between η^1 and η^3 type bonding, by the appearance or absence, respectively of a bond just above 1600 cm^{-1}. The conversion of a η^1-bonded allyl group into a η-arrangement is of greater interest from a mechanistic view point and has been more generally studied in detail by n.m.r. technique (cf. Chapter 5). An example of such a rearrangement which has been followed by i.r. and ^{1}Hn.m.r. spectroscopy is illustrated below:

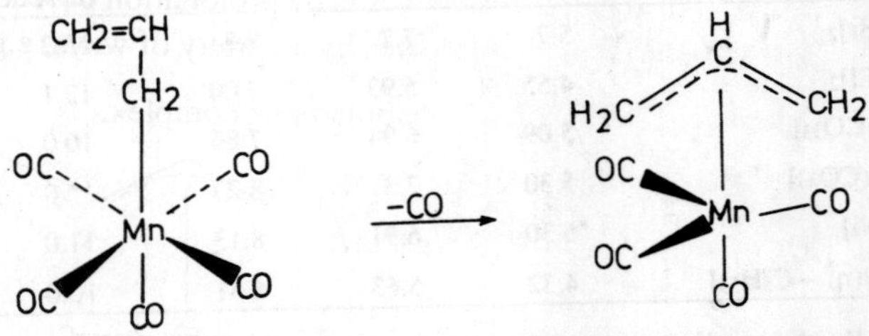

^{1}H n.m.r. : 1.778(2)doublet (αCH$_2$)
4.778(2) triplet (=CH$_2$)
6.158(1) multiplet (=CH—)

I.r.: 1620 cm^{-1}

(a)

^{1}H n.m.r. : 1.728(2) doublet (*exo* H*)
J = 14 Hz
2.698(2) doublet
(*endo* H**) J = 14 Hz
4.708(1) 9 lines

I.r.: 1505 cm^{-1}

(b)

Fig. 4.41 Convergence of σ-bonded allyl group into a π-arrangement

The loss of a strongly π-bonded CO group from the coordination sphere enhances the electron density on the metal atom resulting in a stronger back-donation of 'd' electron from the metal to the allylic π-system and contributes to its greater stability of 4.41b. In i.r. spectrum, this change is indicated by the replacement of the medium νC =C band due to 'free' double bond at 1620 cm^{-1} in 4.41a by a weaker band at 1505 cm^{-1} in 4.41b.

(b) *Proton magnetic resonance spectra*. The proton magnetic resonance spectra of transition metal-allylic complexes fall essentially into three classes: (1) the σ-bonded allyls, M—CH_2—CH= CH_2, give spectra which are expected from their formulation; (2) the π-allylic complexes, e.g., (π—C_3H_5)$Mn(CO)_4$, show spectra typical of A_2M_2X system (shown in Fig. 4.39) and (3) the fluxional allylic complexes, in which allyl groups have equivalent terminal hydrogens and exhibit spectra typical of A_4X system (Fig. 4.40). Proton n.m.r. studies of many η^3-allyl metal complexes have shown that the η^3-allyl ligands are stereochemically non-rigid. At lower temperature, the η^3-allyl ligand appears to be static (AA′BB′X pattern), while at higher temperature the spectrum collapses to a dynamic spectrum (A_4X pattern) owing to a rapid intramolecular rearrangement that causes AA′ (*anti*) and BB′ (*syn*) protons to be equivalent. A careful examination of the possible mechanisms, which could lead to the observed proton equilibration, has led to the accepted conclusion that the η^3-allyl group becomes σ-bonded to metal in the crucial transition state, i.e., a σ-allyl group. In this transition state, rotation about the M—C bond followed by reformation of the η^3 from accounts for the stereochemical non-rigidity. Equilibria of the type η^3-allyl ⇌ σ-allyl also are presumably crucial to the importance of metal-allyl complexes in organic synthesis and catalysis.

The ^{1}Hn.m.r. data for some typical η^3-allyl complexes are listed in Table 4.8.

Table 4.8 ^{1}H n.m.r. Data for Some η^3 Allyl Metal Complexes

Complex	H_a	H_b	H_c	J_{ac} (cps)	J_{ab} (cps)
[η^3—C_3H_5NiBr]$_2$	5.7	7.7	8.8	13.3	7.0
[η^3—C_3H_5PdCl]$_2$	4.55	5.93	7.00	12.1	6.9
[η^3—$C_3H_5Co(CO)_3$]	5.09	6.94	7.86	10.0	6.0
[η^3—$C_3H_5Mn(CO_4)$]	5.30	7.3.	8.23	[illegible].0	10.0
[(η^3—C_3H_5)$_2$Ni]	6.30	6.91	8.13	11.0	3.0
[(η^3—C_3H_5)Pd(η^5—C_5H_5)]	4.32	5.63	6.81	10.0	5.0

Chemically the π-allyl group often behaves as an electron-rich ligand, but some of these reactions involve prior dissociation of one terminus to form an intermediate σ-allyl group. This kind of process is often facile and reversible, leading to fluxional rearrangement (σ-π rearrangement) on the ^{1}H n.m.r. time-scale:

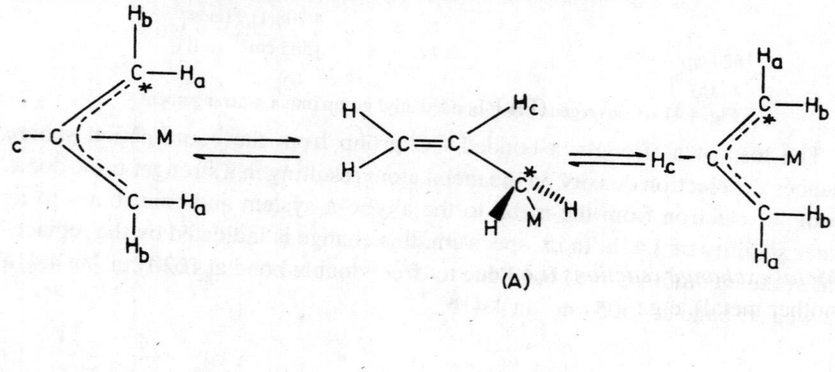

(A)

Many η^3-allyl complexes, especially homoleptic ones, exhibit such fluxional behaviour (Chapter 5). In the σ-allyl form, there is "free rotation" about the single carbon-carbon bond, so that this facile η^3 to η^1 rearrangement provides a mechanism for exchanging the *syn* and *anti* groups. It may be pointed out that since the η^1-allyl group can not act as a π-acid, it is more electron-rich than the η^3-allyl group. The transformation η^3 to η^1 (π to σ) allyl also creates a vacant coordination site on the metal atom as in A. The presence of this arrangement clearly accounts for the wide range of reactivities exhibited by metal allyl complexes.

4.3.4.3 Chemical Properties

Transition metal η^3-allyl complexes are susceptible to a wide range of chemical reagents and these reactions in general may be classified as:

(a) *Insertion reactions* (i.e. reactions in which an organic or inorganic substrate is inserted into the metal-carbon bond), e.g.:

$$Ni(\eta^3—C_3H_5) + CO_2 \longrightarrow \underset{\text{intermediate}}{[Ni(OCOC_3H_5)(\eta^3—C_3H_5)]}$$

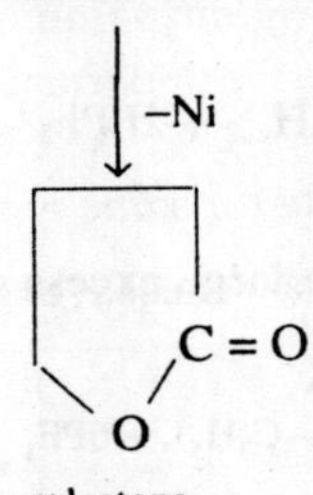

γ-lactone

$$[\{PdCl(\eta^3 - C_3H_5)\}_2] + CO \xrightarrow{HCl}$$

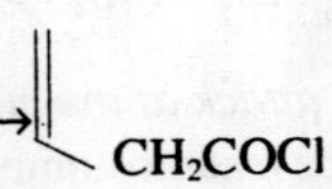

$$2[Pt(PPh_3)_2(\eta^3—C_3H_5)]^+Cl^- + 2CO \xrightarrow{CHCl_3}$$

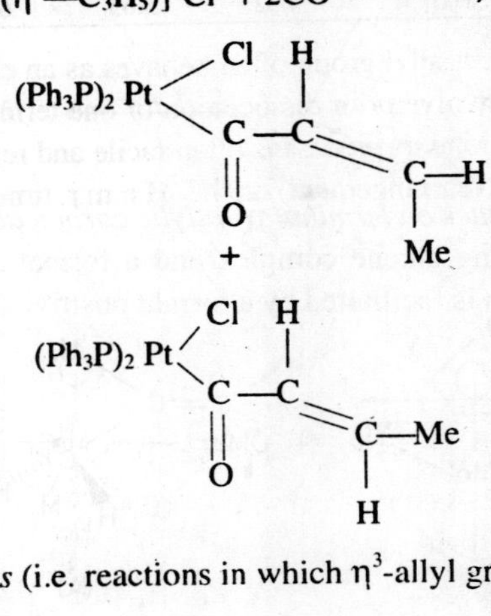

(b) *Metal exchange reactions* (i.e. reactions in which η^3-allyl group is transferred to another metal), e.g.:

$$[\{PdX(\eta^3—C_3H_5)\}_2] + 2Hg \longrightarrow 2Pd + 2Hg(\eta^1—C_3H_5)X$$

or

$$Hg(\eta^1—C_3H_5)_2 + HgX_2$$

$$[\{PdX(\eta^3—C_3H_5)\}_2] + NaCo(CO)_4 \longrightarrow (Co(CO)_3)(\eta^3—C_3H_5)$$

$$\downarrow PPh_3$$

$$Co(CO)_2(PPh_3)(\eta^3—C_3H_5)$$

(c) *Halide exchange reactions* (i.e. reactions in which halide ions of the heteroleptic allylic complexes are better leaving groups) e.g.:

$$[\{Pd(\mu—X)(\eta^3—C_3H_5)\}_2] + 2Tl^+C_5H_5^- \longrightarrow 2Pd(\eta^3—C_3H_5)(\eta^5—C_5H_5) + 2TlX$$

(d) *Displacement reactions* in which neutral donor ligands such as tertiary phosphines, tertiary arsines or isocyanides (in limited amounts) react with η^3-allyl complexes of the transition metal to give firstly a change of bonding from η^3 to η^1, e.g.:

$$Pt(\eta^3–2–MeC_3H_4)_2 + 2EPh_3 \longrightarrow Pt(EPh_3)_2(\eta^1—CH_2—CMe{=}CH_2)$$

(E = P, As)

However, with a large excess of some neutral ligands, the η^3-allyl group is displaced, e.g.:

$$Cr(\eta^3—C_3H_5)_3 + 6PF_3 \longrightarrow Cr(PF_3)_6$$

$$Co(CO)_3(\eta^3—C_3H_5) + NO \longrightarrow Co(CO)_3(NO)$$

$$Co(\eta^3—C_3H_5)_2 + 3CO \longrightarrow (OC)_3Co—$$ 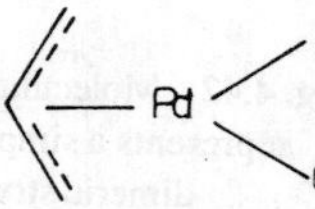

(e) *Nucleophilic attack at the metal* involving interaction of η^3-allylic complexes with donor ligands leads to compounds which show dynamic behaviour in solution, which can be followed by n.m.r. spectroscopy:

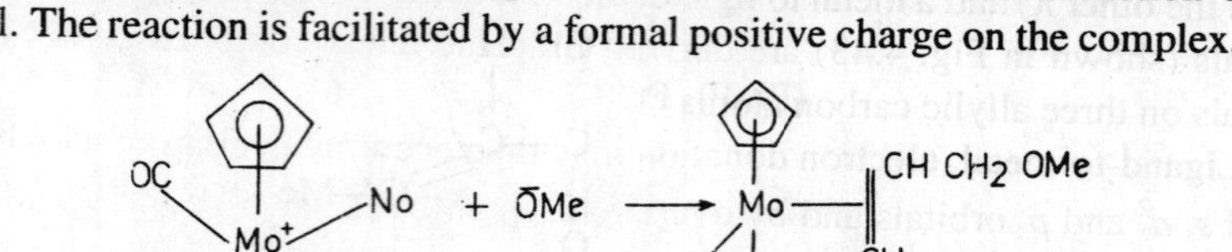

(f) *Nucleophilic attack on terminal η^3-allylic carbon atom from* the back side leads to the corresponding alkene complex and a formal 2-electron reduction of the metal. The reaction is facilitated by a formal positive charge on the complex, e.g.:

4.3.4.4 *Structure and Bonding*

Out of the structures of a large number of transition metal allyl complexes, which have been elucidated by X-ray studies only a few are illustrated below:

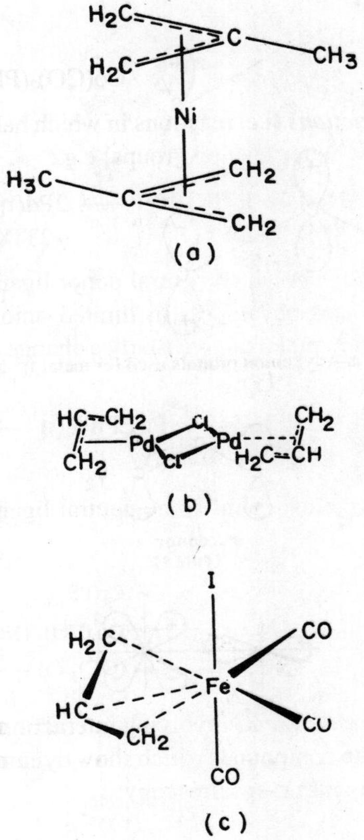

Fig. 4.42 Molecular structure of some typical (η^3-allyl) metal complexes, Figure 4.42a represents a simple monomeric homoleptic molecule, Figure 4.42b is typical of a dimeric structure, and Figure 4.42c represents a heteroleptic complex.

The bonding in η^3-allyl groups involves two ligand-to-metal donor bonds (one σ and the other π) and a metal to ligand, $d\pi$—$p\pi$ bond. The three ligand molecular orbitals (shown in Fig. 4.43) are derived from the linear combinations of the p_z orbitals on three allylic carbon atoms

Ligand-to-metal, electron donation may then occur by overlap of ψ_1 with the metal s, d_z^2 and p_z orbitals and by overlap of ψ_2 with the metal p_y and d_{yz} orbitals. The metal-to-ligand back-bond then derives from d_{xz} overlap with ψ_3 .

The extent to which these orbitals are, infact, involved in bonding depends on their relative energies, their radical diffuseness and the actual extent of orbital overlap.

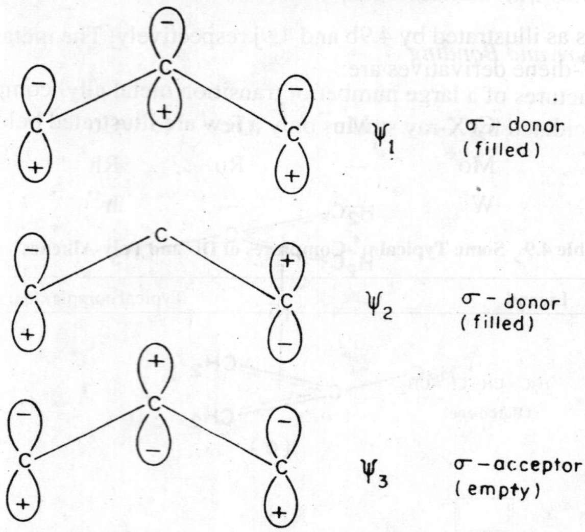

Fig. 4.43 π-Allyl anion orbitals used for metal η^3-allyl bonding.

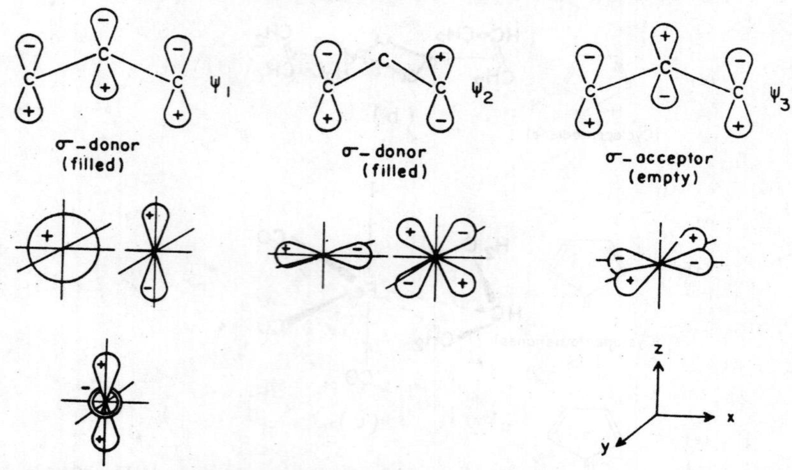

Fig. 4.44 π-Allyl molecular orbitals and the metal orbitals undergoing effective overlap to produce substantial metal-ligand bonding.

4.3.5 η^4-Diene Complexes

Compared to mono-alkene (η^2) and allyl (η^3) ligands, dienes (or dialkenes) can obviously function as four electron donor ligands towards metals. A number of complexes formed by such ligands are illustrated in Table 4.9. In fact, in some cases like 4.9h and 4.9i, the ligands function as 4-electron donors (η^4) by employing only 2 out of the 3 or 4 alkene bonds available respectively in the ligand molecules. Another interesting feature of the chemistry of η^4-diene complexes is that although cyclobutadiene  and trimethylenemethyl 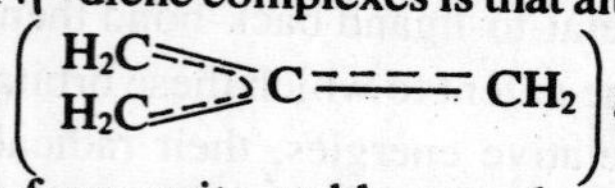 groups cannot be isolated in the free state, yet they form quite stable complexes with transition metals as illustrated by 4.9b and 4.9j respectively. The metals known to form isolable η^4-diene derivatives are:

Ti*	V	Cr	Mn**	Fe	Co	Ni
—	Nb*	Mo	—	Ru	Rh	Pd
—	—	W	—	—	Ir**	Pt

Table 4.9 Some Typical η^4-Complexes of Di- and Poly-Alkenes

	Ligand	Typical complex(es)
	$H_2C=CH-CH=CH_2$ (Butadiene)	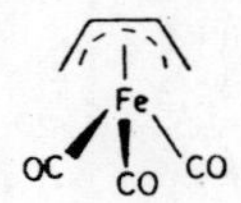
(b)	 (Cyclobutadiene)	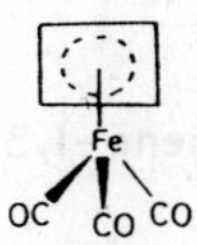
(c)	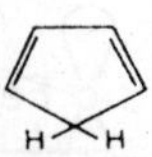 (Cyclopentadiene)	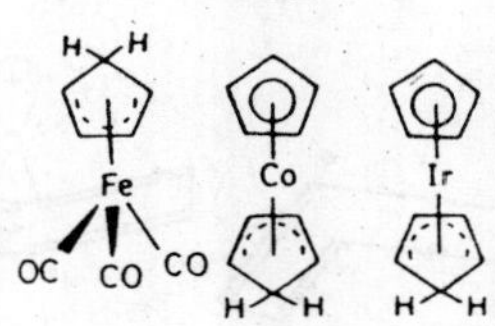
(d)	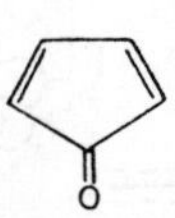 (Cyclopentadienones)	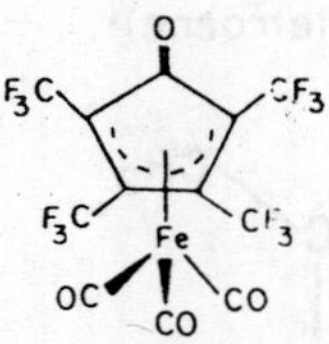
(e)	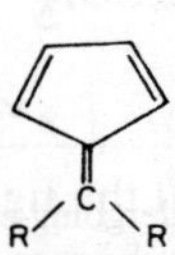 (Fulvenes)	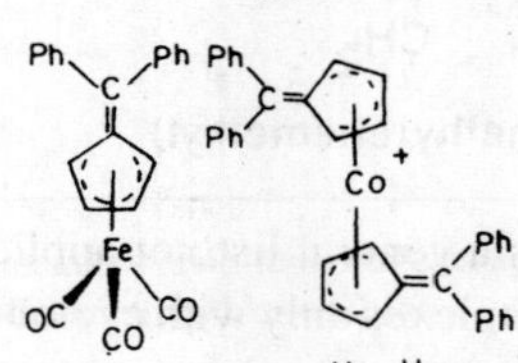
(f)	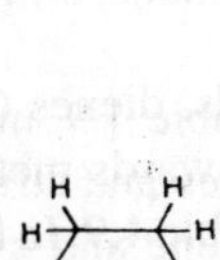 (Cyclohexadiene-1,3)	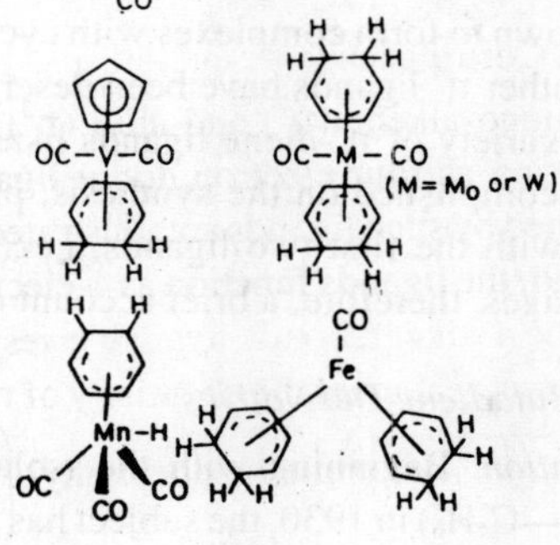

(g) 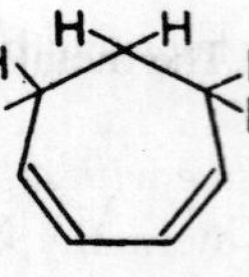

(Cyclohexadiene -1,3)

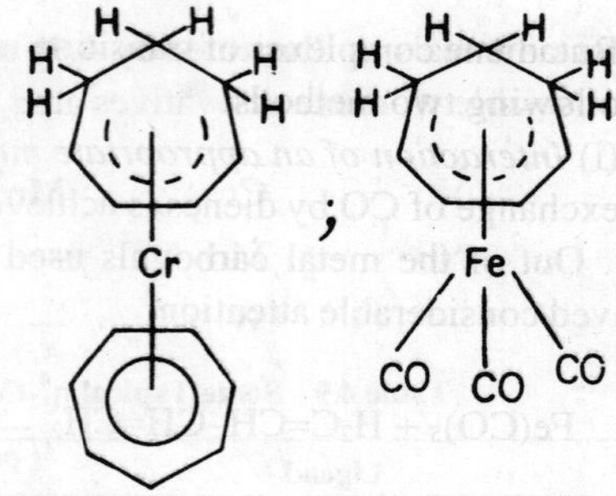

(h) 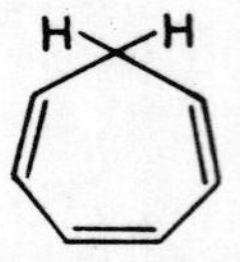

(Cycloheptatriene-1,3,5)

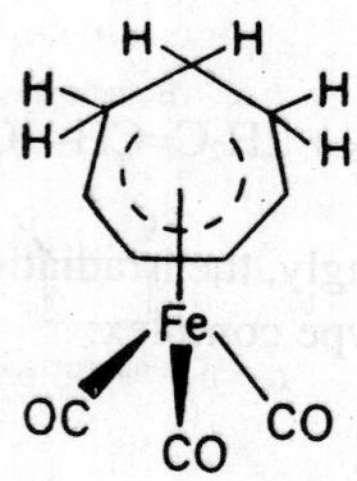

(i) 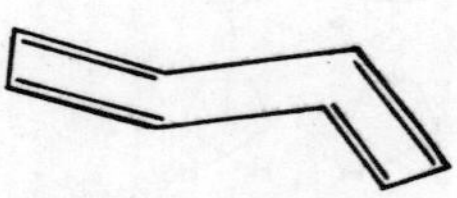

(Cyclooctatetraene)

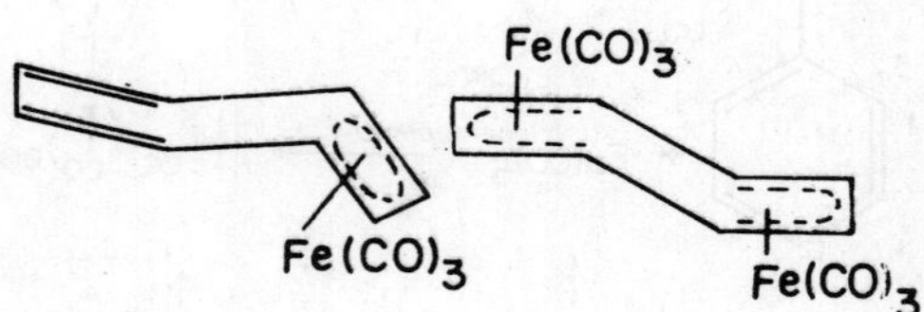

(j) 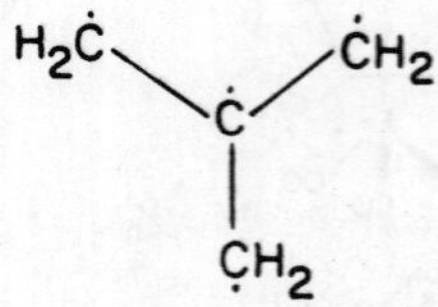

(Trimethylenemethyl)

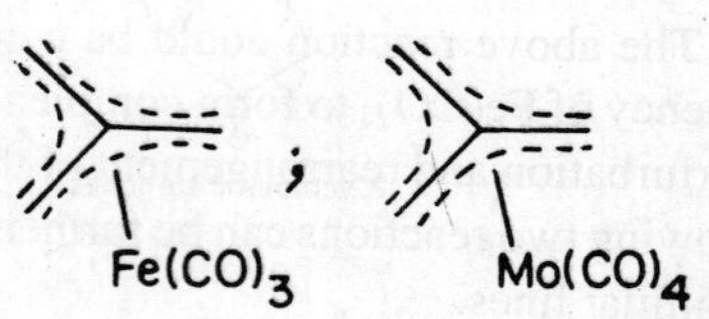

The above is a general list not applicable to all the ligands; for example Ti* and Nb* from complexes only with cyclobutadiene, whereas Mn** and Ir** have not yet been known to form complexes with cyclobutadiene, although their complexes with some other η^4 ligands have been described.

Of the variety of η^4-diene ligands listed in table 4.9, major amount of work has been accomplished on the synthesis, properties and structural aspects of the complexes with the first two ligands, i.e., butadiene and cyclobutadiene. In the following pages, therefore, a brief account of these only is being presented.

4.3.5.1 η^4-Butadiene Derivatives

(a) *Preparation.* Beginning with the isolation of the first butadiene complex, $Fe(CO)_3(\eta^4—C_4H_6)$ in 1930, the subject has grown at a fast pace particularly during the last two decades.

Butadiene complexes of transition metals have generally been synthesized by the following two methods:

(i) *Interaction of an appropriate metal carbonyl derivative with a butadiene* The exchange of CO by dienes is achieved photochemically, thermally, or chemically. Out of the metal carboynls used for such reactions, iron carbonyls have received considerable attention:

$$Fe(CO)_5 + H_2C{=}CH{-}CH{=}CH_2 \xrightarrow[\text{pressure}]{} Fe(CO)_3(\eta^4{-}C_4H_6) + 2CO$$

Irradiation of $Fe(CO)_5$ and butadiene together yields a bis(η^4-butadiene) complex:

$$Fe(CO)_5 + 2H_2C{=}CH{-}CH{=}CH_2 \xrightarrow{\text{U.V.}} Fe(CO)(\eta^4{-}C_4H_6)_2 + 4CO$$

Interestingly, the irradiation of a $Fe(CO)_5$—vinylbenzene mixture also yields a butadiene type complex:

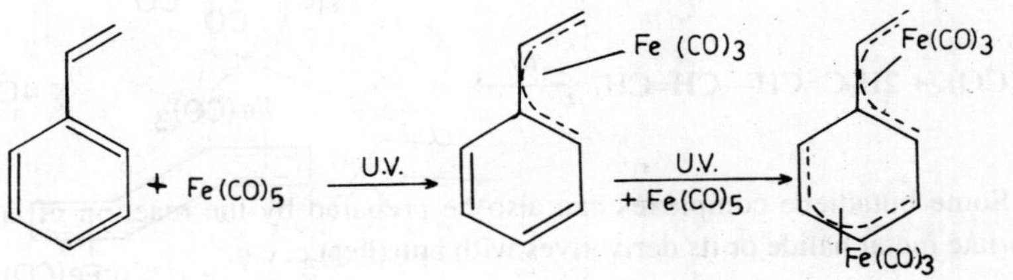

The above reaction could be considered as an example in which the strong tendency of $Fe(CO)_3$ to form complexes with η^4-butadiene type of ligands leads to a perturbation and rearrangement cf the electron cloud in an aromatic system. The following two reactions can be further cited as illustrations which can be explained on similar lines.

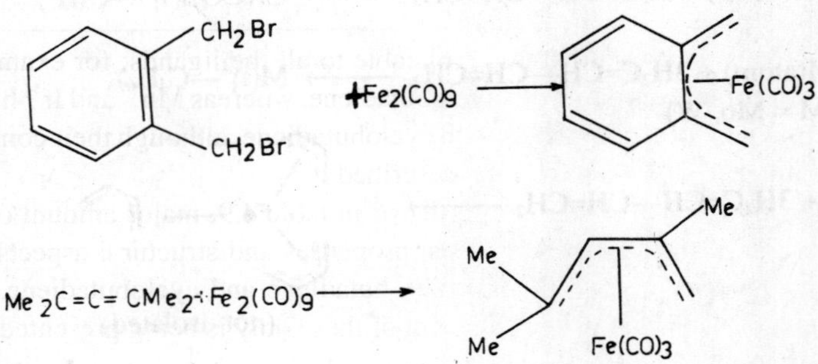

Interaction of bromodicarbonylcyclopentadienyl iron $FeBr(CO)_2(\eta^5{-}C_5H_5)$, with butadiene in the presence of aluminium bromide results in the formation of a cationic complex, $[Fe(CO)_2(\eta^4{-}C_4H_6)(\eta^5{-}C_5H_5)]^+$:

$$FeBr(CO)_2(\eta^5—C_5H_5) + H_2C{=}CH—CH{=}CH_2 \xrightarrow[-CO]{AlBr_3} [Fe(CO)(\eta^4—C_4H_6)(\eta^5—C_5H_5)]^+[AlBr_4]^-$$

Reactions of a few more metal carbonyls yielding butadiene complexes are represented by the following reactions:

$$Mo(CO)_6 + H_2C{=}CH—CH{=}CH_2 \xrightarrow{h\nu} Mo(CO)_4(\eta^4—C_4H_6) + 2CO$$

$$Mo(CO)_6 + 2H_2C{=}CH—CH{=}CH_2 \xrightarrow{h\nu} Mo(CO)_2(\eta^4—C_4H_6)_2 + 4CO$$

$$Mn(CO)_3(\eta^5—C_5H_5) + H_2C{=}CH—CH{=}CH_2 \xrightarrow{h\nu} Mn(CO)(\eta^4—C_4H_6)(\eta^5—C_5H_6) + 2CO$$

$$Co(\eta^5—C_5H_5)_2 + H_2C{=}CH—CH{=}CH_2 \rightarrow Co(\eta^4—C_4H_6)(\eta^5—C_5H_5) + C_5H_5$$

$$Co_2(CO)_8 + 2H_2C{=}CH—CH{=}CH_2 \xrightarrow{h\nu}$$

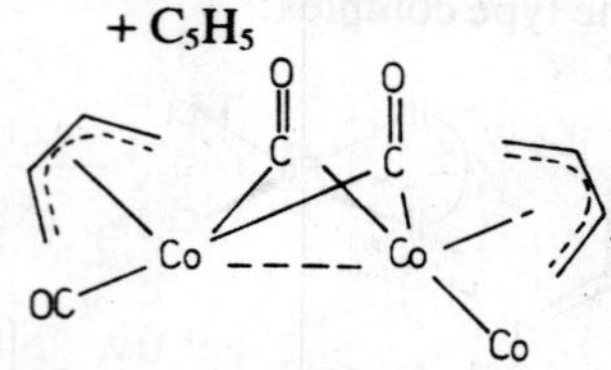

$+ 4CO$

Some butadiene complexes can also be prepared by the reaction of an appropriate metal halide or its derivatives with butadience, e.g.:

$$RhCl_3 + H_2C{=}CH—CH{=}CH_2 \longrightarrow RhCl(\eta^4—C_4H_6)_2$$

$$Na_2IrCl_4 + H_2C{=}CH—CH{=}CH_2 \longrightarrow IrCl(\eta^4—C_4H_6)_2$$

(ii) *Metal-atom vapour synthesis*. The application of this route can be illustrated by the following reactions:

$$Cr(atom) + H_2C{=}CH—CH{=}CH_2 \xrightarrow{+4CO} Cr(CO)_4(\eta^4—C_4H_6)$$

$$M(atom) + 3H_2C{=}CH—CH{=}CH_2 \longrightarrow M(\eta^4—C_4H_6)_3$$

(M = Mo, W)

$$Fe(atom) + 3H_2C{=}CH—CH{=}CH_2 \longrightarrow$$

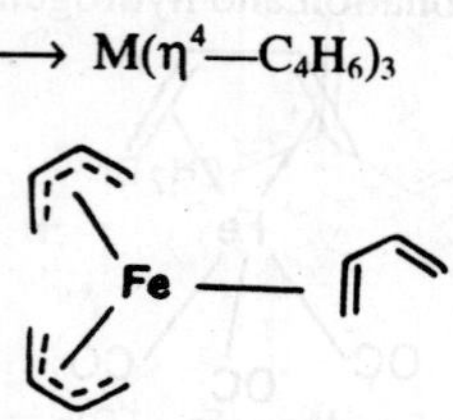

(not isolated)

↓ L

$Fe(L)(\eta^4—C_4H_6)_2$

(L = CO or PR_3)

(b) *Chemical properties*. The chemical properties of η^4-diene complexes in general can be represented by the following types of reactions:

(i) *Nucleophilic addition to η^4-diene complexes* (nucleophilic attack at an olefinic carbon centre). Unsaturated hydrocarbons, e.g., ethylene (cf. Section 4.3.2), butadiene, or benzene (cf. Section 4.3.7) do not normally undergo nucleophilic addition or substitution reactions. However, when these molecules are coordinated to electron withdrawing metal centres, they are attacked by a wide range of nucleophiles such as H^-, R^-, CN^-, MeO^-, R_3N, etc. The nucleophilic attack can occur in three different ways: (a) the nucleophile displaces the diene; (b) the nucleophile attacks the metal but displaces ligands other than the diene; and (c) the nucleophile attacks at an olefinic carbon of the diene. These general types of reactions are illustrated below with the help of diene-palladium(II) complexes:

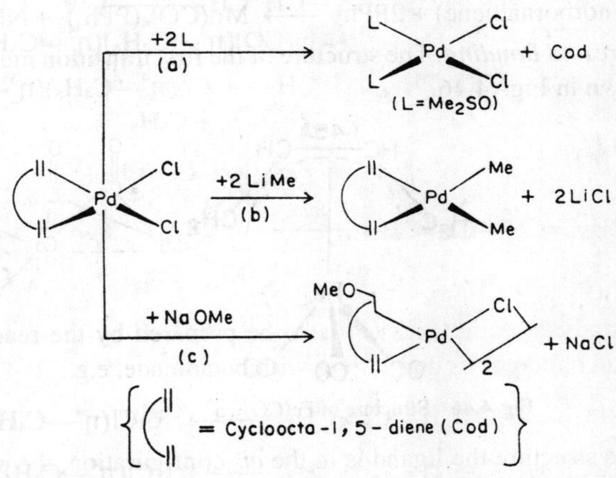

Fig. 4.45 Different types of diene-palladium(II) complexes

(ii) *Electrophilic additions*. In contrast with the ready susceptibility of coordinated alkenes to undergo attack by nucleophiles, attack by electrophiles is relatively uncommon. Electrophilic reactions, however, may be illustrated by direct protonation and hydrogen abstraction, e.g.:

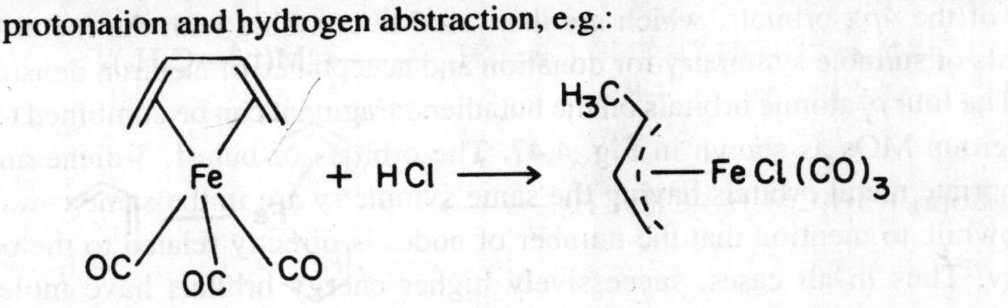

It may, however, be noted that the above type of reactions lead to a change in the hydrocarbon ligand and its mode of bonding.

(iii) *Reactions involving substituent groups*, e.g.:

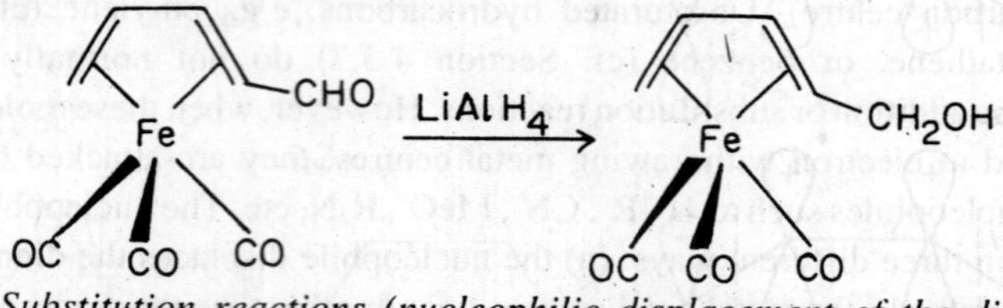

(iv) *Substitution reactions (nucleophilic displacement of the diene) as illustrated below:*

$$Fe(\eta^4\text{—}C_4H_6)(CO)_3 + 2PPh_3 \longrightarrow Fe(CO)_3(PPh_3)_2 + C_4H_6$$

$$Mo(CO)_4(\eta^4\text{—norbornadiene}) + 2PPh_3 \longrightarrow Mo(CO)_4(PPh_3)_2 + NBD$$

(c) *Structure and Bonding*. The structure of the first transition metal butadiene complex is shown in Fig. 4.46.

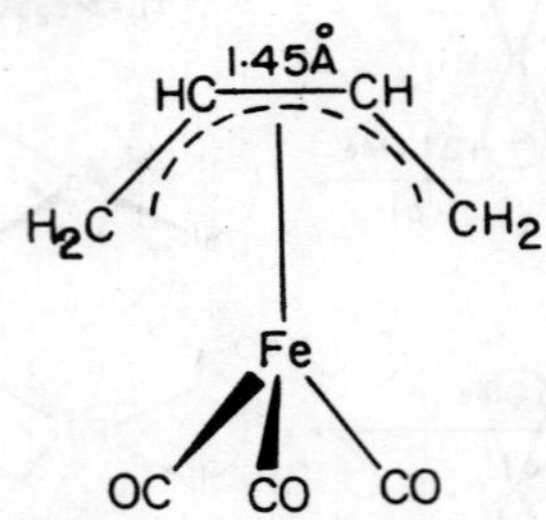

Fig. 4.46 Structure of $Fe(CO)_3(\eta^4\text{—}C_4H_6)$

In the above structure the ligand is in the *cis* configuration, the plane of four carbon atoms being nearly parallel to that of the carbonyl groups. The metal atom is equidistant from each of the four carbon atoms in the diene and the carbon-carbon bond lengths are all equal.

The bonding in η^4-butadiene transition metal complexes can be described by constructing molecular orbitals of the diene system with the appropriate combinations of the $4p\pi$ orbitals, which are then used to combine with the metal-based orbitals of suitable symmetry for donation and acceptance of electron density.

The four p_z atomic orbitals on the butadiene fragment can be combined to give four group MOs as shown in Fig. 4.47. The orbitals of buta-I, 3-diene and the appropriate metal orbitals having the same symmetry are in the same rows. It is worthwhile to mention that the number of nodes is directly related to the orbital energy. Thus in all cases, successively higher energy orbitals have molecular orbital nodes symmetrically placed and increasing in number by one in the order of decreasing stability.

Bonding molecular orbitals in metal-butadiene derivatives are formed when orbitals on the metal and diene have similar energies and their wave function, ψ, match in the bonding region. The closer the match, the more stable will be the resulting bonding molecular orbital (BMO). If ψ_1 and ψ_2 interactions with metal

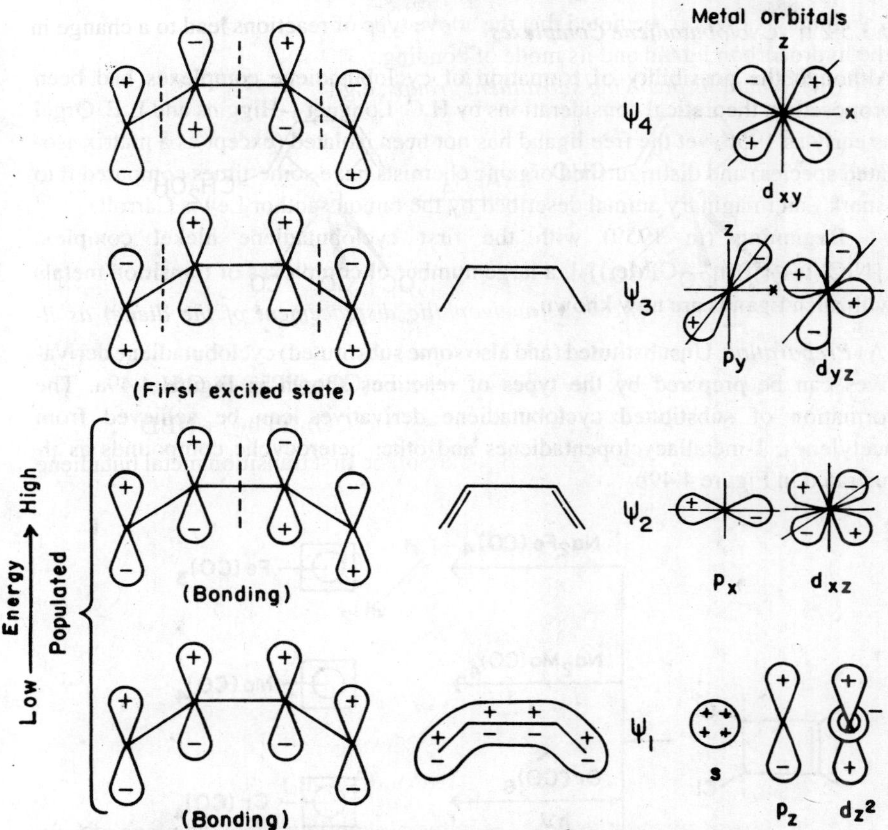

Fig. 4.47 Orbitals used in the bonding of metal η^2-butadiene complexes.

orbitals are important, then the geometry of butadiene moiety is pretty well like 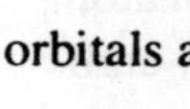 (Figure 4.48a) and 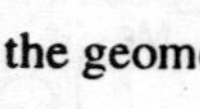 (Figure 4.48b) respectively, but if ψ_3 interactions are important, then we have the first excited-state geometry of butadiene fragment (Figure 4.48c).

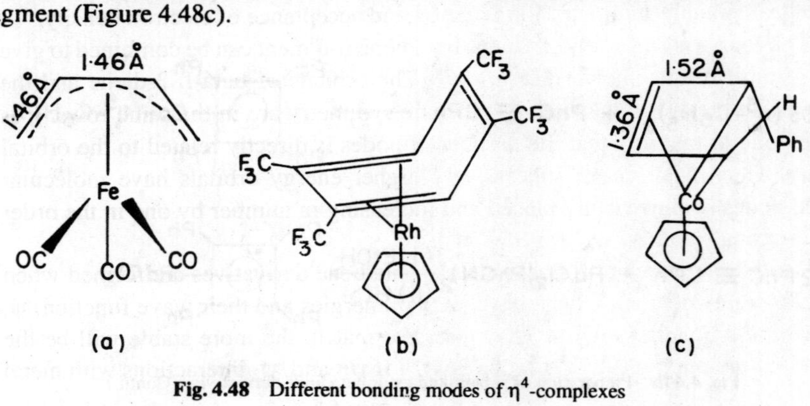

Fig. 4.48 Different bonding modes of η^4-complexes

4.3.5.2 η^4-Cyclobutadiene Complexes

Although the possibility of formation of cyclobutadiene complexes had been proposed on theoretical considerations by H.C. Longuet—Higgins and L.E. Orgel as early as 1956, yet the free ligand has not been isolated (except as a matrix-isolated species) and distinguished organic chemists have some-times compared it to 'snark', an imaginary animal described by the famous author Lewis Carroll.

Beginning (in 1959) with the first cyclobutadiene nickel complex, $[\{NiCl(\mu—Cl)(\eta^4—C_4Me_4)\}_2]$, a large number of complexes of transition metals with such ligands are now known.

(A) *Preparation.* Unsubstituted (and also some substituted) cyclobutadiene derivatives can be prepared by the types of reactions shown in Figure 4.49a. The formation of substituted cyclobutadiene derivatives can be achieved from acetylenes, 1-metallacyclopentadienes and other heterocyclic compounds as illustrated in Figure 4.49b:

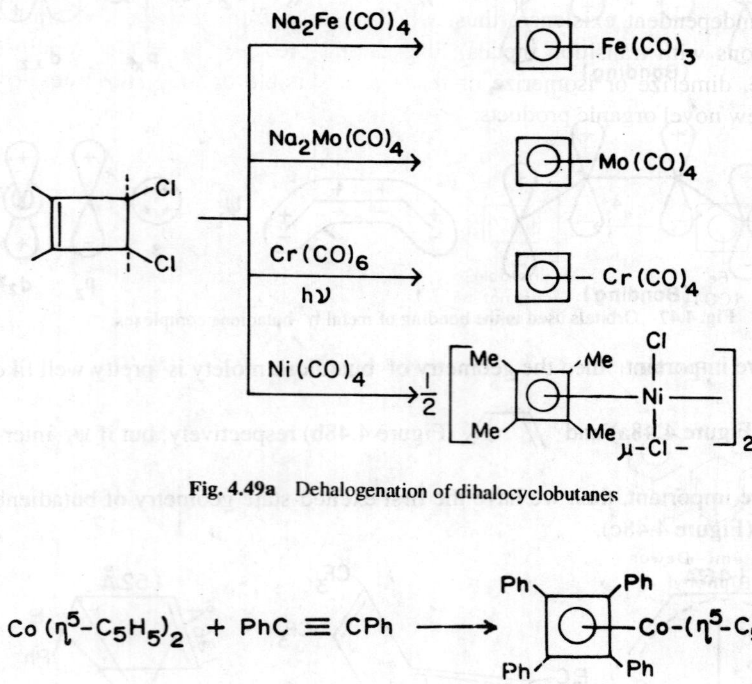

Fig. 4.49a Dehalogenation of dihalocyclobutanes

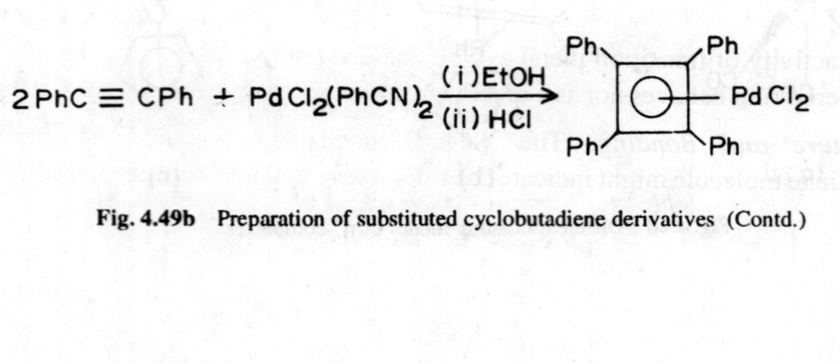

Fig. 4.49b Preparation of substituted cyclobutadiene derivatives (Contd.)

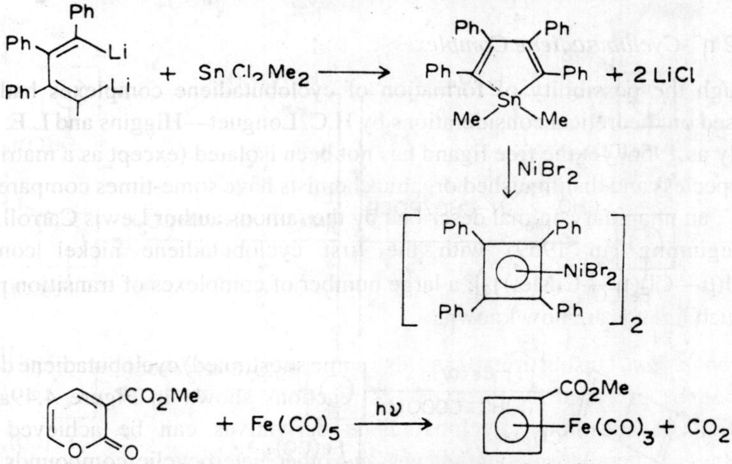

Fig. 4.49b Preparation of substituted cyclobutadiene derivatives.

(B) *Chemical Reactivity*. As indicated earlier, cyclobutadiene appears to be too labile for independent existence, thus, when released from their captive traps (combinations with transition metals), these highly reactive molecules rapidly decompose, dimerize or isomerize or react with suitable organic substrates to generate new novel organic products:

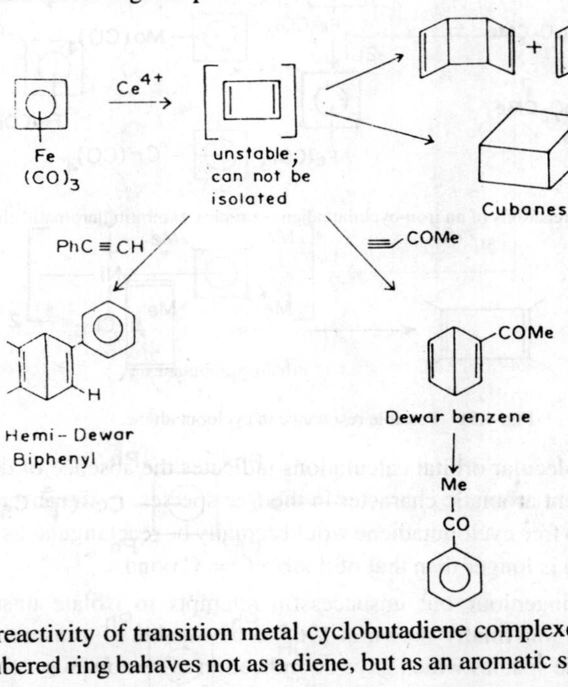

The reactivity of transition metal cyclobutadiene complexes reveals that the four membered ring bahaves not as a diene, but as an aromatic species (Fig. 4.50).

(C) *Structure and Bonding*. The valence bond representation of free cyclobutadiene molecule might indicate considerable resonance stabilization (Fig. 4.51).

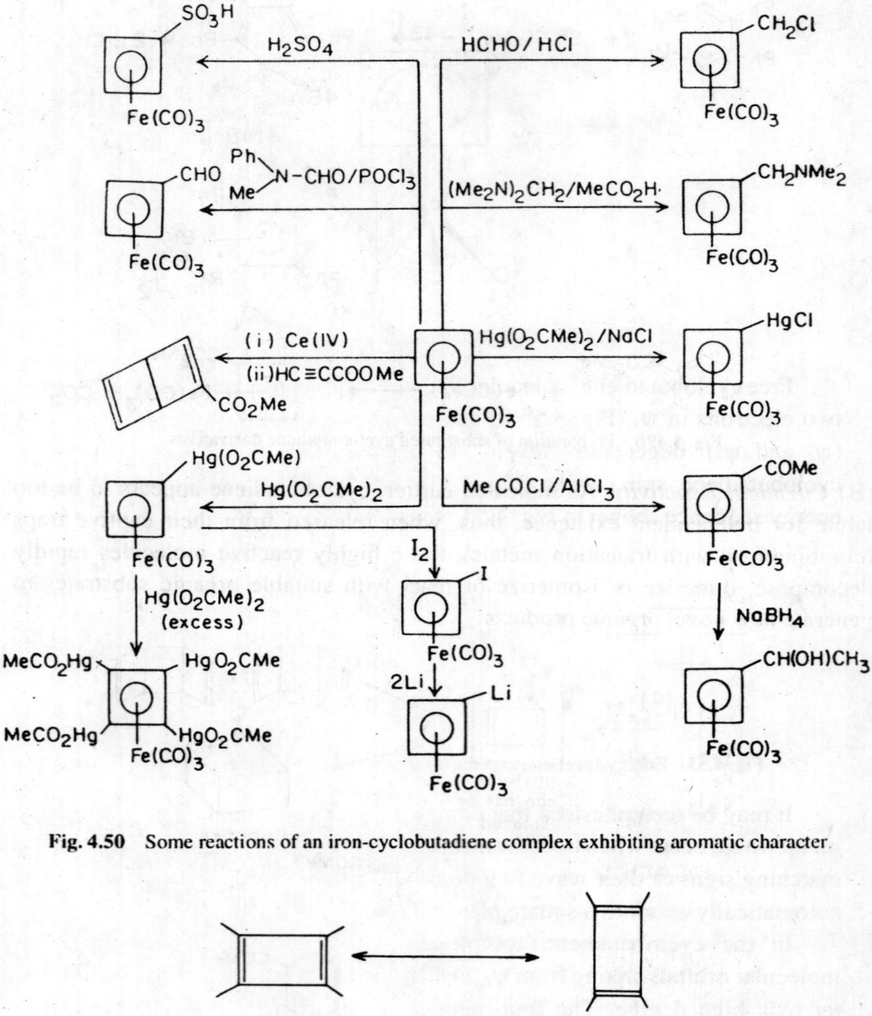

Fig. 4.50 Some reactions of an iron-cyclobutadiene complex exhibiting aromatic character.

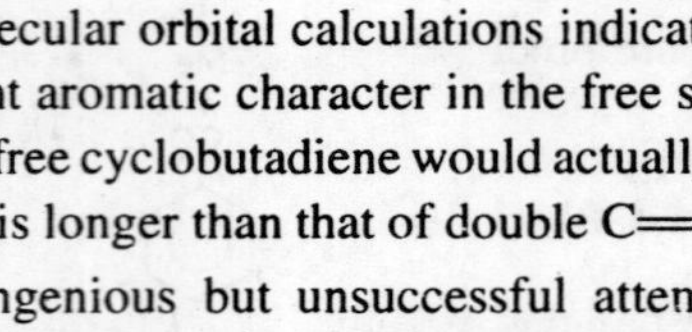

Fig. 4.51 Possible resonance in cyclobutadiene.

However, molecular orbital calculations indicates the absence of delocalization and consequent aromatic character in the free species. In the absence of this delocalization, the free cyclobutadiene would actually be reactangular as the single C—C bond length is longer than that of double C═C bond.

After many ingenious but unsuccessful attempts to isolate unsubstituted cyclobutadiene, it was finally shown in 1965 to be capable of transient existence as a highly reactive species which gave no indication of reasonance stabilization.

However, it has been conclusively shown that on complexation with transition metals, cyclobutadiene assumes a square configuration as illustrated by the structure of the iron carbonyl complex (Fig. 4.52).

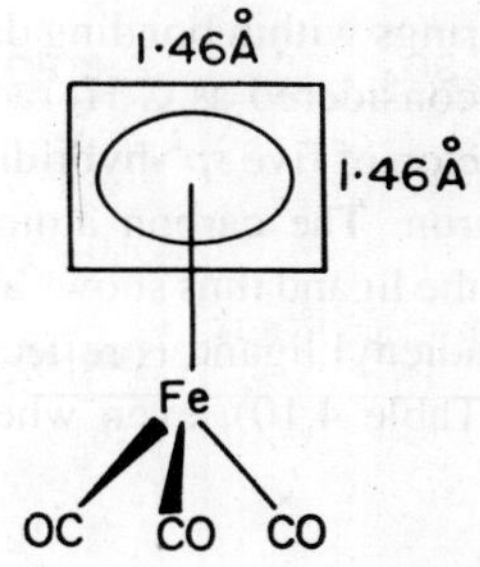

Fig. 4.52 Cyclobutadiene complex of iron carbonyl

Free cyclobutadiene molecules in their square planar forms would incorporate two electrons in ψ_1 (Fig. 4.53 a) and only one unpaired electron in each of two (ψ_2 and ψ_3) degenerate levels. In the alternative rectangular forms of cyclobutadiene, spin pairing (Fig. 4.53b) will effectively bring the orbitals in the energy sequence shown in Fig. 4.53.

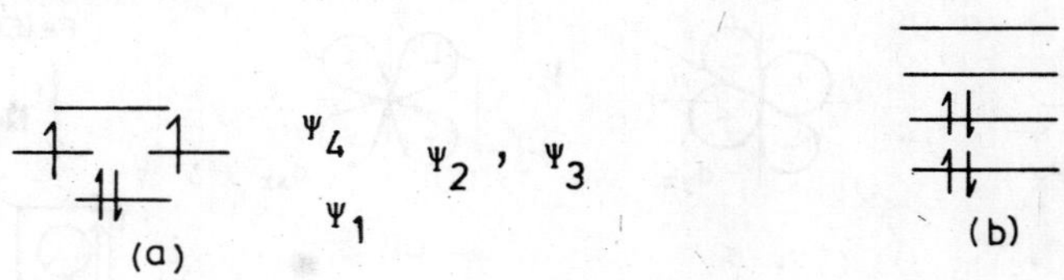

Fig. 4.53 Energy-level diagram for (a) square, and (b) rectangular cyclobutadiene.

It may be reemphasized that complexation of cyclobutadiene with transition metal would occur when both the metal and ligand have similar energies along with matching signs of their wave functions. On complexation, cyclobutadiene ligand automatically assumes a square planar shape (Fig. 4.52).

In the cyclobutadiene complexes, the electrons are accommodated in molecular orbitals arising from ψ_2 and ψ_3 (Fig. 4.53), which retain ligand character to a high degree. The four-membered ring ligated cyclobutadiene should, therefore, be considered preferably as a six-electron aromatic system, rather than as a two π-electron aromatic system. Thus, the overlaps between metal and ψ_2 as well as ψ_3 orbitals contribute predominantly to the bonding in the complex, as illustrated in Fig. 4.54.

4.3.6 η^5-Cyclopentadienyl d-Block Metal Complexes

The name 'cyclopentadienyl' in this section will be employed in a general sense to include not only unsubstituted η^5—C_5H_5 (abbreviated sometimes as Cp) but also substituted η^5—$C_5H_{5-n}R_n$ derivatives (represented as Cp*).

Although the potassium derivative of cyclopentadienide anion was described as early as 1901 by Thiele, yet an exciting phase in metal cyclopentadienyl chemistry began in early 1950's with the discovery of 'ferrocene' (cf. Section 1.3.2), which was shown to have a novel 'sandwich' structure with all the carbon atoms of both the cyclopentadienyl rings within bonding distance of the metal atom. The two cyclopentadienyl rings (considered as C_5H_5 radicals, with 5π-electrons each distributed over a planar skeleton of five sp^2-hybridized carbon atoms) complete a noble-gas configuration of iron. The carbon atoms of the ring are equidistant and lie in the same plane and the ligand thus shows aromaticity. The strong affinity of metals towards cyclopentadienyl ligand is reflected by the formation of many other "metallocence" (see Table 4.10), even when the 18-electron rule is not obeyed.

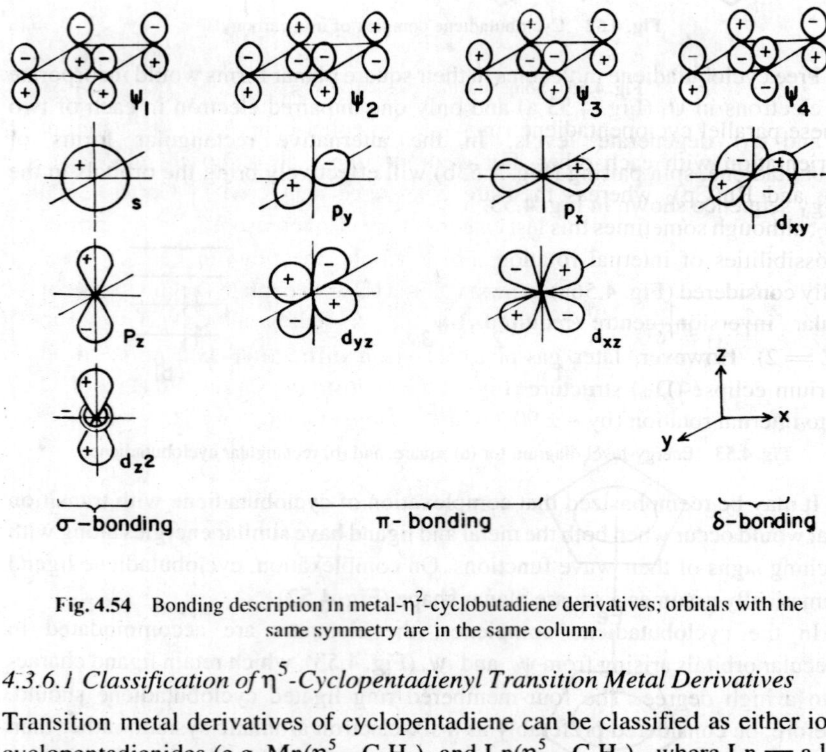

Fig. 4.54 Bonding description in metal-η^2-cyclobutadiene derivatives; orbitals with the same symmetry are in the same column.

4.3.6.1 Classification of η^5-Cyclopentadienyl Transition Metal Derivatives

Transition metal derivatives of cyclopentadiene can be classified as either ionic cyclopentadienides (e.g. $Mn(\eta^5—C_5H_5)_2$ and $Ln(\eta^5—C_5H_5)_3$, where Ln = a lanthanoid metal) or covalent cyclopentadienyls (e.g. almost all other transition metal cyclopentadienyl derivatives shows covalent character). Although these structural extremes are common, other structures also have been found, depicting dynamic structural aspects (see Section 5.4.2). Alternatively, these derivatives can be of three main types: (i) $M(Cp)_2$, symmertrical "sandwich" type derivatives represented by $Fe(\eta^5—C_5H_5)_2$ (ferrocene), $Co(\eta^5—C_5H_5)_2$, $Mn(\eta^5—C_5H_5)_2$, etc., often called 'metallocenes' (see Figure 4.55). The two rings of metallocenes and their derivatives may be parallel, as in the following three examples of iron, ruthenium, and titanium:

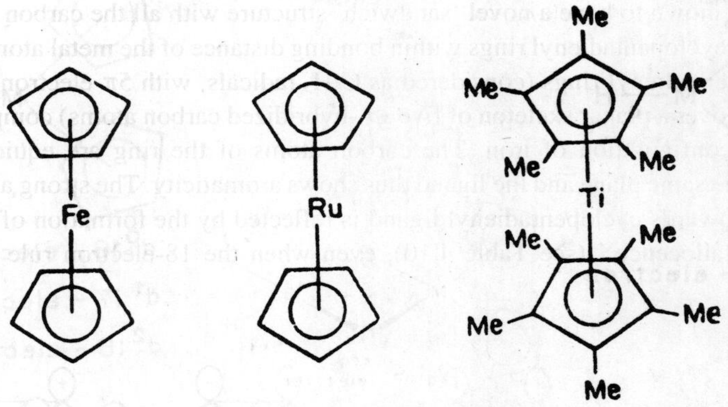

Fig. 4.55 Some metallocene derivatives

These parallel cyclopentadiene rings may be either eclipsed or staggered in their orientation with each other. For example, these are eclipsed in $V(Cp)_2$, $Cr(Cp)_2$ and $Ru(Cp)_2$ whereas they are considered staggered in $Co(Cp)_2$ and $Fe(Cp)_2$, although sometimes this last category may also represent eclipsed species with possibilities of internal rotation. For example, the rings in $Fe(Cp)_2$ were generally considered (Fig. 4.56a) to be staggered (D_{5d}) to explain the indication of molecular inversion centre required by the crystallographic space group (C^5_{2h}, Z == 2). However, later gas-phase elecrron diffraction data point to an equilibrium eclipse (D_{5h}) structure (Fig. 4.56b) with rather a low barrier (~4 kJ mol^{-1}) to internal rotation (by ~ ± 90°) of the cyclopentadiene rings.

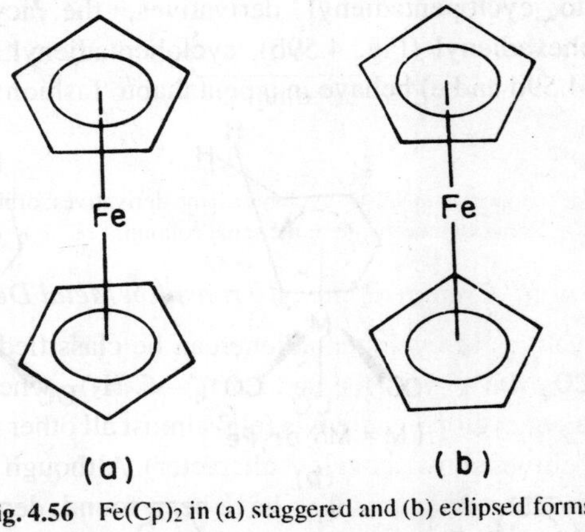

Fig. 4.56 $Fe(Cp)_2$ in (a) staggered and (b) eclipsed forms.

(ii) $(Cp)_2ML_x$, "bent metallocene" derivatives, in which the plane of two cyclopentadienyl rings are at an angle and L represents some other ligand such as hydride, alkyl(aryl), halide, alkene, carbon monoxide or even η^1—cyclopentadienyl groups and 'x' may have values generally from one to three (Fig. 4.57).

(iii) $(Cp)ML_x$, 'half-sandwich" compounds, where L == ligands like H^-, halide, CO, PR_3 etc., and x may have values from one to four. Compounds

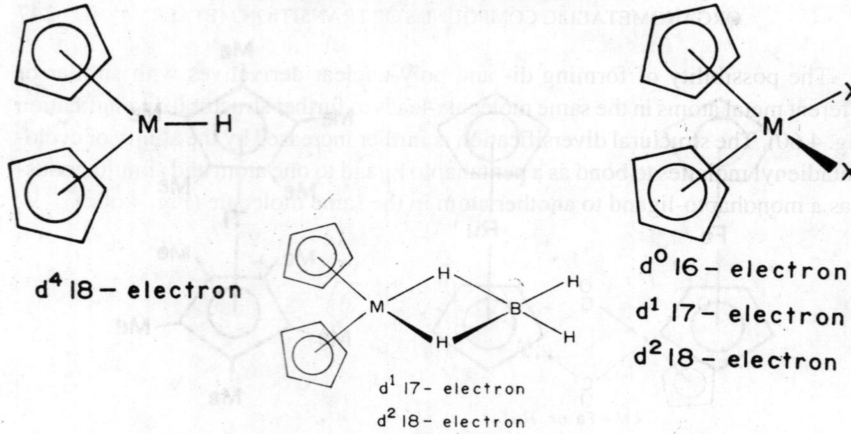

Fig. 4.57 Examples of 'bent' metallocene' derivatives.

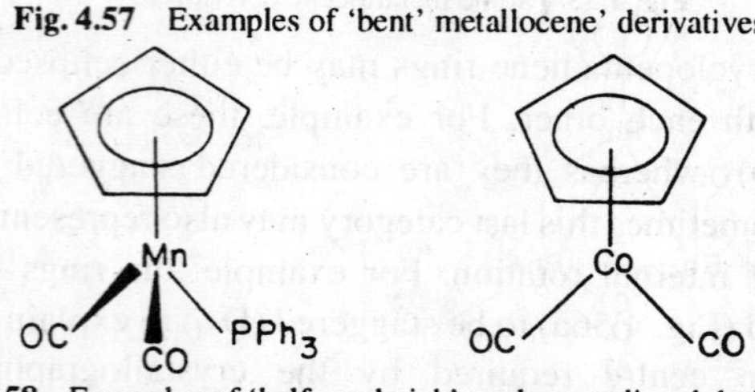

Fig. 4.58 Examples of 'half-sandwich' cyclopentadienyl derivatives.

$Mn(CO)_2(PPh_3)(\eta^5—C_5H_5)$, $Co(CO)_2(\eta^5—C_5H_5)$, are typical examples (Fig. 4.58) of this class of compounds.

In addition to cyclopentadienyl derivatives, the acyclic pentadienyl (Fig. 4.59a), cyclohexadienyl (Fig. 4.59b), cycloheptadienyl (Fig. 4.59c) and other ligands (Fig. 4.59d and e) behave in a pentahapto fashion:

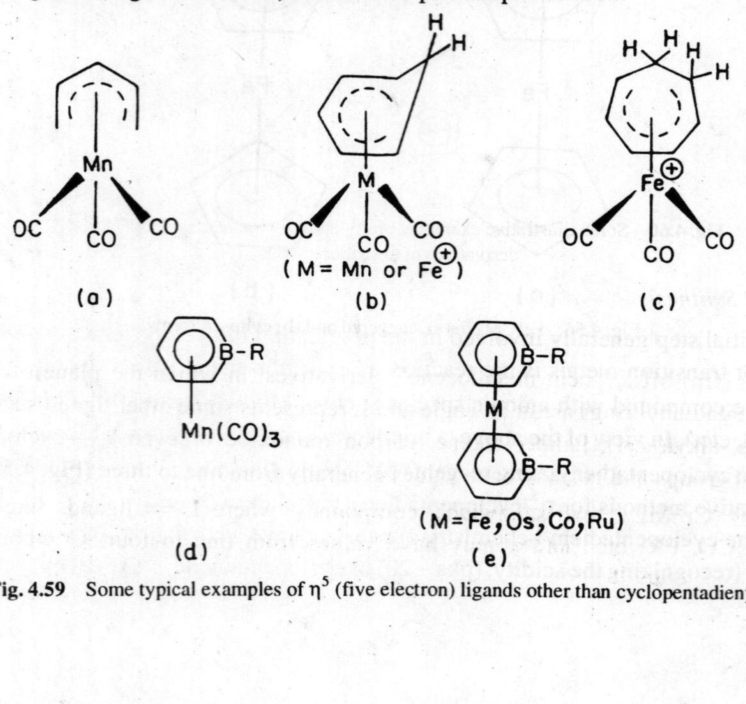

Fig. 4.59 Some typical examples of η^5 (five electron) ligands other than cyclopentadienyl.

The possibility of forming di- and poly-nuclear derivatives with similar or different metal atoms in the same molecule-leads to further structural diversification (Fig. 4.60). The structural diversification is further increased by the ability of cyclopentadienyl moieties to bond as a pentahapto ligand to one atom and simultaneously as a monohapto-ligand to another atom in the same molecule (Fig. 4.60c).

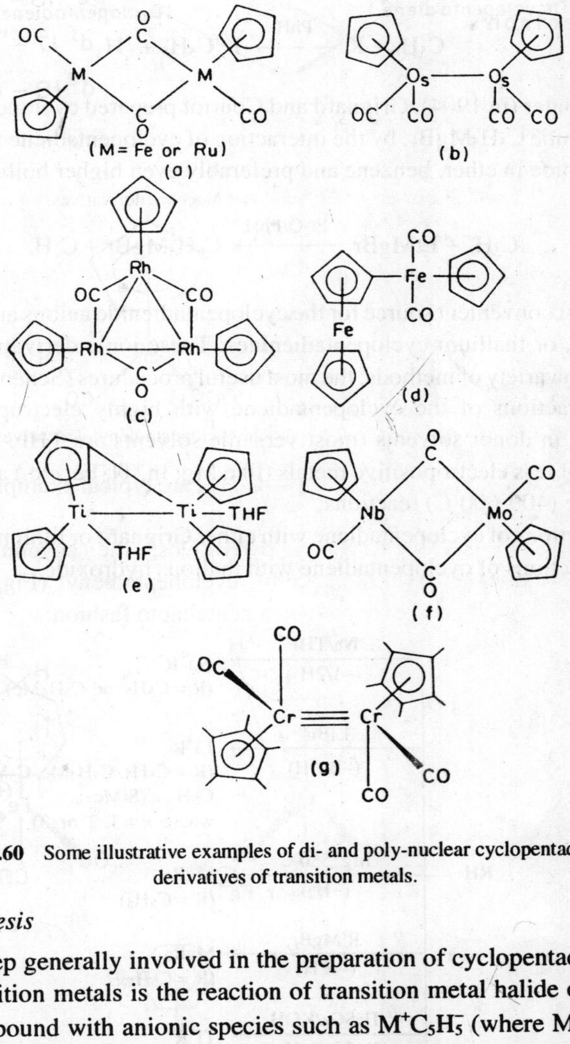

Fig. 4.60 Some illustrative examples of di- and poly-nuclear cyclopentadienyl derivatives of transition metals.

4.3.6.2 Synthesis

The initial step generally involved in the preparation of cyclopentadienyl derivatives of transition metals is the reaction of transition metal halide or some other suitable compound with anionic species such as $M^+C_5H_5$ (where M = Li, Na, K, Mg, Tl, etc.). In view of the above, a brief account of the preparative routes to such anionic cyclopentadienylating reagents $M^+C_5H_5$ is illustrated first, followed by the preparative methods for η^5-cyclopentadienyl metal derivatives.

The cyclopentadienyl chemistry appears to have been started in 1901, when Thiele (recognizing the acidity, (pka ~ 20) of cyclopentadiene, obtained by thermal cracking of Diels-Alder Dimer, dicyclopentadiene) used it in the preparation of potassium cyclopentadienide:

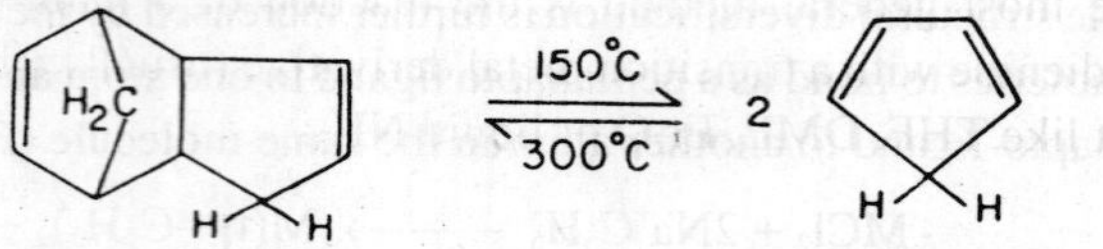

(Dicyclopentadiene) (Cyclopentadiene),

$$C_5H_6 + K \xrightarrow{PhH} K^+C_5H_5^- + \tfrac{1}{2}H_2$$

Three years later (in 1904), Grignard and Courtot prepared cyclopentadienylmagnesium bromide C_5H_5MgBr, by the interaction of cyclopentadiene with ethylmagnesium bromide in ether, benzene and preferably even higher boiling solvents:

$$C_5H_6 + EtMgBr \xrightarrow{Et_2O/PhH} C_5H_5MgBr + C_2H_6$$

The most convenient source for the cyclopentadienide anions are alkali metal-, magnesium-, or thallium-cyclopentadienides. These ionic derivatives have been prepared by a variety of methods; the most useful procedures (Scheme 4.1) include:

(i) Interactions of the cyclopentadiene with highly electropositive metals (usually Na) in donor solvents (most versatile solvents are THF, or DME). The derivatives of less electropositive metals (like Mg, In, Tl, Fe, etc.) are prepared by vapour phase (400-600°C) reactions;

(ii) Reactions of cyclopentadiene with either Grignard or lithium reagents; and

(iii) Reactions of cyclopentadiene with thallous hydroxide.

RH →

- $\xrightarrow[(-1/2H_2)]{Na/THF}$ Na^+R^- ($R = C_5H_5$ or C_5H_4Me)
- $\xrightarrow[(-Bu^nH)]{LiBu^n}$ Li^+R^- ($R = C_5H_5$, C_5H_4Me, C_5Me_5 or $C_5H_{5-n}(SiMe_3)_n$ where $n = 1, 2$ or 3)
- $\xrightarrow[(-H_2)]{Mg.\ 550°C}$ MgR_2 ($R = C_5H_5$)
- $\xrightarrow[(-R'H)]{R'MgBr}$ MgR_2 ($R = C_5H_5$)
- $\xrightarrow[(-K_2SO_4,\ -H_2O)]{Tl_2SO_4/KOH}$ Tl^+R^- ($R = C_5H_5$, C_5H_4Me, or C_5Me_5)

Scheme 4.1 Preparative routes to some anionic cyclopentadienylating reagents.

These ionic derivatives are air- and mositure-sensitive (except $Ti^+C_5H_5^-$ which is an air stable yellow solid), and therefore, these have to be isolated with extreme care or can be used *in situ* with a metal chloride for synthesis of the desired product.

The methods used for the synthesis of transition metal cyclopentadienyls can be grouped as below:

(i) The most general method is the reaction of a main group ionic metal cyclopentadienide with a transition metal derivative (usually a halide or carbonyl) in a solvent like THF, DME, Et_2O or liquid NH_3:

$$MCl_2 + 2Na^+C_5H_5^- \longrightarrow M(\eta^2\text{–}C_5H_5)_2 + 2NaCl$$

(M = Cr, Fe, Co, Ni)

$$MnBr_2 + 2Na^+C_5H_5^- \xrightarrow{\text{liq. } NH_3} Mn(\eta^5\text{–}C_5H_5)_2 + 2NaBr$$

$$MCl_4 + 2M'Cp^* \longrightarrow MCl_2(\eta^5\text{–}Cp^*)_2 + 2M'Cl$$

$$MCl_2(\eta^5\text{–}C_5H_5)_2 + 2M'C_5H_5 \longrightarrow M(\eta^5\text{–}C_5H_5)_2(\eta^1\text{–}C_5H_5)_2 + 2M'Cl$$

(M = Ti, Hf; M′ = Li, Na, or Tl; Cp* = C_5H_5, C_5H_4Me, C_5H_4Et, $C_5H_4Pr^i$, $C_5H_4Bu^t$, $C_5H_4SiMe_3$, $C_5H_3(SiMe_3)_2$ or C_5Me_5).

$$W(CO)_6 + Na^+C_5H_5^- \xrightarrow{\text{THF}} Na^+\,(\eta^5\text{–}C_5H_5)W(CO)_3 + 3CO$$

$$VCl + \underset{\text{(excess)}}{C_5H_5MgBr} \longrightarrow V(\eta^5\text{–}C_5H_5)_2$$

$$Ru(acac)_3 + C_5H_5MgBr \longrightarrow Ru(\eta^5\text{–}C_5H_5)_2$$

$$MoCl_5 + Na^+C_5H_5^- \longrightarrow Mo(\eta^5\text{–}C_5H_5)_2(\eta^1\text{–}C_5H_5)_2$$

The cyclopendtadienide anion derived from more electropositive metals is a potential nucleophile and strong reducing agent. Thus easily reducible metal ions generally undergo simultaneous reduction in some of the above reactions (in such reactions involving reduction also, the chemical equations are not balanced as the oxidation products of cyclopentadiene are not fully identified). Incidentally in a few of these cases hydrido complexes are obtained, the solvent being the source of hydrido ligand:

$$ReCl_5 + Na^+C_5H_5 \xrightarrow[\text{reflux}]{\text{THF}} ReH(\eta^5\text{–}C_5H_5)_2$$

$$RhCl_3 + C_5H_5MgBr \longrightarrow Rh(\eta^5\text{–}C_5H_5)_2^+ + Rh_3H(\eta^5\text{–}C_5H_5)_4$$

In addition to the above the cyclopentadienylthalium finds increasing synthetic applicability, because in this case cyclopentadienide anion is a weak nucleophile and a poor reducing agent and side-reactions (leading to reduction products) are often minimized by its use:

$$Tl^+C_5H_5^- + RuCl_2(\eta^6\text{–}C_6H_6) \longrightarrow (\eta^6\text{–}C_6H_6)Ru^+(\eta^5\text{–}C_5H_5) + Cl^- + TlCl$$

$$Tl^+C_5H_5^- + MoCl_2(NO)_2 \longrightarrow MoCl(\eta^5\text{–}C_5H_5)(NO)_2 + TlCl$$

(ii) The reaction of metal halides with cyclopentadiene in the presence of a base (in excess) like diethylamine, piperidine, or even potassium hydroxide (in some specific cases involving non-hydrolyzable products):

$$MCl_2 + 2C_5H_6 + 2Et_2NH \longrightarrow M(\eta^5{-}C_5H_5)_2 + 2Et_2NH_2^+Cl^-$$

(M = Fe, Co, or Ni)

$$FeCl_2{\cdot}4H_2O + 2C_5H_6 + 3KOH \xrightarrow[-2KCl]{-6KOH{\cdot}H_2O} Fn(\eta^5{-}C_5H_5)_2$$

(iii) In some cases, the reaction of cyclopentadienes (or even dicyclopentadiene), or fulvenes with the metal or its carbonyl gives the complex, e.g.:

$$Mg_{(s)} + 2C_5H_{6(g)} \xrightarrow{heat} Mg(\eta^5{-}C_5H_5)_2 + H_2$$

$$2C_5H_6 + Fe\ (atoms) \longrightarrow Fe(\eta^5{-}C_5H_5)_2 + H_2$$

$$Fe(CO)_5 + C_{10}H_{12} \longrightarrow Fe(\eta^5{-}C_5H_5)(CO)_2$$

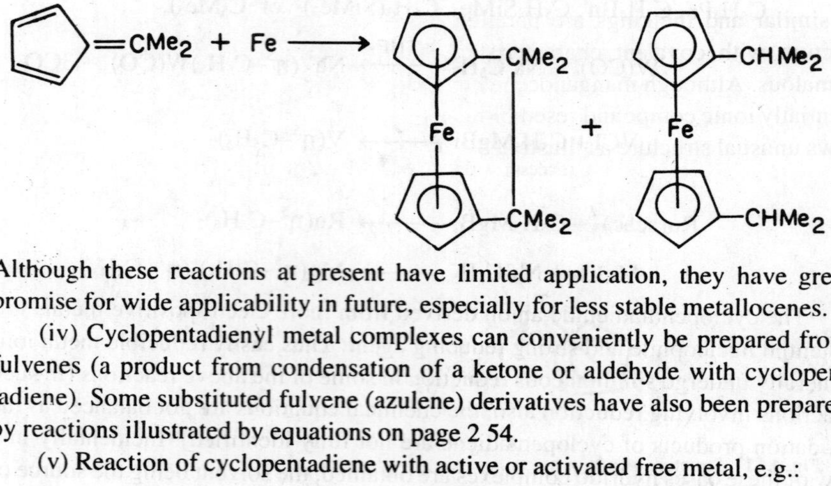

Although these reactions at present have limited application, they have great promise for wide applicability in future, especially for less stable metallocenes.

(iv) Cyclopentadienyl metal complexes can conveniently be prepared from fulvenes (a product from condensation of a ketone or aldehyde with cyclopentadiene). Some substituted fulvene (azulene) derivatives have also been prepared by reactions illustrated by equations on page 2 54.

(v) Reaction of cyclopentadiene with active or activated free metal, e.g.:

$$nC_5H_6 + M \longrightarrow M(\eta^5{-}C_5H_5)_n + n/2\ H_2$$

4.3.6.3 Properties

As mentioned above, cyclopentadienyl derivatives of transition metals are available in various oxidation states as exemplified by $[CpMo(CO)_3]^-$, $CpMn(CO)_3$, $Fe(Cp)_2$, $[Co(Cp)_2]^+$, $TiCl_2(Cp)_2$, $NbBr_3(Cp)_2$ with central metals, in 0, 1, 2, 3, 4 and 5 oxidation states respectively.

(A) Physical characteristics. Metallocenes from the 3*d* transition series are generally paramagnetic (with certain exceptions such as $Fe(\eta^5{-}C_5H_5)_2$, $[Co(\eta^5{-}C_5H_5)_2]^+$ and $Ti(\eta^2{-}C_5H_5)_2$. These are isomorphous in many cases and have melting points which are remarkably constant at or near 173°C (Table 4.10). However, only ferrocene is air-stable, the others being oxidatively labile. The stability order is Ni

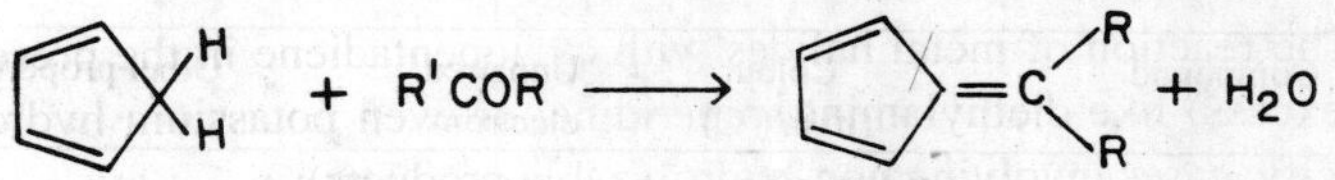

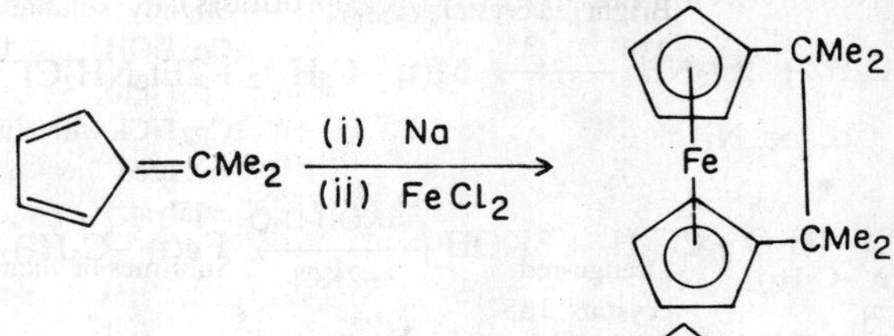

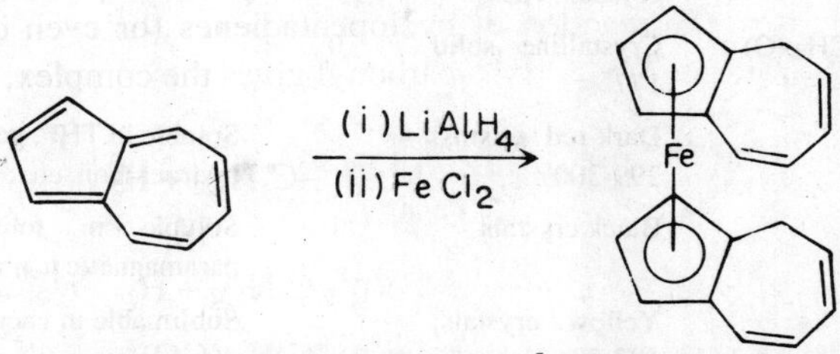

> Co > V >> Cr. In typical metallocenes, M(η^5–C_2H_5)$_2$, all the C—C bond lengths are similar and the rings are parallel; these have either eclipsed or staggered structure with covalent characteristics. The manganocene and titanocene are anomalous. Although manganocene has a sandwich structure, yet it behaves as an essentially ionic compound resembling Mg(η^5–C_5H_5)$_2$. Titanocene as THF adduct shows unusual structure as illustrated below (Fig. 4.61):

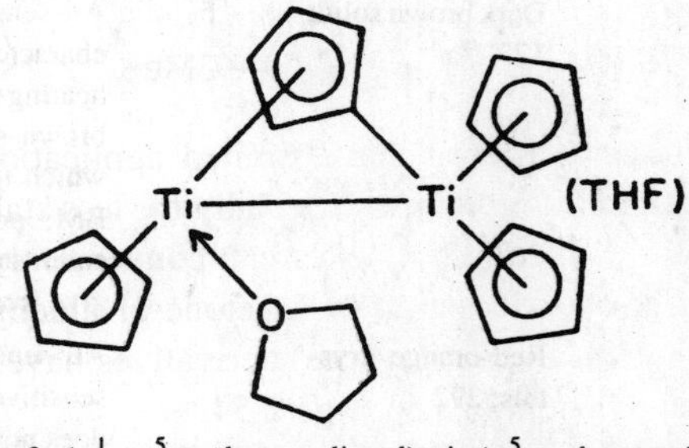

Fig. 4.61 Structure of μ(η^1 : η^5-cyclopentadienyl) tris (η^5-cyclopentadienyl) dititanium.

Some 4*d* and 5*d* transition metals such as Zr, Hf, Nb, Ta, Mo, W, Ru, Os do form ferrocene analogues, but some of these are highly reactive and exist as intermediates in reactions showing "carbene" like behaviour (cf. Section 4.5).

Table 4.10 Some Di-η^5-Cyclopentdienylmetal Compounds

Compound	Colour m.p. (°C)	Unpaired electrons	Other properties
ScCl(η^5–C_5Me_5)$_2$	White	0	Monomeric in the solid state
(η^5–C_5Me_5)$_2$Y(μ–Cl)YCl (η^5–C_5Me_5)$_2$	Yellow	0	Dimeric in the solid state
"Ti(η^5–C_5H_5)$_2$"	Dark green solid 200° (decomp)		Actually exists as [(C_5H_5) (C_5H_4) TiH]$_2$
Ti(η^5–C_5Me_5)$_2$	Orange crystals	2	Paramagnetic, μ_{eff} = 2.6 BM

Compound	Colour m.p. (°C)	Unpaired electrons	Other properties
$TiCl_2(\eta^5-C_5H_5)_2$	Bright red crystal; 230°	0	Slightly soluble in water giving Cp_2TiOH^+; LiC_6H_5 gives Cp_2TiPh_2; reducible to Cp_2TiCl, with aluminium alkyls; it is used as a polymerization catalyst.
$Ti(CH_2SiMe_3)_2(\eta^5-C_5H_5)_2$	Orange-red crystals; 185°	0	Sublimes in vacuo
$Cp_2Zr(-CH_2C_6H_4CH_2-O)$	Crystalline solid 177°	0	
$V(\eta^5-C_5Me_5)_2$	Dark-red prisms; 299-300°		Soluble in THF, pentane, aromatic hydrocarbon, etc.; paramagnetic.
$NbCl_2(\eta^5-C_5H_5)_2$	Black crystals	1	Soluble in toluene, CH_2Cl_2; paramagnetic $\mu_{eff} = 1.63$ BM
$MoH_2(\eta^5-C_5H_5)_2$	Yellow crystals; 183-5°		Sublimable in vacuo at 50°C
$[Mo(\eta^5-C_5H_5)_2]_x$	Red-brown solid		Sparingly soluble in THF
$[W(\eta^5-C_5H_5)_2]_x$	Red solid		
$W(\eta^5-C_5H_5)_2H_2$	Yellow crystals; 163°	0	Moderately stable in air; soluble in benzene, etc.; soluble in acids giving $Cp_2WH_3^+$
$Mn(\eta^5-C_5H_5)_2$	Dark brown solid; 173°	5	Air-sensitive; hydrolyses with characteristic cracking sound; heating under N_2 to 158°C turns the brown solid to a pale pink solid; which melts at 173°C; $\mu_{eff} = 5.86$ BM; polymeric structure in the solid state; reacts with $FeCl_2$ to give ferrocene.
$Mn(\eta^5-C_5Me_5)_2$	Red-orange crystals; 292		Air-sensitive in solution, but less sensitive to oxygen as a solid; it does not react with $FeCl_2$.
$Fe(\eta^5-C_5H_5)_2$	Orange solid; 174°	0	Air stable; sublimes at > 100°C; oxidizes to $Fe(\eta^5-C_5H_5)_{2^+}$ diamagnetic.
$Fe(\eta^5-C_5H_5)(\eta^5-C_5H_4SiMe_3)$	Dark orange-red mobile liquid; 23°		
$Co(\eta^5-C_5H_5)_2^+$	Yellow in aqueous solution	0	Forms a number of stable salts; thermally stable to 400°C
$Co(\eta^5-C_5Me_5)_2$	Black crystals; 300°	1	Paramagnetic.
$Ni(\eta^5-C_5H_5)_2$	Green, 173° (decom.)	2	Fairly air stable; sublimes, 80-90°C; paramagnetic $\mu_{eff} = 2.86$ BM.

(a) *Infrared studies*. Infrared spectroscopy has been quite useful in diagonising the mode of bonding in a particular metallocene. For example, for different modes of bonding: ionic bonding, centrally σ-bonding, and centrally π-bonding can be distinguished by infrared studies.

The cyclopentadienide group in ionically bonded complexes would have D_{5h} symmetry and should be expected to show for infrared active bands (Table 4.11).

Table 4.11 Characteristic Infrared Frequencies of Cyclopentadienide Group

3100-3000 cm^{-1}	ν(C–H)
1500-1400 cm^{-1}	ν(C–C) or ring deformation
1010-1000 cm^{-1}	δ(CH)
700-600 cm^{-1}	π(CH)

In centrally σ-derivatives, the local symmetry of the ring is reduced to C_{5v} and the number of infrared active bands is increased from four to seven (Table 4.12). Additionally a band due to metal ring stretching frequency can be seen in the far infrared region.

Table 4.12 Assignments of Infrared Stretching Frequencies in σ-Bonded (Centrally) Cyclopentadienyl Complexes

(i) 3100-3020 cm^{-1}	ν(CH)
(ii) 2970-2900 cm^{-1}	ν(CH)
(iii) 1450-1410 cm^{-1}	ν(CC)
(iv) 1120-1110 cm^{-1}	ν(CC)
(v) 1010-990 cm^{-1}	δ(CH)
(vi) 820-740 cm^{-1}	π(CH)
(vii) 790-710 cm^{-1}	π(CH)

Centrally π–bonded complexes which represent one of the largest groups of metal π-complexes, exhibit infrared spectra in regions almost similar to those of the centrally σ-bonded derivatives discussed above (local symmetry, C_{5v}), but these compounds are characterized by metal-ring vibrations as well as other skeletal vibrations in the far i.r. region. In addition, while still exhibiting the seven bands of the centrally σ-bonded derivatives, the two deformation bands (vi) and (vii) appear generally at higher frequencies than those of the σ-bonded derivatives. I.r. studies have been made on a large number of π-bonded metallocenes, out of which the absorptions observed in two typical cases, i.e., ferrocene and nickelocene are listed in Table 4.13:

Table 4.13 Infrared Assignments of Ferrocene and Nickelocene

Compound	ν(CH)	ν(CH)	ν(CC)	ν(CC)	δ(CH)	π(CH)	π(CH)	Antisym-metry ring tilt	M–R str.
$Fe(\eta^5\text{–}C_5H_5)_2$	3086	2909	1408	1104	1001	854	814	490	478
$Ni(\eta^5\text{–}C_5H_5)_2$	3052	2891	1421	1109	1002	839	772	?	355

(b) Nuclear magnetic resonance studies. A single resonance peak in the ^{1}Hn.m.r. spectrum is observed (showing equivalence of all the five ring hydrogens) due to the cyclopentadienyl ring protons. However, splitting in this peak result on coupling with a n.m.r. active metal nucleus such as $^{195}Pt(I = 1/2)$ or $^{103}Rh(I = 1/2)$ or with protons directly attached to the metal, e.g. $ReH(\eta^5\text{–}C_5H_5)_2$. The appearance of a singlet for substituents in unsymmetrical cyclopentadienyl complexes is due to ring whizzling of a cyclopentadienyl group about the metal-ring axis (see Fig. 4.62):

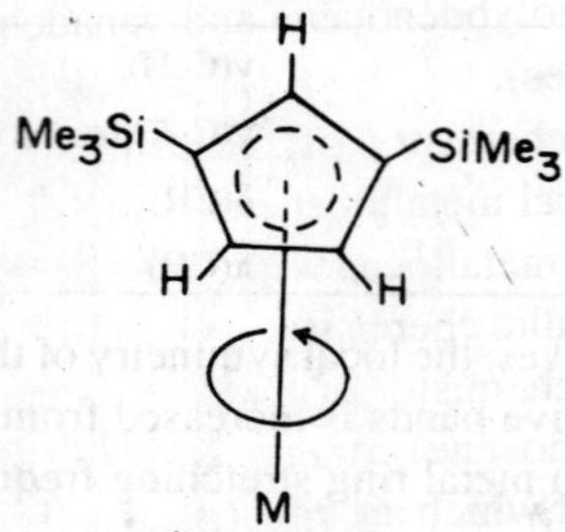

Fig. 4.62

In these cases, interchange of ring hydrogen positions is generally fast enough on the n.m.r. time-scale (10^{-2}-10^{-9} s) to give only a time average signal. When hydrogens of the C_5H_5 ligand are substituted by Me, Pr^i, Bu^t, or Me_3Si groups, more complex spectra are obtained.

(B) *General Characteristics and Chemical Reactivity.* Some general properties of neutral metallocenes and their derivatives have been listed in Table 4.10. Metallocenes of the type $M(\eta^5\text{–}C_5H_5)_2$ are generally intensely coloured, volatile, air-sensitive solids, the majority of which are soluble in organic solvents. Amongst the metallocenes that are known, ferrocene is unique in depicting greater oxidative and thermodynamic stability; the stability order for 3*d* metallocenes is: Fe > Ni > Co > V >> Cr > Ti. All the C–C bond lengths in η^5 – C_5H_5 complexes are equivalent and resemble those of benzene which is in conformity with the aromatic character of the C_5H_5 ring. Theoretically metallocene are expected to undergo aromatic substitution reactions such as Friedel-Crafts acylation, sulphonation and metallation, etc. However, most cyclopentadienyl complexes do not survive under the reaction conditions involved in aromatic substitution and similar reactions. Such reactions have been widely studied only in some special cases, prominent amongst which are ferrocene, ruthenocene, etc. In general, all metallocenes are more reactive towards electrophilic reagents than benzene, indicating a higher availability of electrons in the metallocene systems.

In addition to their aromaticity, metallocenes beginning with ferrocene have attracted special attention due to their covalent characteristics, except in the cases of cyclopentadienides of strongly electropositive inner transition metals; e.g., lanthanides and also in the special case of manganese (explained on the basis of the stability of half-filled d^5 configuration). This special ionic behaviour of $Mn(\eta^5\text{–}C_5H_5)_2$ is reflected in its high air-sensitivity and reactivity towards $FeCl_2$

yielding ferrocene. It may be worth noting that the permethylated manganese derivative, $Mn(\eta^5\text{-}C_5Me_5)_2$, behaves quite differently, in being much less air-sensitive and unreactive towards $FeCl_2$.

From the point of their reactivity, therefore, metallocenes and their derivatives can be generally classified into the following three main categories with typical examples of each class given in parenthesis: (i) covalent and oxidatively as well as thermodynamically stable (e.g. ferrocene; ruthenocene); (ii) covalent and highly reactive (e.g. titanocene; molybdenocene and vanadenocene) and (iii) ionic (e.g. manganocene; lanthanocenes).

We now present the chemistry of these three categories of metallocenes together with that of a typical metallocene derivative: zircocene dichloride, highlighting the potentialities of metallocenes and their derivatives in generating newer and fascinating oeganometallic chemistry.

The most remarkable chemistry of ferrocene (and other similar compounds such as ruthenocene and ormocene) ermerge out from the aromaticity of cyclopentadienyl rings, some facts of which are illustrated in Scheme 4.2.

Compared to rather stable covalent later transition metal derivatives typified by $Fe(Cp)_2$, the metallocenes, $M(Cp)_2$ and $M(Cp^*)_2$ of earlier transition metal (M = Ti, Zr, V, Cr, Mo etc.) in their low bivalent states are coordinatively unsaturated and highly reactive as depicted in Scheme 4.3.

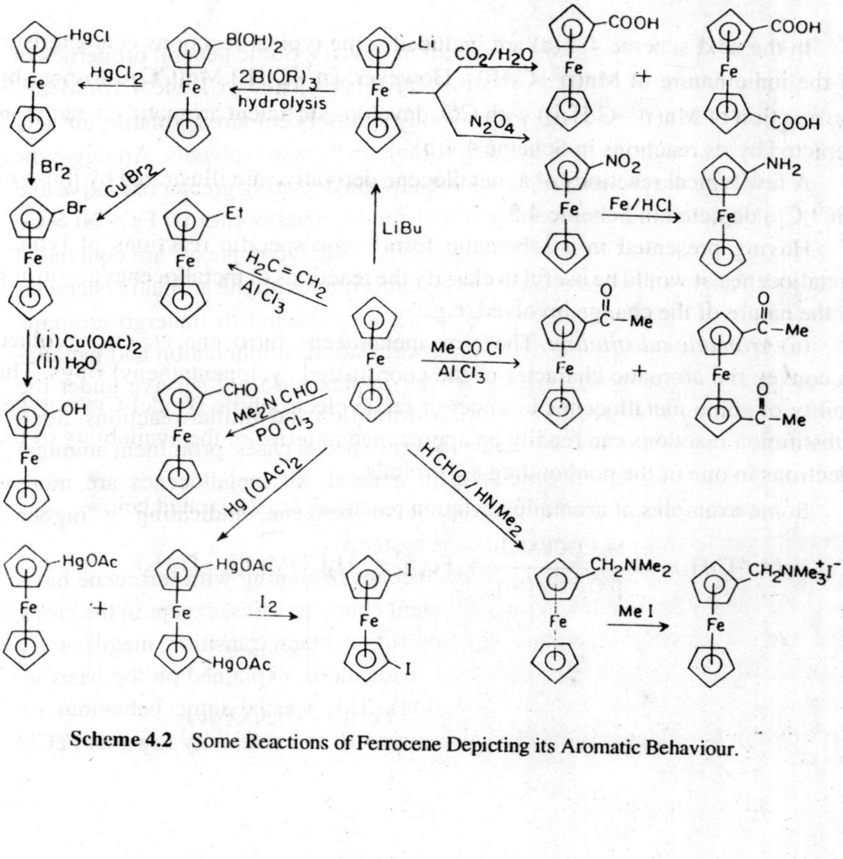

Scheme 4.2 Some Reactions of Ferrocene Depicting its Aromatic Behaviour.

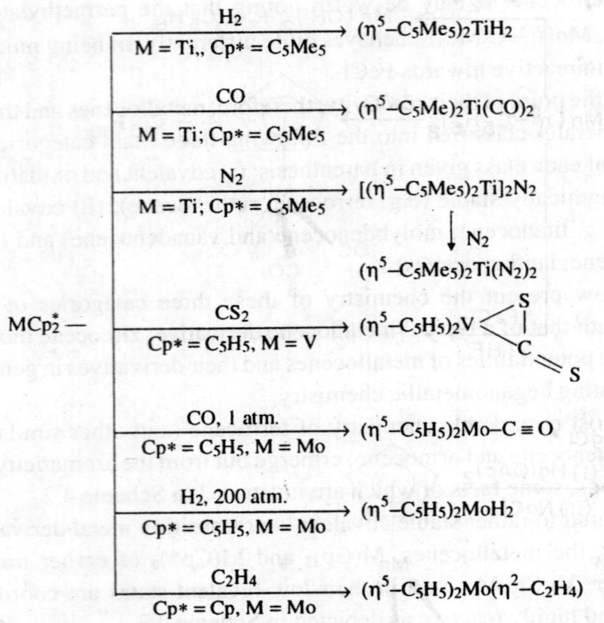

Scheme 4.3 Some Typical Reactions of Covalent Highly Reactive Metallocenes.

In the next scheme 4.4 (a) are included some typical reactions characteristic of the ionic nature of $Mn(\eta^5{-}C_5H_5)_2$. However, $(\eta^5{-}C_5H_5)\ Mn(CO)_3$, formed by the reaction of $Mn(\eta^5{-}C_5H_5)_2)$ with CO, develops sufficient aromatic character as depicted by its reactions in Scheme 4.4(b):

A few typical reactions of a metallocene derivative are illustrated by those of $TiCl_2Cp_2$ depicted in Scheme 4.5.

Having presented in a schematic form some specific reactions of typical metallocenes, it would be useful to classify the reactions of metallocenes according to the nature of the change involved, e.g.:

(a) *Aromatic substitution*. The terms metallocene, ferrocene, etc., were coined to convey the aromatic character of the coordinated cyclopentadienyl rings. The ability of some metallocenes to undergo facile electrophilic (Fridel-Crafts type) substitution reactions can readily be appreciated in terms of the availability of the electrons in one of the nonbonding e_{2g} orbitals.

Some examples of aromatic acylation reactions are illustrated below:

$$Fe(\eta^5{-}C_5H_5)_2 \xrightarrow[\text{0°C (very fast)}]{\text{MeCOCl/AlCl}_3} Fe(\eta^5{-}C_5H_4COMe)(\eta^5{-}C_5H_5)$$

$$\downarrow \text{Slower}$$

$$Fe(\eta^5{-}C_5H_4COMe)_2$$

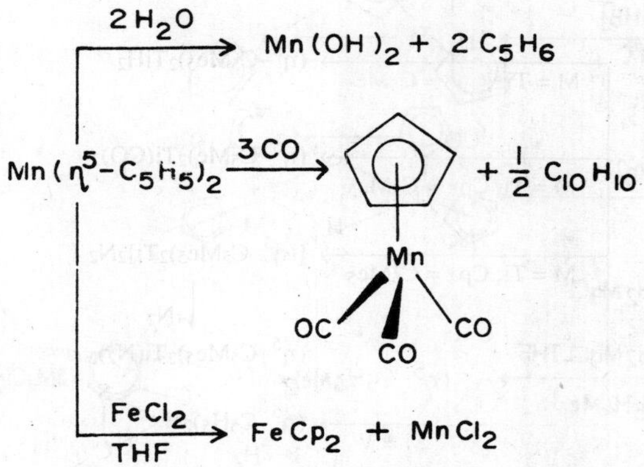

(a)

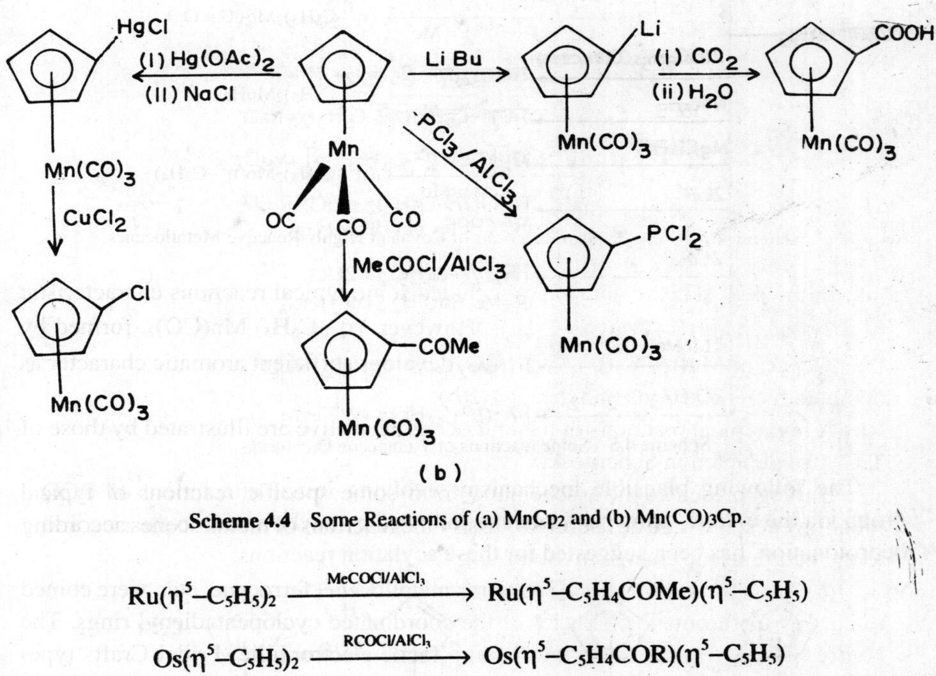

(b)

Scheme 4.4 Some Reactions of (a) $MnCp_2$ and (b) $Mn(CO)_3Cp$.

$$Ru(\eta^5\text{–}C_5H_5)_2 \xrightarrow{MeCOCl/AlCl_3} Ru(\eta^5\text{–}C_5H_4COMe)(\eta^5\text{–}C_5H_5)$$

$$Os(\eta^5\text{–}C_5H_5)_2 \xrightarrow{RCOCl/AlCl_3} Os(\eta^5\text{–}C_5H_4COR)(\eta^5\text{–}C_5H_5)$$

The first stage of acylation of ferrocene is extremely fast, its rate has been estimated to more than 10^6 times faster than that of benzene. The second stage of acylation of ferrocene is much slower than the first, but even then it is 200 times faster that of benzene. Amongst the iron group metallocenes, the rate of acylation follows the order:

$$Fe(\eta^5\text{–}C_5H_5)_2 > Ru(\eta^5\text{–}C_5H_5)_2 > Os(\eta^5\text{–}C_5H_5)_2 > C_6H_6$$

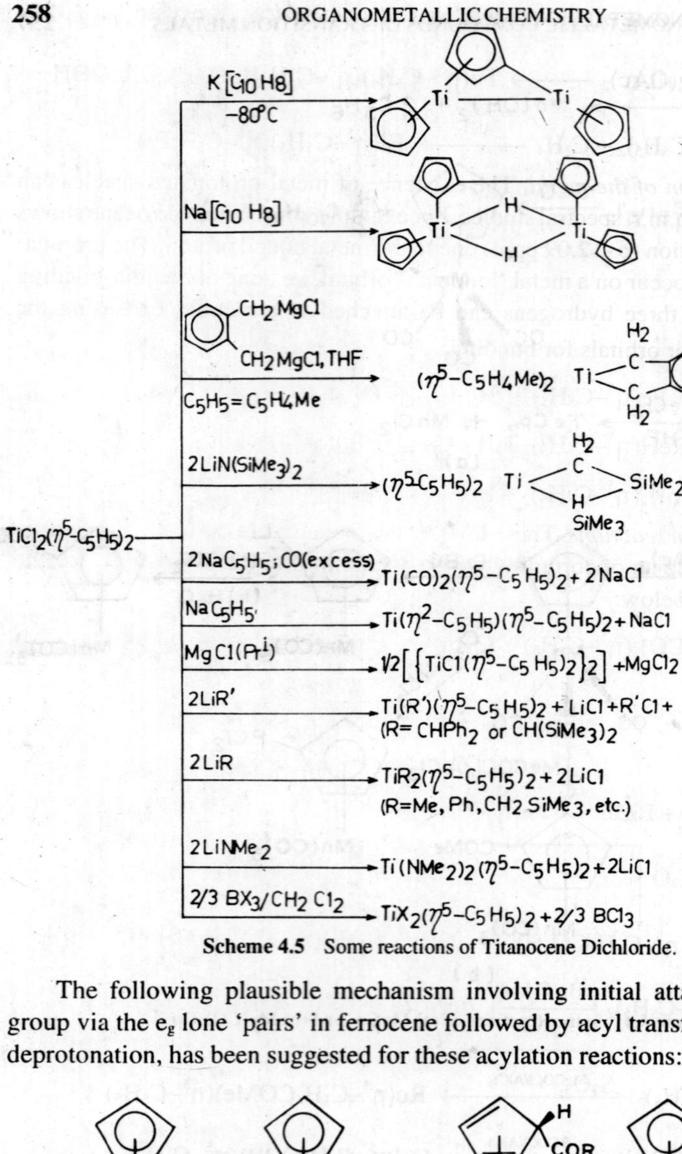

Scheme 4.5 Some reactions of Titanocene Dichloride.

The following plausible mechanism involving initial attachment of $R\overset{+}{C}O$ group via the e_g lone 'pairs' in ferrocene followed by acyl transfer to the ring and deprotonation, has been suggested for these acylation reactions:

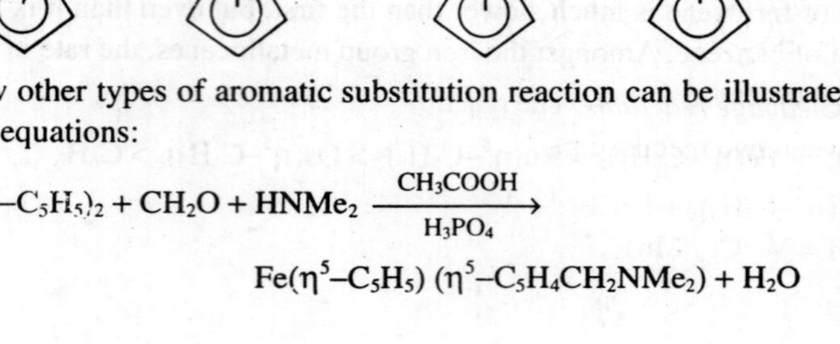

A few other types of aromatic substitution reaction can be illustrated by the following equations:

$$Fe(\eta^5{-}C_5H_5)_2 + CH_2O + HNMe_2 \xrightarrow[H_3PO_4]{CH_3COOH}$$

$$Fe(\eta^5{-}C_5H_5)\ (\eta^5{-}C_5H_4CH_2NMe_2) + H_2O$$

$$Fe(\eta^5\text{–}C_5H_5)_2 + Hg(OAc)_2 \longrightarrow Fe(\eta^5\text{–}C_5H_5)(\eta^5\text{–}C_5H_4HgOAc) + AcOOH$$

$$Fe(\eta^5\text{–}C_5H_5)_2 + C_2H_4 \xrightarrow{AlCl_3} Fe(\eta^5\text{–}C_5H_5)(\eta^5\text{–}C_5H_4Et)$$

(b) *Protonation of the metal.* The existence of metal-protonated species can be revealed by the n.m.r. spectral studies. For example, protonated ferrocene shows a high field absorption (δ -12.09 ppm) due to the metal bound proton. The protonation is believed to occur on a metal 'lone pair' orbital, i.e., one of the non-bonding e_{2g} orbitals. Upto three hydrogens can be attached to a metallocene using the η' and δ molecular orbitals for bonding:

$$Fe(\eta^5\text{–}C_5H_5)_2 + H^+ \rightarrow [FeH(\eta^5\text{–}C_5H_5)_2]^+$$

$$ReH(\eta^5\text{–}C_5H_5)_2 + H^+ \rightarrow [ReH_2(\eta^5\text{–}C_5H_5)_2]^+$$

$$MoH_2(\eta^5\text{–}C_5H_5)_2 + H^+ \rightarrow [MoH_3(\eta^5\text{–}C_5H_5)_2]^+$$

(c) *Metallation reactions.* These involve the replacement of ring hydrogens by metals, and provide an opportunity to introduce other functional groups into the ring as illustrated below:

$$Mn(CO)_3(\eta^5\text{–}C_5H_5) + LiBu \rightarrow Mn(CO)_3(\eta^5\text{–}C_5H_4Li)$$

$$\downarrow CO_2;\ H_2O$$

$$Mn(CO)_3(\eta^5\text{–}C_5H_4CO_2H)$$

$$Fe(\eta^5\text{–}C_5H_5)_2 + LiBu \rightarrow Fe(\eta^5\text{–}C_5H_5)(\eta^5\text{–}C_5H_4Li) + Fe(\eta^5\text{–}C_5H_4Li)_2$$

CO_2/H_2O $\downarrow$ $\qquad$ $\downarrow$ $B(OBu)_3;\ H_2O$

$$Fe(\eta^5\text{–}C_5H_5)(\eta^5\text{–}C_5H_4CO_2H) \qquad Fe(\eta^5\text{–}C_5H_5)(\eta^5\text{–}C_5H_4B(OH)_2)$$

$$Fe(\eta^5\text{–}C_5H_5)_2 \xrightarrow[KCl]{Hg(OAc)_2} Fe(\eta^5\text{–}C_5H_5)(\eta^5\text{–}C_5H_4HgCl)$$

$S_2O_3^{2-}$ $\downarrow$ $\qquad$ $\downarrow$ I_2

$$Fe(\eta^5\text{–}C_5H_5I)(\eta^5\text{–}C_5H_5)$$

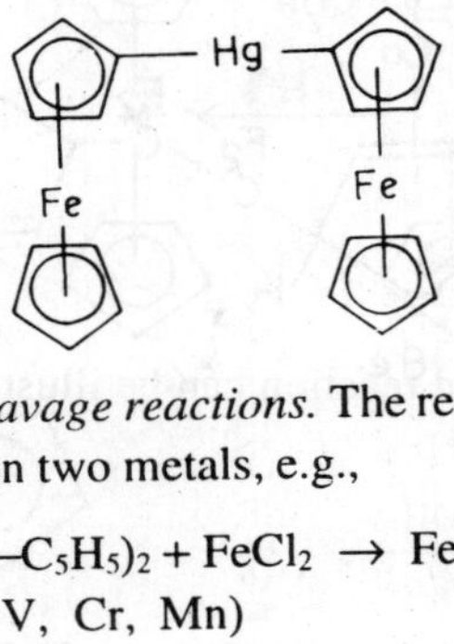

(d) *Cleavage reactions.* The reactions either involve the transfer of a $\eta^5\text{–}C_5H_5$ ring between two metals, e.g.,

$$M(\eta^5\text{–}C_5H_5)_2 + FeCl_2 \rightarrow Fe(\eta^5\text{–}C_5H_5)_2 + MCl_2$$

$(M = V,\ Cr,\ Mn)$

$$Li(\eta^5\text{–}C_5H_5) + W(CO)_6 \rightarrow Li[W(CO)_3(\eta^5\text{–}C_5H_5)] + 3CO$$

$$Ni(\eta^5\text{–}C_5H_5)_2 + Ni(CO)_4 \rightarrow [\{Ni(CO)(\eta^5\text{–}C_5H_5)\}_2] + 2CO$$

$$Ni(\eta^5\text{–}C_5H_5)_2 + Fe(CO)_5 \rightarrow [(\eta^5\text{–}C_5H_5)Ni(CO)_2Fe(CO)(\eta^5\text{–}C_5H_5)] + 2CO$$

or involve the protonolysis of the metal-cyclopentadienyl bond, e.g.:

$$Mn(\eta^5\text{–}C_5H_5)_2 + 2H_2O \rightarrow Mn(OH)_2 + 2C_5H_6$$

(e) *Reduction reactions*. These may involve reductions either the metal and/or the cyclopentadienly group; the cyclopentadienyl ring is either expelled as the C_5H_5 anion or becomes coordinated to the metal in a *tetrahapto* fashion, e.g.:

$$Fe(\eta^5\text{–}C_5H_5)_2 \xrightarrow{Li/EtNH_2} Fe + 2\,C_5H_5^-$$

$$Fe(\eta^5\text{–}C_5H_5)_2 \xrightarrow[\text{Rh catalyst}]{H_2} \text{no hydrogenating}$$

(cf. C_6H_6 hydrogenation in 5 min.)

$$Co^+(\eta^5\text{–}C_5H_5)_2 + NaBH_4 \longrightarrow Co(\eta^5\text{–}C_5H_5)(\eta^4\text{–}C_5H_6)$$

$$Co(\eta^5\text{–}C_5H_5)_2 + NaBH_4 \longrightarrow Co(\eta^5\text{–}C_5H_5)(\eta^4\text{–}C_5H_6)$$

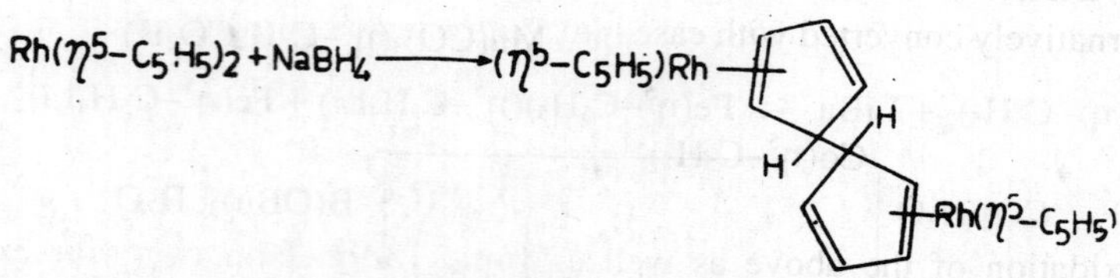

(f) *Addition reactions.* In view of comparative stability of 18 electron systems, cobaltocene (19e) and Nickelocene (20e) reactions in which the hapticity of one of the rings is changed in manner such as to afford an 18-electron system:

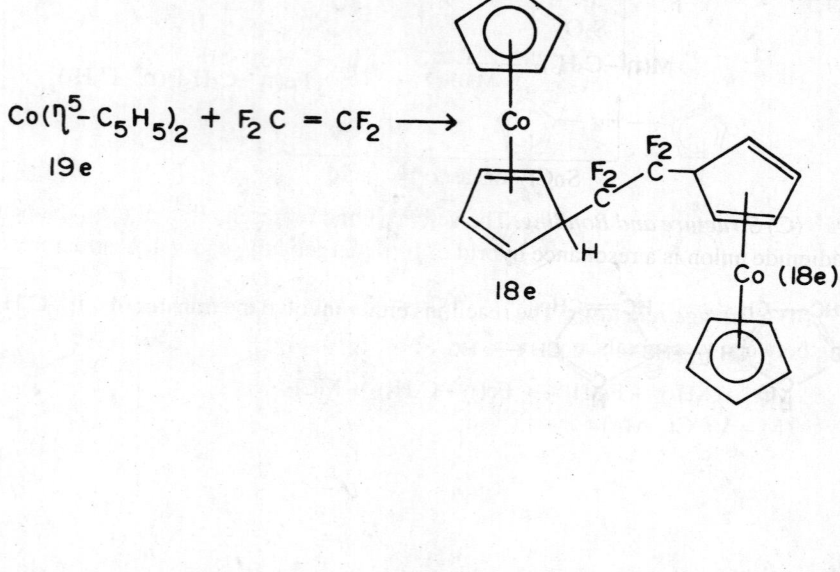

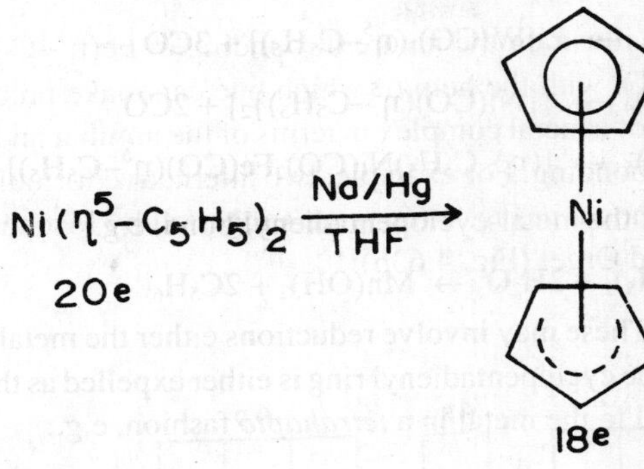

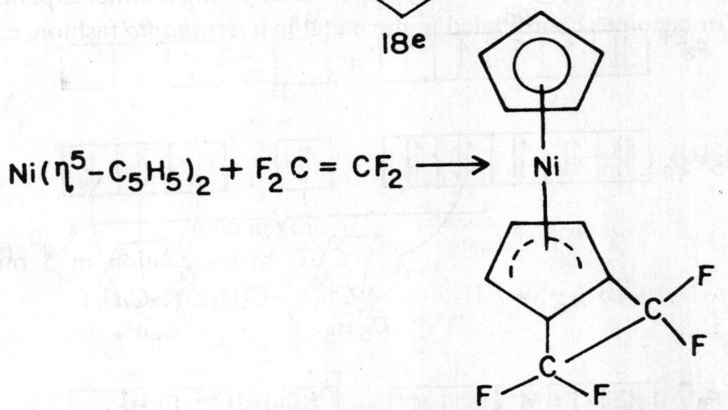

(g) *Electron transfer reactions*. The 19e cobaltocene and 20e nickelocene can be alternatively converted with ease into 18e systems by loss of electron(e), e.g.,

$$Co(\eta^5\text{–}C_5H_5)_2 \xrightarrow{[O]} [Co(\eta^5\text{–}C_5H_5)_2]^+$$

Oxidation of the above as well as some 18e metallocenes (for example, ferrocene, ruthenocene and osmocene) can be effected electrolytically, photochemically or chemically using a variety of inorganic/organic oxidants; these reactions are explained to occur due to the availability of e_g electrons in metallocene systems. Generally such reactions are reversible:

$$M(\eta^5\text{–}C_5H_5)_2 \underset{M'BF_4}{\xrightarrow{-e}} [M(\eta^5\text{–}C_5H_5)_2]^+[BF_4]^-$$

$$[M(\eta^5\text{–}C_5H_5)_2]^+[BF_4]^- \xrightarrow[\text{SnCl}_2 \text{ or ascorbic acid}]{+e} M(\eta^5\text{–}C_5H_5)_2$$

(C) *Structure and Bonding*. The velence bond representation of the cyclopentadienide anion is a resonance hybrid of five energetically equivalent structures:

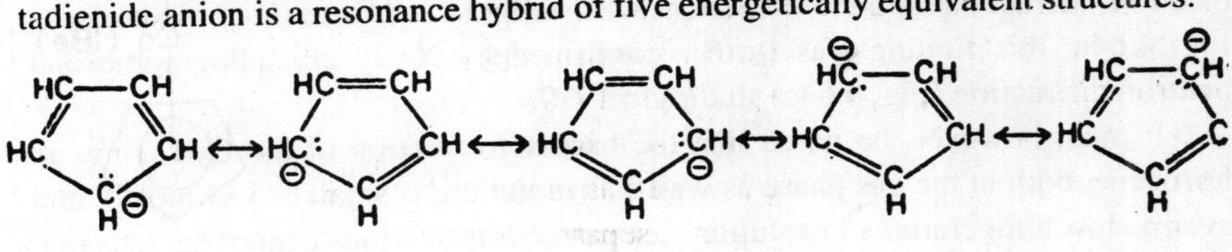

The simplest approach to illustrate the nature of bonding in $Fe(\eta^5\text{–}C_5H_5)_2$ is the valence bond theory (VBT), with the help of which one can make prediction about the molecular geometry of a metal complex in terms of the number and types of metal orbitals available for bonding. For example, two alternative formulations of the bonding in ferrocene (shown in Fig. 4.63) were suggested by Fischer (Fig. 4.63a) as well as by Dunitz and Orgel (Fig. 4.63b):

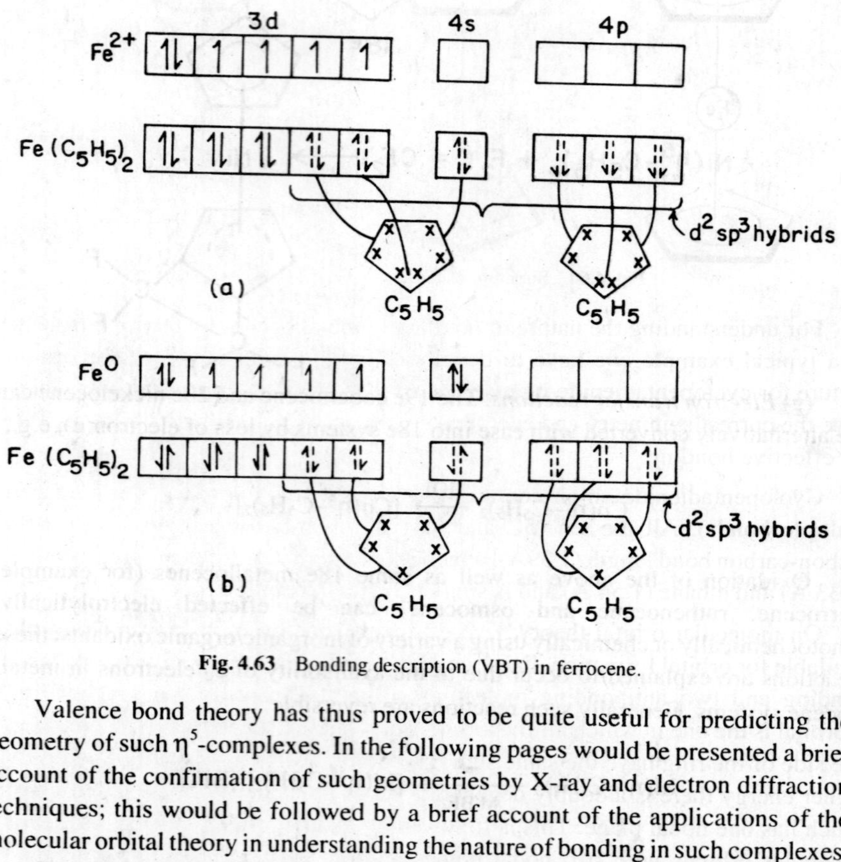

Fig. 4.63 Bonding description (VBT) in ferrocene.

Valence bond theory has thus proved to be quite useful for predicting the geometry of such η^5-complexes. In the following pages would be presented a brief account of the confirmation of such geometries by X-ray and electron diffraction techniques; this would be followed by a brief account of the applications of the molecular orbital theory in understanding the nature of bonding in such complexes.

Crystallographic studies in 1956 led to the authentication of sandwich structure of ferrocene, which was shown to posses staggered (D_{5d}) conformation of two planar pentahapto cyclopentadienyl rings (Fig. 4.64b). By contrast, gas phase electron diffraction data are consistent with eclipsed structure (D_{5h}) as shown in Fig. 4.64a; this finding was further confirmed by X-ray crystallographic and neutron diffraction (Fig. 4.64c) studies in 1979.

It may, however, be noted that the barrier to rotation of the C_5H_5 rings in ferrocene, both in the gas phase as well as in solution, is small (~5 kJ mol^{-1}), and even at low temperatures in solution, separate conformations cannot be detected.

Thus the two different conformations (staggered or eclipsed) in the solid state possibly arise from packing considerations.

Ruthenocene and osmocene have similar molecular structures with eclipsed parallel cyclopentadienyl rings. Other metallocenes such as $V(\eta^5\text{–}C_5H_5)_2$, $Cr(\eta^5\text{–}C_5H_5)_2$, $Co(\eta^5\text{–}C_5H_5)_2$ or $Ni(\eta^5\text{–}C_5H_5)_2$ have been shown to possess staggered (D_{5h}) conformation.

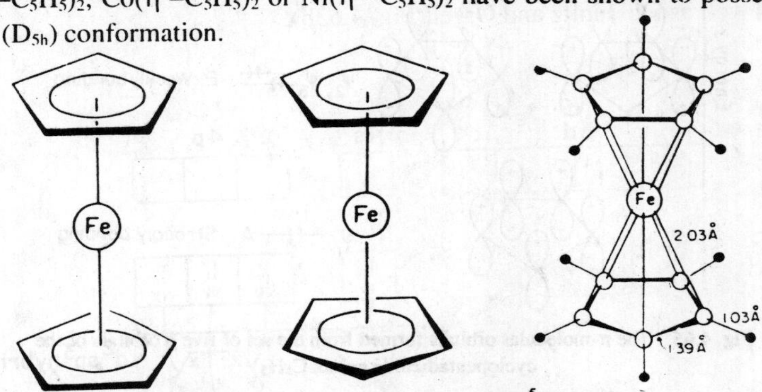

Fig. 4.64 Structure of ferrocene, $Fe(\eta^5\text{–}C_5H_5)_2$

For understanding the nature of bonding in metallocenes by taking ferrocene as a typical example, we have to develop, firstly, a suitable molecular orbital picture for cyclopentadienide ligand (s), and then select the metal orbitals which have the correct symmetry and energies to overlap with the ligand group orbitals for effective bonding.

Cyclopentadienide anion contains delocalized structure (i.e., non-localized multiple bonds). All the ten (5C and 5H) atoms are coplanar, with the same carbon-carbon bond length (1.39Å), which is intermediate between those of ethene (1.33 Å) and ethane (1.54 Å) and is similar to that in benzene.

On molecular orbital theory, five p orbitals (one on each carbon atom) are available for orbital formation, linear combinations of which can give rise to three bonding and two antibonding molecular orbitals. The most strongly bonding π-orbital is the one in which all five orbitals overlap (i.e. all the wave functions on one side of the ring have the same sign) and this is designated as 'A'. At slightly higher energy there is a doubly degenerate set of molecular orbitals (E_1) each of which has one nodal plane. This is followed by another doubly degenerate set of molecular orbitals with two nodal planes and yet higher energy. A schematic representation of the π-molecular orbitals of cyclopentadienide anion is given in Fig. 4.65, with their energies increasing from bottom to top.

To construct the ten ligand group orbitals of ferrocene from two cyclopentadienide anion, the five group molecular orbitals ($\psi_1 - \psi_5$) of one cyclopentadienide anion ($C_5H_5^-$) are combined in pairs with a similar set from the second C_5H_5 in symmetric (German gerade, g even) and antisymmetric (ungerade, u odd) fashion with respect to inversion. The resulting combinations (i.e. ligand group orbitals) and matching atomic orbitals of iron for ferrocene are shown in Fig. 4.65a. In an LCAO molecular orbital treatment of bonding in metallocenes, the atomic

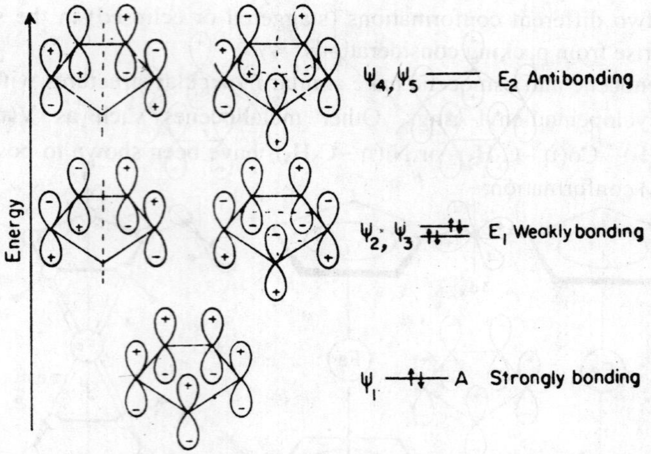

Fig. 4.65 The π-molecular orbitals formed from the set of five *p* orbitals of the cyclopentadienide anion, $C_5H_5^-$

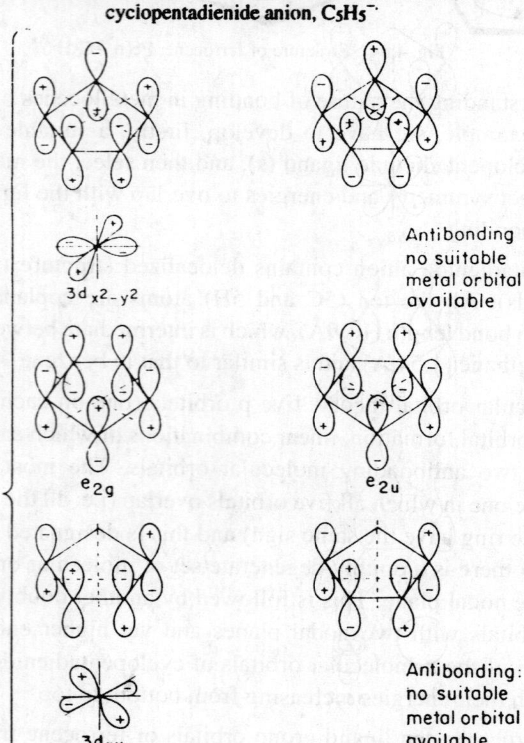

Fig. 4.65a: continued to top of page 265

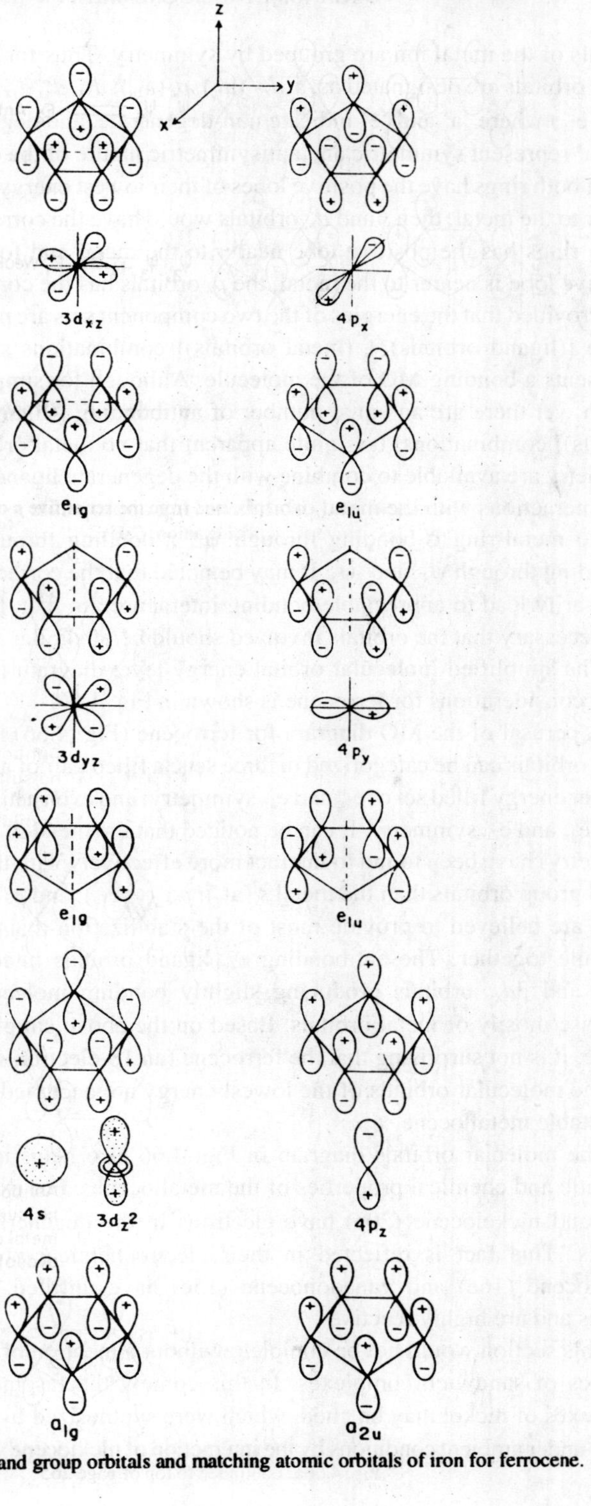

Fig. 4.65a Ligand group orbitals and matching atomic orbitals of iron for ferrocene.

orbitals of the metal ion are grouped by symmetry. Thus for ferrocene, the metal (iron) orbitals are designated as: s, d_{z^2} (a_{1g}), p_z (a_{2u}); d_{xz}, d_{xz} d_{yz} (e_{1g}); p_x, p_y (e_{1u}); d_{xy}, $d_{x^2-y^2}$ (e_{2g}) where 'a' and 'e' indicate non-degenerate, and degenerate where as 'g' and 'u' represent symmetric and antisymmetric nature of the orbitals.

If both rings have the positive lobes of their lowest energy orbitals on the sides nearer to the metal, then s and d_{z^2} orbitals would have the correct symmetry; if one of the rings has the positive lobe nearer to the metal and for the other ring, the negative lobe is nearer to the metal, the p_z orbitals has the correct symmetry.

Provided that the energies of the two component sets are not too far apart, each of the [(ligand orbitals) + (metal orbitals)] combinations shown in Fig. 4.65a represents a bonding MO of the molecule. Although for simplicity, these are not shown, yet there are an equal number of antibonding [(ligand orbitals) – (metal orbitals)] combinations. It is quite apparent that no metal orbitals of appropriate symmetry are available to combine with the degenerate ligand sets labelled as e_{2u}.

Interactions with the metal orbitals having appropriate symmetry for overlap, lead to metal-ring σ-bonding through ψ_1, π-bonding through ψ_2 and ψ_3, and δ-bonding through ψ_4 and ψ_5. It may be noted that the correct symmetry will not necessarily lead to appreciable bonding interaction; in fact, for this purpose it is also necessary that the orbitals involved should be of similar energies.

The simplified molecular orbital energy-level diagram that results from the above considerations for ferrocene is shown in Fig. 4.66.

A perusal of the MO diagram for ferrocene (Fig. 4.66) indicates that ligand group orbitals can be categorized in three sets: a filled pair of a_{1g} and a_{2u} symmetry; a higher energy filled set of e_{1g} and e_{1u} symmetry; and even a higher energy unfilled set of e_{2g} and e_{2u} symmetry. It can be noticed that the metal d_{xz} and d_{yz} orbitals (e_{1g} symmetry) have been found to interact more effectively with the cyclopentadienyl ligand group orbitals than the metal s (a_{1g}); p_x, p_y (e_{1u}); and p_z (a_{2u}) orbitals. These bonds are believed to provide most of the stabilization that holds the ferrocene molecule together. The antibonding e_{2g} ligand orbitals interact with the metal $3d_{x^2-y^2}$ and $3d_{xy}$ orbitals producing slightly bonding molecular orbitals which comprise mostly of metal orbitals. Based on the above simple molecular orbital picture, it is not surprising that the ferrocene (an 18-electron system) in which all the nine molecular orbitals of the lowest energy are occupied by electrons, is the most stable metallocene.

The moleclar orbitals diagram in Fig. 4.66 also helps us to rationalize the magnetic and chemical properties of the metallocenes. For example, cobaltocene (19e) and nickelocene (20e) have electrons in the degenerate, antibonding e^*_{1g} orbitals. This fact is reflected in their electrochemical properties. Similarly, chromocene (16e) and vanadonocene (15e) have unfilled bonding molecular orbitals and are highly reactive.

This section would not be complete without a mention of the interesting new varieties of sandwich complexes. In this context the fascinating 'tripledecker' complexes of nickel may be cited, which were synthesized by Werner and Salzer (1972) under ambient conditions by the interaction of nicklocene with carbonium ions:

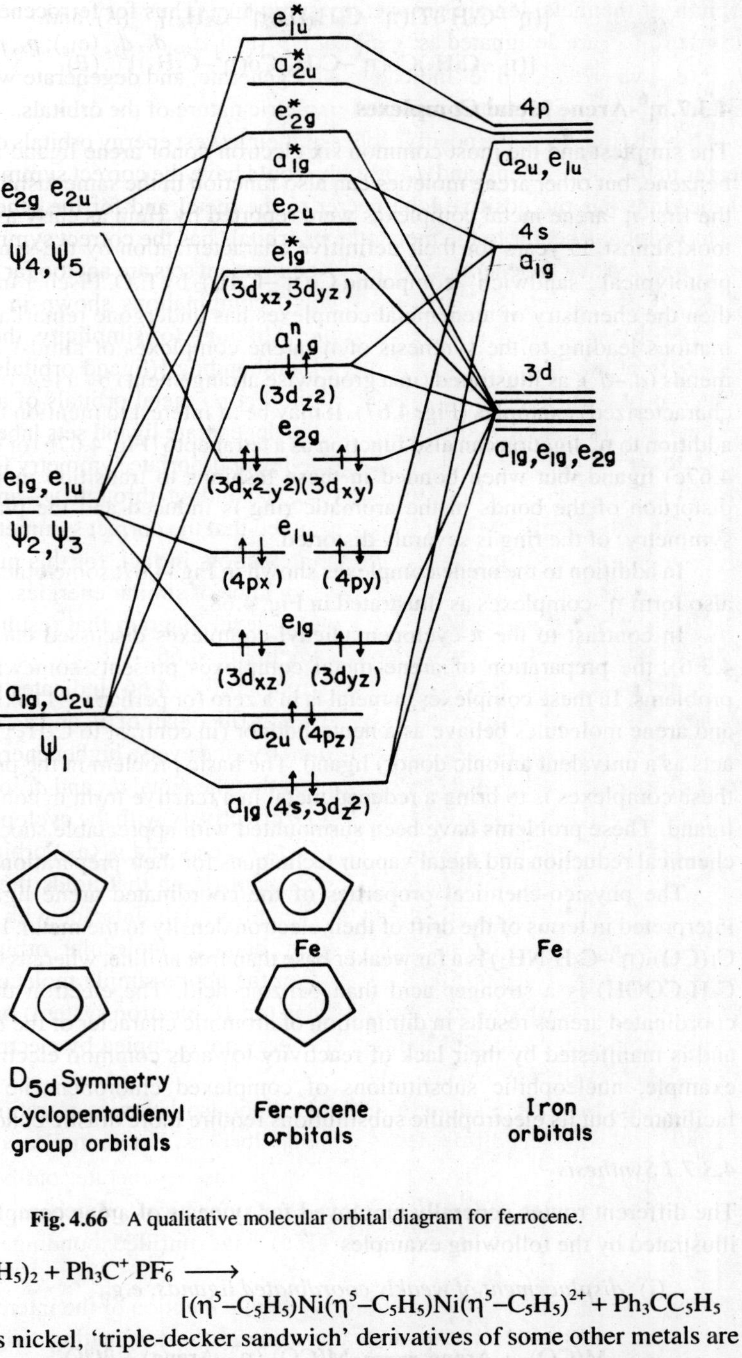

Fig. 4.66 A qualitative molecular orbital diagram for ferrocene.

$$2Ni(\eta^5-C_5H_5)_2 + Ph_3C^+PF_6^- \longrightarrow$$
$$[(\eta^5-C_5H_5)Ni(\eta^5-C_5H_5)Ni(\eta^5-C_5H_5)^{2+} + Ph_3CC_5H_5$$

Besides nickel, 'triple-decker sandwich' derivatives of some other metals are also known as illustrated by the following cyclooctatetraene and mixed derivatives of titanium (*A*) and cobalt (*B*) respectively:

$$[(\eta^8\text{–}C_8H_8)Ti(\eta^8\text{–}C_8H_8)Ti(\eta^8\text{–}C_8H_8)]^{2-}\ (A)\ \text{and}$$

$$[(\eta^5\text{–}C_5H_5)Co(\eta^8\text{–}C_8H_8)Co(\eta^5\text{–}C_5H_5)]^{2+}\ (B).$$

4.3.7. η^6-Arene Metal Complexes

The simplest and the most common six electron donor arene ligand is, of course, benzene, but other arene moieties can also function in the same fashion. Although the first η^6-arene metal complexes were reported by Hein as early as 1919, yet it took almost 35 years for their definitive characterization by the synthesis of the prototypical "sandwich" compound $Cr(\eta^6\text{–}C_6H_6)_2$ by E.O. Fischer in 1955. Since then the chemistry of arenemetal complexes has undergone remarkable transformations leading to the synthesis of η^6-arene complexes of almost all transition metals (d^2–d^8), as illustrated (in a groupwise arrangement) by a few typical (X-ray characterized) examples (Fig. 4.67). It may be of interest to mention that arenes in addition to η^6-ligation can also function as a tetrahapto (Fig. 4.67b) or dihapto (Fig. 4.67c) ligand, but when bonded in these fashions to transition metal species a distortion of the bonds in the aromatic ring is induced and the planarity (D_{6h}) symmetry) of the ring is severaly distorted.

In addition to the arene complexes shown in Fig. 4.67, some other ligands can also form η^6-complexes as illustrated in Fig. 4.68.

In contrast to the π-cyclopentadienyl complexes discussed earlier (Section 4.3.6), the preparation of arene metal complexes presents somewhat different problems. In these complexes, a metal is in a zero (or perhaps + 1) oxidation state, and arene molecules behave as a neutral donor (in contrast to C_5H_5 groups which acts as a univalent anionic donor) ligand. The basic problem in the preparation of these complexes is to bring a reduced metal in a reactive form in contact with the ligand. These problems have been surmounted with appreciable success by using chemical reduction and metal vapour techniques for their preparations.

The physico-chemical properties of the coordinated arene ligands can be interpreted in terms of the drift of their electron density to the metal. For example, $Cr(CO)_3(\eta^6\text{–}C_6H_5NH_2)$ is a far weaker base than free aniline, whereas $Cr(CO)_3(\eta^6\text{–}C_6H_5COOH)$ is a stronger acid than benzoic acid. The electron deficiency of coordinated arenes results in diminution of aromatic character of the benzene ring and is manifested by their lack of reactivity towards common electrophiles. For example, nucleophilic substitutions of complexed chlorobenzene are greatly facilitated, but its electrophilic substitutions require more drastic conditions.

4.3.7.1 Synthesis

The different routes generally employed for sytnesis of arene complexes can be illustrated by the following examples:

(i) *displacement of weakly coordinated ligands, e.g.,*

$$M(CO)_6 + \text{Arene} \xrightarrow{h\nu} M(CO)_3(\eta^6\text{–Arene}) + 3CO$$

(M = Cr, Mo or W)

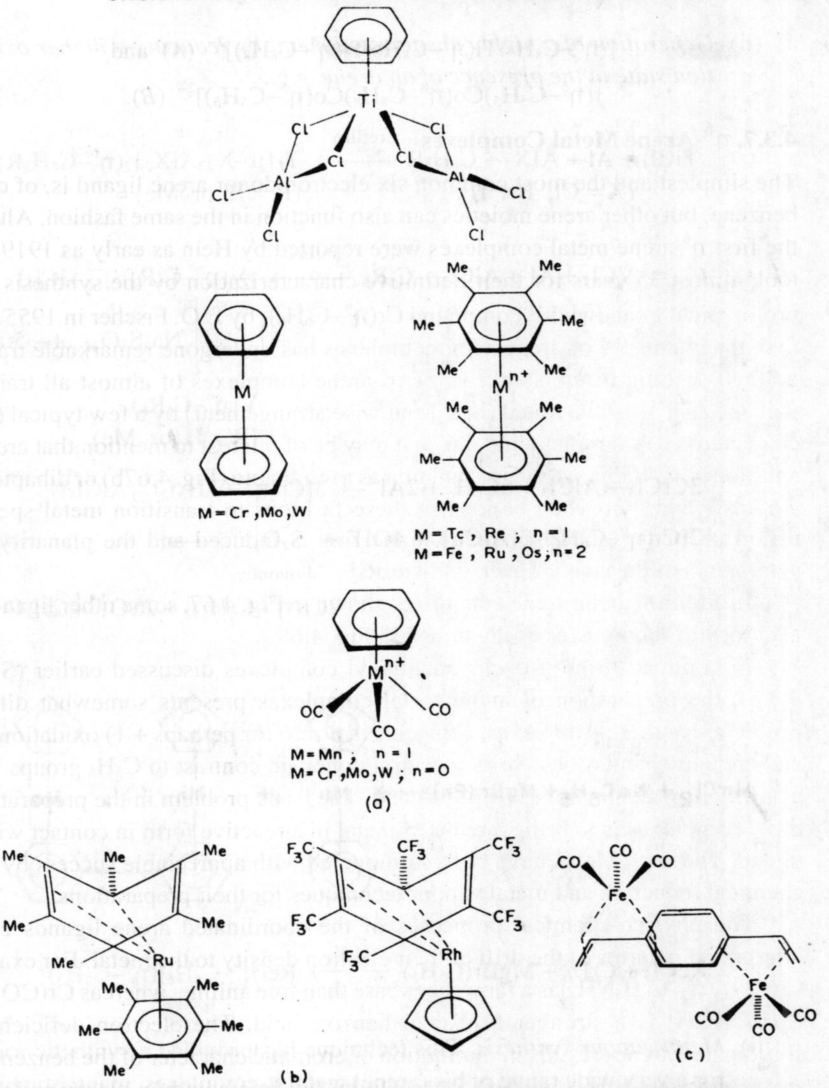

Fig. 4.67 Transition metal arene complexes exhibiting varied structural and ligation patterns (a) hexahapto-, (b) tetrahapto-, and (c) dihaptro-bonding modes.

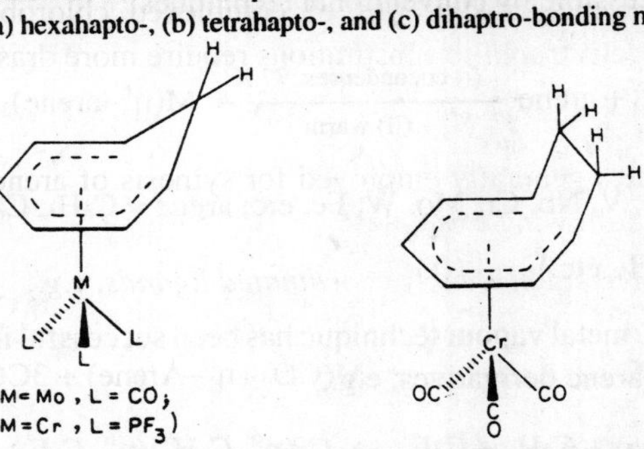

Fig. 4.68 Some η^6-complexes derived from ligands other than benzene.

(ii) *Generation of a highly unsaturated metal by reducing a higher oxidation state in the presence of an arene, e.g.,*

$$\underset{(X = Cl,\ Br,\ I)}{TiCl_4 + Al + AlX_3 + C_6H_5R} \xrightarrow{\text{reflux}} \underset{(R = H \text{ or } Me)}{Ti\{\mu\text{–}X)_2AlX_2\}_2(\eta^6\text{–}C_6H_5R)}$$

$$VCl_4 + Al + AlCl_3 + C_6R_6 \xrightarrow{\text{reflux}} [V(\eta^6\text{–}C_6R_6)_2]^+[AlCl_4]^-$$

$$\downarrow Na_2S_2O_4,\ \text{at} - 50°C$$

$$\underset{(R = H \text{ or } Me)}{V(\eta^6\text{–}C_6R_6)_2}$$

$$3CrCl_3 + AlCl_3 + 6C_6H_6 + 2Al \rightarrow 3[Cr(\eta^6\text{–}C_6H_6)_2]^+\ [AlCl_4]^-$$

$$2[Cr(\eta^6\text{–}C_6H_6)_2]^+[AlCl_4]^- + 4OH^- + \underset{\text{dithionite}}{S_2O_4^{2-}} \longrightarrow$$

$$2Cr(\eta^6\text{–}C_6H_6)_2 + 2SO_3^{2-} + 2H_2O$$

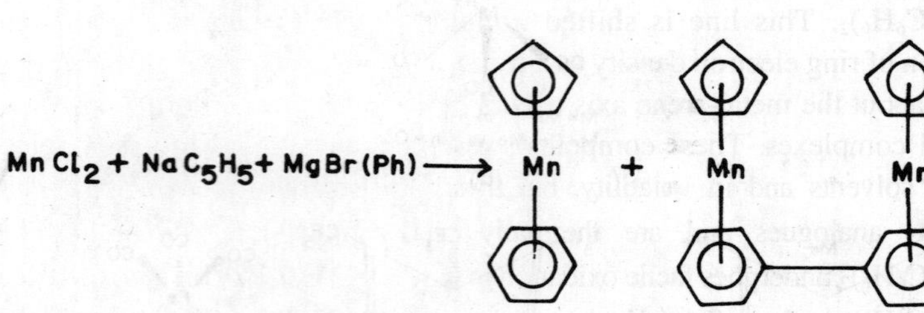

$$ReCl_5 + C_6H_6 + MgBr(C_5H_5) \xrightarrow{\text{U.V.}} Re(\eta^6\text{–}C_6H_6)(\eta^5\text{–}C_5H_5)$$

(iii) *Metal vapour synthesis*. This technique has provided a synthetic route for a very wide range of bis (arene) metal π-complexes, many of which were inacessible by conventional techniques:

$$M_{(a)} + \text{arene} \xrightarrow[\text{(ii) warm}]{\text{(i) cocondenses, 77 K}} M(\eta^6\text{–arene})_2$$

(M = Ti, V, Nb, Cr, Mo, W, Fe, etc; arene = C_6H_6, C_6H_5Me, $Me_3C_6H_3$, etc.)

Similarly, metal vapour technique has been successful in the preparation of mixed arene derivatives, e.g.,

$$Cr(a) + C_6H_6 + C_6F_6 \rightarrow Cr(\eta^6\text{–}C_6H_6)(\eta^6\text{–}C_6F_6)$$

$$Mn(a) + C_5H_5R + arene \xrightarrow[\text{(ii) warm}]{\text{(i) cocondense, 77K}} (\eta^6\text{–arene})Mn(\eta^5\text{–}C_5H_4R)$$

(arene = PhMe, or $Me_3C_6H_3$; R = H or Me)

(iv) *Cyclization of alkynes, e.g.,*

$$MnCl_2 + MeC \equiv CMe \xrightarrow[\text{(ii) } H_2O]{\text{(i) MgBr(Ar)}} Mn(\eta^6\text{–}C_6Me_6)_2$$

(v) *Grignard method.* Through the formation of a σ-bonded derivative e.g.,

$$CrCl_3(thf)_3 + MgBr(Ph) \rightarrow CrPh_3(thf)_3$$

$$\downarrow \text{(i) warm; (ii) hydrolysis/}H_2O$$

$$Cr(\eta^6\text{–}C_6H_6)_2$$

4.3.7.2 General Characteristics and Chemical Reactivity

Although in view of their paramagnetic nature, the p.m.r. spectra of bis (η^6-arene) complexes of transition metals are rather broad, yet a narrow line is observed for $Cr(\eta^6\text{–}C_6H_6)_2$. This line is shifted upfield relative to benzene, consistent with depletion of ring electron density as a result of complexation. Ring whizzing of arene groups about the metal-arene axis occurs in a similar manner to that of η^5-cylopentadienyl complexes. These compounds resemble metallocenes in their solubility in organic solvents and in volatility, but these are generally more reactive than their $\eta^5\text{–}C_5H_5$ analogues and are thermally and aerially less stable. For example, $Cr(\eta^6\text{–}C_6H_6)_2$ undergoes facile oxidation to $[Cr(\eta^6\text{–}C_6H_6)_2]^+$. This ease of oxidation of $Cr(\eta^6\text{–}C_6H_6)_2$ is also reflected by its reaction with alkyl halides, generating alkyl radicals which undergo usual disproportionation/addition reactions:

$$Cr(\eta^6\text{–}C_6H_6)_2 \xrightarrow{C_2H_5I} [Cr(\eta^6\text{–}C_6H_6)_2]^+ + C_2H_5^{\bullet}$$

$$\downarrow$$

$$(C_2H_6 + C_2H_4 + C_4H_{10})$$

Amongst the bis (η^6-arene) complexes, the chromium derivative appears to exhibit comparatively the higher thermal stability. This could be explained on the basis of its 18-electron system. However, not only the corresponding complexes of later first row transition metals, but also those of molybdenum and tungsten are less stable. Alkyl substitution in the arene ring increases the stability of the metal derivatives which can be easily understood on the basis of the enhancement in the electron density of coordinated arene system due to the inductive (+I) effect of the alkyl group. Even for such ligands like $\eta^6\text{–}C_6Me_6$, strong reducing agents like alkali metals are required to produce the 20 and 21 electron systems as in $Fe(\eta^6\text{–}C_6Me_6)_2$ and $Co(\eta^6\text{–}C_6Me_6)_2$ derivatives respectively.

The metal-arene bonds in bis (η^6-arene) metal complexes appear to be less stable and tend to be more easily cleaved than those in the corresponding η^5-cyclopentadienyl metal complexes.

As stated earlier (Sec. 4.3.7), (η^6-arene) derivatives depict weaker aromatic character and thus these do not undergo the reactions (such as Friedel-Crafts acylation, sulphonation, or nitration) characteristic of benzene and its derivatives. The loss of aromaticity enhances competition for the incoming electrophile by relatively electron-rich metal centre. However, (η^6-arene) complexes do undergo substitution, addition, metallation, mercuration, nitration, expansion, reactions in limited and specific cases. There should be much less competitions for nucleophiles and therefore nucleophilic substitution is enhanced compared with free arenes. Unlike the cyclopentadienyl group, arene ligands are readily displaced from complexes by other neutral ligands, such as carbon monoxide, phosphines, 2, 2′-bipyridyl, etc.

The typical reactions of (η^6-arene) metal derivatives are illustrated in the following pages.

(i) *Arene replacement reactions.* Bis (arene) transition-metal complexes undergo a variety of arene exchange processes, e.g.,

$$Cr(\eta^6\text{–}C_6H_3Me_3)_2 + 2C_6H_6 \xrightarrow{\text{reflux}} Cr(\eta^6\text{–}C_6H_6)_2 + 2Me_3C_6H_3$$

On the other hand, arene displacement by non-aromatic ligands is less facile and requires drastic reaction conditions, e.g.,

$$Cr(\eta^6\text{–}C_6H_6)_2 + 6L \xrightarrow[\text{high pressure (300 atm)}]{200^\circ C} CrL_6 + 2C_6H_6$$

$$(L = CO \text{ or } PF_3)$$

$$Mo(\eta^6\text{–}C_6H_6)_3 + 6PF_3 \xrightarrow[600 \text{ atm}]{100^\circ C} Mo(PF_3)_6 + 2C_6H_6$$

$$Cr(\eta^6\text{–}C_6H_6)_2 + C_6H_5Me + 3CO \xrightarrow{\text{pressure}} Cr(CO)_3(\eta^6\text{–}C_6H_5Me) + C_6H_6$$

On the other hand, reaction with a monodentate tertiary phosphines and phosphites generally causes smooth release of only one arene ligand, e.g.,

$$Mo(\eta^6\text{–arene})_2 + 3PR_3 - Mo(PR_3)_3(\eta^6\text{–arene}) + \text{arene}$$

However, the naphthalene ligand is very readily displaced from $Cr(\eta^6\text{–}C_{10}H_8)_2$ by a variety of other ligands as shown in Scheme 4.6.

(ii) *Redistribution reactions*

$$(\eta^6\text{–}C_6H_6)_2 + Cr(CO)_6 \rightarrow 2Cr(CO)_3(\eta^6\text{–}C_6H_6)$$

(iii) *Reactions at the arene ligands*

Nucleophilic substitution/addition reactions: Benzene and its derivatives generally undergo electrophilic addition or substitution reactions. However, when

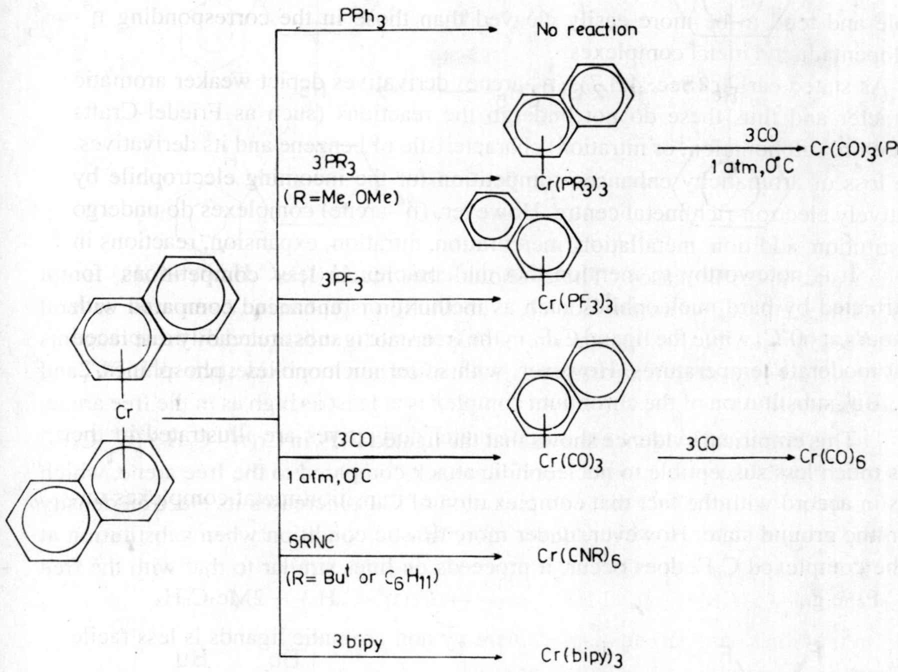

Scheme 4.6 Reactions of Bisnaphthalene Chromium(O).

these molecules are coordinated to electron withdrawing transition metal centres they become much more susceptible to be attacked by nucleophiles such as H^-, R^-, CN^-, MeO^-, R_3N, etc. This can be interpreted in terms of lowered electron density on the rings, arising from drift of electrons from the ring towards metal centre. Some examples illustrating these reactions are:

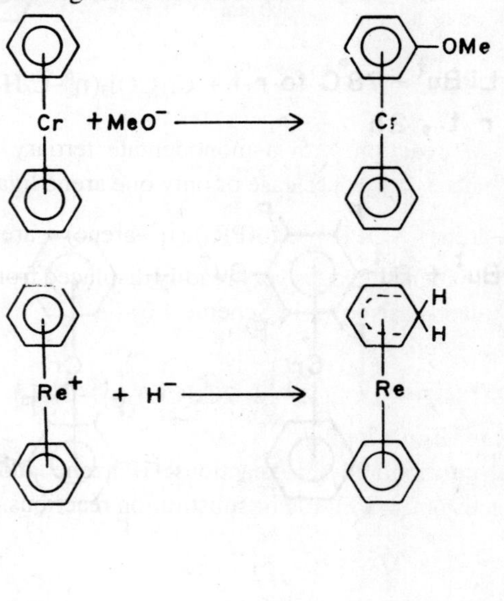

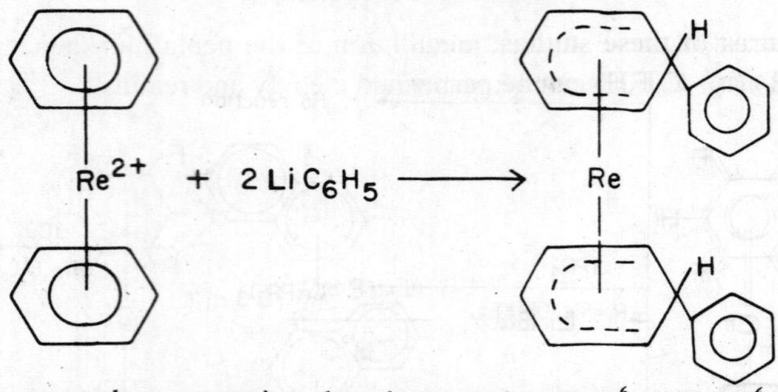

It is noteworthy to mention that the complex $Cr(\eta^6{-}C_6H_6)(\eta^6{-}C_6F_6)$ is not affected by hard nucleophiles such as methoxy or piperidine even after several hours at 60°C, while the ligand C_6F_6 in the free state is substituted by these reagents at moderate temperatures. However, with softer nucleophiles such as $LiBu^n$ and $LiBu^t$, substitution of the chromium complex is at least as high as in the free arene.

This empirical evidence shows that the ligated C_6F_6 in $Cr(\eta^6{-}C_6H_6)$ $(\eta^6{-}C_6F_6)$ is much less susceptible to nucleophilic attack compared to the free arene, which is in accord with the fact that complexation of C_6F_6 increases its electron density in the ground state. However, under more drastic condition when substitution at the complexed C_6F_6 does occur, it proceeds on lines similar to that with the free C_6F_6, e.g.,

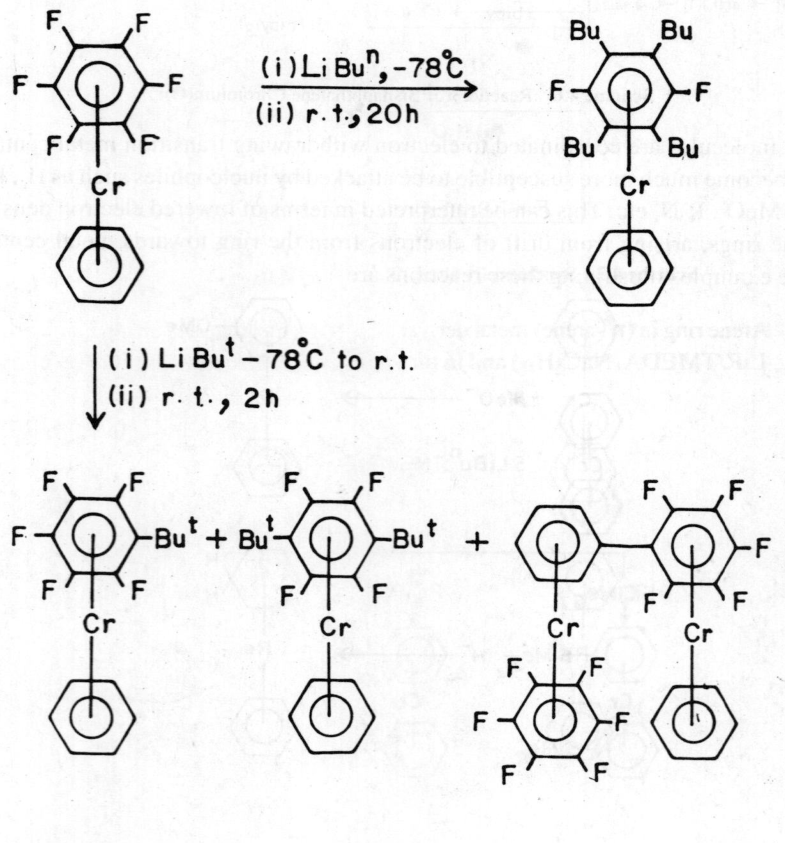

In contrast of these studies, metallation of the pentafluorobenzene complex $Cr(\eta^6-C_6H_6)(\eta^6-C_6F_5H)$ can be performed cleanly and readily:

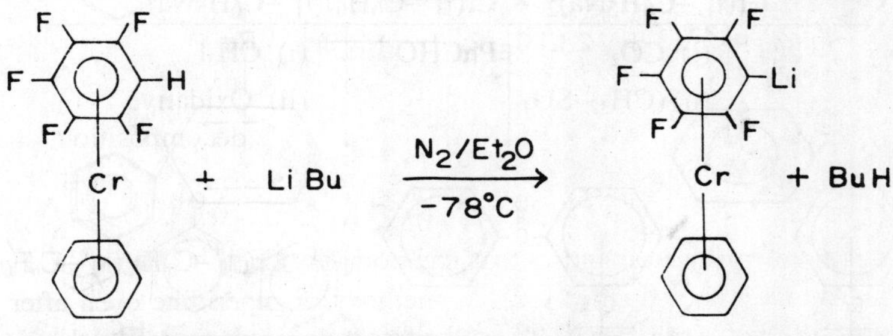

In *situ*, treatment of these monolithiated species with various reagents has provided convenient routes to a range of mono-functionalized bis(arene) chromium(O) complex intermediates (Scheme 4.7):

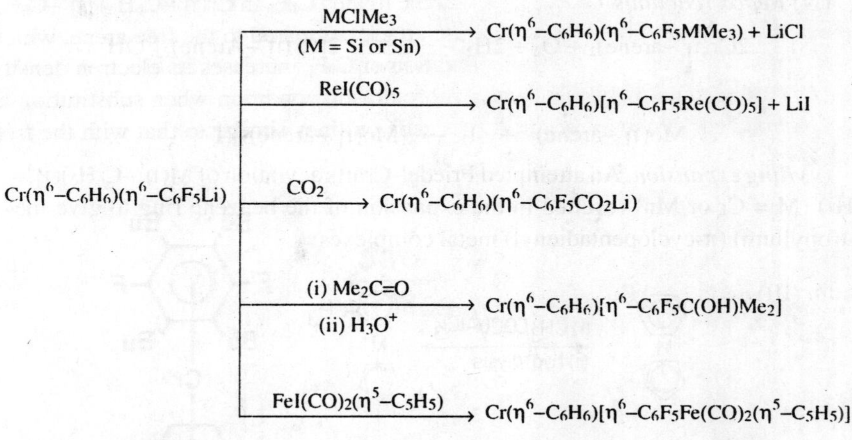

Scheme 4.7 Synthetically Useful Reactions of $Cr(\eta^6-C_6H_6)(\eta^6C_6F_5Li)$.

Arene ring in (η^6-arene) metal derivatives can be metallated with alkali metal alkyls (e.g. LiR/TMEDA, NaC_5H_{11}) and in turn converted into useful organic compounds:

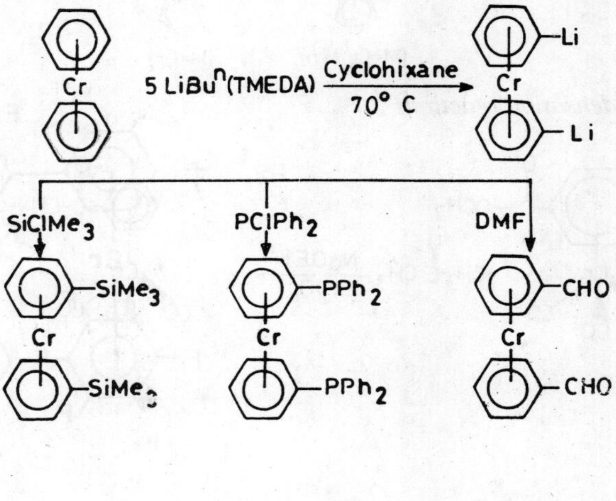

$$Cr(\eta^6\text{–}C_6H_6)_2$$
$$\downarrow NaC_5H_{11},\ \text{hexane}$$
$$Cr(\eta^6\text{–}C_6H_5Na)_2 + Cr(\eta^6\text{–}C_6H_6)(\eta^6\text{–}C_6H_5Na)$$

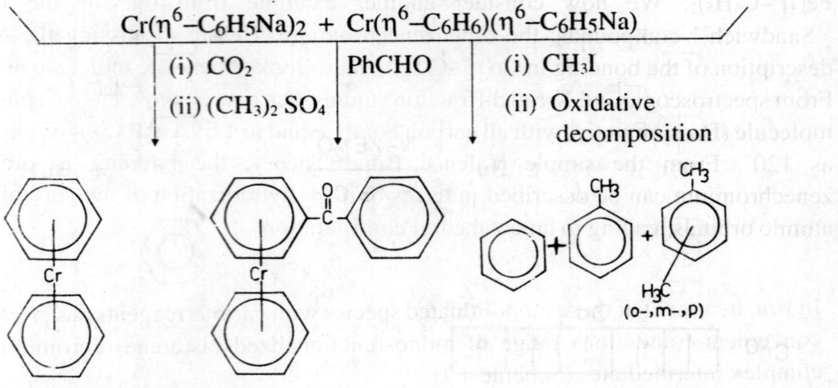

(iv) *Redox reactions*

$$2Cr(\eta^6\text{–arene})_2 + O_2 + 2H_2O \underset{Na_2S_2O_4}{\rightleftharpoons} 2[Cr(\eta^6\text{–Arene})_2]^+OH^-$$

$$Mo(\eta^6\text{–arene})_2 + \tfrac{1}{2}I_2 \rightarrow [Mo(\eta^6\text{–arene})_2]^+I^-$$

(v) *Ring expansion.* An attempted Friedel-Crafts acylation of $M(\eta^6\text{–}C_6H_6)(\eta^5\text{–}C_5H_5)$ (M = Cr or Mn) resulted in the expansion of the benzene ring to give the (π-tropylium) (π-cyclopentadienyl) metal complexes:

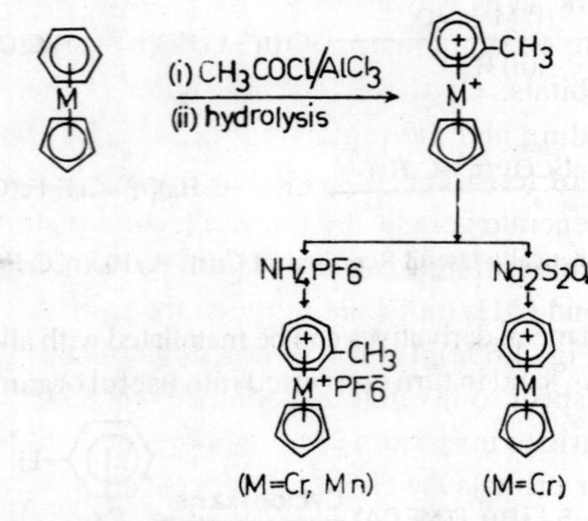

(vi) *Condensation reactions*

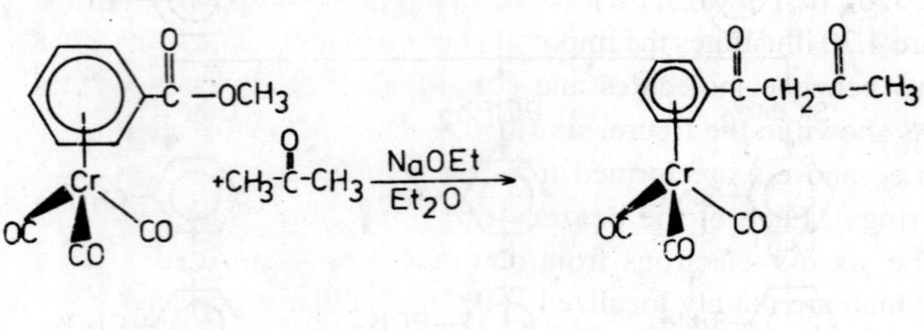

4.3.7.3 Structure and Bonding

In the preceding section, we have presented bonding description in ferrocene $Fe(\eta^5\text{–}C_5H_5)_2$. We now consider another example from one of the first "Sandwich" compounds, the dibenzenechromium. Before discussing the MO description of the bonding in $Cr(\eta^6\text{–}C_6H_6)_2$, let us look at benzene molecule itself. From spectroscopic and X-ray diffraction studies, benzene is known to be a planar molecule (D_{6h} symmetry, with all carbon bonds equal to 1.39Å and C–$\hat{C}$–C angles as 120°. From the simple Valence Bond Theory, the bonding in dibenzenechromium can be described in terms of d^2sp^3 hybridization of the chromium atomic orbitals leading to an octahedral configuration:

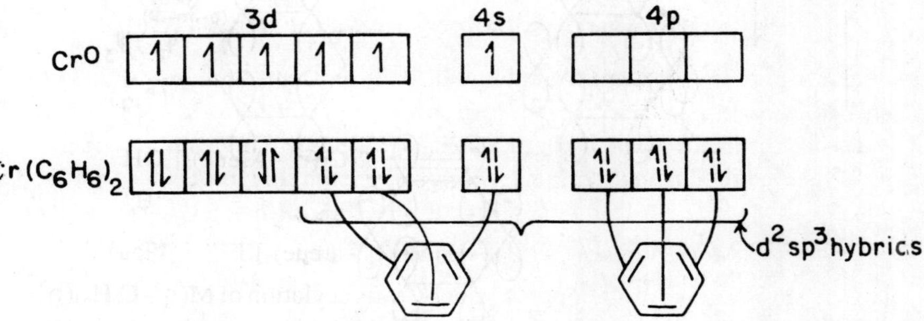

In the MO description the six p_π atomic orbitals on the planar C_6H_6 molecule can be combined in much the same way as we did for $C_5H_5^-$ group in the preceding section 4.3.6. Thus there are six combinations of the six component p_π orbitals, to produce six molecular orbitals. Of these, six molecular orbitals three can be properly described as bonding and the remaining three as anti-bonding orbitals (Fig. 4.69). As in the case of ferrocene, one can place one ring on the top of the other in such a way as to generate gerade and ungerade combinations.

These six group molecular orbitals can themselves be combined in pairs with a similar set from the second C_6H_6 molecule in much the same way as described already for ferrocene (cf. Fig. 4.65a). If may be noted that one of these combinations is symmetric with respect to inversion (*gerade*, even, g, whereas the other combination is antisymmetric to inversion, i.e., *ungerade*, odd, u). These resulting combined group molecular orbitals (twelve in number) can in turn be combined with metal orbitals having similar symmetry properties (Fig. 4.70). Their symmetry labels are s, d_{z^2} (a_{1g}); p_z(a_{2u}); p_x, p_y(e_{1u}); d_{zx}, d_{zy}(e_{1g}); and d_{xy}, $dx^2 - y^2$ (e_{2g}).

Figure 4.70 illustrates the important bonding interactions between the orbitals of the two benzene molecules and chromium metal valence orbitals in $Cr(\eta\text{–}C_6H_6)_2$. As shown in the figure, six filled bonding MO s (one each a_{1g} and a_{2u} and two each e_{1g} and e_{1u}) are formed from the 12-π-electrons (6 each from the two benzene rings). Each of the benzene rings, therefore, functions as a 6-electron donor. The six d^5s^1-electrons from chromium(O) seem to fill in the next three orbitals which are mainly localized at the metal. The system is further stabilized

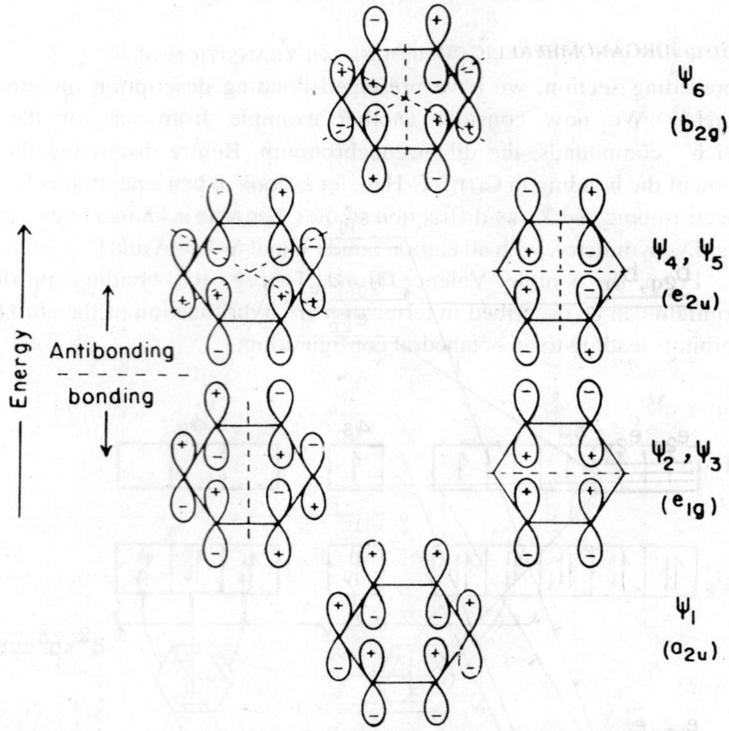

Fig. 4.69 Schematic representation of the π-molecular orbitals of benzene.

by the back-bonding of two of the $e_{2g}(d_{x^2-y^2}$ and $d_{xy})$ orbitals overlapping with the empty antibonding benzene orbitals.

The non-bonding nature of the metal a_{1g} and e_{2g} orbitals in $Cr(\eta^6-C_6H_6)_2$ provides a facile explanation for the observation that it is possible to synthesize a wide range of bis-arene complexes of transition metals with 16 (e.g. $Ti(\eta^6-C_6H_6)_2$) and 17 (e.g. $V(\eta^6-C_6H_3Me_3)_2$) valence electrons.

4.3.8 Derivatives of Heptahapto (η^7) Ligands

Cycloheptatriene, C_7H_8, is an isomer of norbornadiene. Whereas norbornadiene functions only as a σ or/and π-alkene ligand (cf. Sec. 4.3.2.1), cycloheptatriene exhibits an intriguing flexibility in its ligating patterns. Although cycloheptariene is a six electron donor similar to benzene, but it differs in the localization of its π-electrons as illustrated by the molybdenum complex, $Mo(CO)_3(\eta^6-C_7H_8)$ (Fig. 4.71); the alternation in bond length in the free cycloheptariene is retained in the complexes.

The complex (Fig. 4.71) can be oxidized (hydride ion abstraction) to form tropylium complexes, which could be reduced back to the parent complex with $NaBH_4$:

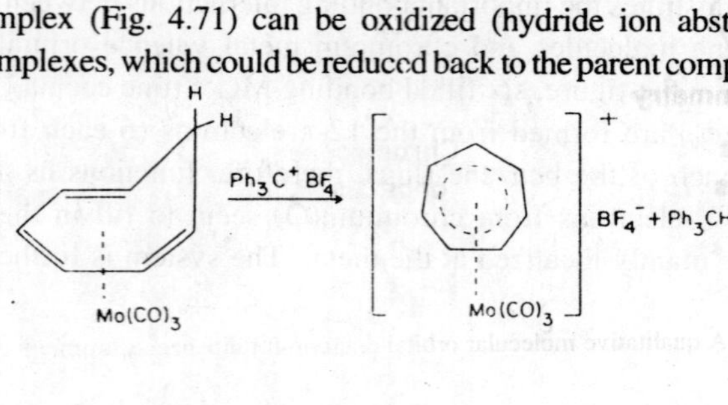

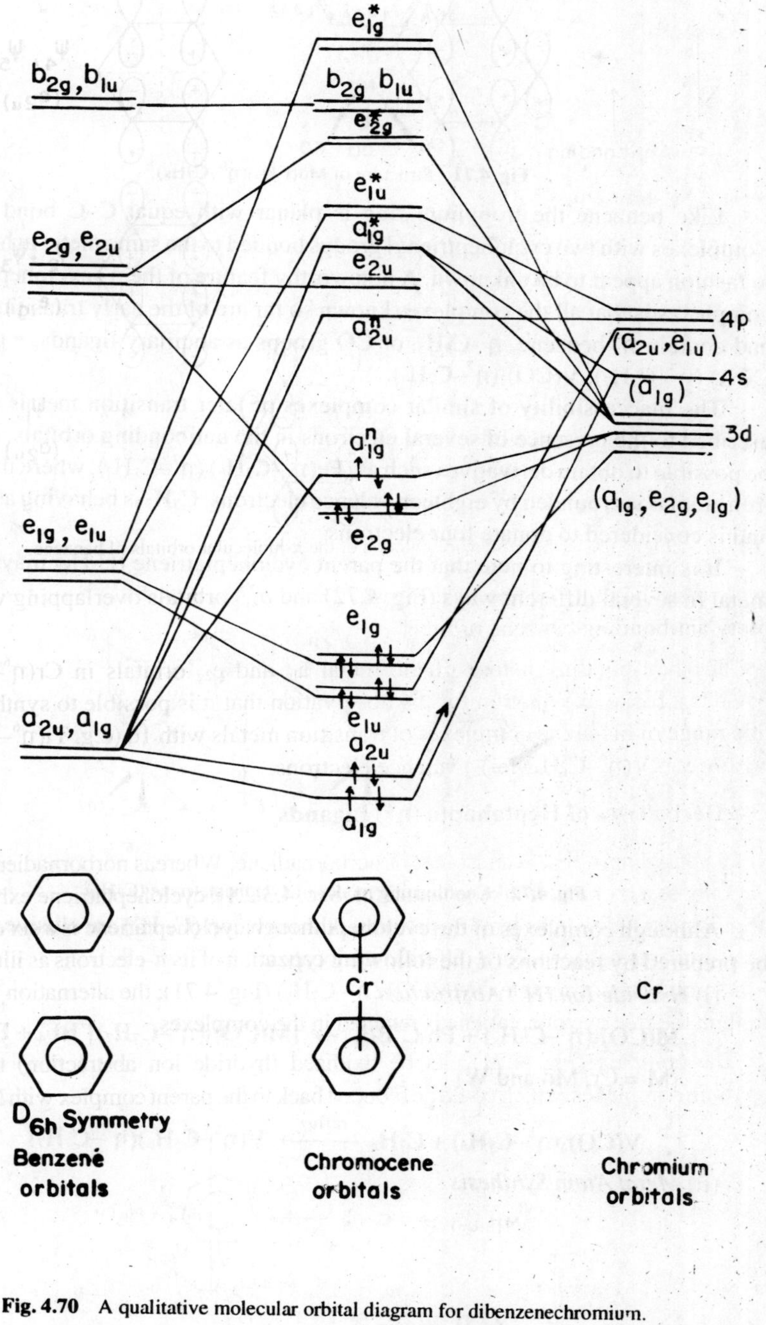

Fig. 4.70 A qualitative molecular orbital diagram for dibenzenechromium.

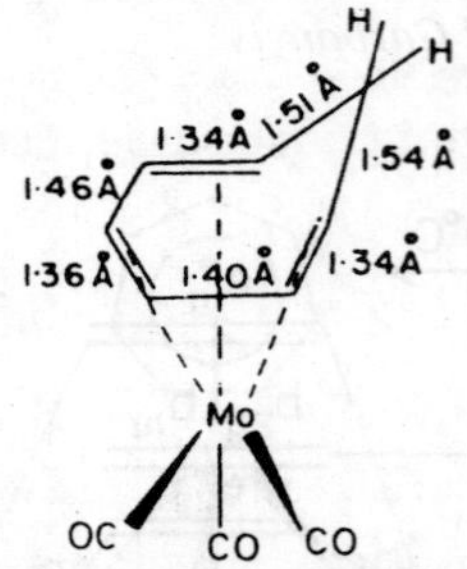

Fig. 4.71 Structure of $Mo(CO)_3(\eta^6-C_7H_8)$.

Like benzene the tropylium ring is planar with equal C–C bond lengths. Complexes with two cycloheptrienyl groups bonded to the same metal in heptahapto fashion appear to be unknown. A noteworthy feature of the η^7-cycloheptatrienyl complexes is that all the complexes known so far are of the early transition metals and contain η^6-benzene, η^5-C_5H_5, or CO groups as ancillary ligands, e.g., $V(\eta^5-C_5H_5)(\eta^7-C_7H_7)$, $V(CO)_3(\eta^7-C_7H_7)$.

The inaccessibility of similar complexes of later transition metals could be ascribed to the presence of several electrons in the antibonding orbitals. It should be possible to obtain derivatives such as $(Fe(\eta^3-C_3H_5)(\eta^7-C_7H_7)$, where the central iron atom is surrounded by eighteen valence electrons; C_7H_7 is behaving as trihapto and is considered to donate four electrons.

It is interesting to note that the parent cycloheptatriene (C_7H_8) may bind the metal in several different ways (Fig. 4.72).

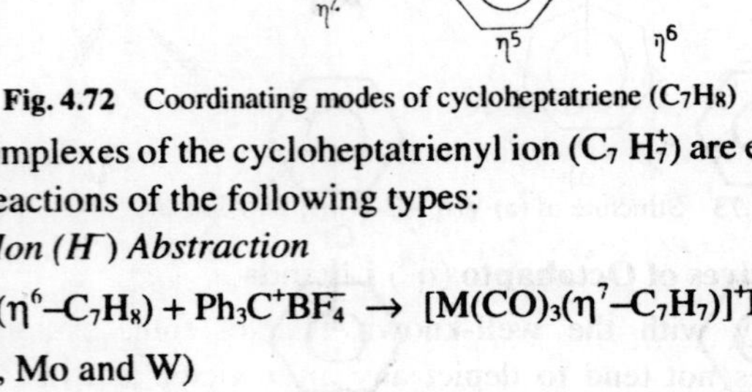

Fig. 4.72 Coordinating modes of cycloheptatriene (C_7H_8)

Although complexes of the cycloheptatrienyl ion ($C_7H_7^+$) are elusive, they can be prepared by reactions of the following types:

(i) *Hydride Ion (H^-) Abstraction*

$$M(CO)_3(\eta^6-C_7H_8) + Ph_3C^+BF_4^- \rightarrow [M(CO)_3(\eta^7-C_7H_7)]^+BF_4^- + Ph_3CH$$

(M = Cr, Mo and W)

$$V(CO)_4(\eta^5-C_5H_5) + C_7H_8 \xrightarrow{\text{reflux}} V(\eta^5-C_5H_5)(\eta^7-C_7H_7)$$

(ii) *Metal Atom Synthesis*

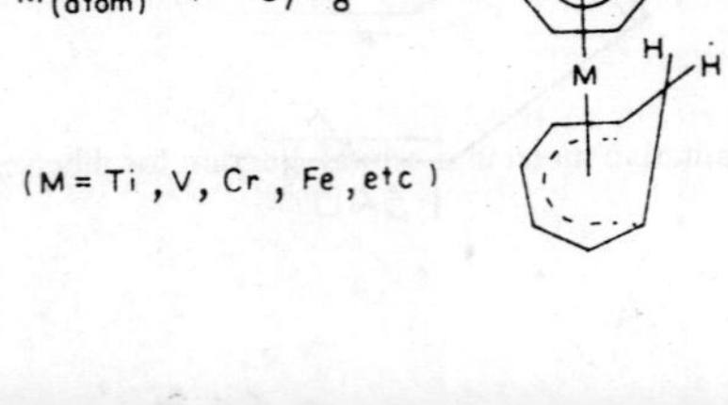

(iii) *Reaction with Metal Carbonyls*

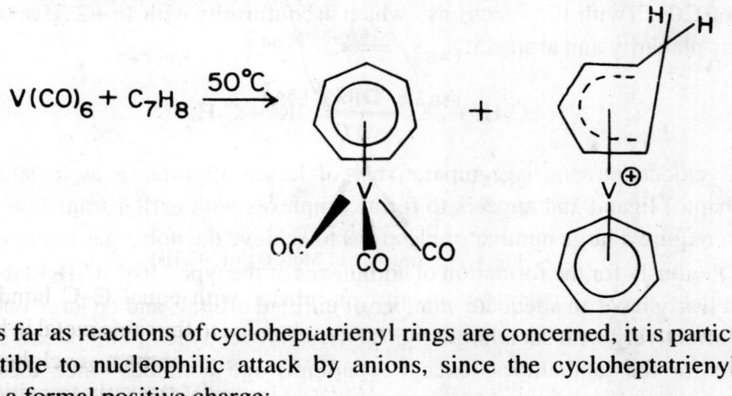

As far as reactions of cycloheptatrienyl rings are concerned, it is particularly susceptible to nucleophilic attack by anions, since the cycloheptatrienyl ring carries a formal positive charge:

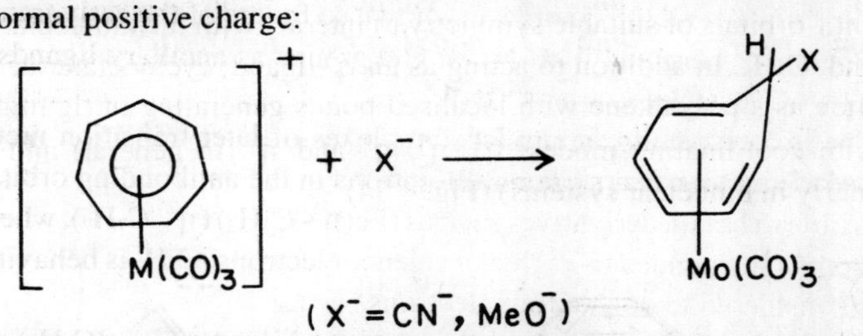

The bonding in cycloheptatrienyl complexes appears to be similar to that in η^5–C_5H_5 and η^6–C_6H_6 complexes. For example, both the complexes $V(\eta^5–C_5H_5)(\eta^7–C_7H_7)$ and $V(CO)_3(\eta^7–C_7H_7)$ feature symmetrical planar C_7 rings as illustrated in Fig. 4.73; the rings are eclipsed in the former compound:

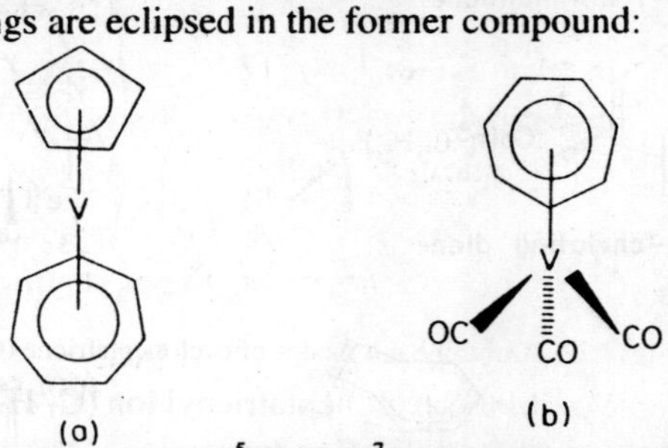

Fig. 4.73 Structure of (a) $V(\eta^5–C_5H_5)(\eta^7–C_7H_7)$, and (b) $V(CO)_3(\eta^7–C_7H_7)$.

4.3.9 Derivatives of Octahapto (η^8) Ligands

In conformity with the well-known Hückel rule, cyclooctatetraene (C_6H_8) molecule does not tend to depict any aromaticity and in its free state, it is a non-planar molecule, in which the adjacent C–C bond lengths correspond alternatively to the characteristic single and double bonds:

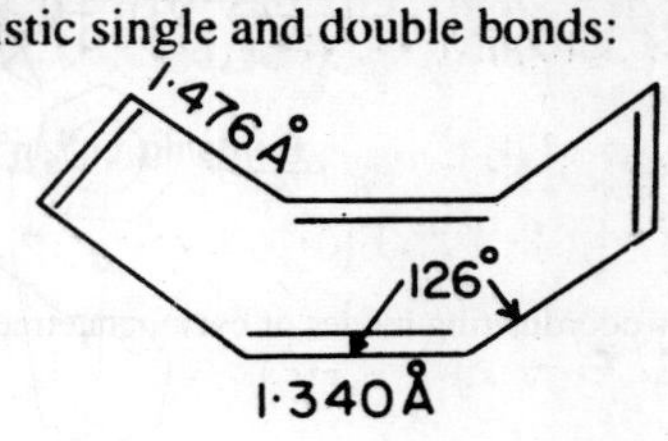

The reaction of cyclooctatetraene with an electropositive metal produces the dianion $C_8H_8^{2-}$ (with 10π-electrons), which in conformity with 4n + 2 (Hückel) rule, depicts planarity and aromacity:

$$C_8H_8 + 2K \xrightarrow[-30^\circ C]{THF} 2K^+ + C_8H_8^{2-}$$

Cyclooctatetraene is comparatively of lesser importance as a delocalized (octahapto) ligand and appears to form complexes with earlier transition metals, which require a large number of electrons to achieve the noble gas configuration.

Obviously for the formation of complexes of the type $M(\eta^8-C_8H_8)_2$, the metal must clearly have an adequate number of unfilled orbitals and be large enough to bond effectively with such large ring. Thus, sandwich type of complexes are only known for the lanthanides (except scandium) and actinides since these have sufficient f-orbitals of suitable symmetry to interact with the molecular orbitals of the ligand, C_8H_8. In addition to acting as an η^8-ligand, cyclooctatetraene behaves more often as a polyalkene with localised bonds generating intriguing structural types with coordinating modes: η^2, η^4, η^6 and η^8 (in general) and η^3 and η^5 (particularly in binuclear systems) (Fig. 4.74):

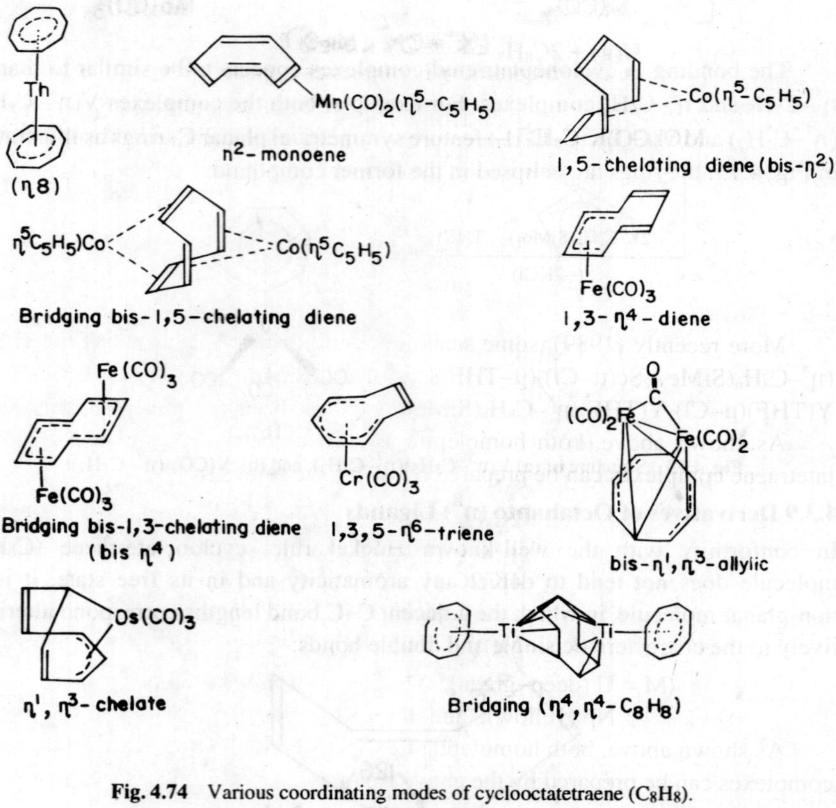

Fig. 4.74 Various coordinating modes of cyclooctatetraene (C_8H_8).

The synthetic routes generally employed for cyclooctatetraene planar complexes can be illustrated by the following typical reactions:

$$Ti(OBu)_4 + AlEt_3 + C_8H_8 \rightarrow Ti(C_8H_8)_2$$

$$NbCl_5 + 2K^+C_8H_8^{2-} \xrightarrow{THF} NbCl_2(THF)(\eta^8\text{–}C_8H_8) + 2KCl$$

$$(\eta^5\text{–}C_5H_5)\,CrCl_2 + C_8H_8 \xrightarrow[THF]{MgBr\,(i\text{–}Pr)} Cr(\eta^5\text{–}C_5H_5)(\eta^6\text{–}C_8H_8)$$

$$LnCl_3 + 4K^+2[C_8H_8]^{2-} \xrightarrow{THF} K^+[Ln(\eta^8\text{–}C_8H_8)_2]^- + 3KCl$$

(Ln = Ce, Pr, Nd, Sm)

$$Ln(atom) + C_8H_8 \xrightarrow[at\ -196^\circ C]{cocondensation} \xrightarrow[with\ THF]{extraction} [Ln(\eta^8\text{–}C_8H_8)(THF)_2]^+[Ln(\eta^8\text{–}C_8H_8)_2]^-$$

(Ln = La, Ce, Nd, or Er)

The metal complexes of cyclooctatetraene can be broadly classified into two types of derivatives with (i) planar, and (ii) non-planar C_8H_8 rings.

4.3.9.1 Derivatives with Planar C_8H_8 Rings

$$U(g) + 2C_8H_8 \xrightarrow[at\ -196^\circ C]{co\text{–}condensation} U(\eta^8\text{–}C_8H_8)_2$$

$$MCl_4 + 2K^+C_8H_8^{2-} \xrightarrow[0^\circ C]{THF} M(\eta^8\text{–}C_8H_8)_2 + 2KCl$$

(M = Th, U, Np and Pu)

$$\xrightarrow[(-2KCl)]{2K^+C_8H_6(SiMe_3)_2^{2-},\ THF} + M(\eta^8\text{–}C_8H_6(SiMe_3)_2)_2$$

(M = Th, U)

More recently (1989), some scandium and yttrium derivatives of the types, $(\eta^8\text{–}C_8H_6(SiMe_3)_2Sc(\mu\text{–}Cl)_2(\mu\text{–}THF)Sc(\eta^8\text{–}C_8H_6(SiMe_3)_2)$ and $(\eta^8\text{–}C_8H_6(SiMe_3)_2)Y(THF)(\mu\text{–}Cl)_2Y(THF)(\eta^8\text{–}C_8H_6(SiMe_3)_2$ have also been prepared by similar routes.

As shown above, both homoleptic as well as heteroleptic planar cyclooctatetraene complexes can be prepared by the above routes.

$$2C_8H_8 + 4K$$
$$\downarrow THF, -30^\circ C$$
$$4K^+2C_8H_8^{2-}$$
$$\downarrow MCl_4,\ THF,\ 0^\circ C$$
$$M(\eta^8\text{–}C_8H_8)_2 + 4KCl$$

(M = U (deep–green), Th (colourless), Pa (yellow) Np (yellow), and Pu (cherry red))

As shown above, both homoleptic as well as heteroleptic planar octatetraene complexes can be prepared by the above routes.

Complexes of planar cyclooctatetraene can be further sub-divided into two general types: (i) $M(C_8H_8)_n$ and (ii) $M(C_8H_8)X$. Structures of both types have been elucidated by X-ray diffraction technique.

Complexes of the type $M(\eta^8 - C_8H_8)_2$ (M = Th, Pa, U and Pu), have been shown by X-ray crystallography to have eclipsed parallel planar rings (Fig. 4.75):

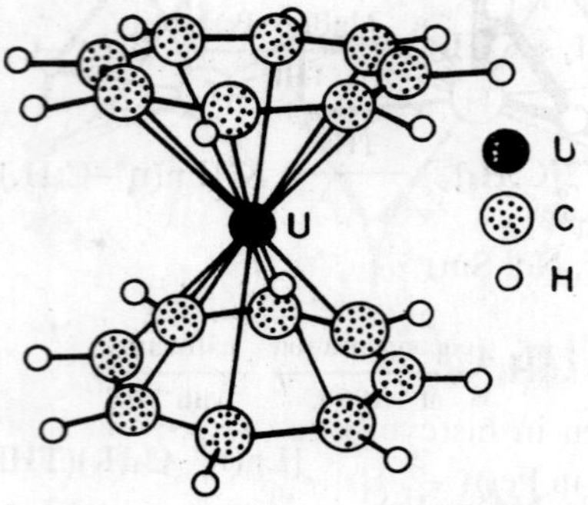

Fig. 4.75 Structure of $U(\eta^8\text{–}C_8H_8)_2$ (D_{8h} symmetry).

The structure of a rather unusual pair of complex ions, such as $[Nd(C_8H_8)(THF)_2]^+$ and $[Nd(C_8H_8)_2]^-$, with the same lanthanoid metal is depicted as below:

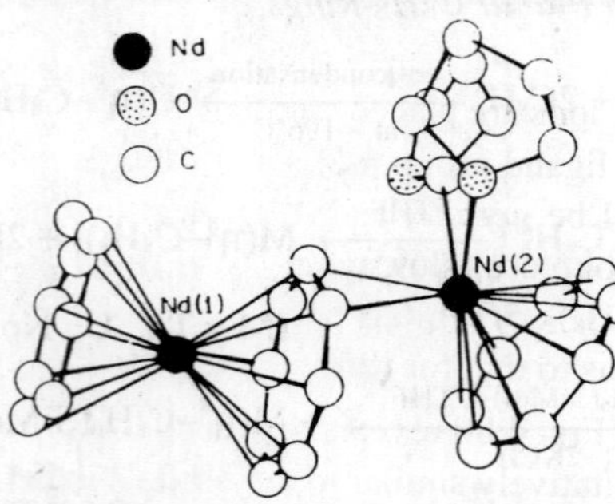

Fig. 4.76 Structure of the $[Nd(C_8H_8)(THF)_2]^+[Nd(C_8H_8)_2]^-$ ion pair.

An X-ray diffraction study for the cerium complex $[Ce(\eta^8\text{–}C_8H_8)Cl(THF)_2]_2$ (which belongs to the family of $M(\eta^8\text{–}C_8H_8)X$ compounds, demonstrates its dimeric nature (Fig. 4.77).

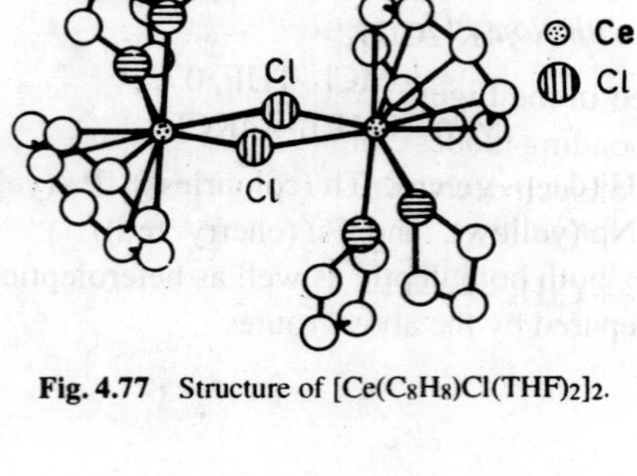

Fig. 4.77 Structure of $[Ce(C_8H_8)Cl(THF)_2]_2$.

One of the elusive example of η^8-bonding to a *d*-block element is in the intriguing complex $Ti_2(C_8H_8)_3$ (Fig. 4.78), which has been shown to contain two planar cyclooctatetraene (η^8) and a third central puckered ring bridging the two titanium atoms in a bis-η^4-mode.

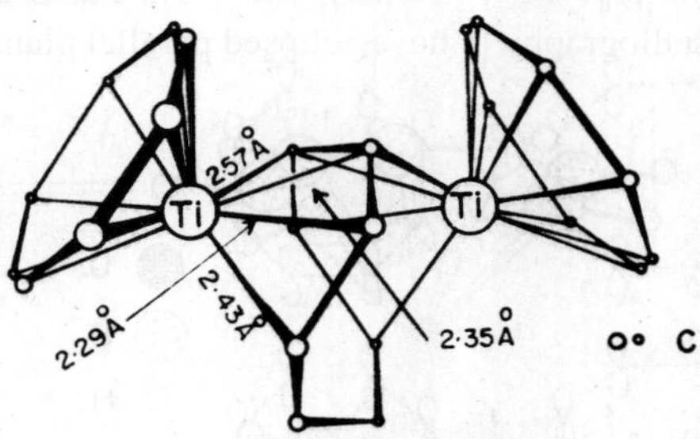

Fig. 4.78 Structure of $Ti_2(C_8H_8)_3$.

The bonding pattern in bis(cyclooctatetraenyl) actinides, $An(\eta^8–C_8H_8)_2$, is basically similar to that in $Fe(\eta^5–C_5H_5)_2$ and $Cr(\eta^6–C_6H_6)_2$ and can, therefore, be understood by constructing first the ligand group MO's and grouping them according to their symmetry properties followed by identification of ligand group orbitals with symmetries suitable to overlap with the f-orbitls of the metal. In cyclooctatetraene dianion, eight *p* orbitals combine to produce eight molecular orbitals.

When two of $C_8H_8^{2-}$ ions are placed parallel with one another and with carbon atoms eclipsed, sixteen ligand group molecular orbitals will be generated. Out of these sixteen, eight will be *gerade*, and the seven *ungerade f* orbitals match all except the b_{1u} ligand group orbital in symmetry. Based on the supposition that in the derivative, $U(\eta^8–C_8H_8)_2$, 5*f* orbitals participate in bonding, a simple molecular orbital scheme analogous to that for ferrocene can be suggested (Fig. 4.80).

An examination of Fig. 4.80 reveals that in addition to a_{1g}, a_{2u}, e_{1u}, e_{1g} and e_{2g} orbitals, which are qualitatively similar to the orbitals in $Fe(\eta^5–C_5H_5)_2$, the e_{2u} and e_{3u} orbitals are formed by appropriate combinations with the 5*f* orbitals of the metal. The first six of the these comprise of a total of ten $C_8H_8^{2-}$ orbitals, which together donate (from a formal consideration) twenty electrons to the metal. Two of the 5*f* electrons $U^{4+}(5f^2)$, $Np^{4+}(5f^3)$ and $Pu^{4+}(5f^4)$ in the metals will, therefore, occupy the doubly degenerate e_{3u} orbital and the remaining electrons in the metals (none in U^{4+}, one in Np^{4+} and two in Pu^{4+}) would be unpaired; these expectations have been verified experimentally.

4.3.9.2 Derivatives with Non-Planar C_8H_8 Rings

As already illustrated in the Figure 4.74, complexes of neutral cyclooctatetraene with η^2, η^4, and η^6 bonding modes exhibit varying degrees of non-planarity. Some typical preparations of such derivatives are described below:

$$Fe(CO)_5 + C_8H_8 \xrightarrow{\text{octane, 125°C}} Fe(CO)_3(\eta^4–C_8H_8) + 2CO$$

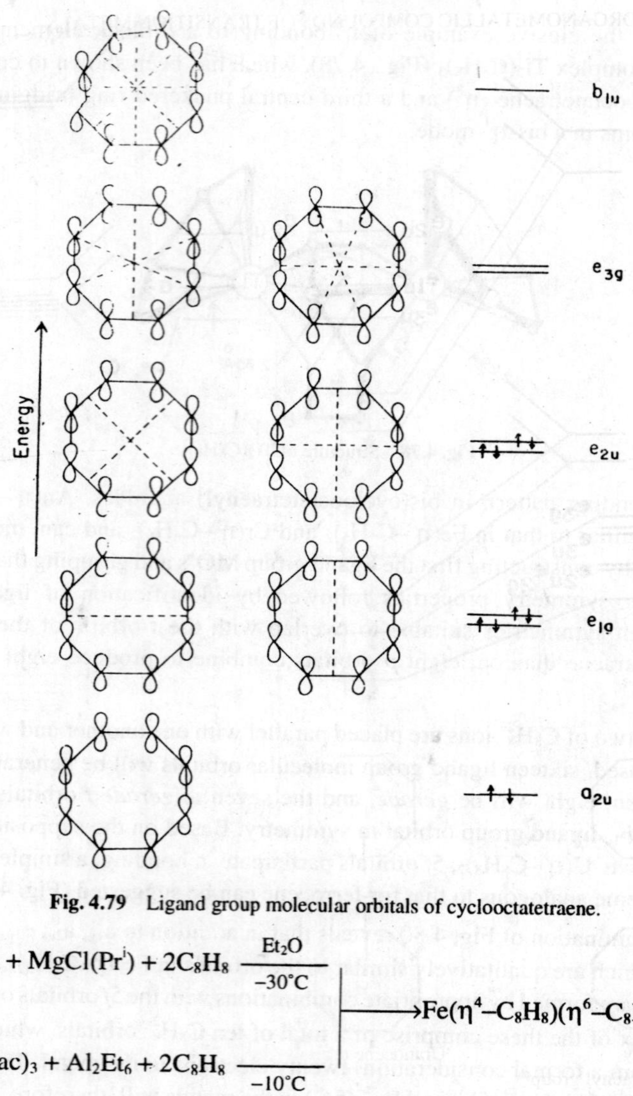

Fig. 4.79 Ligand group molecular orbitals of cyclooctatetraene.

$$FeCl_3 + MgCl(Pr^i) + 2C_8H_8 \xrightarrow[-30^\circ C]{Et_2O} Fe(\eta^4\text{–}C_8H_8)(\eta^6\text{–}C_8H_8)$$

$$Fe(acac)_3 + Al_2Et_6 + 2C_8H_8 \xrightarrow[-10^\circ C]{} Fe(\eta^4\text{–}C_8H_8)(\eta^6\text{–}C_8H_8)$$

$$Co(CO)_2(\eta^5\text{–}C_5H_5) + C_8H_8 \xrightarrow{h\nu} Co(\eta^4\text{–}C_8H_8)(\eta^5\text{–}C_5H_5) + 2CO$$

η^4 (isolated double bond) rigid

η^4 (Conjugated double bond) fluxional

In addition, a number of binuclear derivatives, with cyclooctatetraene acting as a bridging ligand have been synthesized, e.g.:

$$2CrCl_3 + 3MgCl(Pr^i) + 3C_8H_8 \xrightarrow{Et_2O,\ h\nu} Cr_2(C_8H_8)_3$$

$$Cr(g) + 2C_8H_8(g) \xrightarrow[-196^\circ C]{\text{co–condensation}} Cr_2(C_8H_8)_3$$

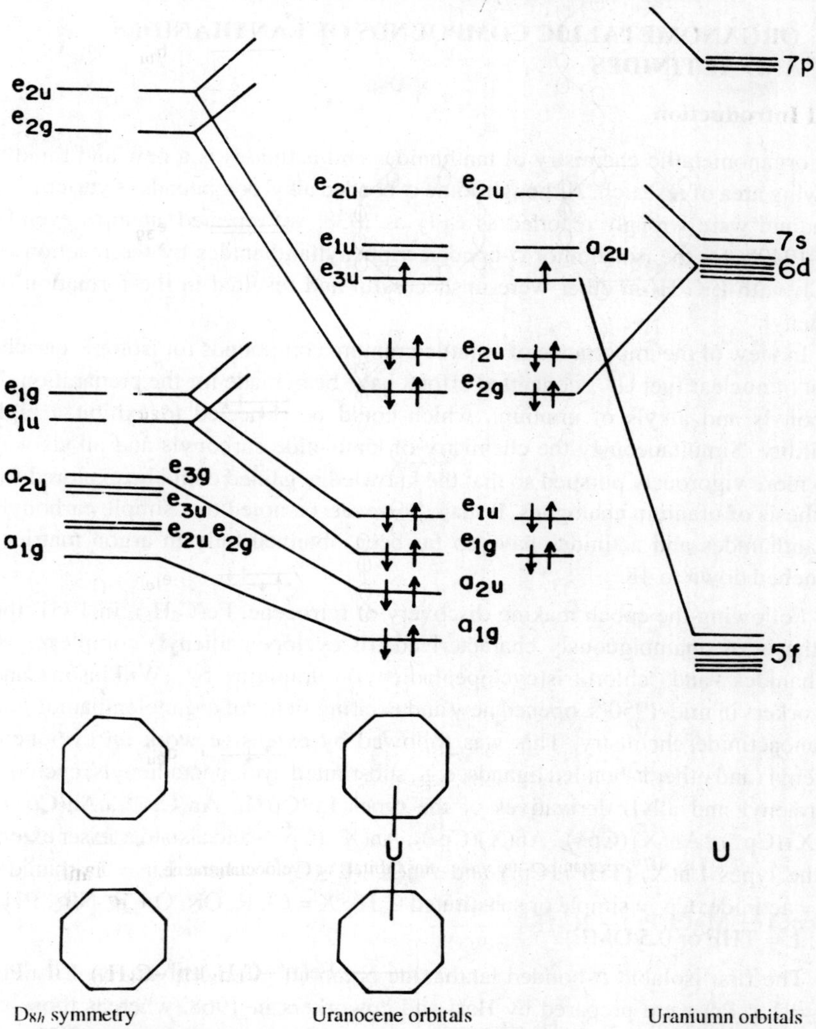

Fig. 4.80 A qualitative molecular orbital scheme for $U(\eta^8 - C_8H_8)_2$.

In both the above two complexes, the terminal rings are bonded in η^4 fashion and the bridging ring is η^5 bonded to both the metal centres and the metal nuclei themselves are triply bonded.

An interesting feature of cyclooctatetraene bridging is the occurrence of syn- and anti-coordination modes in the extensively studied derivatives of the type, $\{(\eta^5\text{–}C_5H_5)M\}_2C_8H_8$ synthesis (M = V, Cr, Fe, Ru, Co, Rh). The synthesis of which is illustrated by the following preparation:

$$2CrCl_2 + 2Na^+C_5H_5^- + 2K^+C_8H_8^{2-} \xrightarrow[\text{(ii) 150°C}]{\text{(i) 25°C}} \{(\eta^5\text{–}C_5H_5)Cr\}_2C_8H_8$$

4.4 ORGANOMETALLIC COMPOUNDS OF LANTHANIDES AND ACTINIDES

4.4.1 Introduction

The organometallic chemistry of lanthanides and actinides is a new and rapidly growing area of research. Although some σ-bonded alkyl compounds of yttrium and scandium were wrongly reported as early as 1938, yet repeated attempts even in mid-1940's for the isolation of σ-bonded triphenyllanthanides by the reaction of $LnCl_3$ with LiC_6H_5 in ether were unsuccessful and resulted in the formation of biphenyl.

In view of the importance of volatile uranium compounds for isotopic enrichment of nuclear fuel U^{235}, concerted efforts have been made for the preparation of carbonyls and alkyls of uranium, which could be expected to exhibit a high volatility. Simultaneously the chemistry of lanthanide carbonyls and alkyls was also more vigorously pursued so that the knowledge gained could be explored for synthesis of uranium analogues. It may, however, be noted that simple carbonyls of lanthanides and actinides have so far been obtained only in argon matrices quenched down to 4K.

Following the epoch making discovery of ferrocene, $Fe(C_5H_5)_2$ in 1951, the synthesis of unambiguously characterized tris(cyclopentadienyl) complexes of lanthanides and chlorotris(cyclopentadienyl) uranium by Wilkinson and coworkers in mid-1950's, opened new and exciting fields of organolanthanide and organoactinide chemistry. This was followed by extensive work on cyclopentadienyl (and other π-bonded ligands, e.g., substituted cyclopentadienyls, cyclooctatetraenyl and allyl) derivatives of the types $Ln(Cp^*)_3$, $An(Cp^*)_3$, $An(Cp^*)_4$, $Ln(X)(Cp^*)_2$, $An(X)(Cp^*)_2$, $An(X)(Cp^*)_3$, $An(X)_2(Cp^*)_2$ and also to a lesser extent of the types $Ln(X)_2(THF)_3(Cp^*)$ and $An(X)_3(L)_2(Cp^*)$ where Ln = lanthanide; An = actinide; Cp^* = simple or substituted C_5H_5; X = Cl, R, OR, O_2CR, NR_2, BH_4, etc.; L = THF or 0.5 DME).

The first isolable σ-bonded lanthanide organometallic derivatives, $LiLnPh_4$ (Ln = La, Pr) were prepared by Hart and coworkers in 1968, whereas those of actinides of the type $(C_5H_5)_3An–R$ were prepared independently by three different research groups (those of Tsutsui, Marks and Mazzei) in 1973.

The first homoleptic derivatives $ScPh_3$ and YPh_3 were synthesized in 1970 by the reactions of anhydrous metal chlorides with three equivalents of phenyllithium. Relatively stable compounds of the type $LnR_3(thf)_3$ have been since obtained by Lappert and coworkers by use of the group (R*) such as $–CH_2CMe_3$, $–CH_2SiMe_3$, or $–CH(SiMe_3)_2$. All the three of these ligands are bulky as well as free from a β-hydrogen atom. Although steric bulk and β-hydrogen elimination processes both have been well established (Section 4.2.1) in determining the stability of σ-bonded organometallic of transition metals, it is not quite clear as to which of the above two factors plays a more dominant role in the chemistry of organometallics of inner transition metals, i.e. lanthanides and actinides. It is, therefore, not surprising that in spite of strenuous efforts to synthesize homoleptic alkyls and aryls of actinides, the only simple alkyl of the actinide series showing some stability at ambient temperatures is tetrabenzylthorium, $Th(CH_2C_6H_5)_4$, synthesized in 1974.

In spite of some differences, not yet fully understood, the development of organometallic chemistry of '*f*' elements shows a close parallel to that of '*d*' elements in general. Conventionally, the two most important factors governing the stability of *f*-block organometallics are optimization of electrostatic interactions and kinetic saturation of the coordination sphere of the metal either by enhancing its coordination states or by bringing about steric congestion with the help of bulky ligands in order to block sterically the possible decomposition pathways (cf. Section 4.2.1). For example, the increase in the stability of σ-alkyls like $(\eta^5\text{—}Cp^*)_2LnR$ and $(\eta^5\text{—}C_5H_5)_3AnR$ by the presence of cyclopentadienyls or other polyhapto ligands can be easily understood on the former ground (i.e. coordinative saturation) whereas the extraordinary stability of species like $[Ln\{CH(SiMe_2)_2\}_3]$ reflects the role of steric factors.

In view of the stronger electropositive character of *f*-metals, a greater ionic character could be expected in all their compounds including the organometallics. On the other hand, evidence for covalency in these compounds has been adduced from a variety of physico-chemical techniques including photoelectron spectroscopy; the extent of comparative derivatives of 4*f* and 5*f* elements has not been yet fully clarified, although some interesting comparisons have been made. For example, investigations on relative covalencies in the lanthanide series $GdCl_3$, $GdCpCl_2$. 3THF; $[GdCp_2Cl]_2$, $GdCp_3$, $[GdCp_2Me]_2$ and $[GdCp_2(C\equiv CPh)]_n$ have indicated a prominent ionic character and only a very marginal 4*f*-orbital covalent character in the different metal-ligand bonds present. By contrast, in the analogous complexes of actinides with ligands like cyclooctatetraenyl, cyclopentadienyl and indenyl, much larger 5*f*-orbital contribution has been adduced to explain their comparatively higher covalent character.

As the investigations on organometallic chemistry of lanthanides and actinides (mainy U, Th, Pu, and Np) have been concentrated on cylopentadienyls, σ-bonded alkyl and aryl derivatives (more recently since early 1970's), and cyclooctatetraenyls (already described in Section 4.3.9), a brief account of these is being presented here. For sake of convenience, the account of *f*-element organometallics in this introductory text-book would be limited to the following types in the order: simple alkyls and aryls followed by cyclopentadienyls.

4.4.2 Simple Alkyls and Aryls

Alkyls and aryls of lanthanides and actinides may be prepared by metathetic reactions in tetrahydrofuran or ether. For example, the treatment of anhydrous scandium and yttrium chlorides with phenyllithium can be represented by the following equation:

$$Sc/YCl_3 + 3LiPh \rightarrow Sc/YPh_3 + 3LiCl$$

With bulky alkyl groups, thermally stable neutral alkyls of lanthanides (including Sc and Y) have been isolated:

$$LnCl_3 + 3LiR^* \xrightarrow{THF/Et_2O} LnR^*_3(thf)_3 + 3LiCl$$

$$(R^* = CH_2CMe_3, CH_2SiMe_3, CH_2Me_2Ph)$$

Another interesting derivative with a sterically more demanding 'Bu^t' group (which contains even β-hydrogen atoms) can be prepared in spite of a lower (five) coordination state of the metal:

$$Er(OBu^t)_3 + 3LiBu^t \xrightarrow{TMEDA} Li(Bu^t)_3\,(TMEDA) + 3LiOBu^t$$

The stability of the organolanthanides and actinides thus appears to depend on a balance of three factors: absence of β-hydrogen on the ligand; steric bulk of the ligand; and the coordinative saturation of the metal; the relative role of each is still being explored by more penetrating physicochemtical studies.

In view of the smaller size of scandium and yttrium, monomeric trisalkyls ScR^*_3 and YR^*_3 ($R^* = CH_2CMe_3, CH_2SiMe_3, CH_2Me_2Ph$) could also be synthesized:

$$Sc/YCl_3 + 3LiR^* \xrightarrow{Et_2O/C_5H_{12}} Sc/Y\,R^*_3 + 3LiCl$$

Another mode of stabilization of neutral σ-bonded organometallics is by the use of alkyl and aryl ligands containing built-in chelating groups, like (dimethylamino) o-benzyl or (dimethylamino) methylphenyl:

$$LnCl_3 + 3Li[\,o-(CH_2)\,C_6\,H_5\,NMe_2\,] \xrightarrow{THF} Ln\,[\,o-(CH_2)\,C_6\,H_4N\,Me_2]_3 + 3LiCl$$

(Ln = Sc, Er, Yb, Lu)

The reaction of anhydrous thorium tetrachloride with benzyllithium in tetrahydrofuran yielded the first reasonably stable (at ambient temperatures) σ-bonded homoleptic organometallic of actinide series:

$$ThCl_4 + 4LiCH_2\,C_6\,H_5 \xrightarrow[-20^\circ C]{THF} Th\,(CH_2\,C_6\,H_5)_4 + 4LiCl$$

As discussed earlier (cf. Section 4.2.1), bulkier ligands and attainment of higher coordination numbers tend to stabilize the derivatives by steric and coordinative saturation of the metal. Thus, anionic ("ate") complexes such as $[LnPh_4]^-$, $[Ln(CMe_3)_4]^-$, $[LnMe_6]^{3-}$, $[UR_6]^{2-}$ and $[UR_8]^{3-}$ are thermally more stable than their analogous neutral complexes. This strategy has been successfully exploited in *f*-element organometallics. For example, although simple triphenyls, $LnPh_3$, of larger lanthanides (e.g., La, Pr) have not been isolated, yet when the same reaction was carried out with 4 mols of LiPh, the tetraphenylate products could be isolated:

$$LnCl_3 + 4LiPh \xrightarrow{THF} LiLnPh_4 + 3LiCl$$

The stability of such phenylate derivatives could further be enhanced with a bulkier substituted phenyl group such as 2,6-$Me_2C_6H_3$, the lutecium derivative of which was incidently the first σ-bonded organolanthanide fully characterized by

X-ray diffraction by Hart and coworkers in 1972; it can be prepared by the simple metathetic reaction:

$$LuCl_3 + 4LiR \xrightarrow{THF} [Li(thf)_4]^+ [LuR_4]^- + 3LiCl$$

As indicated earlier, simple alkyls, LnR_3 (R = Me, Et, Pr, Bu, etc.) appear to be too labile to be isolated at ambient temperatures. However, fairly stable alkylate derivatives can be synthesized by using sterically demanding ligands {e.g. Bu^t, CH_2Bu^t, CH_2SiMe_3, $CH(SiMe_3)_2$} and enhancing the coordination state of the metal:

$$LuCl_3 + 4LiBu^t \xrightarrow{THF} [Li(thf)_4]^+ [Ln(Bu^t)_4]^- + 3LiCl$$

$$LnCl_3 + 4LiCH_2EMe_3 \xrightarrow{THF} [Li(thf)_4]^+ [Ln(CH_2EMe_3)_4]^- + 3LiCl$$

(E = C or Si)

$$LnCl_3 + 3LiCH(SiMe_3)_2 \xrightarrow{THF/Et_2O} [Li(thf)_4]^+ [LnCl\{CH(SiMe_3)_2\}_3]^- + 2LiCl$$

$$[Ln(OC_6H_3Bu^t_2-2,6)_3] + 3LiCH(SiMe_3)_2 \xrightarrow{pentane} Ln\{CH(SiMe_3)_2\}_3 + 3Li(OC_6H_3Bu^t_2-2,6)\downarrow$$

Ln = Y, La, Sm, Lu

$$UCl_4 + \underset{(excess)}{LiR} \xrightarrow{THF} [Li(thf)_4]_2^+ [UR_6]^{2-}$$

$$U_2(OEt)_{10} + \underset{(excess)}{16LiR^*} \xrightarrow{dioxane} 2[Li(dioxane)]_3^+ [UR_8]^{3-} + 10LiOEt$$

($R^* = CH_2CMe_3, CH_2SiMe_3$)

Of the simple alkyls, the methyls could be expected to be more stable due to the absence of β-hydrogens which are known to facilitate the decomposition of transition metal alkyls. This factor coupled with an attainment of higher coordination state of the metal has been exploited in the synthesis of comparatively stable methyl derivatives as in $[LnMe_6]^{3-}$, $[UMe_6]^{2-}$, and $[UMe_8]^{3-}$:

$$LnCl_3 + 6LiMe + 3\ TMEDA \xrightarrow[-20^\circ C]{Et_2O} [Li(TMEDA)]_3^+ [LnMe_6]^{3-} + 3LiCl$$

$$LnCl_3 + 6LiMe + 3\ DME \xrightarrow[-78^\circ C]{Et_2O} [Li(DME)]_3^+ [LnMe_6]^{3-} + 3LiCl$$

$$LnCl_3 + 10LiMe + 2\ TEED \xrightarrow[-78^\circ C]{Et_2O} [Li(TEED)(Et_2O)_2]_2^{3+} [LnMe_6]^{3-} + 6\ LiCl$$

$$UCl_4 + \underset{\text{(excess)}}{6LiMe} \xrightarrow{\text{THF}} [Li(THF)_4]_2^+ [UMe_6]^{2-} + 4LiCl$$

$$U_2(OEt)_{10} + \underset{\text{(excess)}}{16LiMe} \xrightarrow{\text{dioxane}} 2[Li(dioxane)]_3^+ [UMe_8]^{3-} + 10LiOEt$$

All those complexes are extremely sensitive to oxygen and moisture and some of them are pyrophoric. However, most of these are stable at ambient temperatures in the absence of air.

4.4.3 Cyclopentadienyl Derivatives

Cyclopentadienyls of lanthanides and actinides can be divided into two main types, i.e. (i) homoleptic, e.g., $Ln(C_5H_5)_3$ (Sc, Y, lanthanide metal), $Np(C_5H_5)_3$, $Pu(C_5H_5)_3$, $Cf(C_5H_5)_3$, $Th(C_5H_5)_4$, $U(C_5H_5)_4$, $Np(C_5H_5)_4$, and (ii) heteroleptic, e.g., $[\{LnCl(C_5H_5)_2\}_2]$, $LnCl_2(C_5H_5) \cdot 3THF$, $ThCl(C_5H_5)_3$, $UCl(C_5H_5)_3$, $ThCl_2(Cp^{**})_2$, $UCl_2(Cp^{**})_2$, $UCl(Cp^{**})_2$, $BkCl(C_5H_5)_2$, $UCl_3(C_5H_5)$ (where Cp** = substituted cyclopentadienyls with considerable steric bulk) etc., as well as derivatives in which 'Cl' has been replaced by moieties like R, OR, NR_2 or BH_4.

The simple bis (η^5-cyclopentadienyl) halides of the lanthanides and actinides of the types, $[\{LnX(\eta^5\text{–}C_5H_5)_2\}_2]$, $[\{AnX(\eta^5\text{–}C_5H_5)_2\}_2]$ and $AnX_2(\eta^5\text{–}C_5H_5)_2$ are only known for heavier lanthanides and actinides, but similar derivatives of early lanthanides, as well those of thorium and uranium can be stabilized only when the cyclopentadienyl ligands are of sterically demanding types. For example, synthesis of early lanthanocene chlorides, thorocene- and uranocene-chloride with two ancillary cyclopentadienyl ligands has been possible only with Cp** ligands such as $C_5H_3(SiMe_3)_2$ (Lappert and Singh, 1981), 1,3–$(CH_2)_3(C_5H_4)_2$ (John and Tsutsui, 1980), and C_5Me_5 (Marks and coworkers, 1978); this appears to be the consequence of steric saturation of the larger lanthanide, uranium and thorium atoms. These bis-cyclopentadienyl derivatives of the lanthanides can also be of three types: i.e. neutral, e.g., $[\{LnXCp_2\}_2]$; anionic, e.g., $[LnX_2Cp_2]^-$; and heterobimetallic, e.g., $Cp_2Ln(M\text{–}X)_2Li(L)_2$ where L is Lewis base like THF, Et_2O or $(L)_2$ = TMEDA or DME.

As metallocenes of the types, $[\{LnX(\eta^5\text{–}C_5H_5)_2\}_2]$, $[\{LnX(\eta^5\text{–}Cp^{**})_2\}_2]$ and $AnX_2(\eta^5\text{–}Cp^{**})_2$ (An = Th, U) have undergone mushroom growth during the past decade and continue to generate exciting chemistry, these will attract our major attention. This will be followed by a brief mention of other heteroleptic cyclopentadienyl complexes of 4*f*- and 5*f*- elements containing three (e.g., $AnX(\eta^5\text{–}C_5H_5)_3$) and one (e.g. $LnX_2(THF)_3(\eta^5\text{–}C_5H_5)$ and $AnX_3(THF)_2(\eta^5\text{–}C_5H_5)$) cyclopentadienyl ligands, and homoleptic derivatives, e.g., $Ln(C_5H_5)_3$, $An(C_5H_5)_3$, $Th(C_5H_5)_4$, and $U(C_5H_5)_4$.

In addition to the above, some lanthanides like Sm, Eu, and Yb form cyclopentadienyl derivatives in their +2 oxidation state, whereas Ce forms derivatives in +4 oxidation state; both of these types are also dealt with very briefly in the last section.

4.4.3.1 Bis (η^5–Cyclopentadienyl) Derivatives of the f-elements

(i) *Preparation of Halide Precursors.* The general method of preparation of bis (cyclopentadienyl) halide complexes of the lanthanides and actinides is the interaction of anhydrous lanthanide/actinide halides with ionic cyclopentadienides of electropositive elements like lithium, sodium, potassium, thallium, and magnesium in an organic solvent like benzene, tetrahydrofuran or diethyl ether:

$$LnCl_3 + 2MCp^* \rightarrow 1/2[\{LnCl(\eta^5\text{–}Cp^*)_2\}_2] + 2MCl$$

(Ln = Sc, Y and later lanthanide metal, when $Cp^* = C_5H_5$ or C_5H_4Me; Ln = Sc, Y, early and later lanthanides, when $Cp^* = Cp^{**}$ such as $C_5H_3(SiMe_3)_2$, 1, 3-$(CH_2)_3(C_5H_4)_2$; M = Li or Na)

$$AnCl_4 + 2LiCp^{**} \rightarrow AnCl_2(\eta^5\text{–}Cp^{**})_2 + 2LiCl$$

(An = Th, U; $Cp^{**} = C_5H_3(SiMe_3)_2$, (C_5Me_5))

$$AnCl_4 + 2MgCl(C_5Me_5)(THF) \xrightarrow[100°C]{PhMe} AnCl_2(\eta^5\text{–}C_5Me_5)_2 + 2MgCl_2(THF)$$

(An = Th, U)

$$2UCl_4 + 2Li_2\ 1,3\text{–}(CH_2)(C_5H_4) \rightarrow [LiU_2Cl_5(THF)_2\{CH_2(C_5H_4)_2\}_2] + 3LiCl$$

Another interesting reaction used for the synthesis of bis (cyclopentadienyl) uranium dichloride can be illustrated as below:

$$UCl_4 + 2SnBu_3(C_5Me_4Et) \xrightarrow[120°C]{PhMe} UCl_2(\eta^5\text{–}C_5Me_4Et)_2 + 2SnBu_3Cl$$

$$4BkCl_3 + 5Be(C_5H_5)_2 \rightarrow 2Bk(\eta^5\text{–}C_5H_5)_3 + [\{BkCl(\eta^5\text{–}C_5H_5)_2\}_2] + 5BeCl_2$$

For lanthanides there are two other alternative approaches also: (i) ligand redistribution reaction, e.g.,

$$LnCl_3 + 2Ln(C_5H_5)_3 \xrightarrow{THF} 3LnCl(\eta^5\text{–}C_5H_5)_2(THF)$$

(Ln = Later lanthanide metal)

$$2LaCp_3^{**} + LaCl_3 \xrightarrow{THF} 2LaCl(\eta^5\text{–}Cp^{**})_2(THF)$$

$(Cp^{**}) = C_5H_3(SiMe_3)_2$

and (ii) cleavage of cyclopentadienyl group, e.g.,

$$2Ln(\eta^5\text{–}C_5H_5)_3 + 2NH_4Cl \xrightarrow{PhH} [\{LnCl(\eta^5\text{–}C_5H_5)_2\}_2] + 2C_5H_6 + 2NH_3$$

(Ln = Sm, Yb)

$$Ln(\eta^5\text{–}C_5H_5)_3 + HCl \xrightarrow{THF} LnCl(\eta^5\text{–}C_5H_5)_2(THF) + C_5H_6$$

$$Ln(\eta^5\text{–}Cp^{**})_3 + HCl \xrightarrow{THF} LaCl(\eta^5\text{–}Cp^{**})_2(THF) + Cp^{**}H$$

$(Cp^{**} = C_5H_3(SiMe_3)_2)$

Reductive reactions, as illustrated below, have also been utilized for the synthesis of bis (pentamethylcyclopentadienyl) uranium monochloride:

$$6U(R)(Cl)(\eta^5\text{–}C_5Me_5)_2 + \underset{(1\ atm)}{3H_2} \xrightarrow[25^\circ C]{PhMe} 2[\{UCl(\eta^5\text{–}C_5Me_5)_2\}_3) + 6RH$$

$(R = Me, CH_2SiMe_3)$

$$3UCl_2(\eta^5\text{–}C_5Me_5)_2 + 3LiBu^t \rightarrow [\{UCl(\eta^5\text{–}C_5Me_5)_2\}_3] + 3LiCl + \text{organic products}$$

$$UCl_2(\eta^5\text{–}C_5Me_5)_2 \xrightarrow[PhMe]{Na/Hg} 1/3[\{UCl(\eta^5\text{–}C_5Me_5)_2\}_3] + NaCl$$

(ii) *General Properties, Reactions and Structures*. All these types of bis (cyclopentadienyls) are usually thermally stable, sublimable, and soluble in polar solvents but they are unstable with respect to oxygen and moisture. The metal atoms in compounds like $LnCl(\eta^5\text{–}Cp^*)_2$ and $UCl(\eta^5\text{–}C_5Me_5)_2$ display an apparent tendency to increase their coordination numbers. Thus, members of the former class exist as dimeric species (Fig. 4.81a), while the latter is trimeric (Figure 4.81b). In both the cases, an essentially tetrahedral disposition of two of these $(\eta^5\text{–}Cp^*)$ rings and two bridging chloride groups around the metal appears to be general. However, the bulky C_5Me_5 ligand with smaller yttrium and scandium forms an unsymmetrical chloride bridged dimer $(\eta^5\text{–}C_5Me_5)_2Y(\mu\text{–}Cl)YCl(\eta^5\text{–}C_5Me_5)_2$ (Fig. 4.81c) and monomeric derivative $Sc(\eta^5\text{–}C_5Me_5)_2Cl$ (Fig. 4.81d), respectively. As the size of Y^{3+}(0.88Å) is similar to those of Ho^{3+}(0.89Å) and Er^{3+}(0.88Å), it is likely that bis$(\eta^5\text{–}C_5Me_5)_2$ chlorides of these two metals and also of smaller lanthanides i.e., Tm^{3+}, Yb^{3+} and Lu^{3+} should attain unsymmetrical chloride bridged structural patterns. While derivatives of the type $AnCl(\eta^5\text{–}Cp^*)_2$ are generally associated, compounds of the form $AnCl_2(\eta^5\text{–}Cp^*)_2$ are monomeric (Fig. 4.81e). The observed lack of association in the latter compounds can be explained on the basis of steric congestion around the metal atoms.

Replacement of 'Cl' in derivatives $[\{(Cp^*)_2LnCl\}_2]$ by alkoxy, alkyl, amido, aryl, aryloxy, PR_2 or BH_4 groups, etc., provides the most versatile route for generating novel and exciting chemistry in this field as illustrated by the following reactions:

$$[\{(\eta^5\text{–}Cp^*)_2LnCl\}_2] + 2LiR \rightarrow \underset{(A)}{(\eta^5\text{–}Cp^*)_2Ln(\mu\text{–}R)_2Ln(\eta^5\text{–}Cp^*)_2} + 2LiCl$$

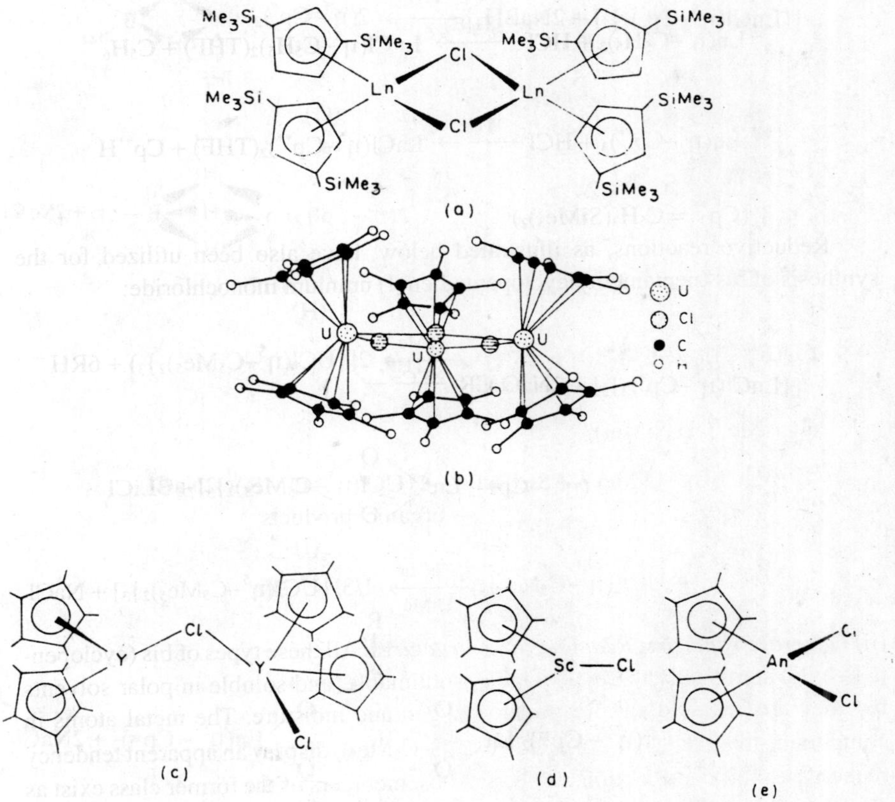

Fig. 4.81 Some structural patterns of bis (η^5-cyclopentadienyl) chlorides of lanthanides and actinides.

$$[\{\eta^5\text{–}Cp^*)_2LnCl\}_2] + 2LiAlMe_4 \xrightarrow{PhMe} (\eta^5\text{–}Cp^*)_2Ln(\mu\text{–}Me)_2AlMe_2 + 2LiCl$$

(B)

$$[\{(\eta^5\text{–}Cp^*)_2LnCl\}_2] + 2LiR \xrightarrow{THF} 2(\eta^5\text{–}Cp^*)_2LnR{\cdot}THF + 2LiCl$$

(C)

$$[\{(\eta^5\text{–}Cp^*)_2LnCl\}_2] + 2Mg(CH_2\text{–}CH{=}CH_2)Br \xrightarrow[(-2MgBrCl)]{THF/Et_2O}$$

$$2(\eta^5\text{–}Cp^*)_2Ln\text{–}CH_2\text{–}CH = CH_2$$

(E)

or

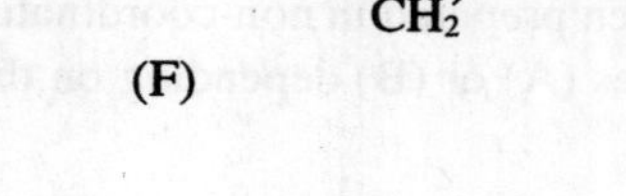

(F)

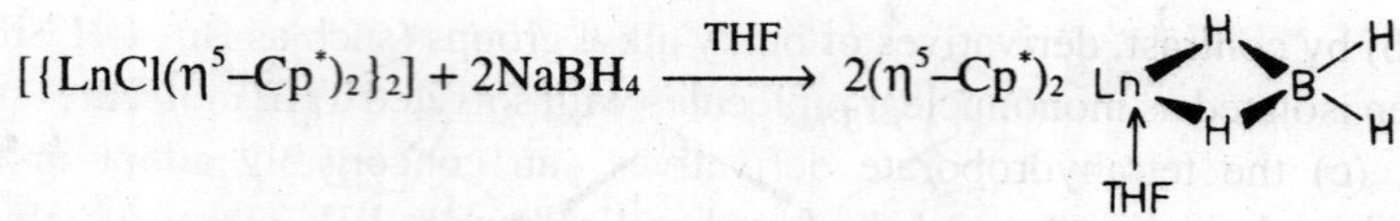

(G)

or

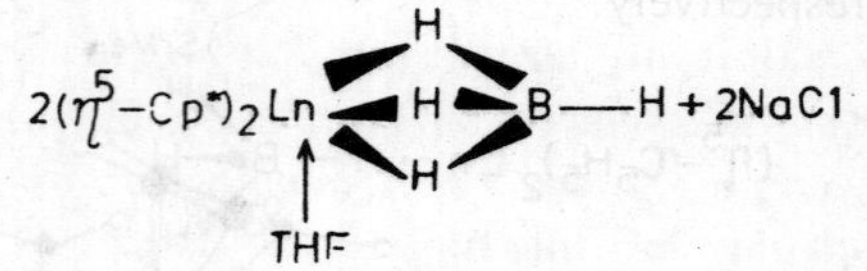

(H)

$$[\{LnCl(\eta^5\text{–}Cp^*)_2\}_2] + 2NaO_2CR \xrightarrow{THF}$$

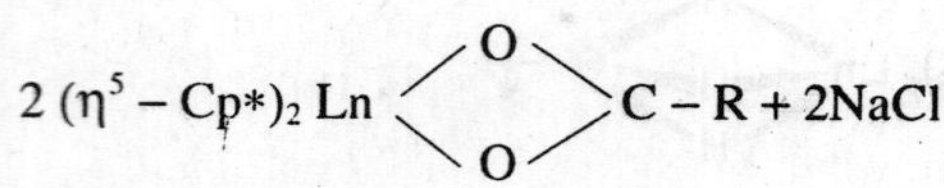

(I)

or

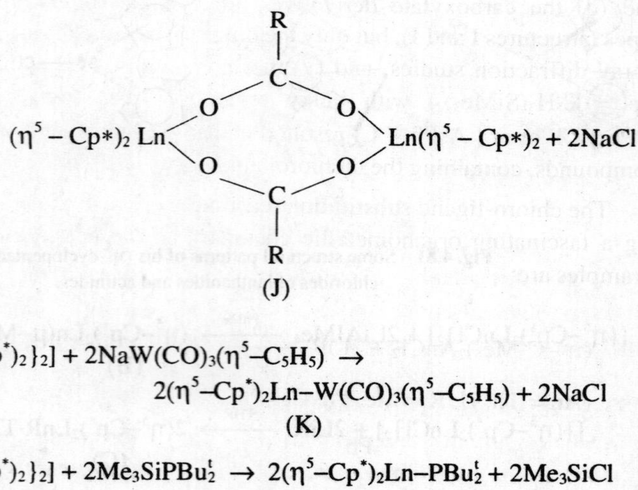

(J)

$$[\{LnCl(\eta^5\text{–}Cp^*)_2\}_2] + 2NaW(CO)_3(\eta^5\text{–}C_5H_5) \rightarrow$$
$$2(\eta^5\text{–}Cp^*)_2Ln\text{–}W(CO)_3(\eta^5\text{–}C_5H_5) + 2NaCl$$

(K)

$$[\{LnCl(\eta^5\text{–}Cp^*)_2\}_2] + 2Me_3SiPBu^t_2 \rightarrow 2(\eta^5\text{–}Cp^*)_2Ln\text{–}PBu^t_2 + 2Me_3SiCl$$

(L)

$$\tfrac{1}{2}[\{LnCl(\eta^5\text{–}Cp^*)_2\}_2] + \overset{+}{M}{}^*Cl^- \xrightarrow{CH_2Cl_2} [M^*]^+[LnCl_2(\eta^5\text{–}Cp^*)_2]^-$$

(M)

($M^* = PMePh_3, PPh_4, AsPh_4$, etc.)

A few salient features of these chloro-ligand substitution reactions in lanthanocene chemistry are: (a) mixed alkyl (alkenyl) cyclopentadienyl complexes when prepared in non-coordinating solvents exhibit alkyl (alkenyl) bridged structures (A) or (B) depending on the alkylating agent, i.e., LiR or $[AlMe_4]^-$ respectively; (b) by contrast, derivatives of bulky alkyl groups (such as Bu^t, CH_2SiMe_3) have been isolated as mononuclear molecules with solvated THF molecule (structure C); (c) the tetrahydroborate derivatives can conceivably adopt in their mononuclear derivatives any of the four bonding modes I-IV out of which only types II and III have been actually observed in complexes (G) and (H) respectively:

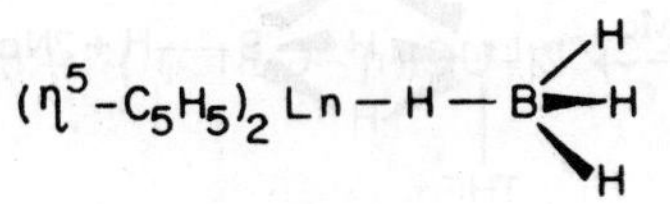

(I)

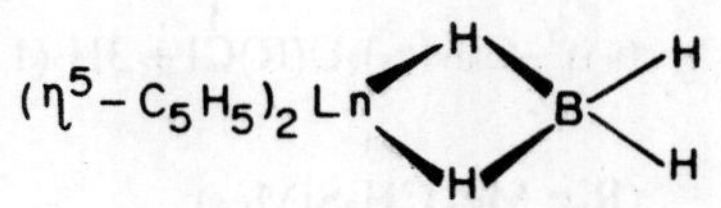

(II)

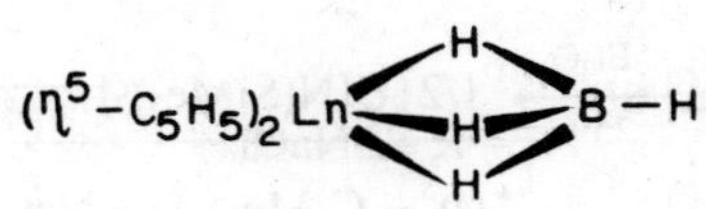

(III)

$(\eta^5-C_5H_5)_2Ln^+ \quad BH_4^+$

(IV)

and (d) the carboxylate derivatives are capable of exhibiting two structural types (structures I and J), but only the dimeric structure (J) has been recognized by X-ray diffraction studies; and (v) the interaction of $[\{LnCl(\eta^5\text{–}Cp^*)_2\}_2]$ (where $Cp^* = C_5H_3(SiMe_3)_2$) with bulky organic halides such as $[PMePh_3]^+Cl^-$ or $[PPh_4]^+Cl^-$ or or $[AsPh_4]^+Cl^-$ resulted in a new class of lanthanide organometallic compounds, containing the dichlorolanthanocene(III) anion (structure M).

The chloro-ligand substitution reactions have also proved crucial in developing a fascinating organometallic chemistry of the actinides. Some illustrative examples are:

$$(\eta^5\text{–}C_5Me_5)_2AnCl_2 + 2LiR \xrightarrow{\text{ether}} An(R)_2(\eta^5\text{–}C_5Me_5)_2 + 2LiCl$$

(An = Th, U; R = Me, Ph, CH_2Ph, CH_2SiMe_3, CH_2CMe_3)

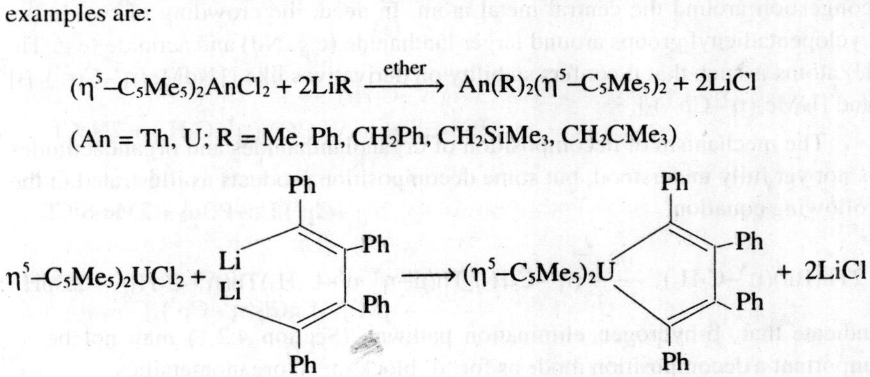

$$(\eta^5\text{–}C_5Me_5)_2AnCl_2 + An(R)_2(\eta^5\text{–}C_5Me_5)_2 \xrightleftharpoons{\text{PhMe}} 2An(R)(Cl)(\eta^5\text{–}C_5Me_5)_2$$

(An = Th, U; R = Me, Ph, CH_2Ph, CH_2CMe_3, CH_2SiMe_3)

$$ThCl(NMe_2)(\eta^5\text{–}C_5Me_5)_2 + LiMe \rightarrow Th(NMe_2)Me(\eta^5\text{–}C_5Me_5)_2 + LiCl$$

$$(\eta^5\text{–}C_5Me_5)_2AnCl_2 + nLiNR_2 \xrightarrow[25^\circ C]{Et_2O} (\eta^5\text{–}C_5Me_5)_2AnCl_{2-n}(NR_2)_n + nLiCl$$

(An = Th, U; R = Me, Et; n = 1, 2)

$$(\eta^5\text{–}C_5Me_5)_2AnCl_2 + LiR \rightarrow (\eta^5\text{–}C_5Me_5)_2An(R)Cl + LiCl$$

$$6(\eta^5\text{–}C_5Me_5)_2U(R)Cl + 3H_2(1\ atm) \xrightarrow[25^\circ C]{PhMe} 2[\{UCl(\eta^5\text{–}C_5Me_5)_2\}_3] + 6RH$$

(R = Me, CH_2SiMe_3)

$$UCl_2(\eta^5\text{–}C_5Me_5)_2 \xrightarrow[PhMe]{Na/Hg} [\{UCl(\eta^5\text{–}C_5Me_5)_2\}_3] + NaCl$$

$$1/3\{UCl(\eta^5\text{–}C_5Me_5)_2\}_3 + NaN(SiMe_3)_2 \xrightarrow[25^\circ C]{Et_2O} 1/2\{U[N(SiMe_3)_2](\eta^5 - C_5Me_5)_2\}_2 + NaCl$$

$$1/3\ \{UCl(\eta^5\text{–}C_5Me_5)_2\}_3 + LiCH(SiMe_3)_2 \xrightarrow[25^\circ C]{Et_2O} 1/2\{U[CH(SiMe_3)_2](\eta^5\text{–}C_5Me_5)_2\}_2 + LiCl$$

(ii) *Properties of σ-Bonded Heteroleptic Organometallics of Lanthanides and Actinides*

(a) *Thermal stability and modes of decomposition*. The alkyl and aryl derivatives of f-metals are extremely air-sensitive. Their stability increases with greater steric congestion around the central metal atom. In deed, the crowding of two bulky cyclopentadienyl groups around larger lanthanide (e.g. Nd) and actinide (e.g. Th, U) atoms is such that it confers stability on derivatives like $[\{NdMe(\eta^5\text{–}Cp^{**})_2\}_2]$ and $[UMe_2(\eta^5\text{–}Cp^{**})_2]$.

The mechanism of decomposition of organolanthanides and organoactinides is not yet fully understood, but some decomposition products as illustrated in the following equation,

$$2Th(Bu)(\eta^5\text{–}C_5H_5)_3 \xrightarrow{170^\circ C} (\eta^5\text{–}C_5H_5)_2Th(\mu\text{–}\eta^5, \eta^1\text{–}C_5H_4)Th(\eta^5\text{–}C_5H_5)_3 + 2BuH$$

indicate that β-hydrogen elimination pathway (Section 4.2.1) may not be as important a decomposition mode as for 'd' block metal organometallics.

(b) *Reactions with electrophiles (protic reagents)*. In general the σ-An-C bond is more reactive than η^5 An–C bond towards electrophiles as illustrated by the following equations:

$$(\eta^5\text{–}C_5H_5)_3ThR + R'OH \rightarrow (\eta^5\text{–}C_5H_5)_3Th(OR') + RH$$

$$(\eta^5\text{–}C_5H_5)_3UR + AlR'_3 \rightarrow (\eta^5\text{–}C_5H_5)_3U(R') + AlR'_2R$$

$$(R = Et, Bu, Ph, OEt;\ R' = Me, Et, Bu^t)$$

In the case of thorium derivatives, no cleavage of η^5 bonds is observed till cleavage is complete, but in the case of uranium derivatives, the two occur almost simultaneously:

$$(\eta^5\text{–}C_5H_5)_3UR + R'OH \longrightarrow \begin{cases} (\eta^5\text{–}C_5H_5)_3UOR' + RH \\ (\eta^5\text{–}C_5H_5)_2U(OR')\,R + C_5H_6 \end{cases}$$

(c) *Hydrogenolysis.* The σ-bonded organo-lanthanides and -actinides exhibit a wide range of stabilities which depend upon the nature of the supporting ligands present, the nature of the organic ligands as well as on the size of the concerned metal atom.

The cleavage of organotransition metal-carbon bond by hydrogen plays an important role in many catalytic processes. For example, the terminating steps in hydroformylation (cf. Section 6.4.2.3) and homogeneous hydrogenation (cf. Section 6.4.2.1) processes require cleavage of metal acyl or alkyl by hydrogen. Similarly the Fischer-Tropsch heterogeneous reaction (cf. Section 6.4.2.8) requires several steps in which M–H bonds are formed, including the final release of hydrocarbon or oxygenate product from the metal surface. Although interstitial and binary hydrides of lanthanides and actinides have been known for many years, yet the first organo-lanthanide hydrides were reported by Marks and coworkers in early 1980's:

$$Er(CH_2SiMe_3)(\eta^5\text{–}C_5H_4Me)_2 + 2H_2 \xrightarrow[70\text{–}80^\circ C]{PhMe} [\{ErH(\eta^5\text{–}C_5H_4Me)_2\}_2] + 2SiMe_4$$

Another attempt to make lanthanide hydrides by hydrogenolysis of $Ln(CMe_3)(THF)(\eta^5\text{–}C_5H_5)_2$ complexes proved crucial to yield the first crystallographically characterized organolanthanide hydrides (A):

$$(\eta^5\text{–}C_5H_5)_2(THF)Ln\text{–}CMe_3 + H_2 \longrightarrow$$

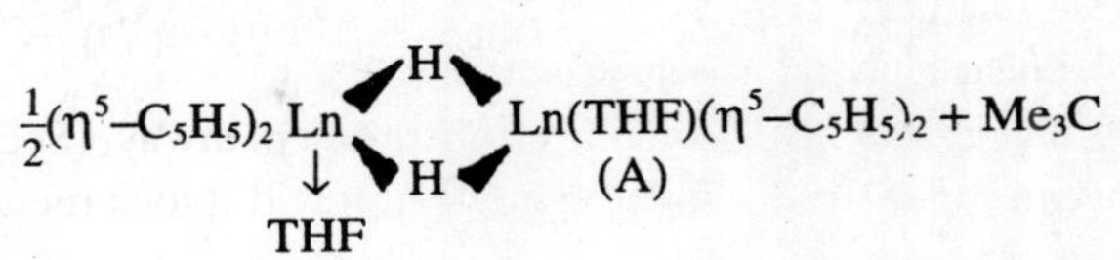

$$(Ln = Y, Lu)$$

The bis(pentamethylcyclopentadienyl) thorium and uranium dialkyls are highly reactive towards hydrogen. In contrast to the tricyclopentadienyls, these compounds readily undergo hydrogenolysis:

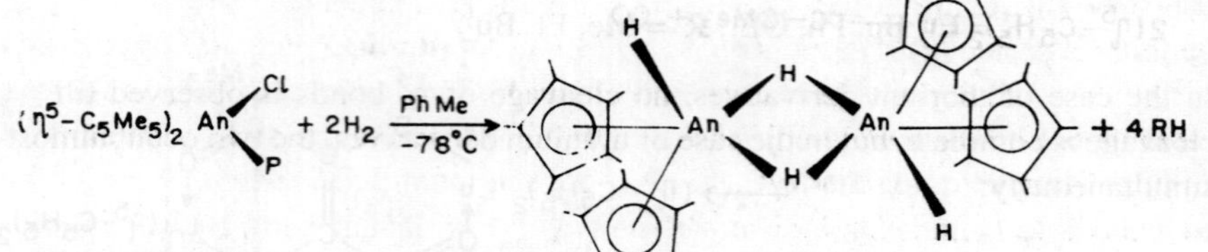

(An = Th, U) (B)

The resulting organoactinide dihydrides (B) show by single crystal neutron diffraction studies that hydrogen atoms occupy both terminal and bridging position in the dimer. While the thorium hydrido derivatives do not tend to lose hydrogen, the uranium analogues readily (and reversibly) lose hydrogen:

$$n/2[\{UH_2(\eta^5-C_5Me_5)_2\}_2] \rightleftharpoons [\{UH(\eta^5-C_5Me_5)_2\}n] + n/2\ H_2$$

This difference may be explained on the basis of greater facility with which uranium can shuttle between different oxidation states. The hydridic nature of both the thorium and uranium compounds is evidenced by the rapid and quantitative reduction of methyl chloride:

$$AnH_2(\eta^5-C_5Me_5)_2 + 2CH_3Cl \rightarrow AnCl_2(\eta^5-C_5Me_5)_2 + 2CH_4$$

(d) *Insertion of carbon monoxide.* The classical reaction pattern of σ-bonded transition metal (M) organometallics with carbon monoxide is 'migratory insertion' and formation of acyl complexes (cf. Section 4.2.3.3)

$$M-CH_3 + CO \rightleftharpoons \underset{}{\overset{CH_3}{\overset{|}{M}}} \leftarrow CO \rightleftharpoons M-\overset{CH_3}{\overset{|}{C}}=O$$

This type of acyl formation is an important step in a number of catalytic processes such as hydroformylation (cf. Section 6.4.2.3). In lanthanides and actinides, the initial products undergo a facile conversion in which the acyl ligand appears to be dihapto bonded due to special affinity of these metals towards oxygen; e.g.,

$$(\eta^5-C_5H_5)_2(THF)Lu-CMe_3 + CO \xrightarrow{PhMe} (\eta^5-C_5H_5)_2Lu-\underset{\underset{O}{\|}}{C}-CMe_3$$

$$\downarrow$$

$$(\eta^5-C_5H_5)_2Lu \leftarrow :\overset{\overset{O}{|}}{C}-CMe_3 \quad (\text{O also bonded to Lu})$$

In this behaviour, '4f' derivatives show similarity to those of some earlier transition (d) metals. Further reaction with CO forms a bimetallic enedionediolate complex:

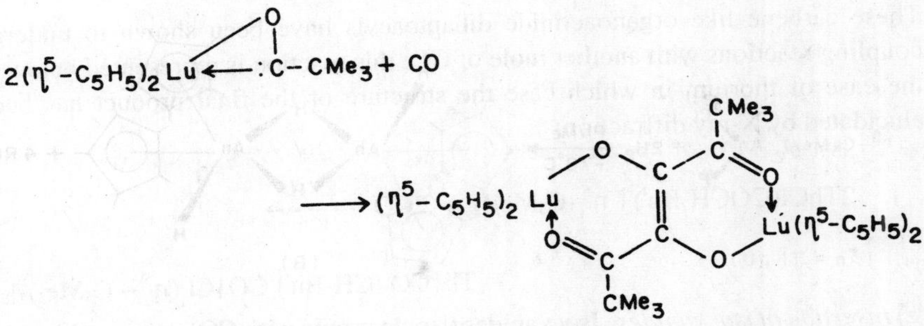

With tetravalent methylactinide (Th and U) analogoues the over-al, reactions can be represented as below:

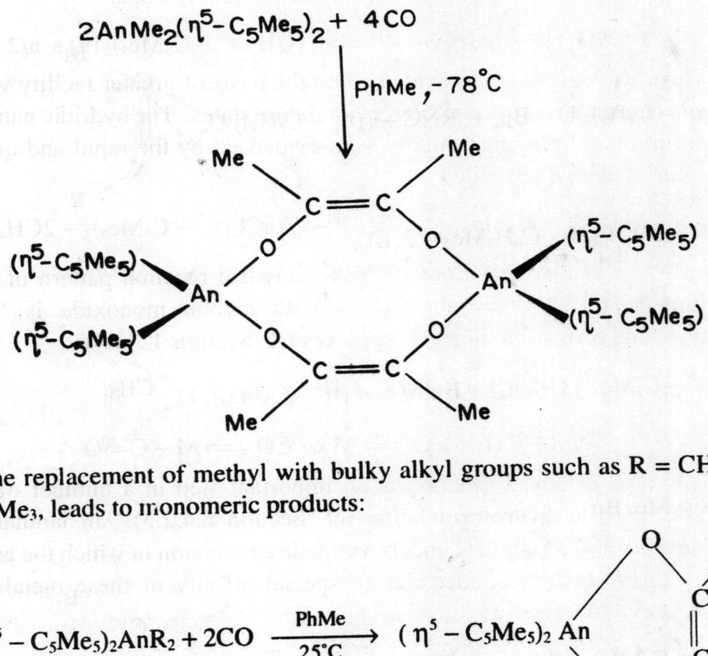

The replacement of methyl with bulky alkyl groups such as R = CH_2CMe_3, CH_2SiMe_3, leads to monomeric products:

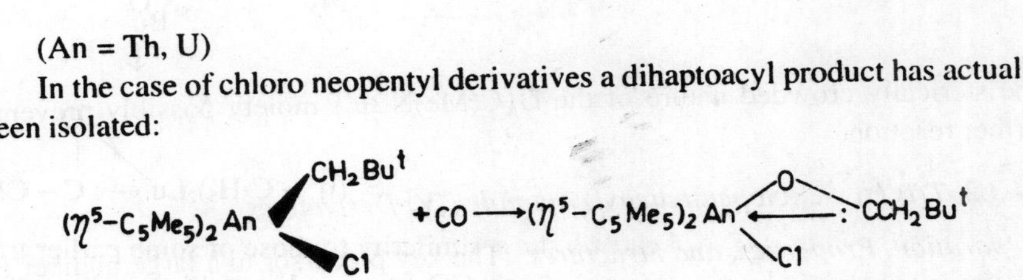

(An = Th, U)

In the case of chloro neopentyl derivatives a dihaptoacyl product has actually been isolated:

These carbene like organoactinide dihaptoacyls have been shown to undergo coupling reactions with another mole of CO; this reaction is represented below in the case of thorium in which case the structure of the final product has been elucidated by X-ray diffraction:

$$2ThCl(COCH_2Bu^t)(\eta^5\text{–}C_5Me_5)_2 + 2CO \xrightarrow[12h]{25^\circ C} [Th\{CO(CH_2Bu^t)CO\}Cl(\eta^5 - C_5Me_5)_2]_2$$

(e) *Insertion of isocyanides.* Isocyanides (isoelectronic with CO) react readily with $U(R)(\eta^5\text{—}C_5H_5)_3$, $U(R)(Cl)(C_5Me_5)_2$, and $U(Me)_2(\eta^5\text{—}C_5Me_5)_2$ to give iminoalkyl insertion products; with $U(R)_2(\eta^5\text{—}C_5Me_5)_2$ only one insertion reaction occurs. These reactions are illustrated below:

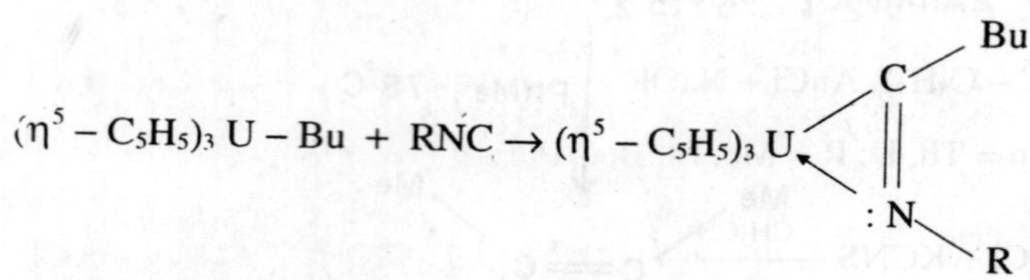

$(R = Bu^t, C_6H_{11}, C_6H_3Me_2 - 2, 6)$

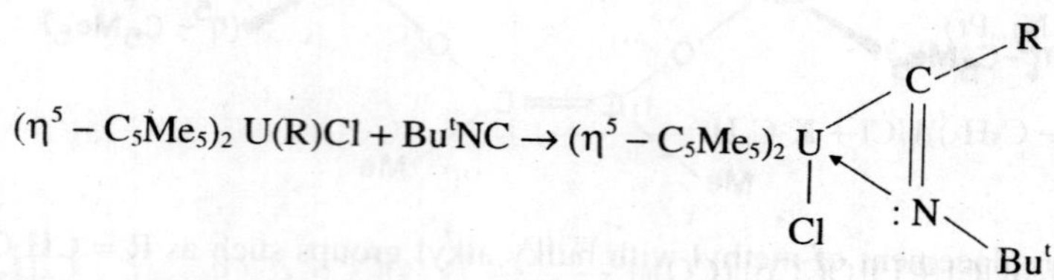

$(R = Me, Bu)$

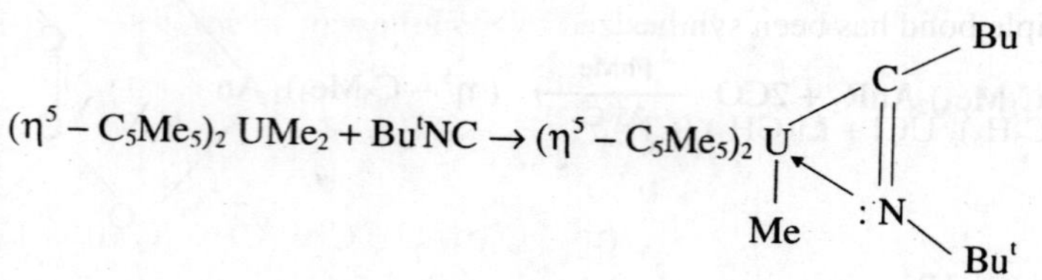

The sterically crowded nature of the $U\{C(Me)NBu^t\}$ moiety possibly prevents further reaction.

4.4.3.2 *Tris (η^5–Cyclopentadienyl) Actinide (IV) Derivatives*

Preparation, Properties, and Structures. The key precursor to a large number of tris (η^5-cyclopentadienyl) actinide (IV) compounds, $AnX(\eta^5\text{–}C_5H_5)_3$ is the corresponding chloro-derivative, although other halide complexes are also known.

For the synthesis of chloro-precursos, the following simple metathetic reactions have proved to be of great utility in most cases:

$$\mathrm{AnCl_4 + 3M'C_5H_5 \rightarrow AnCl(\eta^5\text{–}C_5H_5)_3 + 3\,M'Cl}$$

$$\mathrm{(An = Th, U, Np;\ M' = Li, Na, K\ or\ Tl)}$$

Replacement of Cl in $AnCl(\eta^5\text{–}C_5H_5)_3$ by other halogens, or by alkoxy, alkyl, aryl, alkenyl, alkynyl, BH_4 or some other anionic groups, provides the most extensive synthetic routes for new and fascinating organometallics of the actinides. Some typical examples are mentioned as follows:

$$\mathrm{(\eta^5 - C_5H_5)_3\,UCl + NaF \rightarrow UF(\eta^5 - C_5H_5)_3 + NaCl}$$

$$\mathrm{(\eta^5 - C_5H_5)_3\,AnCl + \underset{or\ LiR}{Mg(R)X} \rightarrow AnR(\eta^5 - C_5H_5)_3 + \underset{or\ LiCl}{MgClX}}$$

(An = Th, U, Np; R = Me, Pr^n, Pr^i, Bu^n, Bu^t, CH_2Bu^t, CH_2SiMe_3, Ph, CH_2Ph, allyl, etc.)

$$\mathrm{(\eta^5 - C_5H_5)_3\,AnCl + NaOR \xrightarrow{PhH} An(OR)\,(\eta^5 - C_5H_5)_3 + NaCl}$$

(An = Th, U; R = Me, Et, Bu, Bu^t, etc.)

$$\mathrm{(\eta^5 - C_5H_5)_3\,UCl + KCNS \xrightarrow{CH_3CN} U(NCS)(CH_3CN)(\eta^5 - C_5H_3)_3 + KCl}$$

$$\mathrm{2(\eta^5 - C_5H_5)_3UCl + K_2M(CN)_4 \xrightarrow{H_2O} [U(\eta^5 - C_5H_5)_3]_2\,[M(CN)_4] + 2KCl}$$

(M = Ni, Pt)

$$\mathrm{2(\eta^5 - C_5H_5)_3UCl + K_2C_8H_8 \xrightarrow[-\,30^\circ C]{THF} [\,U(\eta^5 - C_5H_5)_3]_2\,[C_8H_8] + 2KCl}$$

$$\mathrm{(\eta^5 - C_5H_5)_3UCl + Li[OCCo_3\,(CO)_9] \xrightarrow{THF} U\,\{OCCo_3\,(CO)_9\}\,(\eta^5 - C_5H_5)_3 + LiCl}$$

The first X-ray crystallographically characterized compound with a uranium carbon multiple bond has been synthesized by the following reaction:

$$\mathrm{(\eta^5 - C_5H_5)_3\,UCl + Li(CH_2)\,(CH_2)P(CH_3)\,(C_6H_5) \xrightarrow[-\,40^\circ C]{THF} (\eta^5 - C_5H_5)_3U = CHP(CH_3)_2(C_6H_5) + LiCl}$$

The ligand, BH_4^-, forms covalent complexes with a number of actinide metal ions and exhibits tridentate ligation, e.g.,

$$\mathrm{(\eta^5\text{–}C_5H_5)_3\,AnCl + Na^+[BH_4]^- \xrightarrow{THF} (\eta^5\text{–}C_5H_5)_3}$$ 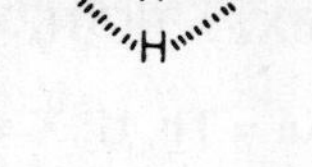

(An = Th, U, etc.)

All these compounds are thermally stable, mostly sublimable, but extremely air-sensitive solids. Their bonding is evidently more covalent than that in organolanthanides presumably because of the involvement of 5f orbitals.

The overall geometry of tris(cyclopentadienyl) actinide (IV) compounds, $AnX(\eta^5—C_5H_5)_3$, in most cases (with a few exceptions) is similar. The coordination sphere around the actinide atom (An) is roughly tetrahedral, with three η^5-bonded cyclopentadienyl (Cp) ligands and a η^1-bonded anionic group giving approximately C_{3v} geometry. Two representative structural types are illustrated in Figure 4.82.

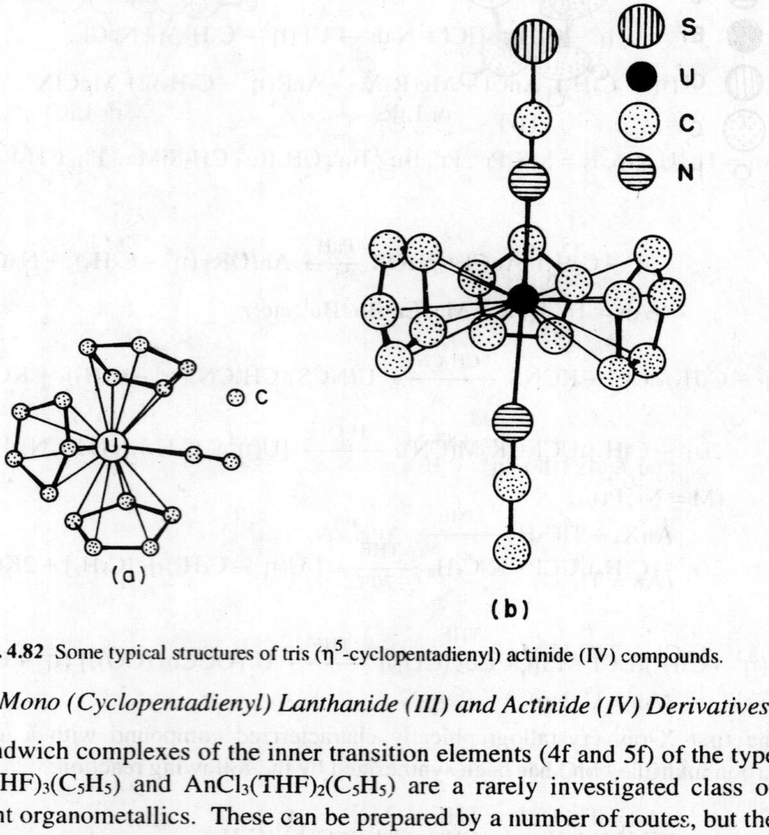

Fig. 4.82 Some typical structures of tris (η^5-cyclopentadienyl) actinide (IV) compounds.

4.4.3.3 Mono (Cyclopentadienyl) Lanthanide (III) and Actinide (IV) Derivatives

Half-sandwich complexes of the inner transition elements (4f and 5f) of the type $LnCl_2(THF)_3(C_5H_5)$ and $AnCl_3(THF)_2(C_5H_5)$ are a rarely investigated class of f-element organometallics. These can be prepared by a number of routes, but the most convenient and common amongst those are the metathetic reactions as illustrated by the following reactions:

$$LnCl_3 + MC_5H_5 \xrightarrow{THF} LnCl_2\,(THF)_3\,(\eta^5 - C_5\,H_5) + MCl$$

(Ln = Sm, Eu, Gd, Dy, Ho, Er, Lu; M = Na or Tl)

$$AnX_4 + TlC_5H_5 \xrightarrow{THF} AnX_3(\eta^5 - C_5H_5)(THF)_2 + TlCl$$

(An = Th, U; X = Cl, Br)

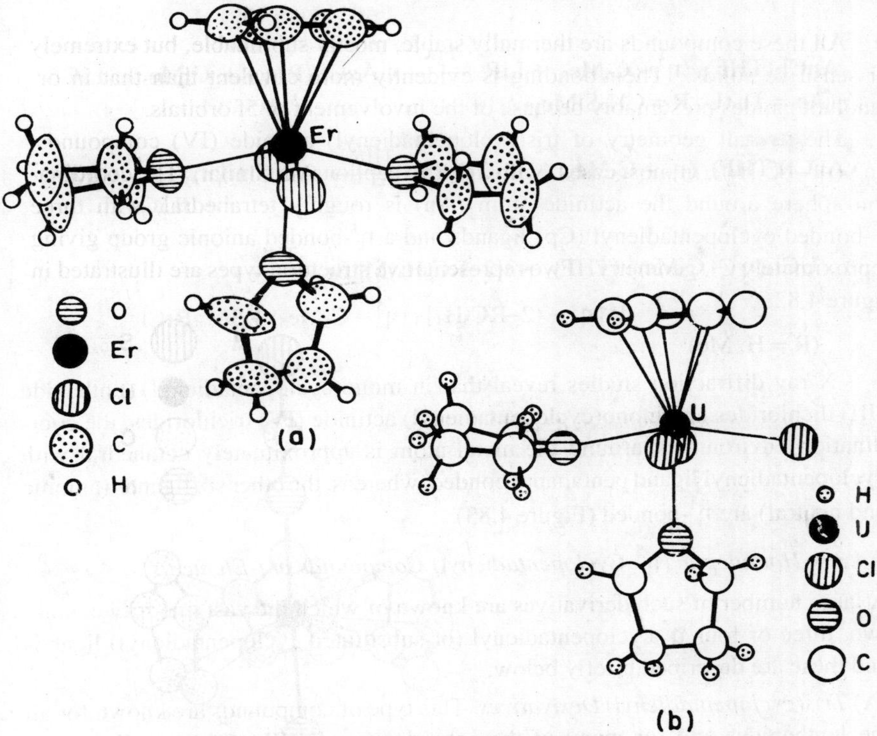

Fig. 4.83 Molecular structures of
(a) $ErCl_2(THF)_3$ ($\eta^5 - C_5H_5$), and (b) UCl_3 $(THF)_2$ ($\eta^5 - C_5H_4Me$).

$$AnX_4 + TlC_5H_5 \xrightarrow{DME} AnX_3(\eta^5 - C_5H_5)\ (DME) + TlCl$$

(An = Th, U)

$$AnCl_4 + TlC_5H_4Me \xrightarrow{THF} AnCl_3(\eta^5 - C_5H_4Me)\ (THF) + TlCl$$

(An = U, Np)

$$AnCl_4 + MC_5Me_5 \xrightarrow{L} AnCl_3(\eta^5 - C_5Me_5)\ (L)_2 + MCl$$

(An = Th, U; L = THF, 1/2 DME)

A large variety of interesting compounds can be prepared from these chloride precursors. A few typical examples are described as follows:

$$UCl_3\ (THF)_2\ (\eta^5 - C_5H_5) + Na(acac) \xrightarrow{THF} U\ (acac)Cl_2(\eta^5 - C_5H_5) + NaCl$$

$$AnX_3(THF)_2\ (\eta^5 - C_5H_5) + KHB(pz)_3 \xrightarrow{THF} AnX_2\ (\eta^5 - C_5H_5)\ (HB(pz)_3) + KX$$

(An = Th, U; X = Br, Cl)

$$AnCl_3(THF)_2(\eta^5-C_5Me_5) + LiR \xrightarrow{Et_2O} AnCl_2(R)(\eta^5-C_5Me_5) + LiCl$$
(An = Th, U; R = CH_2SiMe_3)

$$AnCl_3(THF)_2(\eta^5-C_5Me_5) + 3LiCH_2Ph \xrightarrow{THF} An(CH_2Ph)_3(\eta^5-C_5Me_5) + 3LiCl$$

$$UCl_3(\eta^5-C_5Me_5)(THF)_2 + 3(2\text{-}RC_3H_4)MgBr \xrightarrow{THF} U\{\eta^3-(2\text{-}RC_3H_4\}_3(\eta^5-C_5Me_5) + 3MgBrCl$$
(R = H, Me)

X-ray diffraction studies reveal that in mono(cyclopentadienyl) lanthanide (III) dichlorides and mono(cyclopentadienyl) actinide (IV) trichlorides, the coordination environment around the metal atom is approximately octahedral with cyclopentadienyl ligand pentahapto bonded, whereas, the other six ligands (anionic and neutral) are η^1-bonded (Figure 4.83)

4.4.3.4 Homoleptic (η^5-Cyclopentadienyl) Compounds of f-Elements

A large number of such derivatives are known of which the vast majority contain two, three or four η^5-cyclopentadienyl (or substituted cyclopentadienyl) ligands and these are described briefly below:

(A) *Tris(cyclopentadienyl) Derivatives.* This type of compounds are known for all the lanthanides and for many of the actinides (e.g. U(III). Cf(III)). The most general method for their preparation is by ligand exchange reactions, although thorium (III) and U(III) derivatives, can also be prepared by reductive routes:

$$LnCl_3 + 3NaC_5H_5 \xrightarrow{THF} Ln(\eta^5-C_5H_5)_3(THF) + 3NaCl$$

$$\downarrow -THF$$

$$Ln(\eta^5-C_5H_5)_3$$

(Ln = Sc, Y, La, Ce, Pr, Nd, Sm, Gd, Dy, Er, Yb)

$$LnCl_3 + 3KC_5H_5 \xrightarrow[Et_2O]{PhH\ or} Ln(\eta^5-C_5H_5)_3 + 3KCl$$

$$2LnF_3 + 3Mg(C_5H_5)_2 \longrightarrow 2Ln(\eta^5-C_5H_5)_3 + 3MgF_2$$
(Ln = Ce, Nd, Sm)

$$UCl_3(THF)_n + 3NaC_5H_5 \xrightarrow{THF} U(\eta^5-C_5H_5)_3(THF) + 3NaCl$$

$$\downarrow -THF$$

$$U(\eta^5-C_5H_5)_3$$

$$UCl_3 + 3KC_5H_5 \xrightarrow{PhH} U(\eta^5-C_5H_5)_3 + 3KCl$$

The tris(cyclopentadienyl) compounds of promethium (III) and other actinides (III) are best prepared by the reactions:

$$2PmCl_3 + 3Be(C_5H_5)_2 \xrightarrow{65°C} 2Pm(\eta^5 - C_5H_5)_3 + 3BeCl_2$$

$$2AnCl_3 + 3Be(C_5H_5)_2 \xrightarrow{65°C} 2An(\eta^5 - C_5H_5)_3 + 3BeCl_2$$

The synthesis of tris(cyclopentadienyl) compounds sometimes results in the isolation of a MCp_3 base complex from which base free species can be easily obtained. All these compounds are thermally robust, but they are extremely air and moisture sensitive.

Although lanthanide(III) and actinide(III) ions are typical 'a' class metals (hard acids), their organometallic compounds form adducts with soft bases (e.g. phosphines and sulphides) as well as with hard bases (e.g. carboxyl and nitrosyl groups, ammonia and tetrahydrofuran).

Some typical reactions and structures of the resulting products are illustrated as follows:

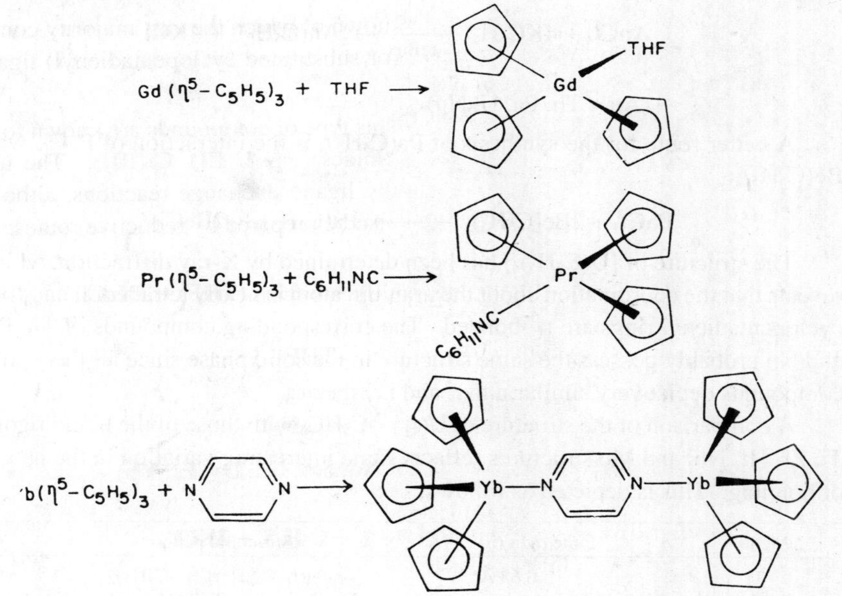

In all these compounds, X-ray structures have revealed approximate tetrahedral environment around the lanthanide metal by the three pentahapto bonded C_5H_5 rings and one η^1-bonded Lewis base ligand.

The protolysis of the metal-ring bonds in tris(cyclopentadienyl) lanthanide(III) and actinide(III) can be represented by the following equations:

$$3Ln(C_5H_5)_3 + 3H_2O \longrightarrow 3Ln(OH)_3 + 3C_5H_6$$

$$3An(C_5H_3)_3 + 3H_2O \longrightarrow 3An(OH)_3 + 3C_5H_6$$

$$Ln(C_5H_5)_3 + HCl \longrightarrow 1/2[\{LnCl(\eta^5 - C_5H_5)_2\}_2] + C_5H_6$$

$$U(C_5H_5)_3 + HCN \longrightarrow U(CN)\,(C_5H_5)_2 + C_5H_6$$

These reactions can be taken as evidence for significant ionic nature in the metal-cyclopentadienyl ring bonding.

Structural studies of only a few of these homoleptic derivatives have been carried out so far, but these have revealed some interesting features; for example in $[Sm(C_5H_5)_3]_n$, each C_5H_5 group is pentahapto towards one metal atom, but some of these are simultaneously η^1- or η^2-bonded towards an adjacent metal atom, producing a chain-type (cf. structure of $Ca(C_5H_5)_2$). By contrast, $[Nd(C_5H_4Me)_3]_4$ depicts a simpler tetrameric structure in which out of the three η^5-bonded methyl cyclopentadienyl rings to each metal, one is also bonded in a monohapto manner to the adjacent neodymium atom resulting in an over-all tetrameric structure.

(B) *Tetrakis (cyclopentadienyl) Derivatives.* These are conveniently prepared by treating $AnCl_4$ with KC_5H_5 in benzene:

$$AnCl_4 + 4KC_5H_5 \xrightarrow[\text{reflux}]{\text{PhH}} An(C_5H_5)_4 + 4KCl$$

(An = Th, Pa, U, Np)

A better route for the synthesis of $Pa(C_5H_5)_4$ is the interaction of $PaCl_4$ with $Be(C_5H_5)_2$:

$$PaCl_4 + 2Be(C_5H_5)_2 \longrightarrow Pa(C_5H_5)_4 + 2BeCl_2$$

The structure of $[U(C_5H_5)_4]$ has been determined by X-ray diffraction, which reveals that the coordination about the uranium atom is nearly tetrahedral and four cyclopentadienyl rings are η^5-bonded. The corresponding compounds of Th, Pa, and Np probably possess the same structure in the solid phase since all these four compounds depict very similar n.m.r. and i.r. spectra.

A comparison of the structure of $U(\eta^5—C_5H_5)_4$ with those of the homologous Ti, Zr, Hf, Nb, and Mo structures reflects some interesting variation in the nature of bonding. This is depicted as follows:

Metal	Ionic radii (M^{4+}) (Å)	Structural pattern
Ti	0.68 Å	$Ti(\eta^1–C_5H_5)_2(\eta^5–C_5H_5)_2$
Zr	0.74 Å	$Zr(\eta^1–C_5H_5)(\eta^5–C_5H_5)_3$
Hf	0.75 Å	$Hf(\eta^1–C_5H_5)_2(\eta^5–C_5H_5)_2$
Nb	0.74 Å	$Nb(\eta^1–C_5H_5)_2(\eta^5–C_5H_5)_2$
Mo	0.70 Å	$Mo(\eta^1–C_5H_5)_2(\eta^5–C_5H_5)_2$
U	0.93 Å	$U(\eta^2–C_5H_5)_4$

(C) *Bis(η^5-cyclopentadienyl) Lanthanide(II) Complexes.* In contrast to the reasonably extensive literature associated with trivalent lanthanide organometallics discussed in earlier sections, the divalent organolanthanides compounds are rather sparse. Although divalent oxidation states are feasible for a number of lanthanides (e.g. Nd, Pm, Sm, Eu, Dy, Tm, Yb), yet well defined, isolable divalent organometallics are at present accessible under normal chemical conditions only for $Sm^{2+}(4f^6)$, $Eu^{2+}(4f^7)$, and $Yb^{2+}(4f^{14})$ ions. Of these, as expected from their electronic configurations, samarium (II) organometallics are the most reactive. Concerted efforts by a number of chemists in this novel area in recent years have generated a variety of new classes of complexes (Fig. 4.84) and reactivity patterns (Scheme 4.8). In view of the magnitude of the standard electrode potential, E° for Ln^{3+}/Ln^{2+} (Sm^{3+}/Sm^{2+} – 1.50 V; Eu^{3+}/Eu^{2+}, — 0.43 V; Yb^{3+}/Yb^{2+}, — 1.15 V) systems, these divalent organometallics are strong reducing agents and undergo one electron oxidation reactions cleanly and in quantitative yields.

(a) *Preparation.* General methods for the preparation of divalent (Ln^{2+}) lanthanide cyclopentadienyl complexes include:

(i) metathetic reactions, e.g.:

$$SmI_2 + 2KC_5Me_5 \xrightarrow{THF} Sm(\eta^5 - C_5Me_5)_2(THF)_2 + 2KI$$

$$EuCl_3 + 3NaC_5Me_5 \xrightarrow{THF/Et_2O} Eu(\eta^5 - C_5Me_5)_2(THF)(Et_2O) + 3NaCl + \text{reduced organic products}$$

$$YbCl_2 + 2NaC_5Me_5 \xrightarrow{THF} Yb(\eta^5 - C_5Me_5)_2(THF)\,0.5\,C_6H_5Me + 2NaCl$$

$$YbBr_2 + 2KC_5Me_5 \xrightarrow{THF \text{ or } Et_2O} Yb(\eta^5 - C_5Me_5)_2(L)_2 + 2KBr$$

(ii) direct method, i.e. reaction of the lanthanide metal (Sm, Eu, Yb) with cyclopentadiene or its substituted derivatives, e.g.:

$$Ln + 3C_5H_6 \xrightarrow[(-C_5H_8)]{NH_3} Ln(C_5H_5)_2(NH_3)_x \xrightarrow[\text{vacuum}]{420^\circ C} Ln(C_5H_5)_2$$

(Ln = Eu or Yb)

$$Yb + 2C_5Me_5H \xrightarrow{THF/NH_3} Yb(C_5Me_5)_2(NH_3)(THF)$$

$$\underset{(\text{vapour})}{Sm + 2C_5Me_5H} \xrightarrow[-120^\circ C]{THF} Sm(C_5Me_5)_2(THF)_2$$

(iii) reductive method, e.g.:

$$1/2\,[\{YbCl(\eta^5 - C_5H_5)_2\}_2] + 2Na \xrightarrow{THF} Yb(\eta^5 - C_5H_5)_2(THF)_x + 2NaCl$$

$$1/2\,[\{YbCl(\eta^5 - Cp')_2\}_2] \xrightarrow[THF]{Na/Hg} Yb(\eta^5 - Cp')_2(THF)_2 + 2NaCl$$

($Cp' = C_5H_4SiMe_3$)

$$1/2\ [\{LnCl(\eta^5 - Cp'')_2\}_2] \xrightarrow[THF]{Na/Hg} Ln(\eta^5 - Cp'')_2\,(THF)_n$$

$$(Cp'' = C_5H_3(SiMe_3)_2\ ;\ Ln = Eu,\ n = 1;\ Ln = Yb,\ n = 0)$$

$$Yb(C_5H_5)_3\,(THF) + Na \xrightarrow{THF} Yb(\eta^5 - C_5H_5)_2(THF)_x + NaC_5H_5$$

$$1/2\ [\{YbCl(\eta^5 - C_5H_5)_2\}_2] + 2Yb \xrightarrow{THF} Yb(\eta^5 - C_5H_5)_2(THF)_x + YbCl_2$$

$$Sm(C_5H_5)_3 \xrightarrow{KC_{10}H_8/THF} Sm(\eta^5 - C_5H_5)_2(THF)_x + KC_5H_5$$

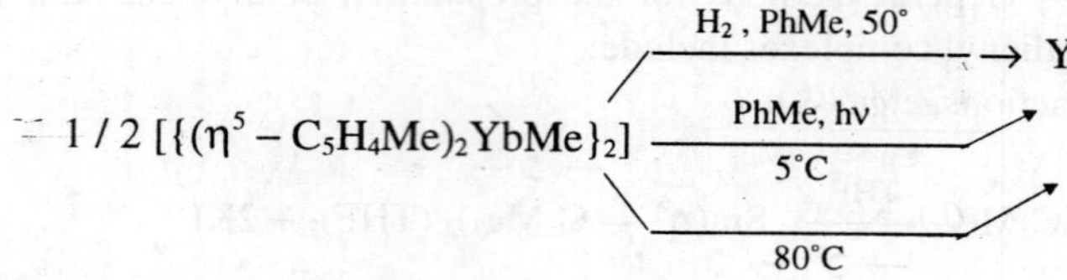

(b) *Properties.* These complexes are highly oxygen and moisture sensitive coloured solids and are strong reducing agents. The oxygen sensitivity of these compounds $Ln(\eta^5—C_5H_5)_2$ (Ln = Sm, Eu, or Yb) increase in the sequence: Eu < Yb < Sm. Magnetic moment (the magnitudes of μ eff., are those expected for Ln^{2+} ions), infrared (the spectra are very similar to those of alkaline earth bis-cyclopentadienides), X-ray power diffraction ($Eu(C_5H_5)_2$ and $Sr(C_5H_5)_2$ are isostructural) and ^{151}Eu mössbauer spectral measurements on these derivatives are indicative of the presence of symmetrical five membered rings with predominant ionic bonding.

The chemistry of substituted cyclopentadienyl derivatives of samarium and ytterbium has been explored in recent years. Some typical examples are illustrated in the following schemes 4.8a and b:

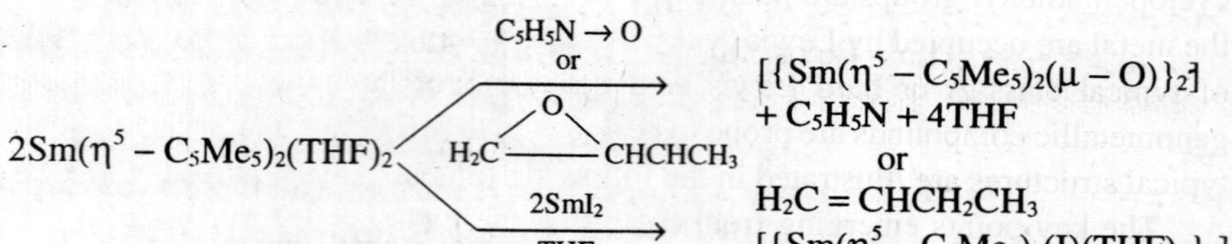

(a)

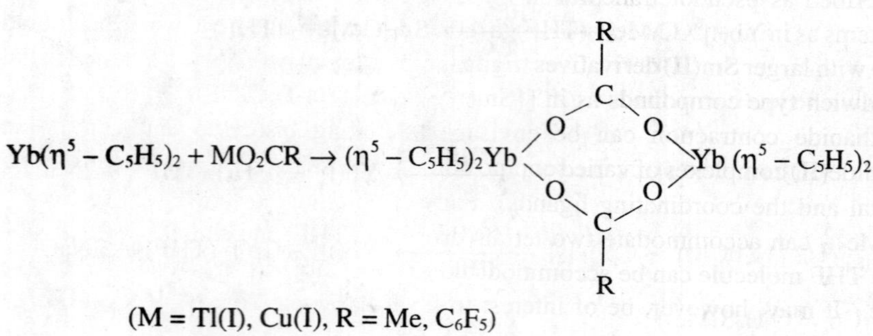

(M = Tl(I), Cu(I), R = Me, C_6F_5)

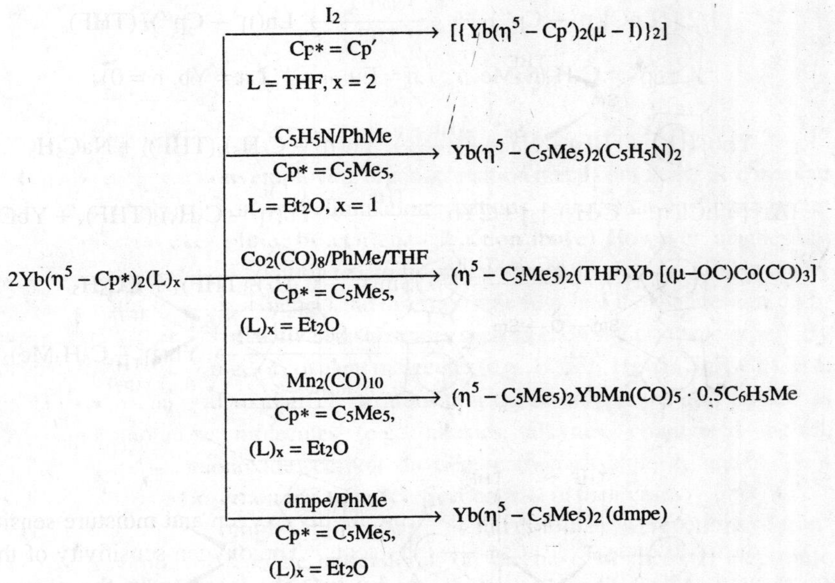

(b)

Scheme 4.8 Some Reactions of (a) $Sm(\eta^5—C_5Me_5)_2\ (THF)_2$, and (b) $Yb(\eta^5—Cp^*)_2\ (L)_x$ (where $Cp^* = C_5H_5$, C_5Me_5, $C_5H_4SiMe_3(Cp')$; L = THF or Et_2O, X = O, 1, 2)

(c) *Structure.* In the earlier sections, coordinative unsaturation of $LnCp_2^*$ (formally in six coordination state) coupled with lanthanide contraction and selection of the coordination ligands have been exploited in the synthesis of some interesting divalent organolanthanide derivatives. In all these derivatives, cyclopentadienyl groups are pentahapto bonded and vacant coordination sites on the metal are occupied by Lewis bases. An interesting point to note is that in spite of typical class A or hard Lewis acid behaviour of lanthanide ions, their organometallic compounds are prone to be attacked to soft donor ligands also. Some typical structures are illustrated in the following figures.

The key points emerging from these structural studies (as illustrated in Fig. 4.84) are that (i) the coordination geometry around the lanthanide atom may be described as pseudotetrahedral in most cases; (ii) in some sterically congested systems as in $Yb(\eta^5–C_5Me_5)_2(THF)$, trigonal planar arrangement is more favoured; (iii) with larger Sm(II) derivatives trigonal bipyramidal geometry is attained in half sandwich type compounds as in $[\{Sm(\eta^5–C_5Me_5)\ (\mu–I)\ (THF)_2\}_2]$; (iv) effect of lanthanide contraction can be envisaged isolating bis(cyclopentadienyl) lanthanide(II) complexes of varied composition depending upon both the nature of the metal and the coordinating ligands. For example, larger Sm(II) ion in $Sm(\eta^5–C_5Me_5)_2$ can accommodate two tetrahydrofuran (THF) molecules, whereas only one THF molecule can be accommodated in the analogous smaller Yb(II) derivative. It may, however, be of interest to note that even $Yb(\eta^5–C_5Me_5)_2(THF)$ is

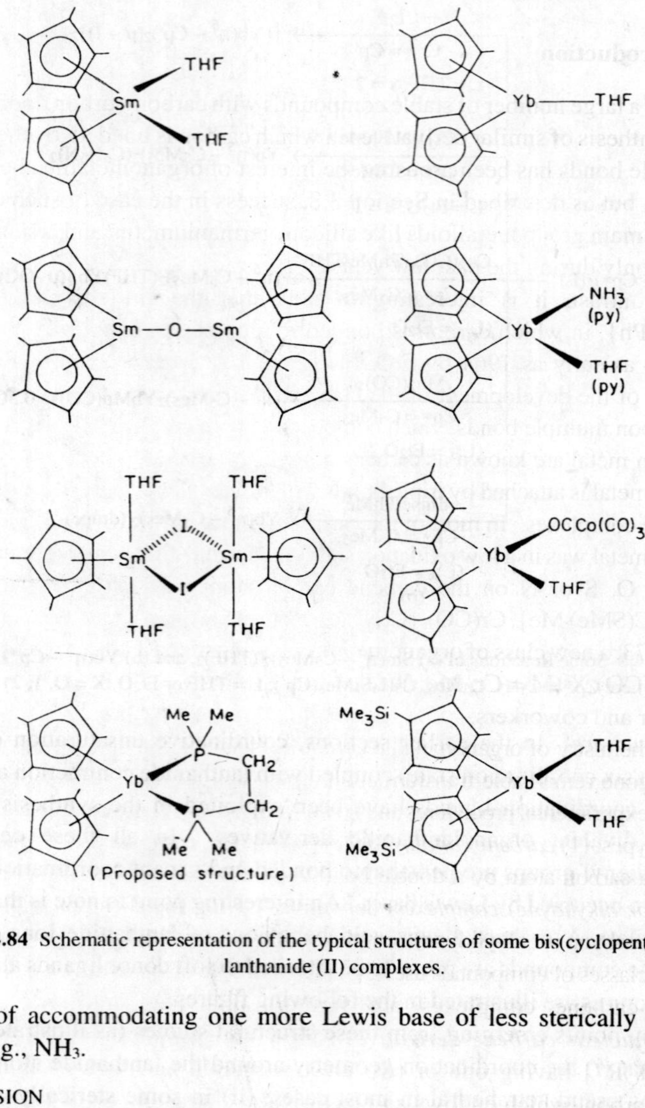

Fig. 4.84 Schematic representation of the typical structures of some bis(cyclopentadienyl) lanthanide (II) complexes.

capable of accommodating one more Lewis base of less sterically demanding nature, e.g., NH_3.

CONCLUSION

Our discussion shows that the chemistry of *f*-element organometallics, although in its infancy, is moving at a fast pace. The bonding, particularly in the +3 oxidation states of the metals, is essentially ionic, but the tetravalent derivatives of early actinides show appreciable covalent character with the involvement of '5*f*' as well as '6*d*'-orbitals. Evidence is also accumulating of the potential applications of both 4*f*- and 5*f*-organometallics as homogeneous as well as supported catalysts. The future of this rather new branch is thus quite bright, and it can be expected to emerge into a unique chapter of organometallic chemistry in the years to come.

4.5 MULTIPLE METAL-CARBON BONDED COMPOUNDS

4.5.1 Introduction

In view of a large number of stable compounds with carbon-carbon double or triple bonds, synthesis of similar derivatives in which carbon is bonded to other elements by multiple bonds has been arousing the interest of organometallic chemists for a long time, but as described in Section 3.8, success in the case of such derivatives even with main group metalloids like silicon, germanium, tin, and arsenic has been achieved only during the past few years.

By contrast, it is interesting to note that the first derivative $W(CO)_5$ {C(OMe)Ph}, in which carbon is bonded to a transition metal by a double bond was made as early as 1964 by E.O. Fischer and A. Maasböl and this marked the beginning of the development of the chemistry of compounds containing transition metal-carbon multiple bonds. Such compounds in which divalent carbon is bound to a transition metal are known as carbene complexes; likewise compounds in which a transition metal is attached by triple bonds with monovalent carbon moiety are termed as carbyne complexes. In most of the earlier examples of the carbene complexes, the transition metal was in a low oxidation state and the derivatives incorporating a hetero atom like O, S or N on the carbene carbon atom e.g., $Cr(CO)_5$ {C(OMe)Me}, $W(CO)_5${C(SMe) Me}, $Cr(CO)_5${C(NMe_2)Me}, etc. were comparatively more stable.

In 1973, a new class of organometallics having metal-carbon triple bonds, e.g., M(≡CR) $(CO)_4X$ (M = Cr, Mo, W; R = Me, Et, Ph; X = Cl, Br, I) was discovered by Fischer and coworkers.

The chemistry of organometallic compounds having multiple metal carbon bonds has undergone remarkable transformation during the last three decades, incorporating ingenious experimental procedures and novel compounds. These complexes are generally of two types: (1) *carbene (or alkylidene) complexes:* compounds, in which a metal is bound to a carbon atom by a double bond, are known as carbene complexes; and (ii) *carbyne (or alkylidyne) complexes*: derivatives having metal-carbon triple bonds are termed as carbyne (or alkylidyne) complexes (the systematic (IUPAC) nomenclature of these two classes of compounds uses the suffix -ylidene or -ylidyne respectively.

The carbene complexes are further subdivided into two categories: (i) *heteronuclear carbene derivatives*: these are complexes of the general type $L_nM[:C(R')R'']$ having one or two electronegative heteroatoms bonded to the carbene carbon, which endows an electrophilic character on the carbene carbon. These are generally in low oxidation states and comparatively more stable and continue to be called as Fischer type (after the discoverer); and (ii) *homonuclear carbene derivatives*: these are compounds with the metals in high oxidation states and further in these, the carbene which is bound to hydrogen and carbon atoms only is nucleophilic in nature. Discovered by R.R. Shrock in 1974, these can be more appropriately termed as alkylidene complexes in view of the homonuclear nature of the ligands involved. It may be noted that some homonuclear carbenes such as $(CO)_5W[:CPh_2]$ exhibit high electrophilic character, and therefore these are included in the Fischer type carbenes.

The carbene complexes are, therefore, subdivided into the following two categories:

	Fischer Type	Schrock Type
(i)	Discovered by Fischer (1963)	Discovered by Schrock (1976)
(ii)	The :C(X)Y ligand has generally at least one heteroatom attached to the carbene carbon, i.e., X or Y or both = halogen, OR, etc., e.g., $(OC)_5Cr\{C(OEt)Me\}$, $(OC)_5Cr\{C(NEt_2)Cl\}$,	In the :C(X)Y ligand, both X and Y are alkyl groups/hydrogen atoms, e.g., $Cp_2(CH_3)Ta(:CH_2)$, $(Me_3CCH_2)_3\,Ta(:CHCMe_3)$.
(iii)	The carbene carbon is electrophilic in nature e.g., $(OC)_5\,Cr\{:C(OMe)R\}$ + HNR′R″ → $(OC)_5\,Cr\{:C(NR'R'')R\}$ (R = alkyl or aryl)	The carbene carbon is nucleophilic in nature e.g., $Ta(CH_2CMe_3)_3\left(:C\begin{matrix}H\\CMe_3\end{matrix}\right)$ + HCl → $TaCl(CH_2CMe_3)_4$
(iv)	The metal M in M:C(X)Y complex is generally in a low oxidation state, e.g., {W(I)} in $(OC)_5\,W{:}C\begin{matrix}OMe\\Ph\end{matrix}$	The metal M in M:C(R)R is comparatively high oxidation state, e.g., {Ta(V)} in $Ta(CH_2CMe_3)_3\left(:C\begin{matrix}H\\Bu^t\end{matrix}\right)$

Note: In view of the electrophilic nature of carbene carbon in derivatives like $(OC)_5\,W\,(:CPh_2)$, these are also sometimes included in the first category.

4.5.2 Transition Metal-Carbene Complexes

4.5.2.1 Carbene Ligands

Carbene (also called carbenoid) ligands are divalent carbon species, :Ċ(X)Y, with only six valence electrons on the carbene carbon, which is covalently bonded to two other groups (X and Y), thus leaving two non-bonded electrons on the carbon atom. These two electrons may be spin paired (with opposing spins, ↓↑), in which case the carbene (such as $:CCl_2$) is in the singlet state (Fig. 4.85a), whereas these unbonded electrons may have parallel spins in which case the carbene (e.g. $:C(C_6H_5)_2$) assumes a triplet state (Fig. 4.85b):

These two electronic configurations should be reflected in the differing geometries and chemical reactivity, as well as in the important physical charac-

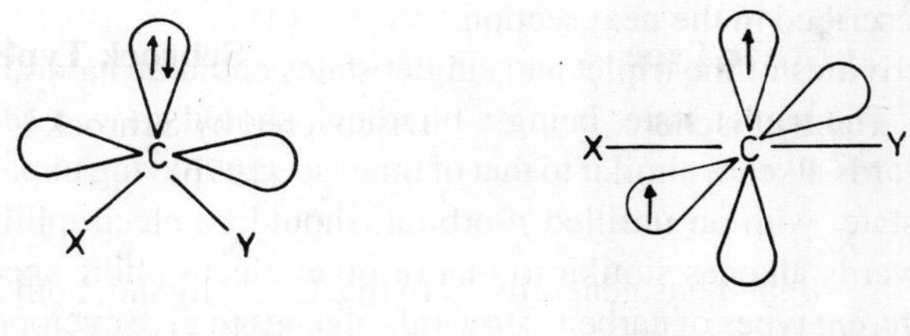

(a) singlet state, sp^2 hybridization

(b) triplet state sp hybridization

Fig. 4.85 Molecular orbital (MO) diagrams for carbenes.

teristic that whereas the singlet is diamagnetic (similar to carbonium species), the triplet would be paramagnetic (with two unpaired electrons, similar to bi-radical species). Electron spin resonance (e.s.r.) spectroscopy should, therefore, be an appropriate technique for the study of carbenes in the triplet state.

For the geometries of the two types of carbenes, a qualitative picture can be obtained from the hybridization at the carbon atom. The bonding in the singlet assumes sp^2 hybridization with the electron pair in the sp^2 hybridized orbital and with an unoccupied p orbital. The $X\hat{C}Y$ angle could be expected to be contracted slightly (provided X and Y are small) from the ideal 120° angle due to lone pair-bond pair repulsion being greater than bond pair-bond pair repulsion. The sp hybridization in triplet state would tend to make the carbene (B type) linear, with the two unpaired electrons occupying two mutually orthogonal p-orbital.

On the basis of MO calculations, the $H\hat{C}H$ angle for methylene ($:CH_2$) can be predicted as ~ 135° and ~ 105° respectively for the triplet and singlet states, with the triplet being slightly more stable (~ 10 kcal or 42 kJ/mole lower in energy) than the singlet. All available physical and chemical evidence is consistent with the triplet as the ground state in which $H\hat{C}H$ angle is estimated to be 125–140° by e.s.r. spectroscopy.

The nature of substituents X and Y could be expected to have the effect of perturbing relative energies of singlet and triplet states. In general, the dialkylcarbenes are ground-state triplets. Lone pair donors in the substituents X and Y should tend to stabilize the singlet state in preference to the triplet state by π-donation into the emply p-orbital.

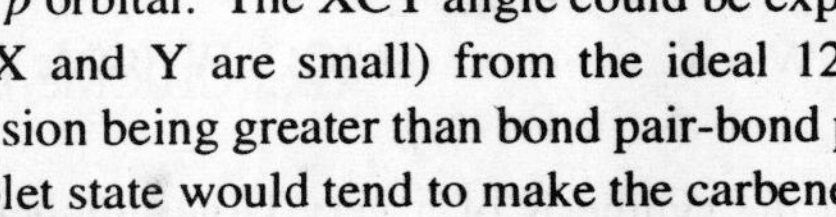

(X = F, Cl, Br, or NR_2)

The dihalocarbenes have, for example, all been shown to have singlet ground state with $X\hat{C}Y$ angle equal to 100 ± 10°. As far as chemical properties are concerned, the stabilization of singlet state by electron release from adjacent heteroatoms may be so pronounced as to cause the electrophilic properties of a carbene to vanish; this has been of special significance in the discovery of Fischer type carbene complexes described in the next section.

The reactivities of the triplet and singlet states could be naturally expected to be different. The triplet state, being a biradical, should show a selectivity in its reactions towards alkenes similar to that of other species having unpaired electrons. The singlet state, with an unfilled *p*-orbital, should be electrophilic and exhibit reactivity towards alkenes similar to that of other electrophilic species (for reactivities of different types of carbenes towards alkenes to give cyclopropane derivatives, any good organic text may be referred to):

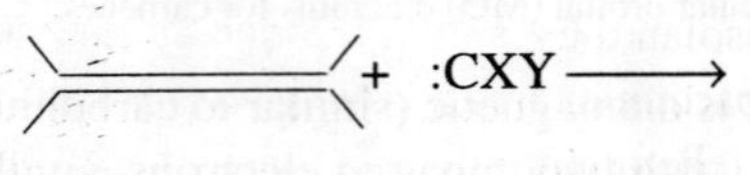

4.5.2.2 Nature of Transition Metal-Carbene Complexes

As mentioned in the preceding section, the nature of carbene ligand with an electron-donating hetero-atom in the substituent X (e.g. OMe) was exploited by E.O. Fischer in 1964 for the synthesis of the first transition metal carbene complex:

$$(OC)_6W + :C(OMe)(Ph) \rightarrow (OC)_5W\{(OMe)Ph\}$$

This was followed by the synthesis of a very large number of similar carbene complexes, now generally known as those of Fischer type. In all these, the carbene ligands appears to function as neutral two electron donor ligands similar to carbon monoxide $\{(C = O \longleftrightarrow \overset{-}{C} \equiv \overset{+}{O})\}$ with the transition metals.

From a study of various physico-chemical properties, it appears that out of the three main canonical forms (A, B and C) for such Fischer type of complexes, the first two (A and B) forms are predominant:

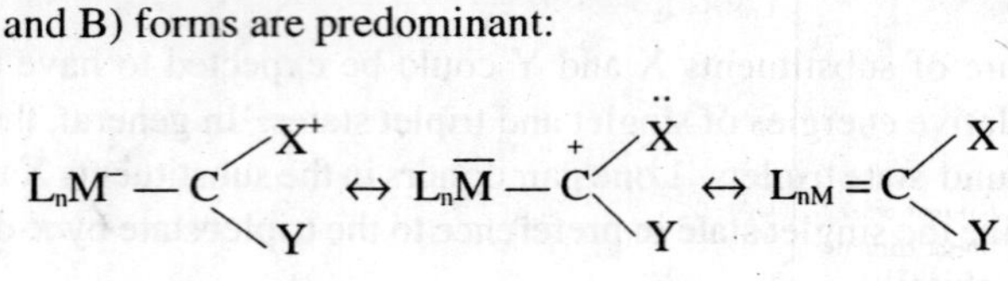

Fig. 4.86 Resonance forms of Fischer type carbene (with X having an electron donor hetero-atom) complexes.

As discussed in later sections dealing with structural aspects (Section 4.5.2.3c) and chemical reactions (section 4.5.2.3B) of these Fischer type complexes, the predominance of forms (A) and (B) is reflected in (i) the absence of a bond order significantly by higher than one in M—$C_{carbene}$ bond, (ii) considerable double bond nature of $C_{carbene}$ ——→ X bond, and (iii) the electrophilic character of $C_{carbene}$ in such complexes.

A large number of such Fischer type complexes are now known for d^5—d^8 metals (usually in 0 or + 1 formal oxidation states, assuming the carbene to be a neutral two electron ligand). Although complexes with carbenes having no heteroatom substitutents had been postulated as intermediates in a number of reactions, yet only three compounds, i.e., $Cr(CO)_5$(2,3-diphenylcyclopropenylidene), $W(CO)_5(CPh_2)$, and $Fe(\eta^5-C_5H_5)(CO)_2$(benzocyclobutenylidene) had been isolated prior to 1974. These three were also similar to Fischer-type complexes in being 18-electron species with the metal in formally 0 or + 1 oxidation state; these are also characterized by having at least one stabilizing resonance form and in showing reactivity at $C_{carbene}$ with nucleophiles.

In 1974, a ten-electron tantalum complex, $Ta(CH_2CMe_3)_3(CHCMe_3)$ (in formally +3 oxidation state) was isolated by Shrock. Since then, marked intensification of research activity in this particular area by many workers has generated novel and fascinating organometallic chemistry.

In the following two sections is being presented a brief account of Fisher type carbene complexes followed by that of Schrock type carbene (alkylidene) derivatives.

4.5.2.3 Fischer Type Carbene Complexes

(A) *Synthesis of Carbene Complexes (Fischer Type).* Transition metal carbene complexes (Fischer Type) have generally been obtained from three classes of precursors (starting materials), which are depicted as (i), (ii) and (iii) in Scheme 4.9.

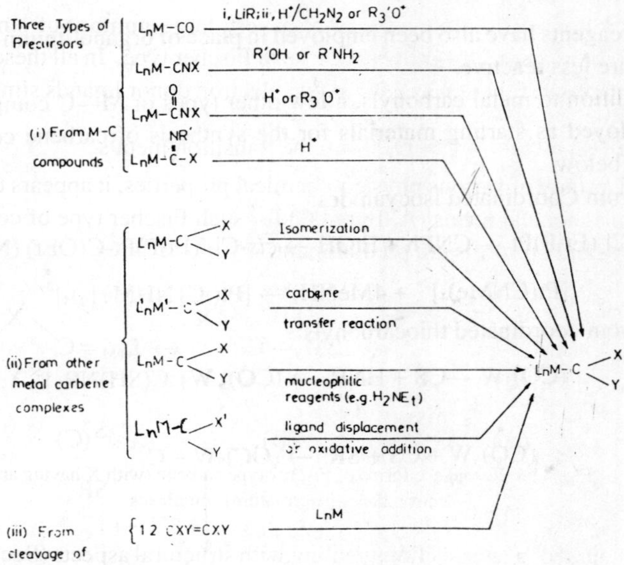

Scheme 4.9 Main preparative routes for Fischer type transition metal-carbene derivatives.

(i) *Synthesis from metal-carbon compounds.* The most useful general route for the synthesis of carbene complexes from non-carbene precursors involves the nucleophilic attack on the carbonyl carbon of a metal carbonyl. Of the two reactions represented in the following diagram, the route on the left was employed for the synthesis of the first tungsten carbene complex in 1964 by Fischer and Maasböl. The synthesis was later improved (as shown by the route on the right) by employing oxonium salts:

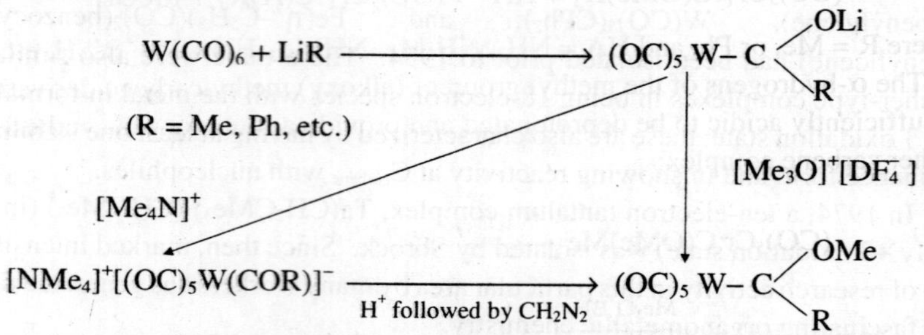

The synthesis has later been extended to a number of other transition metals (e.g., Cr, Mo, Mn, Fe, Ru and Re) and to other ligands such as:

R′(R″S)C— , R′(R″$_2$N)C— ,

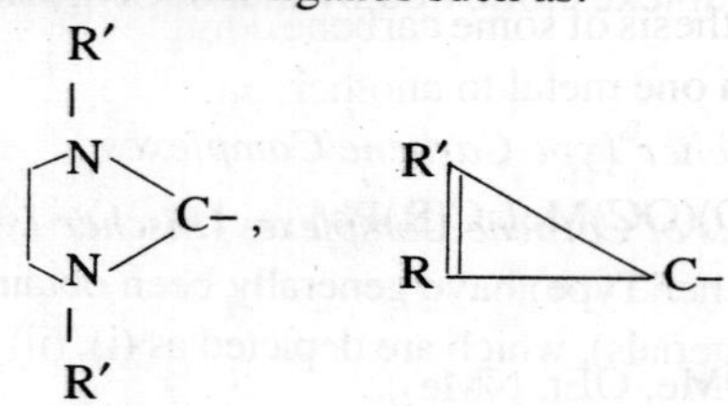

Grignard reagents have also been employed in place of organolithium compounds, but these are less reactive.

In addition to metal carbonyls, a few other types of M—C compounds have been employed as starting materials for the synthesis of carbene complexes as illustrated below:

(a) From Coordinated isocyanides:

$$cis\text{-}Cl_2(Et_3P)Pt — CNPh + EtOH \rightarrow cis\text{-}Cl_2(Et_2P)Pt\text{-}C(OEt)(NHPh)$$

$$[Pt(CNMe)_4]^{2+} + 4MeNH_2 \rightarrow [Pt(:C\{NHMe\}_2)_4]^{2+}$$

(b) From coordinated thiocarbonyls:

$$(CO)_5W — CS + HNR_2 \rightarrow (CO)_5 W\{:C(SH)NR_2\}$$

$$(CO)_5W - CS + SR^- \rightarrow (OC)_5W - C(=S^-)(SR) \xrightarrow{+R'I\,(-I^-)} (CO)_5W\{:C(SR)SR'\}$$

(R = Me, Et, Pr^i, Bu^n, Bu^t, R′ = Me, Et, Pr^i, Bu^n)

(c) From coordinated carbon disulphide derivatives:

$$Os(\eta^2\text{–}CS_2)(CO)_2(PPh_3)_2 + 2MeI \rightarrow OsI_2(CO)_2(PPh_3)_2\{:C(SMe)_2\}$$

$$Pt(\eta^2\text{—}CS_2)(PPh_3)_2 + 2MeI \rightarrow PtI(PPh_3)_2\{:C(SMe)_2\}I$$

(ii) *Synthesis using metal carbene precursors.* This route for the synthesis of new carbene complexes can be illustrated by the following general equations:

$$(OC)_5Cr\{:C(OMe)R\} + HA \rightarrow (OC)_5Cr\{:C(A)R\} + MeOH$$

(Where R = Me, or Ph, and HA = NH_3, NH_2Me, $NHMe_2$, $HN{:}CPh_2$, R′SH, etc.).

The α-hydrogens of the methyl group of (alkoxy) methylcarbene derivatives are sufficiently acidic to be deprotonated and provide a route for the synthesis of another carbene complex:

$$(CO)_5Cr{:}C(OMe)Me \xrightarrow{NaOMe}$$

$$\xrightarrow{Me_3O^+BF_4^-} (OC)_5Cr\{:C(OMe)CH_nMe_{3-n}\}$$
$$(n = 1 \text{ to } 3)$$

The synthesis of some carbene complexes can be accessible via carbene ligand transfer from one metal to another, e.g.:

$$Cp(NO)(OC)Me\{:C(R)Ph\} + Fe(CO)_5 \xrightarrow{h\nu} Mo(CO)_2(NO)Cp + (OC)_4Fe\{:C(R)Ph\}$$

(R = OMe, OEt, NMe_2)

$$(OC)_5W\{:C(X)Ph\} + Au(H)Cl_4 \xrightarrow{heat} Cl - Au\{:C(X)Ph\} + WCl_2(CO)_4$$

(X = OMe, NH_2, NMe_2)

Synthesis of new transition metal carbene complexes can be achieved via displacement of either neutral or anionic ligands from parent transition metal carbene derivatives:

$$(PPh_3)_2Cl_3Rh\{:C(NMe_2)H\} + 2PEt_3 \rightarrow (PEt_3)_2Cl_3Rh\{:C(NMe_2)H\} + 2PPh_3$$

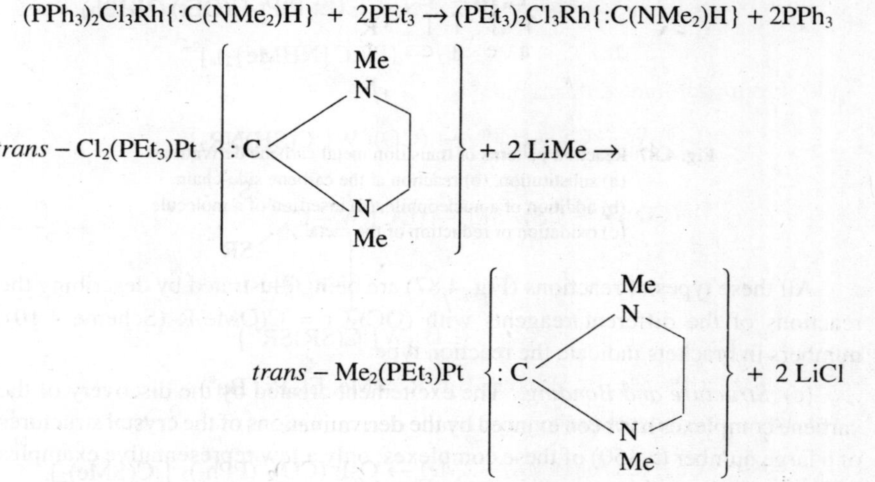

(iii) *From cleavage of electron-rich alkenes.* This is a remarkable, quite general route to the synthesis of diamino-substituted carbenes, and involves interaction of various metal derivatives with highly electron rich alkenes such as the bis(imidazolidizidene) mono, bis, neutral, and cationic carbene derivatives involving a majority of the transition metals have been prepared by this procedure:

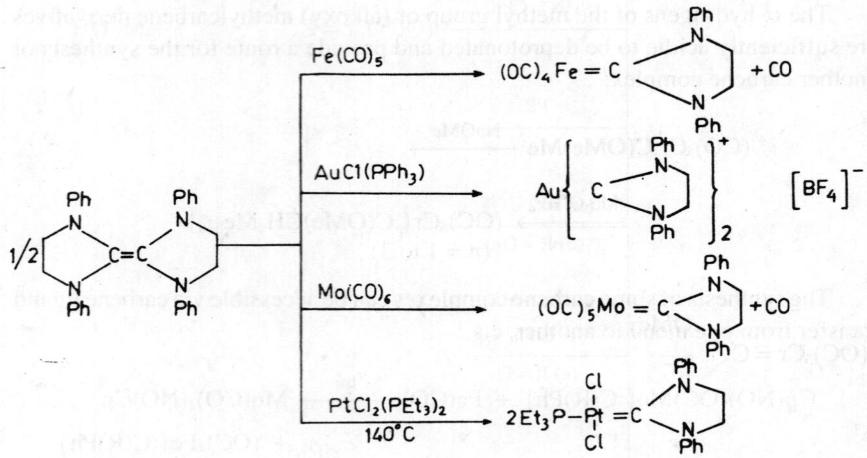

(B) *Reactions of Carbene Complexes (Fischer Type).* There are several different ways in which transition metal carbene derivatives may undergo chemical transformation as illustrated in Fig. 4.87.

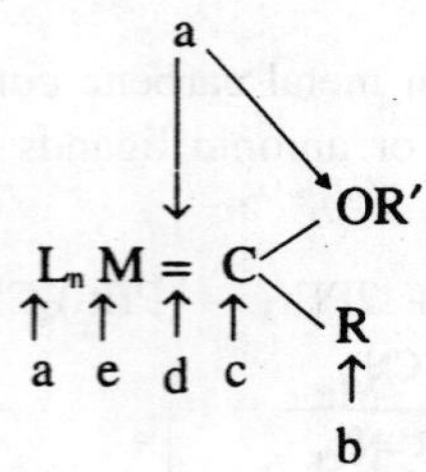

Fig. 4.87 **Reaction patterns of transition metal carbene derivatives: (a) substitution, (b) reaction at the carbene side-chain, (b) addition of a nucleophile, (d) insertion of a molecule (c) oxidation or reduction of the metal.**

All these types of reactions (Fig. 4.87) are being illustrated by describing the reactions of the different reagents with $(OC)_5Cr = C(OMe)R$ (Scheme 4.10); numbers in brackets indicate the reaction type:

(c) *Structure and Bonding.* The excitement created by the discovery of the carbene complexes has been evinced by the determinations of the crystal structures of a large number (> 150) of these complexes; only a few representative examples of Fischer type complexes are shown in Fig. 4.88.

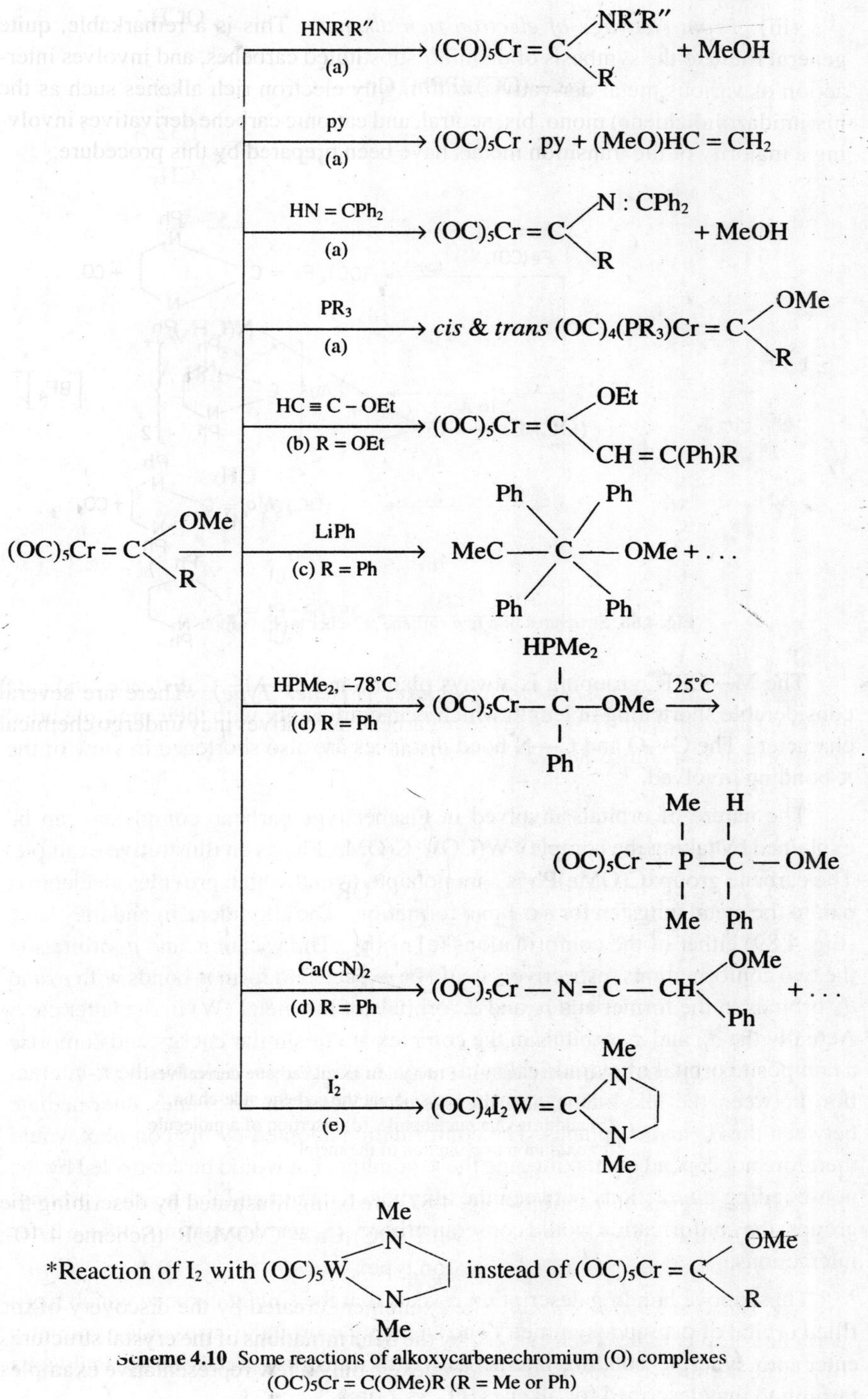

Scheme 4.10 Some reactions of alkoxycarbenechromium (O) complexes $(OC)_5Cr = C(OMe)R$ (R = Me or Ph)

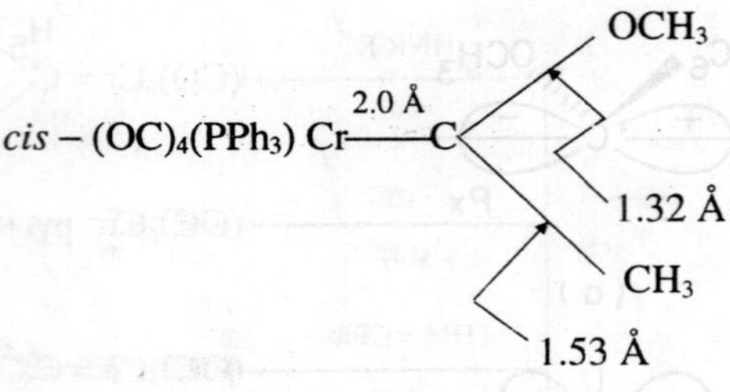

(a)

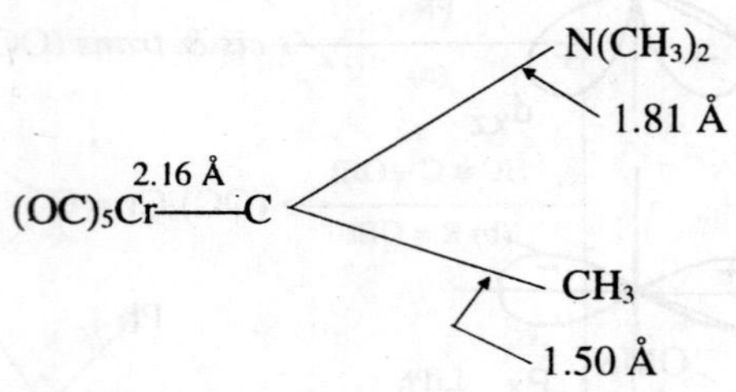

(b)

Fig. 4.88 Structures of a few carbene (Fischer type) complexes.

The M—CRR′ grouping is always planar, and the M—C distances indicate considerable shortening in length which is in conformity with their multiple bond character. The C—O and C—N bond distances are also shortened in view of the π-bonding involved.

The nature of orbitals involved in Fischer type carbene complexes can be explained by taking the complex $W(CO)_5\{C(OMe)Ph\}$ as an illustrative example. The carbene group: C(OMe)Ph is a monohapto ligand which provides an electron pair to the metal tungsten for a σ-bond formation. The alkylidene ligand may have (Fig. 4.89) either of the conformations (a) or (b). The vacant p_x and p_y orbitals in the two conformations respectively could be expected to form π-bonds with p_x and d_{xz} orbitals in the former and p_y and d_{yz} orbitals of the metal (W) in the latter case. Actually the d_{xz} and d_{yz} orbitals in the complex are of similar energy and comprise a composite orbital of cylindrical symmetry, making it possible for the π-interaction between the alkylidene and the metal to occur in all planes intermediate between the d_{xz} and d_{yz} planes. The conformation adopted by the complex would therefore not depend on maximizing the π-bonding, but would be controlled by the non-bonding interactions between the alkylidene ligand and the nearby carbonyl groups; the conformation would consequently be staggered to minimize these latter interactions.

Thus, above bonding description reveals that the singlet carbene which has a filled orbital of σ-pseudosymmetry and an empty *p*-orbital of π-pseudo-symmetry enter into synergic bonding interactions with transition metals in an analogous fashion to that described for alkenes (cf. Section 4.3.2.3):

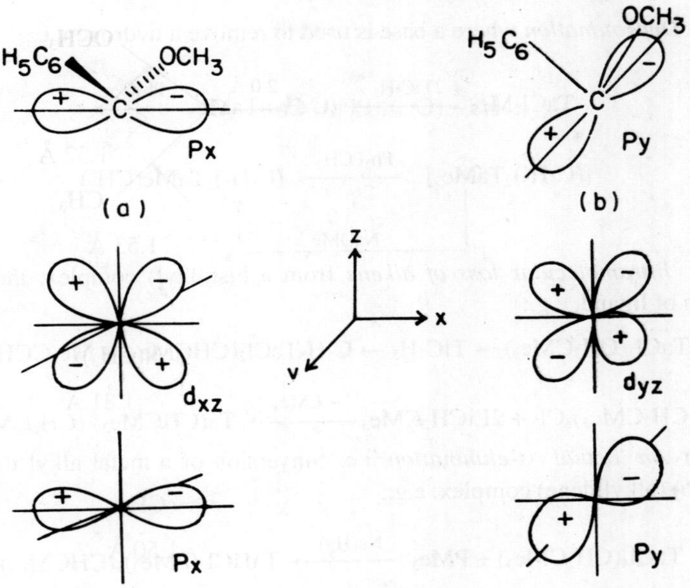

Fig. 4.89 Schematic illustration of the atomic orbitals of both : C(OMe)Ph and tungsten that are available for π-bonding.

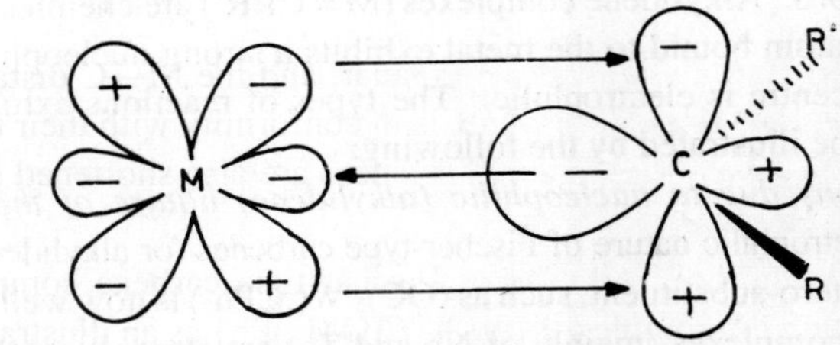

Fig. 4.90 Synergic mode of bonding in metal carbene (alkylidene) complexes.

4.5.2.4 Homonuclear (or Schrock Type) Carbene (Alkylidene) Complexes

The developments in the chemistry of Fischer type complexes in which the ligand has a hetero-atom substituent led to a continuous search for complexes with carbenes having only alkyl(aryl/hydrogen) substituents on the carbene carbon. This was further accentuated by the fact that most of the carbene intermediates postulated in catalytic (e.g. alkene metathesis; Fischer-Tropsch synthesis and possibly in Ziegler-Natta polymerization) processes were of the latter and not of the former (Fischer) type. Success in the isolation of first stable derivative of this class was, however, achieved in 1974 by Schrock with tantalum (which incidentally is not particularly reactive in catalytic reactions).

The first isolable alkylidene complex of this type was $(Me_3CCH_2)_3Ta(:CHCMe_3)$; this has been followed the development of a fascinating chemistry, a brief account of which is being presented in the following pages:

(A) *Synthesis.* Three general types of reaction routes are accessible for transformation of an alkyl to an alkylidene derivative.

(i) *Deprotonation* where a base is used to remove a hydrogen, e.g.:

$$TaCl_2Me_3 \xrightarrow{2TiC_5H_5} (C_5H_5)_2TaMe_3 \xrightarrow{Ph_3C^+}$$

$$[(C_5H_5)_2TaMe_2]^+ \xrightarrow{Ph_3PCH_2} (C_5H_5)_2TaMe(CH_2)$$

$$(C_5H_5)_2TaMe_3 \xrightarrow{NaOMe} (C_5H_5)_2TaMe(CH_2)$$

(ii) *Intramolecular loss of alkene* from a bis(alkyl) complex, induced by addition of ligand, e.g.:

$$TaCl_3(CH_2CMe_3)_2 + TlC_5H_5 \rightarrow C_5H_5TaCl_2(CHCMe_3) + Me_3CCH_3$$

$$Ta(CH_2CMe_3)_3Cl_2 + 2LiCH_2CMe_3 \xrightarrow{-CMe_4} Ta(CHCMe_3)\,(CH_2CMe_3)_3$$

(iii) *An 'actual' α-elimination,* i.e. conversion of a metal alkyl to a metal (hydrido) (alkylidene) complex, e.g.:

$$TaCl_4(CH_2CMe_3) + PMe_3 \xrightarrow{Na(Hg)} TaHCl_2(PMe_3)_3(CHCMe_3)$$

$$C_5H_5TaCl_3(CH_2CMe_3) + PMe_3 \xrightarrow{Na(Hg)} C_5H_5TaHCl(PMe_3)\,(CHCMe_3)$$

(B) *Reactions.* Alkylidene complexes (M = CRR′) are chemically very labile and the carbon atom bound to the metal exhibits a strong nucleophilic character, whereas metal centre is electrophilic. The types of reactions exhibited by such complexes can be illustrated by the following:

(i) *Reactions due to nucleophilic (alkylidene) nature of the (alkylidene) carbon.* The electrophilic nature of Fischer-type carbenes (or alkylidenes) including those with no hetero-substituent, such as $(OC)_5\,W(:CPh_2)$ is now well documented. The alkylidene complexes (mainly of Nb and Ta) are strongly nucleophilic, and therefore, exhibit greater tendency towards protonation reactions, e.g.:

$$(Me_3CCH_2)_3\,Ta(:CHCMe_3) + HCl \xrightarrow{\text{low temp.}} TaCl(CH_2CMe_3)_4$$

With Lewis acids such as Al_2Me_6 also, these form addition complexes, e.g.:

$$(C_5H_5)_2TaMe(:CH_2) + 1/2\ Al_2Me_6 \rightarrow TaMe(CH_2AlMe_3)\,(C_5H_5)_2$$

Interactions with organic halides have been shown to proceed via $S_N{}^2$ mechanism, e.g.:

$$TaMe(:CH_2)\,(\eta^5 - C_5H_5)_2 \xrightarrow{MeI} TaIMe(\eta^2 - C_2H_4)\,(\eta^5 - C_5H_5)_2$$

$$TaMe(:CH_2)\,(\eta^5 - C_5H_5)_2 \xrightarrow{SiBrMe_3} TaMe(CH_2SiMe_3)\,(\eta^5 - C_5H_5)_2Br$$

Nucleophiles such as phosphine ligands add (when possible because of coordinative unsaturation) only to the metal centre, and not (as in Fischer type complexes) to the alkylidene fragment e.g.:

$$Ta(CH_2CMe_3)_3\,(CHCMe_3) + 2PMe_3 \xrightarrow{-\,CMe_4} Ta(CH_2CMe_3)\,(=CHCMe_3)_2\,(PMe_3)_2$$

The ability of alkylidene complexes to react with carbonyl compounds is due to the combined effect of both the nucleophilicity of alkylidene carbon and oxophilicity of the metal centre. Some illustrative examples are shown in Scheme 4.11:

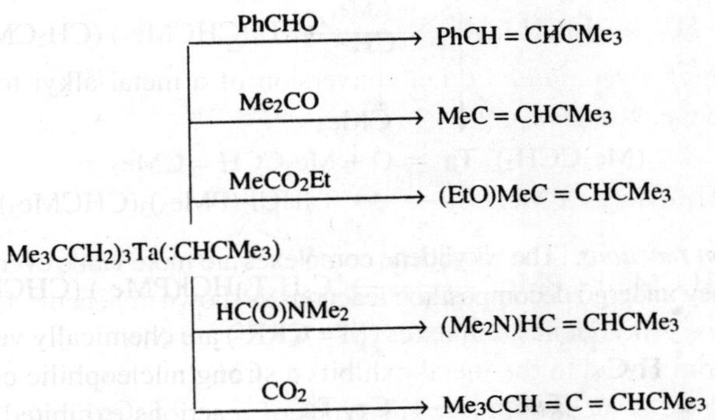

Scheme 4.11 Some reactions of a tantalum alkylidene complex $Ta(CH_2CMe_3)_3\,(=CHCMe_3)$

(ii) *Reactions with unsaturated organic compounds.* In view of their proposed role in alkene metathesis reactions (cf. Chapter 6), the interactions of unsaturated compounds with alkylidene complexes assumes special significance. Although alkenes do react with coordinatively unsaturated complexes e.g., $MCl_2(:CHCMe_3)\,(\eta^5\text{—}C_5H_5)$, yet these do not appear to be accompanied by metathetic type transformations and merely result in the formation of a new alkene probably through an insertion reaction

$$TaCl_2(CHCMe_3)\,(\eta^5-C_5H_5) + H_2C=CH_2 \rightarrow (\eta^5-C_5H_5)\,Cl_2Ta\langle\square\rangle + Me_3CCH_2CH=CH_2$$

The failure to isolate alkylidene intermediates postulated in the mechanism of metathesis reactions is ascribed to their instability. However, complexes of the type $MX_n\,(OR)_{3-x}\,(CHR)L$ exhibit more interesting behaviour:

$$TaCl(OBu^t)_2(CHCMe_3)\,(PMe_3) + PhCH=CH_2 \xrightarrow{PMe_3} TaCl(OBu^t)_2\,(CHPh)\,(PMe_3)_2$$

Other examples of the addition of alkylidenes to unsaturated organic molecules include:

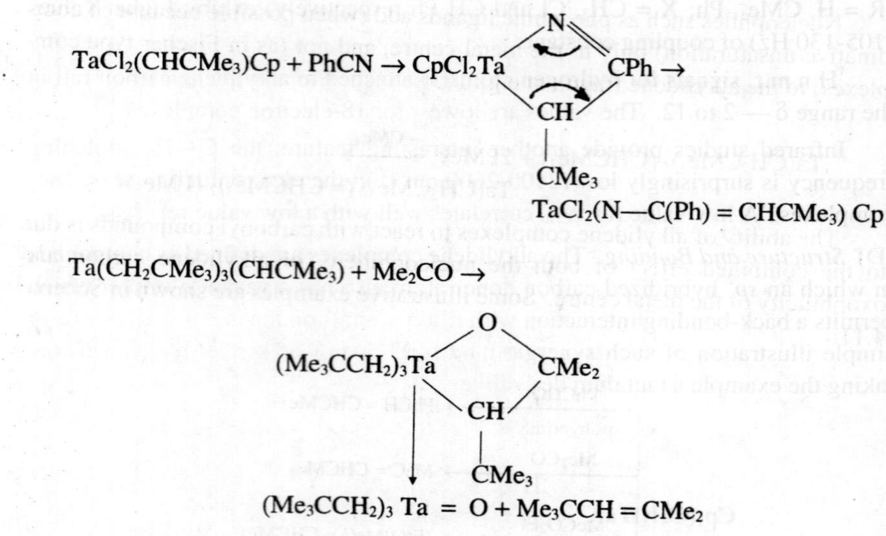

(iii) *Decomposition reactions.* The alkylidene complexes are more labile even at room temperature, that they undergo decomposition reactions as shown in Scheme 4.12.

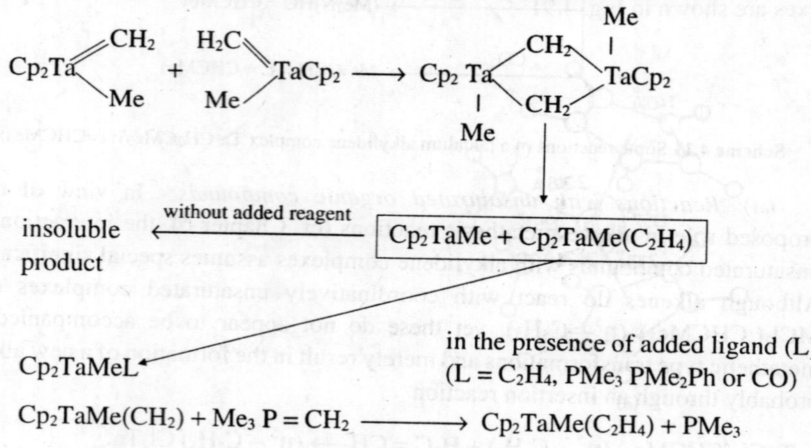

Scheme 4.12

(C) *Spectroscopic properties.* The ^{13}C n.m.r. spectroscopy is a useful tool for characterizing alkylidene complexes since the alkylidene carbon resonance is invariably found at low field (200-300 ppm downfield of $SiMe_4$). Such a large chemical shift appears to be characteristic of many types of carbon atoms which are multiply bonded to some extent to a transition metal. Another distinguishing feature of alkylidene complexes is low $^1J_{CH\alpha}$ coupling constant values (75-100 Hz) found for example, in neopentylidene and benzylidene complexes having less than 18 valence electrons (e.g. $TaCl(CH_2CMe_3)(CHCMe_3)$ $(\eta^5\text{–}C_5Me_5)$ = 14e, $^1J_{CH\alpha}$ = 76 Hz). By contrast 18 electron complexes of the type, $MX(CHR)$ $(\eta^5\text{–}C_5H_5)_2$

(R = H, CMe_3, Ph; X = CH_3, Cl and CH_2Ph respectively) exhibit higher values (105-130 Hz) of coupling constants.

^{1}H n.m.r. signals for hydrogen atoms(s) attached to alkylidene carbon fall in the range δ — 2 to 12. The values are lowest for 18-electron complexes.

Infrared studies provide another interesting feature; the C—H_α stretching frequency is surprisingly low (2400-2600 cm^{-1}) in the electron deficient (< 18e) complexes. A low value for $\nu_{CH\alpha}$ correlates well with a low value for $^1J_{CH\alpha}$.

(D) *Structure and Bonding*. The alkylidene complexes are defined as compounds in which an sp^2 hybridized carbon donor atom also has a vacant *p*-orbital which permits a back-bonding interaction with filled orbitals on the metal (Fig. 4.90). A simple illustration of such synergic mode of bonding is further described here taking the example a tantalum derivative:

$$\underset{\uparrow\ sp^2\ \text{hybridized}}{Cp_2(CH_3)Ta{:}C\begin{smallmatrix}H\\ H\end{smallmatrix}} \rightleftharpoons Cp_2(CH_3)Ta = C\begin{smallmatrix}H\\ H\end{smallmatrix}$$

It may be noted that the true M—C bond order in such complexes would usually be substantially less than two. Two typical structures of alkylidene com-lexes are shown in Fig. 4.91.

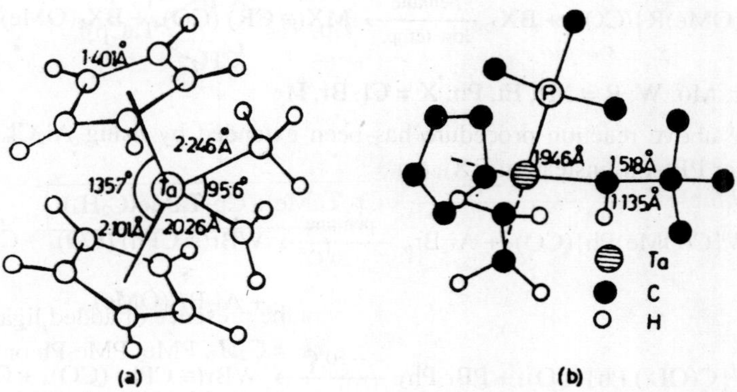

Fig. 4.91 Some typical structures of alkylidene complexes.

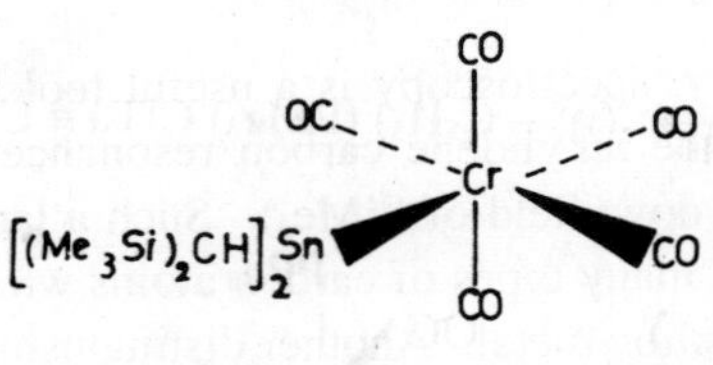

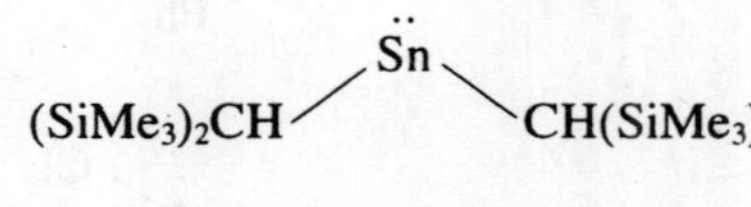

Before passing on to carbyne complexes, attention may be drawn to an interesting complex (Fig. 3.40) described in Section 3.7.1; the tin containing ligand in this complex can be considered an inorganic carbene:

Another example of the same type is the germylene chromium (O) complex:

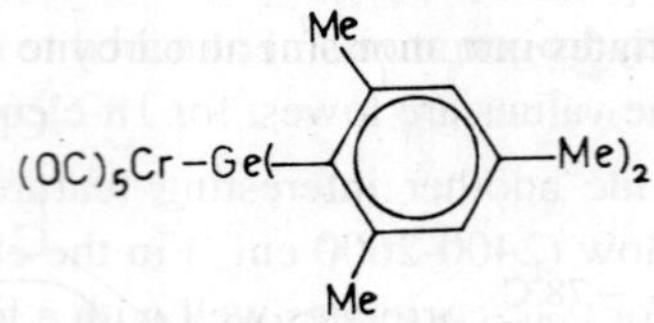

4.5.3 Transition Metal-Carbyne Complexes

The carbyne ligand, a monovalent carbon species $[:C{-}R]^+$, with two empty *p*-orbitals is capable of accepting electron density in a typical π-acceptor fashion, from a metal as well as functioning as a σ-donor. Since the carbyne has two empty *p*-orbitals, two sets of metal π-electrons can be accepted leading to a formal bonding of the type, $[(OC)_5Cr \equiv CR]^+$.

Alternatively, carbynes could be supposed to contribute three electrons to the metal; pairing with metal electrons in these cases leads to the formation of a metal-carbon triple bond: $M \equiv CR$.

(A) *Synthesis.* The first carbyne complex was accidentally synthesized (in 1973) when an attempt to replace methoxy group from a complexed carbene by halide ion (from boron trihalide) resulted instead into a carbyne complex:

$$M\{C(OMe)R\}(CO)_5 + BX_3 \xrightarrow[\text{low temp.}]{\text{pentane}} MX(\equiv CR)(CO)_4 + BX_2(OMe) + CO$$

(M = Cr, Mo, W; R = Me, Et, Ph; X = Cl, Br, I)

The above reaction procedure has been extended by using Al_2Cl_6, $GaCl_3$, Al_2Br_6, or PBr_2Ph_3 instead of BX_3, e.g.,

$$W\{C(OMe)Ph\}(CO)_5 + Al_2Br_6 \xrightarrow[-30^\circ C]{\text{pentane}} WBr(\equiv CPh)(CO)_4 + CO + Al_2Br_5(OMe)$$

$$W\{C(OLi)Ph\}(CO)_4 + PBr_2Ph_3 \xrightarrow{-50^\circ C} WBr(\equiv CPh)(CO)_4 + CO + LiBr + P(O)Ph_3$$

Another synthetic route of greater utility uses carbene precursors:

$$(\eta^5-C_5H_5)Cl_2Ta = C(H)(Ph) \xrightarrow[\text{(ii) } Ph_3P = CH_2]{\text{(i) } PMe_3} (\eta^5-C_5H_5)(PMe_3)ClTa \equiv CPh$$

The interaction of $MoCl_5$ and WCl_6 with neopentyllithium ($LiCH_2CMe_3$) yields the dimeric species $[M(CH_2CMe_3)_3(\equiv C—CMe_3)]_2$ with bridging-carbyne ligands. The dimer dissociates into monomeric carbyne complexes:

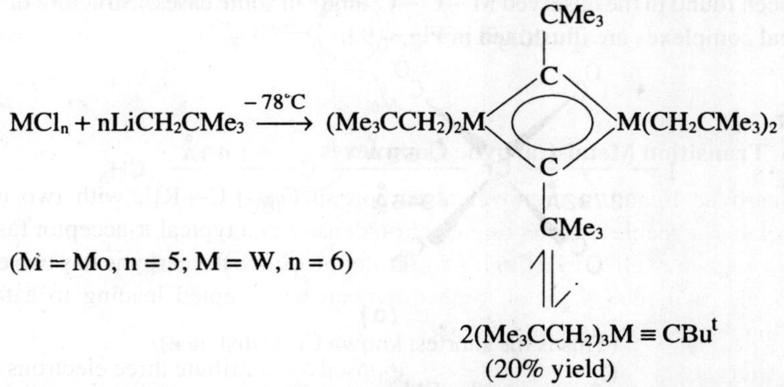

(M = Mo, n = 5; M = W, n = 6)

A higher yield route to $W(\equiv CCMe_3)(CH_2CMe_3)_3$ consists of the reaction between $WCl_3(OMe)_3$ with six equivalents of bis(neopentyl) magnesium:

$$WCl_3(OMe)_3 + 6Mg(CH_2CMe_3)_2 \xrightarrow{\text{ether}} W(\equiv CCMe_3)(CH_2CMe_3)_3$$

The reaction of WCl_4 with $Hg(CH_2SiMe_3)_2$ or $MgX(CH_2SiMe_3)$ yields a bridged carbyne:

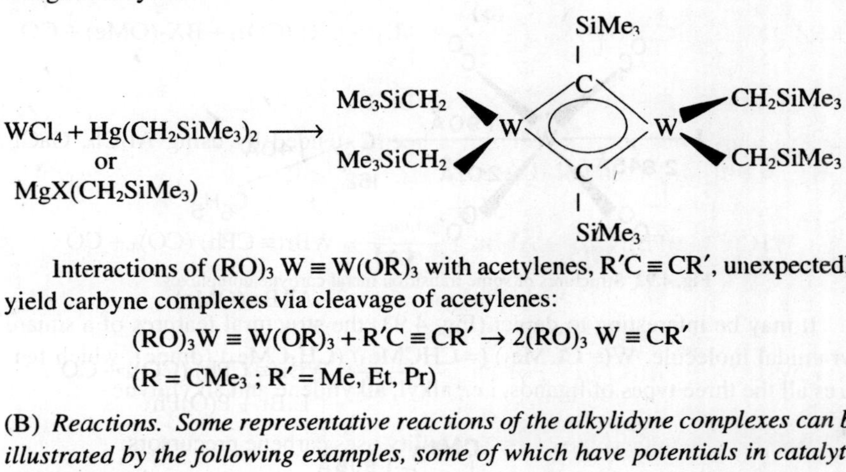

Interactions of $(RO)_3W \equiv W(OR)_3$ with acetylenes, $R'C \equiv CR'$, unexpectedly yield carbyne complexes via cleavage of acetylenes:

$$(RO)_3W \equiv W(OR)_3 + R'C \equiv CR' \rightarrow 2(RO)_3W \equiv CR'$$

$$(R = CMe_3\ ;\ R' = Me, Et, Pr)$$

(B) *Reactions. Some representative reactions of the alkylidyne complexes can be illustrated by the following examples, some of which have potentials in catalytic processes:*

$$W(CCMe_3)(OCMe_3)_3 + \underset{\text{excess}}{RC \equiv CR} \rightarrow W(\equiv CR)(OCMe_3)_3 + RC \equiv CCMe_3$$

(R = Pr or Ph)

(D) *Structure.* The metal-carbon bond in these carbyne complexes,, M≡CR, could be expected to significantly short with the bond angle M – Ĉ – C equal to ~ 180°. While the X-ray structural studies have generally confirmed the former, variation has been found in the observed M – Ĉ – C angle in some cases. Structure of a few typical complexes are illustrated in Fig. 4.92:

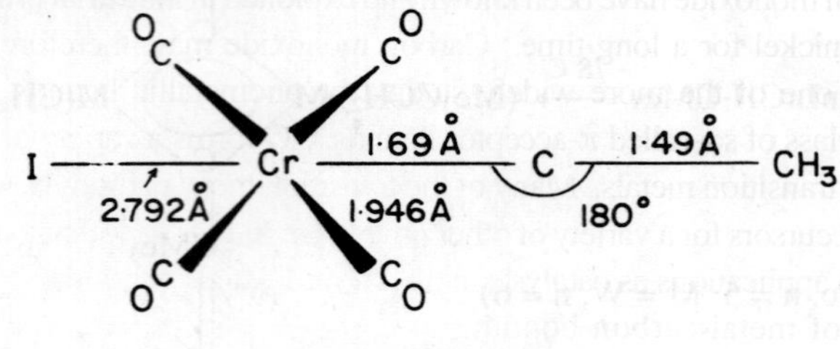

(a)

(This is the shortest known Cr–C distance)

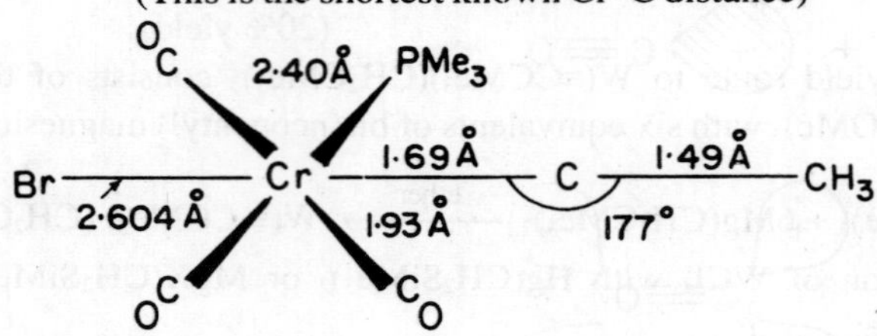

(b)

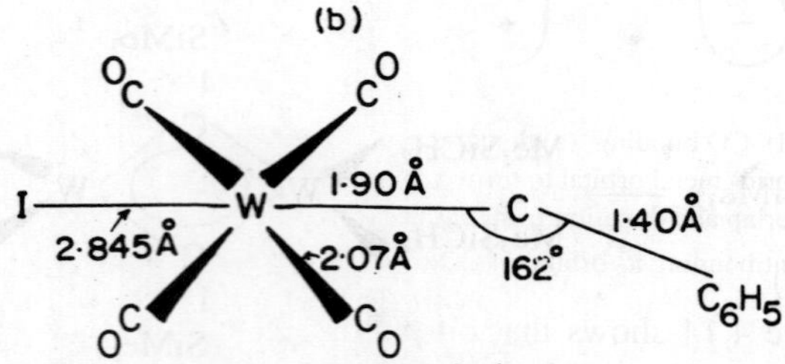

(c)

Fig. 4.92 Structures of some transition metal carbyne complexes.

It may be interesting to depict (Fig. 4.93) the structural features of a square pyramidal molecule, W(≡ CCMe$_3$) (= CHCMe$_3$) (CH$_2$CMe$_3$) (dmpe), which features all the three types of ligands, i.e., alkyl, alkylidene and alkylidyne:

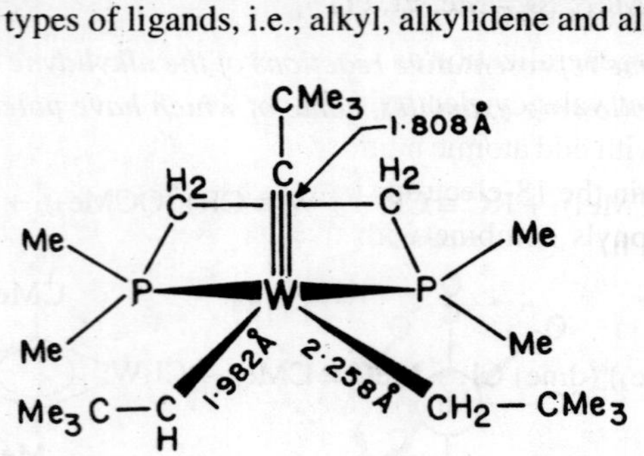

Fig. 4.93 Structure of W (≡ CCMe$_3$) (= CHCMe$_3$) (CH$_2$CMe$_3$) (dmpe).

4.6 METAL CARBONYLS $M_x(CO)_y$

4.6.1 Introduction

Another type of compounds which belong to the general class of multiple bonded metal-carbon derivatives is that of carbonyls. These interesting derivatives of transition metals with carbon monoxide have been known and exploited in industrial processes such as metallurgy of nickel for a long time. Carbon monoxide may, therefore, in a sense considered to be one of the more widely studied organometallic ligands. Being the prototype of the class of so-called π-acceptor ligands, CO forms a variety of derivatives mainly with d^3-d^8 transition metals. Many of the transition metal carbonyls are of special significance as precursors for a variety of other novel compounds including organometallics and find wide applications as catalysts in diverse industrial processes.

The nature of metal-carbon bonding in carbonyls is similar to that in metal alkene or -carbene complexes, as illustrated schematically in Fig. 4.94.

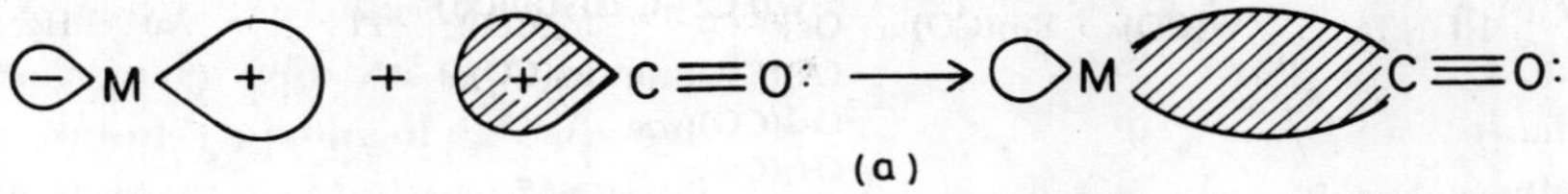

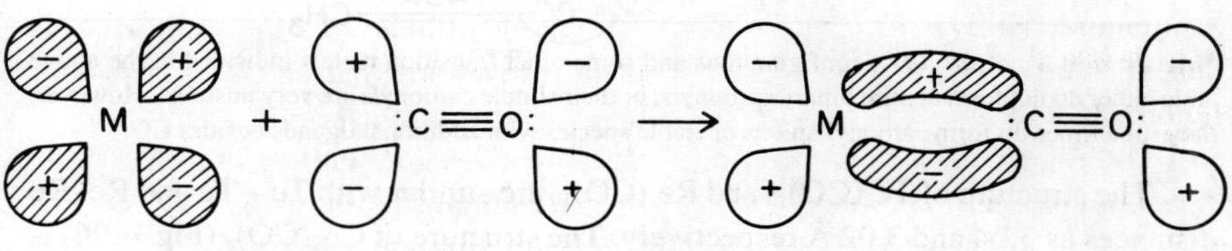

Fig. 4.94 Metal (M)–CO bonding: (a) donation and overlap from the lone-pair on C onto a vacant (hybrid) metal orbital to form a σ M ← C bond and simultaneous (b) π overlap and donation from d_{xz} or d_{yz} orbital on M into a vacant antibonding π_p^* orbital of CO to form π M → C bond.

A look at Table 4.14 shows that only those elements which have an even atomic number from mononuclear carbonyls and in all of these, i.e., $Cr(CO)_6$, $Mo(CO)_6$, $W(CO)_6$, $Fe(CO)_5$, $Ru(CO)_5$, $Os(CO)_5$ and $Ni(CO)_4$; the number of valency electrons around the central metal atom adds to $18(d^{10}s^2p^6)$, i.e., all these metals attain the effective atomic number of the next noble gas, provided each ligated CO molecule is calculated to donate 2 electrons to the metal; $V(CO)_6$ is the only exception to this generalization as vanadium itself has an odd atomic number and the total number of valency electrons also in $V(CO)_6$ add upto 17 only.

Other elements with odd atomic numbers (i.e., ^{25}Mn, ^{43}Tc, ^{75}Re, ^{27}Co, ^{45}Rh and ^{77}Ir) are unable to attain the 18-electron status in the above manner. These in their simplest binary carbonyls combine with that number of CO ligands, e.g., as in

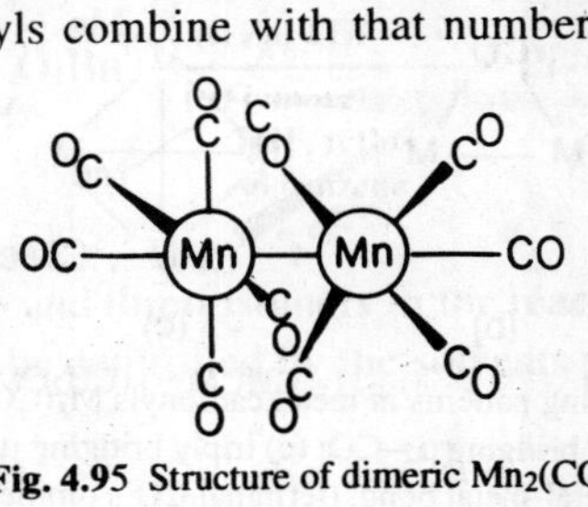

Fig. 4.95 Structure of dimeric $Mn_2(CO)_{10}$.

$Mn(CO)_5$ and $Co(CO)_4$ which makes their valency electrons equal to 17 and then these tend to fall in line with the 18-electron rule by dimerization. The structure of $Mn_2(CO)_{10}$ is represented by Fig. 4.95:

Table 4.14 The known Neutral Binary Metal Carbonyls

III d^1s^2	IV d^2s^2	V d^3s^2	VI d^4s^2	VII d^5s^2	 d^6s^2	VIII d^7s^2	 d^8s^2	I d^9s^2	II $d^{10}s^2$
Sc	Ti	$V(CO)_5$	$Cr(CO)_6$	$Mn_2(CO)_{10}$	$Fe(CO)_5$ $Fe_2(CO)_9$ $Fe_3(CO)_{12}$	$Co_2(CO)_8$ $Co_4(CO)_{12}$ $Co_6(CO)_{16}$	$Ni(CO)_4$	Cu	Zn
Y	Zr	Nb	$Mo(CO)_6$	$Tc_2(CO)_{10}$ $Tc_3(CO)_{12}$	$Ru(CO)_5$ $Ru_2(CO)_9$ $Ru_3(CO)_{12}$ $Os(CO)_5$ $Os_2(CO)_9$	$Rh_2(CO)_8$ $Rh_4(CO)_{12}$ $Rh_6(CO)_{16}$ $Ir_2(CO)_8$	Pd	Ag	Cd
La	Hf	Ta	$W(CO)_6$	$Re_2(CO)_{10}$	$Os_3(CO)_{12}$ $Os_5(CO)_{16/19}$ $Os_6(CO)_{18/20}$ $Os_7(CO)_{21}$ $Os_8(CO)_{23}$	$Ir_4(CO)_{12}$ $Ir_6(CO)_{16}$	Pt	Au	Hg

*Metals with d^1, d^2, d^9, (d^{10}) configurations and some other transition metals indicated in the above table either do not form neutral binary carbonyls, or their simple carbonyls are very unstable. However, they sometimes do form carbonyl anions or stable species with additional ligands besides CO.

The structure of $Tc_2(CO)_{10}$ and $Re_2(CO)_{10}$ are similar with Tc – Tc and Re–Re distances as 3.04 and 3.02 Å respectively. The structure of $Co_2(CO)_8$ (Fig. 4.96) is also similar in solution (a), but in the solid state (b), it shows a bent Co – Co bond and two bridging carbonyl groups:

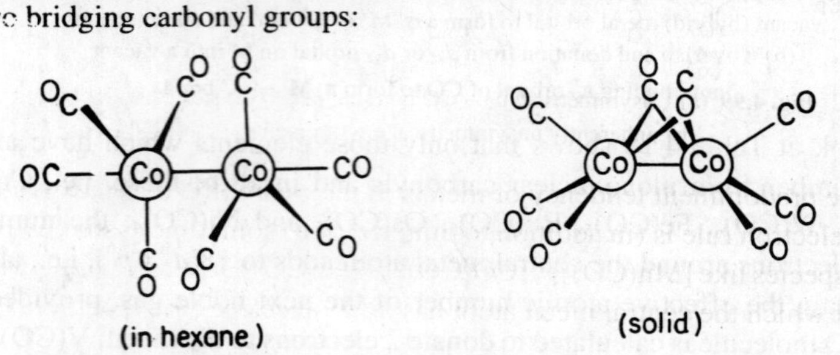

Fig. 4.96 Structure of $Co_2(CO)_8$ in (a) solution and (b) solid state.

This brings us to other patterns in which CO can bridge two metal atoms. The common bonding patterns of metal carbonyls have been summarized in Fig. 4.97:

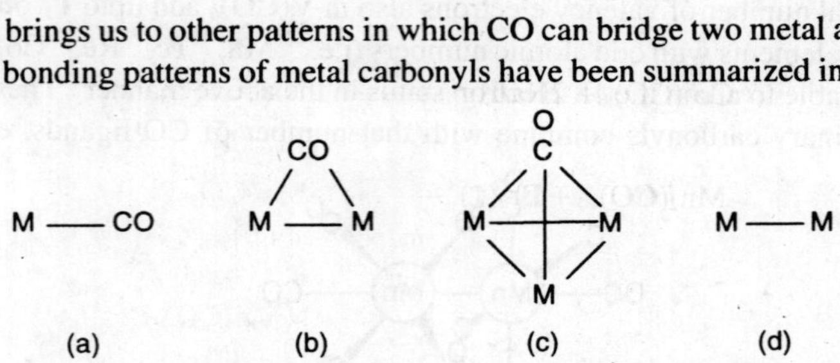

Fig. 4.97 General bonding patterns in metal carbonyls $M_x(CO)_y$: (a) terminal CO; (b) symmetrical bridging μ_2–CO; (c) triply bridging μ_3–CO, and (d) metal-metal bond. (terminal CO's omitted)

In addition to the above, two other types of bonding (Fig. 4.98) have also been observed in metal carbonyls. For example, the unsymmetrical (or semi-bridging) CO position arises in general as intermediates in the transition from structure 4.97(a) with non-bridging CO's to 4.97(b) with μ_2–CO groups (cf. Sec. 5.4.6), as well as in the side-on bonded (Fig. 4.98ii). It may also arise when the formation of one side of the bridge is sterically hindered as in $(\eta^5\text{–}C_5H_5)_2Mo_2(CO)_4$ (RC≡CR) and alternatively in inherently unsymmetrical environments as depicted in Fig: 4.99(a) and (b) respectively.

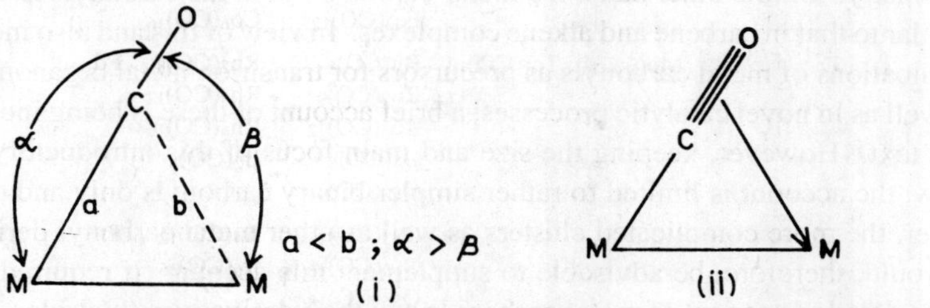

Fig. 4.98 (i) Unsymmetrical bridging CO group, and (ii) side-on bonding of CO, observed in $(CO)_2Mn(Ph_2PCH_2PPh_2)_2(\mu — CO)Mn(CO)_2$.

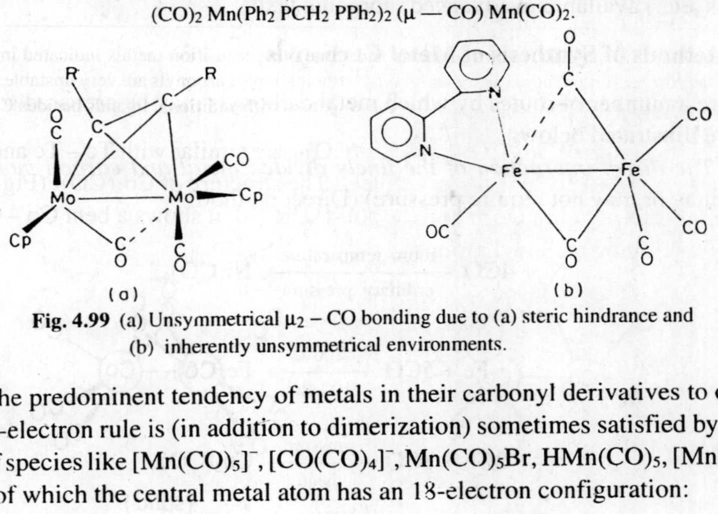

Fig. 4.99 (a) Unsymmetrical μ_2 – CO bonding due to (a) steric hindrance and (b) inherently unsymmetrical environments.

The predominent tendency of metals in their carbonyl derivatives to observe the 18-electron rule is (in addition to dimerization) sometimes satisfied by formation of species like $[Mn(CO)_5]^-$, $[CO(CO)_4]^-$, $Mn(CO)_5Br$, $HMn(CO)_5$, $[Mn(CO)_6]^+$ in all of which the central metal atom has an 18-electron configuration:

$$Mn_2(CO)_{10} + 2Li \xrightarrow{\text{THF}} 2Li^+[Mn(CO)_5]^-$$

$$Co_2(CO)_8 + 2Na/Hg \xrightarrow{\text{THF}} 2Na^+[Co(CO)_4]^-$$

$$Mn_2(CO)_{10} + Br_2(l) \xrightarrow{40^\circ C} 2Mn(CO)_5Br$$

$$2Mn(CO)_5Br \xrightarrow[\text{in petroleum ether } 120^\circ C]{} [\{Mn(CO)_4Br\}_2] + 2CO$$

$$Mn(CO)_5Cl + AlCl_3 + CO \longrightarrow [Mn(CO)_6]^+[AlCl_4]^-$$

Even elements like Ta which do not tend to form simple neutral binary carbonyls can give carbonyl anions under special conditions:

$$TaCl_5 + 6CO + Na\ (excess) \xrightarrow[300\ atm.]{diglyme\ 100^\circ C} [Na(diglyme)_2]^+ [Ta(CO)_6]^-$$

Elucidation of the nature of bonding in metal carbonyls has contributed richly to a clearer understanding of many classes of coordination compounds and therefore, these continue to be a significant portion of teaching and research of inorganic chemistry. On the other hand, the metal-carbon bond in the carbonyls is basically similar to that in carbene and alkene complexes. In view of this and also increasing applications of metal carbonyls as precursors for transition metal organometallics as well as in novel catalytic processes, a brief account of these is being included in this text. However, keeping the size and main focus of this introductory text in view, the account is limited to rather simpler binary carbonyls only and does not cover, the more complicated clusters as well as other metal carbonyl derivatives. It would, therefore, be advisable to supplement this account (if required) with a more detailed account of metal corbonyls and their derivatives (hydrides, halides, nitrosyls, etc.) available in advanced inorganic texts.

4.6.2 Methods of Synthesis of Metal Carbonyls

There are a number of routes by which metal carbonyls may be prepared some of these are illustrated below:

1. *The direct interaction of the finely divided metal and carbon monoxide* (which may or may not require pressure) (Direct method).

$$Ni + 4CO \xrightarrow[ordinary\ pressure]{room\ temperature} Ni(CO)_4$$

$$Fe + 5CO \xrightarrow[100\ atm]{200^\circ C} Fe(CO)_5$$

$$2Co + 8CO \xrightarrow[heat]{pressure} Co_2(CO)_8$$

$$W + 6CO \xrightarrow{pressure} W(CO)_6$$

$$Mo + 6CO \xrightarrow[heat]{pressure} Mo(CO)_6$$

The reactions with nickel and iron proceed readily and are industrially important. However, special conditions of temperature, catalysts, and pressure are unusually critical if a satisfactory yield of metal carbonyl is to be obtained in other cases.

2. *Reduction of metal salts (or complexes) in the presence of carbon monoxide under pressure (reductive carbonylation)*:

(a) By use of suitable reducing agent (such as triethylaluminium in ether, zinc and magnesium powder in pyridine, sodium in diglyme, hydrogen, aluminium, copper, lithium aluminium hydride, sodium benzophenone ketyl, and Grignard reagent):

$$VCl_3 + 3Na + 6CO \xrightarrow[\text{diglyme}]{100^\circ C,\ 150\ atm.} V(CO)_6 + 3NaCl$$

$$MoCl_5 + 5Na + 6CO \xrightarrow[\text{diglyme}]{} Mo(CO)_6 + 5NaCl$$

$$*CrCl_3 \xrightarrow[\text{ether}]{Et_3Al\ /\ CO} Cr(CO)_6$$

$$*CrCl_3 \xrightarrow{Al/AlCl_3/CO} Cr(CO)_6$$

$$*CrCl_3 \xrightarrow{RMgX/CO} Cr(CO)_6$$

$$CrCl_3 + 6CO + LiAlH_4 \xrightarrow[\text{ether}]{115^\circ C,\ 70\ atm.} Cr(CO)_6 + LiCl + AlCl_3$$

$$CoCO_3 + CO + H_2 \xrightarrow[\text{pet. ether}]{150^\circ C,\ 300\ atm.} Co_2(CO)_8$$

$$RuI_3 + 5CO + 3Ag \xrightarrow{175^\circ C,\ 250\ atm.} Ru(CO)_5 + 3AgI$$

$$2CoS + 8CO + 4Cu \xrightarrow{200^\circ C,\ atm.} Co_2(CO)_8 + 2Cu_2S$$

(In some cases*, composition to which the reducing agents are oxidised in the process are not fully classified).

For manganese carbonyl, sodium benzophenone ketyl may be employed with advantage:

$$Na + Ph_2CO \rightarrow Ph_2CONa$$

$$MnCl_2 + 2Ph_2CONa \rightarrow Mn(Ph_2CO)_2 + 2NaCl$$

$$2Mn(Ph_2CO)_2 + 10CO \xrightarrow[\text{THF}]{200^\circ C,\ 200\ atm.} Mn_2(CO)_{10} + 4Ph_2CO$$

(b) Use of carbon monoxide itself as reducing agent:

$$OsO_4 + 9CO \xrightarrow{100^\circ C,\ 50\ atm.} Os(CO)_5 + 4CO_2$$

$$Re_2S_7 + 17CO \xrightarrow{250^\circ C,\ 200\ atm.} Re_2(CO)_{10} + 7COS$$

In place of carbon monoxide; $Fe(CO)_5$ can be used to convert certain metal salts to carbonyls:

$$2MoCl_5 + 5Fe(CO)_5 \longrightarrow 2Mo(CO)_6 + 5FeCl_2 + 13CO$$

$$WCl_6 + 3Fe(CO)_5 \xrightarrow{100^\circ C} W(CO)_6 + 3FeCl_2 + 9CO$$

(3) Preparation of polynuclear metal carbonyls by photo-or thermal-decomposition of mononuclear metal carbonyls:

$$2Fe(CO)_5 \xrightarrow{h\nu} Fe_2(CO)_9 + CO$$
$$\downarrow \Delta$$
$$2/3\ Fe_3(CO)_{12} + CO$$

4.6.3 Physical Properties

Some properties of the simple metal carbonyls are collected in Table 4.15. The notable features are: (i) all mononuclear carbonyls are colourless (except $V(CO)_6$ which is black), while polynuclear derivatives are coloured (except $Re_2(CO)_{10}$ which is colourless), the colour becoming darker as the nuclearity of the metal atom increases, e.g., $Fe(CO)_5$ (yellowish liquid), $Fe_2(CO)_9$ (orange solid), and $Fe_3(CO)_{12}$ (very dark green-black solid); (ii) all the known metal carbonyls are solids, except $Fe(CO)_5$, $Ru(CO)_5$ $Os(CO)_5$ and $Ni(CO)_4$ which are liquids at ambient temperatures; (iii) some carbonyls such as $Fe(CO)_5$ and $Ni(CO)_4$ are very toxic, and (iv) all carbonyls with the exception of $Fe_2(CO)_9$, are more or less soluble in organic solvents.

Table 4.15 Some Properties of Mononuclear and Binuclear Metal Carbonyls

Compound	Colour	M.P. (°C)	Structure	Comments
$V(CO)_6$	black-green crystals	70 (dec.)	O_h	paramagnetic
$Cr(CO)_6$	white crystals		O_h	diamagnetic, stable to air: dec. 180–200°C
$Mo(CO)_6$	white crystals		O_h	diamagnetic, stable to air: dec. 180–200°C
$W(CO)_6$	white crystals		O_h	diamagnetic, stable to air: dec. 180–200°C
$Mn_2(CO)_{10}$	golden yellow crystals	154-155	D_{4d}	Mn—Mn bond = 2.93Å
$Tc_2(CO)_{10}$	white crystals	160	D_{4d}	
$Re_2(CO)_{10}$	white crystals	177	D_{4d}	
$Fe(CO)_5$	yellow liquid	–20	D_{3h}	
$Ru(CO)_5$	colourless liquid	–22	D_{3h}	unstable with respect to light catalyzed decomposition to $Ru_3(CO)_{12}$
$Os(CO)_5$	colourless liquid	–15	D_{3h}	very unstable with respect to $Os_3(CO)_{12}$
$Fe_2(CO)_9$	golden-yellow crystals	100–120 (dec.)	D_{3h}	Fe—Fe bond = 2.46 Å
$Os_2(CO)_9$	orange yellow crystals	64–67	D_{3h}	thermally- and photolytically-sensitive
$Co_2(CO)_8$	orange-red crystals	51(dec.)	C_{2v} (solid) D_{3d} (solution)	Co—Co bond = 2.54Å

Table 4.15 (*Contd.*)

Compound	Colour	M.P. (°C)	Structure	Comments
$Ni(CO)_4$	colourless liquid	–25	T_d	
$Pd(CO)_4^{**}$				
$Pt(CO)_4^{**}$				

** Matrix isolation sample

The carbonyls vary greatly in their resistance to atmospheric oxidation: $Fe(CO)_5$ and $Ni(CO)_4$ are readily oxidized, forming explosive mixtures with air and $Co_2(CO)_8$ decomposes in air under ambient conditions forming a carbonate, whereas $Cr(CO)_6$, $Mo(CO)_6$ and $W(CO)_6$ are quite stable in air. The stability of the latter derivatives, however, appears to be determined by kinetic rather than thermodynamic factors. Mean bond dissociation energies $\bar{D}$(M—CO)/kJ mol^{-1} increase in the sequence $Cr(CO)_6$ 109, $Mo(CO)_6$ 151, $W(CO)_6$ 176 and these are in the same range as those of $Mn_2(CO)_{10}$ 100, $Fe(CO)_5$ 121, $Co_2(CO)_8$ 138 and $Ni(CO)_4$ 147.

In the next sub-section (4.5.3.1) is presented a brief account of the applications of infrared spectroscopy which has proved to be the most powerful tool in the elucidation of characteristic features of metal carbonyls; it may be pertinent to mention that the applications of nuclear magnetic resonance, which has provided valuable information on the structures as well as fluxionality of metal carbonyls are described in Chapter 5.

4.6.3.1 Infrared Studies

The applications of infrared spectroscopy in elucidation of structural features of organometallic compounds has already been discussed in Section (3.6.4.1). The potential of this technique would be further illustrated with respect to metal carbonyls.

The CO stretching frequencies appear as sharp and intense bands in a region where vibrational modes of most other ligands are generally absent. In view of the relative ease of observing and interpreting the νCO absorptions of metal carbonyls, their i.r. spectroscopy has proved to be the most effective technique in the study of metal carbonyl chemistry mainly in three ways: (i) the frequency range in which the bands are observed, (ii) the number of bands and to some extent, (iii) the relative intensities of the observed bands.

(i) The frequency range in which CO absorptions occur in a metal carbonyl generally provides an effective criterion for differentiating between the terminal and bridging CO groups; these absorptions occur in the frequency ranges 1850-2125 cm^{-1} and 1700-1860 cm^{-1} respectively.

For comparison, it may be worthwhile to point out that free carbon monoxide absorbs at 2143 cm^{-1} and the CO group in ketones, R_2CO absorbs in the range 1715 ± 10 cm^{-1}. The position of νCO, therefore, is merely a reflection of the bond order between C and O atoms and hence, while trying to differentiate between the terminal and bridging CO groups in a metal carbonyl, care should be taken to keep in mind any collateral factors which could affect the bond order and consequently the νCO in an appreciable manner. When some of the CO groups in a carbonyl are replaced by donor (e.g, nitrogen) ligands with low or negligible π-acceptor ability, or in the presence of a negative charge on the metal atom {such as in $V(CO)_6^-$}, the back-donation from metal to carbon would be enhanced bringing about a lowering in C—O bond order as illustrated in Fig. 4.100:

$$\overset{\delta-}{M} \longleftarrow :\overset{\delta+}{C} \equiv O \rightleftharpoons M = C = O$$

Fig. 4.100 Lowering of C—O bond order by enhanced back donation from metal to carbon.

The lower C—O bond order would be reflected in a decrease in the observed νCO. For example, whereas $Cr(CO)_6$ shows absorptions at 2100, 2000 and 1985 (exact values may vary with phase and solvent), $Cr\{NH(CH_2CH_2NH_2)_2\}(CO)_3$, shows νCO absorptions at ~ 1900 and 1760 and in the spectra of $V(CO)_6$, νCO occurs at 1860 cm^{-1}.

The applications of the above simple facts would be discussed in the actual elucidation of the structural features of some typical carbonyls. Thus $Fe_2(CO)_9$ shows two CO absorptions in 2000-2100 cm^{-1} (terminal) range and one at 1830 cm^{-1} which may be clearly ascribed to a bridging CO group (in conformity with structure 4.101a). However, $Os_3(CO)_{12}$ shows CO absorptions only in the 2000-2070 cm^{-1} (terminal) range again in conformity with the structure 4.101b, which does not have any bridging carbonyl groups:

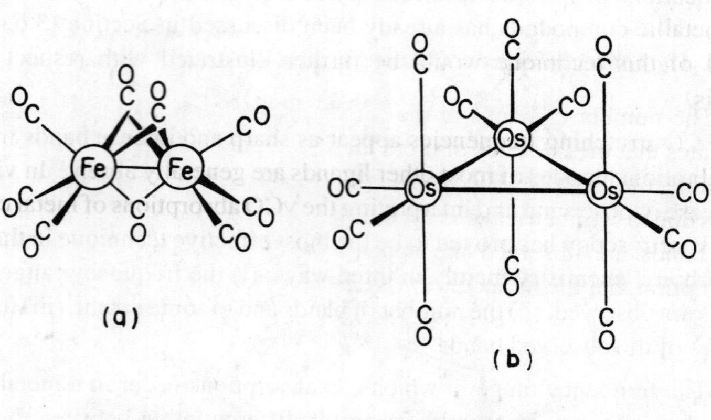

Fig. 4.101 Structures of (a) $Fe_2(CO)_9$; (b) $Os_3(CO)_{12}$.

(ii) The number of absorption bands appearing in i.r. spectrum of any compound (e.g. a metal carbonyl in the present context) gives an idea about the symmetry of the molecule and in the case of a metal carbonyl about the arrangement of CO group around the metal atom(s).

The mode of application of this criterion to actual molecules can be illustrated by taking $M(CO)_4(PPh_3)_2$ species as an example. Compounds of the above composition can obviously exist in two geometrical forms *cis* and *trans* and these can be characterized by the presence of the number of bands assignable to CO stretching frequencies. For example, Fig. 4.98 shows that in the *cis* form four bands are infrared active whereas in the *trans* form two bands are infrared active, but their frequencies are the same. Thus the *cis* compounds $M(CO)_4(PPh_3)_2$ exhibit four infrared bands at 2016, 1915, 1900, 1890 cm^{-1}, whereas the trans-analogues exhibit one strong band at 1890 cm^{-1}:

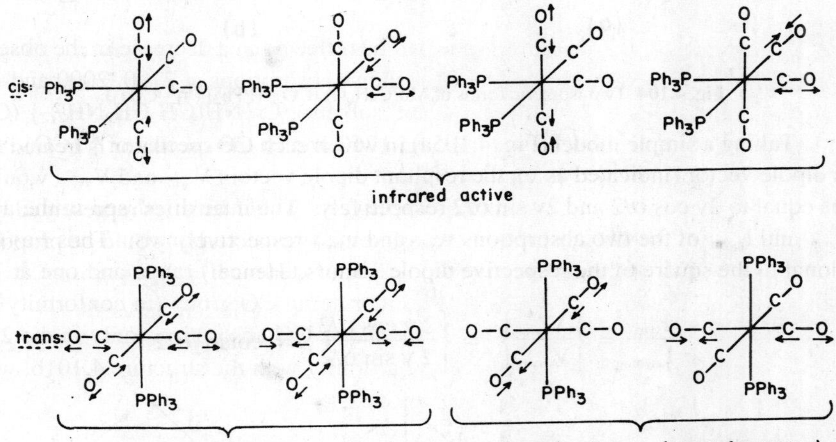

Fig. 4.102 Infrared active CO stretching vibrations of *cis* and *trans* $M(CO)_4(PPh_3)_2$.

The number of bands in the carbonyl stretching region for some molecules indicates that the presence of more than one conformational isomer. For example, the molecule $Fe(\eta^1{-}C_5H_5)\,(CO)_2(\eta^5{-}C_5H_5)$ would be expected to give only two bands due to symmetrical and asymmetric modes. However, in the actual spectrum four bands are observed the appearance of which has been explained by the assumption that the molecule exists in the following two conformations:

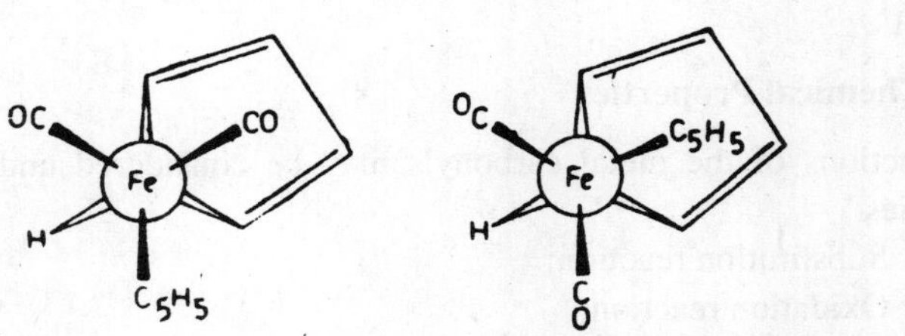

Fig. 4.103 Two conformational isomers of $Fe(\eta_1{-}C_5H_5)\,(CO)_2\,(\eta^5{-}C_5H_5)$.

(iii) The relative intensities of different CO stretching modes can provide further information about the bond angles α and β in the cis and trans isomeric ions shown in Figure 4.104a and b.

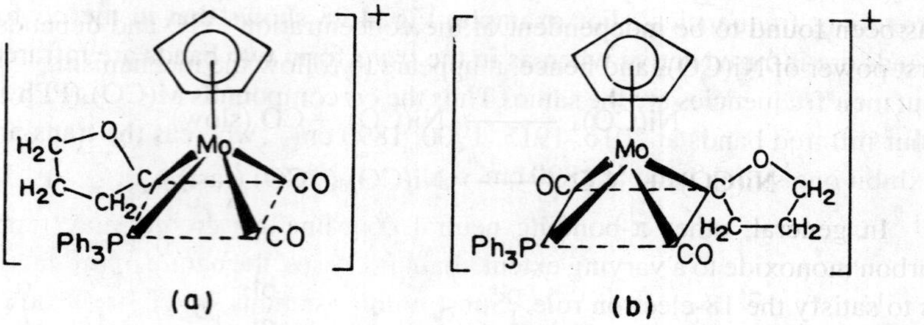

Fig. 4.104 Two isomeric forms of $Mo(C_4H_6O)(CO)_2(PPh_3)(\eta^5-C_5H_5)$.

Taking a simple model (Fig. 4.105a) in which each CO oscillator is treated as a dipole vector (indicated as ν), the resultant dipole vectors $V_{sym.}$ and $V_{asym.}$ would be equal to 2ν cos α/2 and 2ν sin α/2 respectively. The intensities represented by $I_{sym.}$ and $I_{asym.}$ of the two absorptions $\nu_{sym.}$ and $\nu_{asym.}$ respectively would be proportional to the square of the respective dipole vectors. Hence:

$$\frac{I_{sym.}}{I_{asym.}} = \left\{\frac{V_{sym.}}{V_{asym.}}\right\}^2 = \left\{\frac{2\,\nu\cos\alpha/2}{2\,\nu\sin\alpha/2}\right\}^2 = \cot^2\alpha/2$$

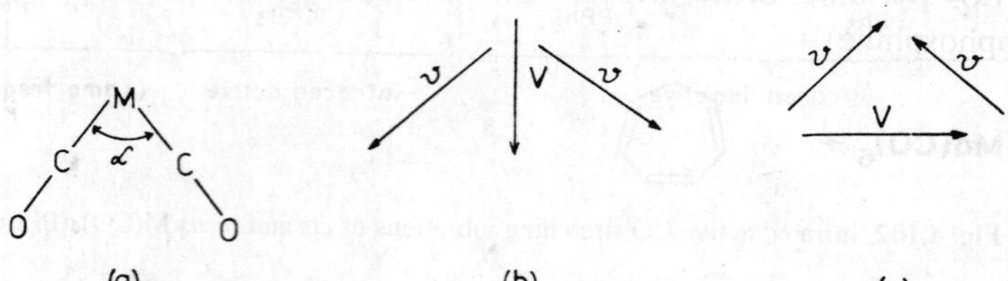

Fig. 4.105 Vector summation of dipole vectors of individual CO oscillators.

For the two isomers 4.104a and b, the ratios of the intensities of $\nu_{sym.}$ and $\nu_{asym.}$ has been found to be 1.44 and 0.32 respectively in their i.r. spectra. It follows therefore, that $\cot^2\alpha/2$ for the *cis* isomer (4.104a) = 1.44, from which α can be calculated as 79° and $\cot^2\alpha'/2$ for the *trans* isomer (4.104b) = 0.32 and hence α′ = 121°.

4.6.4 Chemical Properties

The reactions of the metal carbonyls may be considered under the following categories:

(1) Substitution reaction;
(2) Oxidation reaction;
(3) Reduction reaction; and
(4) Lewis base behaviour.

4.6.4.1 Substitution Reactions

The rate of exchange of some metal carbonyls has been studied with labelled ^{14}CO. $Ni(CO)_4$ and $Co_2(CO)_8$ exchange rapidly while $Fe(CO)_5$, $Cr(CO)_6$ and $Mn_2(CO)_{10}$ exchange very slowly with ^{14}CO. The rate of interchange with $Ni(CO)_4$ has been found to be independent of the concentration ^{14}CO and depends on the first power of $Ni(CO)_4$ and hence, it appears to follow the mechanism:

$$Ni(CO)_4 \longrightarrow Ni(CO)_3 + CO \text{ (slow)}$$

$$Ni(CO)_3 + {}^{14}CO \longrightarrow Ni(CO)_3({}^{14}CO) \text{ (fast)}$$

In general, other π-bonding neutral coordinating groups may replace the carbon monoxide to a varying extent. In many cases, the nature of products is such as to satisfy the 18-electron rule. Substitution reactions which have been studied include:

$$Ni(CO)_4 + 2PPh_3 \longrightarrow Ni(CO)_2(PPh_3)_2 + 2CO$$

$$Ni(CO)_4 + nPF_3 \longrightarrow Ni(CO)_{4-n}(PF_3)_n + nCO$$

(with NF_3 no reaction)

$$Cr(CO)_6 + 2C_5H_5N \longrightarrow Cr(CO)_4(C_5H_5N)_2 + 2CO$$

$$Co_2(CO)_8 + 2NO \longrightarrow 2[Co(CO)_3NO] + 2CO$$

$$Mo(CO)_6 + 3L \longrightarrow Mo(CO)_3(L)_3 + 3CO$$

(where L = pyridine, diene, triphenylphosphine, dichlorophenylphosphine, and trifluorophosphine)

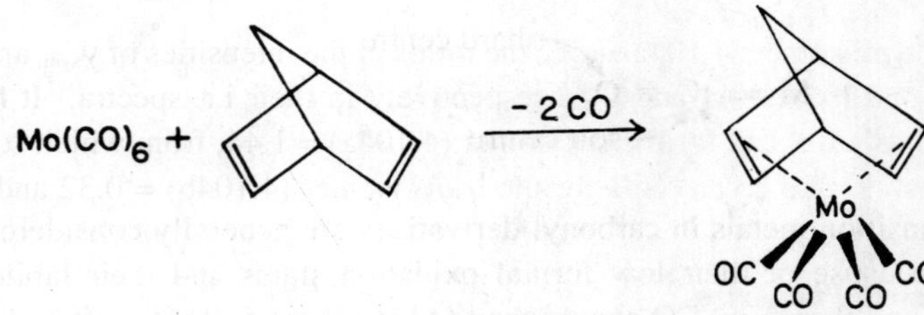

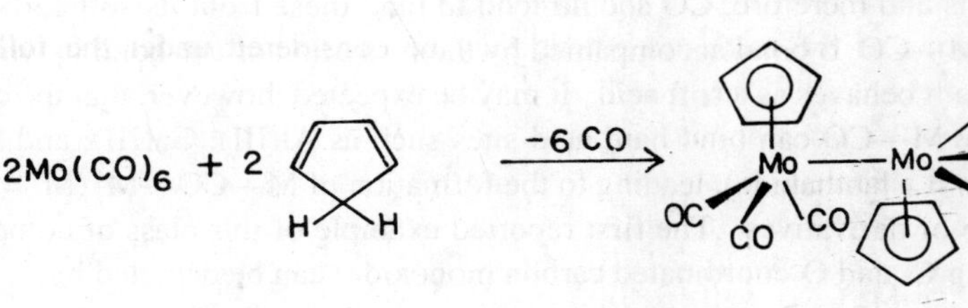

However, in the reactions of metal carbonyls with other substituents the 18-electron rule does not appear to be applicable to the same extent:

$$2Fe(CO)_5 + 3C_5H_5N \xrightarrow{80^\circ} Fe_2(CO)_4(C_5H_5N)_3 + 6CO$$

$$2Fe_3(CO)_{12} + 3C_5H_5N \longrightarrow 3Fe(CO)_3(C_5H_5N) + 3Fe(CO)_5$$

$$Mn_2(CO)_{10} + 2PPh_3 \longrightarrow 2Mn(CO)_4(PPh_3) + 2CO$$

$$Mn_2(CO)_{10} + 2\ o\text{–}C_6H_4(AsMe_2)_2 \xrightarrow{160^\circ} 2[Mn(CO)_3C_6H_4(AsMe_2)_2] + 4CO$$

4.6.4.2 Oxidation Reactions

These reactions may be illustrated by the following examples:

$$Mn_2(CO)_{10} \xrightarrow{Br_2} 2Mn^{I}Br(CO)_5$$

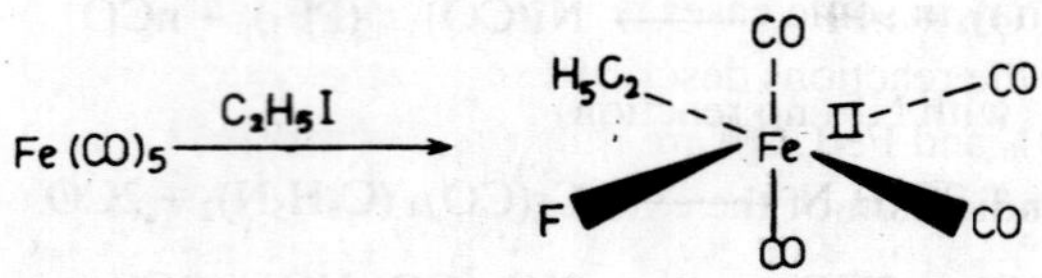

4.6.4.3 Reduction Reactions

These reactions may be represented by the following examples:

$$V(CO)_6 + Na \longrightarrow Na^+[V(CO)_6]^-$$

$$Mn_2(CO)_{10} + 2Na \longrightarrow 2Na^+[Mn(CO)_5]^-$$

4.6.4.4 Lewis Base Behaviour

In terms of the criterion of hard and soft acid and base interactions, the carbon end of carbon monoxide is apparently softer Lewis base site with the oxygen being much harder:

$$M \rightleftharpoons :C \equiv O: \quad (\text{O: hard centre; C: soft centre})$$

The transition metals in carbonyl derivatives are generally considered to be soft acids because of their low formal oxidation states and their labile outer *d*-electrons and therefore, CO should tend to bind these from its soft carbon end forming M—CO σ bond accompanied by the formation of a π-bond in which the carbon atom behaves as a soft acid. It may be expected, however, that the oxygen end of the M—CO can bind hard acid sites such as Al(III), Ga(III), and Ln(III) (where Ln = a lanthanide), leading to the formation of M—CO—M′ (M′ = a hard acid) type of derivatives. The first reported example of this class of compounds (exhibiting C- and O-coordinated carbon monoxide) can be depicted by:

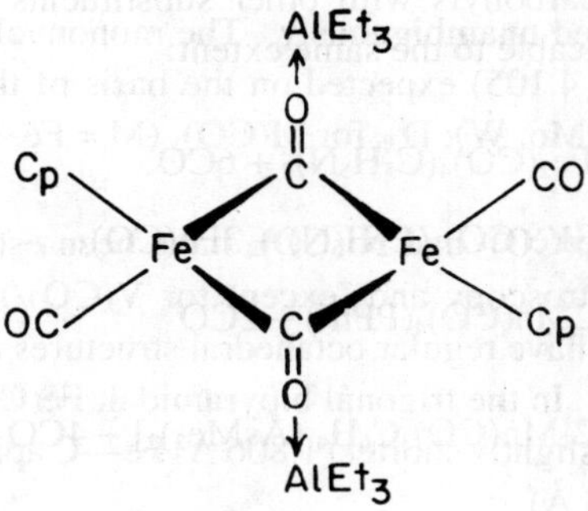

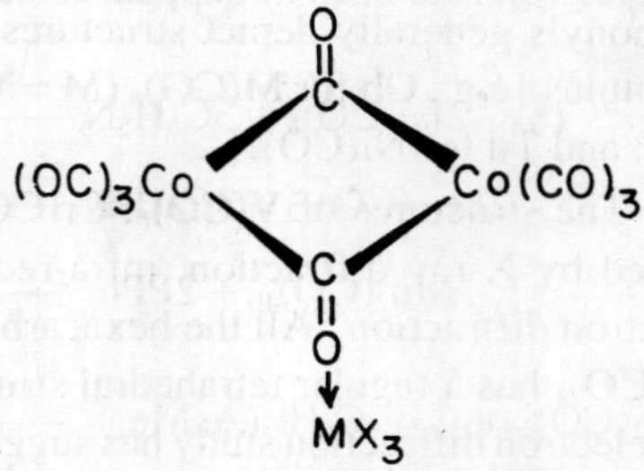

$MX_3 = AlBr_3$ or $(C_5H_5)_3Ln$

As expected, the formation of M—CO—M′ linkages results in a considerable lowering of the C—O stretching frequency of the carbonyl ligand bridging to hard Lewis acid centre, M′. The infrared absorptions of the remaining uncomplexed carbonyl ligands may in some cases tend to increase slightly in frequency.

In addition to the reactions described above, the reactions of the typical metal carbonyls $Mo(CO)_6$ and $Fe(CO)_5$ are illustrated in schemes 4.12 and 4.13 respectively to give an indication of the extensive chemistry associated with individual metal carbonyls:

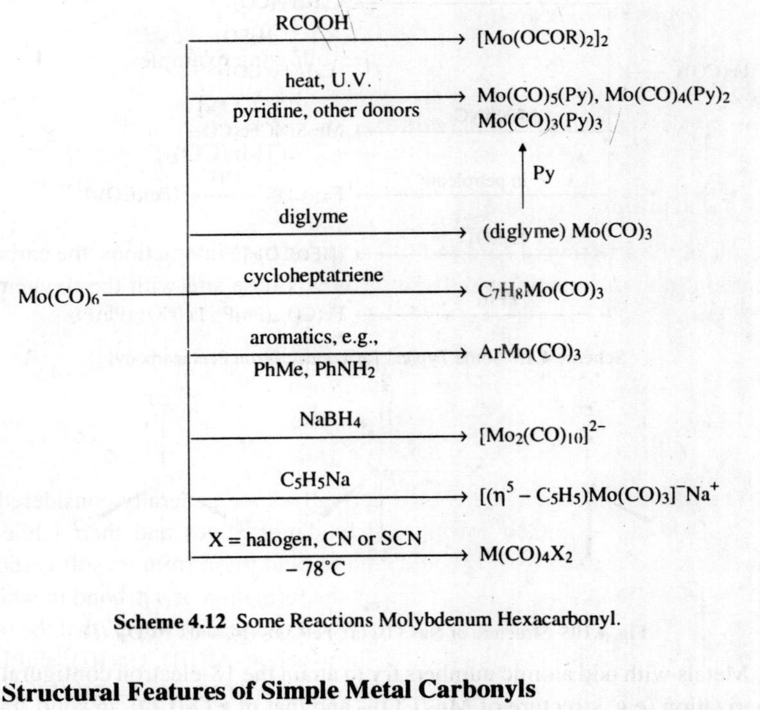

Scheme 4.12 Some Reactions Molybdenum Hexacarbonyl.

4.6.5 Structural Features of Simple Metal Carbonyls

The determination of structures of metal carbonyls has attracted a wide spread interest for many ye^rs. The structure of $Ni(CO)_4$ was first established in 1935, but many others have only recently been defined unambiguously. The mononuclear carbonyls generally depict structures (Fig. 4.105) expected on the basis of their formulas {e.g., Oh for $M(CO)_6$ (M = V, Cr, Mo, W); D_{3h} for $M(CO)_5$ (M = Fe, Ru, Os); and Td for $Ni(CO)_4$}.

The structures of $V(CO)_6$, $Cr(CO)_6$, $Fe(CO)_5$ and $Ni(CO)_4$ have been established by X-ray diffraction, infra-red spectroscopy and (except for $V(CO)_6$) by electron diffraction. All the hexacarbonyls have regular octahedral structures and $Ni(CO)_4$ has a regular tetrahedral structure. In the trigonal bipyramidal, $Fe(CO)_5$ the electron diffraction study has suggested slightly shorter (1.806 Å) Fe—C apical bond compared to the equatorial one (1.833 Å).

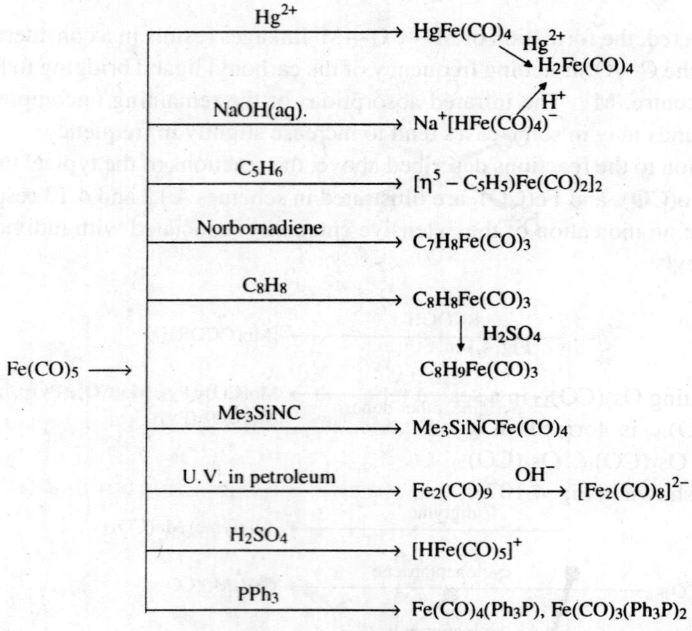

Scheme 4.13 Some Typical Reactions of Iron Pentacarbonyl.

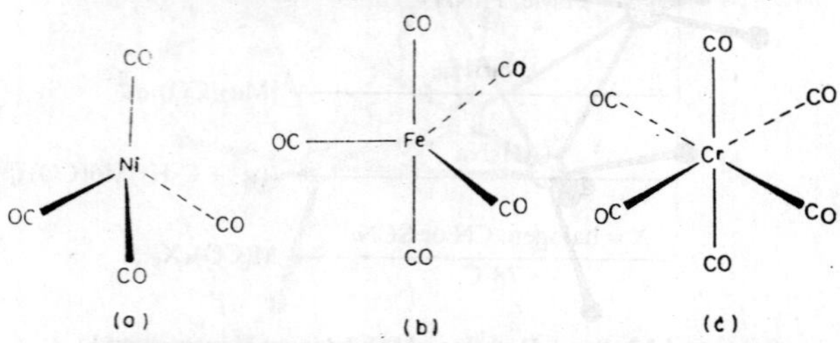

Fig. 4.105 Structure of $Ni(CO)_4$ (a), $Fe(CO)_5$ (b), and $Cr(CO)_6$ (c)

Metals with odd atomic numbers try to attain the 18-electron configuration by dimerization (e.g. structure of $Mn_2(CO)_{10}$ and that of $CO_2(CO)_8$ in solid state and solution, Section 4.6.1). The metal carbonyls and their substitution products often provide interesting examples of fluxional molecules (cf. Chapter 5).

As shown in Table 4.14, the d^7s^2 and d^6s^2 metals in general tend to form di- and poly-nuclear neutral binary carbonyl species. The structure of $Fe_3(CO)_{12}$ has been depicted in Fig. 4.106. It is interesting to record that whereas the structures of $Ru_3(CO)_{12}$ and even some mixed carbonyls like $FeRu_2(CO)_{12}$ and $FeOs_2(CO)_{12}$ are similar to that of $Fe_3(CO)_{12}$ without any bridging carbonyl ligands, that of $Fe_3(CO)_{12}$ does involve bridging CO groups and can be represented by Fig. 4.106. The structures of $Os_3(CO)_{12}$ and $Ru_3(CO)_{12}$ depict D_{3h} symmetry, whereas the $Fe_3(CO)_{12}$ structure is highly unsymmetrical and can be thought as derived from the D_{3h} structure by rotation of one of the CO groups on each of the two iron atoms into bridging positions across the bond between these two atoms, accompanied by rotations of residual $Fe(CO)_3$ moieties containing those metal atoms.

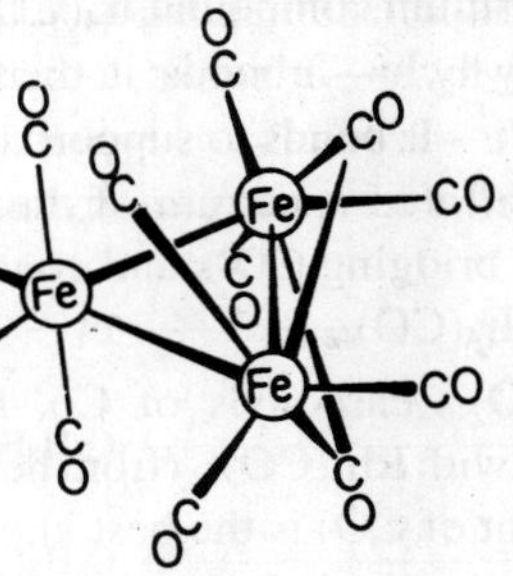

Fig. 4.106 Structure of $Fe_3(CO)_{12}$.

On heating $Os_3(CO)_{12}$ in a sealed tube at 190°C, carbon monoxide is evolved and $Os_6(CO)_{18}$ is formed in good yield along with less well characterized $Os_5(CO)_{16}$, $Os_5(CO)_{19}$, $Os_6(CO)_{20}$, $Os_7(CO)_{21}$ and $Os_8(CO)_{23}$. The structure of $Os_6(CO)_{18}$, shown in Fig. 4.107, is that of a bicapped tetrahedron of Os atoms, with

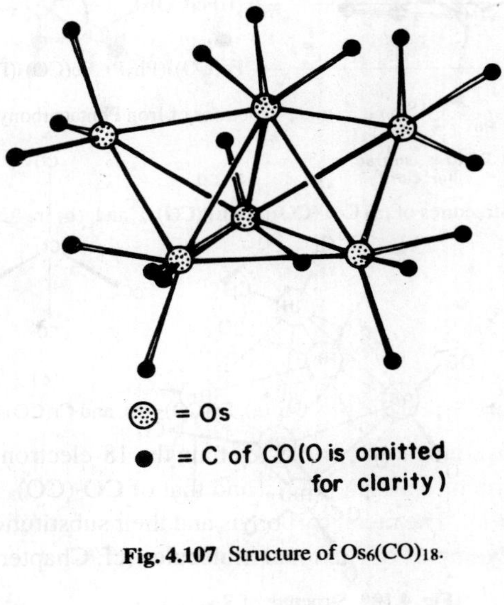

Fig. 4.107 Structure of $Os_6(CO)_{18}$.

the 2 capping atoms being susceptible to nucleophilic attack. $Os_6(CO)_{18}$ is thus itself not octahedral, but it is converted into the octahedral $Os_6(CO)_{18}$ anion by the nucleophilic attack of I^- ions on the two capping atoms.

In view of the odd atomic number of cobalt (as well as rhodium and iridium), these form well defined dimeric units, $M_2(CO)_8$. Some interesting facts of the structure of $Co_2(CO)_8$ have already been presented in Section 4.5.1. These crystalline air-sensitive orange-red crystals of $Co_2(CO)_8$ are converted into black $Co_4(CO)_{12}$ by heating in an inert atmosphere. The corresponding $M_4(CO)_{12}$ are the most stable carbonyls of rhodium and iridium and can be obtained by heating MCl_3 with copper metal under 200 atm. of CO. The structure of these, shown in Fig. 4.108 differ in that whereas the cobalt and rhodium carbonyls each incorporate 3 bridging carbonyls, the iridium compound $Ir_4(CO)_{12}$ consists of a tetrahedron of Ir atoms held together solely by Ir—Ir bonds; it, therefore, appears that iridium atoms may be too large for the Ir—Ir bonds to support a CO bridge. It may be of interest to record here that the infrared spectrum of the mixed derivatives $Co_2Ir_2(CO)_{12}$ indicates the presence of bridging CO's and consequent similarity in structure to $Co_4(CO)_{12}$ rather than to $Ir_4(CO)_{12}$.

Of the three $M_6(CO)_{16}$ carbonyls of Co, Rh and Ir, the dark-brown Rh compound, formed along with $Rh_4(CO)_{12}$ (from heating of $RhCl_3$ with copper under high pressure environment of CO) is the best known. The structure of $Rh_6(CO)_{16}$ (Fig. 4.109) consists of an octahedral array of $Rh(CO)_2$ units with the remaining 4 CO's bridging 4 faces of the octahedron.

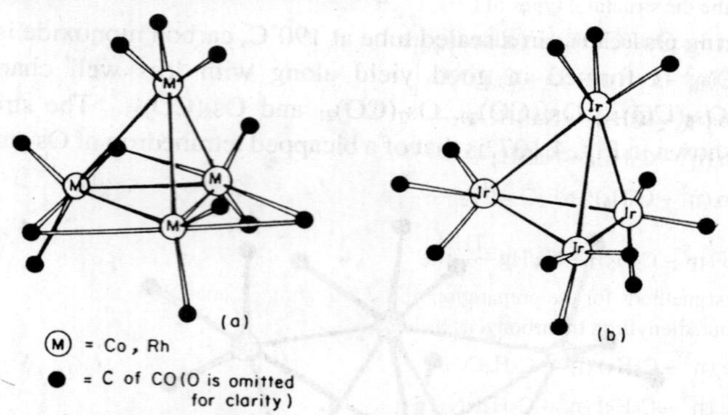

Fig. 4.108 Structures of (a) $Co_4(CO)_{12}$; $Rh_4(CO)_{12}$ and (b) $Ir_4(CO)_{12}$.

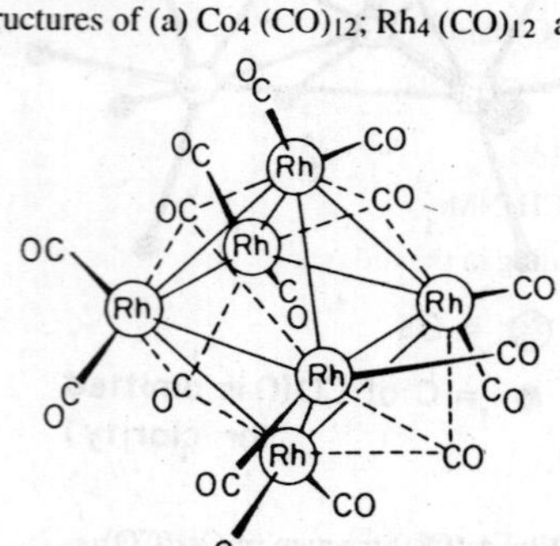

Fig. 4.109 Structure of $Rh_6(CO)_{16}$

Study Questions

1. Why were transition metal alkyls discovered rather late? What are the factors determining the stability of transition metal alkyls?
2. Discuss briefly the followings:
   (a) Classification of organotransition metal compounds
   (b) The eighteen electron rule
3. Comment on each of the following observations:
   (a) For free propene, ν (C=C) is at 1652 cm^{-1}, but in the complex $K[PtCl_3(CH_3CH=CH_2)]$ it is at 1504 cm^{-1}.
   (b) The possibility of distinguishing between η^1 and η^3 type of metal-allylic bonding by the i.r. spectroscopy.
   (c) Nickelocene reacts with tetrafluoroethylene to form Ni(η^3-L) (η^5-C_5H_5), where

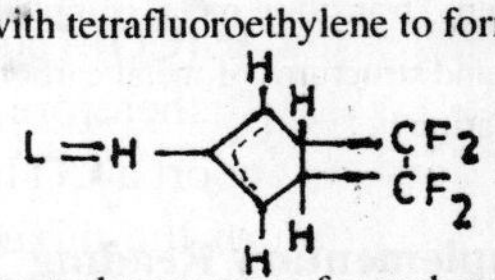

4. Compare proton NMR spectral resonances for each pair of compounds shown below:
   (a) $\eta^1 - C_3H_5$ and $\eta^3 - C_3H_5$ ligands
   (b) Free and complexed ethylene
   (c) Free and compled alkynes having at least one acetylenic proton.
   (d) C_6H_6 and an η^6–C_6H_6 ligand
5. The $\eta^5 - C_5H_5$ ligand is susceptible both to nucleophilic and electrophilic attack. Discuss the relative reactivity of ferrocene and cobalticenium complexes in these types of reactions.
6. Describe the structural types of C_5H_5 coordination geometries.
7. Predict the products of the following reactions:
   (a) $Fe(\eta^5 - C_5H_5)_2 + Li + C_2H_5NH_2 \longrightarrow$
   (b) $Co(\eta^5 - C_5H_5)_2^+ + NaBH_4 \longrightarrow$
   (c) $Rh(\eta^5 - C_5H_5)_2 + NaBH_4 \longrightarrow$
   (d) $Co(\eta^5 - C_5H_5)_2 + F_2C = CF_2 \longrightarrow$
   (d) $Ni(\eta^5 - C_5H_5)_2 + Na/Hg \xrightarrow{THF}$
8. Suggest methods for the preparation of the following compounds starting with ferrocene or cyclobutadienyliron tricarbonyl as the case may be:
   (a) $Fe(\eta^5 - C_5H_5)(\eta^5 - C_5H_4COR)$
   (b) $Fe(\eta^5 - C_5H_5)(\eta^5 - C_5H_4HN_2)$
   (c) $Fe(\eta^5 - C_5H_5)(\eta^5 - C_5H_4Et)$
   (d) $Fe(CO)_3(\eta^4 - C_4H_3 - CH(OH)CH_3)$
   (e) $Fe(CO)_3(\eta^4 - C_4H_3Li)$
   (f) $Fe(CO)_3(\eta^4 - C_4H_3CH_2NMe_2)$
9. Discuss the nature of bonding in the following compounds:
   (a) $[PtCl_3(C_2H_4)]^-$
   (b) $Ni(\eta^2 - C_3H_5)_2$
   (c) $Fe(\eta^5 - C_5H_5)_2$;
   (c) $Cr(\eta^6 - C_6H_6)_2$
10. Show that the metal atoms in $Fe_3(CO)_{12}$, $Os_3(CO)_{12}$, $Fe_2(CO)_9$, and the two forms of $Co_2(CO)_8$ satisfy the effective atomic number (EAN) rule.

11. Sketch the metal and ligand orbitals involved in the following complexes:
    (a) Metal-alkynes
    (b) Metal-carbenes
    (c) Metal-carbonyls
12. Give an account of the chemistry, structure and bonding of the π-allyl complexes of transition metals.
13. Discuss the following:
    (a) Extended π system complexes
    (b) σ - π rearrangement in π-allyl complexes
    (c) Arene transition metal complexes
    (d) Fluxional behaviour of organometallic compounds
    (e) Reactions of cyclopentadienyl transition metal complexes.
14. Discuss the nature of bonding and structures of metal carbonyls. How does infrared spectroscopy helps in structural elucidation?

Supplementary Reading

1. '*Organometallic Compounds*', by The Open University Course Team, The Open University Press, Walton Hall, Milton Keynes, U.K. (1977).
2. '*Metal σ-hydrocarbyls, MR_n, Stoichiometry. Structures, Stabilities, and Thermal Decomposition Pathway*', by P.J. Davidson, M.F. Lappert and R. Pearce, Chemical Reviews, vol. 76, pp. 219-268 (1976).
3. '*σ-alkyl and Aryl Complexes of the Group 4-7 Transition Metals*', by R.R. Schrock and G.W. Parshall, *Chemical Reviews,* vol. 76, pp. 243-268 (1976).
4. '*Bridged Hydrocarbyl or Hydrocarbon Binuclear Transition, Metal Complexes: Classification, Structures and Chemistry*', by J. Holton, M.F. Lappert, R. Pearce and P.I.W. Yarrow, *Chemical Reviews,* vol. 83, pp. 135-201 (1983).
5. '*Principles and Applications of Organotransition Metal Chemistry*' by J.P. Collmn, L.S. Hegedus, J.R. Norton and R.G. Finke, University Science Books, Mill Valley, California (1987).
6. '*Organotransition Metal Chemistry*', by R.F. Heck, Academic Press, New York, (1972).
7. '*Metallo-Organic Chemistry*', by A.J. Pearson, John Wiley Chichester (1985).
8. '*Basic Organometallic Chemistry*', by I. Haiduc and J.J. Zuckerman, Walter de Gruyter, Berlin/New York (1985).
9. '*Advanced Inorganic Chemistry*', 5th edn., by F.A. Cotton and G. Wilkinson, pp. 1021-1051; 1122-1185, John Wilkey, New York (1988).
10. '*Inorganic Chemistry*', by K.F. Purcell and J.C. Kotz, pp. 793-962, W.B. Saunders Company, London (1977).
11. '*Introduction to Metal π-Complex Chemistry*', by M. Tsutsui, M.N. Levy, A. Nakamura, M. Ichikawa and K. Mori, Plenum Press, New York (1970).
12. 'Organometallic Compounds of the Transition Metals, by D.St.C. Black, W.R. Jackson and J.M. Swan, in '*Comprehensive Organic Chemistry*', vol. 3, pp. 1127-1311, edited by D.H.R. Barton and W.D. Ollis, Pergamon Press, Oxford (1979).
13. '*Organometallic Chemistry of Titanium, Zirconium and Hafnium*', by P.C. Wailes, R.S.P. Coutts and H. Weigold, Academic Press, New York, (1974).
14. '*Comprehensive Organometallic Chemistry I*', edited by G. Wilkinson, F.G.A. Stone and E.W. Abel; vols. 3 (Chapters 21-28), 4 (Chapters 29-33), 5 (Chapters 34-36), and 6(Chapter 37-39), Pergamon Press, Oxford (1982).
15. '*Transition Metal Carbene Complexes*', edited by K.H. Dotz, H. Fischer, P. Hofmann, F.R. Kreissl, U. Schubert and K. Weiss, Verlag Chemie, Weinheim (1983).

16. '*Characterization of Organometallic Compounds*', edited by M. Tsutsui, Part I (1969) and Part II (1970), John Wiley, New York.
17. '*Vibrational Spectra of Organometallic Compounds*', by E. Maslowsky, Jr., John Wiley, New York (1977).
18. '*Vibrational Spectra of π-Complexes*', by H.P. Fritz, in *Advances in Organometallic Chemistry.* edited by F.G.A. Stone and R. West, vol. 15, pp. 235, Academic Press, New York (1977).
19. '*Metal Carbonyl Spectra*', by P.S. Braterman, Academic Press, New York (1975).
20. '*Chemistry and Spectroscopy of f-Element Organometallics, Part I: The Lanthanides*', by T.J. Marks, in *Progress in Inorganic Chemistry,* edited by S.J. Lippard, vol. 24, pp. 51-107, John Wiley, New York (1978).
21. '*Chemistry and Spectroscopy of f-Element Organometallics, Part-II: The Actinides*', by T.J. Marks, in *Progress in Inorganic Chemistry,* edited by S.J. Lippard, vol. 25, pp. 224-325, John Wiley, New York (1979).
22. '*Organometallic Lanthanide Chemistry*', by W.J. Evans, in Advances in Organometallic Chemistry, edited by F.G.A. Stone and R. West, vol. 24, pp. 131-177, Academic Press, London (1985).
23. '*Alkylidene Complexes of Nb and Ta*', by R.R. Schrock, *Accounts Chemical Research,* vol. 12, pp. 98-104 (1979).
24. '*Transition Metal-Carbene Complexes*', by D.J. Cardin, B. Cetinkaya, and M.F. Lappert, *Chemical Reviews,* vol. 72, pp. 545-572 (1972).
25. '*The Chemistry of Transition-Metal Carbene Complexes and their Role as Reaction, Intermediates*', by D.J. Cardin, B. Cetinkaya, M.J. Doyle, and M.F. Lappert, *Chemical Society Reviews,* vol. 2, pp. 99-144 (1973).
26. '*Carbene and Carbyne-Metal Complexes: Nobel Lecture*', by E.O. Fischer, *Angewandte Chemie,* vol. 86, pp. 651-663 (1974).
27. '*On the Way to Carbene and Carbyne Complexes*', by E.O. Fischer, in *Advances in Organometallic Chemistry,* edited by F.G.A. Stone and R. West, vol. 14, pp. 1-29, Academic Press, London (1976).
28. '*Metal Vapour Synthesis in Organometallic Chemistry*', by J.R. Blackborow and D. Young, Springer-Verlag, Berlin (1979).
29. '*Fundamental Transition Metal Organometallic Chemistry*', by C.M. Lukehart, Brooks/Cole California (1985).
30. '*Organotransition Metal Chemistry*' by A. Yamamoto, John Wiley, New York (1986).
31. '*The Organometallic Chemistry of the Transition Metals*' by R.H. Crabtree, John Wiley, New York (1988); 2nd ed., John Wiley, New York (1994).
32. '*Organomettalic Chemistry of the Transition Elements*' by F.P. Pruchnik (Translated from Polish by S.A. Duraj), Plenum Press, New York (1990).
33. '*Kinetics and Mechanism of Reactions of Transition Metal Complexes*', by R.G. Wlikins, 2nd ed., V.C.H. Weinheim (1991).
34. '*Organometallics: A Concise Introduction*' by Ch. Elschembroich and A. Salzer, 2nd ed., VCH, Weinheim (1992).
35. '*Comprehensive Organometallic Chemistry II*', edited by E.W. Abel, F.G.A. Stone and G. Wilkinson; Vols. 3 (Chapters 1-3), 4 (Chapters 1-13), 5 (Chapters 1-8), 6 (Chapters 1-10), 7 (Chapters 1-16), 8 (Chapters 1-4), 9 (Chapters 1-9), and 10 (Chapters 1-7), Pergamon Press, New York (1995).
36. '*Synthetic Methods of Organometallic and Inorganic Chemistry*', edited by W.A. Herrmann, Vols. 6, 7 & 8, Thieme Verlag, New York, 1997.

Fluxional Organometallic Compounds (Structural Non-Rigidity)

5.1 GENERAL AND HISTORICAL

In the earlier chapters, the structures of a number of organometallic compounds have been described. In most of the cases, these structures are unique and well-defined. The atoms of the molecule are of course continuously involved in approximately harmonic vibrations about their equilibrium positions, but in other respects the structure may be considered rigid.

As stated in the last-chapter also, alkene complexes of transition metals represent a special class of compounds, in which the molecular vibrations or intramolecular rearrangements change a molecule from one type of configuration into another. Such molecules have, therefore, more than one thermally accessible structures, which under suitable conditions (especially of temperature) are interconvertible from one to another fairly rapidly. This phenomenon may be termed as 'fluxional behaviour' and is generally of two types:

(i) In some cases (e.g., keto-enol equilibrium), the non-equivalence of two or more structures results, in the existence of molecules with different instantaneous identities. Such a process of interconversion is called 'tautomerism' and has been known in the case of organic molecules for a long time, e.g.,

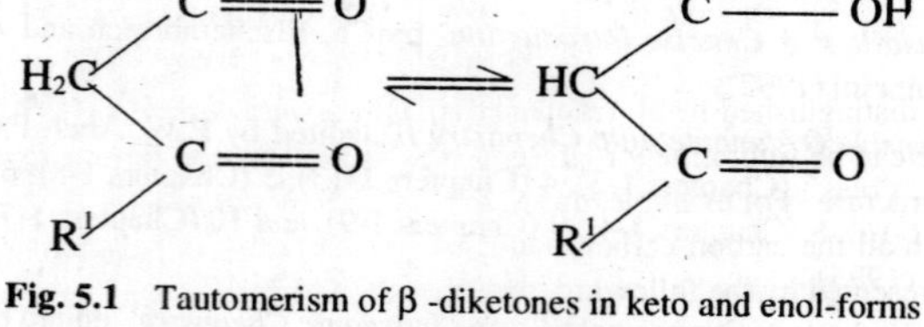

Fig. 5.1 Tautomerism of β -diketones in keto and enol-forms

(ii) In other cases, the two or more configurations are equivalent (in structure, bonding and energy content). Such molecules are called fluxional* (name suggested by Doering and Roth, 1963).

An alternative term 'Stereochemical non-rigidity' has been suggested (by Muetterties, 1965) for the latter type of 'fluxional molecules'. However, F.A. Cotton, a prominent worker in this field, has suggested (1968) that the term 'stereochemical non-rigidity' be reserved as a general term for molecules of both the above types, i.e., tautomeric and fluxional, which undergo rapid reversible intramolecular rearrangements.

The development of the concept of 'fluxionality' in molecules has an interesting history. In view of the novelty of the idea and involvement of prominent organometallic chemists, the historical development of the concept is being presented in a little more detail and examples of even some simple inorganic molecules like PF_5 have been included for elucidating the concept before presenting the features of a few 'fluxional' organometallic compounds. Although the n.m.r. non-equivalence of the fluxional atoms in PF_5 was reported by Gutosvsky and coworkers in 1953, yet a fuller understanding of the accompanying phenomenon had to wait till the commercial n.m.r. spectrometers became available in late fifties.

The first observations leading to the concept of fluxional behaviour of organometallic molecules appear to have been made in 1956 by Piper and Wilkinson, when they observed in the respective proton n.m.r. spectra, only a singlet for $Hg(\eta^1-C_5H_5)_2$; two singlets for the two cyclopentadienyl ligands at the room temperature in $Fe(\eta^1-C_5H_5)(CO)_2(\eta^5-C_5H_5)$ and a singlet for all five ring protons in $Cu(\eta^1-C_5H_5)P(C_2H_5)_3$:

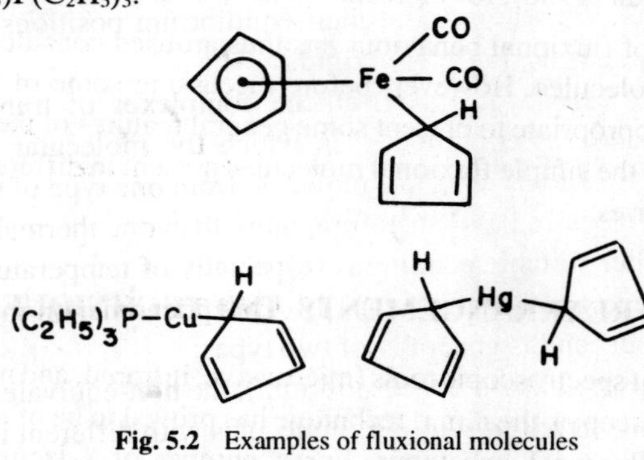

Fig. 5.2 Examples of fluxional molecules

These workers explained the appearance of single resonances for all protons in presumably σ-bonded cyclopentadienyl rings in all these compounds to be due to a rapid shifting of metal-carbon bond among the five possible positions*.

* This concept is to be distinguished from 'resonance' in which *only one type of actually more stable structure of a molecule* can be figuratively represented by more than one form, each of which is less stable than the actual structure. For example, due to the inability in representing the actual structure of benzene molecule (with all the carbon-carbon bonds of equal lengths, in between single and double bond-lengths), it is represented by the following two resonating forms:

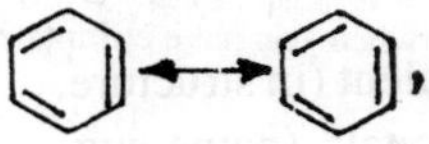, each of which is less stable than the actual structure.

Although the explanation based on fluxionality of the molecules is well accepted now, yet the same had aroused considerable scepticism* in the beginning.

The next observation relevant to this field was the observation of a single ^{13}C peak in the n.m.r. spectrum of $Fe(CO)_5$, which according to the two plausible structures, trigonal bipyramid (tbp) and square pyramid (sp), was expected to give two signals in the ratios of 2/3 or 1/4 respectively. The observation about the appearance of only a single peak was confirmed by Nyholm and coworkers (1962) who believed in the 'sp' structure contrary to the 'tbp' structure assumed by Cotton and coworkers. Incidentally, the structure in the solid state has been established (1974) by X-ray crystallography to be of the 'tbp' type.

The next significant development in the field was the synthesis in 1959 in four laboratories of cyclooctatetraene complex of iron carbonyl, $Fe(CO)_3$ (η^4–C_8H_8), which again gave only a single line in the p.m.r. spectrum. Dickens and Lipscomb (1962) showed that the structure in the crystal involved coordination of two adjacent double bonds as had been shown in the case of the corresponding well-known compound, (1, 3-butadieno)$Fe(CO)_3$ (described by Reihlen and coworkers, 1930). To account for the single proton resonance, Lipscomb and Dickens suggested a dynamical effect amounting to permutation of the carbon atoms of the ring relative to the $Fe(CO)_3$ groups. However as Cotton remarked later (1975) that it was doubtful if any one had correlated this observation with that of Piper and Wilkinson (1956), for molecules shown in Fig. 5.2.

The concept of fluxional behaviour has thus aroused considerable interest in organometallic molecules. However, before discussing some of these in greater detail, it may be appropriate to present some general features of the rates and types rearrangements in the simple fluxional molecules present in different coordination states and geometries.

5.2 RATES OF REARRANGEMENTS AND TECHNIQUES OF STUDY

Out of the different spectroscopic tools (microwave, infrared, and nuclear magnetic resonance spectroscopy), the n.m.r. technique has proved to be of maximum utility in the study of fluxional behaviour. For a number of systems, the speed of rearrangement is fast enough at the ambient temperatures for the n.m.r. techniques

* Professor F.A. Cotton, who was working in the same Department of Harvard University, has recorded the reactions of Professor G. Wilkinson's research group on this unexpected observation in the words: 'I can well remember the consternation and cudgeling in brains that this produced in Wilkinson's little research group. After the initial obvious suspicions that the spectrometer was not working properly or had been ineptly operated were invalidated by careful rechecking. Wilkinson himself produced the idea that the ring was perhaps moving so that the metal-to-carbon bond switched rapidly enough from one carbon atom to the next to cause all resonances to be seen as one. He himself has pointed out that this idea came to him as a kind of analogy with motion-picture projection. Be that as it may, that idea seemed rather bizarre and I doubt if any one in the group, the boss included, would have been prepared to bet his life insurance on it. However, no one thought of anything else which could explain in facts. Following Holmsian (cf. A. Conan Doyle, '*The Sign of Four*' principle, ''when you have eliminated the impossible, whatever remains, however, improbable, must be the truth''.

not to be able to detect the same, but at lower temperatures the rearrangements become slow enough to be detected by n.m.r. spectroscopy.

The choice of a technique suitable for the study of a fluxional system would obviously depend on the rate of rearrangements in a specific case. From this point of view, the spectroscopic techniques could be divided into two broad groups:

(a) In some fast techniques like electronic and vibrational spectroscopy and gas phase electron diffraction, the act of observing the molecule is completed in $< 10^{-11}$ sec. Processes of rearrangement are generally not fast enough to influence these observations. Thus by these techniques, (i) no fluxional behaviour would be expected to be observed if the different configurations are equivalent, and (ii) for tautomeric or isomeric molecules, both forms will register their presence in the spectrum, but no indication of the interconversations could be possible.

(b) The rearrangements involved in stereochemically non-rigid behaviour are rate processes with activation energies of different magnitudes. When these activation energies are in the range of 25-100 kJ mol^{-1}, the rate of rearrangements can be brought generally between temperatures of +150° to −150°C into a range in which the signals due to different fluxional species can be detected separately by n.m.r. technique. Thus by proper choice of temperature, many such rearrangements can be controlled so that they are slow enough at lower temperatures to allow signals due to different fluxional species appearing distinctly, and rapid enough at higher temperatures for the signals from the different configurations or environments to be averaged into a single line at a mean position.

Normally, studies of fluxional molecules are carried out with the following two main objectives:

(i) establishing definitely the instantaneous structure (for which information can be achieved by recognising the limiting spectrum) of the molecule; and

(ii) to study the pathway of rearrangement.

These targets can be conveniently achieved by n.m.r. line shape analysis and this technique will, therefore, be discussed in some detail.

5.2.1 Rate Processes and NMR Spectra

We shall discuss in this section the influence or rate processes on n.m.r. spectra. If a nucleus is exchanging between two molecular configurations, A and B, then we have equilibrium:

$$\underset{n_A}{A} \rightleftharpoons \underset{n_B}{B}$$

where n_A and n_B are the mole fractions of A and B. An equilibrium of the type can be generally characterized by two parameters:

(i) The equilibrium position is determined by the free energy of the process (ΔG°), i.e.,

$$n_A / n_B = \exp(-\Delta G/RT)$$

and $$n_A + n_B = 1$$

(ii) The free energy of activation (ΔG°) determines the rate of interconversion, i.e., the rate of reaction (k) for the conversion $A \rightarrow B$ is given by

$$k = \frac{RT}{Nh} \exp(-\Delta G^\circ/RT)$$

Suppose a nucleus in the molecule, in the states A and B, exhibits chemical shift δ_A and δ_B respectively with J_A and J_B coupling in Hz. There may be three types of situations (depending on the rate of exchange) which should be considered in this connection:

(a) *Slow exchange*: When the rate of interconversion of A and B is slow (on the n.m.r. time scale), two separate signals representing two non-equivalent sites (due to two separate species A and B) are observable at ν_A and ν_B with couplings J_A and J_B;

(b) *Fast exchange*: Due to rapid nucleus migration from one site (A) to another (B), the n.m.r. spectrum exhibits one signal in which the chemical shifts and couplings would be the weighted average of the values in A and B and can be represented by:

$$\nu_{av.} = n_A\nu_A + n_B\nu_B$$

and coupling $$J_{av.} = n_AJ_A + n_BJ_B$$

(c) *Intermediate rates of exchange* In these cases broad signals are observed. It may be noted that there are certain other factors which could cause broadening of the n.m.r. signals, such as the presence of quadrupolar nuclei and paramagnetic moieties (as an impurity), and poor sampling. Besides these easily recognizable features, the presence of broad signals, therefore, is an indication of an intermediate rate process.

Consider the simplest $Ti(\eta^1\text{–}C_5H_5)_2(\eta^5\text{–}C_5H_5)_2$, which gives rise in its n.m.r. spectrum near the lower temperature limit two equal intensity signals which collapse, on raising the temperature (to ~ 36°) of recording the spectrum, into one signal with no coupling present. The temperature at which the two separate signals just merge into one is called the 'coalescence temperature' (Tc). At this temperature, the rate 'k' for exchange between two uncoupled equally populated sites (A and B), is given by the equation:

$$k = \frac{\pi(\nu_A - \nu_B)}{\sqrt{2}}$$

where $(\nu_A - \nu_B)$ is the separation (in Hz) of the two signals in the absence of exchange.

Taking the logarithm of both sides in the equation $k = \frac{RT}{Nh} e^{-\Delta G^{\neq}/TR}$, and rearranging, gives

$$-\Delta G^{\neq} = RT \log \frac{RT\sqrt{2}}{Nh\pi} - RT \log K$$

Substituting $k = \frac{\pi \Delta \nu}{\sqrt{2}}$ and collecting the constant terms gives

$$\Delta G^{\neq} = \log \left\{ \frac{R\sqrt{2}}{Nh\pi} + \log \frac{T}{\Delta \nu} \right\}$$

Finally, converting to $\log_{10}$ and substituting values for the constant terms gives the expression:

$$\Delta G^{\neq}_{Tc} = 4.57\, T \left(9.97 + \log_{10} \frac{T}{\Delta \nu} \right)$$

5.3 SOME SIMPLE EXAMPLES OF NONRIGID MOLECULES IN DIFFERENT COORDINATION GEOMETRIES

Although most of the coordination polyhedra are essentially rigid molecules, more and more examples of nonrigidity, specially fluxionality, in such polyhedra are coming to light. In fact, with 5 and 7 as well as higher coordination states, non-rigidity appears to be quite a common feature.

5.3.1 Four-Coordinate Systems

A common feature of fluxional behaviour in such systems is the inversion of pyramidal molecules represented by the following figure:

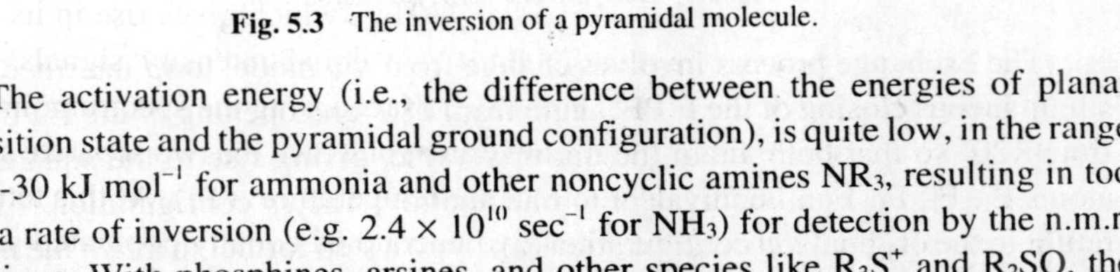

Fig. 5.3 The inversion of a pyramidal molecule.

The activation energy (i.e., the difference between the energies of planar (transition state and the pyramidal ground configuration), is quite low, in the range of 24-30 kJ mol^{-1} for ammonia and other noncyclic amines NR_3, resulting in too high a rate of inversion (e.g. 2.4×10^{10} sec^{-1} for NH_3) for detection by the n.m.r. technique. With phosphines, arsines, and other species like R_3S^+ and R_2SO, the activation energies are higher than 100 kJ mol^{-1}, and inversions are slow enough to allow separation of enantiomers in cases of molecules such as RR′R″P, RR′R″As, and RR′SO.

Amongst four-coordinate transition metal complexes, planar-tetrahedral fluxional changes are quite common; e.g.,

$$\underset{\text{Planar}}{Ni(R_3P)_2X_2} \xrightarrow[10^5\ sec^{-1}]{45\ kJ\ mol^{-1}} \underset{\text{Tetrahedral}}{Ni(R_3P)_2X_2}$$

(room temp.)

5.3.2 Trigonal-Bipyramidal Molecules

It was in such systems, e.g., PF_5 and $Fe(CO)_5$ that fluxional behaviour was first detected. In ^{19}F and ^{13}C n.m.r. spectra respectively of the above two molecules, only one resonance peak was observed instead of two in the ratio of 3 : 2 (i.e. ratio of axial to equatorial sites).

From theoretical considerations, it can be concluded that when two nuclei, the resonance frequencies, of which differ by $(\nu_1 - \nu_2) = \Delta\nu$/sec., change site at a frequency higher than $(\nu_1 + \nu_2)$/sec., then only one resonance peak would be observed at the mean $(\nu_1 + \nu_2)/2$ position of the two frequencies characteristic of two types of sites, between which interchange is occurring.

In both PF_5 and $Fe(CO)_5$ the dissociative or bimolecular mechanism appears to be ruled out in view of the persistence of $^{31}P - ^{19}F$ and $^{57}Fe - ^{13}C$ coupling. Although the probability of 1– for –1 exchange in which one equatorial and one axial ligands are involved in each step cannot be ruled out, the available experimental evidence appears to favour a 2– for –2 process in which axial ligands exchange simultaneously with two equatorial ones. The most probable mechanism for this type of exchange was suggested by Berry in 1960 as follows (Fig. 5.4):

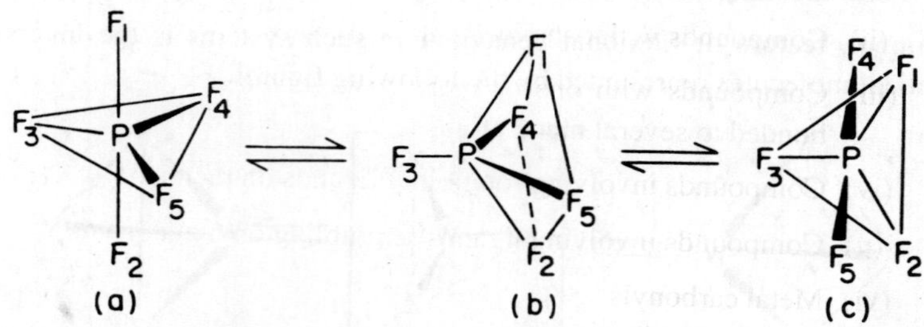

Fig. 5.4 Berry Mechanism (Pseudorotation).

The exchange process involves change from *tbp* model to *sp* intermediate by simultaneous closing of the F^1PF^2 angle from 180° and opening of the F^4PF^5 angle from 120° so that both attain the midway value, giving rise to the square set of atoms F_1, F_2, F_4, F_5 all equivalent to one another. The *sp* configuration (b) could return to the original *tbp* configuration (a), or go a step further to attain the new *tbp* configuration (c) in which the molecules have effectively rotated round the F_3P axis through 90°. In view of this apparent rotation, Berry mechanism is termed 'pseudorotation'.

The first unambiguous evidence for a rapid intramolecular rearrangement in a d^8 pentacoordinate transition metal complex was reported by Udovich and Clark in 1969. The variable low-temperature ^{19}F n.m.r. spectrum of the compound

$F_3CCo(CO)_3PF_3$ indicated two equilibrating isomers which exchange ligands between axial and equatorial sites.

5.3.3 Systems with Coordination Numbers 6 or Higher

The octahedral configuration is rather rigid and fluxional or tautomeric rearrangements generally do not occur in these except through metal ligand rupture. Amongst the few instances in which fluxional changes have been noticed can be cited: iron and ruthenium complexes with the composition $MH_2(PR_3)_4$ where M = Fe or Ru. The *cis* and *trans* isomers of the former complex have separate well-resolved signals at –50°, but with increasing temperature these broaden and collapse until at 60°C, a sharp line is observed indicating rapid exchange between the two isomeric forms.

For higher coordination states, fluxionality appears to be a normal feature, e.g. ReF_7, IF_7; ReH_9^{-2}; $ReH_8PR_3^-$ and $ReH_7(PR_3)_2$ all appear to be fluxional molecules with interchange processes which have been studied by n.m.r. spectroscopic techniques.

5.4 CLASSIFICATION OF FLUXIONAL ORGANOMETALLIC COMPOUNDS

Fluxional organometallic molecules, which have been studied so far, can be divided into the following six major classes:

(i) Compounds with acyclic alkenes;

(ii) Compounds with σ-bonded cyclic alkenes;

(iii) Compounds with one or more cyclopolyenes π-bonded or σ and π-bonded to several metal atoms;

(iv) Compounds involving rotation of ligands on metals;

(v) Compounds involving ligands scrambling on metals; and

(vi) Metal carbonyls

5.4.1 Compounds with Acyclic Alkenes

As mentioned in section 4.3.2, the η^3-allyl or "π-allyl" group is an ubiquitous ligand. A large number of homoleptic as well as heteroleptic π-allyl complexes of both transition and non-transition metals such as:

$Pt(\eta^3\text{–}C_3H_5)(\eta^5\text{–}C_5H_5)$; $Co(CO)_3(\eta^3\text{–}C_3H_5)$; $Mn(CO)_4(\eta^3\text{–}C_3H_5)$;

$Zr(\eta^3\text{–}C_3H_5)_4$; $Cr(\eta^3\text{–}C_3H_5)_3$; $Ni(\eta^3\text{–}C_3H_5)_2$;

$Mo(CO)(\eta^3\text{–}CH_2C_6H_5)(\eta^5\text{–}C_5H_5)$; $MgBr(\eta^3\text{–}C_3H_5)$; and

$Zn(\eta^3\text{–}C_3H_5)_2$ are known. In these complexes, the instantaneous structure may be represented either by localized σ-bonding (Fig. 5.5a) or delocalized π-bonding (Fig. 5.5b).

In the σ-complex (5.5a), the two hydrogen atoms on α- and γ-C atoms would be magnetically equivalent to each other. However, for η^3-allyl complexes (5.5b)

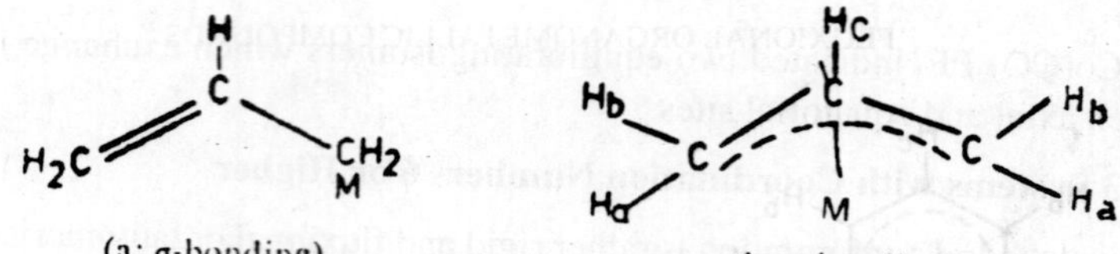

Fig. 5.5 σ and π-bonded complexes.

which are rigid on the ^{1}H n.m.r. time-scale, the *syn* and *anti* hydrogens (H_b and H_a) both exhibit a doublet due to the coupling with the central hydrogen (H_c) which shows a multiplet, Typical n.m.r. parameters observed are: H_c, δ 6.5–4; H_b, δ 5–2.5; H_a, δ 3–1; J_{bc}, ~ 7Hz; Jac ~ 11Hz, and J_{ab} ~ OHz.

The π -allyl group often behaves as an electron-rich moiety, but in some of these reactions there appears to be prior dissociation of one terminus to form an intermediate σ-allyl moiety. This kind of system is often reversible and fairly facile, leading to fluxional rearrangement on the ^{1}H n.m.r. timescale.

In the σ-allyl system, there is 'free rotation' about the carbon-carbon single bond, so that facile η^3 to η^1 rearrangement provides a mechanism for exchanging the *syn* and *anti* groups. It may be noted that the transformation η^3 to η^1 (π to σ) allyl creates a vacant coordination site on the metal as shown in the Fig. 5.6. The rearrangement of π to σ form undoubtedly accounts for the wide range of reactivities shown by metal allyl complexes.

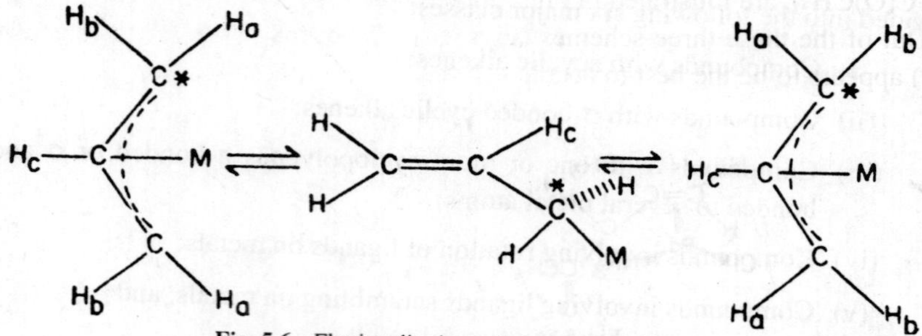

Fig. 5.6 Fluxionality in π-allyl metal complexes.

The complex $Mn(CO)_4(\eta^3\text{–}C_3H_5)$ can exhibit a rigid ^{1}H n.m.r. spectrum such as that shown in Fig. 5.7.

The spectrum shown is of AA′BB′X type where, in the case of the η^3-allyl-metal complex, X is the proton on the central carbon atom, A and A′ are protons (quite different from X) and are located *anti* to X, and B and B′ are protons again quite different from X but similar to A and A′ and located *syn* to X. At elevated temperatures or in the presence of a base, however, this spectrum collapses to a dynamic A_4X spectrum, the chief feature of which is the change of the doublets for the A and B protons to a single doublet presumably owing to a rapid (on the n.m.r. time scale) intramolecular rearrangement that causes the A/A′ (*anti*) and B/B′ (*syn*) protons to have the same chemical shift.

The mechanism of η^3-allyl rearrangement has been extensively examined, and a number of pathways for the scrambling of *syn* and *anti* protons have been

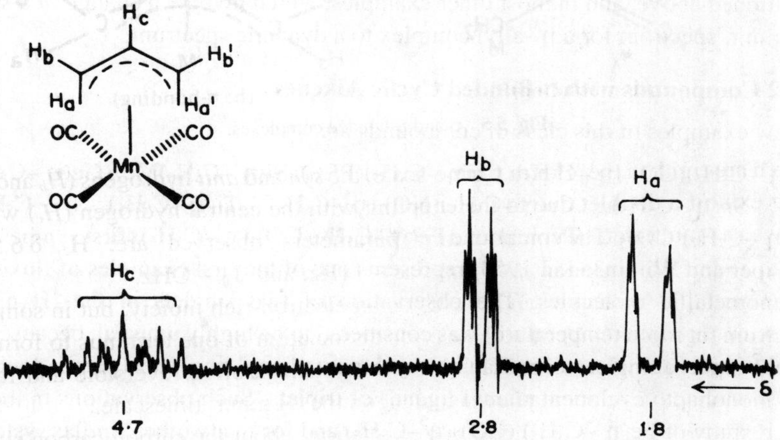

Fig. 5.7 ^{1}H n.m.r. spectrum of (η^3-allyl) tetracarbonylmanganese.

proposed. For example, three of the transition states of intermediates that can lead to *syn/anti* proton equilibration in the complex, $PdCl(NR_3)(\eta^3{-}CH_2{\cdot}CCH_3{\cdot}CH_2{\cdot}C(O)CH_3)$, are illustrated in Figs. 5.8a to 5.8c.

Out of the these three schemes (Figs. 5.8a, 5.8b, 5.8c), the last mechanism (5.8c) appears to be the best to account for proton equilibration in the compound

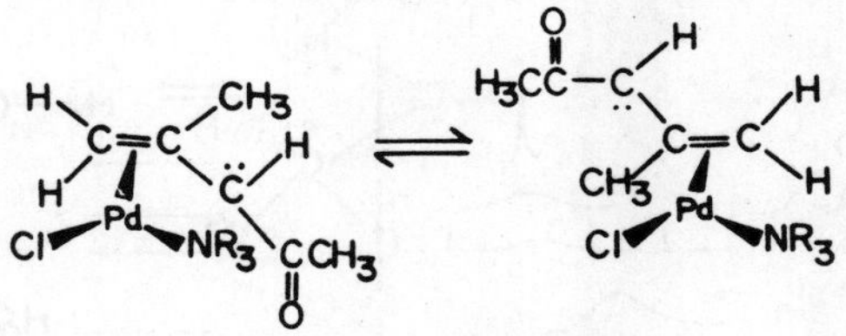

Fig. 5.8a Transition states.

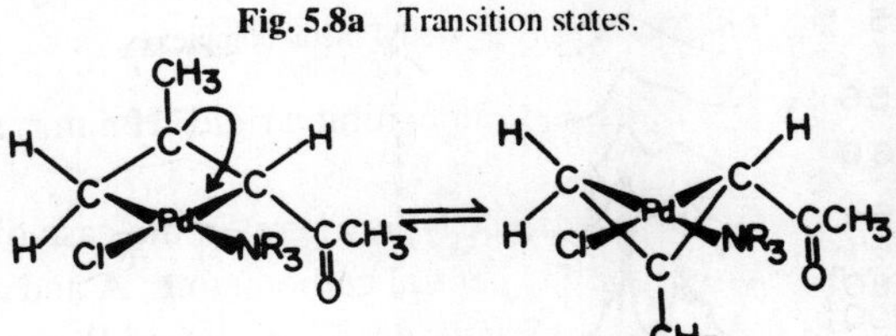

Fig. 5.8b Transition states.

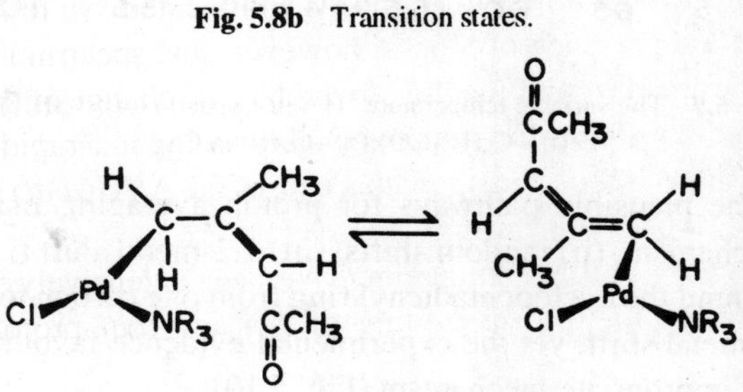

Fig. 5.8c Transition states.

mentioned above, and in most other examples, which involve a change of a static ^1H n.m.r. spectrum for a η^3-allyl complex to a dynamic spectrum.

5.4.2 Compounds with σ-Bonded Cyclic Alkenes

A few examples of this class of compounds are:

$Fe(\eta^1{-}C_5H_5)(CO)_2(\eta^5{-}C_5H_5)$; $Cu(\eta^1{-}C_5H_5)(PEt_3)$; $Si(\eta^1{-}C_5H_5)Me_3$; $Ge(\eta^1{-}C_5H_5)_2Me_2$, $Sn(\eta^1{-}C_5H_5)_4$; $Cr(\eta^1{-}C_5H_5)(NO)_2(\eta^5{-}C_5H_5)$: $Ti(\eta^1{-}C_5H_5)_2$ $(\eta^5{-}C_5H_5)_2$; $Re(\eta^1{-}C_7H_7)(CO)_5$. The compound $Fe(\eta^1{-}C_5H_5)(CO)_2(\eta^5{-}C_5H_5)$ (first synthesized by Piper and Wilkinson in 1953) represents one of the first examples of fluxional organometallic molecules. The observation of two singlets in the ^1H n.m.r. spectrum (at room temperature) was considered to be highly unusual, because one would expect a singlet (for pentahapto cyclopentadienyl ring) and a pair of doublets (for monohapto cyclopentadienyl ligand) or triplets. Such observations in the ^1H n.m.r. study of $Fe(\eta^1{-}C_5H_5)$ $(CO)_2(\eta^5{-}C_5H_5)$ and for many other similar systems, led to a conjecture about a mechanism for rearrangement reactions which could render all the monohapto bonded cyclopentadienyl protons magnetically equivalent. The variable temperature ^1H n.m.r. spectrum of this molecule (Fig. 5.9) over the range + 30°C to –100°C provides valuable information. Some characteristic features of the spectrum are: (i) the nature of the spectrum over the range –80°C to –100°C is almost unaltered (the limiting spectrum) and provides evidence regarding the instantaneous structure in solution, and (ii) recognition of rearrangement pathway.

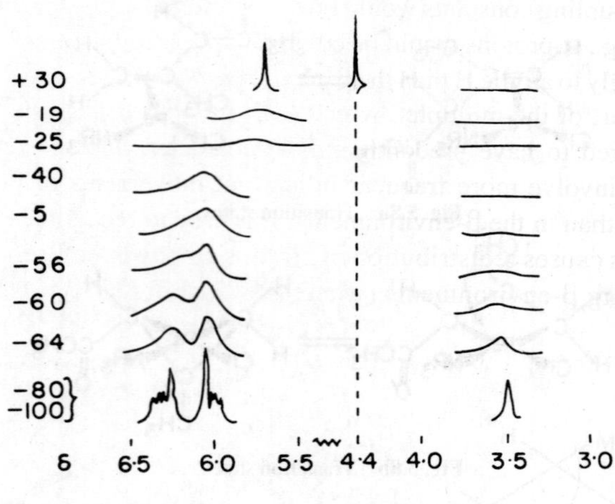

Fig. 5.9 The variable temperature ^1H n.m.r. spectra (at 60 MHz) of $Fe(\eta^1{-}C_5H_5)(CO)_2(\eta^5{-}C_5H_5)$ in CS_2.

Although the plausible pathways for proton averaging may involve (i) a dissociative mechanism, (ii) random shifts, (iii) 1,2-metal shift (i.e. movement of a metal atom around the cyclopentadienyl ring from one carbon to the carbon α to it), or (iv) 1, 3-metal shift, yet the experimental evidence favours the 1, 2-metal shift as the most appropriate mechanism (Fig. 5.10):

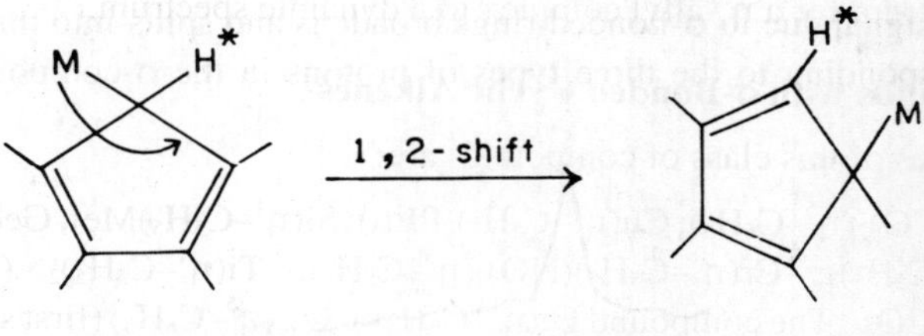

Fig. 5.10 Proton averaging by 1,2-metal shift mechanism.

For the recognition of the rearrangement pathway, generally the following points have been considered to be of great value:

(i) The unsymmetrical collapse of the low field pattern (δ 5.8–6.4) eliminates possibilities like random site exchange or a rearrangement process making all the sites equivalent. This type of observation can in fact lead to only two possibilities, e.g., 1, 2-shifts or 1, 3-shifts (but not a mixture of both). A close examination of the spectrum (Figure 5.9) permits one to distinguish between the latter two pathways provided the two low-field complex multiplets can be assigned;

(ii) Multiplets in the low-field may be described as AA′BB′ resonances, the downfield side of which is broadened more by interaction with allylic H compared to the upfield side; this may be due to α-H atoms. It may be noted that assignment of α- and β-character on the basis of chemical shifts is highly risky. However, magnitudes of coupling constants would be fairly safe to indicate α- or β-character of the protons, e.g., α-protons would be coupled (with a coupling constant value ≈ 2Hz) more strongly to allylic H than the β-proton ($J < 0.5$ Hz). In view of the above, the downfield part of the multiplet, which collapses faster than the upfield half, may be considered to have predominantly α-character, and the rearrangement pathway would involve more frequent in and out movements of protons in the α-environments than in the β-environments. It is easy to recognize that a succession of 1,2-shifts causes redistribution of protons among α-environments, twice as often as among β-environments (whereas 1, 3-shifts have just the opposite effect).

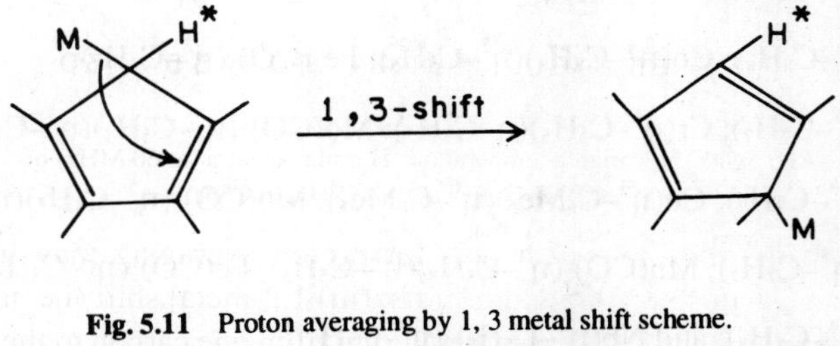

Fig. 5.11 Proton averaging by 1, 3 metal shift scheme.

A second example of study where [1]H n.m.r. has been of assistance is that of metal complex, $(C_2H_5)_3PCuC_5H_5$. From the room temperature down to ~ –21°C, proton spectrum of this compound depicts a singlet only. As the temperature is lowered, the signal due to σ-bonded rings broadens and splits into three separate signals corresponding to the three types of protons in the σ-compound (Figure 5.12).

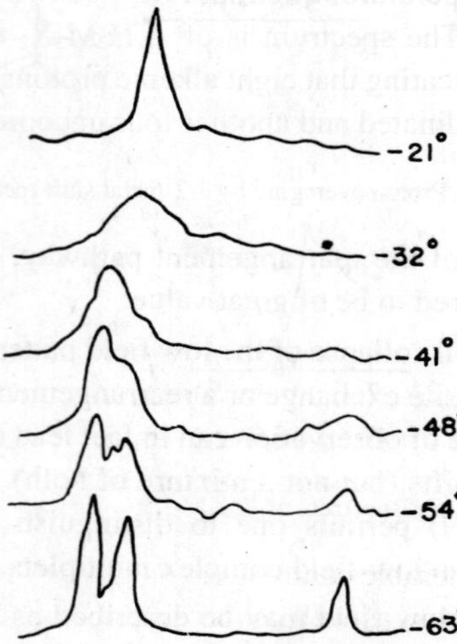

Fig. 5.12 The ^{1}H n.m.r. spectrum at several temperatures of $Cu(\eta^1{-}C_5H_5)P(C_2H_5)_3$.

It is worthwhile to mention that the spectral behaviour of the molecule $Cu(\eta^1{-}C_5H_5)P(C_2H_5)_3$ at –63° is consistent with structure described in section 5.1. However, at intermediate temperatures the spectral behaviour is quite different from that of $(\eta^5{-}C_5H_5)Fe(CO)_2(\eta^1{-}C_5H_5)$, e.g., the upfield portion of the AA′BB′ multiplet collapses more rapidly, thus providing evidence for 1, 3-shifts. An alternative explanation involving 1, 2-shifts also seems to be plausible.

5.4.3 Compounds with π-Bonded Cyclic Alkenes

A few compounds* of cyclic alkenes, in which free alkene ligands has more π-electrons available than the number required to satisfy the 18 Electron or Effective Atomic Number Rule, are listed below:

$Fe(CO)_3(\eta^4{-}C_8H_8)$; $Mo(CO)_3(\eta^6{-}C_8H_8)$; $Cr(CO)_3(\eta^6{-}C_8H_8)$;

$W(CO)_3(\eta^6{-}C_8H_8)$; $Co(\eta^4{-}C_8H_8)(\eta^5{-}C_5H_5)$; $Fe_2(CO)_5(\eta^3{-}C_8H_8)$;

$Fe(CO)_3(\eta^5{-}C_7H_7)$; $Cr(\eta^5{-}C_5H_5)(\eta{-}C_7H_7)$; $Mo(CO)_2(\eta^5{-}C_5H_5)(\eta^3{-}C_7H_7)$;

$Co(CO)_3(\eta^3{-}C_7H_7)$; $Cu(\eta^4{-}C_6Me_6)(\eta^6{-}C_6Me_6)$; $Mo(CO)_2(\eta^5{-}C_5H_5)(\eta^3{-}C_7H_7)$;

$Mn(CO)_3(\eta^5{-}C_7H_7)$; $Mn(CO)_2(\eta^2{-}C_8H_8)(\eta^5{-}C_5H_5)$; $Fe(CO)_2(\eta^2{-}C_8H_8)(\eta^5{-}C_5H_5)$;

$Ru(CO)_3(\eta^4{-}C_8H_8)$; and $Nb(\eta^5{-}C_5H_5)_2(\eta^3{-}C_7H_9)$.

---

* There is potentially far greater variety in this class of compounds but due to the π-bonding of the ligands, their derivatives are of course known mainly with transition metals.

It has been shown by n.m.r. spectroscopy that in solution these molecules generally undergo rapid intramolecular rearrangement which is recognized by fewer lines in the spectrum than one would expect for the static molecule.

The ^{1}H n.m.r. spectrum (Figure 5.13) of $Fe(CO_3)(\eta^4-C_8H_8)$ consists of only a singlet (δ 5.2) at room temperature which splits to give two broad signals (as shown in Fig. 5.13) at –145°C. The spectrum is of $A_2H_2M_2X_2$ type with two peaks of almost equal intensity indicating that eight alkenic protons can be divided into two equal groups of four coordinated and another four uncoordinated ones:

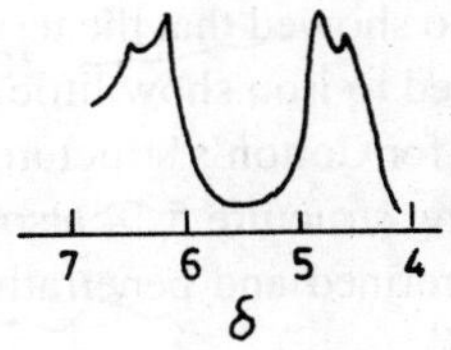

Fig. 5.13 The ^{1}H n.m.r. spectrum of $Fe(CO)_3(\eta^4-C_8H_8)$ at –145°C (60 MHz).

There has been considerable controversy regarding the assignments of the two signals in the above spectrum as well as on the actual instantaneous structure and the rearrangement pathway.

An instantaneous structure (Figure 5.14) at low temperatures as suggested by Kreiter and coworkers (1966), may be represented as below:

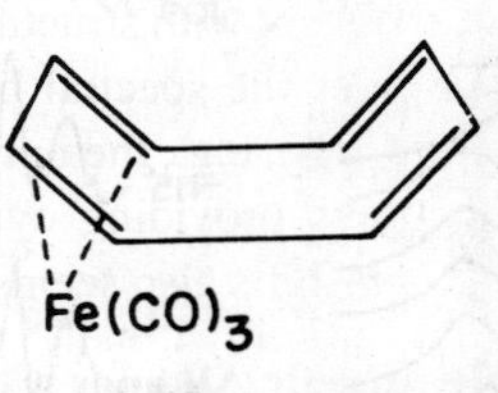

Fig. 5.14 Suggested structure of $Fe(CO)_3(\eta^4-C_8H_8)$.

This structure has been supported by X-ray structural determination confirming that it is of the (1, 3-diene) metal type. Keller and coworkers (1966) have, however, argued in favour of a dynamic exchange (Fig. 5.15):

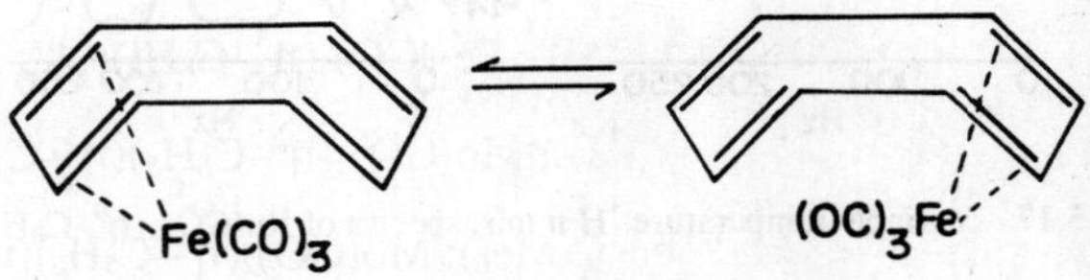

Fig. 5.15 Dynamic exchange in $Fe(CO)_3(\eta^4-C_8H_8)$.

Cotton and coworkers (1966), on the other hand, have suggested a static tub-bonded C_8H_8 structure (Figure 5.16); in which coordinated alkenic protons give rise to the high field signal.

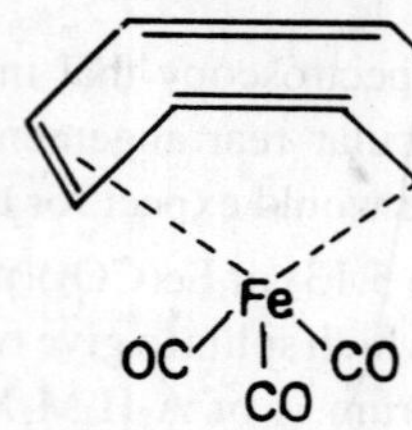

Fig. 5.16 Cotton's suggestion for structure of $Fe(CO)_3(\eta^4-C_8H_8)$.

Cotton's suggestions have received support from a study of the n.m.r. spectrum of methyl-d-cyclooctatetraene—2, 3, 4, 5, 6, 7-d_6-iron tricarbonyl by Anet and coworkers (1967), who showed that the terminal and internal protons of the 1, 3-diene moiety coordinated to iron show little difference in chemical shift. In spite of evidence cited above for Cotton's structure (Fig. 5.16), some more data have been accumulated favouring structure 5.14 also. The story, therefore, is still incomplete and requires more refined and penetrating measurements as well as interpretations.

With the intuition that the behaviour of ruthenium complex would be similar to that of its iron analogue and the rate of rearrangement may be slower, Cotton and coworkers (1969) prepared and studied the complex $Ru(CO)_3(\eta^4-C_8H_8)$. The observed spectra are demonstrated in Figs. 5.17 and 5.18:

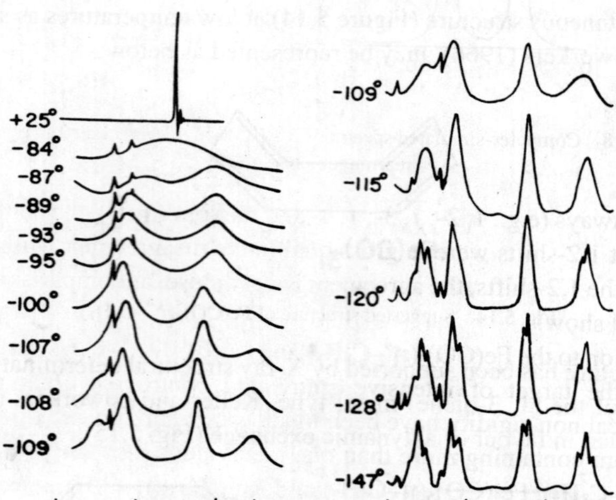

Fig. 5.17 Variable temperature ^1H n.m.r. spectra of $Ru(CO)_3(\eta^4-C_8H_8)$.

The spectra of $Ru(CO)_3(\eta^4-C_8H_8)$ at the lowest temperatures are essentially limiting spectra (even spin-spin hyperfine structure may be seen), and demonstrate fully without anomalies or ambiguities that the instantaneous structure is of the 1,3-diene metal type (Figure 5.14). The technique of computer simulation of spectra was employed to ascertain the rearrangement pathway. Out of various

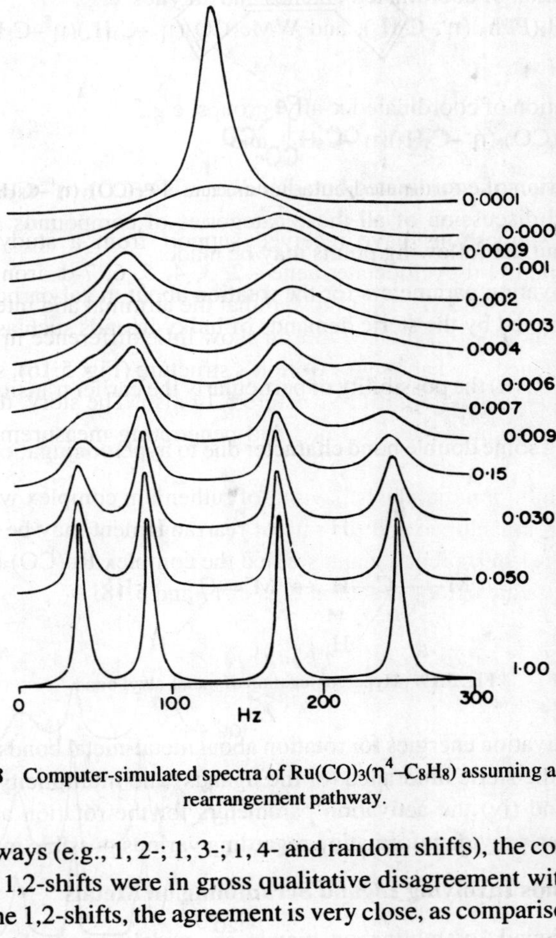

Fig. 5.18 Computer-simulated spectra of $Ru(CO)_3(\eta^4–C_8H_8)$ assuming a 1, 2-shift rearrangement pathway.

possible pathways (e.g., 1, 2-; 1, 3-; 1, 4- and random shifts), the computed spectra for all except 1,2-shifts were in gross qualitative disagreement with the observed spectra. For the 1,2-shifts, the agreement is very close, as comparison of Figs. 5.17 and 5.18 will show.

In addition to the $Fe(CO)_3(\eta^4–C_8H_8)$ and $Ru(CO)_3(\eta^4–C_8H_8)$ molecules which have been the target of intensive study and controversy, n.m.r. studies for stereochemical non-rigidity have been made on several complex products of iron and ruthenium containing more than one metal atom such as $Fe_2(CO)_6(\eta–C_8H_8)$, $Ru_2(CO)_6(\eta–C_8H_8)$, $Fe_2(CO)_5(\eta–C_8H_8)$ and $Ru_2(CO)_5(\eta–C_8H_8)$, etc.

5.4.4 Rotation of Ligands on Metals

The compounds in which fluxionality arises due to rotation of ligands on metals can be grouped in various categories such as:

(a) rotation about metal-carbon σ-bonds, e.g., $ZrMe\{CH(SiMe_3)_2\}(\eta–C_5H_5)_2$,

(b) rotation about metal-carbene bonds, e.g., $TaCl(CHBu^t)(\eta^5–C_5H_5)_2$,

(c) rotation about metal-metal bonds, e.g., $Mo(CO)_3(\eta^5–C_5H_5)_2$,

(d) rotation of coordinated alkenes and alkynes, e.g., $PtCl_2(PPh_3)(\eta^2\text{–}C_2H_4)$, and $WMe(CO)(\eta^2\text{–}C_2H_4)(\eta^5\text{–}C_5H_5)$ respectively,

(e) rotation of coordinated π-allyl groups, e.g., $Mo(CO)_2(\eta^3\text{–}C_5H_5)(\eta^5\text{–}C_5H_5)$, and

(f) rotation of coordinated butadienes, e.g., $Fe(CO)_3(\eta^4\text{–}C_8H_8)$, etc.

A detailed discussion of all these categories of compounds at this stage is unnecessary, but the following points may be made:

(i) the activation parameters for the rotation about metal-carbon σ-bonds are likely to be affected by the steric demands of bulky ligands such as $\text{–}CH(SiMe_3)_2$ and 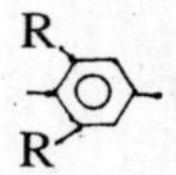 R; (ii) the possibility of particularly the earlier transition metal alkyl bonds assuming some double bond character due to hyperconjugation as illustrated below:

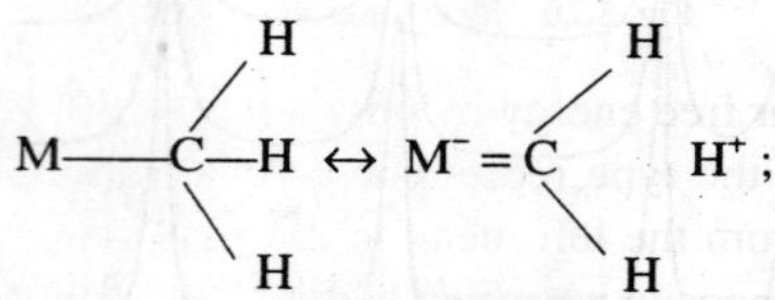

Fig. 5.19 Hyperconiugation of metal-alkyl bond.

(iii) the activation energies for rotation about metal-metal bonds can generally by dictated by the steric hindrance (of the ligands) and multiplicity of the metal-metal bonds, and (iv) the activation parameters for the rotation about metal-π-alkene bonds can provide information regarding various possible mechanisms.

5.4.5 Compounds Involving Ligand Scrambling on Metals

Examples of ligand scrambling on metals are widely known in coordination chemistry for various coordination numbers and stereochemistries and rearrangement processes generally involve a cis ↔ trans isomerism in square-planar and octahedral complexes. For intramolecular rearrangement mechanisms (cf. Section 5.3) such as: tetrahedral ↔ square-planar interconversion of cis ↔ trans isomerism (for four-coordinate complexes) in square planar derivatives, and trigonal bipyramidal ↔ square-pyramidal interconversion (for five-coordinate compounds), are believed to be more probable by the Berry pseudorotation mechanism. There have been insufficient studies for seven-coordinate compounds to lead to any conclusive mechanisms.

For organometallic compounds, mechanistic evidence is weak and it is generally assumed that the compounds are fluxional via the mechanisms found for coordination compounds. This assumption is probably true for compounds involving monodentate ligands, but in the case of π-ligands, other mechanisms (such as rotation of ligands or metals) are also possible.

5.4.6 Metal Carbonyls Involving Bridged-Terminal Exchanges and Scrambling of CO Groups

The fluxional behaviour of $Fe(CO)_5$ has already been discussed (cf. Section 5.3.2) as one of the first observations of Cotton (1958) leading to a clear realisation of the phenomenon itself. It is now generally well known that di- and poly-nuclear metal carbonyls have a pronounced tendency of exhibiting fluxional interchange which has been more specifically termed as 'Carbonyl Scrambling'.

For example, it has been shown in binuclear carbonyl, e.g., $Co_2(CO)_8$, that there are a number of structures of the following types:

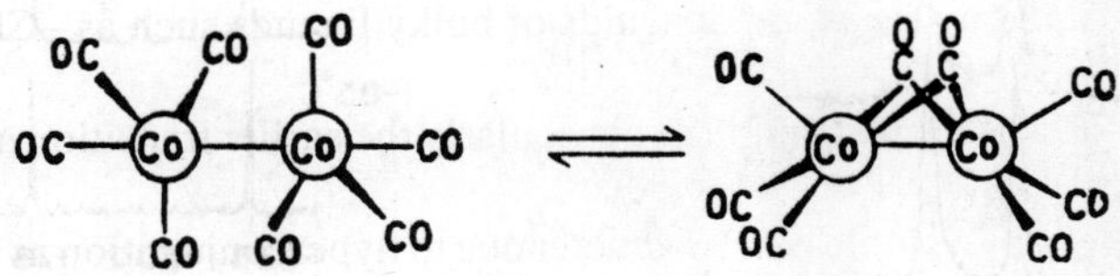

Fig. 5.20 Structure of $Co_2(CO)_8$.

These differ in their free energy by only a few (< 30) kJ mol^{-1}. In such cases, a facile interchange of the type represented in the following diagram occurs in which CO(A) passes from the left metal to the right one, while CO(B) is simultaneously making the opposite journey:

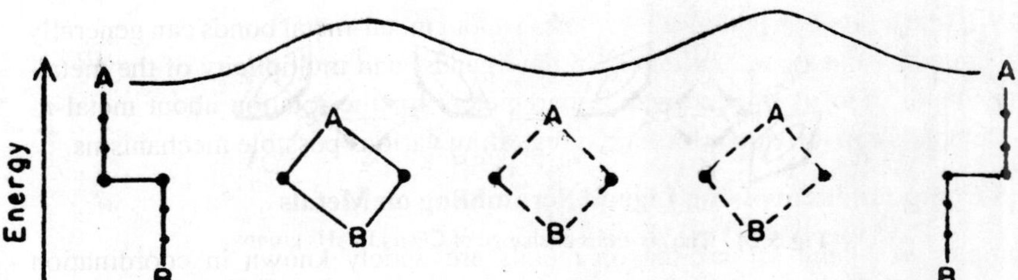

Fig. 5.21 Interchange of CO ligands in metal carbonyls.

In this diagram, the CO bridged structure is slightly less stable than the terminal one, which is the case for $Co_2(CO)_8$. However, the opposite position of bridged ones being slightly more stable than the terminal ones are also known.

Another interesting example of fluxional behaviour was reported (1970) in $(\eta^5\text{–}C_5H_5)_2Fe_2(CO)_4$ molecule, the ^{1}H n.m.r. spectrum of which in solution showed at –70°C, the proton resonances at two positions (Fig. 5.22), but these gradually coalesced when the temperature is raised, giving only one sharp signal at +28°C.

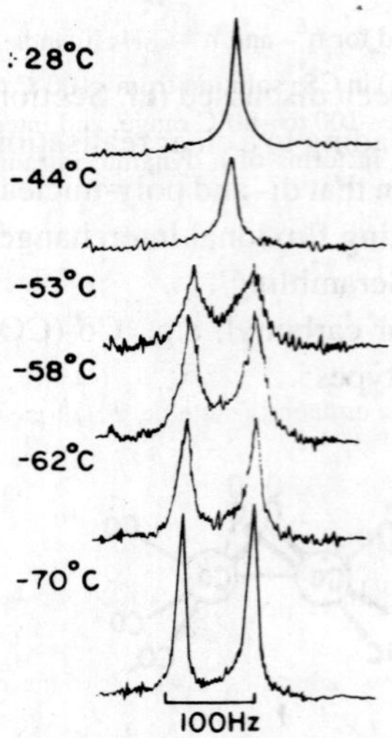

Fig. 5.22 The variable temperature [1]H n.m.r. spectra of the cis- and trans-$[\eta^5\text{–}C_5H_5Fe(CO)_2]_2$.

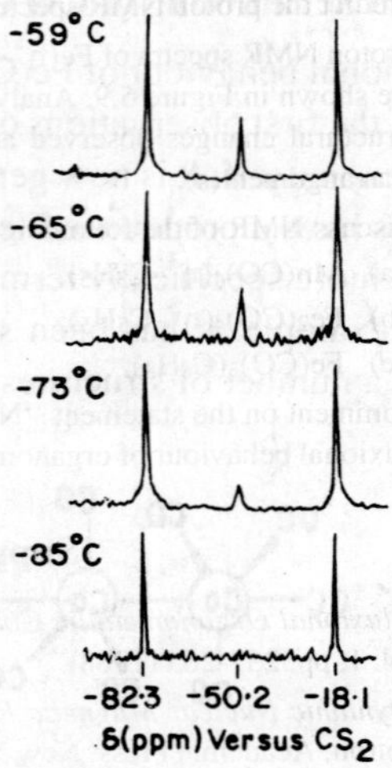

Fig. 5.23 The variable temperature [13]C n.m.r. spectra of the carbonyl groups of $Fe_2(CO)_4(\eta^5\text{–}C_5H_5)_2$.

The [13]C n.m.r. spectrum (Fig. 5.23) of $Fe_2(CO)_4(\eta^5\text{–}C_5H_5)_2$ shows (Fig. 5.24) that the following *cis-trans* exchange is accompanied by the interchange of the CO groups between bridging and terminal positions:

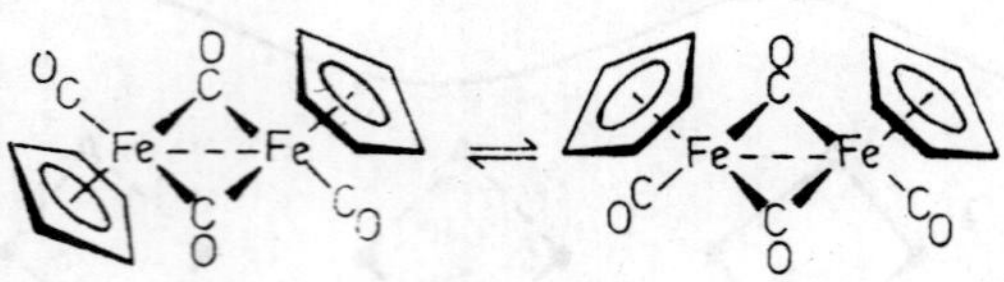

Fig. 5.24 The *cis-trans* exchange of CO and C_5H_5 groups.

It appears that the CO bridges open in a concerted way to give a non-bridged $C_5H_5(CO)_2Fe\text{–}Fe(CO)_2C_5H_5$ intermediate in which rotation about the Fe–Fe bond takes place.

5.5 FUTURE DEVELOPMENTS

The field of organometallic chemistry has provided the largest number of molecules with interesting fluxional behaviour, which has been the main *raison de atre* for including a brief account of only a few of these in a special chapter. Besides the novelty of the concept, the subject could be expected to grow fastly in future years as facilities for solid state and relaxation measurements become more commonly available with sophisticated commercial instruments, which should enable chemists to study even faster processes such as cyclopentadienyl group rotation.

Study Questions

1. Predict the proton NMR spectral resonances expected for η^5– and η^1–C_5H_5 ligands.
2. Proton NMR spectra of $Fe(\eta^1–C_5H_5)(CO)_2(\eta^5–C_5H_5)$ in CS_2 solution from –100°C to +30°C are shown in Figure 5.9. Analyze the spectrum in the –100 to –80°C range, and interpret the structural changes observed at higher temperatures in terms of a dynamic intramolecular rearrangement(s).
3. Discuss NMR of the following fluxional molecules:
   (a) $Mn(CO)_4(\eta^3–C_3H_5)$
   (b) $Fe_2(CO)_4(\eta^5–C_5H_5)_2$
   (c) $Fe(CO)_3(C_8H_8)$.
4. Comment on the statement "NMR spectroscopy is an eminently suitable technique to study fluxional behaviour of organometallic compounds".

Supplementary Reading

1. '*Fluxional Organometallic Compounds*', by F.A. Cotton, *Accounts of Chemical Research*, vol. 1, pp. 257-265 (1968).
2. '*Dynamic Nuclear Magnetic Resonance Spectroscopy*', edited by L.M. Jackman and F.A. Cotton, Academic Press, New York (1975).
3. '*Fluxionality of Organometallic Compounds and Metal Carbonyls*, by F.A. Cotton, *Journal of Organometallic Chemistry*, vol. 100, pp. 29-41 (1975).
4. '*Fluxional and Non-Rigid Behaviour of Transition Metal Organometallic π-Complexes*' by J.W. Faller, in *Advances in Organometallic Chemistry*, edited by F.G.A. Stone and R. West, vol. 16, pp. 221-235, Academic Press, London (1977).
5. '*Non-Rigidity in Organometallic Compounds*', by M.E. Mann, in *Comprehensive Organometallic Chemistry*, edited by G. Wilkinson, F.G.A. Stone, and E.W. Abel, vol. 3, pp. 89-171, Pergamon Press, Oxford (1982).

6

Synthetic and Catalytic Aspects of Organometallic Chemistry

6.1 GENERAL INTRODUCTION

In addition to variety of synthetic applications of some typical main group organometallic reagents (cf. Section 3.5), transition metals and their organometallic derivatives are also finding extensive applications in a large number of industrially and biologically important processes. For example, the well-known Haber process for synthesis of ammonia involves an iron/iron oxide catalyst; the oxidation of ammonia (a process used in the manufacture of nitric acid) is catalyzed by a Pt/Rh catalyst, and the hydrogenation of vegetable oils to solid fats is based on Raney (finely divided) nickel as catalyst. In fact in one guise or another, transition metals are serving as the main targets for catalyst design in the refining of petroleum. Many biologically important reactions also take place with the aid of enzymes and in some of these organometallic species play important roles. For example, a cobalt atom lies at the core of the Vitamin B_{12} coenzymes, which have been shown to play a predominant role in methylation reactions in the environment (Section 7.5.2). Iron atoms are prominently involved in the haemoglobin of blood and in the ferrodoxins which take a significant part in the photosynthetic process. Both molybdenum and iron are essential constituents of nitrogen fixing enzymes in nature. In addition to transition metals, a few main group elements (e.g., Zn, Mg, Ca, K, B, Si and P) play important roles in biological processes and are generally included amongst the 30 elements, now recognized as essential to life (cf. Fig. 7.11). One of the best known examples in this direction can be cited as the role of chlorophyl (a magnesium complex) as catalyst in the conversion of carbon dioxide and water into carbohydrates.

Although the role of organometallic catalysts in the biologically important enzymatic reactions is being recognized more and more, our main endeavour in this chapter would be to concentrate around rather simpler synthetic catalytic procedures which are known or have been postulated to proceed via metal-carbon bond-making and/or breaking pathways.

The simplest definition of a catalyst is a species which facilitates a particular reaction without being consumed in the process; it may be regenerated in its original form but sometimes, in a lower or higher oxidation state also. There are various theories of catalysis, but the most common one involves the formation of intermediate compounds (often transient or/and non-isolable); the ability of transition metals to form labile complexes with molecules such as hydrogen, alkenes, allyls, arenes, and carbon monoxide has led to a flurry of activity in designing various synthetic (catalytic and stoichiometric) processes in which the formation and scission of metal carbon bonds play an important role. It is such synthetic and catalytic processes that we are going to discuss in detail in this chapter.

Another commonly accepted view of the role of a catalyst is that it lowers the activation energy barrier for a thermodynamically allowed (but kinetically slow) reaction resulting in an enhancement of the rate of the process. Further, in cases where alternative reaction paths are feasible, a catalyst may enhance product-specificity by lowering the activation barrier for one path or by raising it for another. Reactions at a catalytic centre or site are believed to involve the intermediate formation of weak catalyst-reactant bonds which have the effect of loosening the bonds in the reactants prior to reaction. Many catalytic reactions taking place in homogeneous (both catalyst and reactants are in one phase) solution have been interpreted in terms of the formation of an activated (or intermediate) complex as depicted in Figure 6.1.

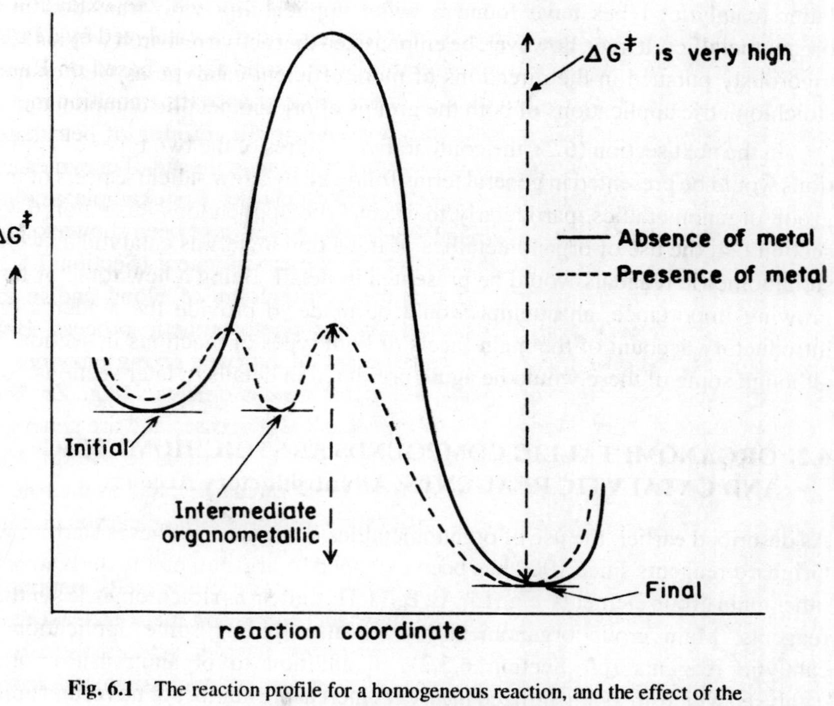

Fig. 6.1 The reaction profile for a homogeneous reaction, and the effect of the presence of a catalyst.

In heterogeneous catalysis, the above steps are generally achieved by the chemisorption of the reactants on the active sites of a metal, facilitating the reaction thereby either by suitable orientation of the molecules or by change in their reactivity brought about by weak bondings effective in chemisorption or in limiting cases even in physical adsorption.

Historically, a large number of processes have been based on heterogeneous catalytic systems. However, of late homogeneous organometallic catalysts have been assuming much greater importance. The main preferential features of the latter are mild (often ambient) conditions required, higher selectivity and controllability as well as faster rates of reactions, whereas the ease of 'product isolation' (Section 6.4.1.3) remains the more attractive feature of the heterogeneous processes. Another factor helpful in a faster growth of homogeneous catalysts is the ease of kinetic and mechanistic studies of such systems (including identification or/and isolation of proposed intermediate product(s) by physico-chemical techniques such as spectroscopy) leading to a better understanding of the fundamentals and consequently improvements in the process itself. Efforts are now been successfully made to combine in some cases the advantages of both the types of systems by the development of polymer supported organometallic catalysts.

Synthetic applications of organometallic compounds can be broadly divided into two main groups, i.e., stoichiometric and catalytic. The former are more important (as discussed in chapter 3) in the main group chemistry, whereas the latter (catalytic) types have found a wider applicability with transition metal organometallics. It may, however, be emphasized that active research work is being vigorously pursued in the directions of more efficient catalytic as well as novel stoichiometric applications of both the groups of organometallic compounds.

In the next section (6.2), the comparative features of the two types of applications would be presented in general terms followed by a few salient aspects of main group organometallics, particularly their catalytic applications in section 6.3. In section 6.4, the use of organometallics of transition metals as catalytic as well as stoichiometric reagents would be presented in detail. Being a new topic of fastly growing importance, an attempt would be made to provide the reader a brief introductory account of the main facets of both types of reactions in section 6.2, although some of these would be again presented in details in later sections.

6.2 ORGANOMETALLIC COMPOUNDS AS STOICHIOMETRIC AND CATALYTIC REAGENTS: An Introductory Account

As described earlier, the use of organometallics in organic syntheses starting with Grignard reagents since 1900 has been extended to organometallic derivatives of other main group elements. e.g., Li, Al, B, Tl, Hg and Sn as stoichiometric synthetic reagents. Main group organometallics are also finding some applications as catalytic reagents (cf. Section 6.3.2), in addition to organotransition metal catalysts, which are being utilized in an ever increasing number of novel and unique synthesis e.g.,

(a) *Hydrogenation of alkenes*

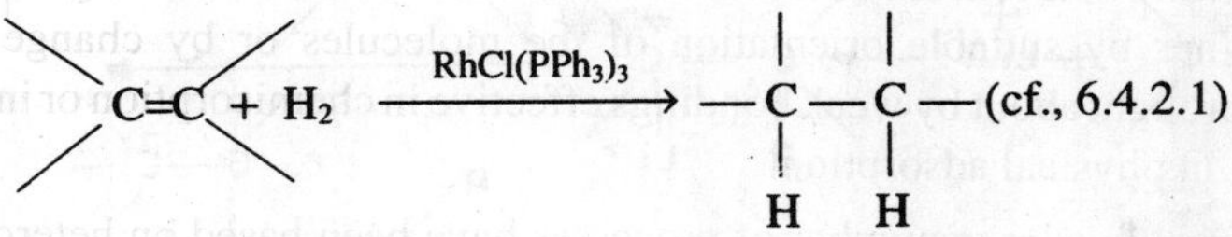

(b) *Hydroformylation of alkenes*

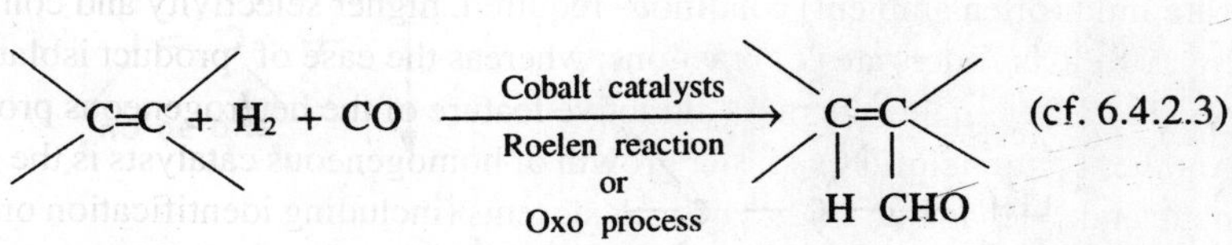

(c) *Oxidation of alkenes*

$$H_2C = CH_2 + 1/2\ O_2 \xrightarrow[\text{Wacker process}]{PdCl_2/CuCl_2} CH_3CHO \qquad (6.4.2.4)$$

(d) *Stereoselective polymerization of alkenes*

$$RCH = CH_2 \xrightarrow[\text{Zeigler–Natta catalysis}]{TiCl_4/Al_2Et_6} \text{``Polythene''} \qquad (\text{cf. } 6.4.2.6)$$

(e) *Polymerization of alkynes*

$$HC \equiv CH \xrightarrow[\text{Reppe synthesis}]{\text{Ni (II) Complexes}} \text{Cyclooctatetraene or Benzene}$$

(f) *Reduction of carbon monoxide by hydrogen*

$$CO + H_2 \xrightarrow[\text{Fischer–Tropsch synthesis}]{\text{Fe, Co, Ni or Ru}} \text{Hydrocarbons or Alcohols}$$

(cf. 6.4.2.8(ii))

In addition to the novel roles of organotransition metal complexes as catalysts, further advances are being continuously made in the applications of main group organometallics especially as 'stoichiometric reagents', but sometimes also as catalytic reagents. The main distinguishing feature between catalytic and stoichiometric reagents used as initiators in synthetic procedures, is that the latter are generally not regenerated as such, but are converted into other species. For example, in the polymerization of alkenes, lithium alkyls initiate the reaction by the following type of mechanism:

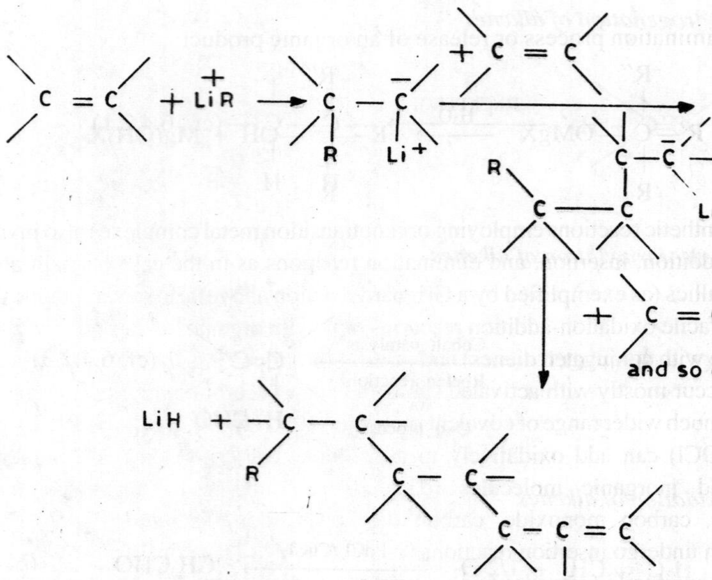

As represented above, the original initiator lithium alkyl is not recovered as such, but is generally converted into lithium hydride at the end of the process. In addition, in some synthetic procedures based on the use of organometallics (e.g., Grignard Reagents), the reagent is required in stoichiometric proportions as it is converted in the process into another derivative which is no longer active. These may be termed as stoichiometric reagents and as stated above, organometallic reagents of main group elements (e.g., Li, Mg, Zn, B, Al, Tl and Sn) function more generally in this manner.

With the above fast developments in the use of organometallic compounds both as stoichiometric reagents as well as catalysts in synthesis, a brief discussion on their comparative mode of functioning would be desirable.

The role of stoichiometric reagents can generally be exemplified by the sequence of the following three typical steps (discussed in detail in section 6.4.1):

(i) interaction (oxidative addition) of the organic substrate (e.g., alkyl halide in the Grignard reaction) with a metal or its complex:

$$Mg + RX \rightarrow R\text{–}Mg - X$$

(ii) reaction of (or on) the coordinated ligand, which, in the case of Grignard reaction, may be the insertion of an unsaturated (C == O, C == N, etc.) moiety into the magnesium-carbon bond:

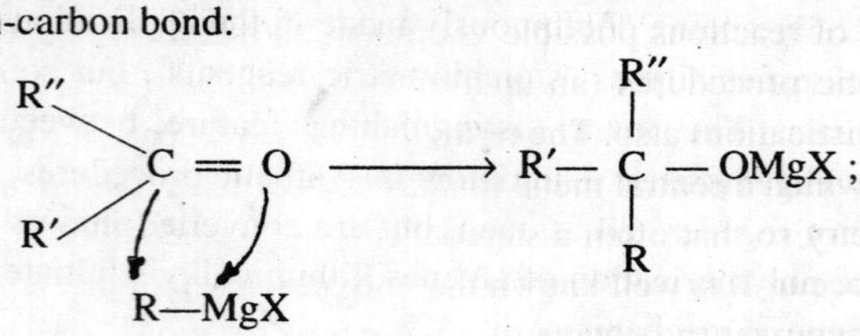

(iii) elimination process or release of an organic product:

$$\mathrm{R'{-}\underset{R}{\overset{R''}{C}}{-}OMgX} \xrightarrow{H_3O^+} \mathrm{R'{-}\underset{R}{\overset{R''}{C}}{-}OH + Mg(OH)X}$$

The synthetic reactions employing organotransition metal complexes also involve oxidative addition, insertion, and elimination reactions as in the case of main group organometallics (as exemplified by a Grignard reaction above). However, magnesium undergoes facile oxidation-addition reactions only with organic halides (and in a very special case with conjugated dienes) and insertion reactions into the magnesium-carbon bonds occur mostly with activated substrates such as carbonyl or nitrile group. By contrast, a much wider range of covalent molecules (e.g., H_2, Cl_2, $HgCl_2$, SnR_3Cl, CH_3I, and CH_3COCl) can add oxidatively to transition metal complexes and numerous organic and inorganic molecules (e.g., alkenes, alkynes, conjugated dienes, isocyanides, carbon monoxide, carbon dioxide, carbon disulphide, and sulphur dioxide) can undergo insertion reactions (cf. Section 6.4) in their cases.

The above facile (complexation, insertion, and elimination) reactions of transition metal complexes can be understood, on the basis of their (i) electronic configurations by virtue of which transition metal atoms act as sinks for, or sources of, electrons thus favouring synergic bonding which facilitates their ability to hold the reacting species in conformations favourable for reaction, and (ii) wider flexibility in oxidation states, coordination number, and stereochemistries. These parameters are obviously responsible for the ability of transition metals to form complexes and to undergo oxidation-reduction (sometimes reversible) reactions. A critical analysis of a variety of synthetic reactions involving transition metal derivatives leads to the following general reaction patterns for such reactions:

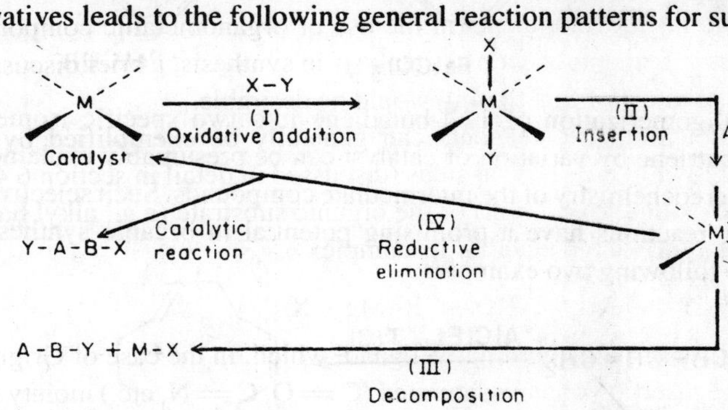

The variety of reactions possible via transition metals are being exploited in template synthetic procedures (an area of synthetic chemistry which is rapidly growing in sophistication) also. The term 'template synthesis' has been suggested for reactions in which a central metal ion has been utilised to hold reactive groups in proper geometry so that often a stereochemically selective reaction (generally multistep) may occur. It is well known that the presence of a metal ion can exert a profound influence over the course of a variety of reactions, and can selectively promote the formation of otherwise inaccessible products. The stabilization of reactive intermediates by complexation with metal ions (or complexes) is also sometimes known as a thermodynamic template effect.

Such reactions are more common in catalytic processes (discussed later in Section 6.4) aided by transition metal complexes: two typical examples are illustrated as follows:

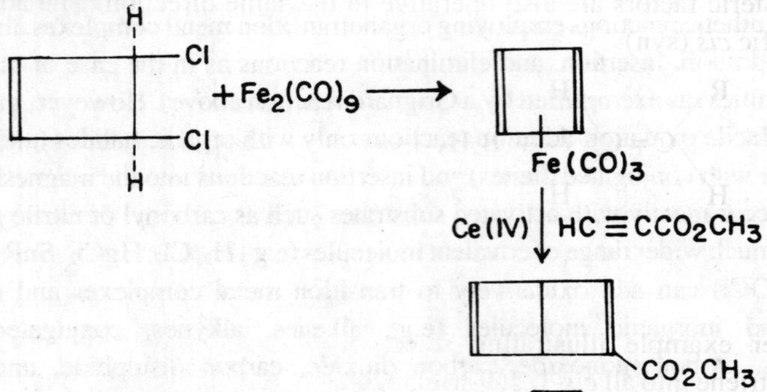

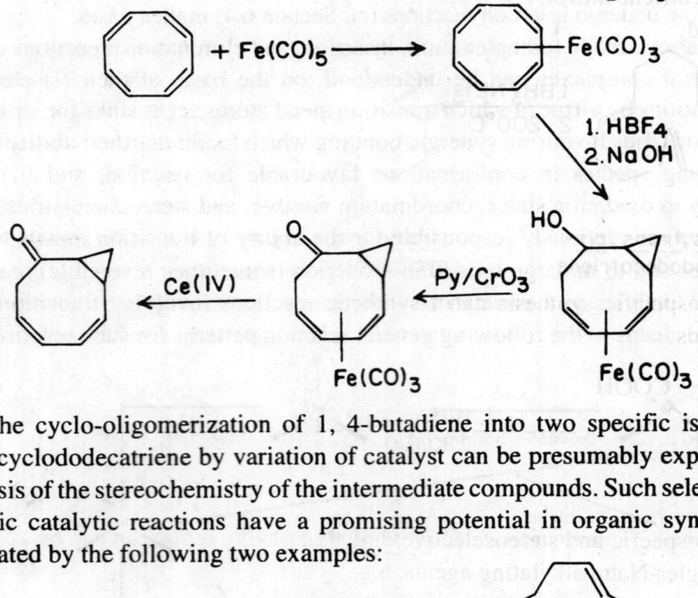

The cyclo-oligomerization of 1, 4-butadiene into two specific isomers of 1,5,3-cyclododecatriene by variation of catalyst can be presumably explained on the basis of the stereochemistry of the intermediate compounds. Such selective and specific catalytic reactions have a promising potential in organic syntheses as illustrated by the following two examples:

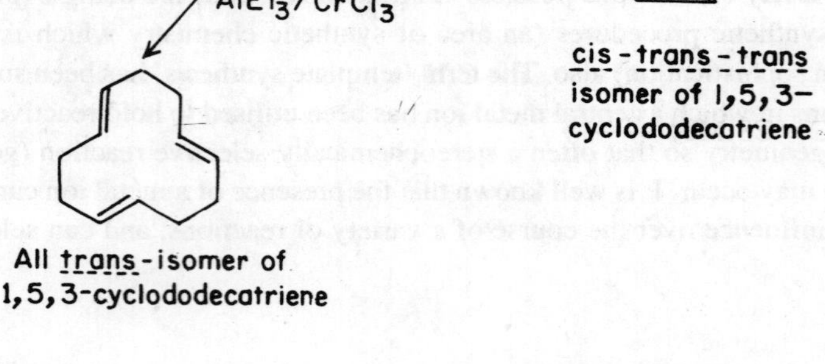

At this stage, it may be emphasized that greater selectivity is also being achieved in stoichiometric syntheses through organometallics of both types. For example, hydroboration (cf. Brown's Nobel Lecture: Appendix 1.3), which has proved to be of phenomenal synthetic utility, is regiospecific, the boron showing preferential attachment to the least substituted carbon atom (anti-Markovnikov rule). This finds ready interpretation in terms of electronic factors and relative bond polarities; steric factors are also operative in the same direction. The addition is stereospecific *cis* (syn):

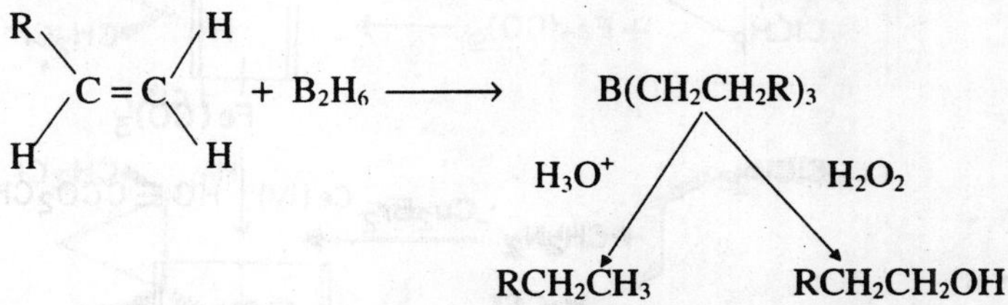

Another example illustrating selectivity is the conversion of 1, 5, 9-cylododecatriene into all *cis*-1, 5, 9-triol mediated by an intermediate organoboron compound:

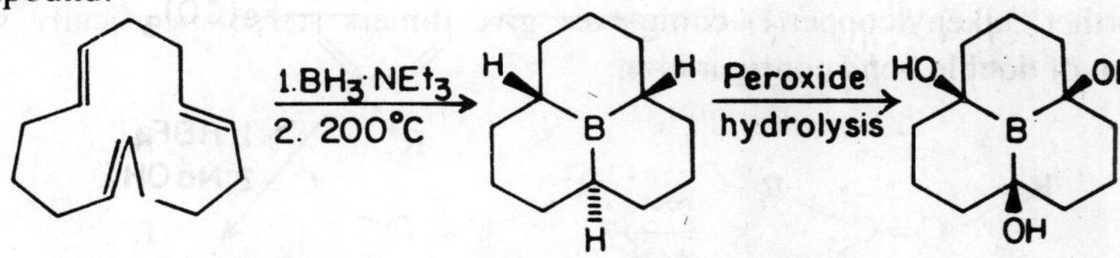

trans,trans,cis-cyclododecatriene — trans,trans,trans cyclododecatriene — cis,cis,cis 1,5,9-triol

Regiospecific synthesis can also be achieved by means of organothallium compounds:

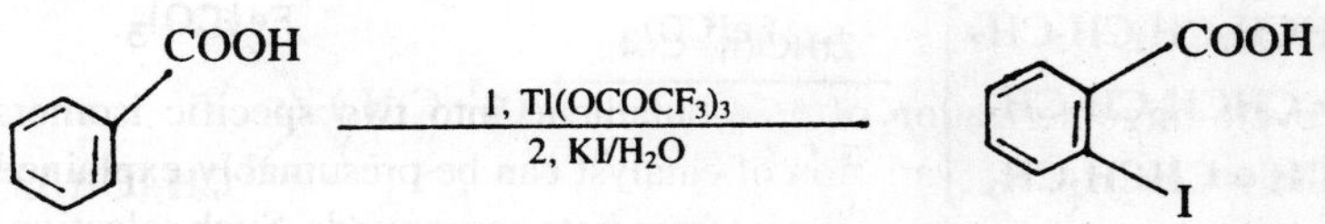

Regiospecfic and stereoselective direct alkylative reduction can be achieved using Ziegler-Natta alkylating agents, e.g.,

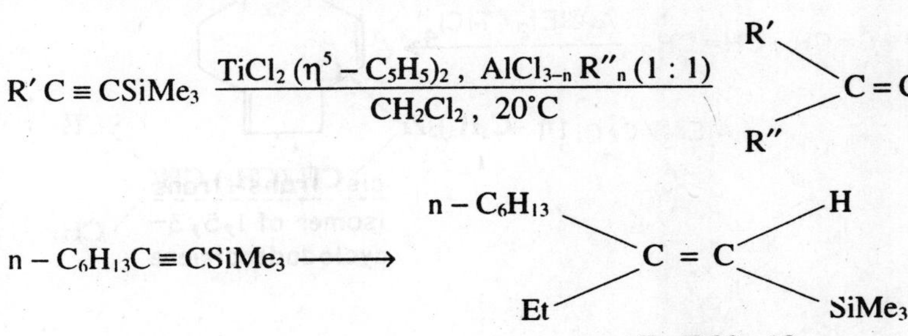

E : Z 90 : 10

Transition metal complexes have also shown great promise for regio- and stereo-specificity as well as selectivity of the reactions. For example, a high degree of stereoselectivity can be readily achieved in the reaction of diazomethane with *cis*- and *trans*-1, 4-dichloro-2-butene in the presence of Cu(I) bromide:

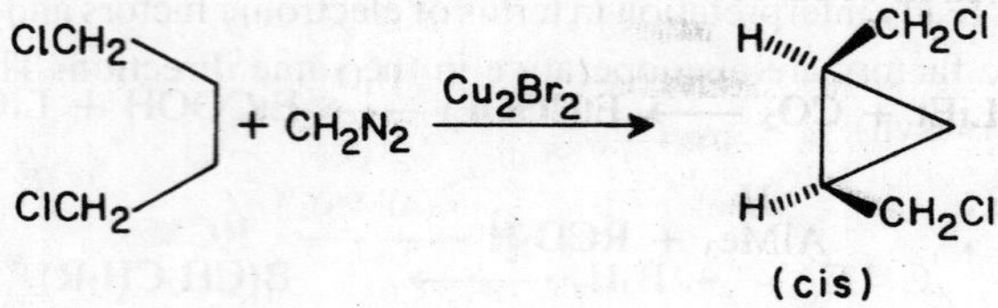

(cis)

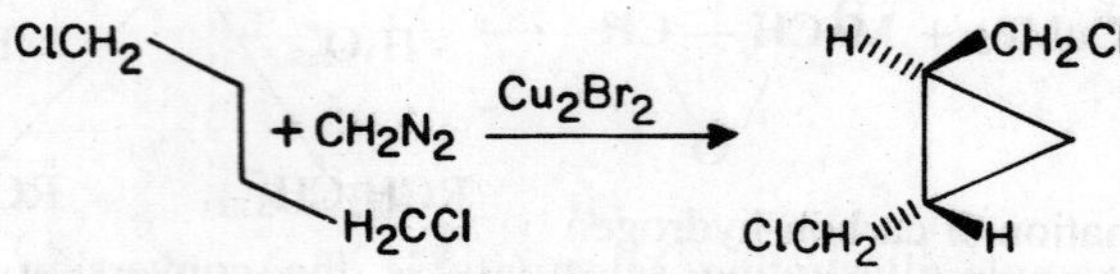

(trans)

Further, alkenylcopper(I) complexes give dimers stereo-specifically with retention of double bond configuration:

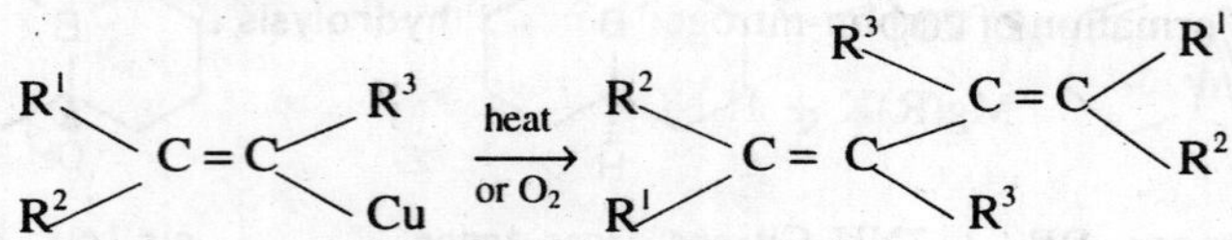

Alkenes react regiospecifically with $ZrHCl(\eta\text{–}C_5H_5)_2$ to form Zr-alkyl bonds with zirconium becoming attached to the least-hindered primary carbon:

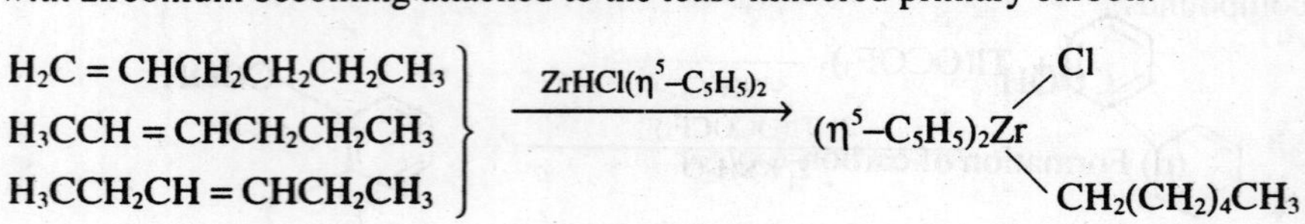

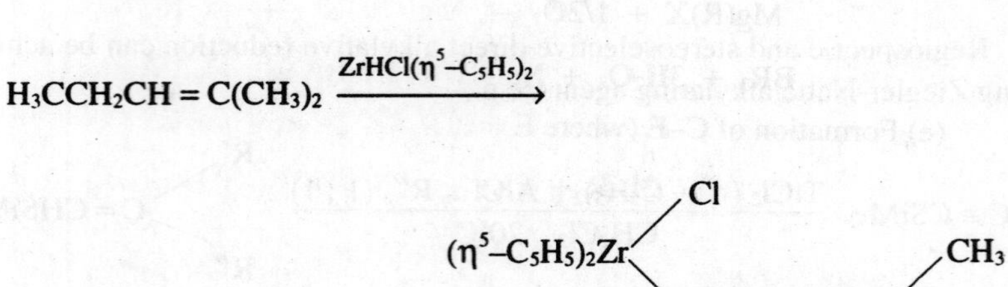

The general characteristics of the use of organometallic compounds in syntheses of both organic and new organometallic derivatives would now be illustrated by their applications in a few types of reactions, which may be classified in general terms as follows:

(a) Formation of carbon-carbon bonds:

$$LiEt + CO_2 \xrightarrow{PhH} EtCOOLi \xrightarrow{H_2O} EtCOOH + LiOH$$

$$AlMe_3 + RCO_2H \xrightarrow{(AlMe_3)_2} RCMe_3$$

$$LiCuMe_2 + \underset{\diagdown\,O\,\diagup}{MeCH - CH_2} \xrightarrow{Et_2O,\ 0^\circ C} \underset{OH}{Me-\overset{|}{C}H-CH_2CH_3}$$

(b) Formation of carbon-hydrogen bonds:

$$LiR + H_2O \rightarrow RH + LiOH$$

$$HgR_2 + HCl \rightarrow RH + HgCl(R)$$

$$B(CH_2CH_2R)_3 + 3CH_3COOH \rightarrow 3RCH_2CH_3 + B(OCOCH_3)_3$$

(c) Formation of carbon-nitrogen bonds:

$$Mg(R)X + H_2NCl \rightarrow RNH_2 + MgCl(X)$$

$$BR_3 + 3NH_2Cl \rightarrow \xrightarrow{hydrolysis} 3RNH_2 + B(OH)_3$$

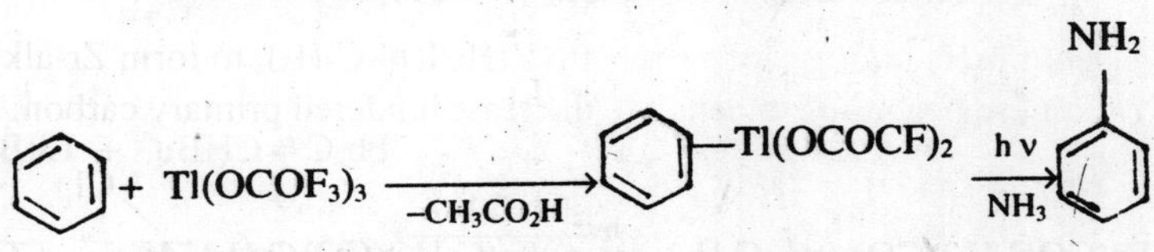

(d) Formation of carbon-oxygen bonds:

$$Mg(R)X + 1/2O_2 \rightarrow Mg(OR)X \xrightarrow{H_3O^+} ROH$$

$$BR_3 + 3H_2O_2 + NaOH \rightarrow NaB(OH)_4 + 3ROH$$

(e) Formation of C–E (where E = P, S, Se, Te etc.) bonds:

$$3LiR + PX_3 \rightarrow PR_3 + 3LiX$$

$$Mg(R)X + SO_2 \rightarrow R-\overset{\overset{O}{\|}}{S}-OMgX \xrightarrow{H_3O^+} R-S(=O)OH$$

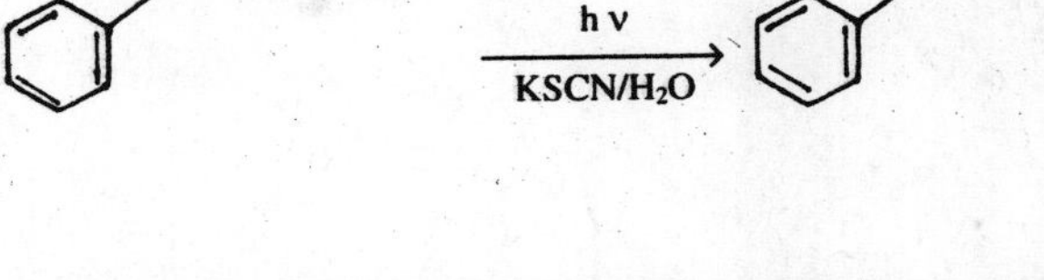

(f) Formation of carbon-halogen bonds:

$$MR + X_2 \rightarrow MX + RX$$
$$(M = \text{alkali metals})$$

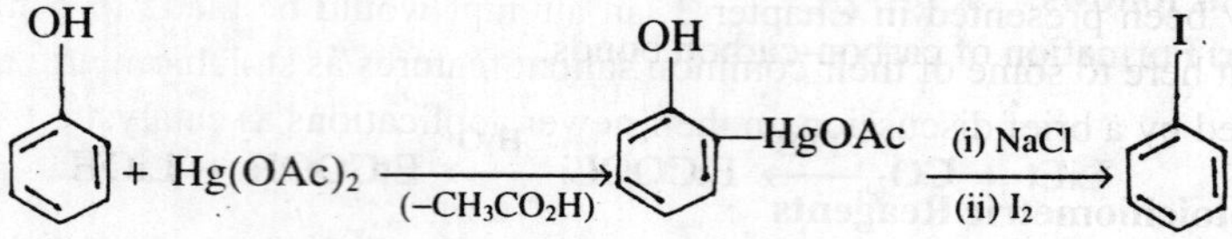

$$ArCOCl + RhCl(PPh_3)_3 \rightarrow ArCl + RhCl(CO)(PPh_3)_2 + PPh_3$$

(g) Formation of new metal (or metalloid)-carbon bonds:

$$BeCl_2 + 2LiPh \xrightarrow{Et_2O} BePh_2 + 2LiCl$$

$$HgMe_2 + Be \xrightarrow{100^\circ C} BeMe_2 + Hg$$

$$2BF_3 + 6Mg(R)X \xrightarrow{Et_2O} 2BR_3 + 3MgF_2 + 3MgX_2$$

$$MnCl_2 + 2dmpe + MgMe_2 \xrightarrow[-78^\circ C]{} MnMe_2(dmpe)_2 + MgCl_2$$

$$WCl_6 + 6LiCH_2CMe_3 \xrightarrow{78^\circ C} (Me_3CCH_2)_3W \equiv CCMe_3 + 6LiCl$$

(h) Elimination, ring opening and rearrangement reactions:

$$Ph_2C = CH_2 \xrightarrow{LIBu^n} Ph_2C(Li) - CH_2Bu^n \xrightarrow{BPh_3} Ph_2C = CHBu^n + Li[BHPh_3]$$

$$Fe(COC_6H_5)(CO)_2(\eta^5-C_5H_5) \xrightarrow{h\nu} Fe(C_6H_5)(CO)_2(\eta^5-C_5H_5) + CO$$

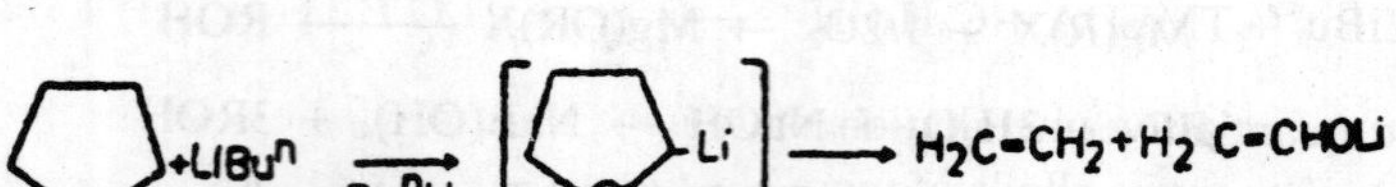

It may be noted that further sub-classification may be made in terms of the substrates employed.

The subject has grown so rapidly in highly diverse directions that it is difficult to present even its main features in a universally acceptable pattern. In this chapter, we are attempting to present a brief account of the use of main group (Section 6.3) followed by transition metal (Section 6.4) organometallics in synthetic processes.

6.3 SYNTHETIC APPLICATIONS OF MAIN GROUP ORGANOMETALLICS

As the synthetic applications of a few typical main group organometallics have already been presented in Chapter 3, an attempt would be made to confine this account here to some of their common salient features as stoichiometric reagents, followed by a brief discussion on their newer applications as catalysts.

6.3.1 Stoichiometric Reagents

The synthetic suitability of an organometallic stoichiometric reagent depends on two key factors, i.e., speed and selectivity of the desired reaction. For example, highly ionic organometallics (which are sources of powerful carbanions or nucleophiles, e.g., organoalkali compounds) are ideal reagents for speedy addition, e.g.:

$$LiCMe_3 + CH_2{=}CH{-}CH{=}CH_2 \rightarrow Me_3CCH_2CH{=}CHCH_2Li \quad (\textit{syn} \text{ and } \textit{anti})$$

$$NaC_6H_5 + LiC_6H_5 \rightarrow Na[Li(C_6H_5)_2]$$

$$LiCH_2CH_3 + Al(CH_2CH_3)_3 \rightarrow Li[Al(CH_2CH_3)_4]$$

$$LiC_6H_5 + Sb(C_6H_5)_5 \rightarrow Li[Sb(C_6H_5)_6]$$

and substitution reactions with suitable substrate, e.g.:

$$LiBu^n + (C_6H_5)_3CH \xrightarrow{\text{hexane}} LiC(C_6H_5)_3 + Bu^nH$$

$$LiBu^n + C_6H_5CH_3 \xrightarrow{\text{hexane}} \text{No substantial metallation}$$

The last reaction in the presence of TMEDA undergoes metallation in quantitative yields (note TMEDA enhances the carbanionic character of $LiBu^n$):

$$LiBu^n + TMEDA + C_6H_5CH_3 \xrightarrow{\text{hexane}} LiCH_2C_6H_5(TMEDA) + Bu^nH$$

The quantitative metallation of toluene is also possible in the absence of TMEDA, if one uses alkyls of more electropositive alkali metals:

$$NaC_5H_{11}{-}n + C_6H_5CH_3 \rightarrow NaCH_2C_5H_5 + n{-}C_5H_{12}$$

However, occasionally selectivity of a reaction is of prime importance and in such cases one has to choose specific metal alkyls. The additions may occur across the ends of the conjugated system (1, 4-additions) or to either one of the double bonds (1, 2-additions).

Organolithium compounds exhibit a much greater tendency to engage in 1, 2-addition and, therefore, these are commonly employed for maximization of this specific type of addition:

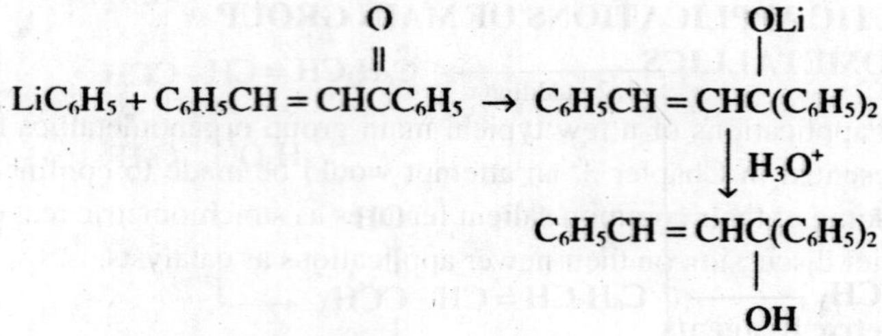

On the other hand, 1, 4-addition may be achieved by using lithium dialkylcuprates:

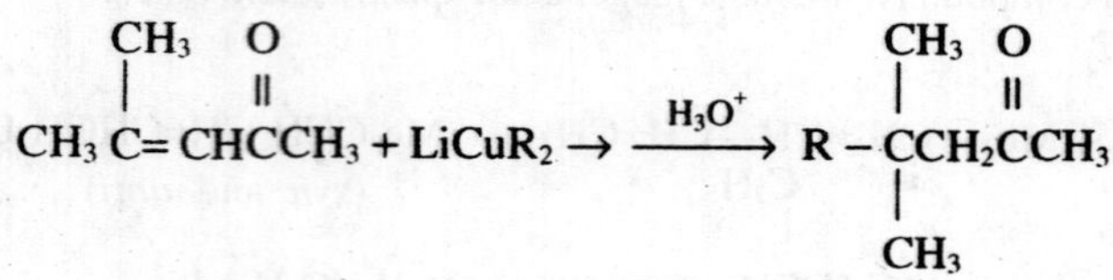

Grignard reagents show variable behaviour depending on the structure of the conjugated system. The most important factor in determining whether the addition is 1, 2 or 1, 4 seems to be steric hindrance. Most α, β-unsaturated aldehydes undergo normal 1, 2-addition to the carbonyl group:

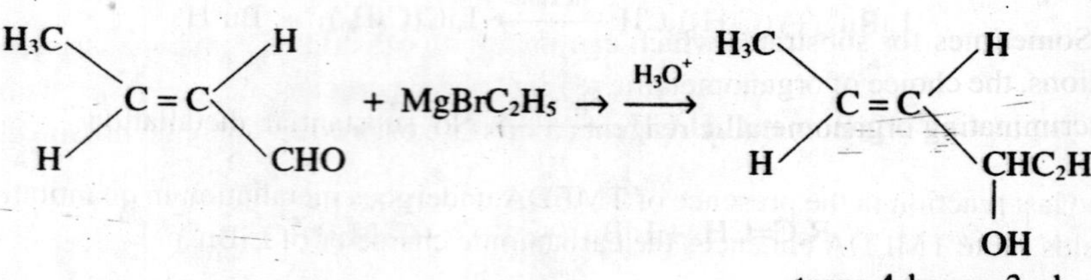

trans-4-hexen-3-ol

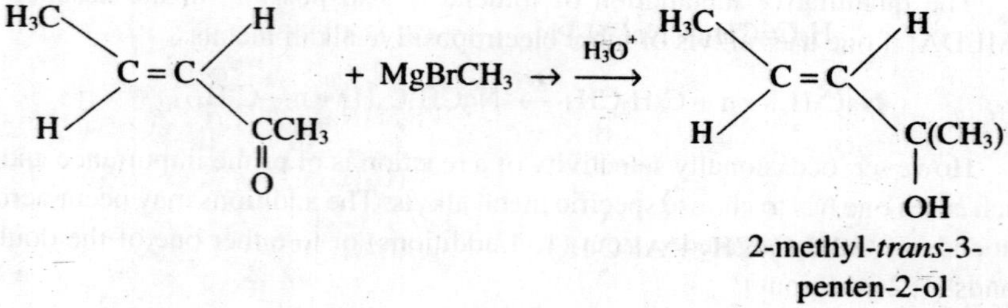

2-methyl-*trans*-3-penten-2-ol

However, many other α, β-unsaturated ketones give substantial amounts also of the 1, 4-adduct:

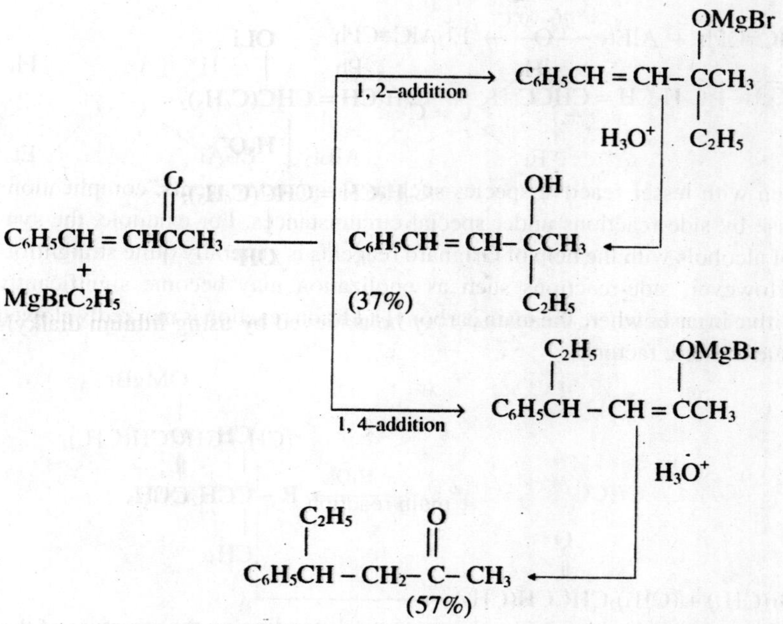

In some cases the 1, 4-adduct is almost the sole product:

$$C_6H_5CH{=}CHCC_6H_5(=O) + MgBrC_6H_5 \rightarrow \xrightarrow{H_3O^+} (C_6H_5)_2CHCH_2CC_6H_5(=O)$$

Sometimes for substrates which can undergo both addition and substitution reactions, the choice of organometallic reagent plays a crucial role, otherwise with indiscriminating organometallic reagents a mixture of products may be obtained, e.g.:

$$R_2C{=}CH_2 + LiBu \rightarrow R_2C(Bu){-}CH_2Li$$

$$Ph_2C{=}CH_2 + NaC_5H_{11} \rightarrow Ph_2CNaCH_2C_5H_{11}$$

$$H_2C{=}CH_2 + NaCH_2Ph \rightarrow PhCH_2CH_2CH_2Na \rightarrow PhCHNaEt$$

$$H_2C{=}CH_2 + Al(C_2H_5)_3 \rightarrow Al[(C_2H_4)_xC_2H_5][(C_2H_4)_yC_2H_5][(C_2H_4)_zC_2H_5]$$

$$HC{\equiv}CH + AlR_3 \rightarrow RCH{=}CHAlR_2$$

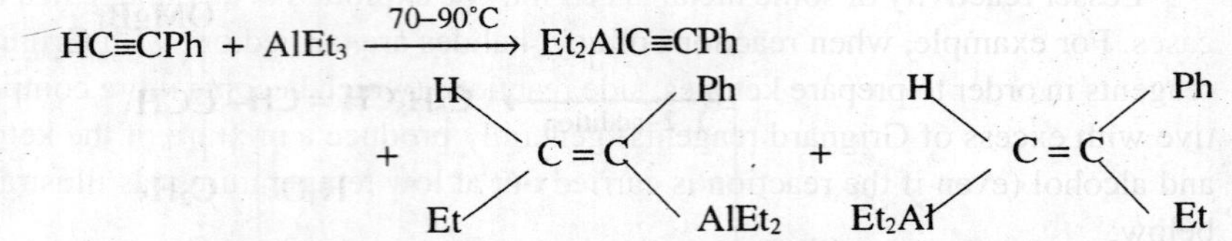

Even with lesser reactive species such as Grignard reagents, complications may arise by side-reactions under special circumstances. For example, the synthesis of alcohols with the help of Grignard reagents is generally quite straightforward. However, side-reactions such as enolization may become significantly competitive in cases where the main carbonyl addition reaction is markedly slowed down due to steric factors:

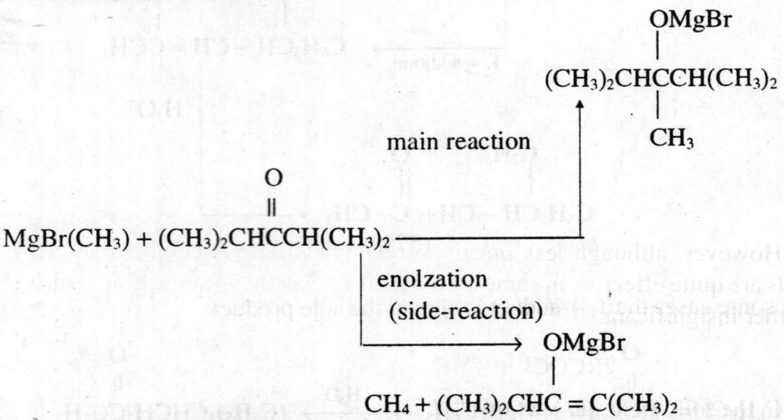

On hydrolyzing the product, the enolate portion would yield the original ketone, thus reducing the yield of the desired alcohol.

In hindered cases, another complication may arise when the Grignard reagent has a β-hydrogen. This is illustrated by the following reaction in which as alkene is a side product:

$$(CH_3)_2CH\overset{\overset{\large O}{\|}}{C}CH(CH_3)_2 + MgBr\{CH(CH_3)_2\} \rightarrow \text{product}$$

$$\xrightarrow{H_2O} (CH_3)_2CH\overset{\overset{\large OH}{|}}{C}HCH(CH_3)_2 + CH_3CH=CH_2$$

In situations like these, alkyllithium reagents are especially useful because they are more reactive than Grignard reagents and can be used at lower temperatures so that the alternative reduction and enolization reactions become less important. A spectacular example of a hindered alcohol which can be synthesized with the help of an organolithium reaction is:

$$(CH_3)_3C-\overset{\overset{\large O}{\|}}{C}-C(CH_3)_3 + LiBu^t \xrightarrow[-70^\circ C]{\text{either}} \xrightarrow{H_2O} Bu^t_3COH$$

Lesser reactivity of some metal alkyls may be exploited to advantage in a few cases. For example, when reactions of acyl halides are carried out with Grignard reagents in order to prepare ketones, side reactions (which become more competitive with excess of Grignard reagents) generally produce a mixture of the ketone and alcohol (even if the reaction is carried out at low temperatures) as illustrated below:

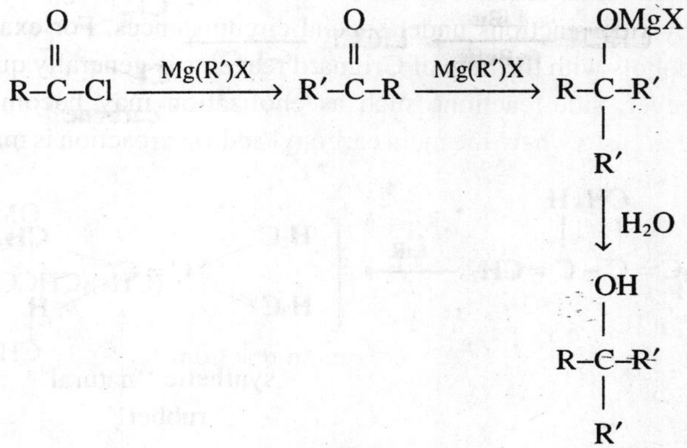

However, although less reactive organometallic compounds like cadmium alkyls are quite effective in some other reactions, yet their interaction with ketones is rather insignificant:

$$2RCOCl + CdR'_2 \rightarrow 2RCOR' + CdCl_2$$

In the following sub-sections, we shall be discussing some salient features of typical main group organometallic reagents, along with a schematic overview of the synthetic potentials of alkyls of lithium and other alkali metals, magnesium, boron, aluminium, thallium, silicon, tin, and mercury (cf. Schemes 6.1 to 6.6).

6.3.1.1 Organolithium and Heavier Alkali Metal Compounds

As a consequence of small size, the lithium cation, Li^+, brings about greater polarization of organic moieties attached to it, thus forming organometallic derivatives which are the most covalent, the most stable and the least reactive amongst its congeners (those of heavier alkali metals). As expected, lithium organometallics are generally (except $LiCH_3$) soluble in hydrocarbon solvents such as benzene, n-hexane, cyclohexane, etc.). It is, therefore, possible to carry out carbanion reactions in non-polar media. In addition, lithium alkyls possess an excellent compromising ability of both reactivity and selectivity. For example, organolithium compounds (n-butyl or phenyllithium) can bring about selective monometallation adjacent to an electronegative group and they are more reactive (compared to their corresponding Grignard reagents) to add to unsaturated linkages such as C=N, C=S, etc. In addition, lithium alkyls can add even

$$CH_3CH_2C{\equiv}CH \xrightarrow{3LiBu/TMEDA} LiC_3CH_3 + 3BuH$$

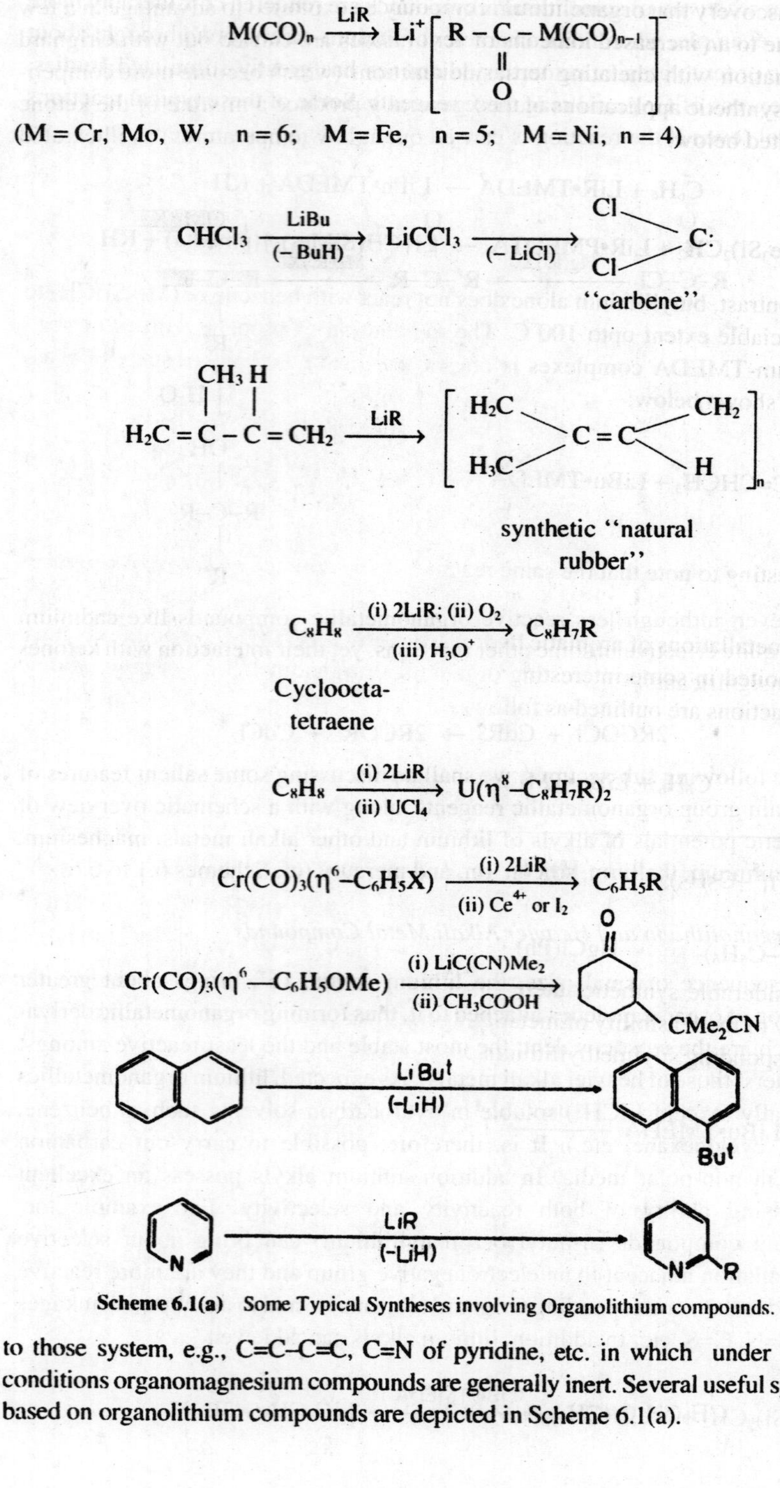

Scheme 6.1(a) Some Typical Syntheses involving Organolithium compounds.

to those system, e.g., C=C–C=C, C=N of pyridine, etc. in which under identical conditions organomagnesium compounds are generally inert. Several useful syntheses based on organolithium compounds are depicted in Scheme 6.1(a).

The discovery that organolithium compounds are rendered considerably more reactive due to an increased ionic character of lithium-carbon bond brought about by coordination with chelating tertiary diamines has greatly stimulated studies, involving synthetic applications of these reagents. Some of these typical reactions are illustrated below:

$$C_6H_6 + LiR{\bullet}TMEDA \rightarrow LiPh{\bullet}TMEDA + RH$$

$$(Me_3Si)_2CH_2 + LiR{\bullet}PMEDTA \rightarrow Li\{CH(SiMe_3)_2\}(pmedta) + RH$$

By contrast, butyllithium alone does not react with benzene or $(Me_3Si)CH_2$ to any appreciable extent upto 100°C. The metallation of propene with n- or sec.-butyllithium-TMEDA complexes is one of the most convenient routes to allyl-lithium as shown below:

$$H_2C{=}CHCH_3 + LiBu{\bullet}TMEDA \rightarrow H_2C{=}CHCH_2Li{\bullet}TMEDA + BuH$$

It is interesting to note that the same reaction in the absence of TMEDA produces a polymer.

The metallations of aromatic hydrocarbons by LiR•TMEDA complexes have been exploited in some interesting organic and organometallic syntheses. Some typical reactions are outlined as follows:

$$C_6H_6 + LiBu{\bullet}TMEDA \xrightarrow{3h,\ r.t.} LiPh{\bullet}TMEDA$$

$$LiPh{\bullet}TMEDA \xrightarrow{TiCl_2(\eta^5\text{-}C_5H_5)_2} TiPh_2(\eta^5\text{-}C_5H_5)_2$$

$$LiPh{\bullet}TMEDA \xrightarrow{HgCl_2} HgCl(Ph)$$

$$LiPh{\bullet}TMEDA \xrightarrow{SiClPh_3} SiPh_4$$

$$LiPh{\bullet}TMEDA \xrightarrow{(i)\ Ph_2C=O\ \ (ii)\ H_3O^+} Ph_3COH$$

$$LiPh{\bullet}TMEDA \xrightarrow{(i)\ CO_2\ \ (ii)\ H_3O^+} PhCO_2H$$

Considerable synthetic interest (see Scheme 6.1a) has been aroused by the discovery of the feasibility of metallating the weekly acidic methylsilanes to form the corresponding silylmethyllithium compounds:

$$SiMe_4 + LiBu{\bullet}TMEDA \xrightarrow{3\ days} LiCH_2SiMe_3 + Me_2NCH_2\text{–}CH_2\text{–}N(Me)\text{–}CH_2Li(TMEDA)$$

$$Me_2NCH_2CH_2N(Me)CH_2Li(TMEDA) \xrightarrow{SiClMe_3} (Me_3Si)_2CH_2 + Me_2N(CH_2)_2N(Me)CH_2SiMe_3$$

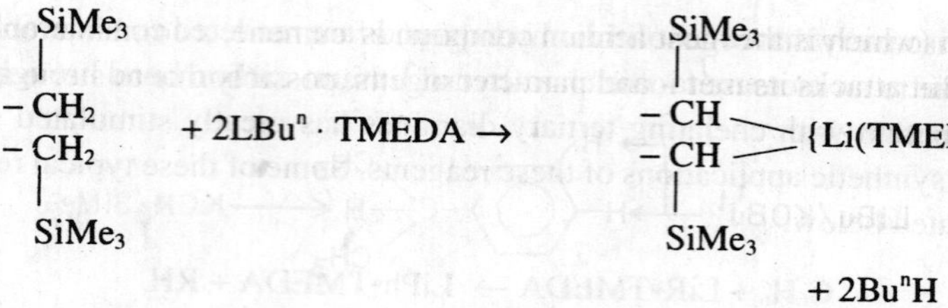

Butyllithium and diphenylacetylene do not react over many hours in hydrocarbon solvents, but a similar reaction in the presence of two equivalents of TMEDA readily produces the desired dilithium reagent resulting from the addition and ortho-metallation. Subsequent reactions with metal halides give the metallocyclic products:

$$PhC \equiv CPh + 2LiBu \cdot TMEDA \xrightarrow[\text{Dt. 24 h}]{\text{hexane}}$$

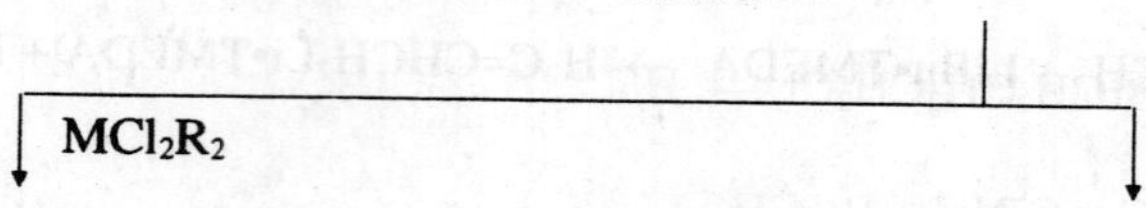

R = Me, M = Si or Sn M = Si, Ge or Sn

R = C_5H_5, M = Ti

Majority of addition and displacement reactions shown here and in scheme 6.1a are probably four-centre concerted processes which permit ligand exchange or insertion reactions:

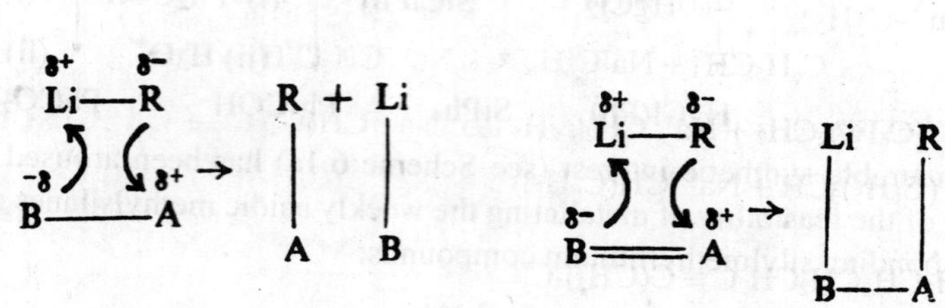

Alkyls of heavier alkali metals (Na → Cs) which are essentially ionic, insoluble (in hydrocarbon solvents), and highly reactive are of limited value in organic synthesis. However, enhanced nucleophilic (or carbanionic) character of the organic moiety in these derivatives has been exploited as a potential metallating agent for less acidic hydorcarbons, e.g.:

$$C_6H_5CH_3 + Na^+ C_5H_{11}^- \rightarrow NaCH_2C_6H_5 + {}^-C_5H_{12}$$

The regio chemistry of the metallation reaction is dependent very much on the nature of the bases used. For example, whereas KCH_2SiMe_3 abstracts the benzylic hydrogen (which is the most acidic hydrogen) of cumene, a combination of LiBu and KOBut attacks its meta- and para-hydrogens, possibly due to steric and other factors:

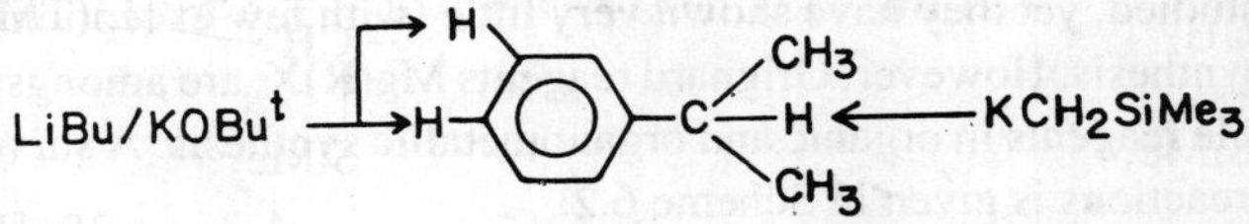

As organosodium and organopotassium compounds are very reactive, they generally metallate alkenes rather than add to them. A few reactions involving sodium organometallics are shown in the Scheme 6.1b; in some of these reactions, solvent appears to play a critical role:

$$RC\equiv CH + NaR' \rightarrow RC\equiv C^-Na^+ + R'H$$

$$\downarrow R''X$$

$$RC\equiv CR'' + Na^+X^-$$

$$HC\equiv C^-Na^+ + BrCH_2CH_2CH_2Cl \rightarrow HC\equiv CCH_2CH_2CH_2Cl + NaBr$$

$$ClCH_2CH_2CH_2C\equiv C^-Na^+ + Br(CH_2)_3Cl \xrightarrow{-\,NaBr} ClCH_2CH_2CH_2C\equiv CCH_2CH_2CH_2Cl$$

$$Me_2CHCOOEt + Na^+C(C_6H_5)_3 \xrightarrow[(-\,(C_6H_5)_3CH)]{} Me_2C^-{-}COOEt \;\; Na^+$$

$$\downarrow MeCOCl$$

$$Me_2C(COMe)(COOEt) + NaCl$$

$$C_6H_6 + Na^+Et^- \rightarrow Na^+C_6H_5 + EtH$$

$$C_6H_5CH_3 + Na^+C_6H_5^- \rightarrow Na^{+\,-}CH_2C_6H_5 + C_6H_6$$

$$(C_6H_5)_2CH_2 + Na^{+\,-}CH_2C_6H_5 \rightarrow Na^{+\,-}CH(C_6H_5)_2 + C_6H_5CH_3$$

$$(C_6H_5)_3CH + Na^{+\,-}CH(C_6H_5)_2 \rightarrow Na^{+\,-}C(C_6H_5)_3 + (C_6H_5)_2CH_2$$

$$CH_3CH_2CH_2CH_2C\equiv C(CH_2)_4C\equiv CH \xrightarrow[NH_3(liq)]{Na}$$

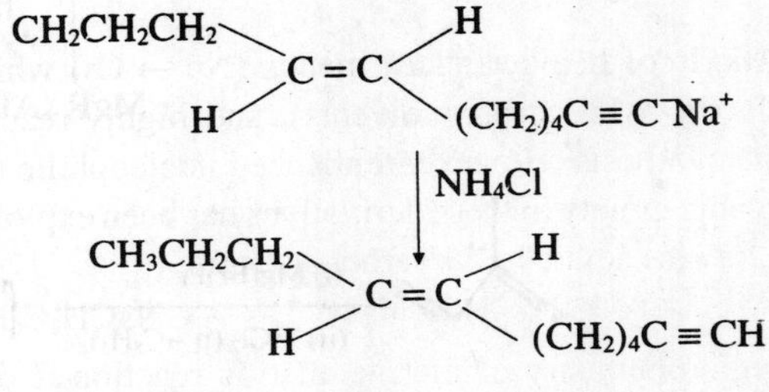

Scheme 6.1b Organic and Organometallic Syntheses involving Sodium Organometallics.

6.3.1.2 Organomagnesium Compounds

Although some reactions of alkyls of beryllium, calcium, strontium, and barium have been studied, yet they have shown very little (with few exceptions) potential in organic synthesis. However, Grignard reagents Mg(R)X, are amongst one of the most versatile reagents in organic and organometallic synthesis. A survey of some interesting reactions is given in Scheme 6.2:

$$MCl_4 + 4Mg(R)X \longrightarrow MR_4 + 4MgClX$$
$$(M = Ti,\ Zr,\ V,\ etc.)$$

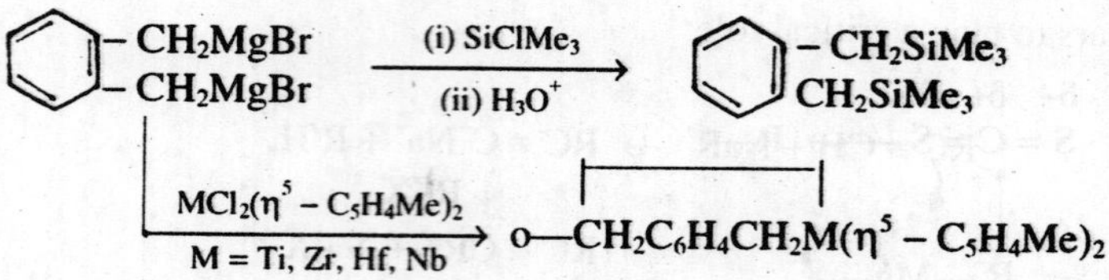

$$GeCl_2 \bullet C_4H_8O_2 + 2MgCl\{CH(SiMe_3)_2\} \rightarrow Ge\{CH(SiMe_3)_2\}_2 + 2MgCl_2 + C_4H_8O_2$$

$$3MgR_2 + BF_3 \bullet OEt_2 \rightarrow 3Mg(R)F + BR_3$$

$$Mg(R)X + ArSO_2C \equiv CR' \rightarrow R'C \equiv CR + Mg(O_2SAr)X$$

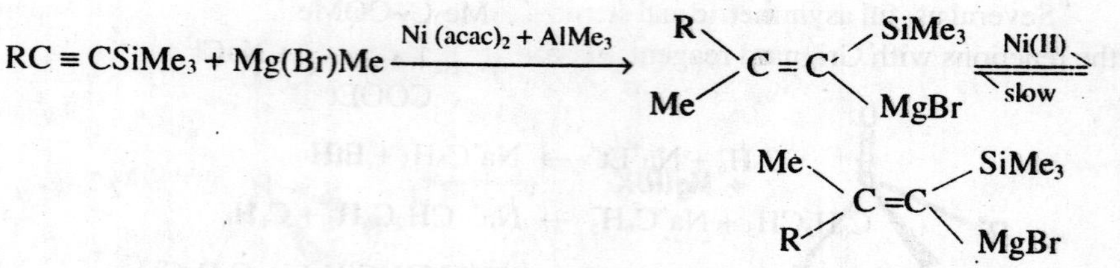

$$HfCl_4 + 2C_8H_8 + 2Mg \rightarrow Hf(\eta^8\text{–}C_8H_8)_2 + 2MgCl_2$$
(cot)

$$2Mg(R)X + Al_2R_6 \xrightarrow{100^\circ C} Mg(AlR_4)_2 + MgCl_2$$
$$\downarrow \text{heat}$$
$$MgR\ (AlR_4) + AlR_3$$

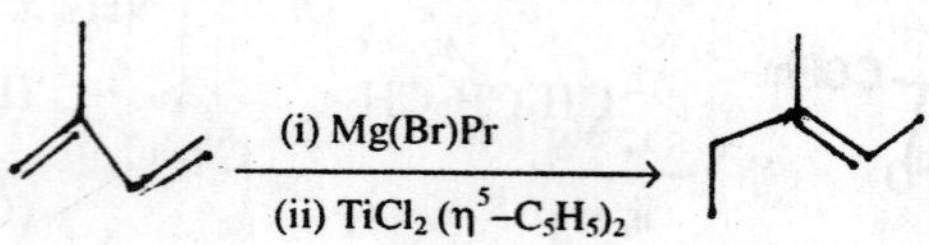

Scheme 6.2 Some Typical Synthetic Applications of Grignard reagents.

In addition to the reactions illustrated in Scheme 6.2, some specific examples are represented below to illustrate the mechanism of Grignard reactions:

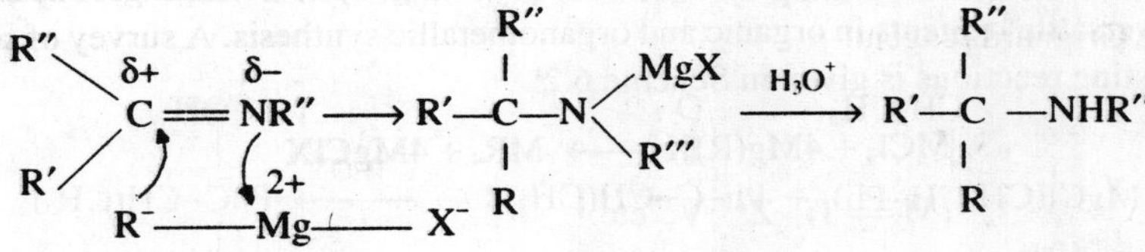

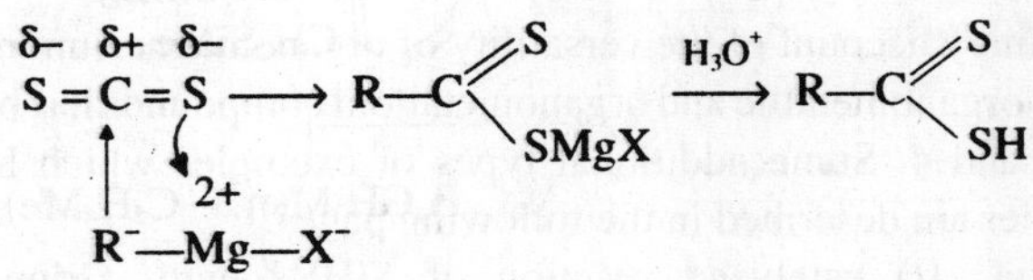

In exceptional cases, some abnormal behaviour of Grignard reagents has also been noted (cf. Section 6.3.1), mainly due to steric factors. For example there is no reaction of methylmagnesium bromide or iodide with hexamethylacetone, $Me_3CC(=O)CMe_3$ and isopropyl methyl ketone, $Me_2CH{-}C(=O){-}Me$, reacts with methylmagnesium iodide but not with t-butylmagnesium chloride.

Several useful asymmetric and stereospecific syntheses have been achieved in the reactions with Grignard reagents. These are represented as follows:

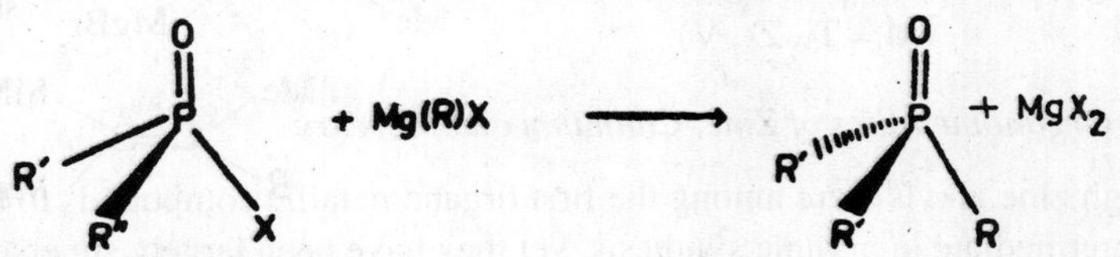

(Chiral substrates)

The above type of reactions generally proceed with inversion of configuration at phosphorous. Although some asymmetric induction can occur even when the chiral centre is remote from the carbonyl group, (as in (s–)–EtCH(Me)CH_2O–C_6H_4–COEt), yet a high degree of stereo-selectivity is possible when the carbonyl group is adjacent to the chiral centre:

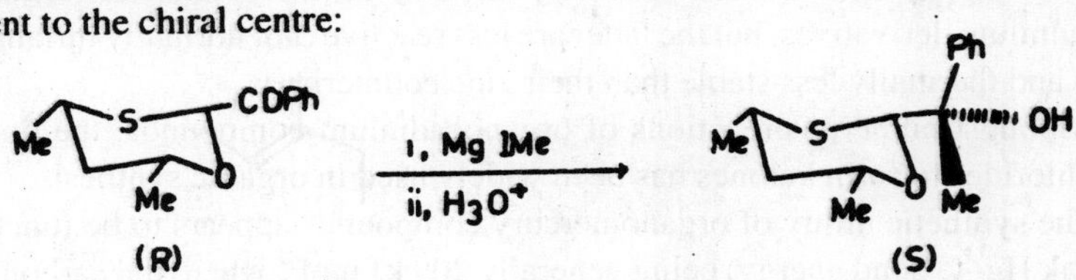

Useful asymmetric synthesis are also accessible by reactions of chiral organomagnesium compounds. For example, the reduction of isopropylphenyl ketone to the corresponding alcohol occurs in 82% optical yield with optically active Grignard reagent:

$$MgCl(CH_2\overset{\displaystyle CH_2CH_3}{\overset{|}{C}}H\text{–}Ph) + Ph\text{–}\overset{\displaystyle O}{\overset{\|}{C}}\text{–}CH(CH_3)_2 \rightarrow \xrightarrow{H_3O^+} Ph\underset{\displaystyle H}{\underset{|}{\overset{\displaystyle OH}{\overset{|}{C}}}}\text{–}CH(CH_3)_2$$

A preliminary account of the versatility of organomagnesium reagents in the preparation of organometallic and organometalloid compounds has been presented in chapters 3 and 4. Some additional types of examples which have not been discussed earlier are described in the following pages.

The nickel (II) catalyzed reaction of $SiHR_3$, with Grignard reagents, $MgR'X$(R' = Me, Ph, CH_2Ph, allyl, vinyl) is an alkylation, which appears to be related mechanistically with hydrosilylation (cf. Section 6.4.2.3) reactions:

$$SiHR_3 \xrightarrow[NiX_2(PPh_3)_2(catalyst)]{Mg(R')X} SiR_3R'$$

The applications of Grignard reagents in the synthesis of transition metal alkyls (especially those like benzyls, neopentyls, trimethylsilylmethyls, neophyls and 1-adamantylmethyls) which do not undergo facile decomposition by β-hydrogen elmination pathways) have been described in Section 4.2.2.1; e.g.:

$$MCl_4 + 4Mg(R)X \rightarrow MR_4 + 4MgClX$$
$$(M = Ti, Zr, V).$$

6.3.1.3 Organometallics of Zinc, Cadmium and Mercury

Although zinc alkyls were among the first organometallic compounds to be used as an intermediate in organic synthesis, yet they have been largely superseded by the Grignard reagents (which are more reactive and convenient to handle), in spite of the fact that many of the reactions for which Grignard reagents are now in use had already been worked out with organozinc compounds. The organozinc compounds are still finding their usefulness in some synthetically important reactions such as Reformatsky reaction, Simmons-Smith synthesis of cyclopropanes, and growing applicability of allylzinc bromides in organic synthesis.

There appears to be close similarity in many of the reactions of organozinc and cadmium derivatives, but the latter are less reactive (not normally inflammable in air) and thermally less stable than their zinc counterparts.

Amongst numerous reactions of organocadmium compounds, the one with acid chlorides to form ketones has been widely used in organic synthesis.

The synthetic utility of organomercury compounds appears to be function of (i) weak Hg–C bond energy, being generally 209 kJ mol^{-1} where the carbon group is a simple alkyl. In fact, due to the weakness of Hg–C bond, organometallics of mercury should be an ideal source for carbon-free radicals and mercury can be replaced by many metals which generate stronger M–C bonds. Thus the preparation of organometallics of other metals (e.g., Li, Na, K, Be, Mg, Ca, B, Al, Sn, Pb, Sb, Bi, Zn, Cd, etc.) is the most important application of organomercury compounds; (ii) the case with which mercury undergoes redox reactions:

$2Hg^{2+} + 2e \rightarrow Hg_2^{2+}$, E = 0.92 V, and (iii) highly covalent and weakly nucleophilic nature of Hg–C bonds, due to small difference (0.6) in electro-negativity between Hg (1.9) and carbon (2.5). For example, unactivated Hg–C bonds are generally not responsive towards typical organic electrophiles, such as organic halides, carbonyl compounds, and expoxides. However, Hg–X bond (where X is an electronegative group) are highly polarizable (''soft''), and these are, therefore, more susceptible to react with ''soft'' organic substrates such as alkenes, acetylenes, and aromatic compounds.

In the present account, we shall deal with the reactions of organo-zinc, –cadmium and –mercury compounds, emphasizing their synthetic aspects.

A. *Organozinc Compounds*

These compounds undergo a wide variety of reactions, such as protonation, redistribution, insertion, halide-carbanion exchange, addition as depicted in Scheme 6.3.

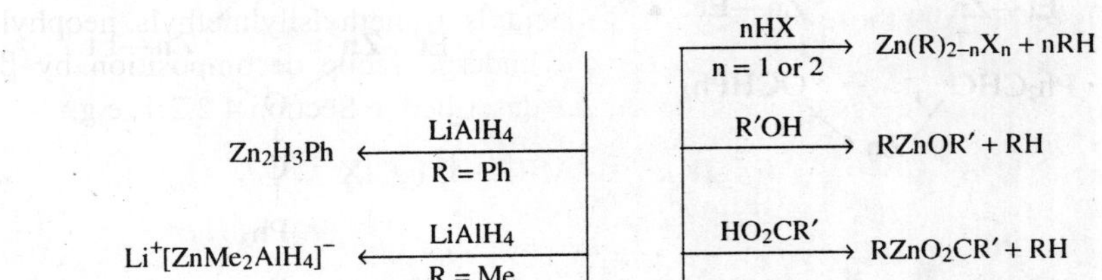

Scheme 6.3 Some Typical Reactions of Organozinc Compounds.

Structurally diorganozinc compounds, ZnR_2 differ fundamentally from the compounds of the type Zn(R)X (X = H, halogen, Or′, NR_2, PR_2, SR′, etc.). The difference arises due to the presence of electronegative atom or group (X) bound directly to zinc, which introduces electrophilicity (on the zinc atom) and a donor site (on the atom bound to zinc) in the same molecule. This enhances associative ability of the molecule, leading to the formation of dimers $(ZnRX)_2$ or oligomers $(ZnRX)_n$ (cf. Section 3.6.2). Higher degrees of association are possible in those cases where the group *X* is H or tends to denote more than one electron pair. A few illustrative structures are shown below.

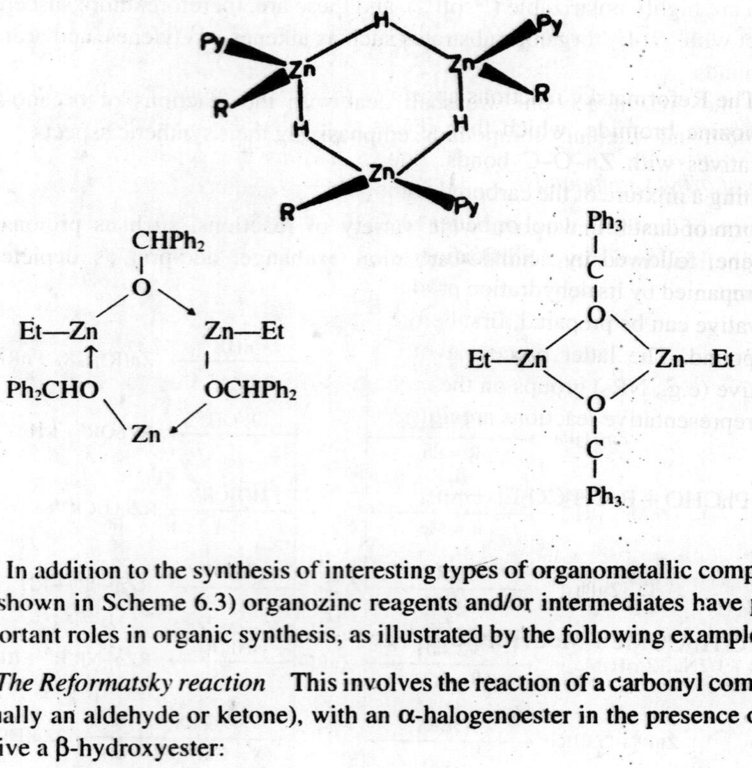

In addition to the synthesis of interesting types of organometallic compounds (as shown in Scheme 6.3) organozinc reagents and/or intermediates have played important roles in organic synthesis, as illustrated by the following examples:

(i) *The Reformatsky reaction* This involves the reaction of a carbonyl compound (usually an aldehyde or ketone), with an α-halogenoester in the presence of zinc to give a β-hydroxyester:

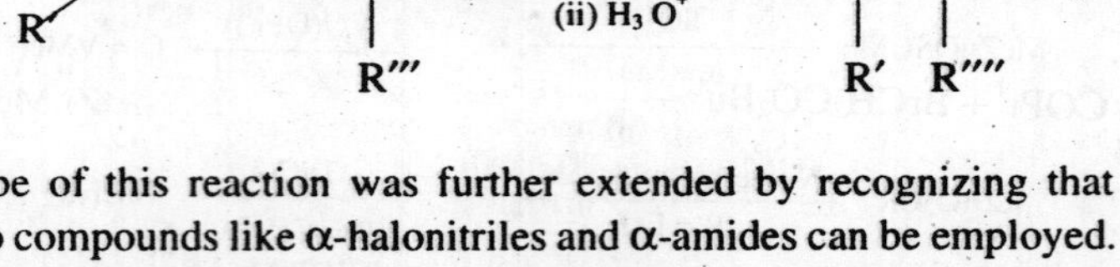

The scope of this reaction was further extended by recognizing that other α-halogeno compounds like α-halonitriles and α-amides can be employed. An ester may be used instead of an aldehyde or ketone as the carbonyl component. In this case the initial product is a β-ketoester, but the latter reaction has found little application in practice:

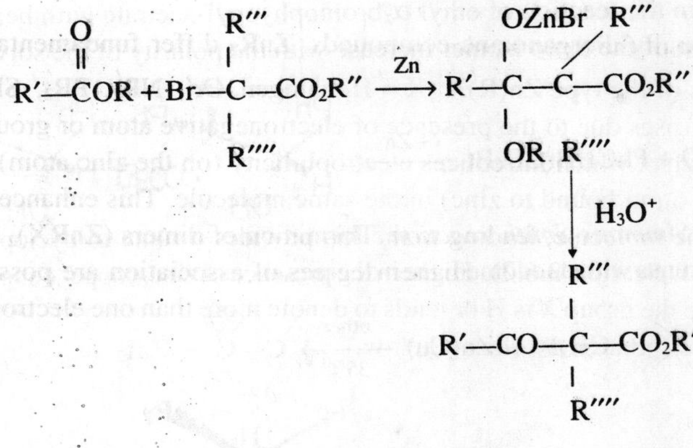

The Reformatsky reactions appear to proceed generally via the formation of organozinc bromide, which then reacts with the carbonyl component to form derivatives with Zn–O–C bonds. Such reactions are generally carried out by warming a mixture of the carbonyl compound, the bromoester and zinc (usually in the form of dust foil, wool or better as Zn–Cu couple) in a solvent such as ether or benzene, followed by acidification which yields the hydroxyester (sometimes accompanied by its dehydration product also). Alternatively, the intermediate zinc derivative can be prepared, first before carrying out its reaction with the carbonyl compound. The latter process avoids the reaction of reducible (e.g., NO_2) or reactive (e.g., NR_2) groups on the carbonyl substrate with the free bromoester. A few representative reactions are illustrated as follows:

$$\mathrm{PhCHO + BrCH_2CO_2Et \xrightarrow[\text{(ii) } H_3O^+]{\text{(i) Zn, ether–benzene}} PhCH(OH)CH_2CO_2Et}$$

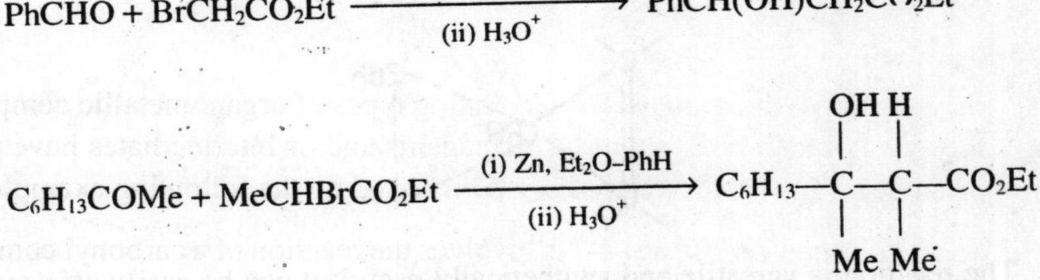

$$\mathrm{PhCHO + BrCMe_2CO_2Bu^t \xrightarrow[\text{(ii) } H_3O^+]{\text{(i) Zn, THF}} PhCH(OH)CMe_2CO_2Bu^t}$$

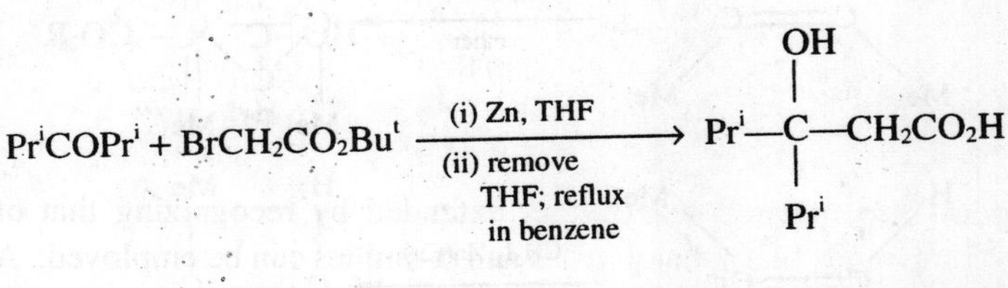

The ratio of erythio and threo-isomers in the reactions involving asymmetric α-haloesters appears to be controlled by the solvents polarity and reaction conditions. Thus, in the reaction of ethyl α-bromophyenyl-acetrate with benzaldehyde the proportion of the threo isomer increase with the polarity of the solvent, e.g.:

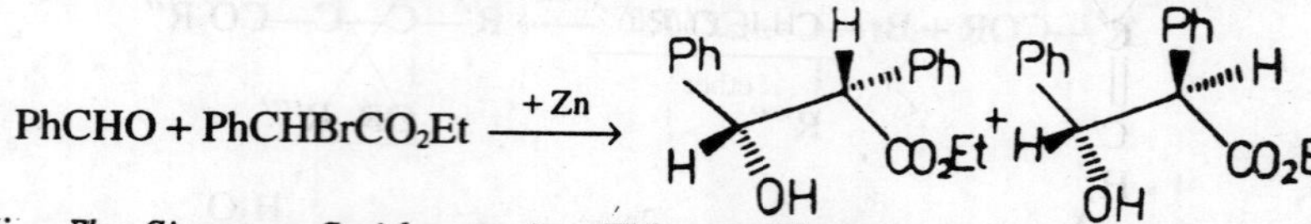

(ii) *The Simmons-Smith reaction* This provides a route to cyclopropanes by reaction of alkenes with duodomethane in the presence of zinc-copper couple or $ZnEt_2$:

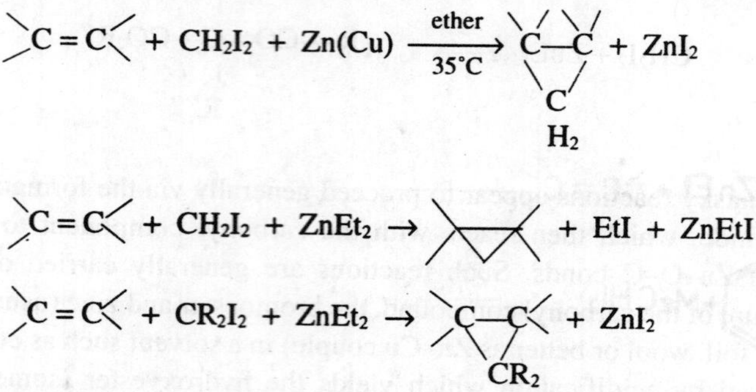

The actual reactive species is an organozinc intermediate of the form ICH_2ZnI or $(ICH_2)_2Zn \cdot ZnI_2$ in the first reaction, ICH_2ZnEt in the second, $EtZnCR_2I$ in the last reaction. The formation of cyclopropanes does not appear to involve free carbenes; nor, apparently does it involve addition of the organozinc intermediate to the double bond followed by elimination of zinc halide. A possible representation of the transition state for concerted mechanism can be represented by the following formula:

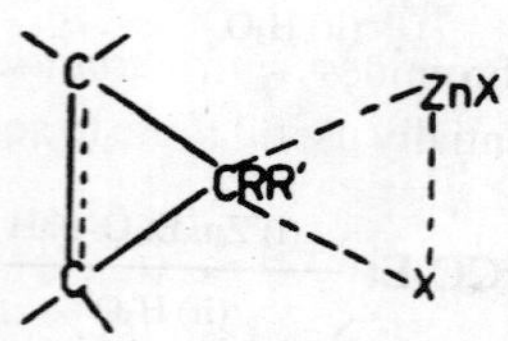

The reaction is versatile and synthetically useful; it can be easily effected in ether under nitrogen. Some examples are shown by the following reactions:

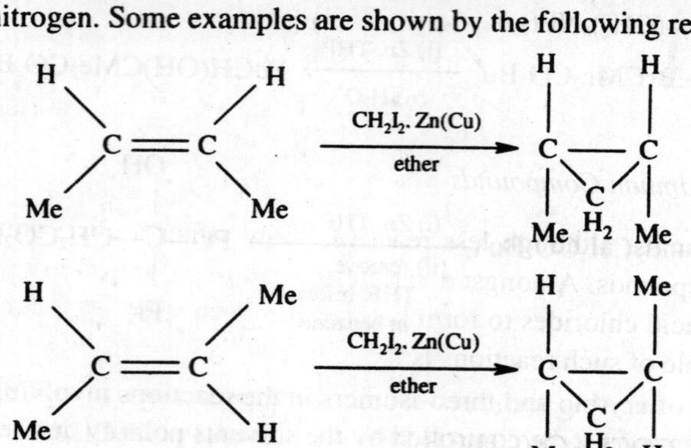

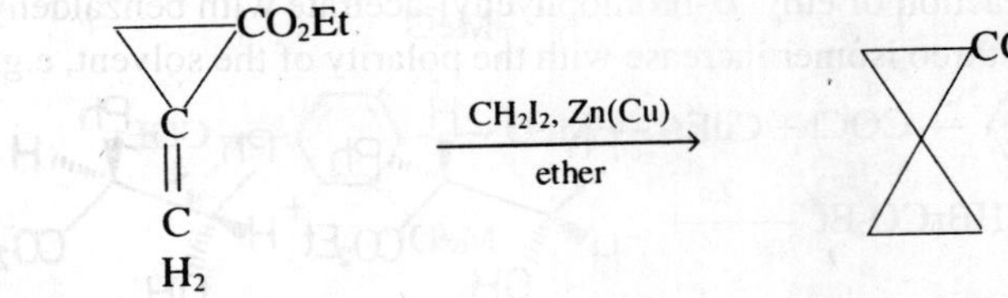

By using diethylzinc in place of Zn(Cu) couple, products in better yields could be obtained, e.g.:

$$CH_2I_2 + ZnEt_2 \longrightarrow ICH_2ZnEt + EtI$$

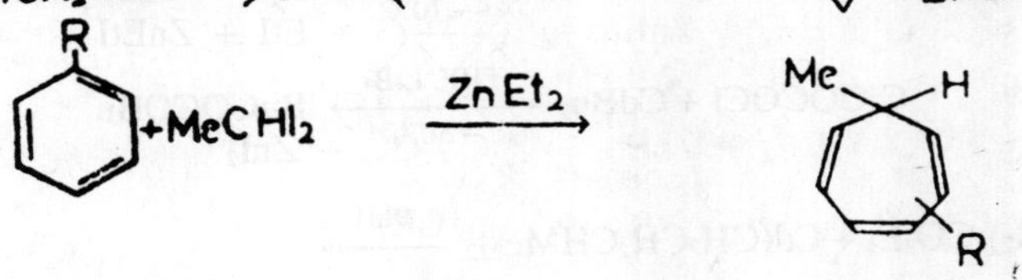

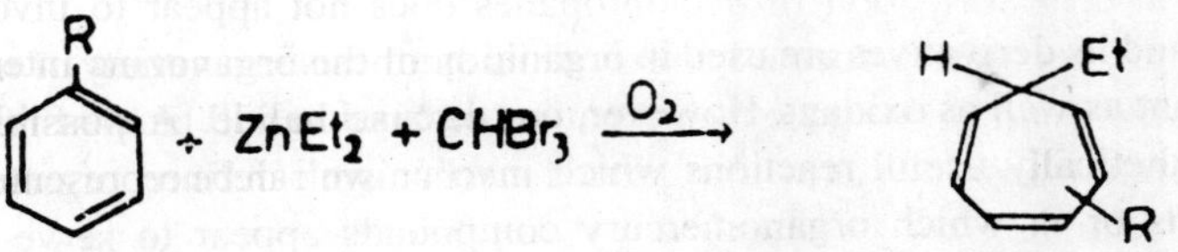

Reactions of alkylzinc bromides with carbon-carbon and carbon-nitrogen multiple bonds provide a potentially useful route to some alkenic compounds, e.g.:

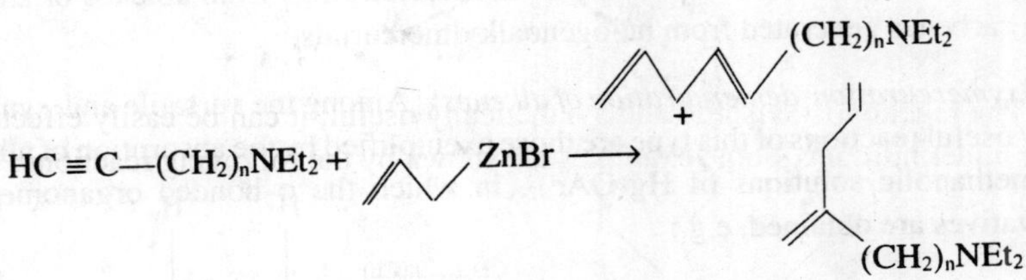

B. *Organocadmium Compounds*

These compounds, although less reactive, show rather a close similarity to organozinc compounds. Amongst a varied types of reactions of diorganocadmium, the one with acid chlorides to form ketones has been widely used in synthesis. A typical example of such reactions is:

$$Me(CH_2)_{16}COCl + Cd(CH_2CH_2Ph)_2 \xrightarrow[\text{reflux}]{\text{PhH}} PhCH_2CH_2CO(CH_2)_{16}Me$$

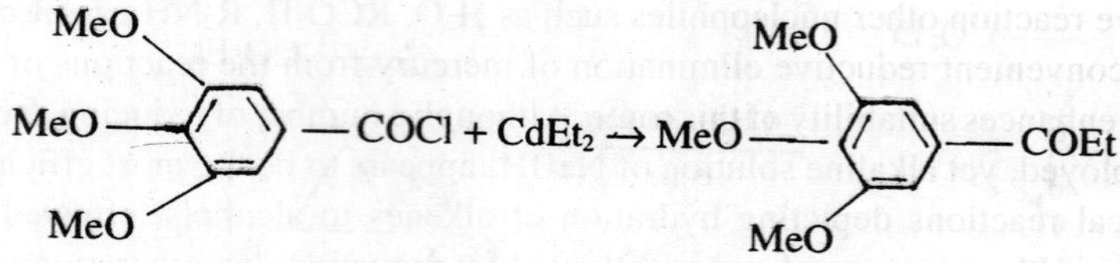

$$2Ph_3CCOCl + CdEt_2 \longrightarrow 2Ph_3CCOEt$$

It is interesting to note that whereas magnesium salts catalyze the reactions of both acid chlorides and carbonyl compounds lithium bromide is more selective in catalyzing only the former reaction. For example, the synthesis of α-oxoesters and of 1, 2-diketones from oxalyl mono- and di-chlorides respectively can be accomplished by the following types of reactions:

$$ClCOCO_2Et + CdBu_2 \xrightarrow[-70^{\circ}C]{THF,\ LiBr} BuCOCO_2Et$$

$$ClCOCOCl + CdBu_2 \xrightarrow[-50^{\circ}C]{THF,\ LiBr} BuCOCOBu$$

$$ClCO(CH_2)_8CO_2Et + Cd(CH_2CH_2CHMe_2)_2 \xrightarrow{PhH} MeCHCH_2CH_2CO(CH_2)_8CO_2Et$$

C. *Organomercury Compounds*

Mercury and its derivatives are used in organic synthesis in a variety of ways such as reductant as well as oxidant. However, our discussion will be confined around those synthetically useful reactions which involve well-defined organomercury compounds or in which organomercury compounds appear to serve as intermediates. Thus, we will discuss mainly two reaction sequences which are at present versatile and of considerable importance in synthesis, typified by (i) the functionalization of carbon atoms by a mercuration-demercuration sequence from alkenes, and (ii) the formation of three membered rings from alkenes or alkynes with carbenes generated from halogenoalkylmercurials.

(i) *Oxymercuration-demercuration of alkenes*: Among the versatile and synthetically useful reactions of this type are those exemplified by the absorption of alkenes by methanolic solutions of $Hg(OAc)_2$, in which the σ-bonded organometallic derivatives are obtained, e.g.:

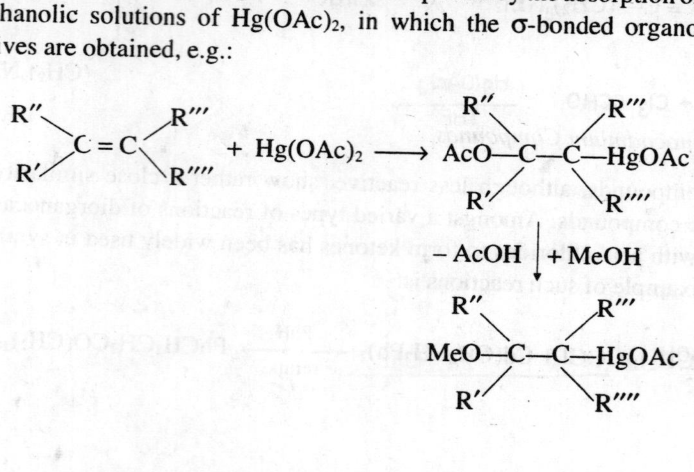

In the above reaction other nucleophiles such as H_2O, RCO_2H, R_2NH can also be used. The convenient reductive elimination of mercury from the reactions of the above type enhances suitability of this route. Although a number of reducing agents can be employed, yet alkaline solution of $NaBH_4$ appears to be the most efficient. Some typical reactions depicting hydration of alkenes to alcohols, effected by mercuration in the presence of water followed by reductive demercuration with sodium borohydride are illustrated as follows:

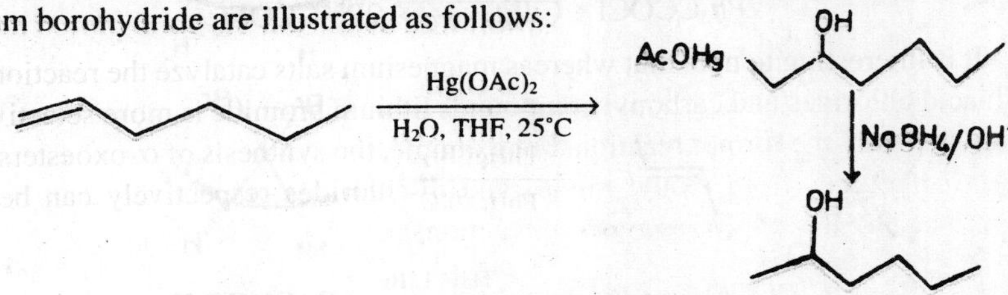

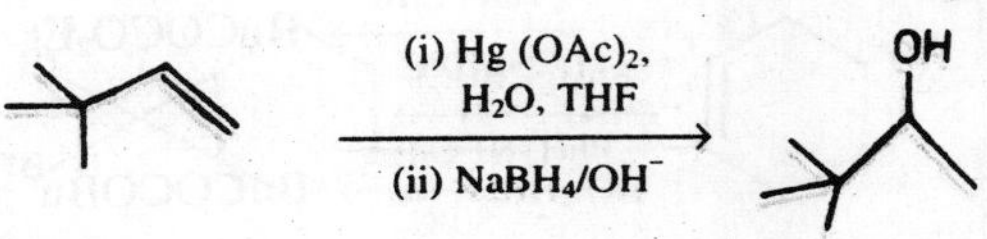

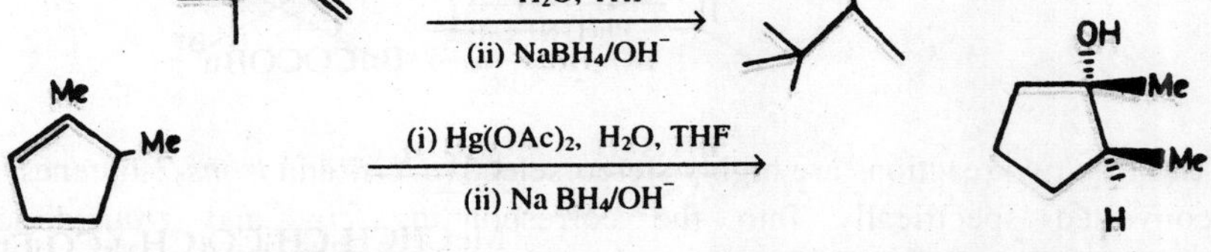

With cyclic alkenes, the stereochemical course of oxymercurations may be influenced by nearly hydroxyl substituents:

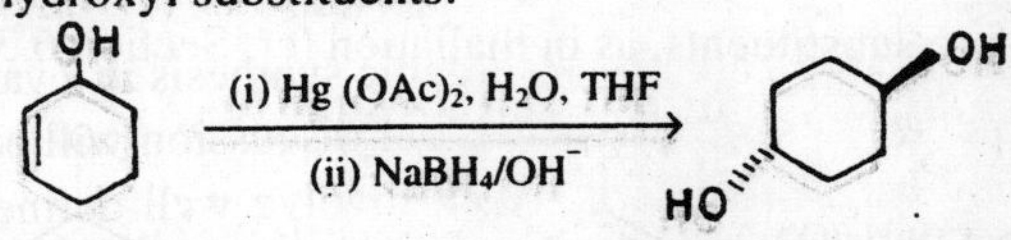

Intramolecular oxymercurations, leading to oxygen heterocycles, may also take place when a hydroxyl substituent in the alkene is suitably placed for attack on the intermediate mercurium ion:

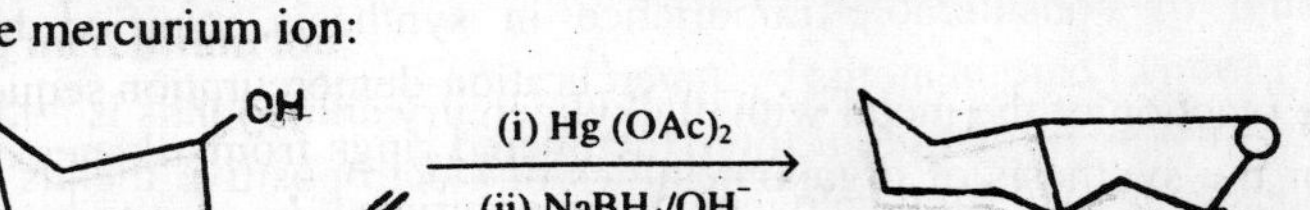

Another useful sequence leads to *cis*-1, 2-cyclohexanediol from 2-cyclohexanol:

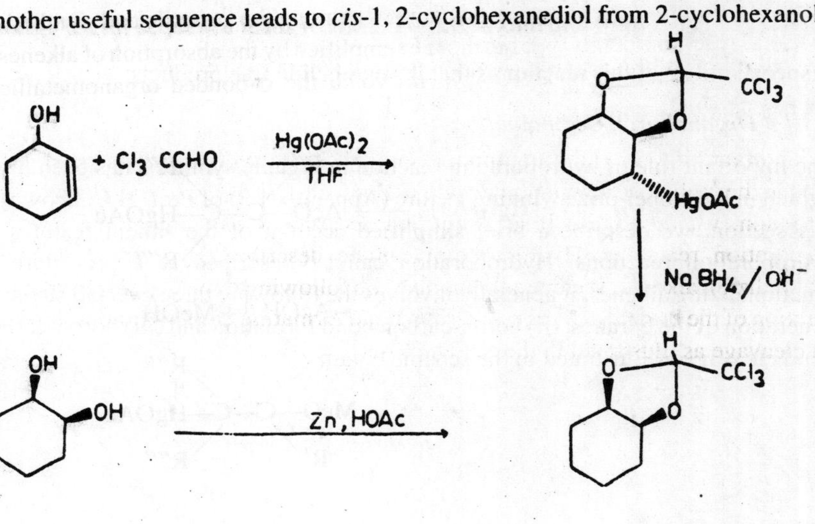

(ii) *Carbene-transfer reactions* α-Halomethylemercury compound which proved to be useful carbene precursors, have been widely employed in the synthesis of gem-dihalogenocyclopropanes, e.g.,

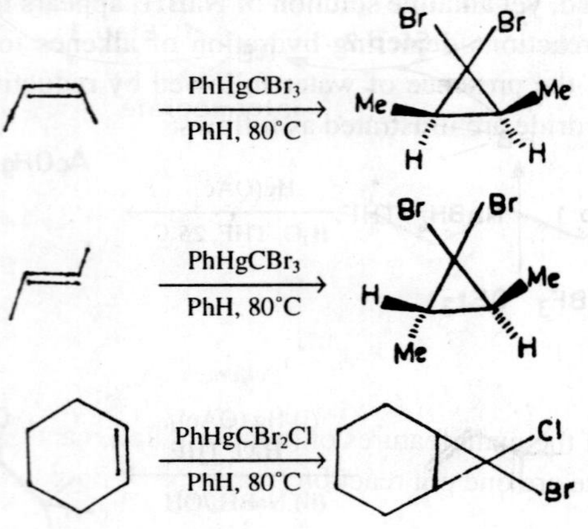

The first two reactions are highly stereo-selective, *cis*- and *trans*-2-butenes being converted specifically into the corresponding *cis*- and *trans*-dimethylcyclopropanes.

In addition to the above, electrophilic metallation of aromatic rings provides an elegant path to substituents, as in thallation (cf. Section 6.3.1.6):

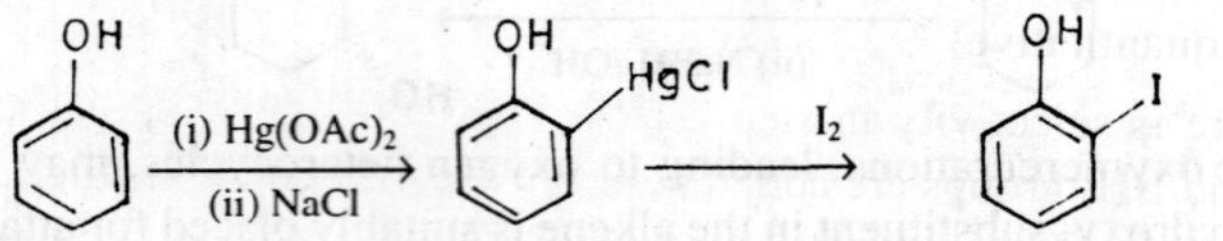

The reaction of the metal with dialkylmercury compounds is a farily general route for the synthesis of organometallics of electropositive metals (cf. Section 3.4.2):

$$nHgR_2 + 2M \rightarrow 2MR_n + n\,Hg$$

A special utility of this reaction is that it gives halide free products.

6.3.1.4 Organoboron Compounds

The important role of hydroboration reaction in organic synthesis has been highlighted in the Nobel-prize winning lecture (Appendix 1.3) of Prof. H.C. Brown. In this section, we describe a brief simplified account of the salient features of hydroboration reactions. Hydroboration can be described as a procedure of functionalizing alkenes; it generally involves the following three over-all steps: (i) generation of the boranes, (ii) boron-carbon bond formation and (iii) boron-carbon bond cleavage as illustrated in the scheme 6.4a:

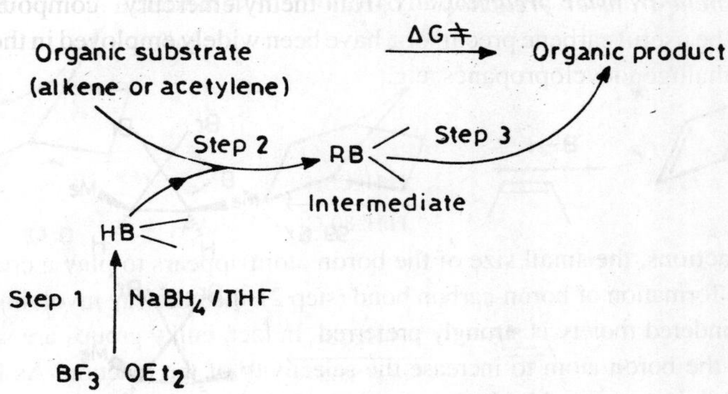

Scheme 6.4a

Some of the main features of hydroboration reactions are:

(a) These are one pot reaction (reagents are prepared in *situ*):

$$NaBH_4 + BF_3 \cdot OEt_2 \xrightarrow{THF} H_3B \cdot THF$$

$$\xrightarrow{>C=C<} RBH_2 \xrightarrow{>C=C<} R_2BH \xrightarrow{>C=C<} BR_3\,;$$

(b) Steps 1, 2 and 3 are all fast and proceed at the room temperature and mild conditions resulting in almost 100% yields (i.e., reactions are essentially instantaneous and quantitative):

(c) There is selectivity in step 2. As a consequence of inductive and steric factors. H and BH_2 groups are added in an anti-Markovnikov manner, e.g.,

$$RCH = CH_2 \xrightarrow{>B-H} RCH_2CH_2B< \xrightarrow{[O]} RCH_2CH_2OH$$

This type of selectivity is illustrated by the actual percentage yields in the hydroboration reactions of the following three alkenes:

$$\underset{6\%}{CH_3CH_2CH_2CH} = \underset{94\%}{CH_2}; \quad CH_3CH_2\underset{1\%}{\overset{CH_3}{C}} = \underset{99\%}{CH_2}; \quad CH_3\underset{2\%}{\overset{CH_3}{C}} = \underset{98\%}{CHCH_3}$$

For similar reasons, the hydroboration reaction involves a *cis* addition of the B–H bonds, e.g.,

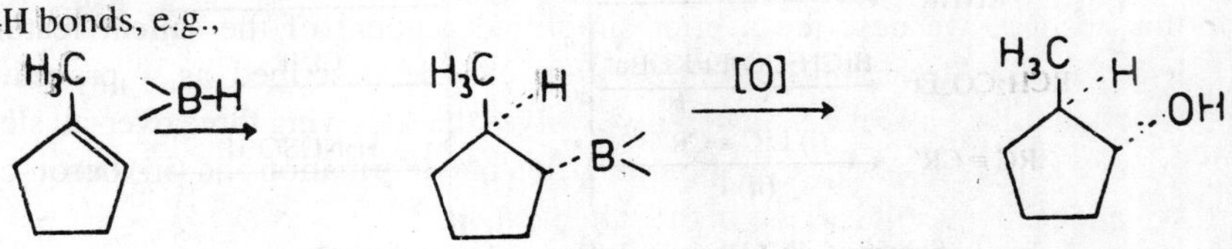

and addition takes place preferentially from the less hindered side of the double bond, e.g.,

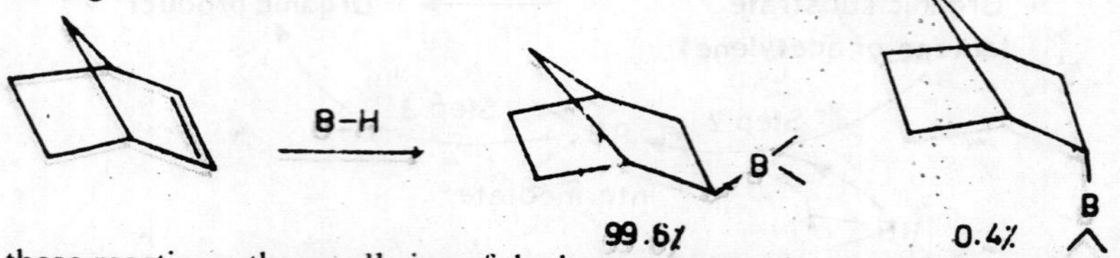

In these reactions, the small size of the boron atom appears to play a crucial role. Thus in the formation of boron-carbon bond (step 2), the pathway involving the least sterically hindered moiety is strongly preferred. In fact, bulky groups are sometimes attached to the boron atom to increase the selectivity of the reaction. As illustrated below, the replacement of hydrogen atoms in BH_3 by sterically demanding alkyl groups leads to preferential attachment (step 2) of the boron species at the less sterically hindered carbon atom of the alkene, leading greater selectivity of the product.

	Bu^nC =	CH_2	$Me_2CH(H)C$ =	$C(Me)H$
BH_3	6%	94%	43%	57%
$(HC(Me)_2—C(Me)H—)_2BH$	1%	99%	3%	97%

(d) The cleavage of boron-carbon bond (step 3) can occur in a variety of ways (Scheme 6.4b), which coupled, with the selectivity of step 2 tends to enhance the practical utility of the hydroboration reactions in synthesis of a specific product in almost quantitative yield:

Product	Reagent		Reagent	Product
RSR′ ←	R′SSR′		H_2O_2-NaOH	→ ROH
RHgOAc ←	$Hg(OAc)_2$		$EtCO_2H$	→ RH
RI ←	$NaOMe/I_2$		$AgNO_3/OH^-$	→ R–R
$ROH + RO_2H$ ←	O_2	BR_3	R′CH = CH_2 (R′ > R)	→ RCH = CH_2
RNHR′ ←	$R'N_3$		$ClNR_2$	→ RNH_2
RCH_2CO_2Et ←	$BrCH_2CO_2Et/KOBu^t$		$ClNR_2$	→ RCl
RC ≡ CR′ ←	(i) LiC ≡ CR″ (ii) I_2		H_2NOSO_3H	→ RNH_2

Scheme 6.4b Varieties in Boron-Carbon Bond Cleavage.

(e) The growing utility of asymmetric hydroboration in organic synthesis, e.g.,

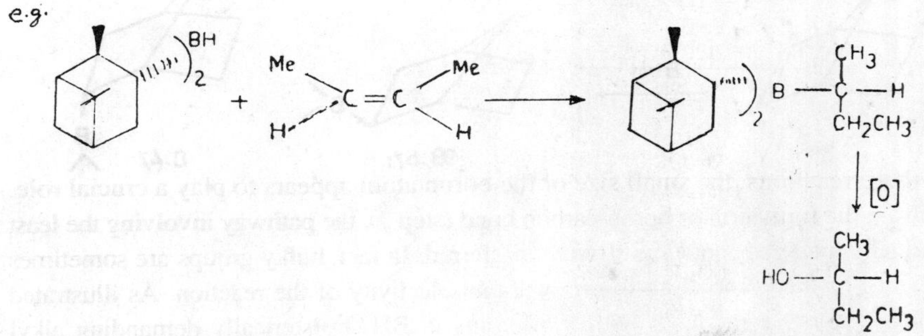

(90% optically pure)

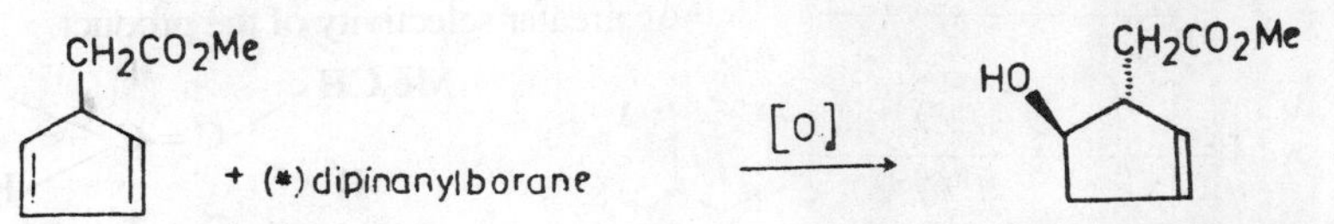

(92% optically pure)

6.3.1.5 Organoaluminium Compounds

The importance of organoaluminium compounds as synthetic reagents (in organic synthesis, organometallic synthesis and catalysis of alkene polymerization) may be understood on the basis of a set of unique properties. (i) the flexible nature of Al-C bond (e.g., the intermediate electronegativity (1.47) of aluminium renders the carbon bonded to aluminium weakly nucleophile only, whereas ready availability of empty orbitals on an aluminium make it strongly electrophilic; thus organoaluminium compounds can act as 'amphophilic' reagents (i.e. derivatives that act as both a nucleophile and an electrophile in a given reaction), (ii) their alkylating as well as reducing ability (particularly towards transition metal salts), (iii) oxophilicity of aluminium, which makes reactions with oxygen and oxygen containing species (e.g. carbonyl) more facile, (iv) ability to form heterometallic bridges, and (v) ability for "ate" complex formation; the formation of 'ate' complexes suppresses their electrophilicity and makes organoaluminates reasonably good and uniquely useful nucleophiles, and (vi) diversified reactivity patterns with alkenes. As a consequence of the above, the various reactions of organoaluminium compounds of synthetic interest can be categorized as (a) C–E (E = an element other than aluminium) bond formation, (b) carbon-carbon bond formation, and (c) reduction.

Synthetic potentialities of organoaluminium compounds are apparent from their reactions discussed briefly in the context of the comparative chemistry of main group III organometallics. A schematic over-all view of the synthetic uses of organoaluminium reagents (including some reactions described in Secion 3.5.3) in organic synthesis is presented in Scheme 6.5, followed by a very brief discussion of the salient features of these reactions. In the Scheme 6.5 the reaction types are indicated by the letters a, b, c in parentheses.

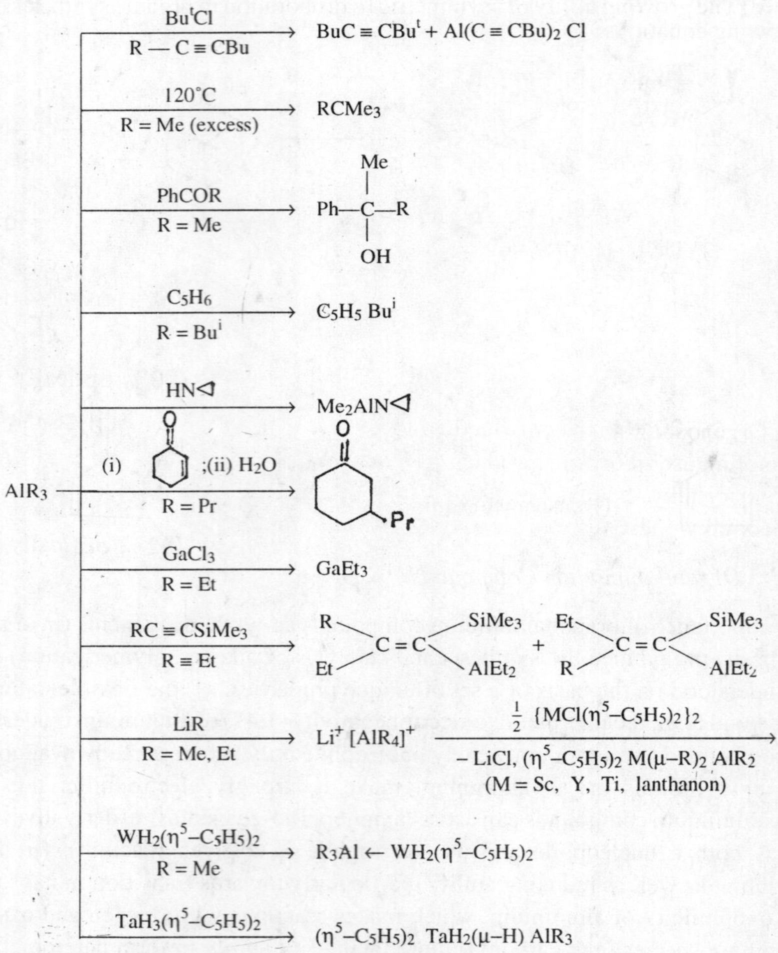

Scheme 6.5 Synthetic Uses of Organoaluminium Compounds

The insertion across Al–C bond and its facile cleavage with numerous inorganic species have been exploited in some useful transformations as depicted in Scheme 6.5. Under suitable reaction conditions, the products can be obtained in high yields.

Although alkylating ability of organoaluminium reagents are considerably lower than those of lithium and Grignard reagents, yet these have proved successful in generating organometallics of main group and transition metals, such as Be, Zn, Hg, B, Ga, Si, Ge, Sn, Pb, Ti, Zr, Ni, Ru, etc. It is important to note that the alkylation reactions of transition metal salts (in which the metal has a high oxidation state) generally proceed with reduction. The most significant reactions of these types are the formation of an active catalyst in the Ziegler-Natta polymerization reaction (Section 6.4.2.6) in which aluminium alkyls appear to bring about both alkylation as well as reducing reaction represented by the following equations:

$$TiCl_4 + AlR_3 \rightarrow RTiCl_3 + R_2AlCl$$

$$RTiCl_3 \rightarrow TiCl_3 + R^{\cdot}$$

$$R^{\cdot} + R^{\cdot} \rightarrow R-R$$

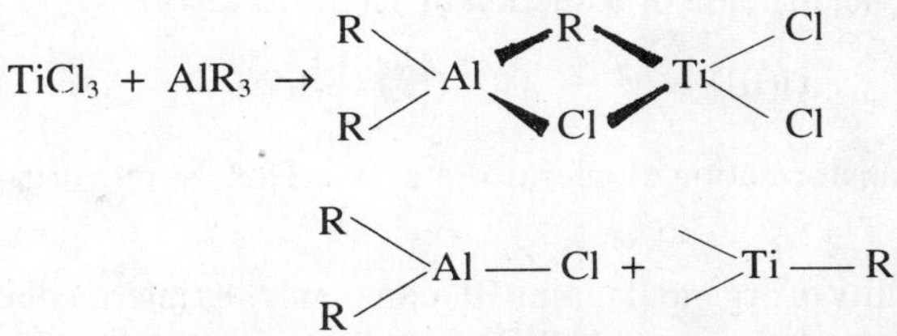

Organoaluminium compounds react with a variety of carbon electrophiles (R^+) generating useful organic products as shown in Scheme 6.5. Typical primary and secondary alkyl halides react sluggishly with AlR_3. By contrast reactions with tertiary alkyl halides are facile to form cross-coupled products:

$$R-C(Me)_2-Cl + AlMe_3 \xrightarrow[-78^{\circ}C]{} R-C(Me)_2-Me + AlClMe_2$$

($R = Me, Et, Pr^i$ and Bu^t)

Another interesting route involving cross-coupling is the reaction of tertiary alkyl halides (R′X) with Al (C ≡ CR)$_3$:

$$Al\,(C \equiv CR)_3 + R'X \xrightarrow{(CH_2Cl)_2,\ 0^{\circ}C} RC \equiv CR' + Al\,(C \equiv CR)_2X$$

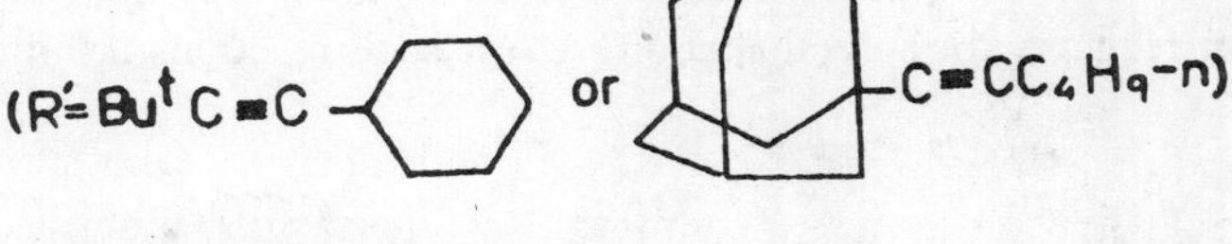

6.3.1.6 Organothallium Compounds

The success achieved by the organometallic chemist in utilizing derivatives of commoner elements has led him to explore the possibility of using derivatives of less common elements like thallium for similar purposes. The most important properties of thallium compounds which have been exploited in organic synthesis are:

(i) Thallium-Carbon bonds are weak, probably of the order of 100–130 kJ mol^{-1}, and cleave under both heterolytic and homolytic conditions.

(ii) Tl^+ is a relatively soft acid, generating tricoordinate species with bidentate ligands (including protic species), which react with organic electrophiles in a regiospecific manner.

(iii) Tl^{3+} is a soft acid cation; the electrophilicity of thallium(III) compounds can, however, be varied substantially and in a controlled manner by the variation of the nature of groups attached to it.

(iv) Thallium(III) is isoelectronic with Hg(II) and Pb(IV) and its redox potential is intermediate between them. Thallium(III), therefore, functions as an efficient 2-electron oxidant.

The thermodynamic ease of reduction of Tl(III) to Tl(I):

$$Tl(III) + 2e \rightarrow Tl(I)\ (E^\circ = -1.25\ V),$$

leads to a facile transformation of intermediates like $Tl(R)X_2$ into organic products and Tl(I) salts.

The applicability of organothallium(III) compounds as intermediates is largely confined to those derived via oxythallation of alkenes and alkynes and thallation of aromatic substrates. Hence the discussion in this section would obviously be centralized around synthetic transformations involving oxythallation and thallation reactions:

(a) *Oxythallation reaction* As stated before Tl(III) undergoes a facile 2-electron reduction to Tl(I). Due to this, oxythallation of alkenes and alkynes is generally quite facile and occurs rapidly under ambient conditions, the monoalkylthallium(III) compounds formed initially tend to decompose readily yielding oxidation products of the organic substrate and Tl(I) salts. In the case of non-aromatic substrates, the utility of Tl(III) salts TlX_3, as oxidants appears to depend mainly on the following three factors, i.e. nucleophilicity of the anion 'X', the solvent and the structure of the alkyl group. The electronegativity of 'X' determines the extent of electrophilicity of Tl(III), enhancing the rate of oxythallation reaction with increasing positive charge on thallium. However, from practical considera-tion, X should preferably be rather a weak nucleophile, facilitating the adduct formation (solvo-thallation) with solvent molecules. For example, trifluoroacete, nitrate, perchlorate and fluoroborate salts of Tl(III) are highly ionic and function as efficient reagents for solvothallation. A few such reactions are illustrated as follows:

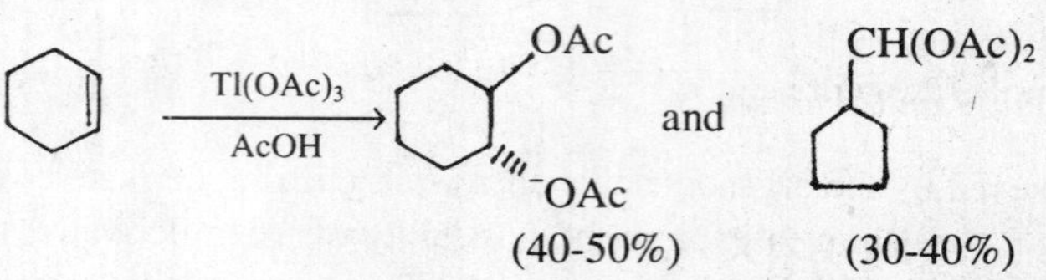

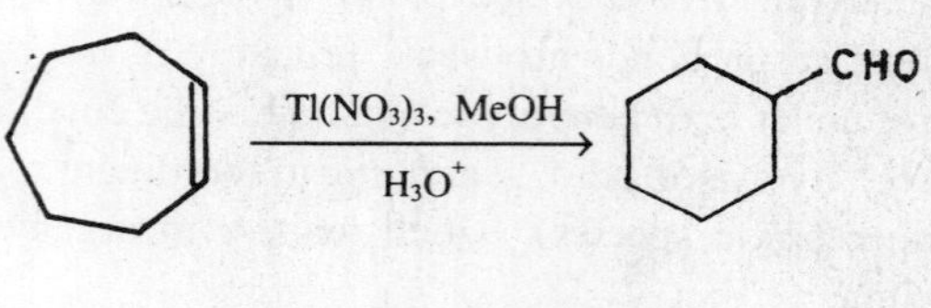

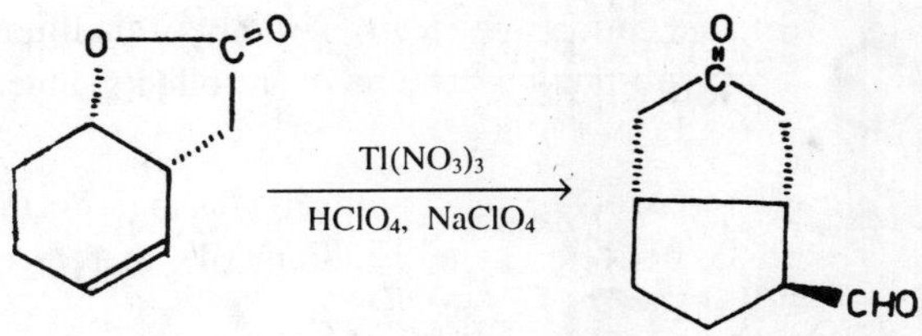

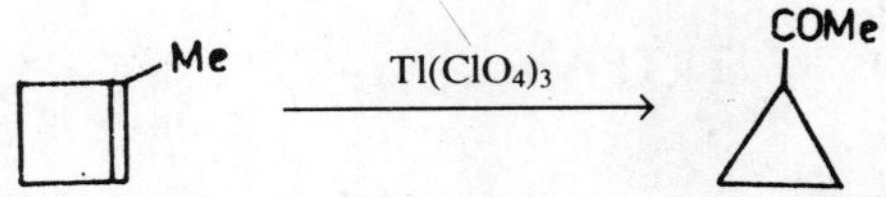

The reactions of alkynes with $Tl(NO_3)_3$ involve oxidation rather than hydration of the triple bond. The nature of the products is strongly dependent on the solvent used and the structure of alkyne, e.g.:

$$RC \equiv CH \xrightarrow[H_2O]{Tl(NO_3)_3} RCO_2H + CH_2O + Tl(I)$$

$$PhC \equiv CPh \xrightarrow[H_3O^+]{Tl(NO_3)_3} Ph{-}\overset{\overset{O}{\|}}{C}{-}\overset{\overset{O}{\|}}{C}{-}Ph$$

$$EtC \equiv CEt \xrightarrow[H_3O^+]{Tl(NO_3)_3} Et{-}\overset{\overset{O}{\|}}{C}{-}\overset{\overset{OH}{|}}{C}HEt$$

$$PhC \equiv CR \xrightarrow[MeOH]{Tl(NO_3)_3} Ph{-}\overset{\overset{R}{|}}{C}HCO_2Me$$

(b) *Thallation reactions* The synthetic utility of aromatic thallation lies in the remarkable degree of orientation control (i.e. regiospecificity) which can be exercised over the process and the ease with which the resultant thallium(III) compounds can be converted into substituted derivatives in which the new substituent enters the ring at the position originally occupied by thallium. A few illustrative examples are represented below:

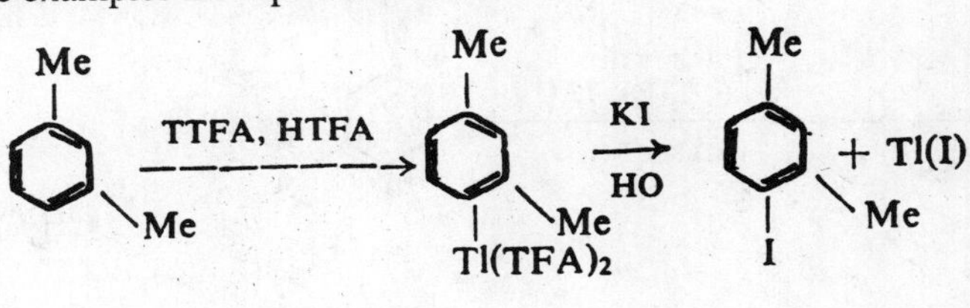

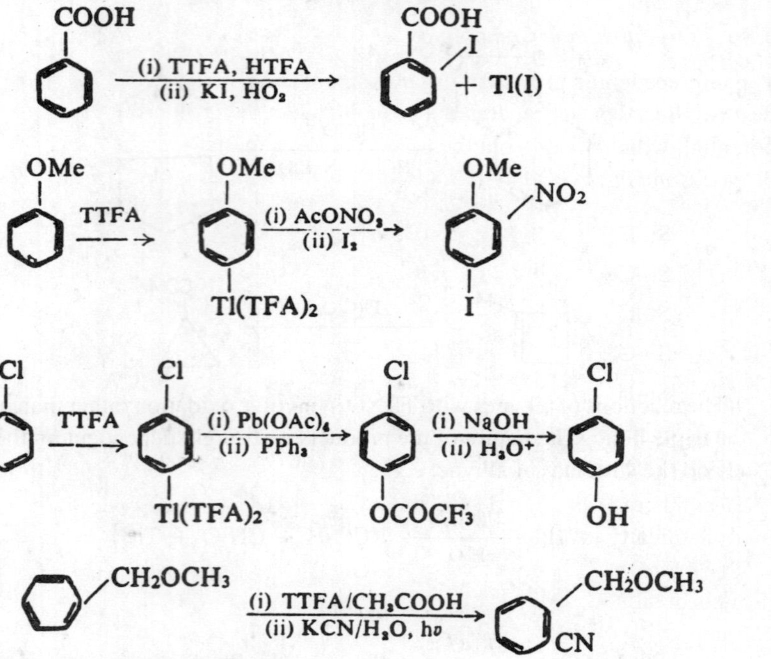

HTFA = Trifluoroacetic acid;

TTFA = Thallium(III) trifluoroacetate

The versatility and control offered by thallation reactions can be illustrated by one other example, which is truly spectacular. Thus, thallation reactions allow 2-phenylethanol to be selectively converted into ortho-, meta- and para-iodo isomers:

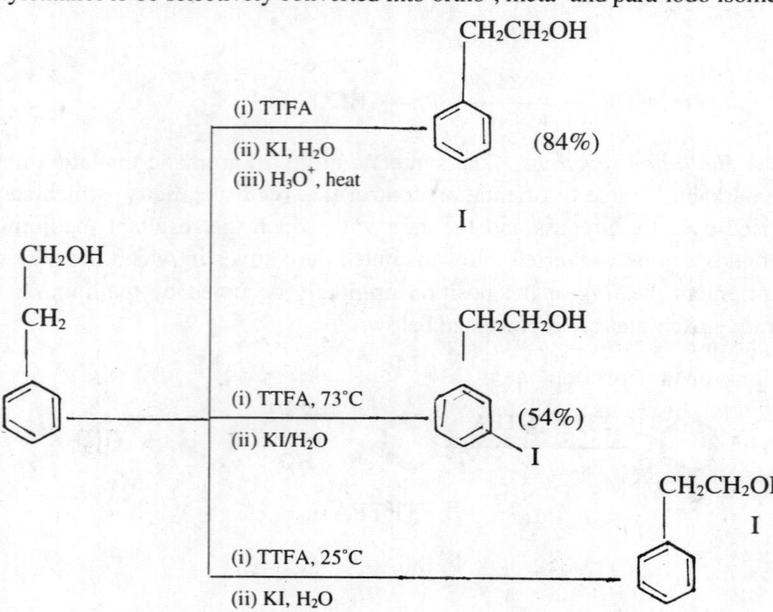

6.3.1.7 *Organosilicon Compounds*

The unprecedented growth of organosilicon chemistry (cf. Section 3.5.4) over the past two decades has generated a growing awareness of its considerable synthetic potential, which arises from the following three main factors:

(i) *Comparative bond dissociation energies* (in kJ mol^{-1})

Si–F	810	C–F	490
Si–Cl	470	C–Cl	340
Si–O	530	C–O	360
Si–C	320	C–C	340
Si–H	340	C–H	420

The Si–F, Si–Cl and Si–O bonds are stronger than the corresponding C–F, C–Cl and C–O bonds, whereas Si–C and Si–H bonds are weaker than C–C and C–H bonds respectively. These characteristics give rise to a wide range of thermodynamically favourable processes.

(ii) *Vacant d-orbitals*. The tendency of π-bond formation between the 3d orbitals of silicon and 2p orbital on a carbon atom appears to stabilize the adjacent carbanion:

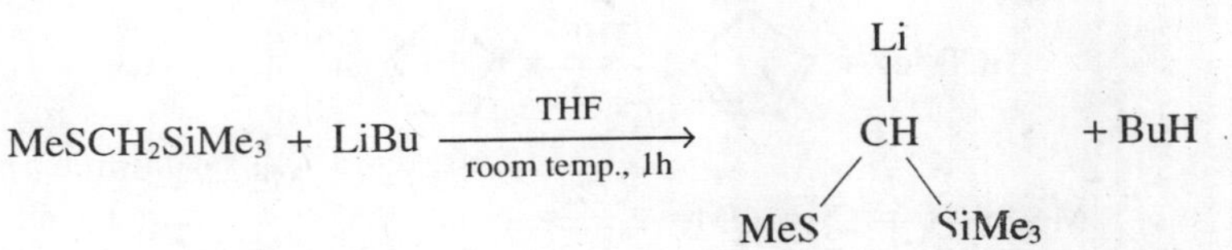

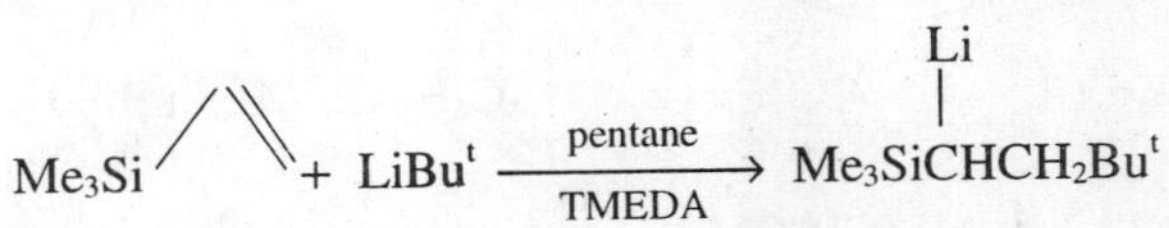

$$(Me_3Si)_2CH_2 + LiBu^t \xrightarrow[-40°C,\ 7.5h]{HMPT} LiCH(SiMe_3)_2 + Bu^tH$$

$$(Me_3Si)_3CH + LiMe \xrightarrow[reflux,\ 6h]{THF} LiC(SiMe_3)_3 + MeH$$

The presence of 'd' orbitals also facilitates the nucleophilic substitution reactions on silicon compared to those on carbon. The ease of nucleophilic substitution reactions at silicon is so dominating that even stronger bonds such as Si–F, Si–O, Si–C, and Si–H are sometimes displaced by suitable reagents. A few illustrative examples are:

$$Et_3Si\text{–}F + AlX_3 \rightarrow Et_3Si\text{–}X + AlFX_2$$

(X = Cl, Br, I)

$$(Me_3Si)_2O + TiCl_4 \rightarrow 2Me_3Si\text{–}Cl + TiOCl_2$$

$$Me_3Si{-}C_6H_5 + HCl \rightarrow Me_3Si{-}Cl + C_6H_6$$

$$Et_3Si{-}H \xrightarrow[\text{room temp.}]{5\%\ Pd/C,\ HX,\ C_6H_6} Et_3Si{-}X \quad (X = F, Cl)$$

(iii) *Relative electronegativity* On the Pauling scale electronegatives of silicon and carbon are 1.8 and 2.5 respectively, thus making silicon-carbon bond polarized in the direction: $Si^{\delta+} - C^{\delta-}$, this renders the silicon atom more susceptible to nucleophilic attack, leading to bond heterolysis, especially when the carbon fragment is a better leaving group, e.g.:

$$Me_3Si{-}Ph + NaOMe \xrightarrow{MeO/HDMSO} Me_3Si{-}OMe + NaPh$$

With the help of the above special features, a large number of synthetically useful reactions can be rationalized in terms of the following general parameters:

(a) Nucleophilic substitution at silicon: The polarizability of the Si–C bond in the direction shown together with the availability of d-orbitals, makes nucleophilic substitution at silicon much easier than the corresponding reactions at carbon, e.g.:

$$Si(CH_2Ph)Me_3 + F^- \longrightarrow SiFMe_3 + {}^-CH_2Ph$$

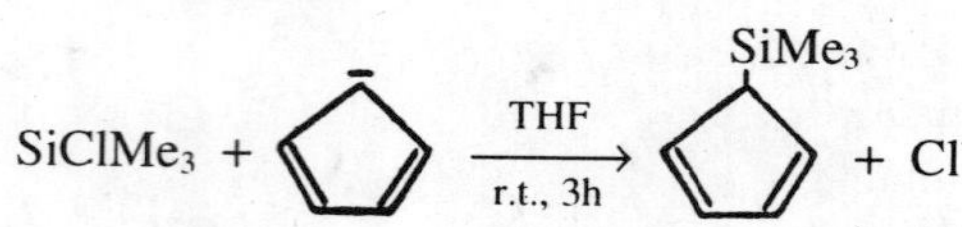

$$Me_3Si{-}I + C_6H_{11}\ddot{O}Me \xrightarrow[50^\circ C,\ 2h]{CCl_4} C_6H_{11}\overset{+}{O}(Me)(SiMe_3)\ I^- \rightarrow C_6H_{11}OSiMe_3 + MeI$$

$$Me_3Si{-}I + Me\ddot{O}CH_2OMe \xrightarrow{25^\circ C} Me_3Si(OMe) + ICH_2OMe$$

It may be pertinent to emphasize that due to weakness of Si=C double bonds, the alternative type of elimination, i.e. $Me_3SiI \xrightarrow{\text{base}} CH_2 = SiMe_2$ does not compete with the facile substitution processes illustrated above.

In fact, the facility of such substitution reactions often makes it possible to add silyl protecting groups to substrates like acetylenes and to remove them again, eg.,

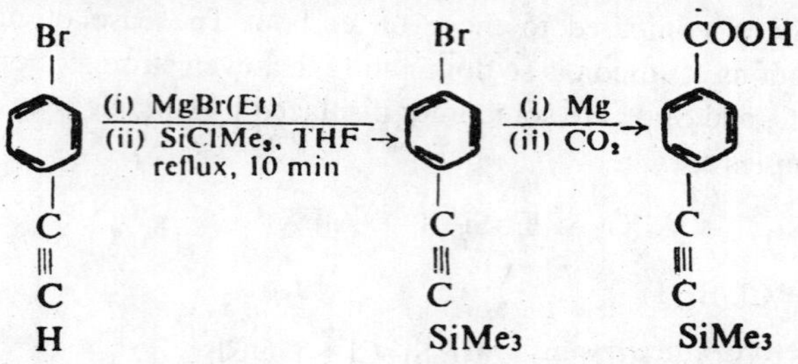

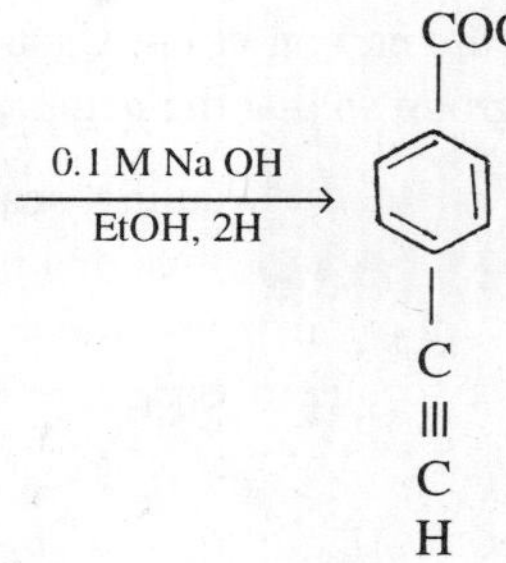

The loss of Me_3Si as a leaving group is extremely facile, but the Me_2Bu^tSi group is comparatively stable (except to acids):

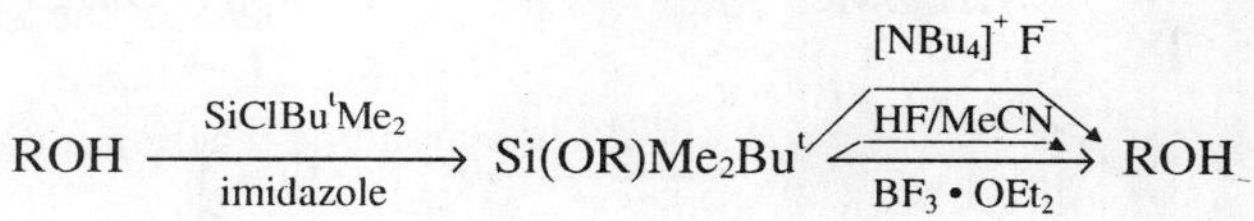

In stereochemical terms these nucleophilic substitution reactions proceed via trigonal bipyramid intermediates. The penta-coordinated organosilicon compounds under some suitable conditions (e.g. if one or more group (X) attached to silicon are strongly electron withdrawing groups such as fluoride atoms) can be stabilized and isolated as reasonably stable moieties:

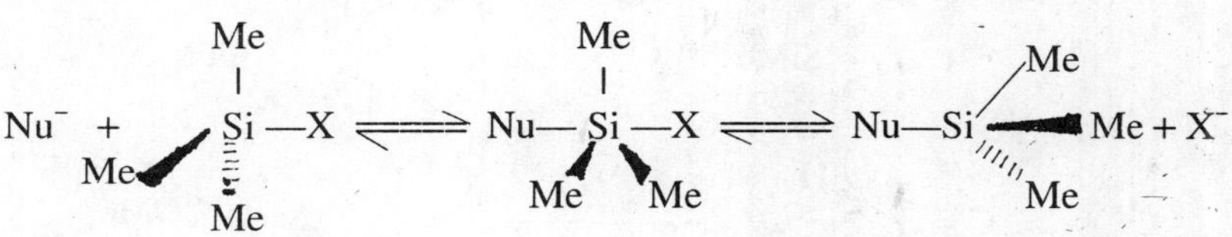

(b) Stabilization of a carbonium ion (i.e. electrophilic character) to the trialkylsilane group (the β-effect) can be illustrated by the greater stability of

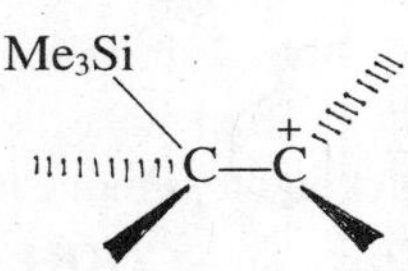

compared to

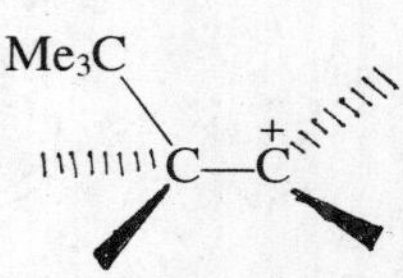

The behaviour of β–functional silanes can be explained in terms of the following factors: (i) inductive effects; (ii) interactions between the d-orbitals of the silicon atom and electrons in π-type orbitals [(π-d) π-bonding] or p-type orbitals [(p-d) π-bonding]; (iii) silicon-carbon hyperconjugation; and (iv) various combinations of these.

As a consequence of these parameters ethynl-, phenyl-, vinyl-, and allylsilanes, and sometimes their homologues, react with electrophiles (E^+) in a well defined and predictable manner. Due to the stabilization of the intermediate cationic species by the Si–C bond, electrophilic attack on ethynyl, phenyl and vinylsilanes is more susceptible at the carbon atom. Usually the outcome of the reaction is elimination of the silyl group so that the net reaction is a site selective substitution.

Typical examples are:

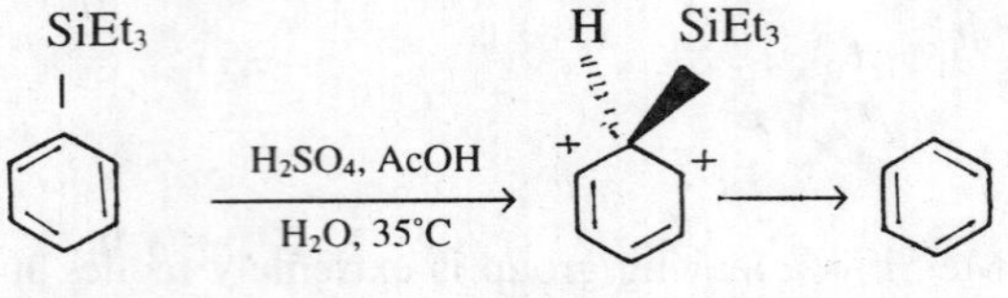

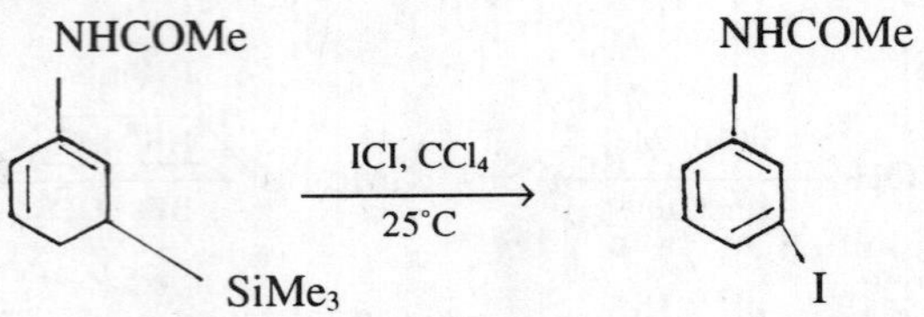

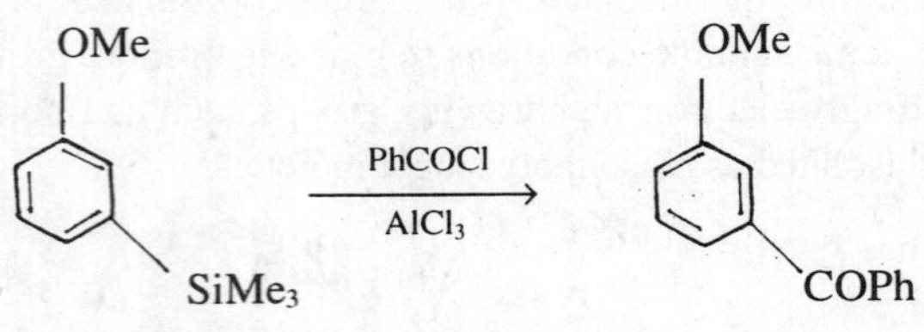

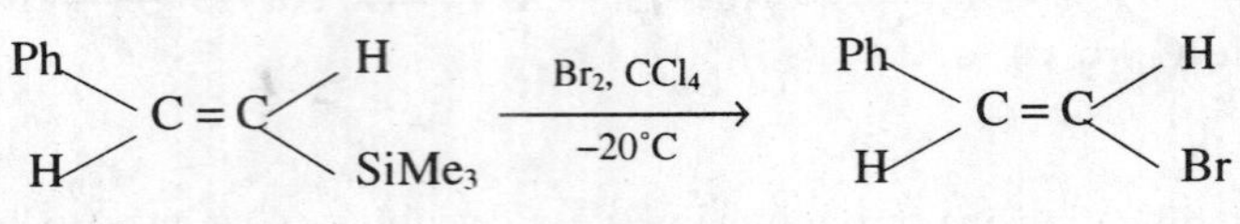

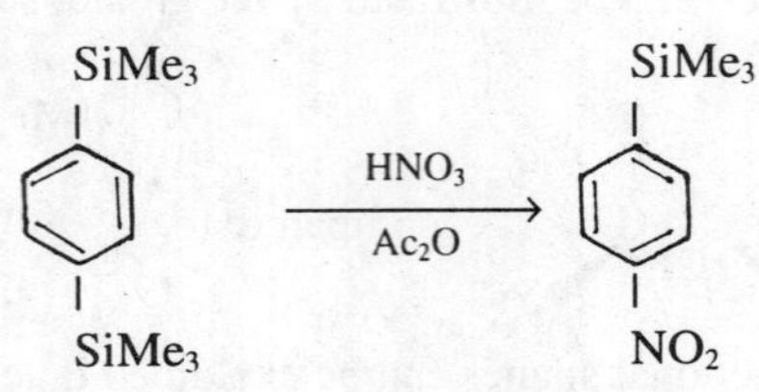

$$Me_3SiC \equiv CSiMe_3 + MeCOCl \xrightarrow[CS_2 \text{ or } PhNO_2]{AlCl_3} Me_3SiC \equiv CCOMe + SiClMe_3$$

(c) The comparatively higher strength of single bonds from silicon to electronegative elements like oxygen, fluorine, and chlorine, affects the direction of some reactions usually in a manner such that a weaker bond to silicon (Si–C, Si–N or Si–H) is replaced by a Si–O or Si-halogen bond:

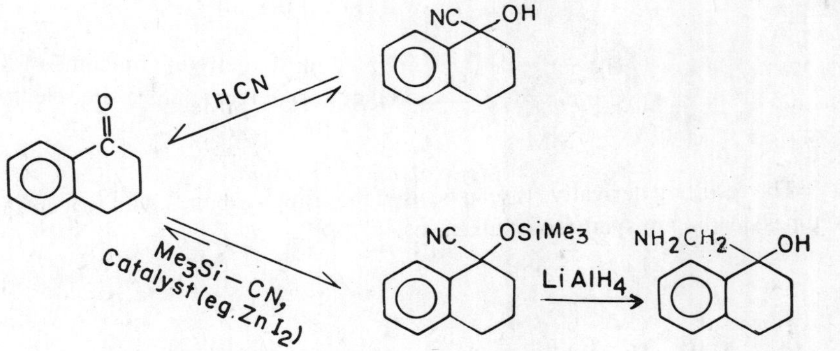

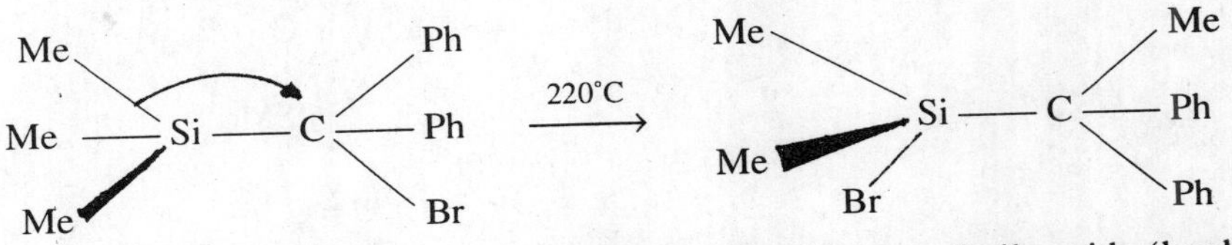

(d) Substitution at silicon bonded to carbon is more facile with 'hard' nucleophiles such as an oxygen, a chloride or a fluoride anion than the corresponding nucleophilic substitution at hydrogen:

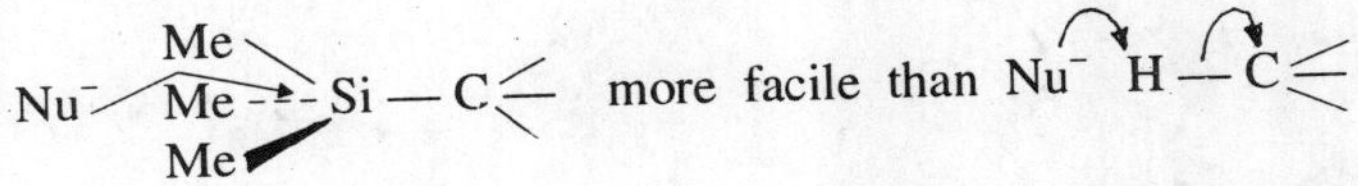

This aspect has been exploited in many applications of carbon bound silicon in organic synthesis:

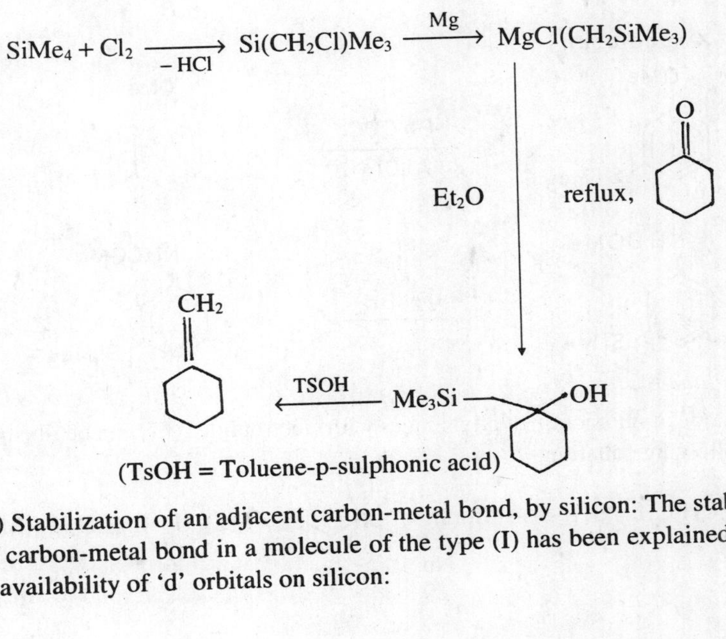

(TsOH = Toluene-p-sulphonic acid)

(e) Stabilization of an adjacent carbon-metal bond, by silicon: The stabilization of carbon-metal bond in a molecule of the type (I) has been explained to be due to availability of 'd' orbitals on silicon:

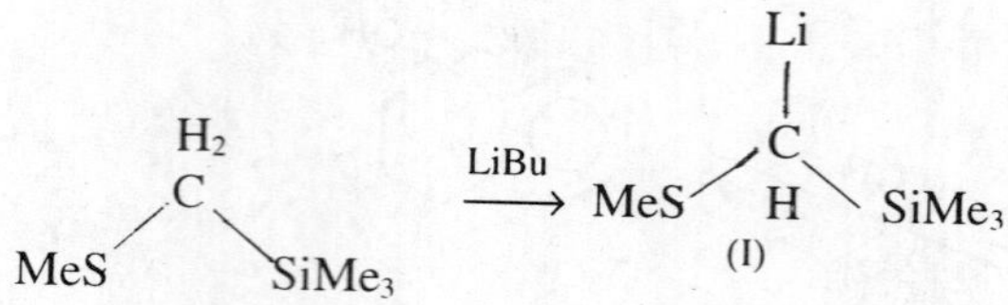

The resulting derivative (I) can be used to form C–C bonds with reagents like ketones, leading to synthesis of alkenes:

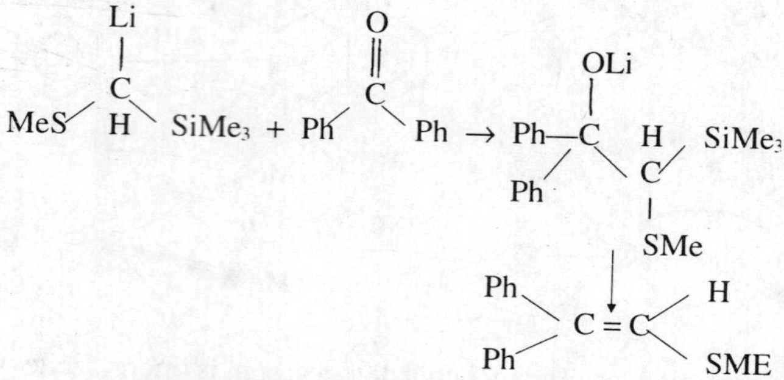

Apart from these synthetically useful reaction types, the utility of some specific organosilicon compounds in interesting organic synthesis can be demonstrated by the following reactions:

Arylsilanes

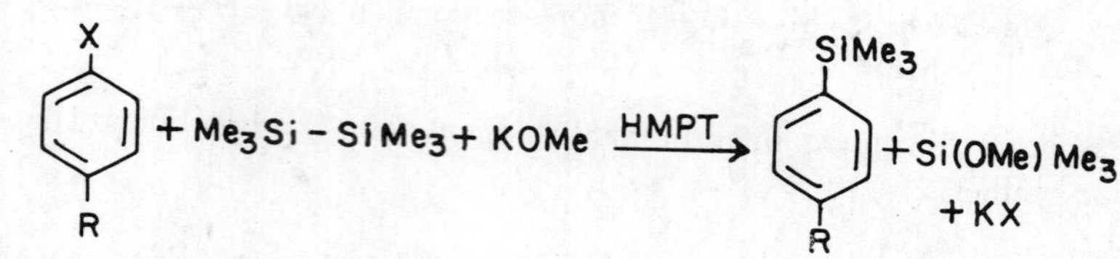

(X = halogen)

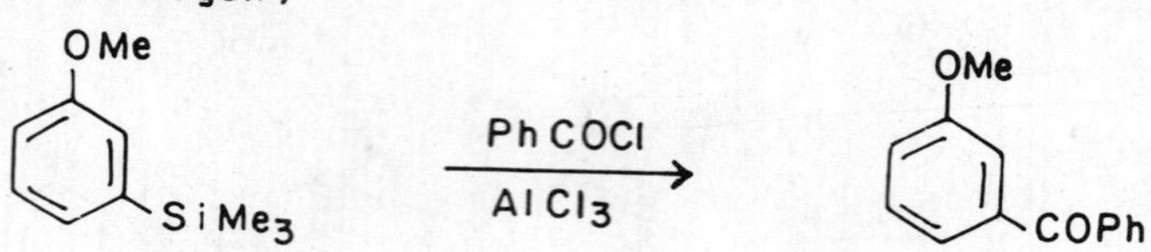

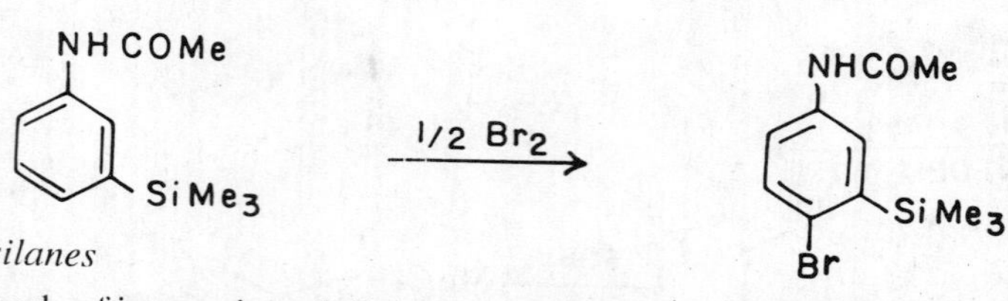

Allylsilanes

The mode of interaction of allylsilanes with electrophiles (E^+) can be illustrated by the following equation:

From synthetic point of view, four typical reactions of allylsilanes are mentioned below:

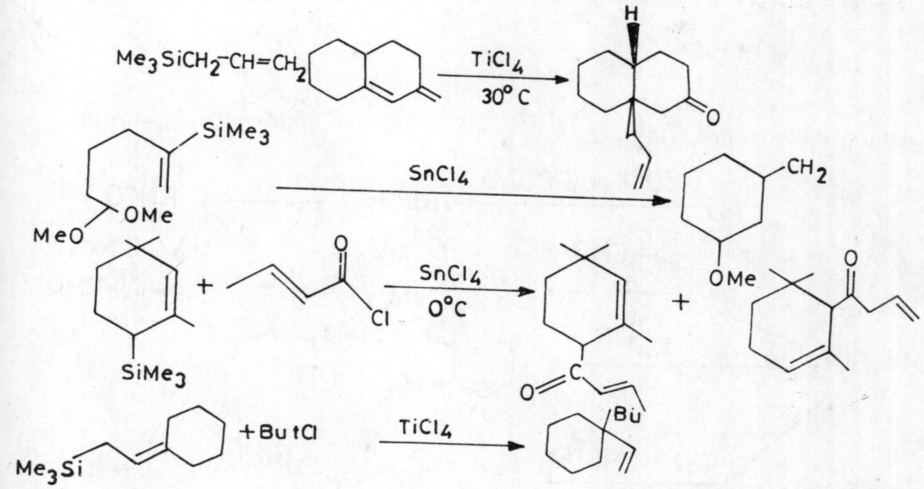

Vinystanes

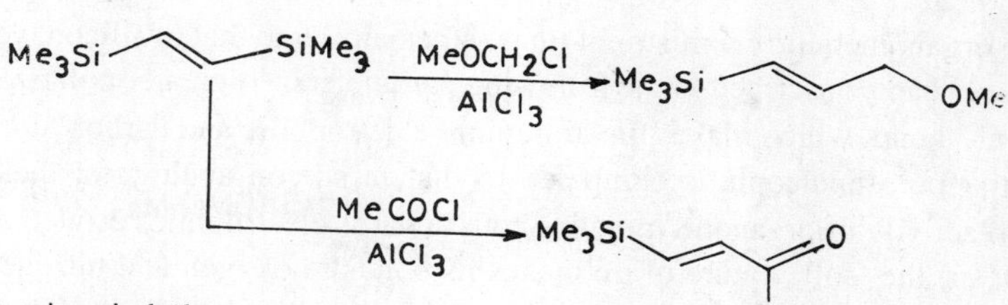

Iodotrimethylsilane

$$\text{MeOCH}_2\text{OMe} + \text{SiIMe}_3 \longrightarrow \text{MeOCH}_2\text{I} + \text{Si(OMe)Me}_3$$

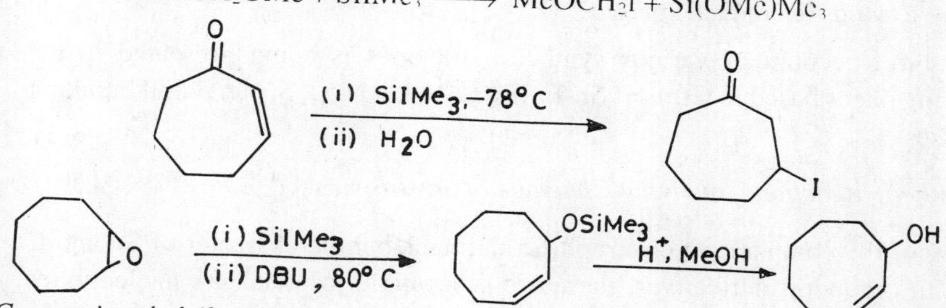

Cyanotrimethylsilane

$$\text{R}'\text{—}\overset{\text{O}}{\overset{\|}{\text{C}}}\text{—R}'' \xrightarrow{\text{Si(CN)Me}_3} \text{R}'\text{—}\underset{\text{CN}}{\underset{|}{\overset{\text{OSiMe}_3}{\overset{|}{\text{C}}}}}\text{—R}''$$

$$\text{R—S—Cl} \xrightarrow{\text{Si(CN)Me}_3} \text{R—S—CN}$$

$$R{-}N{=}C{=}N{-}R \xrightarrow[AlCl_3]{Si(CN)Me_3} R{-}\underset{\underset{SiMe_3}{|}}{N}{-}\overset{\overset{CN}{|}}{C}{=}N{-}R \xrightarrow{MeOH} R{-}\overset{\overset{H}{|}}{N}{-}\overset{\overset{CH}{|}}{C}{=}N{-}R$$

Azidotrimethylsilane

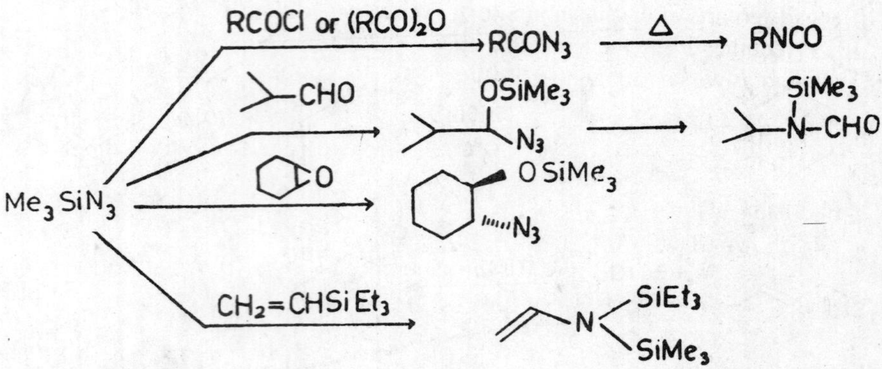

6.3.1.8 Organotin Compounds

Although the organometallic chemistry of tin is rather similar to that of silicon, yet finer differences can be ascribed to the following parameters: (i) greater polarization of $Sn^{\delta+}$–$C^{\delta-}$ bond, which makes tin atom more electrophilic and carbon atom attached to tin more nucleophilic compared to that in silicon analogues; these enhance the reactivity of organotin moieties both towards electrophiles as well as nucleophiles; (ii) the 'soft' nature of tin makes its bonds to oxygen and nitrogen more reactive whereas Sn–S bonds are more stable compared to their silicon counterparts. Another important point to be noted is that Sn–H bond is highly reactive toward free radical species.

A brief account of various synthetic strategies is being presented in the following pages based in terms of Sn-E (E = C, H, Cl, N, O, S, etc.) bond breaking processes.

A. *Synthetic Reactions Involving Cleavage of Sn-C Bonds*

In view of the comparative electrophilic and nucleophilic character of Sn and C atoms in organotin derivatives, these are susceptible to attack by nucleophiles (Nu^-) as well as electrophiles (E^+):

$$\begin{array}{c} H_3C \text{———} Sn(CH_3)_3 \\ \downarrow \; + \; \uparrow \\ Br \text{———} Br \end{array} \xrightarrow{0^\circ C} CH_3Br + SnBr(CH_3)_3$$

$$(\text{cf. } SiMe_4 + Br_2 \xrightarrow[h\nu]{110^\circ C} Si(CH_2Br)Me_3 + HBr)$$

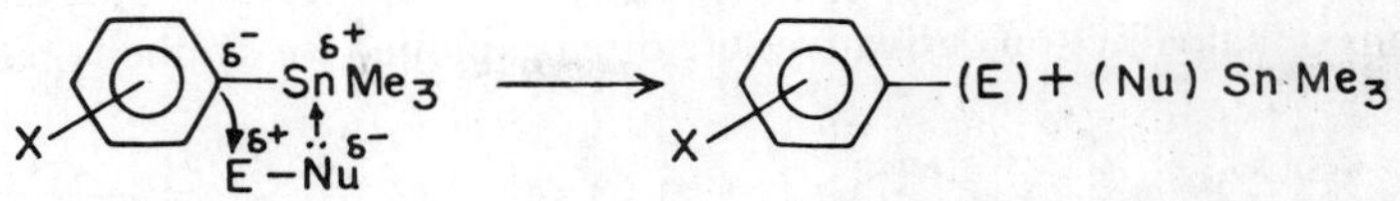

$E - Nu: = RCOCl(AlCl_3, 0°C): RSO_2Cl(AlCl_3, 0°C):$
$ArX(X \neq \text{halogen})$

In the latter reactions, C(phenyl)-Sn bond is preferentially attacked for obvious reasons and the facility of the reactions at this site generally leaves the other sensitive parts of the substrate unaffected.

Another attractive feature of the synthetic utility of these reactions involving alkynyl as well as allyl and benzyl derivatives is their regiospecific nature, e.g.,

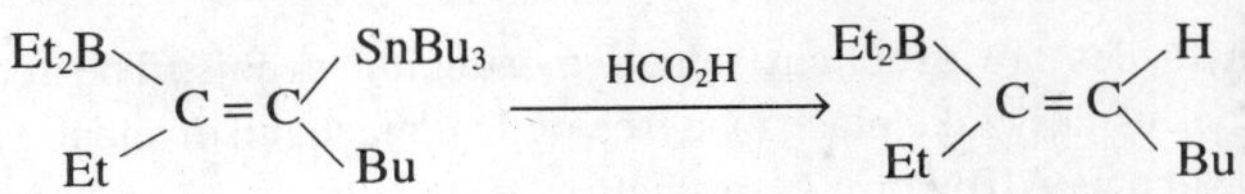

$$RC \equiv SnBu_3 \xrightarrow{SO_2Cl_2} RC \equiv CSO_2Cl$$

$$R_3SnCH_2C \equiv CH \xrightarrow{I_2} ICH = C = CH_2$$

The strongly basic organolithium reagents (LiR) readily cleave Sn-C bonds ($\equiv$Sn—C + LiR → Li—C + $\equiv$Sn—R) by displacing less basic carbanions. The reaction is often stereospecific and is particularly useful for the synthesis of vinyllithium compounds which may be difficult to obtain by other methods:

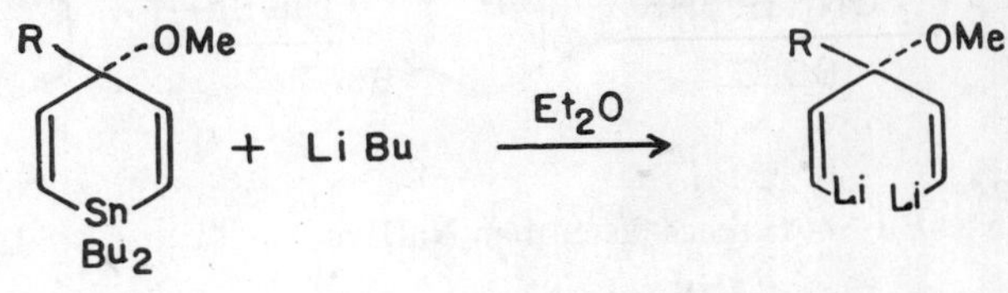

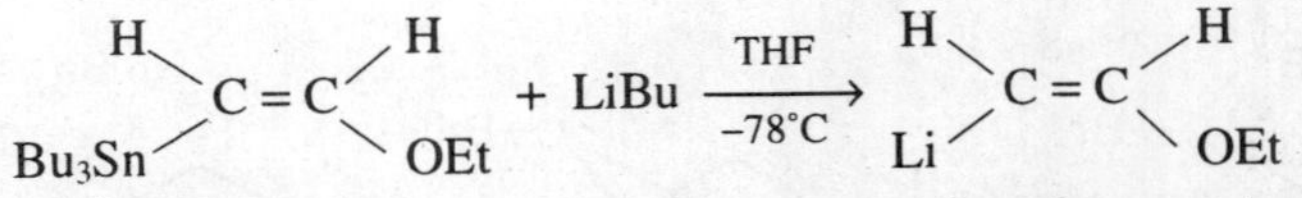

Organotin compounds can undergo Grignard type reactions particularly with carbonyl containing substrate, e.g.,

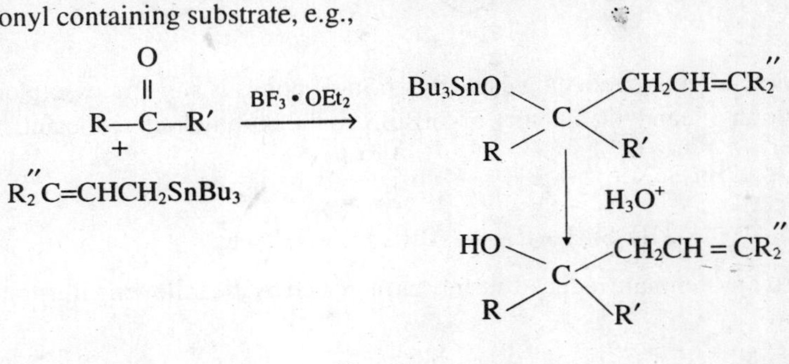

In addition to the above types of synthetically useful reactions, the elimination reactions of haloalkyl-tin derivatives are often exploited for synthetic purposes:

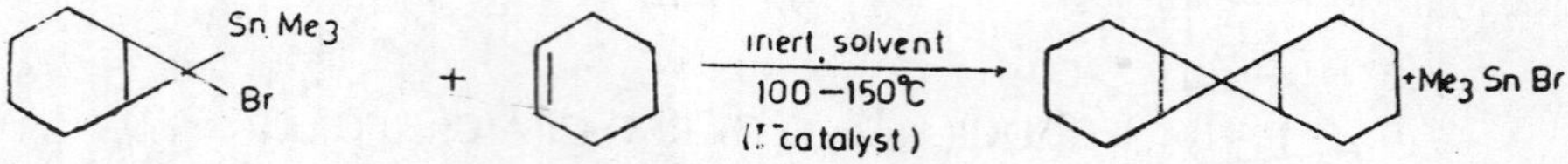

$Bu_3Sn\ O\ CH_2\ CH_2\ CH_2\ Br \xrightarrow[170^\circ C]{heat}$  $+\ Bu_3SnBr$

B. *Reactions of Sn-H Bonds*

Organotin hydrides are excellent selective reducing agents. The reductions of organic halides usually take place by a free radical mechanism and are accelerated by initiators such as AIBN or u.v. irradiation, e.g.:

$$^\bullet R' + R_3Sn - H \rightarrow R'H + {}^\bullet SnR_3$$

$$^\bullet SnR_3 + R''X \rightarrow R_3Sn - X + {}^\bullet R''$$

$$^\bullet R'' + R_3Sn - H \rightarrow R''H + {}^\bullet SnR_3 \text{ etc.}$$

Halide reactivity lies in the order: RI > RBr > RCl > RF, e.g.:

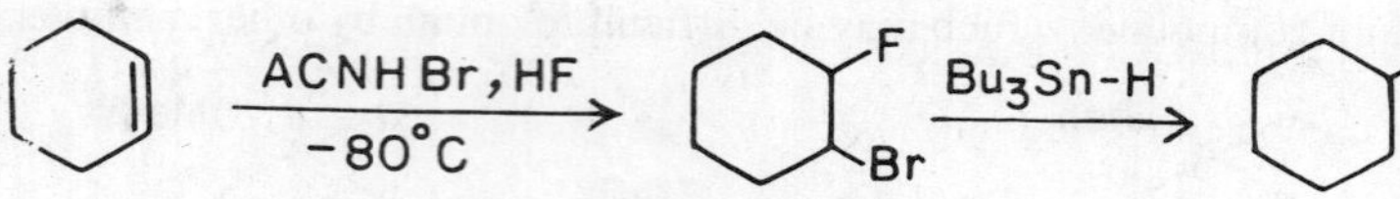

Reagents such as $Bu_3Sn–H$ react faster than $NaBH_4$:

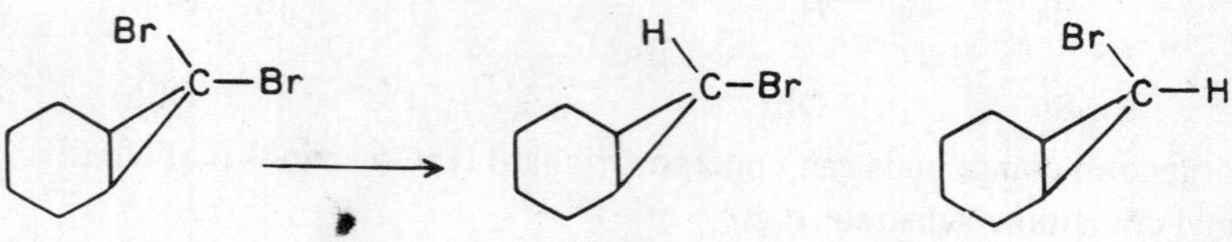

The more effective reducing capacity of Sn–H bonds is sometimes exploited by the addition of catalytic quantity of $SnBu_3X$ to the borohydride reductant:

$$Bu_3SnX + \geq BH^- \rightarrow Bu_3Sn—H + \geq B—X^-$$

$$Bu_3SnH + RX \rightarrow Bu_3SnX + RH, \text{ etc.}$$

Sensitive systems are reduced cleanly as indicated by the following illustrative examples:

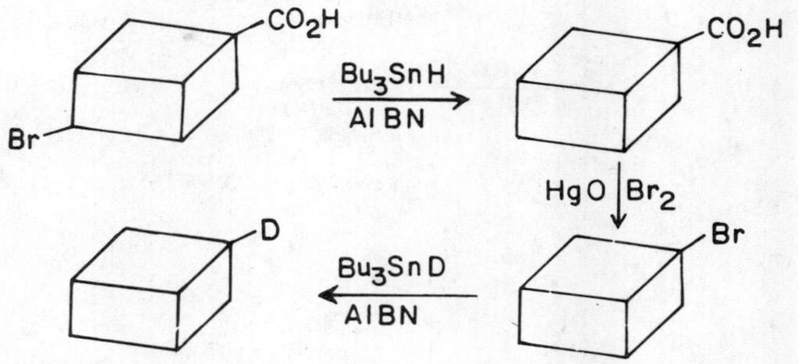

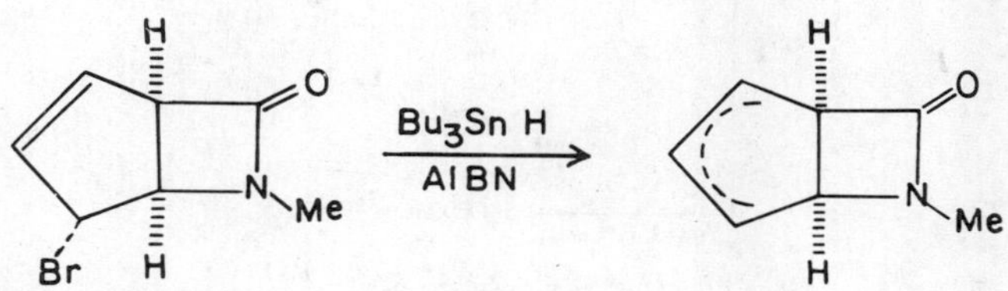

Acyl halides give aldehydes plus esters (a result of acyl radical attack on the product aldehyde):

$$PhCOCl + Bu_3SnH \rightarrow PhCHO + PhCO_2Ph$$

Alcohols can be converted into hydrocarbons *via* orthoformates:

$$ROH \rightarrow RCOCl \xrightarrow{Bu_3SnH} ROCO^{\bullet} \xrightarrow{-CO_2} R^{\bullet} \xrightarrow{Bu_3SnH} RH$$

Reductions of unsaturated substrates ($C \equiv C \rightarrow C = C$) can be illustrated by the reaction:

$H_2C(C \equiv CH)_2 + Bu_2SnH_2 \longrightarrow$ 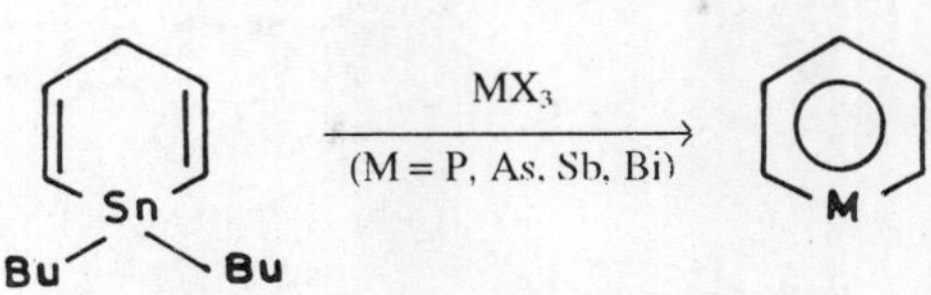

C. *Reactions of Sn–O Bonds*

Tin-oxygen bonds (~398 kJ/mole) are more reactive than Si–O bonds (530 kJ/mole) and add readily to many multiple bonded syntems (Scheme 6.8):

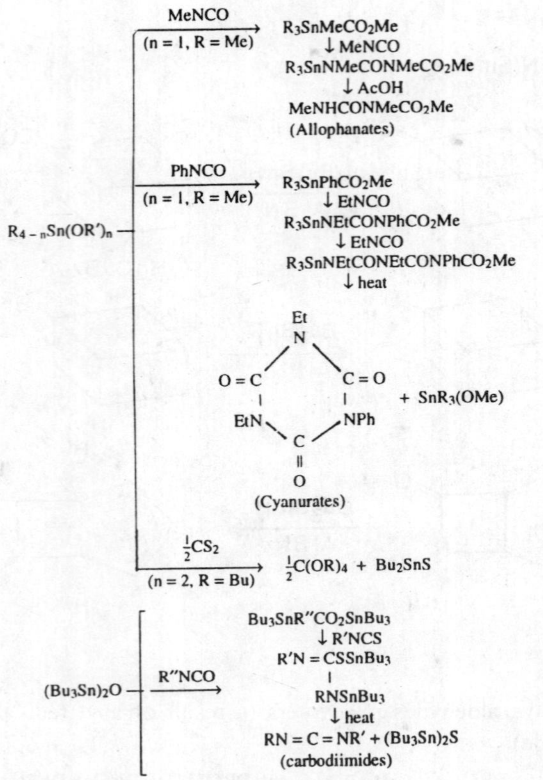

Scheme 6.8 Some Insertion Reactions into Tin-Oxygen Bonds

D. *Reactions of Sn–N Bonds*

Tin-nitrogen bonds are highly reactive and undergo substitution and addition reactions to afford useful products, e.g.,

Substitutions:

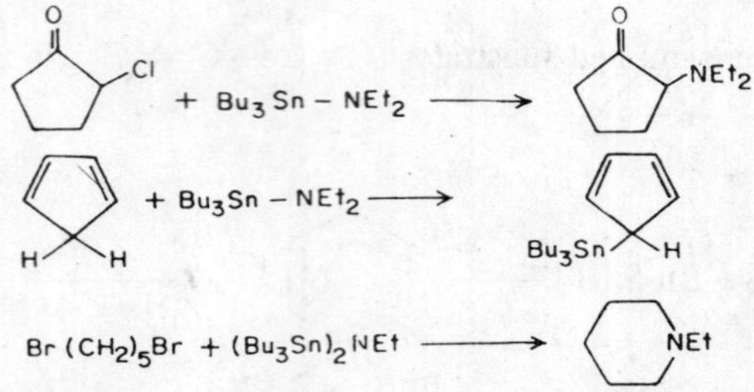

Additions:

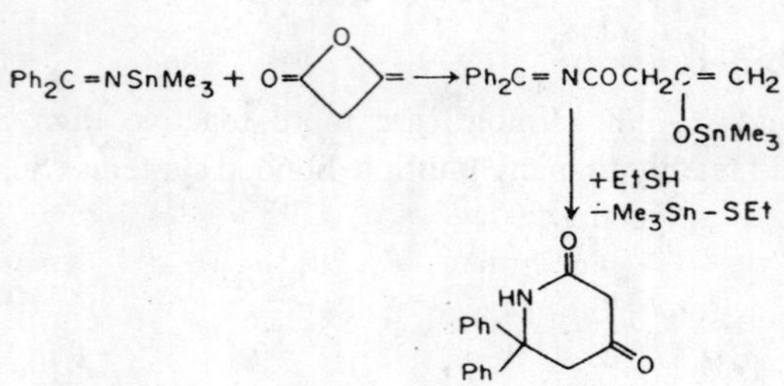

$$MeN(SnMe_3)_2 + PhNCS \xrightarrow{-50^{\circ}C} Me_3SnSC(=NPh)NMeSnMe_3$$

$$\downarrow$$

$$(Me_3Sn)_2S + PhN=C=NMe$$

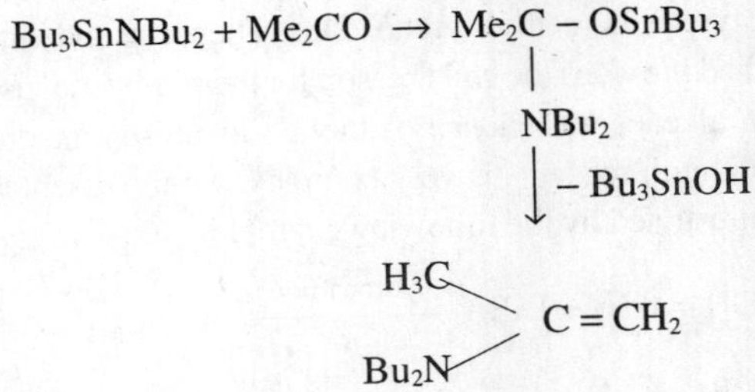

Cycloaddition:

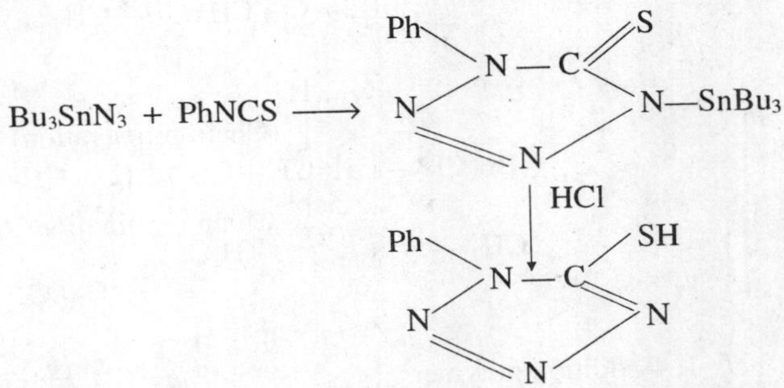

E. *Reactions Involving Sn–S Bonds*. The high thermodynamic stability of the Sn–S bonds over that of the Sn–O (~398 kJ/mole), Sn–N (167–251 kJ/mole), and other Sn-N heteroatom bonds suggest facile elimination of Sn–S bonded compounds from the reaction mixtures containing sulphur and other heteroatom species:

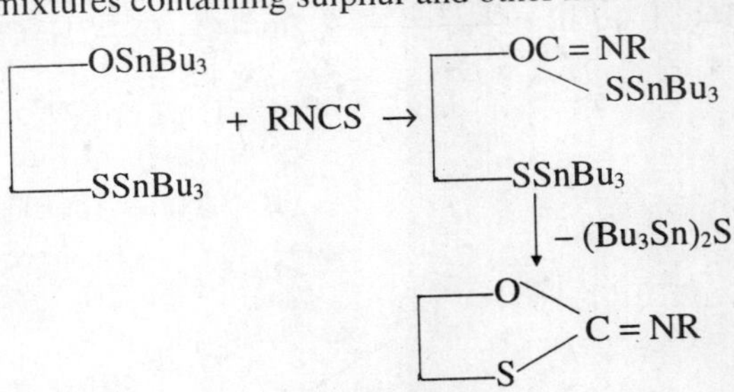

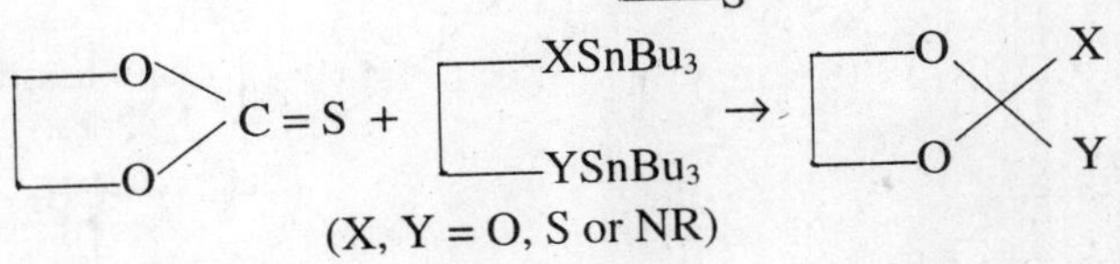

(X, Y = O, S or NR)

6.3.2 Catalytic Reactions

In contrast to the highly versatile synthetic utility of nontransition metal organometallics as stoichiometric reagents, their use as catalytic reagents has so far found limited applications, some of which are being discussed in this section.

The use of organolithium reagents as initiators for polymerization is of considerable industrial importance. For example, ethylene is polymerized by n-butyllithium in the presence of DABCO or TMEDA. The reaction of organolithium compounds with conjugated dienes (e.g. butadiene, isoprene, piperylene, and 2, 3-dimethylbutadiene), or alkenes (ethylene, styrene, α-methylstyrene, divinyl benzene and vinyl toluene) can lead to polymerization, and provides the chief reason for their industrial use for the production of elastomers from alkenes and dienes. Although chemistry of diene polymerization initiated by organolithium reagents is very complex, yet the essential features are quite straightforward as illustrated by the following example:

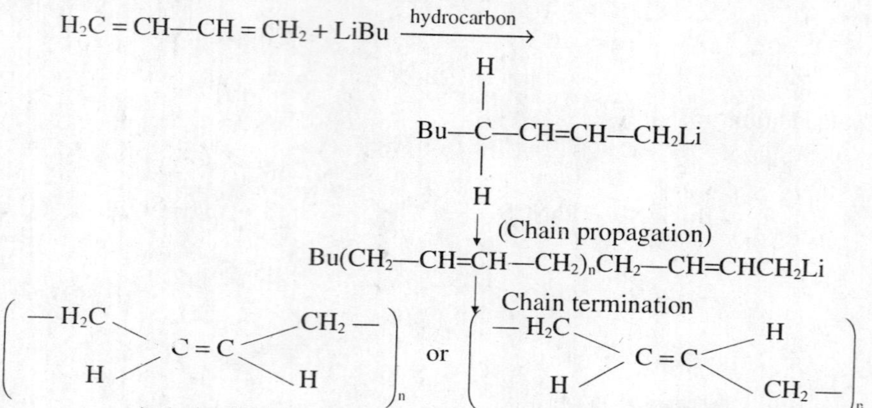

cis-1, 4-addition polymer   *trans*-1, 4-addition polymer

When LiBu brings about polymerization of 1, 3-butadiene in the presence of polar ligands (modifiers), 1, 2-addition appears to be favoured:

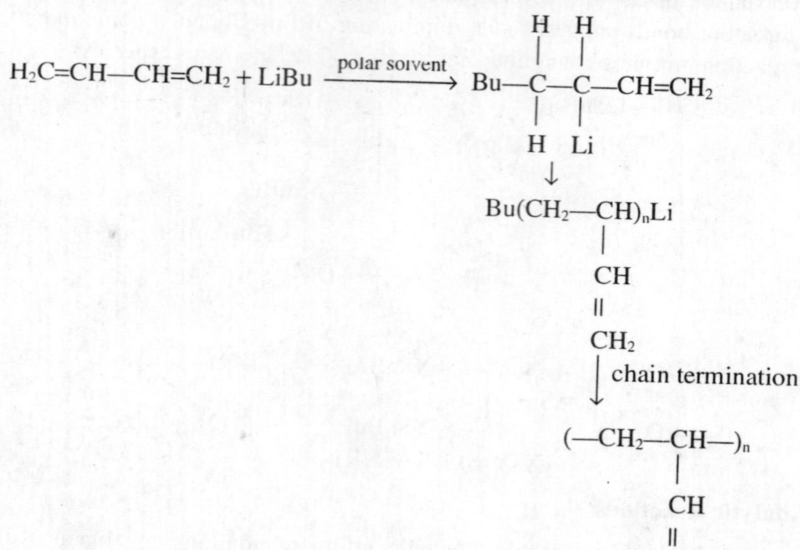

1, 2–addition polyme

The polymer isolated from the type of polymerization shows a high level of 1, 2 (vinyl) content.

Some possible modes of the addition of an alkyllithium compounds to 2-methylbuta-1, 3-diene (isoprene) are shown in Scheme 6.9.

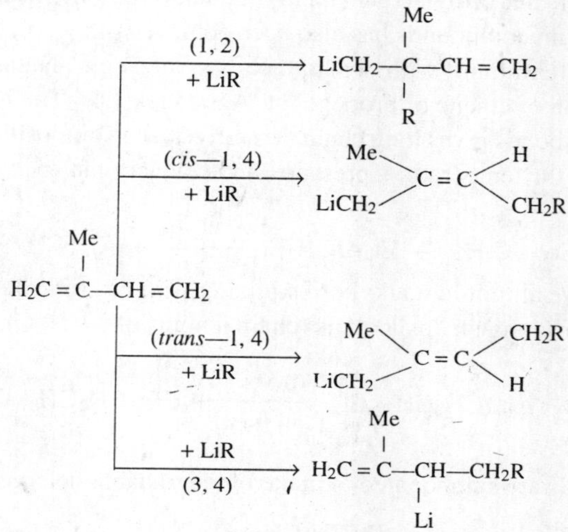

Scheme 6.9 Possible Mode of Addition of LiR to Isoprene.

Out of the above, the *cis*-1, 4-polymerization of isoprene is achieved in an aliphatic hadrocarbon-solvent and is of special significance resulting in a polymer (synthetic rubber) structurally similar to the natural rubber :

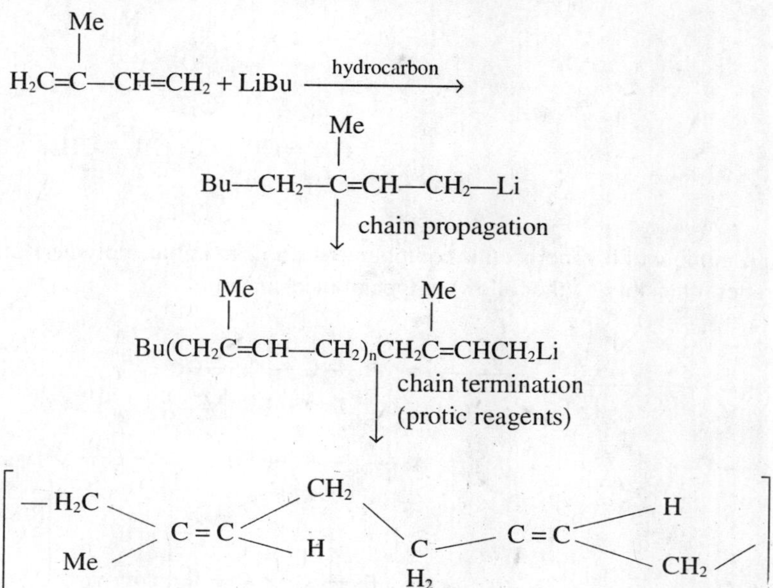

Synthetic rubber

A great deal of work has been done on organosodium initiated polymerizations also. For example, butylsodium has been used for the conversion of butadiene to bunna rubber.

The polymerization of dienes (such as styrene), arylalkenes (e.g. styrene, vinylpyridine), and vinyls (e.g. methylmethacrylate, methacrylonitrile) by organomagnesium compounds has also been achieved.

Triethylaluminium with $TiCl_4$, acts as a Ziegler-Natta catalyst for the polymerization of ethene or propene (cf. Appendex 1.1). The insertion of alkenes into the Al–C bonds gives long chain derivatives, the composition of which can be controlled by the temperature, pressure, and contact time:

$$AlEt_3 + C_2H_4 \rightarrow Et_2AlCH_2Et \xrightarrow{nC_2H_4} Al\{(CH_2CH_2)_nEt\}_3$$

The above aluminium alkyl on oxidizing with air, followed by hydrolysis by water produces long chain alcohols (chain lengths of 14-20 C):

$$Al\{(CH_2CH_2)_nEt\}_3 \xrightarrow[\text{(ii) } H_3O^+]{\text{(i) } O_2} 3Et(CH_2CH_2)_nOH$$

These alcohols are sulphonated to make biodegradable detergents,

$$Et(CH_2CH_2)_nOSO_3^-Na^+.$$

Dimerization of propene in the formation of isoprene is catalyzed by aluminium alkyls:

$$2CH_3CH = CH_2 \xrightarrow{AlR_3} CH_3CH_2CH_2C(CH_3) = CH_2 \xrightarrow{\text{crack}} H_2C = CH—C(CH_3) = CH_2 + CH_4$$

A number of organometallic compounds are able to initiate polymerization of oxiranes, plausibly by the following type of mechanism:

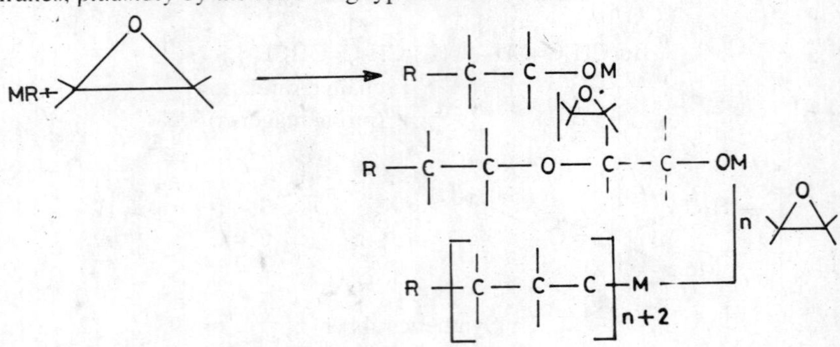

Organozinc compounds, which are often employed as components of Ziegler-Natta catalysts have shown special effectiveness in the polymerizations of oxiranes when these are used as initiators, both the rate as well as extent of polymerization are enhanced by some cocatalysts, like water and alcohols. A synthetically significant feature of the above polymerizations is the formation of an optically active polymer (from a racemic mixture of substituted oxirane) by the use of some optically active co-catalysts.

Organotin compounds also find applications as catalysts in the preparation of polyurethanes and in transesterification reactions in addition to their use in Ziegler-Natta polymerization.

6.4 TRANSITION METAL ORGANOMETALLICS AS CATALYSTS AND SYNTHETIC REAGENTS

6.4.1 Introduction

In a preceding section (6.1), attention has been focused on the growing importance of transition metals in catalysis and synthesis. Transition metal organometallics exhibit a wide variety of promising applications in these directions, and research continues for novel and better methodologies for the use of transition metal organometallics as catalytic as well as stoichiometric reagents in synthesis of organic compounds. Both types of these reactions proceed in general through the three steps described in sub-sections 6.4.1.1 to 6.4.1.3. This account will be followed by a description of a few typical catalytic and stoichiometric processes in sections 6.4.2 and 6.4.3, respectively.

6.4.1.1 Activation Process (Oxidative Addition or Coordination)

This is generally the initiating step in a catalytic process and involves the formation of bonding of some kind (σ, π or other delocalized type) between the transition metal and the organic/inorganic substrate (through oxidative addition or coordination); this is generally a key step in the catalytic or/and stoichiometric process, e.g.:

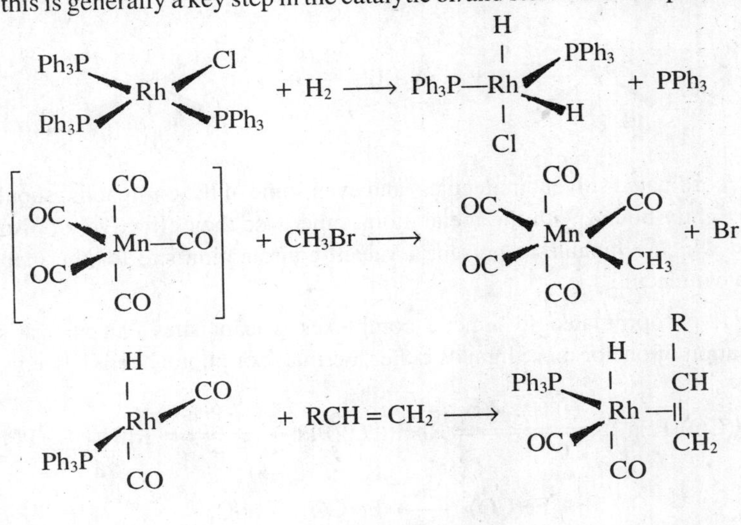

The oxidative addition/coordination reaction of a ligand is generally facilitated by the availability of vacant coordination site(s) on a metal complex in solution (homogeneous catalysis) or a metal surface (heterogeneous catalysis). The coordinative unsaturation may also arise from the dissociation of loosely held solvent molecules, giving rise to potentially vacant site(s). For example, a square planar d^8 (4-coordinate) complex such as $RhCl(PPh_3)_3$ (known as Wilkinson's catalyst) is coordinatively unsaturated and is one of the most active and versatile catalysts known so far. In this connection, it may be noted that coordinative unsaturation is generally guided by the 18-electron rule (or effective atomic number rule). For example, metal complexes with 18 electrons in the valence shell are coordinatively saturated, irrespective of coordination number, and therefore catalytically ineffective, e.g., $IrCl(H)_2(PPh_3)_3(d^6)$, $Fe(CO)_5(d^8)$, etc., whereas species with 16 electrons or less, such as $RhCl(PPh_3)_3$ are coordinatively unsaturated. Column 2, in table 6.1 indicates the number of ligands required to saturate (following 18-electron rule) coordinatively the respective metal ions and consequently, the corresponding complexes with lesser number of ligands would be coordinatively unsaturated.

Table 6.1

Electronic configuration	Saturated coordination number	Examples
d^6	6	Rh^{3+}, Ir^{3+}, Co^{3+}, Ru^{2+}, Mo^0, Pt^{4+}
d^8	5	Fe^0, Ru^0, Os^0, Rh^{1+}, Co^{1+}, Ir^{1+}, Pd^{2+}, Pt^{2+}
d^{10}	4	Ni^0, Pd^0, Pt^0

For coordinatively unsaturated species (e.g. d^8 species) in solution, coordinating solvent molecules occupying the vacant sites could be displaced by reacting molecules:

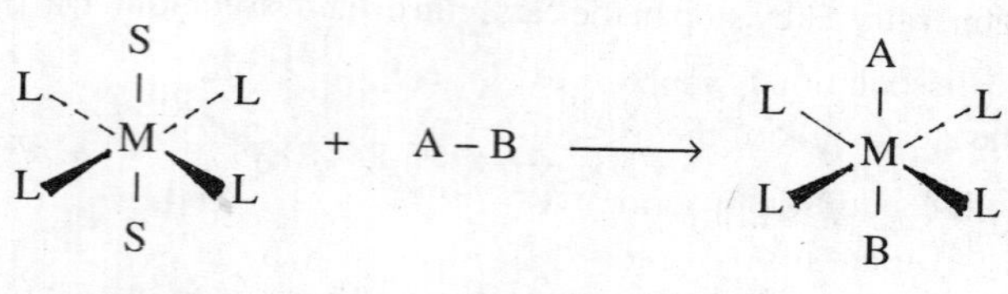

These coordinated solvent molecules (and even some of the co-ligands) should not be too tightly bound with the metal atom, otherwise they effectively poison the catalyst which is unable to present a vacant coordination site for the incoming substrate molecule.

In five coordinated (d^8) metal complexes, vacant sites can also arise by dissociation of one or more ligands either thermally or photochemically e.g.,

$$RhH(CO)(PPh_3)_3 \xrightleftharpoons{(-PPh_3,\ heat)} RhH(CO)(PPh_3)_2 \xrightleftharpoons{(-PPh_3)} RhH(CO)(PPh_3)$$

$$Fe(CO)_5 \xrightarrow{h\nu} Fe(CO)_4 + CO$$

Vacant sites may also be created by the use of ligands with a high *trans*- effect. For example, the reaction

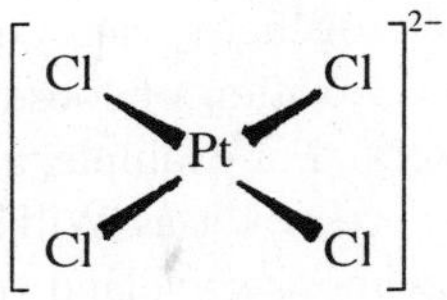
$$+ H_2C=CH_2 \longrightarrow [PtCl_3(C_2H_4)]^- + Cl^-$$

is quite slow, as the ligand chlorine has quite a low *trans*-effect. However, by interaction with $SnCl_2$, the chlorine ligand is converted into $SnCl_3$, which has a strong *trans*-effect, facilitating the following reaction:

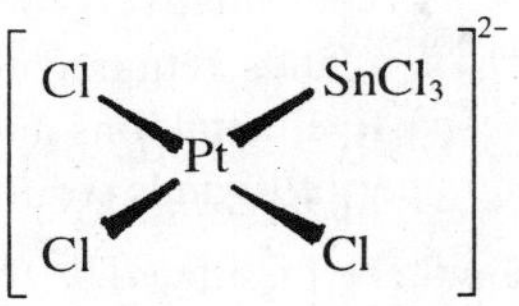
$$+ H_2C=CH_2 \longrightarrow [PtCl_3(C_2H_4)]^- + SnCl_2 + Cl^-$$

Dissociation may also be sometimes prompted by a change in oxidation state. For example, Wilkinson's catalyst, $RhCl(PPh_3)_3$ is by itself not appreciably dissociated in solution. However, when hydrogen is bubbled in its solution, it affords *cis*-$Rh(H)_2Cl(PPh_3)_3$, from which one of the ligated PPh_3 may be dissociated with much greater facility leaving a vacant site:

$$\underset{\text{4-Coordinate}}{Rh^{I}Cl(PPh_3)_3} \underset{-H_2}{\overset{+H_2}{\rightleftharpoons}} \underset{\text{6-Coordinate}}{Rh^{III}(H)_2Cl(PPh_3)_3} \underset{+PPh_3}{\overset{+PPh_3}{\rightleftharpoons}} \underset{\text{5-Coordinate}}{Rh^{III}(H)_2Cl(PPh_3)_2}$$

The above dissociation is supposed to be facilitated by the repulsive tendency arising from the steric bulk of the triphenylphosphine ligand.

The oxidative addition reactions (also, cf. Section 6.1) of organometallic compounds are of great utility in the synthesis of novel and interesting types of compounds; oxidative addition of alkyl halides to transition metal complexes, designated by the following equation, has been extensively studied in view of it involvement in catalytic reactions:

$$ML_n + RX \rightarrow M(R)(X)L_n$$

For example, the addition of methyl iodide to the rhodium(I) complex, $RhI_2(CO)_2^-$ is a key step in the Monsanto acetic acid synthesis; in the initiating reaction rhodium is oxidized from +1 to +3 oxidation state:

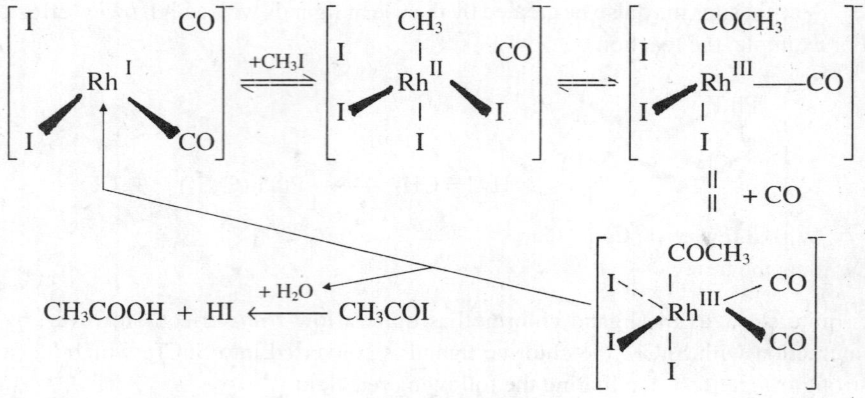

In view of the extensive applicability of oxidative addition reactions in catalysis, some of their essential features are indicated as follows:

(a) *Low Oxidation State of the Metal*

As oxidative addition reaction can be treated as a process undergoing loss of electrons from the electron rich metal centres, the same should be facilitated by the metal atom in the complex being in the lower oxidation state; in fact, any change in the metal complex which enhances the electron density at the central metal leads to an increase in reactivity. Although theoretically metal atoms in metallic (or zero oxidation) state could be expected to be better precursors for such reactions, yet these are of only very limited applicability on energetic grounds. However, the development of metal vapour and reactive metal production techniques have added new dimensions in this particular area.

(b) *Coordinative Unsaturation*

The oxidative addition reactions are generally accompanied by an increase in the coordination number of the central metal atom. It is, therefore, expected that organometallic precursors should be coordinatively unsaturated, e.g. $IrCl(CO)(PPh_3)_2$; $RhCl(PPh_3)_3$, etc. The coordinative unsaturation can be achieved either (i) by the generation of a vacant coordination site or (ii) by the ability of a complex to exist in different coordination numbers.

(c) *Nature of the Metal*

There are many indications that the lighter metals in a given sub-group are less basic or nucleophilic than the heavier elements. For example, while the complex $IrCl(CO)(PPh_3)_2$ (Vaska's compound) reacts completely and irreversibly with CH_3I:

the reaction is reversible for the rhodium analogue:

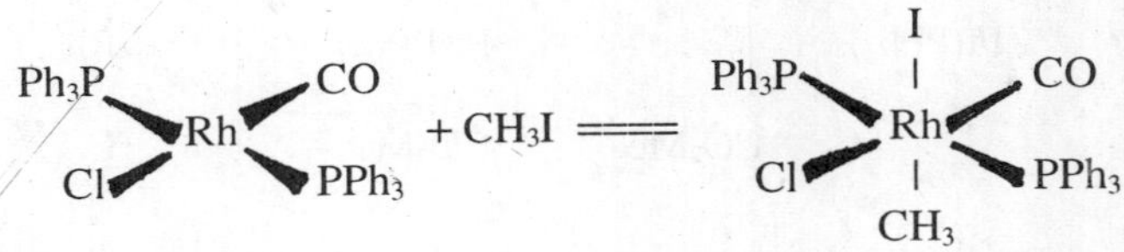

With a few exceptions, the general tendency to undergo oxidative additions parallels the trends indicated below for the Group VIII metals:

Fe(0)	Co(I)	Ni(II)	↓ Increasing tendency for oxidative addition reactions
Ru(0)	Rh(I)	Pd(II)	
Os(0)	Ir(I)	Pt(II)	

← Increasing tendency for oxidative addition reactions

(d) *Electronic Nature of the Ancilliary Ligands*

The presence of ancilliary ligands which enhance the electron density around the metal atom in a complex also influences the course of the reactions. For example, $RhCl(PPh_3)_3$ undergoes facile oxidative addition, but $RhCl(CO)(PPh_3)_2$ is much less active for oxidative addition. The replacement of one strongly donor and weakly acceptor PPh_3 ligand in $RhCl(PPh_3)_3$ by a weakly donor and strongly acceptor CO ligand should result in a lower electron density on the central rhodium atom in the latter complex, $RhCl(CO)(PPh_3)_2$ and thus explain the difference in reactivity of the two complexes, although a steric factor is another parameter to be considered. In a series of isosteric triaryl-phosphines in the Wilkinson complex, the observed order of hydrogenation activity: $(p\text{-}CH_3OC_6H_4)_3P > (C_6H_5)_3P > (p\text{-}FC_6H_4)_3P$ can be explained on the basis of the variation in the donor capability of the three ligands.

(e) *Nature of the Addendum*

The formation of metal-carbon bond by the oxidative additi n of heteronuclear organic substrates such as organic halides is of wide applicability. Oxidative addition reactions involving cleavage of carbon-carbon bond would obviously be much less feasible both on thermodynamic as well as kinetic considerations. However, such cleavage reactions can be brought about under special conditions. For example, sterically crowded addend such as hexaphenylethane are cleaved as it results in the release of a large strain in C_2Ph_6 molecule:

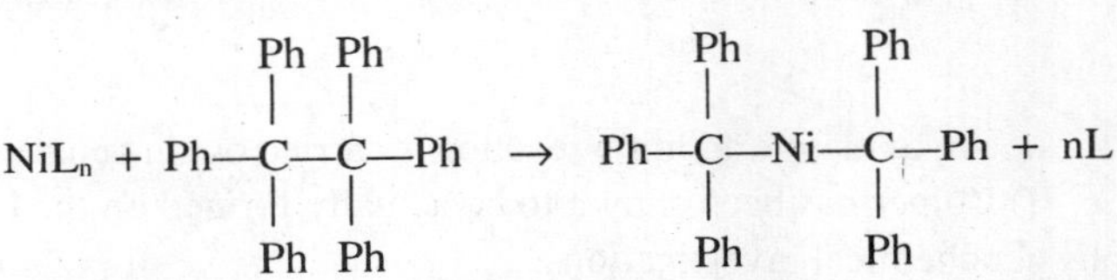

Alternatively, electronic activation of carbon-carbon bonds in molecules like dimethyloxalate and cyanogen can facilitate the cleavage reactions:

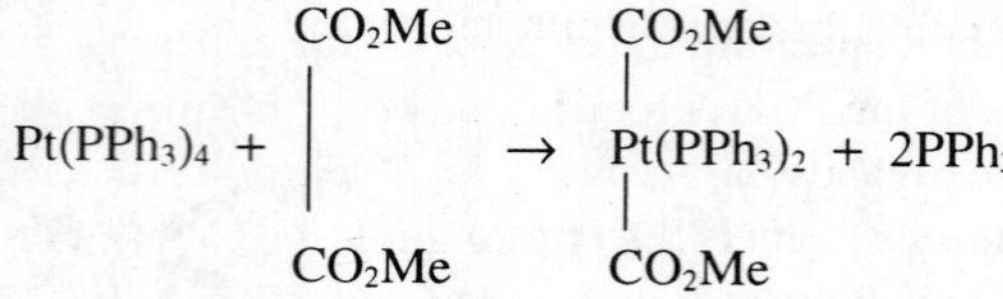

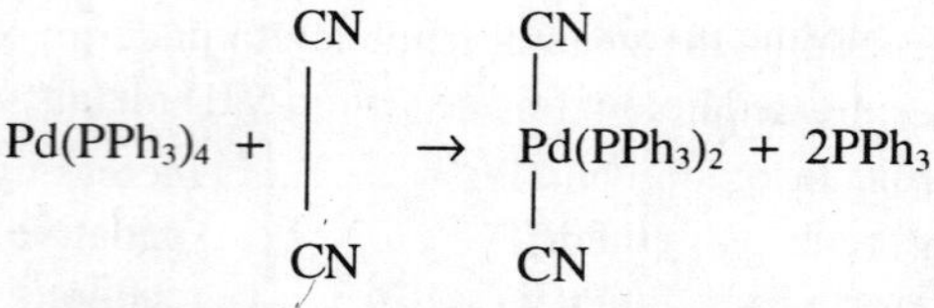

(f) *Stereochemistry and Mechanism of Oxidative Addition Reactions*

A detailed study of the stereochemistry of the product of oxidative addition reactions by physico-chemical techniques such as n.m.r. and i.r. spectroscopy has often provided a valuable insight in the mechanism of such reactions.

For example, the addition of CH_3I to $IrCl(CO)(PMePh_2)_2$ can theoretically yield nine possible stereoisomers (*A* to *I*):

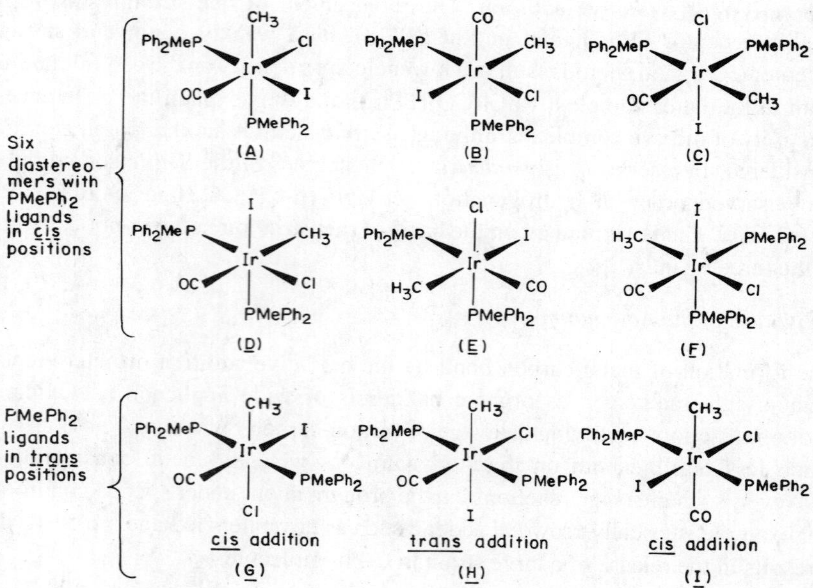

When the above oxidative addition reaction is carried out in benzene medium, only one (i.e. *H*) isomer has been shown to be actually formed on the basis of the following physicochemical investigations:

(i) in the ^{1}H n.m.r. spectrum of the product, the appearance of a triplet due to the methyl protons of the PCH_3 groups establishes unequivocally the presence of the two phosphine ligands in *trans* position to each other; this observation, therefore, eliminates the isomers *A* to *F* and limits the choice to the last three (*G* to *I*) in which the phosphine ligands occupy *trans* positions:

(ii) the i.r. spectrum of the product shows only one band for ν_{Ir-Cl} at 300 cm^{-1}, indicating* that the chlorine is *trans* to carbonyl group.

It may, however, be emphasized that the nature of the solvent also appears to play a significant role in oxidative addition reactions. For example, the use of methanol in place of benzene in the above reaction leads to the formation of *trans* as well as *cis* addition products. On the basis of such observations, it may be concluded that these reactions are kinetically rather than thermodynamically controlled. In fact, the kinetic studies on oxidative addition reaction especially with the help of more recent sophisticated fast reaction techniques have thrown considerable light on the possible reaction pathway.

Following the well established $S_N{}^1$, $S_N{}^2$ and free radical reaction pathways involving alkyl halides in simple organic substitution reactions, the oxidative addition reactions of transition metal moieties with alkyl halides can also be presumed to occur by the following pathways:

(1) *Concerted pathway*

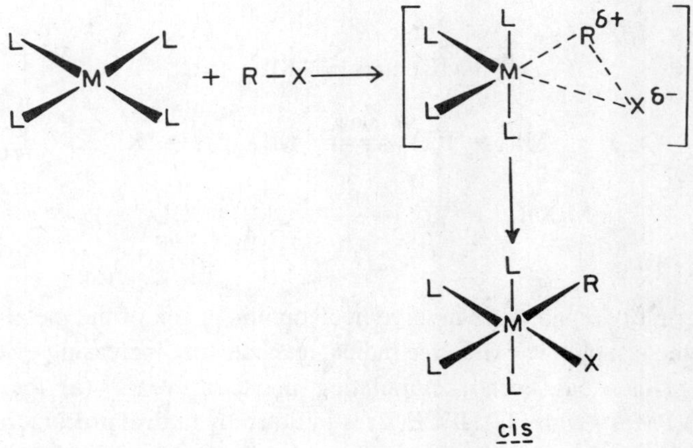

* This conclusion is based on the i.r. spectrum of $IrCl_3(CO)(PMePh_2)_2$, obtained by the oxidative addition of Cl_2 to Vaska's catalyst, $IrCl(CO)(PMePh_2)_2$; this addition product exhibits an absorption at 330 cm^{-1} assignable to the two mutually trans chlorines and another at 310 cm^{-1} due to the chlorine *trans* to the carbonyl group. The conclusion was further supported by the methyl chloride adduct, $IrCl_2(Me)(PMe_2Ph_2)_2$ which exhibited Ir–Cl stretching vibrations at 305 and 257 cm^{-1} corresponding to chlorines *trans* to carbonyl and methyl groups respectively.

(2) *Non-concerted pathway*

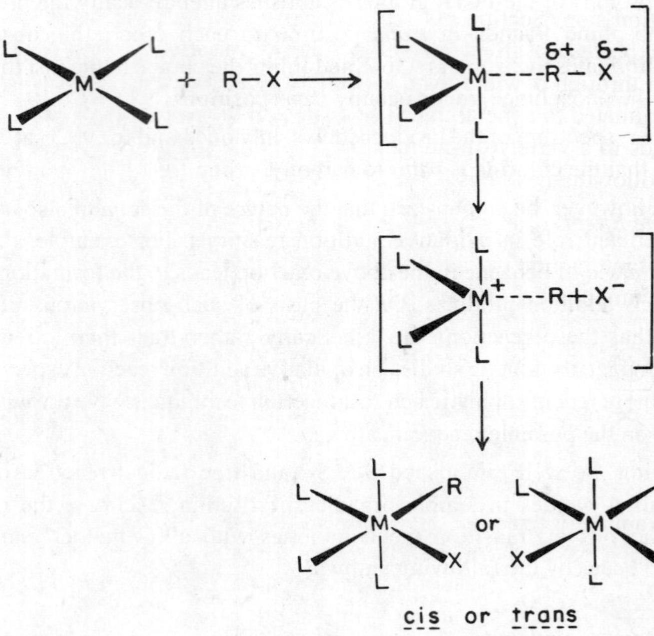

(3) *Free radical pathway*

$$ML_4 \rightleftharpoons ML_3 + L$$

$$ML_3 + RX \overset{slow}{\rightleftharpoons} M(X)L_3 + {}^{\bullet}R'$$

$$M(X)L_3 + {}^{\bullet}R' \xrightarrow{fast} \underset{cis \text{ or } trans}{M(R)(X)L_2} + L$$

The first two pathways involve nucleophilic attack of the metal complex on RX. For the last pathway (i.e. free radical mechanism), increasing evidence on the following lines has been accumulating in recent years: (a) the reaction of $IrCl(CO)(PMe_3)_2$ with $PhCHFCH_2Br$ is initiated by radical precursors such as O_2, AIBN or Bz_2O_2, etc.; (b) the above reaction is inhibited by radical scavengers, e.g., hydroquinone, duroquinone, galvinoxyl, etc.; (c) racemization of the above Ir(I) derivative on addition of a chiral compound, $MeCHBrCO_2Et$; (d) increase in rate for alkyl halides with electronegative substituents in the alkyl moiety; (e) spin trapping experiments involving BuNO; and (f) the ^1H n.m.r. spectrum of a reacting mixture of $Pt(PEt_3)_3$ and Me_2CHI depicts CIDNP effect (CIDNP stands for "chemically induced dynamic nuclear polarization"; such effect is observed only if free radicals are present).

6.4.1.2 *Consequent Changes in the Coordinated Ligand Reactivity*

The activation step (Section 6.3.1.1) of metal-substrate bonding or coordination could be expected to affect markedly (due to variation in intra-ligand charge distribution through σ-withdrawal as well as by π-donation) the reactivity of the ligand coordinated to a metal ion. The change in the reactivity of the coordinated ligands leads to a variety of interactions/reactions, some of which are described under the following headings:

(a) *Template Effect*

This type of interaction involves orientation of the ligated molecules in a manner to facilitate union, e.g.,

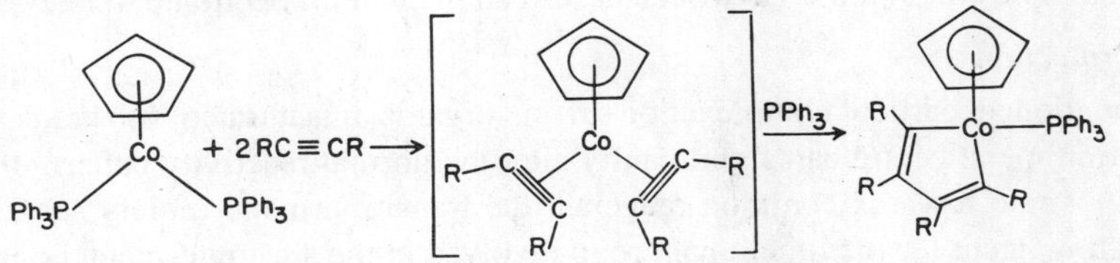

Another example of template effect is the union of two butadiene molecules with Ni(0) to form a bis-π-allyl-nickel (II) complex:

(b) *Insertion Reactions (or Intra-ligand Combination Reactions)*

These insertion reactions (cf. Section 6.4.2) are one of the most important types of reactions mediated by transition metal compounds and can be exemplified by the insertion of an unsaturated moiety into a M-H bond:

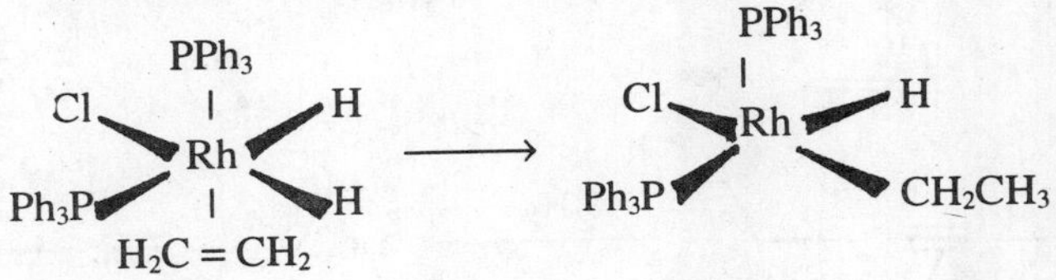

Such a reaction is the vital step in catalytic hydrogenation (cf. Section 6.3.2.1) processes.

In addition to insertion in hydrido complexes, other reactions involve: (i) insertion of an alkene into a metal-alkyl bond to give a longer-chain alkyl group (such a reaction is of greater significance in the Ziegler-Natta polymerization of alkenes (cf. Section 6.4.2.6); (ii) insertion of carbon monoxide into transition metal-carbon bonds. Respective examples are:

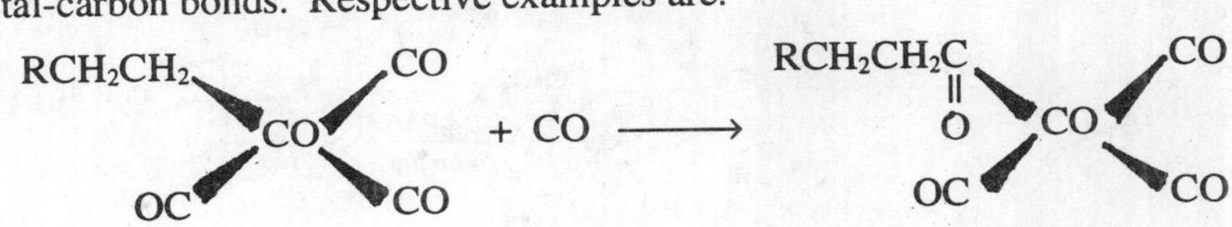

The basic mechanism of an insertion reactions has already been discussed in an earlier Section 4.2.3.3.

(b) *Stabilization Reactive Compounds*

The reactivity of extremely short-lived (highly reactive) species such as cyclobutadiene, 7-norbornadienone, strained acetylenes and dihydropyridines can be sometimes modified to such an extent on coordination to a metal centre as to allow the actual isolation of transition metal derivatives of these highly reactive organic moieties. The coordination process can also lead to the stabilization of a tautomeric form of the ligand which may be thermodynamically unfavourable in the free state, e.g., vinyl alcohol.

Examples of all these have been described in detail in Section 6.4.3.2.5.

(d) *Protection*

As mentioned earlier, complexation of an organic unsaturated substrate to a transition metal centre can significantly alter the normal reactivity pattern of the free supstrate towards common reagents; the transition metal moiety serves as protecting agent for the functional group involved in the substrate-metal bonding and thus allows the chemical reaction at some other reactive site of the ligand e.g.,

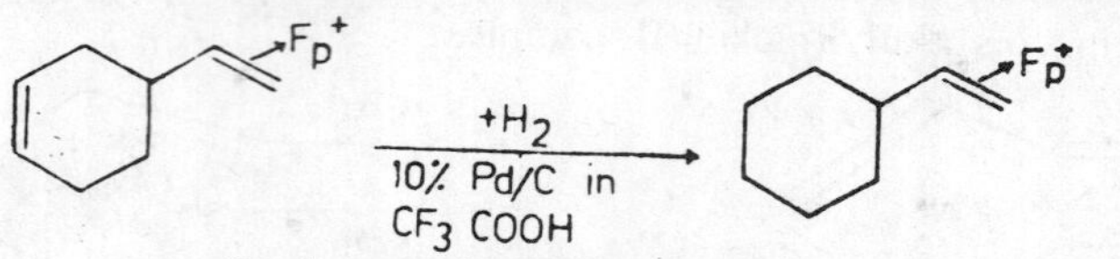

($Fp^+ = Fe(CO)_2(\eta^5—C_5H_5)^+$)

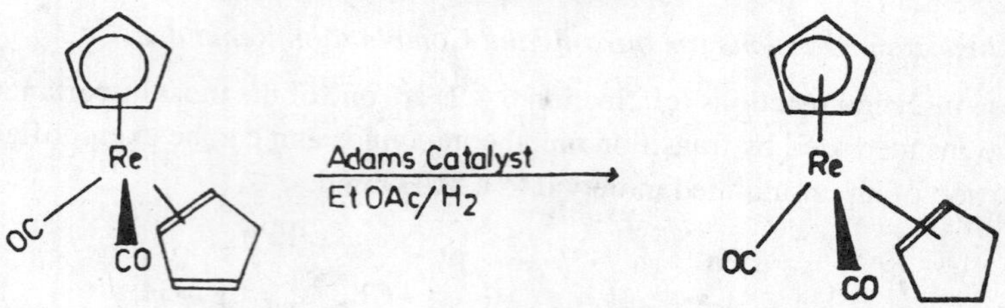

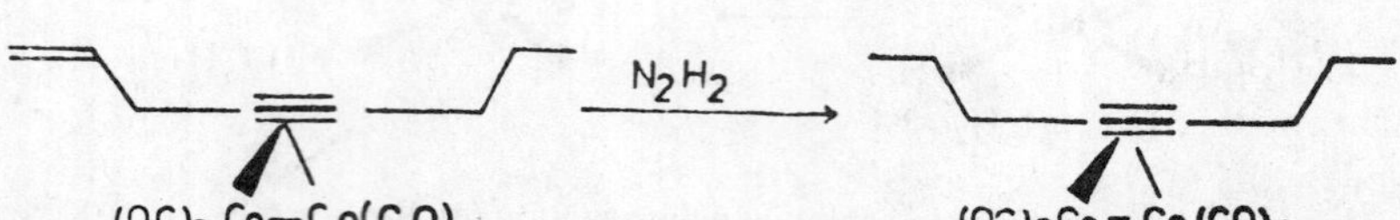

(e) *Steric Control*

The steric inhibition brought about by coordinated bulky groups such as $Fe(CO)_3$ or $Cr(CO)_3$ tends to direct the attack (e.g. deuteration) on the olefinic moiety preferably on a particular exposed side, e.g.,

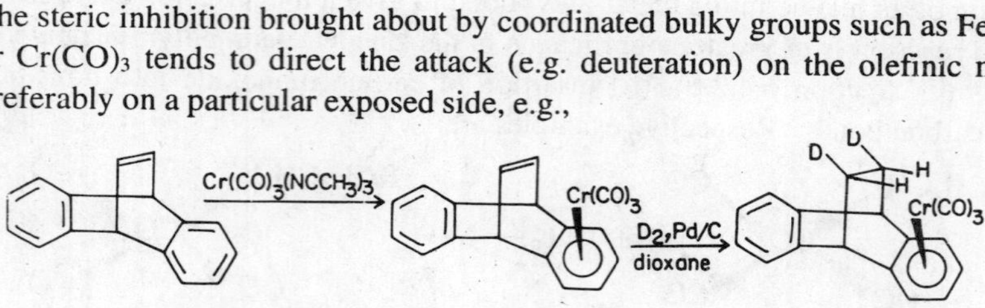

(f) *Facilitation of Nucleophilic Addition Reactions*

On coordination to a transition metal centre, unsaturated hydrocarbons such as ethylene, butadiene or benzene, which normally do not tend to undergo nucleophilic addition or substitution reactions, exhibit pronounced tendency to be attacked by a wide range of nucleophiles such as H^-, R^-, CN^-, MeO^-, etc. Some typical examples of the above are:

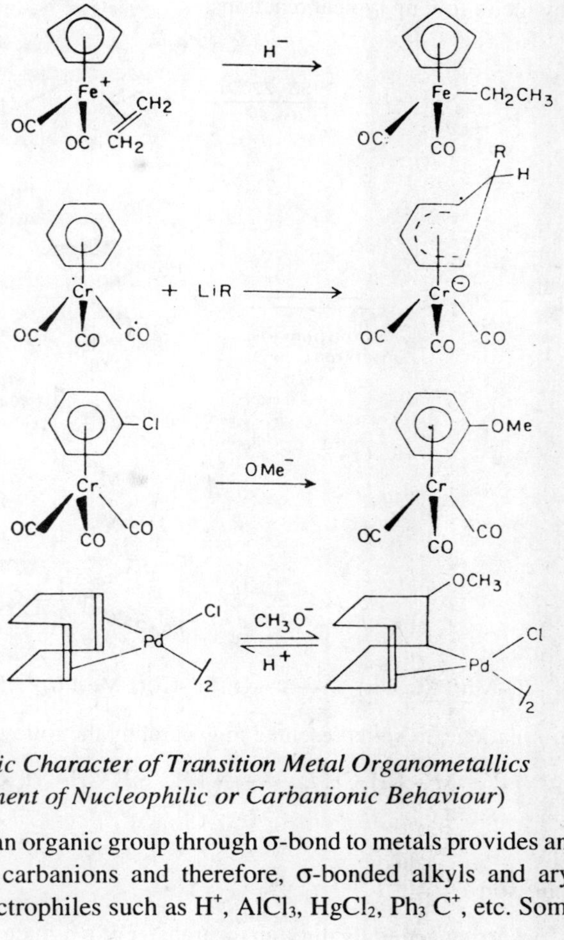

(g) *Nucleophilic Character of Transition Metal Organometallics (Enhancement of Nucleophilic or Carbanionic Behaviour)*

Attachment of an organic group through σ-bond to metals provides an ideal source for generating carbanions and therefore, σ-bonded alkyls and aryls are often attacked by electrophiles such as H^+, $AlCl_3$, $HgCl_2$, Ph_3C^+, etc. Some illustrative example are:

$$ZrCl(R)(\eta^5-C_5H_5)_2 \xrightarrow{H^+Cl^-} RH + ZrCl_2(\eta^5-C_5H_5)_2$$

$$\searrow \xrightarrow{AlCl_3} AlCl_2R + ZrCl_2(\eta^5-C_5H_5)_2$$

$$FeR(CO)_2(\eta^5-C_5H_5) + HgCl_2 \rightarrow Hg(R)Cl + FeCl(CO)_2(\eta^5-C_5H_5)$$

$$ReEt(\eta^5-C_5H_5)_2 + Ph_3C^+BF_4^- \rightarrow [Re(\eta^5-C_5H_5)_2(\eta^2-C_2H_4)]^+BF_4^- + Ph_3CH$$

6.4.1.3 Product Isolation

The final step in any synthetic (catalytic or stoichiometric) process is the liberation (or controlled release) of the desired product which can be effected in a variety of ways, some of which are being summarized below:

(a) *Reductive Elimination*

This process, which is of wide applicability in catalytic procedures, can be represented by the following typical reactions:

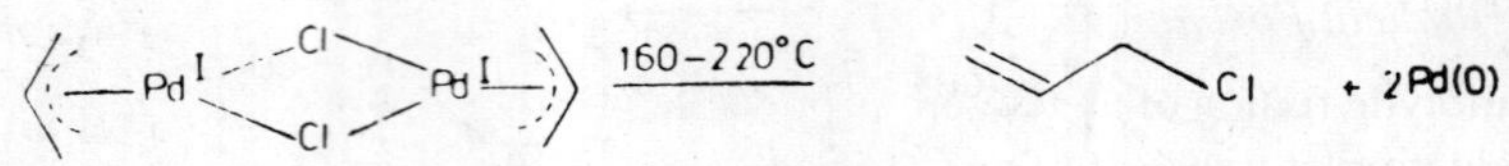

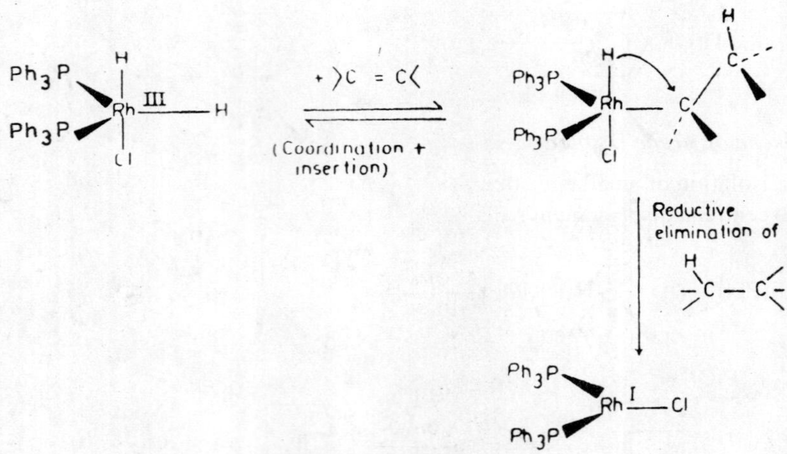

$$Pt^{IV}\ Cl_3Me(PMe_2Ph)_2 \xrightarrow{\text{heat}} Pt^{II}\ Cl_2(PMe_2Ph)_2 + MeCl$$

Reductive elimination can be represented in general by the following equation:

$$\underset{\displaystyle L_n}{R' - \overset{}{M} - CH_2CH_2R} \longrightarrow ML_n + R'CH_2CH_2R$$

(b) β – *Elimination*

This process, occurring generally through the transfer of a β-hydrogen (cf. Section 4.2.1) can be represented by the following general equation:

$$\underset{\displaystyle L_n}{R' - M - CH_2CH_2R} \longrightarrow H_2C = CHR + MH(R')\ L_n$$

Two specific examples of β-elimination are given below:

$$PtCl(CH_2CH_3)\ (PEt_3)_2 \xrightarrow{\Delta} H_2C = CH_2 + PtHCl(PEt_3)_2$$

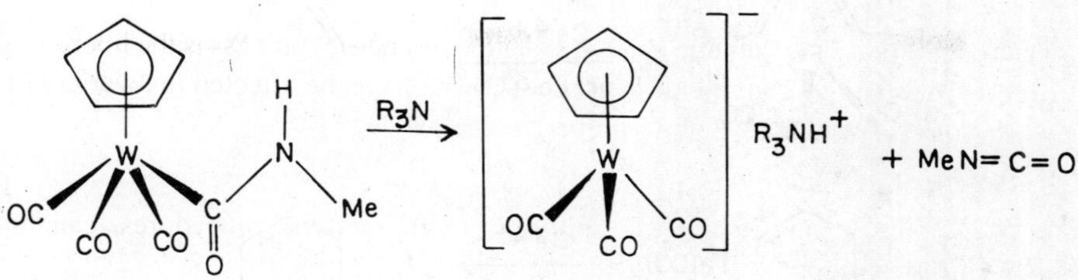

(c) *Radical Formation*

Homolytic fission of metal-carbon bonds under suitable conditions may lead to the production of transient organic radicals which can then undergo their usual (radical) reactions:

$$Ti(OR)_3Ph \xrightarrow{80^\circ C} Ti(OR)_3 + [Ph^{\cdot}\ ] \rightarrow PhH + Ph-Ph$$

$$TiCl_2Ph_2 \xrightarrow{\Delta} Ph-Ph + TiCl_2$$

(d) *Alkene or Arene Displacement by a Competing Ligand*

For the isolation of an alkene, alkadiene, or arene product which is complexed to a metal centre, a displacement reaction with a competitive ligand is often utilized, e.g.

$$Ni(PPh_3)_2\,(\eta^2-C_2\,H_4) + PhCH=CH_2 \rightarrow Ni(PPh_3)_2(\eta^2-PhCH=CH_2) + H_2C=CH_2$$

$$[\ \{PtCl_2\,(\eta^2-C_2H_4)\}_2] + 2PR_3 \rightarrow [\{PtCl_2\,(PR_3)\}_2] + 2H_2C=CH_2$$

$$\xrightarrow{+\,4C_5H_5N} 2PtCl_2\,(C_5H_5N)_2 + 2H_2C=CH_2$$

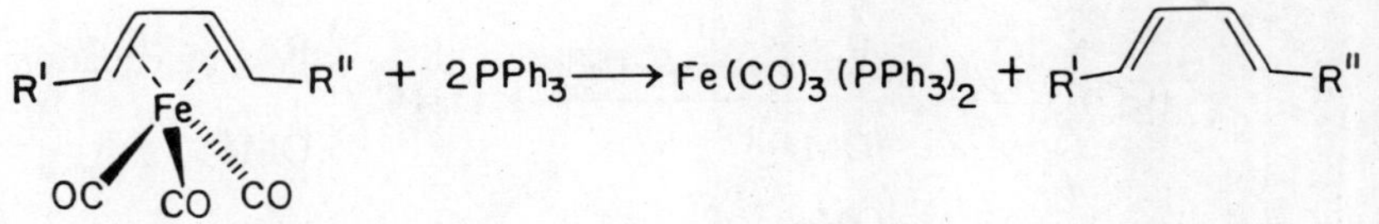

$$M(CO)_3\,(\eta^6-C_6H_6) + 3P(OPh)_3 \longrightarrow M(CO)_3\{P(OPh)_3\}_3 + C_6H_6$$
(M = Cr, Mo, W)

(e) *Electron Transfer from Metal Atom to an Oxidant*

Electron transfer from the central metal atom to a suitable oxidant sometimes results in the liberation of an 'even-electron' ligand such as an alkene or alkadiene. For example:

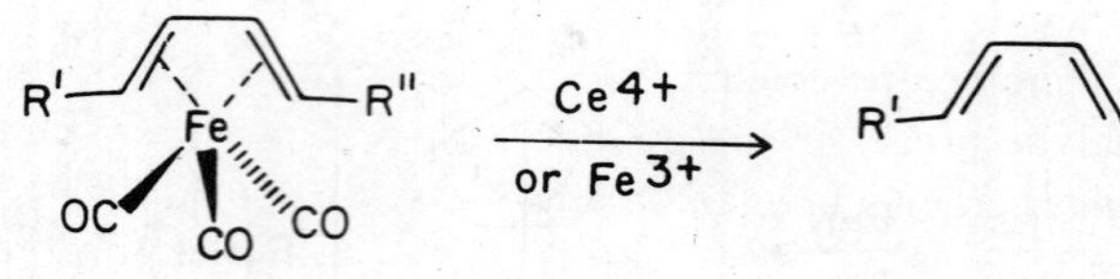

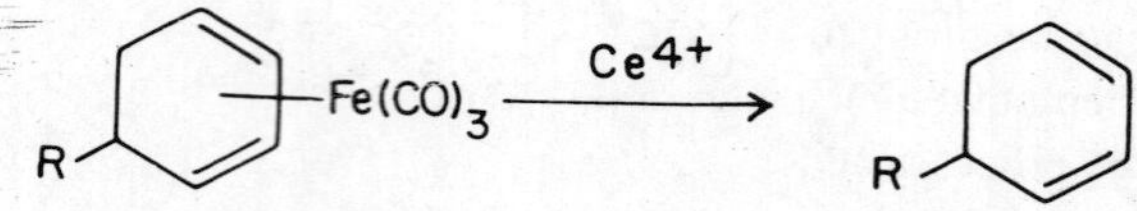

In some cases the liberated organic molecules such as cyclobutadiene and trimethylenemethane are too reactive to be capable of independent existence. However, in these cases, their liberation in the presence of suitable organic substrates can lead to the synthesis of interesting organic compounds, shown as follows:

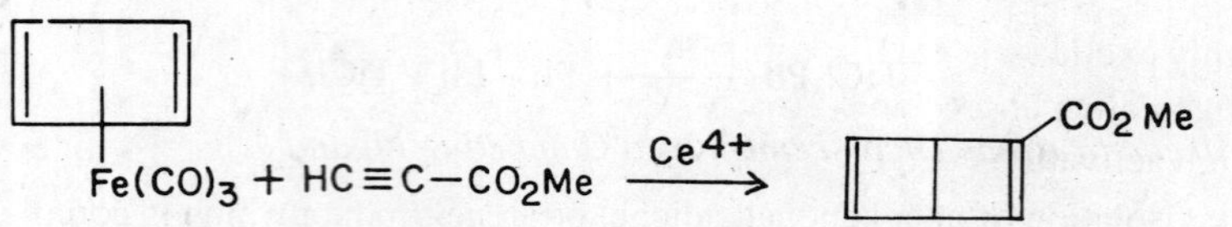

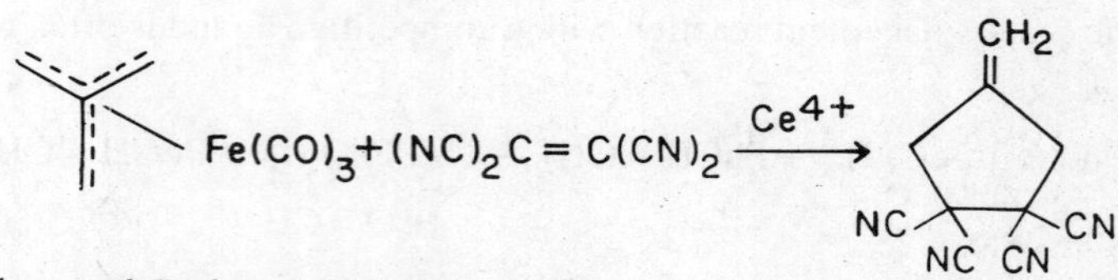

(f) *Release of Carbeniod Ligands*

Reactions with oxygen or some other nucleophile have been often utilized for the release of a carbene fragment from transition metal-carbene complexes. Some of these are illustrated as follows:

$$(CO)_5W=C\begin{matrix}R\\OM\end{matrix} \xrightarrow{O_2 \text{ or } Ce^{4+}} O=C\begin{matrix}R\\OMe\end{matrix}$$

$$(CO)_5Cr=C\begin{matrix}OMe\\CHRR'\end{matrix} \xrightarrow[\Delta]{C_5H_5N} \underset{\substack{|\\OMe}}{HC}=CRR'$$

$$(CO)_5W=C\begin{matrix}Ph\\OMe\end{matrix} + Ph_3P=CH_2 \rightarrow (OMe)PhC=CH_2+W(CO)_5PPh_3$$

6.4.2 Catalytic Processes (involving transition metal organometallics)

Contrary to the applications of main group organometallics (which are generally stoichiometric in nature), an increasing amount of evidence shows that synthetic procedures involving transition metal organometallics are more often catalytic in nature. We have, therefore, reversed (cf. Section 6.3) the order of our presentation here, i.e., the catalytic processes are discussed first followed by stoichiometric reactions, although a certain amount of overlap in the two broad divisions may be unavoidable.

The variety of reactions catalyzed by transition metal organometallics is quite large and these can be grouped in slighly different patterns. In the following sub-sections, an attempt would be made to indicate the specific type of interactions in selected catalytic processes described below under different headings, the first five of which are examples of essentially homogeneous catalysis and the last five deal with homogeneous/heterogeneous processes.

6.4.2.1 Hydrogenation of Alkenes

Although the reaction:

$$H_2C = CH_2 + H_2 \rightarrow H_3C - CH_3$$

is highly exothermic (ΔH° = 136.8 kJ/mole), ethene (ethylene) does not react with hydrogen at an appreciable rate (see Figure 6.2) without an appropriate catalyst; the high activation energy corresponds to negligibly slow reaction rate.

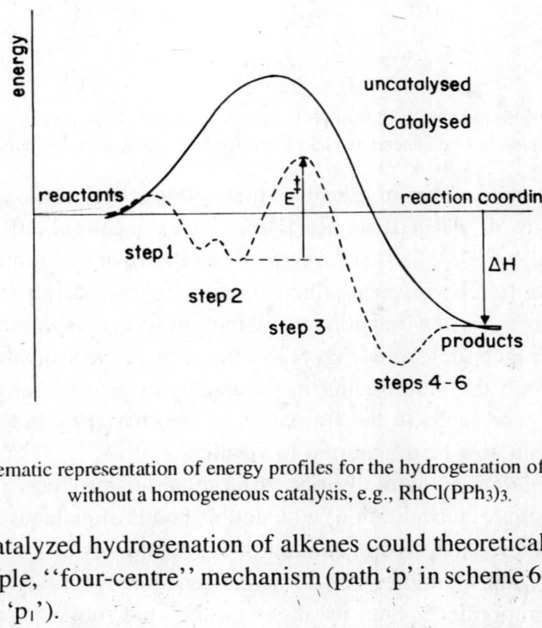

Fig. 6.2 A schematic representation of energy profiles for the hydrogenation of an alkene with and without a homogeneous catalysis, e.g., $RhCl(PPh_3)_3$.

The uncatalyzed hydrogenation of alkenes could theoretically occur by conceptually simple, ''four-centre'' mechanism (path 'p' in scheme 6.10) or a stepwise process (path 'p_1').

However, both the paths are unfavourable, the concerted path ('p') is ''forbidden'' by the Woodward-Hoffman orbital symmetry rules (Fig. 6.3), i.e., electron cannot flow from hydrogen (H_2) to the empty π^* orbital of ethene as there is no net overlap, whereas the stepwise process ('p_1') would necessarily involve high energy hydrogen atoms.

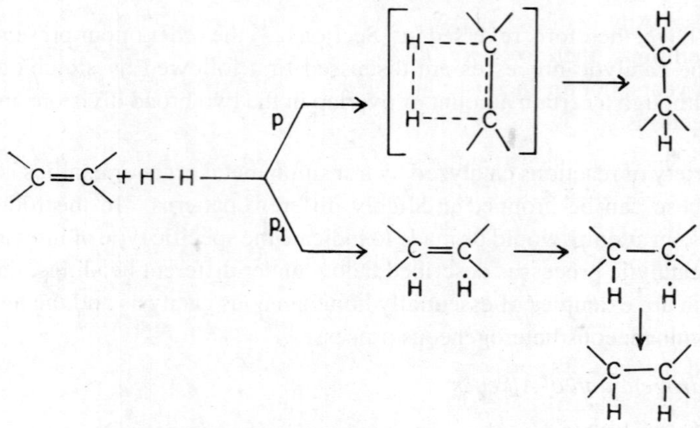

Scheme 6.10 Alternative Pathways for Hydrogenation of alkenes

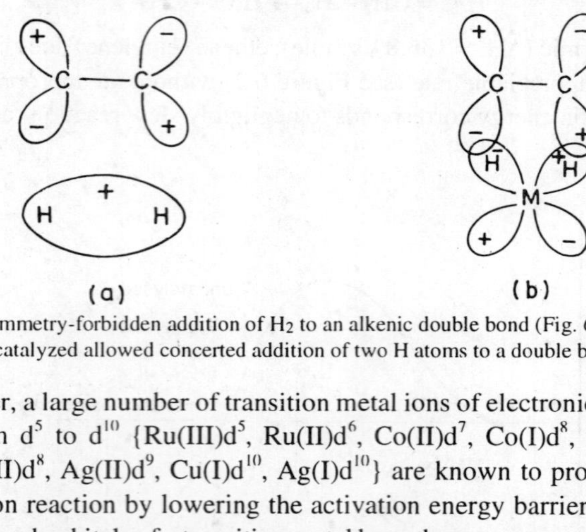

Fig. 6.3 Symmetry-forbidden addition of H_2 to an alkenic double bond (Fig. 6.3a) and metal-catalyzed allowed concerted addition of two H atoms to a double bond.

However, a large number of transition metal ions of electronic configurations varying from d^5 to d^{10} {Ru(III)d^5, Ru(II)d^6, Co(II)d^7, Co(I)d^8, Rh(I)d^8, Ir(I)d^8, Pd(II)d^8, Pt(II)d^8, Ag(II)d^9, Cu(I)d^{10}, Ag(I)d^{10}} are known to promote the above hydrogenation reaction by lowering the activation energy barrier for the process (Fig. 6.2). The d-orbitals of a transition metal have the correct symmetry to interact directly with hydrogen molecule (Fig. 6.3b), the flow of electrons from the filled metal d-orbitals into the anitbonding molecular hydrogen orbital causes H—H bond dissociation and leads to the formation of two metal-hydride bonds. The hydrogen atoms can now be transferred to the alkene.

The hydrogenation of naturally occurring materials/products (general or at selected specific sites of unsaturation) with double bonds often leads to very useful materials. In this direction, hydrogenation of vegetable oils to solid facts can be cited as an important example. These hydrogenation reactions are assuming ever-increasing importance due to their newer and novel applications in petrochemical and drug industries. In these directions, useful heterogeneous catalysts for specific purposes are Adams catalyst (platinum black oxide); Pd/C; Pd/$BaSO_4$; Raney nickel; rhodium; ruthenium, and copper-chromium oxide. Typical hydrogenations (using heterogeneous catalysts) have to be carried out at moderately high pressures of hydrogen (~ 1000/psi) and generally at elevated temperatures. Any catalysts capable of facilitating hydrogenation reactions at ambient conditions, therefore, are obviously very attractive propositions and in more recent years, homogeneous catalytic systems have proved highly successful in these directions.

Out of a large number of homogeneous hydrogenation catalysts or catalyst precursors such as $RhCl(PPh_3)_3$; $RhH(CO)(PPh_3)_2$; $IrCl(CO)(PPh_3)_2$; $RuHCl(PPh_3)_3$; $Ru(H)(Cl)(CO)(PPh_3)_3$ and $PtH(SnCl_3)(PPh_3)_2$, the first one known as Wilkinson's catalyst continues to occupy a position of predominant importance.

The mechanism of homogeneously catalyzed hydrogenations is rather complicated and has been the subject of much speculation and controversy, but the accumulated evidence reveals that metal-hydrogen bond (either present in the parent transition metal complex itself or formed during the catalytic process) appears to be an essential step in the catalytic cycle, followed by coordination and insertion of the alkene and reductive elimination (cf. Section 6.2); the full cycle is illustrated in Figure 6.4.

For example, the chemistry of hydrogenation involving $RhCl(PPh_3)_3$ catalyst can be understood in terms of the following steps (numbered according to their sequence in the cycle illustrated in Fig. 6.4; triphenylphosphine PPh_3 is for convenience symbolized by (P′)

Step 1

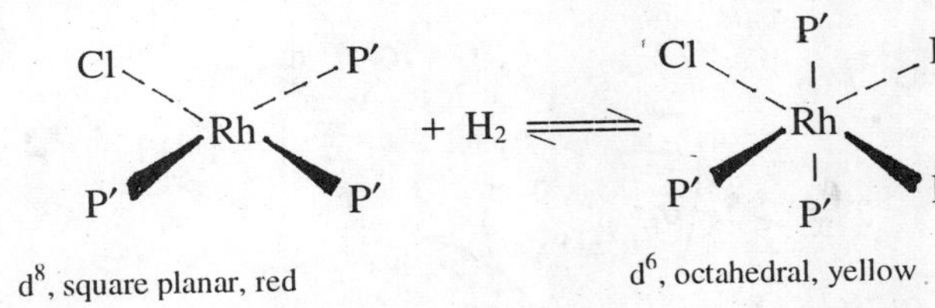

(the above indicates *cis* hydrogen addition and the same has been established by ^{1}H and ^{31}P n.m.r. spectroscopy; the coordination number of rhodium increases from 4 to 6 and its oxidation state from +1 to +3 and hence, the process is known as 'oxidative addition' of hydrogen).

In addition to activation of hydrogen by an oxidative addition process occurring with Wilkinson's catalyst, alternative routes, e.g., homolytic splitting and heterolytic fission, have also been suggested for hydrogen activation; these can be illustrated with the catalysts such as $[Co(CN)_5]^{3-}$ (d^7) and $[RuCl_6]^{3-}$ (d^5) respectively:

$$2\,[Co^{II}(CN)_5]^{3-} + H_2 \xrightarrow[\text{splitting}]{\text{homolytic}} 2\,[Co^{III}H(CN)_5]^{3-}$$

$$[RuCl_6]^{3-} + H_2 \xrightarrow[\text{fission}]{\text{heterolytic}} [RuHCl_5]^{3-} + H^+ + Cl^-$$

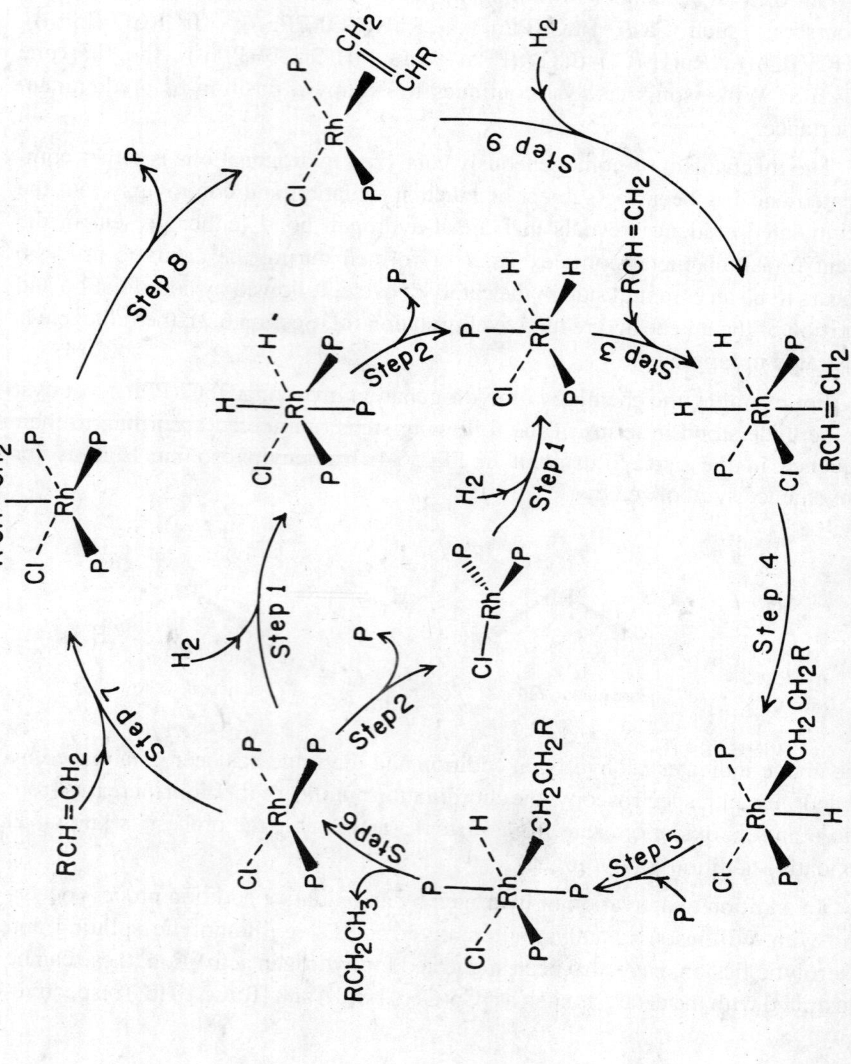

Fig. 6.4 Cycle representing alkene hydrogenation by Wilkinson's Catalyst.

This type of 'heterolytic' activation of hydrogen is more likely with metal complexes in higher oxidation states, which makes the oxidative addition path less feasible:

Step 2

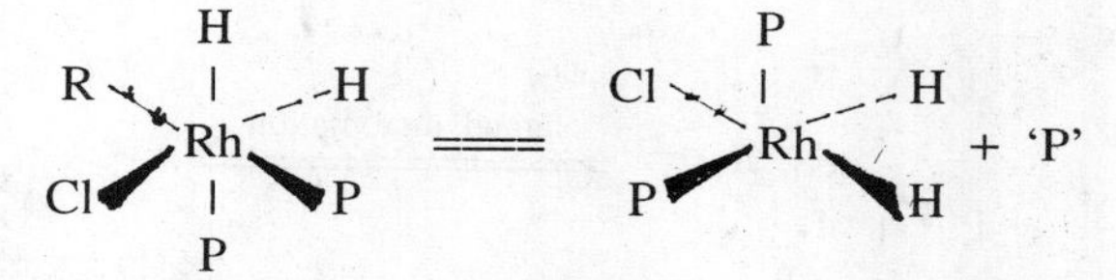

(Phosphine ligand dissociation)

Step 3

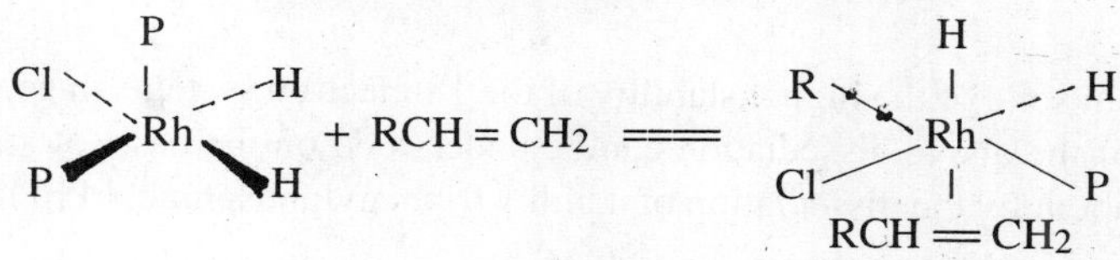

(alkene coordination)

Step 4

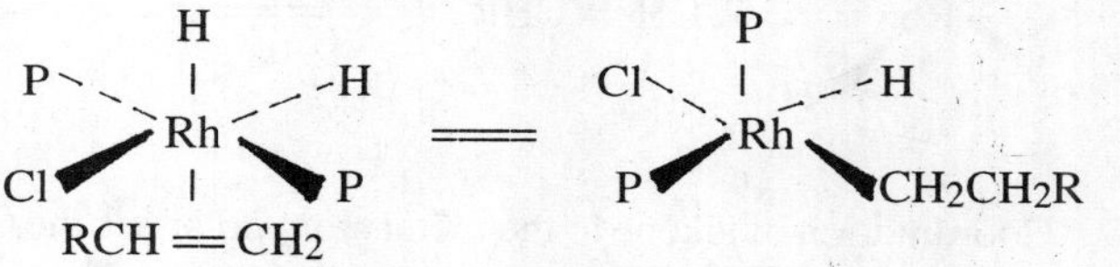

(β-hydrogen transfer; insertion)

Step 5

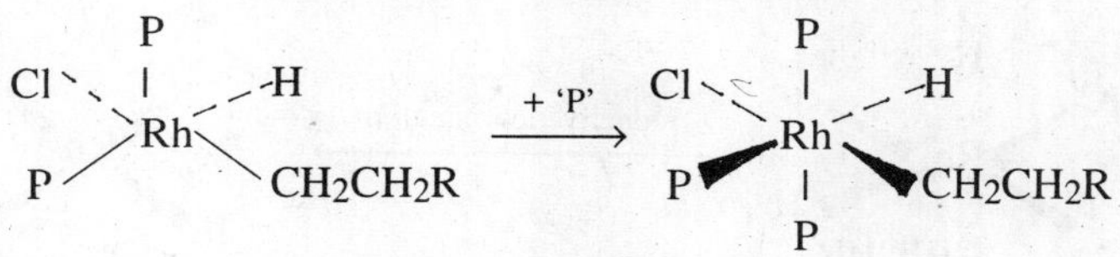

(Phosphine coordination)

Step 6

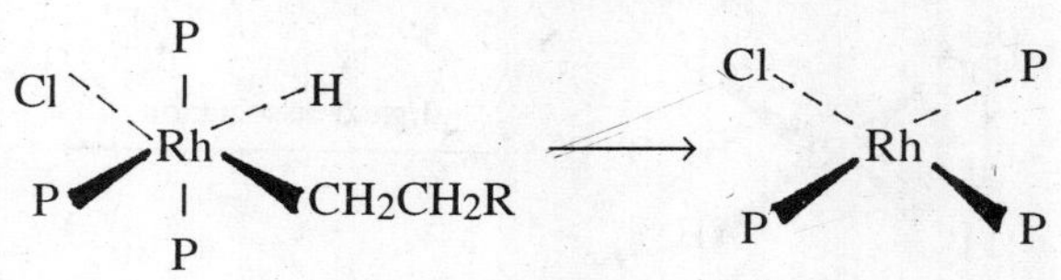

(reductive elimination)

In view of the catalyst being regenerated after a cycle of reactions, it should be possible to represent a catalytic process by a closed loop incorporating various steps of reactions involved. For example, the above reactions can be combined in a loop form as represented in Fig. 6.4.

In addition to the hydrogenation reaction by $RhCl(PPh_3)_3$, the hydridocarbonyltris(triphenylphosphine) rhodium(I) is another effective hydrogenation catalyst for alkenes

$$H_2C = CH_2 + H_2 \xrightarrow{RhH(CO)(PPh_3)_3} CH_3 - CH_3$$

The chemistry of hydrogenation of an alkene with this catalyst can be understood in terms of the following sequence of processes (which could be combined into a cycle as above):

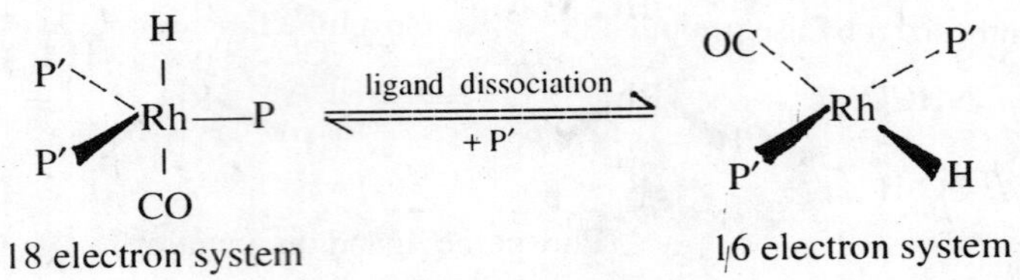

(In spite of the higher stability of the 18-electron systems in general, the tendency of the above dissociation can be understood on the basis of steric relief brought about by the dissociation of a bulky triphenylphosphine, PPh_3 ligand):

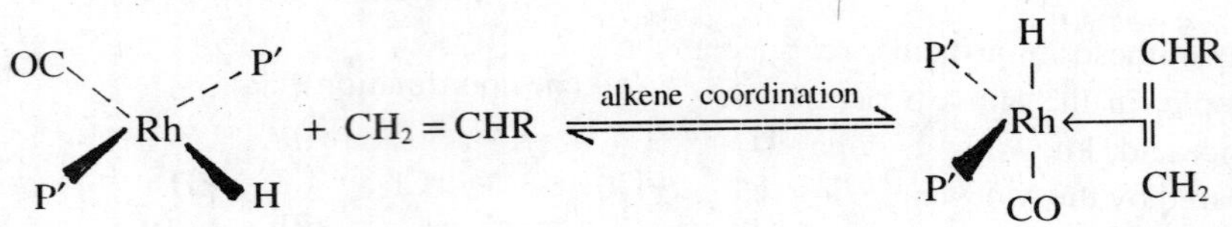

(Coordination of alkene to the metal atom polarises the C—C bond and allows facile migration of the alkyl with its bonding electron pair):

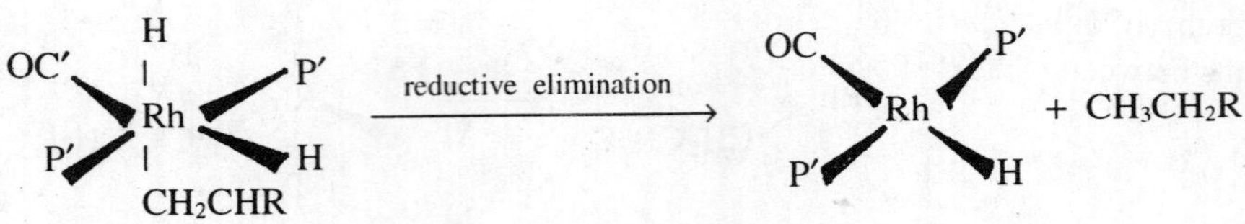

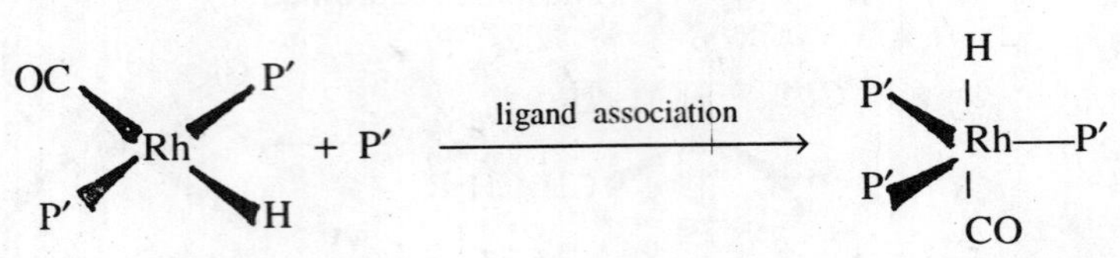

The unique features of above catalytic steps are (i) hydrogenation of alkenes by $RhH(CO)(PPh_3)_3$ is unusual in its being highly selective for the hydrogenation of alk-1-ene compared to alk-2-ene; and (ii) hydrogenation follows anti-Markownikov rule (cf. Section 6.4.2.3).

The hydrogenation of alkenes has been extended to the synthesis of some asymmetric products and also to synthesis involving polymer supported catalysis.

(i) Asymmetric Hydrogenations

Asymmetric synthesis. In general, it can be defined as "a reaction in which an achiral unit in an ensemble of substrate molecules is converted by a reactant into a chiral unit in such a manner that the stereoisomeric products are produced in unequal amounts''. The strategy for asymmetric synthesis often depends on the fact that reagents attack molecules along the less hindered path (cf. hydroboration, Section 6.3.1.4 and Appendix 1.3). An interesting and important application of homogeneous hydrogenation is in the synthesis of chiral organic compounds, e.g., α-amino acids from achiral precursors in optical yields of around 70% using chiral catalysts synthesized by use of optically active phosphine ligands):

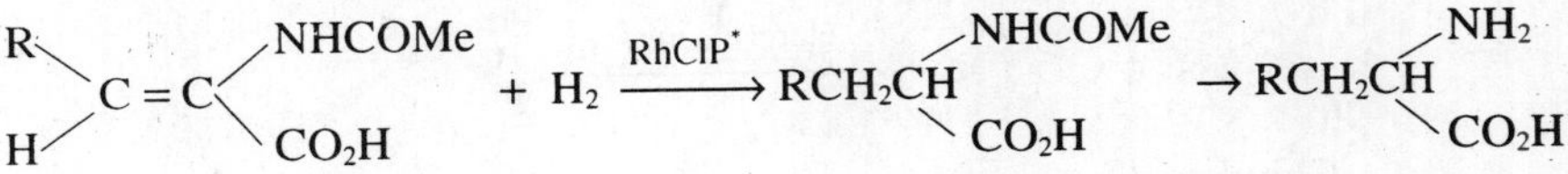

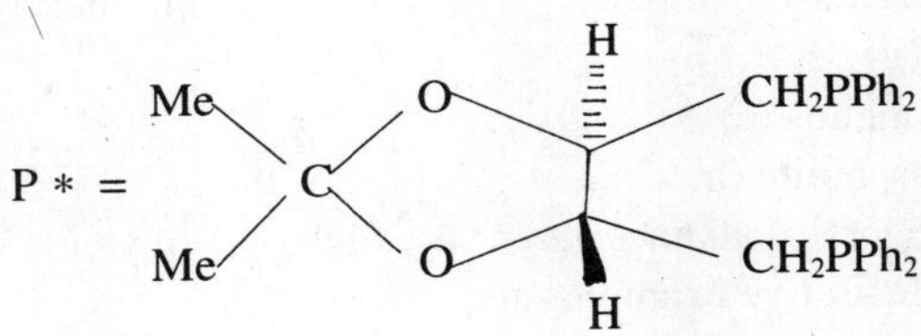

Many of these amino acids are especially useful for their biological properties. For example, in the above reaction when R is 3, 4-dihydroxyphenyl, the resulting amino acid, known as L-dopa, is active against Parkinson's disease and is now prepared by the Monsanto company on an industrial scale.

The use of chiral catalysts in general constitutes a very efficient way of generating centres of optical activity because the catalyst is not consumed. Chiral catalysts can be classified according to their structure as: (i) monophosphines (e.g., Fig. 6.5a-c), the asymmetry being on the phosphorous atom, or in a side chain or on both; (ii) diphosphines (e.g., Fig. 6.5d-h) may contain two, three or four carbon atoms between the two phosphorous atoms.

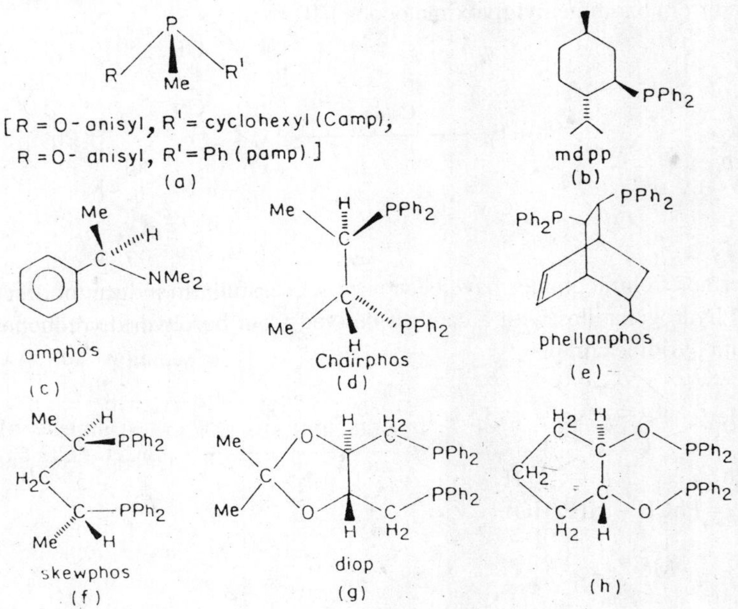

Fig. 6.5 Some main classes of chiral phosphine ligands.

Although many unsaturated systems can be reduced in the presence of chiral rhodium catalysts, yet high optical yields were obtained only with some specific unsaturated systems. In this context, two types of alkenes have to be considered. One in which there is an isolated prochiral double bond and the other in which polar groups are close to the double bond. Optical yields are always low in asymmetric reduction of simple alkenes, irrespective of the chiral catalyst used, e.g.:

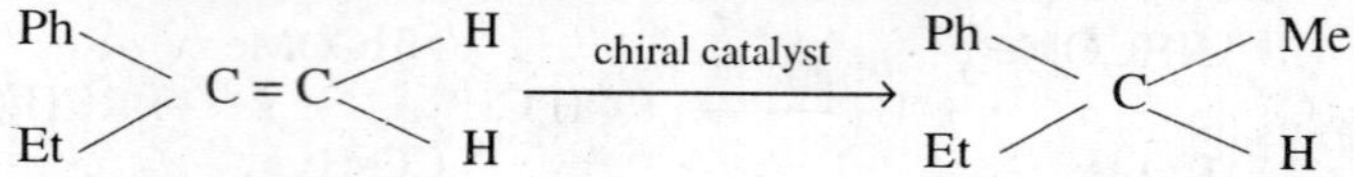

Chiral catalyst	% (enantiomeric excess or e.e.)
diop (g)	15
phellanphos (e)	30
diphosphinite (h)	60

The other category of alkenes gives very high optical yields with many chiral catalysts. A few illustrative examples are:

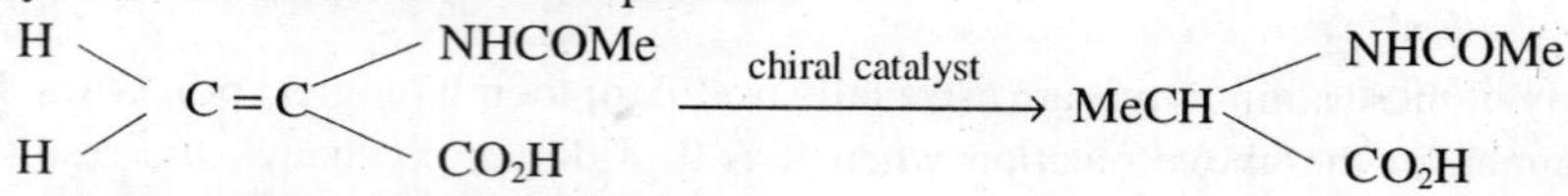

(chiral catalyst = Rh(—) diop) 73% e.e.

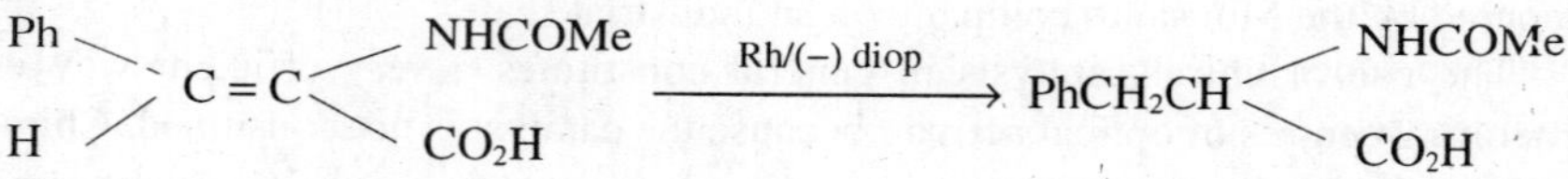

80% e.e.

Although there have been relatively few reports of asymmetric hydrogenation catalysts which employ chiral ligands other than phosphines, yet some successful asymmetric syntheses have also been achieved by adding chiral amines to catalysts derived from bis-dimethylglyoximatocobalt(II) complexes:

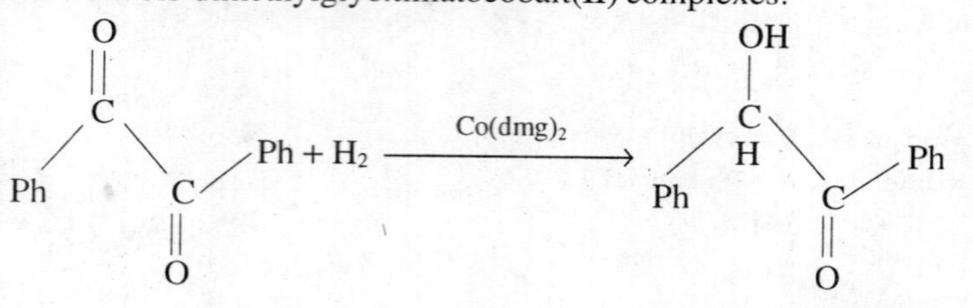

78% e.e.

Furthermore, chiral amides have been used successfully to induce modest asymmetric hydrogenations with a catalyst derived from borohydride reduction of a rhodium pyridine complex:

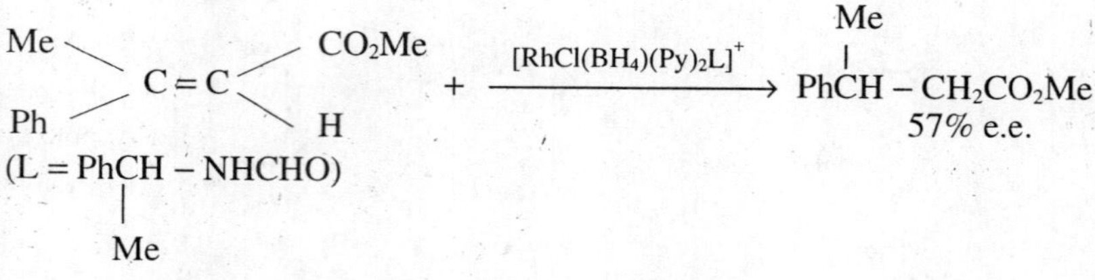

57% e.e.

(ii) *Polymer Supported Catalysts*

Although a homogeneous catalyst possesses many advantages like efficiency, controllability, reconvertibility, and stereoselectivity, the principal disadvantages in its actual application are two-fold:

(a) non-availability of a simple procedure to remove the catalyst at the end of a reaction; and

(b) its high cost.

Separation of the catalyst at the end of the reaction continues to be in fact, a major problem and is one of the reasons why industry has not been so far very enthusiastic to adopt homogeneous catalysts, in spite of their obvious advantages. One of the ways to overcome the above difficulty which is receiving considerable attention is to build the catalyst on an insoluble support, thus combining the advantages of homogeneous and heterogeneous catalysis. The idea is to tie the homogeneous catalysts on a polymer (insoluble) support. This support can be either a natural polymeric material such as cellulose, cotton, wool or a synthetic product such as polystyrene, polyvinylpyridine, polyvinyl alcohol, and the polyacrylates, which are cross-linked by difunctional monomers such as divinyl benzene.

(a) *Synthesis of polymeric catalysts.* The general techniques employed to synthesize functional polymer supports are either by (i) polymerization of the monomer which already contains the desired functional group or by (ii) functionalization of the pre-existing, cross-linked polymer, followed by incorporation of the metal catalyst precursor onto the functionalized polymer support.

Phosphine groups are often used to tie the metal to the solid support. Polystyrene has been most widely employed mainly because of its low cost, easy availability, and its chemical, mechanical, and thermal stability. The general preparative techniques are as follows:

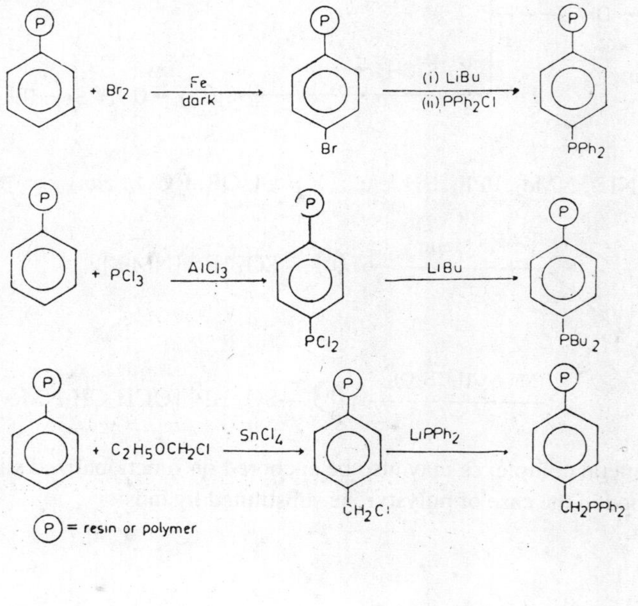

It is possible to replace the three PPh_3 groups of $RhCl(PPh_3)_3$ with resin-bound phosphines. Thus Scheme 6.11 illustrates several ligand replacement reactions used to prepare polymer-anchored homogeneous catalysts. Most commonly, carbonyl substitution or phosphine exchange reactions are employed. Both photochemical and thermal reactions have been used.

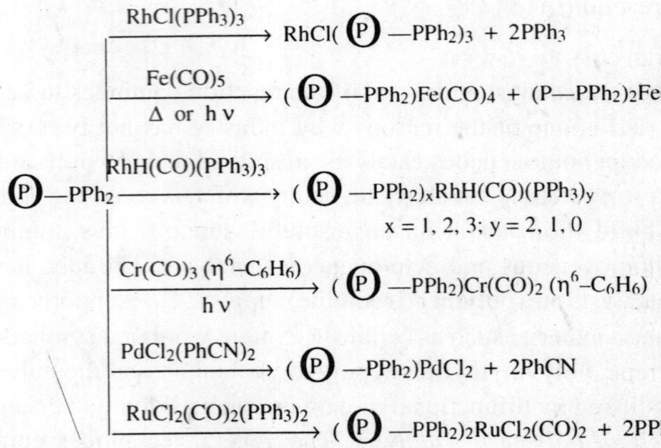

Scheme 6.11 Synthesis of Polymer Supported Catalysts.

Inorganic supports can also be phosphinated easily by utilizing their surface reactive groups such as silanol, sulphonic acid, etc. e.g.

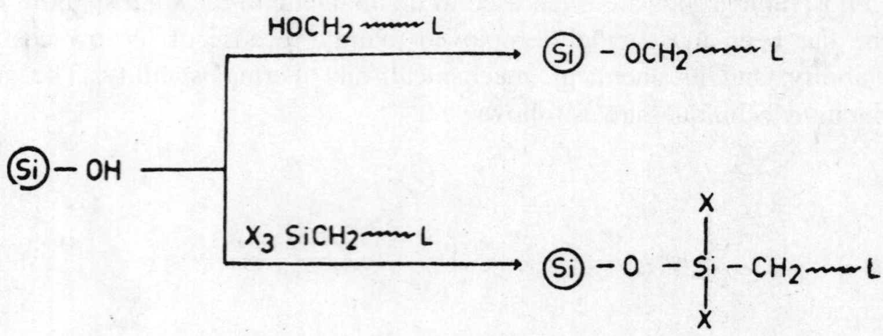

(L = NH_2, NMe_2, PPh_2, SH, etc. ; X = Cl, OR, RCO_2, etc.)

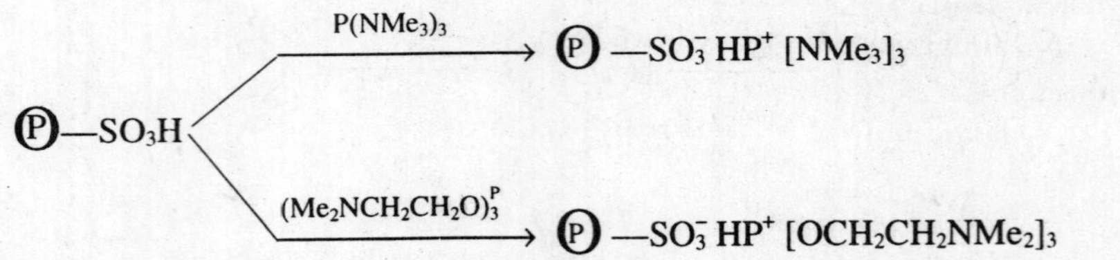

The metal complexes may also be anchored on functionalized silica surfaces as described in the case of polystyrene substituted ligands:

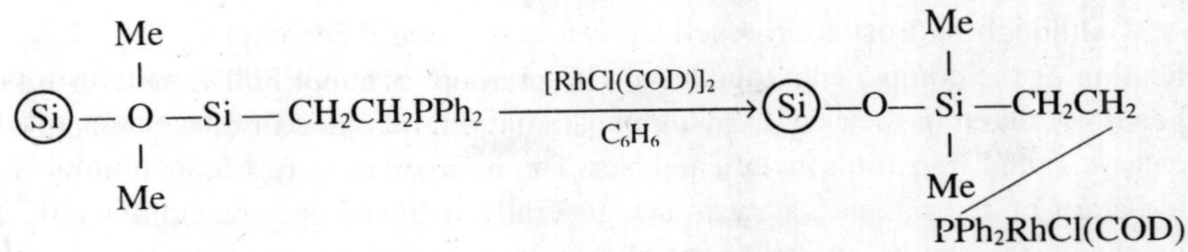

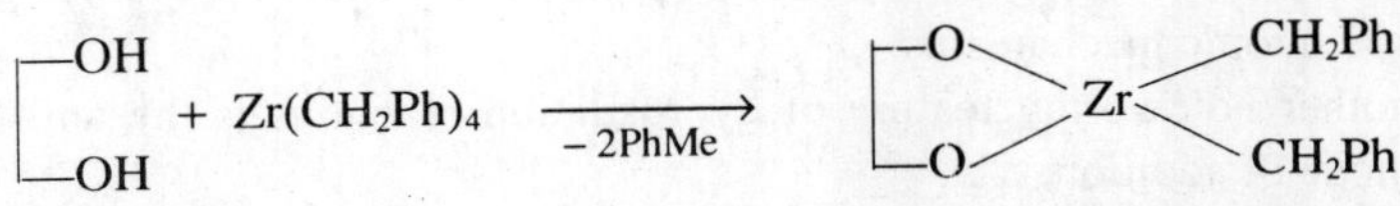

Polymer-bound asymmetric catalysts are also known and these may be classified into two general categories. The first involves support polymers which are inherently chiral, such as cellulose, polypeptides and chiral stereoregular polymers. The second class involves a chiral complex which has been bonded to an inherently chiral polymer support. Only a few examples of the first class have been reported, which involve transition metals at the catalytically active site.

Of a large number of reactions catalyzed by polymer-bound asymmetric catalysts, only a few representative reactions would be described here:

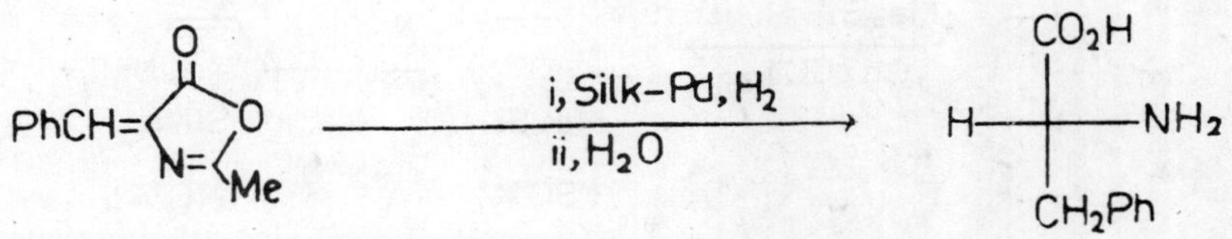

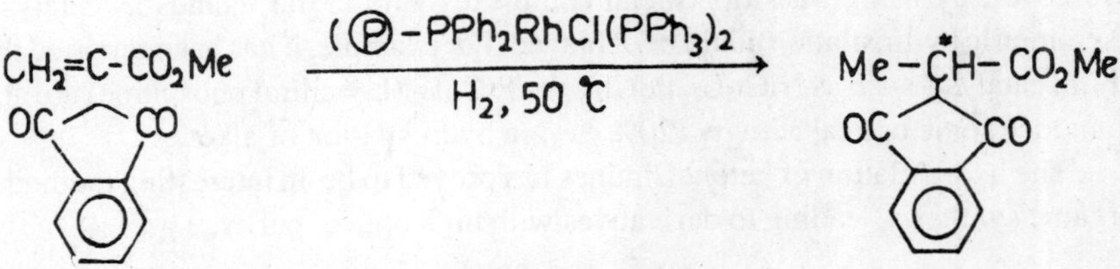

6.4.2.2 Hydrosilation (or Hydrosilylation) Reactions

Insertion reactions of alkenes are not restricted to hydrogen, but occur with a variety of other reagents like compounds containing B—H, Al—H, Si—H or Zr—H bonds. Out of these, the insertion of alkenes in the Si—H bonds has special significance and is termed as hydrosilation; the practical uses of this type of reactions may be illustrated by incorporation of C_2H_4CN groups into rubbers for self-sealing fuel tanks by hydrosilation of acrylonitrile, and in curing of gums into elastomers.

Addition of the Si — H bonds could also occur across C ≡ C triple bonds; e.g.,

$$X_3Si—H + RC\equiv CH \rightarrow RCH = CHSiX_3 \xrightarrow{+X_3Si—H} X_3SiCHRCH_2SiX_3$$

(X = Cl, alkyl, or aryl).

Although hydrosilation reactions can be generally accomplished by simple heating of the components together under pressure at about 300°C, assistance is generally taken in their cases by uv or γ-radiation; peroxide or other free radical catalysts; and transition metal catalysts. The hydrosilation reactions assisted by radiation or fee radical catalysts are generally believed to proceed via radical mechanisms, whereas transition metal catalyzed reactions are presumed generally to occur via ionic mechanisms.

Another noteworthy feature of hydrosilation reactions is the anti-Markovnikov mode of addition, e.g.,

$$X_3Si—H + RCH = CH_2 \rightleftharpoons X_3SiCH_2CH_2R$$

Hydrosilation reactions are often preferred in view of the substituent groups on silicon generally remaining unaffected:

$$Cl_3Si—CH_2SiCl_2 + RCH = CH_2 \xrightarrow{Bz_2O_2} Cl_3SiCH_2SiCl_2{\cdot}CH_2CH_2R$$

$$R_3Si—H + H_2C = CHCH_2NH_2 \xrightarrow{H_2PtCl_6} R_3Si(CH_2)_3NH_2$$

Both regio- and stereo-specific hydrosilation of polyenes are possible, e.g.:

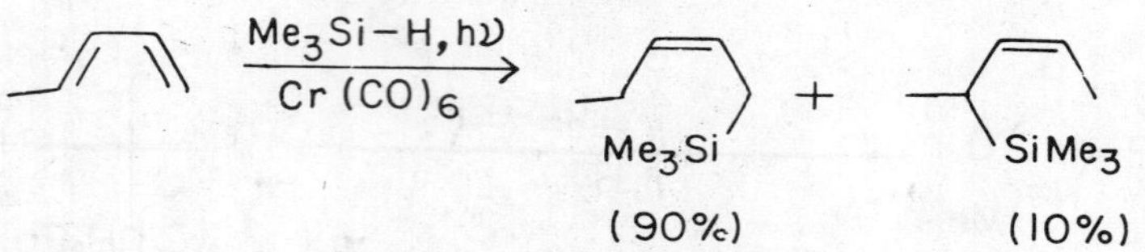

Moreover, by using transition metal complexes with chiral ligands as catalysts, asymmetric hydrosilation could be achieved. For example, it has been reported that chiral catalysts such as $NiCl_2L_2$, $PdCl_2L_2$ or $PtCl_2L_2$ (L = chiral phosphine) are able to induce some optical activity (20% e.e.) in hydrosilation of alkenes.

The hydrosilation of ketones/imines has proved to be an interesting method in organic synthesis, leading to derivatives with high optical purity, e.g.:

$$Me—CO—CO_2Pr + \alpha\text{-}NpPhSiH_2 \xrightarrow{Rh/diop} \underset{85\%\ e.e.(R)}{Me—CHOH—CO_2Pr}$$

$$Me—CO—(CH_2)_2CO_2Bu^t + \alpha\text{-}NpPhSiH_2 \xrightarrow{Rh/diop} \underset{84\%\ e.e.(S)}{Me—CHOH—(CH_2)CO_2Bu^t}$$

$$Ph—CO—Bu^t + \alpha\text{-}NpPhSiH_2 \xrightarrow{Rh/(+)RhCH_2P(Me)Ph} \underset{62\%\ e.e.(S)}{Ph—CHOH—Bu^t}$$

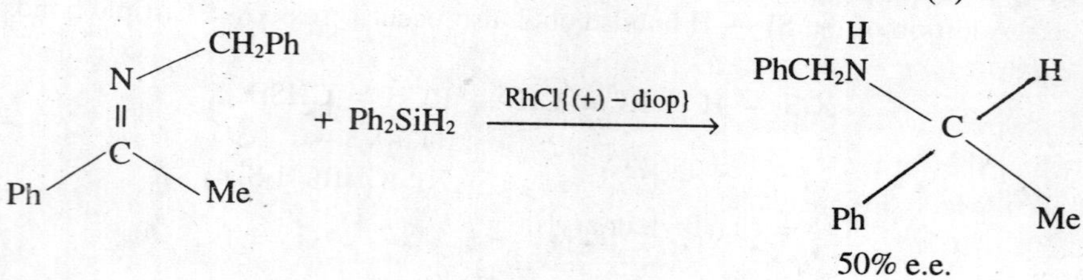

50% e.e.

At the end, an ingenious application of a hydrosilation reactions for producing an asymmetric silicon atom may be mentioned:

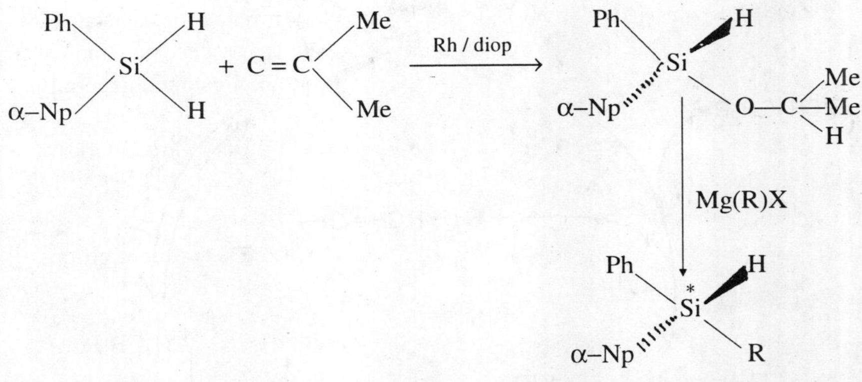

The chemistry of hydrosilation reaction with R_3SiH can be represented as below:

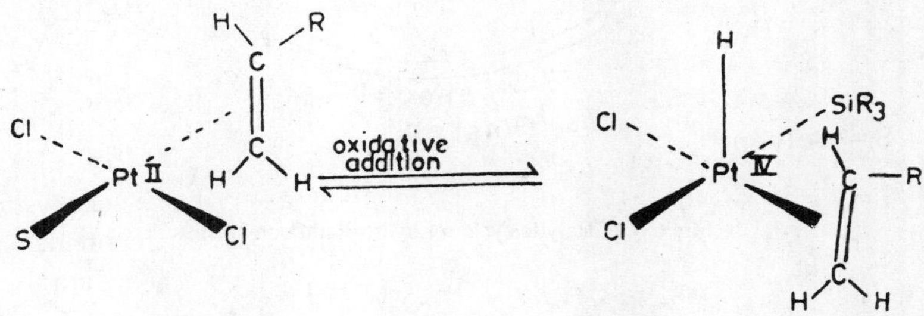

The platinum addition product has not been isolated and is known so far only as a reactive intermediate. The corresponding stoichiometric addition products formed by the reaction of the silanes with the trans $IrCl(CO)(PPh_3)_2$ or $RhCl(PPh_3)_3$ have been actually isolated, and thus provide indirect evidence for the oxidative step envisaged with Speier's catalyst, $IrCl(CO)(PPh_3)_2$:

$$IrCl(CO)(PPh_3)_2 + R_3Si{-}H - IrCl(H)(SiR_3)(PPh_3)_2$$

The catalytic cycle for hydrosilation of alkenes by platinum metal complexes is illustrated in Figure 6.6.

6.4.2.3 Hydroformylation of Alkenes ('Oxo' Process)

The reactions involved in the process are in principle the addition of hydrogen (—H) and the formyl (—CHO) group to alkenes leading to the formation of aldehydes, which are generally precursors for the production of higher alcohols (usually C_6—C_9):

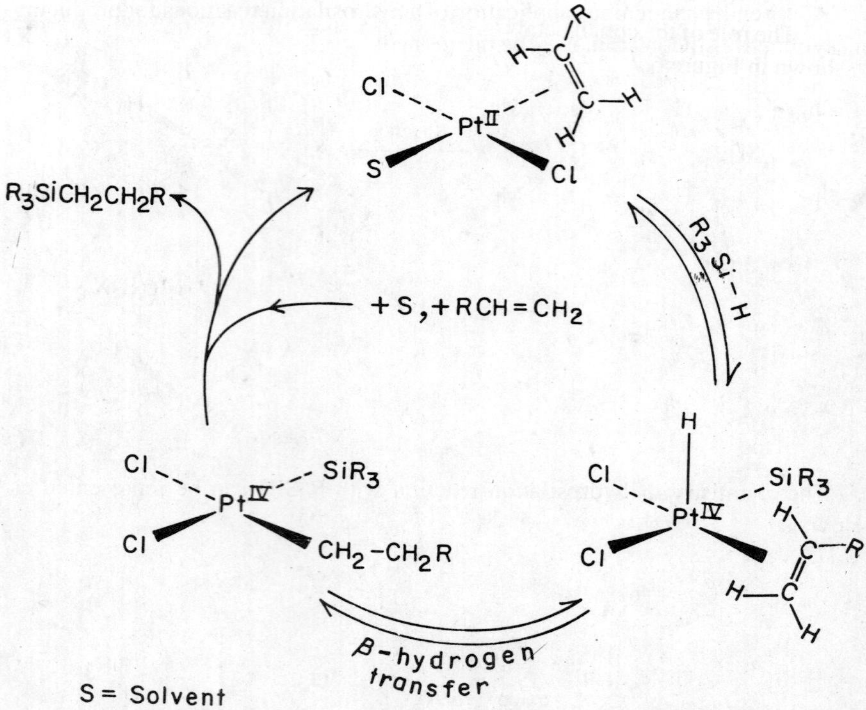

Fig. 6.6 Catalytic cycle for hydrosilation on alkenes

$$RCH=CH_2 + H_2 + CO \rightarrow R{-}\underset{H}{\overset{H}{C}}{-}\underset{CHO}{\overset{H}{C}}{-}H + RCH(OH){-}CH_3$$

Earlier, the reactions used to be catalyzed by cobalt carbonyl $CO_2(CO)_8$, which was not so effective a catalyst as the reaction had to be carried out at high (150–180°C) temperature and pressure (~200 atmospheres). Further the products were generally mixtures of linear and branched-chain isomers in roughly 3 : 1 ratio; the formation of these is directed by anti-Markovnikov and Markovnikov additions respectively, e.g.:

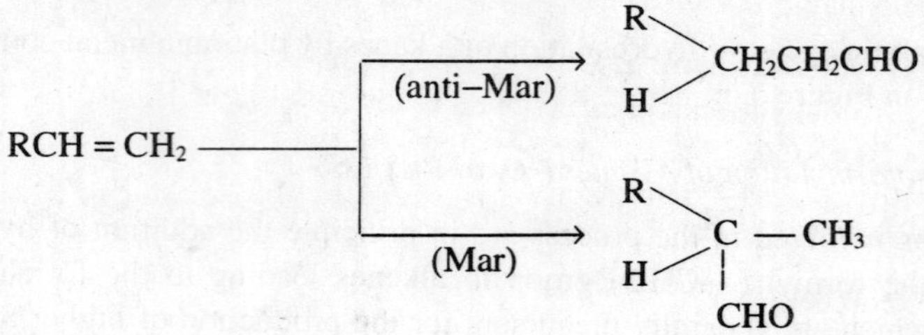

The role of the cobalt catalyst, $CO_2(CO)_8$ can be illustrated by a catalytic cycle shown in Figure 6.7:

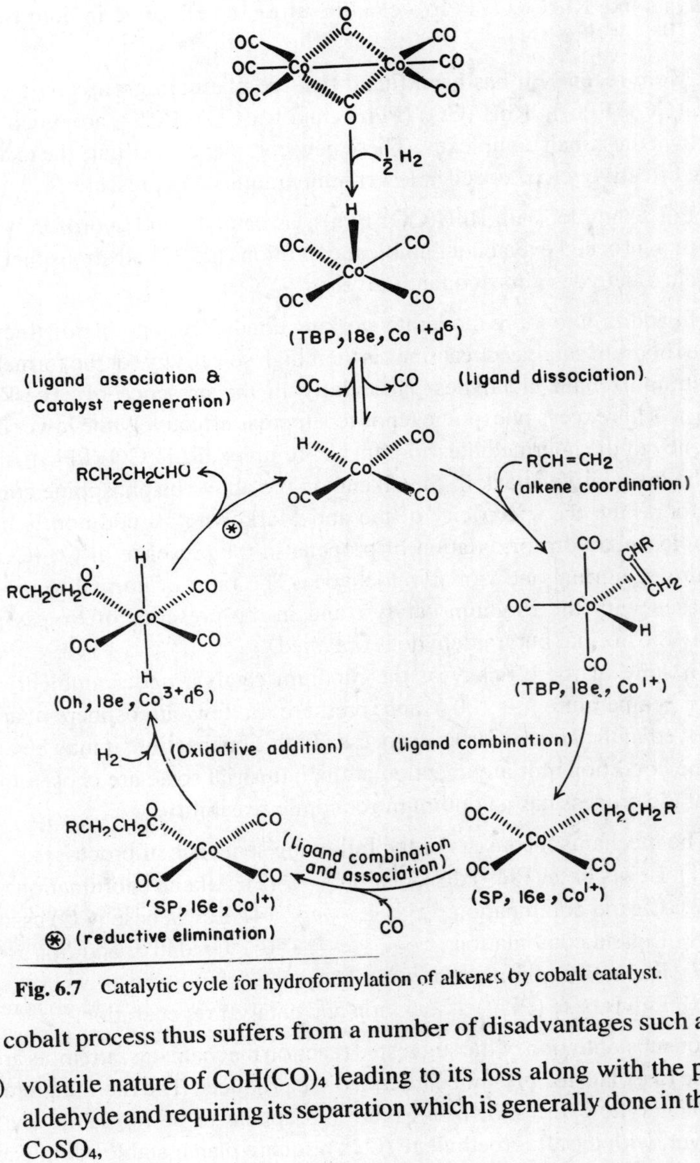

Fig. 6.7 Catalytic cycle for hydroformylation of alkenes by cobalt catalyst.

The cobalt process thus suffers from a number of disadvantages such as:

(a) volatile nature of $CoH(CO)_4$ leading to its loss along with the product aldehyde and requiring its separation which is generally done in the form $CoSO_4$,

(b) loss (~15%) of the alkene by a side hydrogenation reaction,

(c) a small amount of the conversion of aldehydes into alcohols by hydrogenation,

(d) production of generally less useful branched chain aldehydes and

(e) drastic conditions (high temperature and pressure).

A comparative perusal of these hydroformylation and hydrogenation (cf. Section 6.4.2.1) reactions shows their mechanistic similarity. A number of catalysts like $RhH(CO)(PPh_3)_3$ can, therefore, be effective in both types of reactions.

More recently it has been found that the platinum group complexes such as $RhCl(CO)(PPh_3)_3$, $RhH(CO)(PPh_3)_3$ and $Ru(CO)_3(PPh_3)_2$ are much more effective than the cobalt complexes. These new complexes facilitate the reactions much more effectively to proceed at lower temperatures and pressures.

For example, with $RhH(CO)(PPh_3)_3$ as catalyst, the hydroformylation reactions can proceed even under ambient conditions (25°C and atmospheric pressure) to yield aldehydes almost quantitatively.

In addition to more moderate working conditions, another distinct advantage of the rhodium catalyzed reactions is their high selectivity for the formation of only the straight chain aldehydes particularly in the presence of excess phosphine (PPh_3). This excess phosphine tends to suppress effectively the lower dissociative tendency of the intermediate rhodium bis-complex, $RhH(CO)(PPh_3)_2$, thus favouring the associative attack of the alkene on the above bisphosphine complex (Fig. 6.8) for which the specificity of the anti-Markownikoff addition is higher. For example, the hydroformylation of propene in the presence of $Co_2(CO)_8$ yields a mixture of normal and isobutyraldehyde. The ratio of normal to iso-isomers is increased with the rhodium catalyst and in the presence of excess phosphine, mainly the normal butyraldehyde is obtained.

In spite of the efficacy of the rhodium catalyst under ambient conditions, higher temperature (~ 100°) and pressure (a few atmospheres) are actually employed in the Union Carbide process. In this connection, it may be emphasized that the conditions for any reaction at the industrial scale are chosen in a manner so that the process has an optimum economic feasibility.

The mechanism consists of the following sequence of processes:

(1) Lewis base (PPh_3) dissociation; (2) Alkene coordination;
(3) Ligand combination; (4) Lewis base (CO) association;
(5) Ligand combination; (6) Oxidative addition
(7) Reductive elimination;
(8) Lewis base (PPh_3) association and catalyst regeneration.

For substantiation of the suggested reaction mechanisms, attempts are generally made to isolate the presumed intermediate products. The rhodium intermediates of the catalytic cycle shown in Figure 6.8 appear to be too labile for isolation. However, with tetrafluoroethylene (C_2F_4), square planar stable complexes such as $Rh(C_2F_4H)(CO)(PPh_3)_2$ have been isolated. Indirect evidence for the other rhodium intermediates (except dihydridoacyl complex) is based on the characterization of more stable iridium analogues. Although it has not been possible to actually isolate even the iridium analogue of the dihydridoacyl complex, a similar derivative, $IrHCl(COCH_2R)(CO)(PPh_3)_2$ has been characterized.

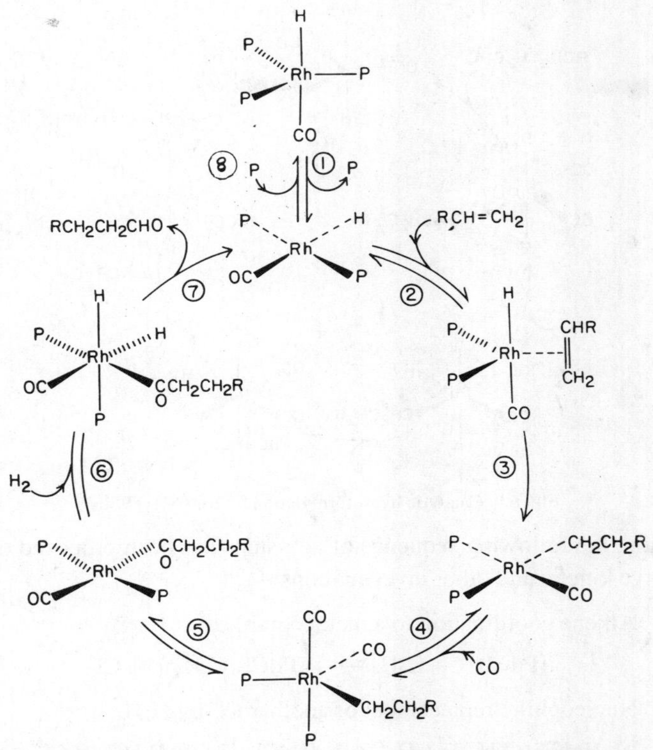

Fig. 6.8 Catalytic cycle for the hydroformylation of alkene (p* represents PPh_3).

In addition to the above mentioned catalysts, a ruthenium complex $Ru(CO)_3(PPh_3)_2$ has also been found effective. Although the basic catalytic steps are similar (as shown in Figure 6.9) to those described in Figure 6.8, yet the mechanism of hydroformylation reaction is of a different pattern.

6.4.2.4 *Oxidation of Alkenes*

One of the first processes to highlight the importance of alkenes in the production of a wide variety of organic compounds was the *Wacker process* developed in Germany in the early 1960's; this is an important large scale industrial process for the production of acetaldehyde, acetone, and methyl ethyl ketone. It has been stated that ''the invention of the Wacker process was a triumph of common sense''. It may be noted that the oxidation of ethylene to acetaldehyde with the help of platinum group metals had been known as for back as 1894:

$$\underset{\text{Zeise's salt}}{[PtCl_3(C_2H_4)]^-} + H_2O \xrightarrow{\text{heat}} CH_3CHO + Pt(O) + 3Cl^-$$

It was, however, more than six decades later that Smidt working at 'Wacker Chemie' laboratories discovered a practical process for the oxidation of ethylene to acetaldehyde using $PdCl_2$ along with $CuCl_2$ as a catalyst. For the mechanism of

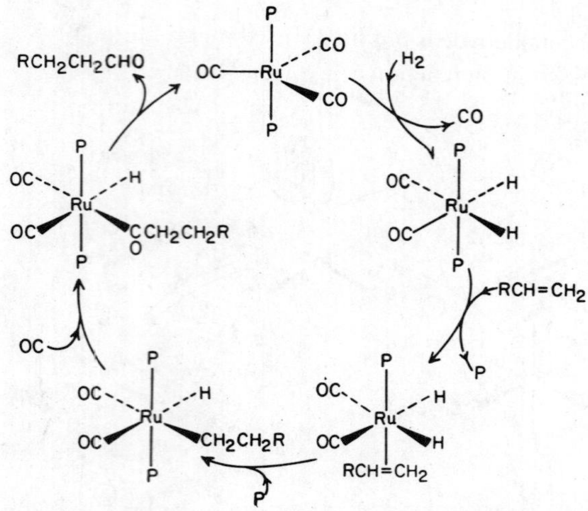

Fig. 6.9 Catalytic hydroformylation by $Ru(CO)_3(PPh_3)_2$.

the reaction, the following sequence of key-steps has been proposed on the basis of extensive kinetic and other investigations:

(a) Alkene coordination to a metal complex,

$$[PdCl_4]^{2-} + C_2H_4 \longrightarrow [PdCl_3(C_2H_4)]^- + Cl^-$$

(b) Nucleophilic replacement of another Cl^- by H_2O,

$$[PdCl_3(C_2H_4)]^- + H_2O \;—\; [PdCl_2(H_2O)(C_2H_4)] + Cl^-$$

(c) Intermolecular attack of H_2O on the coordinated ethene, followed by loss of H^+ to form hydroxometal complex,

$$[PdCl_2(H_2O)(C_2H_4)] \rightleftharpoons [PdCl_2(H_2O)(C_2H_4OH)]^- + H^+$$

(d) β-Hydrogen elimination, and (e) Migratory insertion into Pd-H bond:

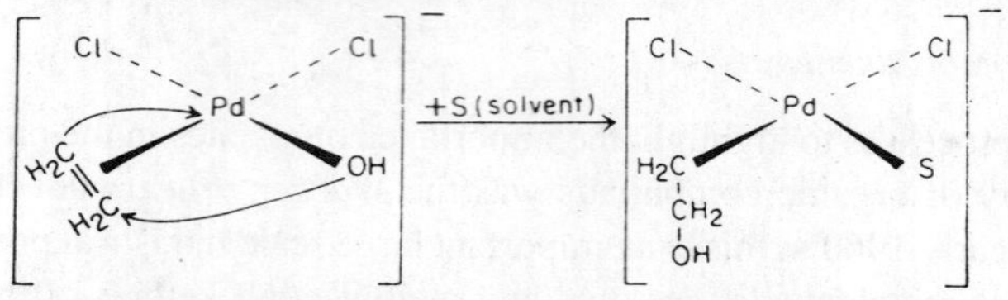

(e) β-Hydrogen elimination

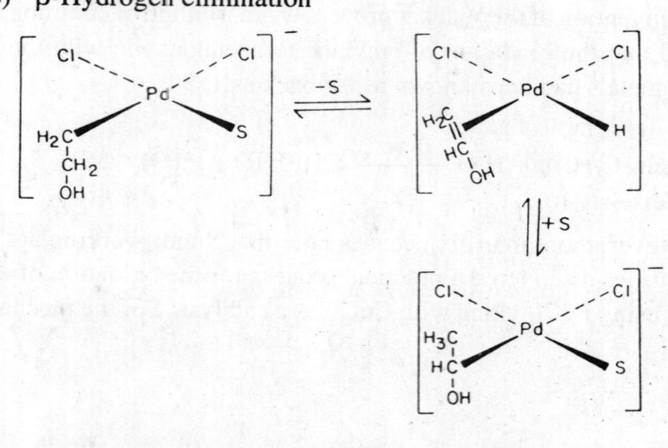

(f) Finally, acetaldehyde is produced in another β-elimination, followed by a base dissociation reaction

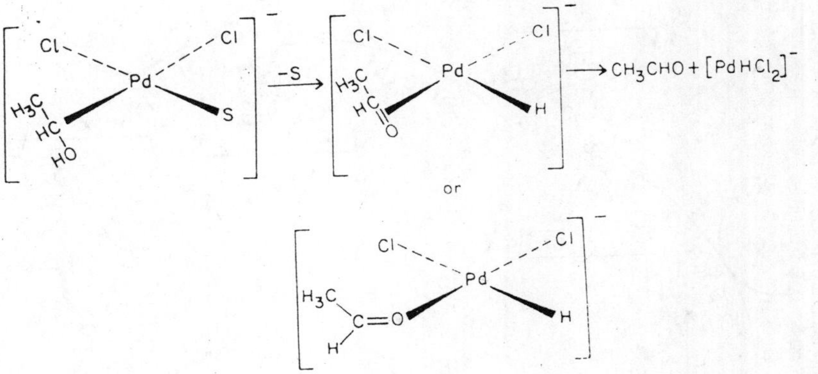

(g) Catalyst regeneration.

Although formation of the species, $[PdCl_3(C_2H_4)]^-$ (an analogue of Zeise's salt) has been postulated in the first step (a) of the above mechanism, it may be pointed out that this derivative has not so far been isolated (In fact, the dimeric complex $[PdCl_2(C_2H_4)]_2$ has been isolated by the reaction of the components in non-aqueous solvents).

The role of copper (II) chloride in the Wacker process appears to be the regeneration of the catalyst $[PdCl_4]^{2-}$ species by oxidation of Pd(O), getting itself reduced to Cu(I) state which is oxidized by oxygen in the presence of HCl back to the +2 oxidation state:

(i) Product (acetaldehyde) formation,

$$[PdCl_4]^{2-} + C_2H_4 + H_2O \rightarrow CH_3CHO + Pd(O) + 2HCl + 2Cl^-$$

(ii) Catalyst, $[PdCl_4]^{2-}$ regeneration,

$$Pd(O) + 2CuCl_2 + 2Cl^- \rightarrow [PdCl_4]^{2-} + 2CuCl$$

(iii) Co-catalyst ($CuCl_2$) regeneration,

$$2CuCl + 2HCl + 1/2O_2 — 2CuCl_2 + H_2O$$

The overall reaction can, therefore, be represented as

$$C_2H_4 + 1/2O_2 \rightarrow CH_3CHO$$

Steps (a) to (f) in the suggested mechanism of the Wacker process can be represented by a cycle (Fig. 6.10) in which the reactants entering and the products leaving the system are indicated at appropriate place. By inclusion of the reactions representing the catalyst and co-catalyst generation the cycle is completed.

6.4.2.5 *Isomerization of Alkenes*

Many transition metal (e.g., molybdenum, iron, cobalt, nickel, rhodium, palladium, iridium and platinum) complexes catalyze double bond migration (i.e., isomerization) in alkenes; out of these, platinum group metal complexes appear to be more

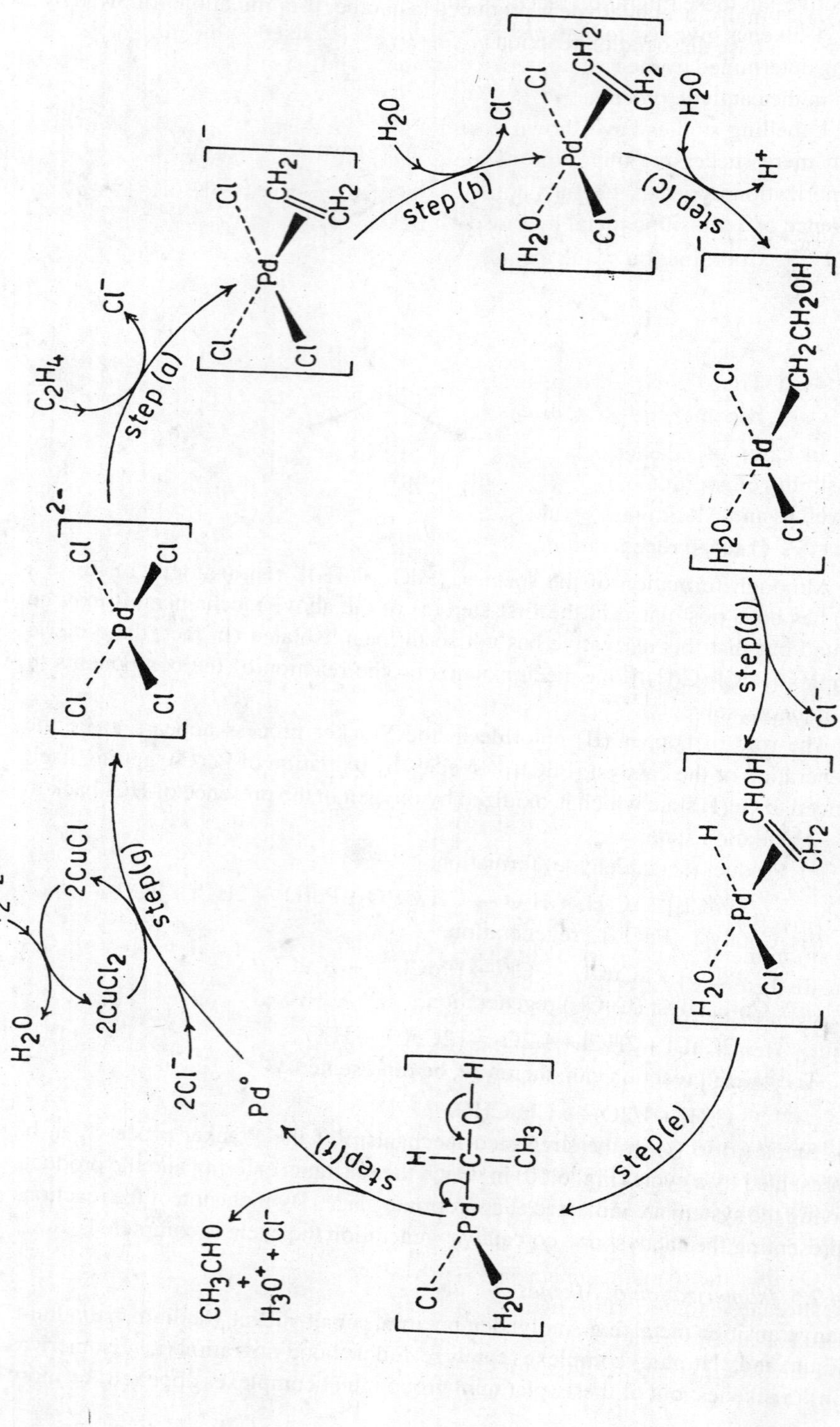

Fig. 6.10 Catalytic cycle of Wacker process.

effective. In these migration reactions, often isomeric mixtures may be obtained; e.g., 1-alkenes give rise to both *cis*- as well as *trans*-2 alkenes, the ratio of the two being determined by their comparative thermodynamic stability as well as the final step in the catalytic process.

Labelling studies have shown that migration of double bond does not result from mere successive migration of hydrogen along the alkene chain. In fact, isomerization appears generally to involve the following types of steps in the presence of a transition metal hydride (which functions as a catalyst):

(i) Coordination of the metal hydride (L_nMH) with the alkene:

$$(L_n)MH + RCH = CH_2 \rightarrow (L_n)(H)M - \begin{matrix} RCH \\ \| \\ CH_2 \end{matrix}$$

(ii) Transfer of the hydrogen atom from the metal to the carbon atom which becomes the β-carbon atom of the new metal alkyl [A or B].

In cases of alkenes other than the simplest member, ethene, there is the possibility of addition of $(L_n)MH$ to the double bond following both Markovnikov as well as anti-Markovnikov rules (just as in the case of addition of alkenes to any other X—H reagent):

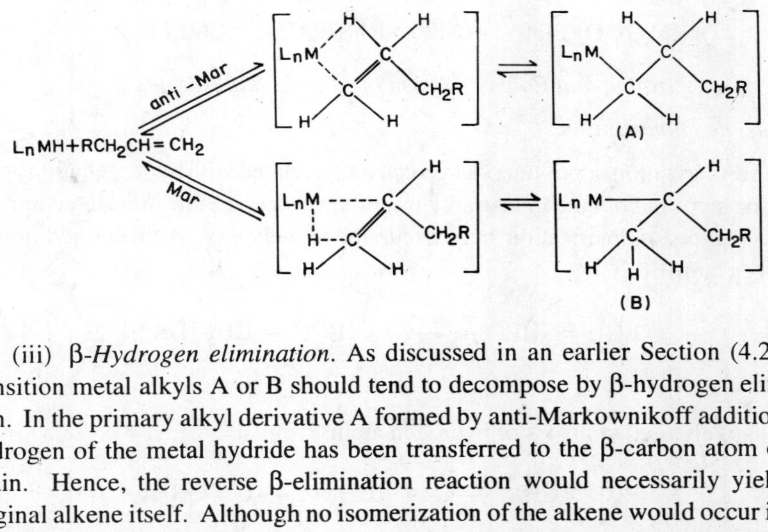

(iii) β-*Hydrogen elimination.* As discussed in an earlier Section (4.2), the transition metal alkyls A or B should tend to decompose by β-hydrogen elimination. In the primary alkyl derivative A formed by anti-Markownikoff addition, the hydrogen of the metal hydride has been transferred to the β-carbon atom of the chain. Hence, the reverse β-elimination reaction would necessarily yield the original alkene itself. Although no isomerization of the alkene would occur in this case, yet in view of the free rotation about the single C—C bond, the same hydrogen (which was transferred from the metal hydride) atom need not necessarily be removed and hence, the possibility of hydrogen exchange does arise (using a deuterium complex L_nMD, the formation of L_nMH also has been observed in conformity with this possibility).

On the other hand, the metal secondary alkyl derivative B can decompose by two alternative routes. If the hydrogen atom is transferred from the CH_3 group, the original 1-alkene would be reproduced. However, the transfer of hydrogen atom from the β-CH_2 group would result in the formation of an 2-alkene. Isomerization of alkenes can thus occur, only if the initial addition is according to Markownikoff rule. However, it should be noted that in this process, the 2-alkene formed may be both *cis* or *trans* isomer.

A brief mention may be made of the extensive investigations carried out to substantiate the above suggested mechanism. The essential role of transition metal-hydrogen bond can be inferred by the observation that although non-hydride complexes such as $(Et_3P)_2NiCl_2$ sometimes do catalyze the isomerization reactions of alkenes, yet they become effective only in the presence of a hydride ion source (such as molecular hydrogen).

Besides possible spectroscopic characterization, the evidence for the first step being the coordination of the alkene with the metal hydride complex can be indirectly inferred from the fact that the reaction of ethylene with $RhHCl_2(PPh_3)_2$ at –25°C to yield $Rh(C_2H_5)Cl_2(PPh_3)_2$ is substantially inhibited by the presence of an excess of PPh_3 (a better coordinating ligand).

The transfer of the hydrogen atom from the metal hydride to the β-position in the alkene has not been unequivocally proved in the case of unsubstituted alkenes due to an extensive scrambling of hydrogen atoms in the over-all readily reversible process. However, indirect evidence for the same has again been provided by the isolation of stable products when fluoroalkenes are used, e.g.,

$$RhH(CO)(PPh_3)_3 + C_2F_4 \rightarrow Rh(CF_2CF_2H)(CO)(PPh_3)_3$$

6.4.2.6 *Dimerization and Polymerization of Alkenes and Alkynes*

(A) *Alkene Dimerization*

Alkene dimerization or polymerization can be correlated with hydrogenation as the initiating steps in both the industrially important processes are considered to be of a similar type. Dimerization may occur reductively (i.e. with the addition of hydrogen atom):

$$2RCH=CHR' \xrightarrow{+2H} RCH_2\underset{\substack{|\\R}}{CH}-\underset{\substack{|\\R}}{CH}CH_2R\,,$$

or oxidatively (i.e., with loss of hydrogen atoms):

$$2RCH=CHR' \xrightarrow{-2H} RCH=\underset{\substack{|\\H}}{C}-\underset{\substack{|\\H}}{C}=CHR',$$

or alternatively by simple addition of one monomer to the other.

Out of a large number of dimerization and oligomerization processes of simple alkenes which have been catalyzed by a wide variety of derivatives of transition metals (e.g. Co, Ni, Ru, Rh and Pd), the most studied reaction is probably the dimerization of ethylene by rhodium trichloride, for which the mechanism depicted in Fig. 6.11 has been proposed on the basis of kinetic and spectroscopic studies. According to this mechanism, the key steps involved in the dimerization of ethylene are:

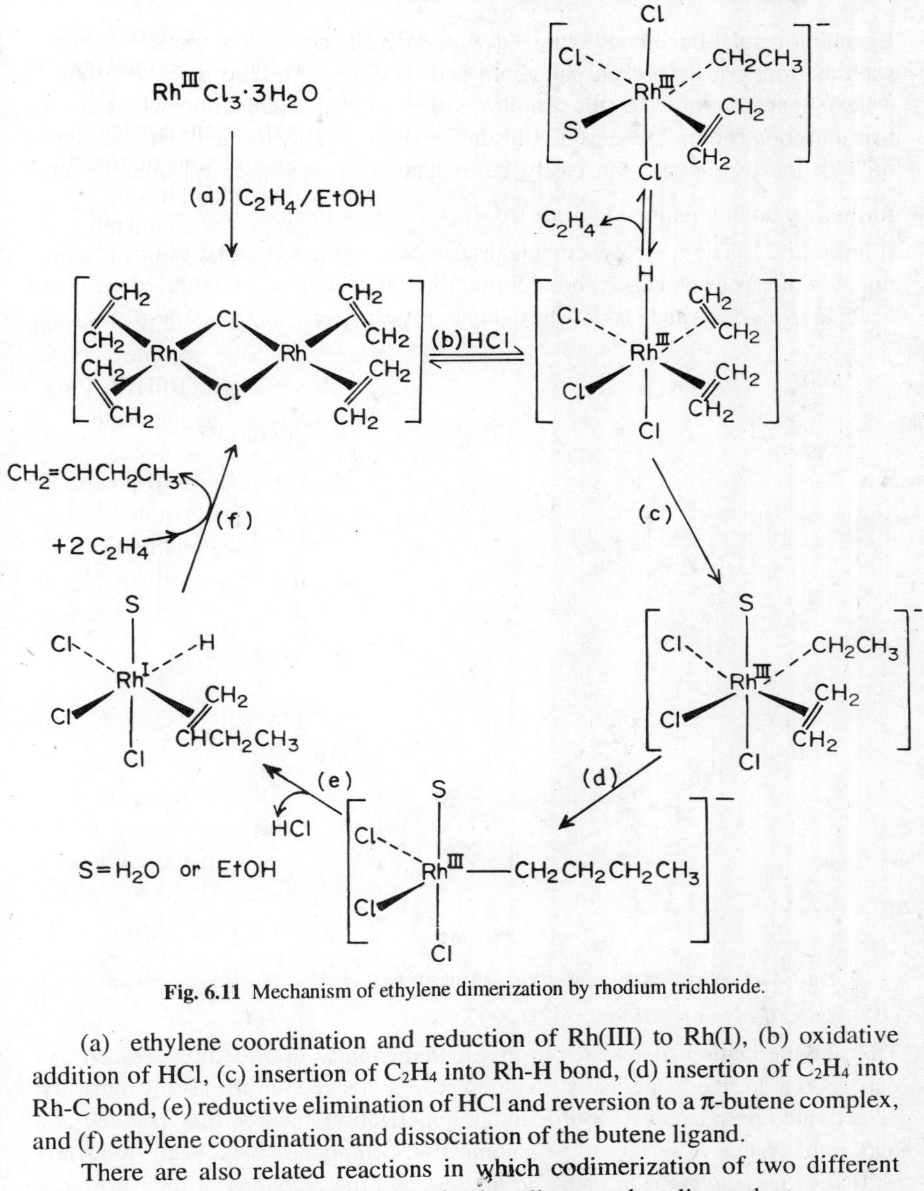

Fig. 6.11 Mechanism of ethylene dimerization by rhodium trichloride.

(a) ethylene coordination and reduction of Rh(III) to Rh(I), (b) oxidative addition of HCl, (c) insertion of C_2H_4 into Rh-H bond, (d) insertion of C_2H_4 into Rh-C bond, (e) reductive elimination of HCl and reversion to a π-butene complex, and (f) ethylene coordination and dissociation of the butene ligand.

There are also related reactions in which codimerization of two different alkenes (e.g., ethylene and butadiene) leads to linear and cyclic products:

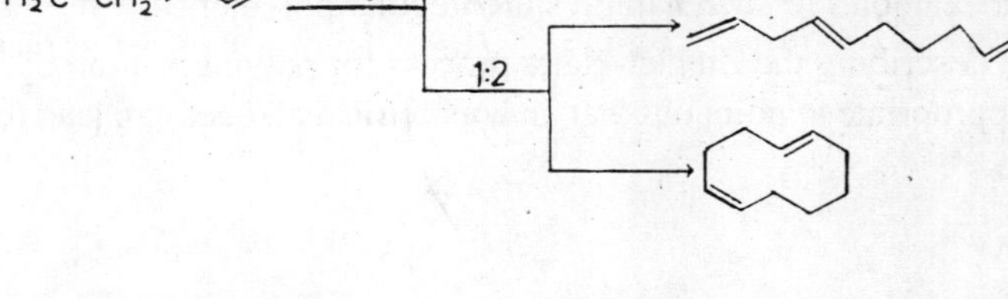

Excellent results have been obtained with catalysts containing transition metals such as iron, cobalt, nickel, palladium and rhodium. The formation of *trans*-1, 4-hexadiene is favoured by the complexes of rhodium and palladium, whereas with iron and cobalt complexes, *cis*-1, 4-hexadiene is preferably formed. The formation of both these products can easily be explained by the η^3-crotyl intermediates formed by adding a metal hydride (MHL_n) to a coordinated butadiene (L_nM 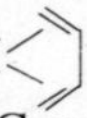) (Figure 6.12). The η^3-crotyl complexes can exist in the *syn*-and *anti*-form. Coordination of ethene and C—C bond formation followed by elimination of the diene leads to *trans*-1, 4- and *cis*-1, 4-hexadiene respectively:

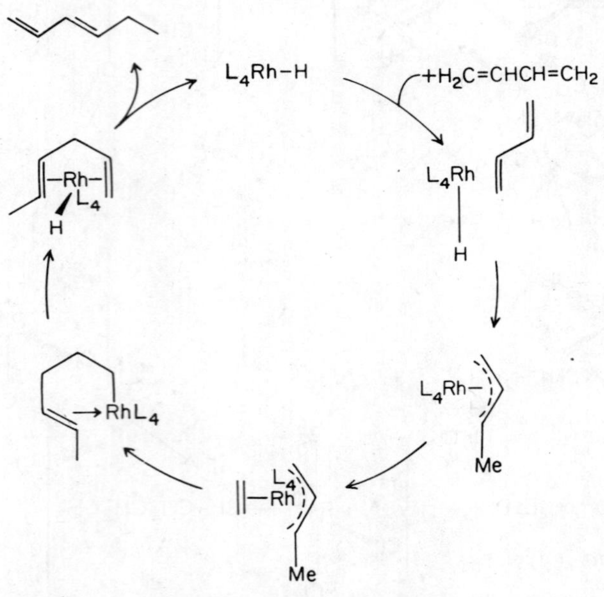

Fig. 6.12 Cooligomerization of ethene and butadiene (L = chloride or a donor solvent).

(B) *Alkene Polymerization*

The polymerization of alkenes to give commercially useful fibres, resins, and plastics is a thermodynamically favoured process. However, the alkene must first be activated by a catalyst, so that the polymerization reaction may proceed at a reasonable rate. A variety of organometallic compounds have been frequently used for this activation or catalytic process and the reactions in their presence generally lead to stereoregular polymers. Stereoregular polymerization leads to the arrangement of the repeating units in a stereoregular arrangement and it may be interesting to point out that nature has been accomplishing the same in an efficient manner all the time. Quartz; the polysaccharides; starch and cellulose; and the polynucleotides: DNA (deoxyribonucleic acid) and RNA (ribonucleic acid) are all examples of such natural stereoregular polymers.

Before describing the Ziegler-Natta process for polymerization of alkenes, it would be appropriate to point out that unsymmetrical alkenes can lead to different ypes of arrangements in the polymer. Catalysed polymerization of ethylene (ethene) produces a stereoregular polyethylene (polythene).

Propylene, on the other hand, may form two stereoregular (either isotactic or syndiotacitc) or a non-regular atactic polvmer (Figure 6.13):

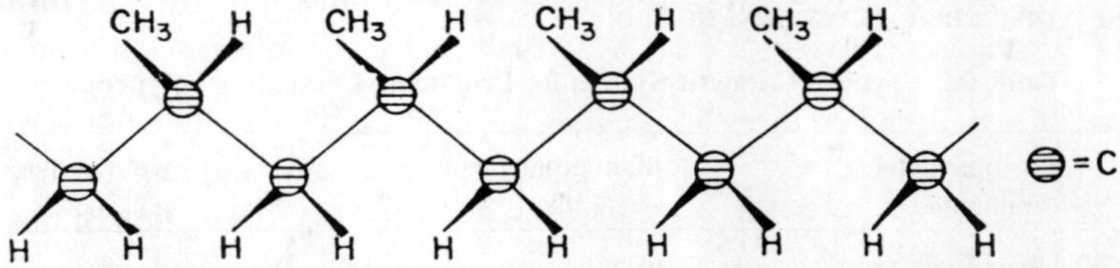

(a) *Isotactic polymer* (In this all methyl groups lie on the same side of the chain, which tends to adopt a helical shape so that the methyl groups do not sterically interfere with one another).

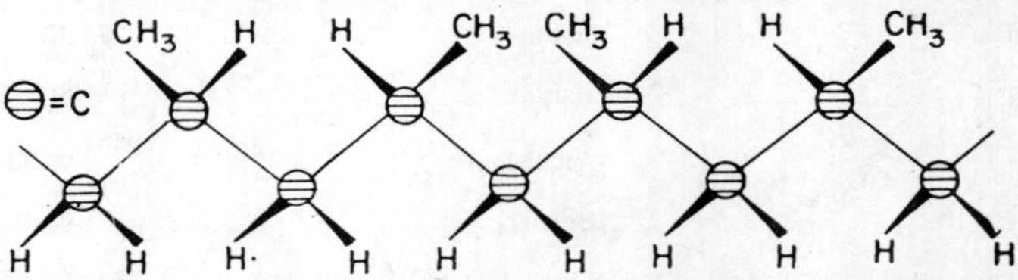

(b) *Syndiotactic polymer* (In this, there is a regular alternation of methyl groups along the chain).

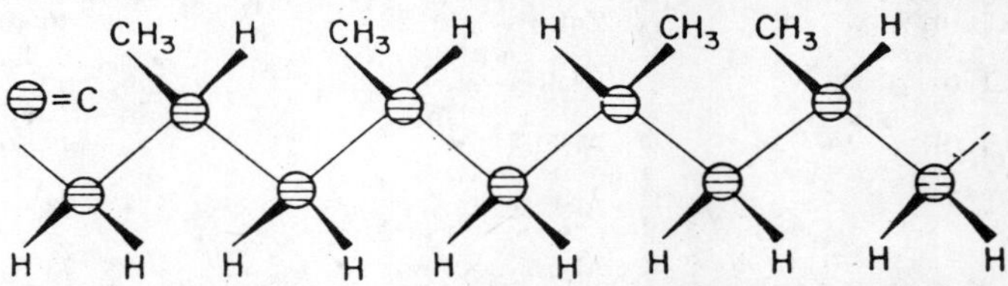

(c) *Atactic polymer* (In this, there is no regularity in the configuration of methyl groups along the chain).

Fig. 6.13 Possible configurations (a, b and c) of polypropylene.

Ziegler-Natta polymerization of alkenes. The original uncatalyzed industrial process

$$nC_2H_4 \xrightarrow[\text{and pressure}]{\text{high temperature}} [C_2H_4]_n$$

for polythene involved the use of high temperatures and pressures and the product was atactic, amorphous, and of low density. However, K. Ziegler (Germany) and G. Natta (Italy) achieved low pressure and room temperature stereoregular catalytic polymerization of ethene and propane in the presence of organometallic mixed catalysts (1955) to give isotactic, high density, harder and tougher, higher melting crystalline polyethylene ("Polythene") and other plastics on an industrial scale, for which they shared Nobel Prize for Chemistry in 1963. A typical Ziegler-Natta catalyst can be made by mixing $TiCl_4$ and Al_2Et_6 in heptane. In general, Ziegler-Natta type of catalytic systems consists of a reducing and/or alkylating agent from Group I-III such as an aluminium alkyl (usually Al_2Et_6 or $\{AlClEt_2\}_2$), and a transition metal compound (usually a halide of early transition metal series, e.g. $TiCl_4$, $TiCl_3$ or VCl_4). In fact, Ziegler cites the following rule: "Take any organometallic combine it with a compound of a transition metal, and you will have a good chance of finding a suitable catalyst for your special polymerization problem". A list of some catalytic systems for producing crystalline polypropene is presented in Table 6.2.

Table 6.2 Typical Catalytic System for Producing Crystalling Polypropene

Transition metal compound	Main group metal alkyl	% yield of the polymer
$TiCl_4$	Al_2Et_6	48
$TiCl_4$	$AlBu^i_3$	60
$TiBr_4$	Al_2Et_6	42
$TiCl_3^*$ (α, γ)	Al_2Et_6	80-92
$TiCl_3$ (α, γ)	$[AlClMe_2]_2$	96-98
$TiCl_3$ (α, γ)	$[BeEt_2]_n$	94-96
$TiCl_3$ (α, γ)	$[MgEt_2]_n$	78-85
$TiCl_3$ (α, γ)	$ZnEt_2$	30-40
$TiCl_3$ (α, γ)	$[NaR]_\infty$	0
$TiCl_3$ (β)	Al_2Et_6	40-50
VCl_4	Al_2Et_6	78
VCl_3	Al_2Et_6	73
$CrCl_3$	Al_2Et_6	36

* Titanium trichloride, $TiCl_3$ has several crystalline forms. The reduction of $TiCl_4$ at 500-1200° by H_2 gives the α-form, whereas the reduction of $TiCl_4$ by aluminium alkyls in inert solvents results in the β-form. The β-form can be converted to the α-variety by heating it at 250-300°. Two other forms, γ and δ are known and these as well as the α-form have layer structures containing $TiCl_6$ groups, whereas the β-form is fibrous with single chains of $TiCl_6$ octahedra sharing edges.

Most Ziegler-Natta catalytic systems are heterogeneous, although several homogeneous catalytic procedures have also been developed (e.g. from $TiCl_2(\eta^5—C_5H_5)_2$ or $TiClEt(\eta^5—C_5H_5)_2$). Recent studies appear to show that the polymerization site is generally the transition metal in an activated form, which is brought about by its reaction with aluminium alkyl (an initiator acting as reducing and alkylating agent). Although the mechanism of the process is not completely understood, yet for many years the accepted view for the functioning of a Ziegler-Natta catalyst has been that due to Cossee (1977). The initiating step, according to Cossee, consists in the replacement of one of the chloride ions coordinated to a titanium(III) ion (which has been initially generated in situ from the parent titanium(IV) chloride by reduction with $Al_2(CH_2CH_3)_6$ at the surface by an alkyl group of the aluminium alkyl:

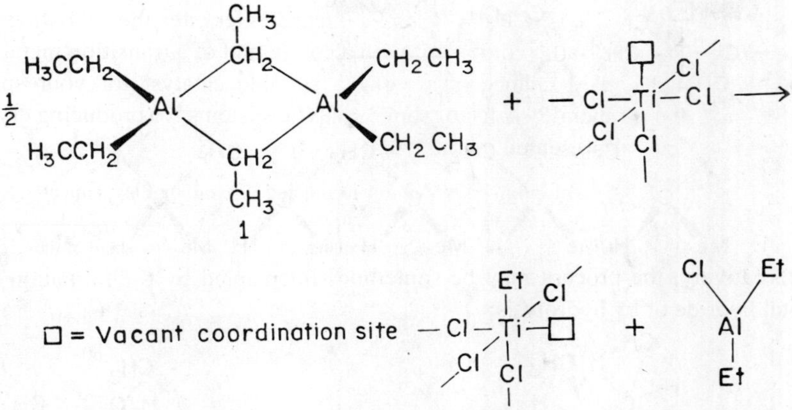

Due to the formation of an empty coordination site exposed at the surface of the alkylated titanium(III) halide, a molecule of propylene (propene) can bind itself to it through its π-electrons (step 2):

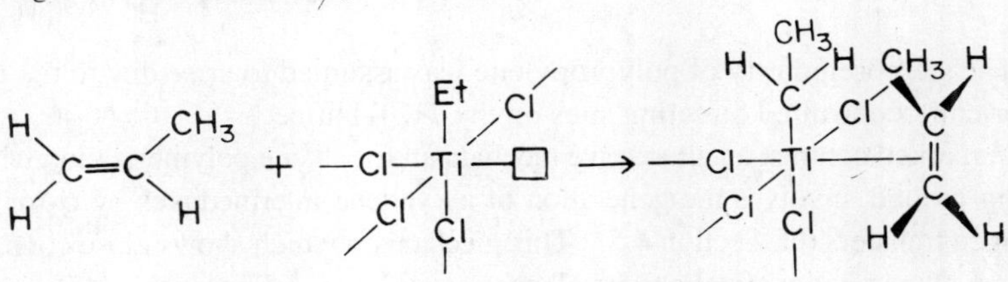

This reaction brings the alkyl group and the alkene in proper stereochemical arrangement, to facilitate migration of the alkyl group to the alkene (i.e., *cis*-insertion in the neighbouring metal-alkyl bond), giving a longer chain (i.e., growing polymer chain) of the alkyl derivative; this results in a fresh vacant site at the surface (step 3):

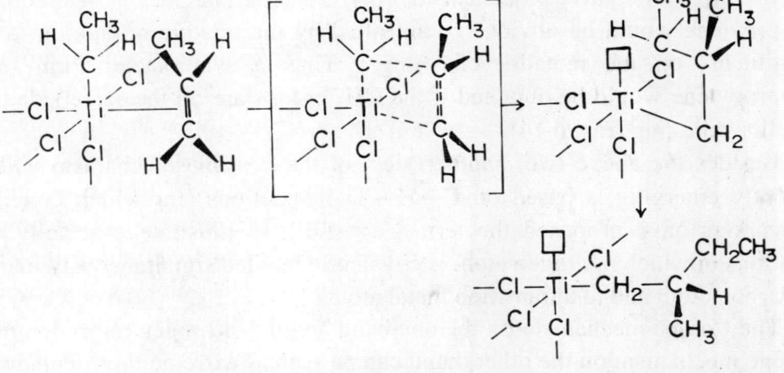

This vacant site can be occupied by another propene molecule and the insertion process is repeated and steps (2) and (3) continued until a chain termination process occurs by the severing of the titanium-alkyl bond:

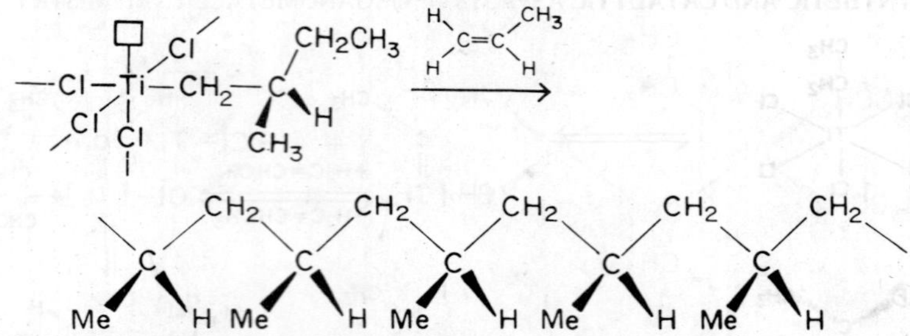

Alternatively, the process may be sometimes interrupted by β-elimination of a metal hydride or by hydrolysis:

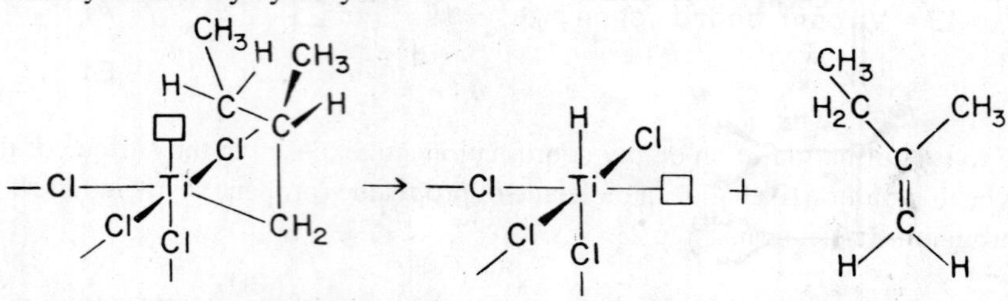

The stereoregularity of polypropylene was assumed to arise due to the nature of sterically controlled directing sites on the $TiCl_3$ lattice.

In marked contrast an alternative mechanism for alkene polymerization which is gaining ground, involves the generation of alkylidene intermediates by α-(or 1, 2-) hydrogen transfers (cf. Section 4.5). This mechanism which shows close connection between alkene polymerization and alkene metathesis (cf. Section 6.4.2.7), provides an explanation for the control of stereochemistry of the polymerization of propene since substituents could lie either as *cis* or *trans* with respect to the metallocyclobutane ring. A scheme for propylene polymerization which involves the following key steps: (i) carbon-carbon bond formation by alkyl migration (Cossee), (ii) 1, 2-hydrogen shift forming an alkylidene hydride, (iii) carbon-carbon bond forming step by metallocycle formation, and (iv) reductive elimination is illustrated in Fig. 6.14.

In this alternative mechanism, the specific nature of stereoregular polyproylene would be obviously controlled by the relative orientations of the substituents on the metallocyclic ring. Thus a syndiotactic form of the polypropylene would be obtained if the CH_3 groups are on the same side of the metallocyclic ring (Fig. 6.14).

Besides the above two, another view of the possible mechanism which is currently emerging is based on C—H→Ti interactions (for which Green and Co-workers have proposed the term ''agostic'', to illustrate especially those situations in which hydrogen atom is covalently bonded simultaneously to both a sp^3 carbon atom and to a transition metal atom).

The Cossee mechanism on the one hand and the hydrogen migration/metal-carbene mechanism on the other, hand can be seen as extreme representations of an alkene insertion reaction involving agostic α-hydrogen interactions. The salient features of the above mechanism are: (i) carbon-carbon bond forming step assisted by the partial migration of an α-hydrogen to the metal giving a C—H→Ti bond;

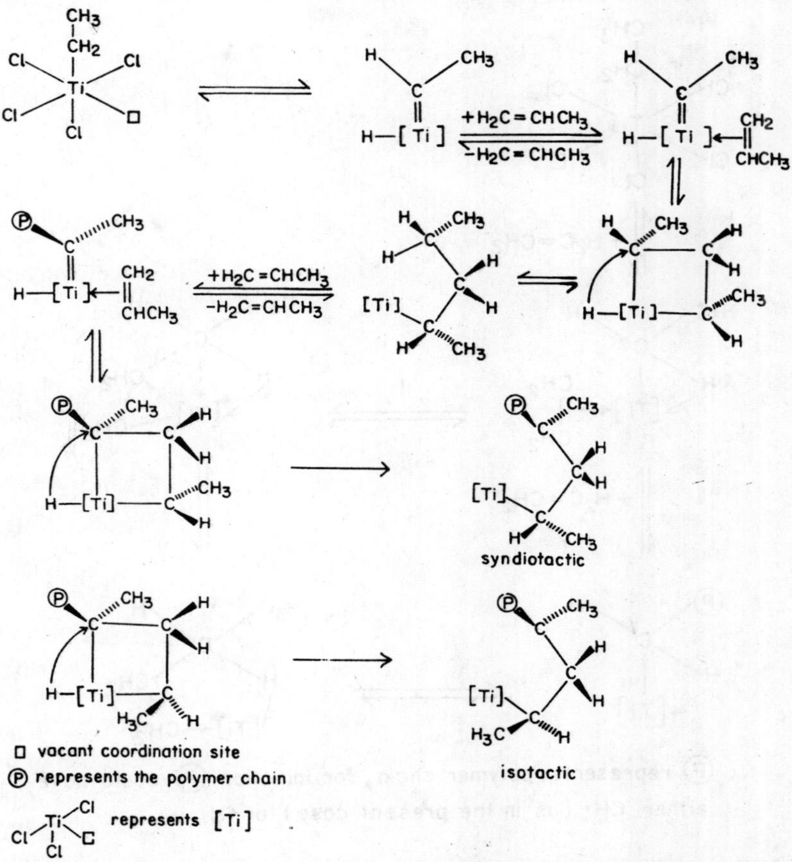

Fig. 6.14 An alternative mechanism of propylene polymerization involving the formation of an alkylidene intermediate.

(ii) the metallocycle intermediate with a γ-CH $\rightharpoonup$ Ti bond; (iii) rearrangement of the agostic hydrogen from the γ- to the α-position; and (iv) addition of the next alkene. In the above figure, the encircled 'P' represents and growing polymer chain. The mechanism based on agostic hydrogen interactions can explain stereoregular polymerization of alkenes on the grounds that (i) the formation of an α-bridging hydrogen by the alkyl group would reduce the steric inhibition to the C—C bond forming step, and (ii) the bidentate nature of the Ti—C(H)—H interaction (Ti—C bridged by H) could provide a means for controlling the stereochemistry (orientation) of the alkyl chain with respect to the incoming alkene.

It may be worthwhile to mention that an interaction of the type Cl_2Ti—$C(H)_2$ (Ti—C bridged by H) has been actually observed in the X-ray structure recorded recently (1982) of $TiCl_3CH_3$ (dmpe).

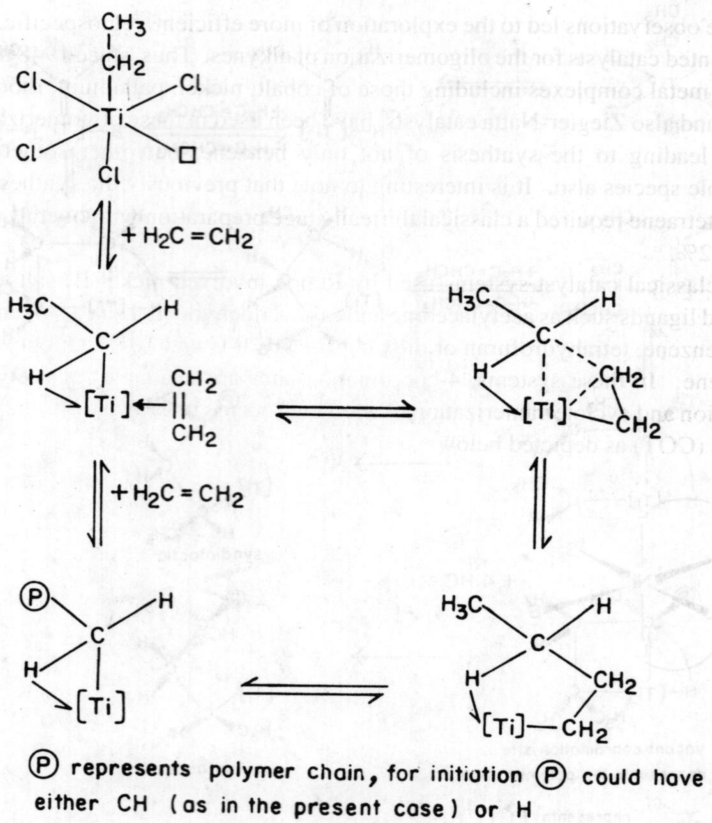

Fig. 6.15 Possible reaction pathway involving agostic hydrogens

C. Oligomerization and polymerization of alkynes

Synthesis of small amount of benzene along with other unsaturated cyclic hydrocarbons (such as styrene and naphthalene) was achieved by Berthelot as early as 1870 by passing acetylene through a hot tube:

$$3HC \equiv CH \rightarrow C_6H_6(g) \text{ (very low yield)}$$

The above conversion is highly exothermic, yet surprisingly in the absence of metal catalysts, it occurs only at high temperatures and in poor yields. In contrast, it was discovered in 1948 by Reppe and coworkers that acetylene could be oligomerized to cyclooctatetraene (COT) in tetrahydrofuran at 60-70°C and 15-20 atm pressure in the presence of nickel(II) cyanide; in this reaction benzene also was a by-product:

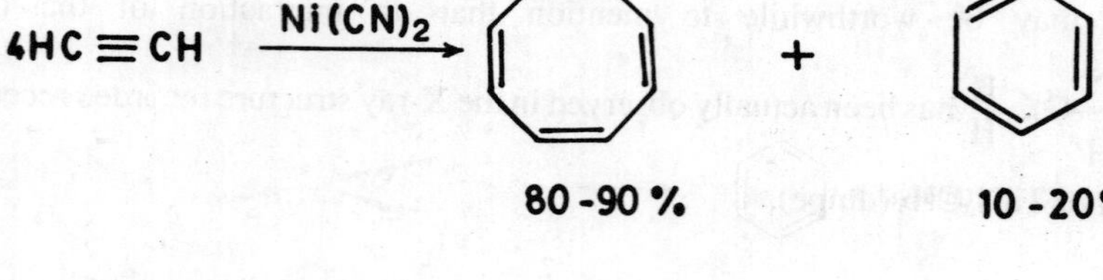

These observations led to the exploration of more efficient, regiospecific, and target oriented catalysts for the oligomerization of alkynes. Thus, since 1948 many transition metal complexes including those of cobalt, nickel, palladium, rhodium and iron, and also Ziegler-Natta catalysts, have been used in these oligomerization reactions leading to the synthesis of not only benzene, but other sometimes inaccessible species also. It is interesting to note that previously the synthesis of cyclooctatetraene required a classical thirteen-stage preparation with overall yield of only 1-2%.

The classical catalyst systems used by Reppe involved nickel(II) salts with weak-field ligands such as acetylacetone and salicylaldehyde (L-L) in a solvent (S) such as benzene, tetrahydrofuran or dioxane at 80-120°C and 10-25 atm pressure of acetylene. In these systems, 4-coordination sites are available for acetylene coordination and cyclotetramerization of acetylene occurs to give mainly cyclooctatetraene (COT) as depicted below:

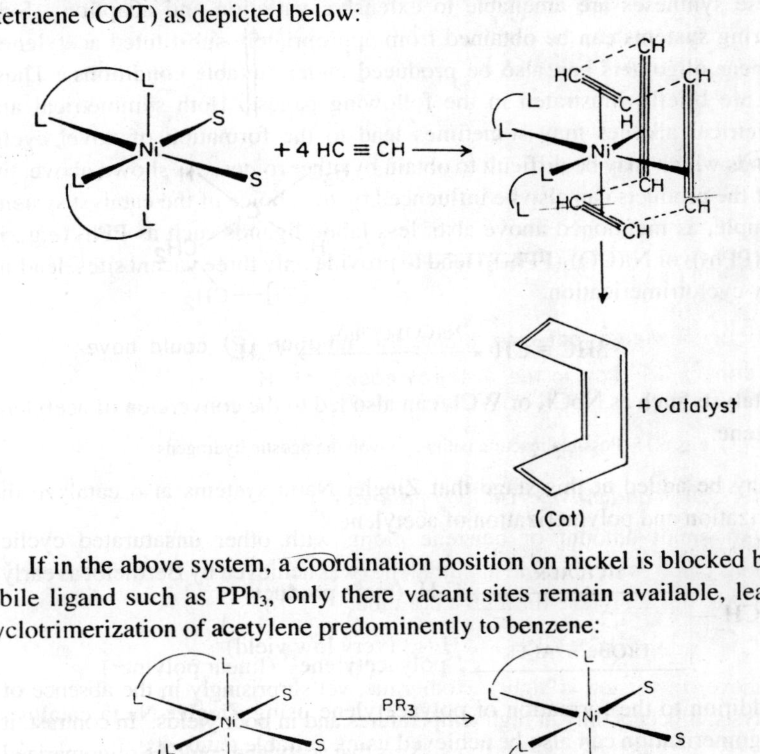

If in the above system, a coordination position on nickel is blocked by a less labile ligand such as PPh_3, only there vacant sites remain available, leading to cyclotrimerization of acetylene predominently to benzene:

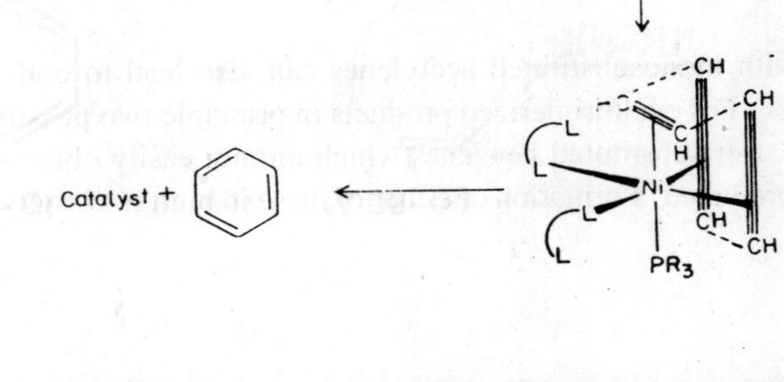

It is worth noting that when PPh_3 is replaced by a bidentate phosphine ligand such as $Ph_2PCH_2CH_2PPh_2$(P—P), no oligomerization appears to take place. This can be explained by preferential release of *trans* sites only leading to intermediates which are not conducive to polymerization reaction(s):

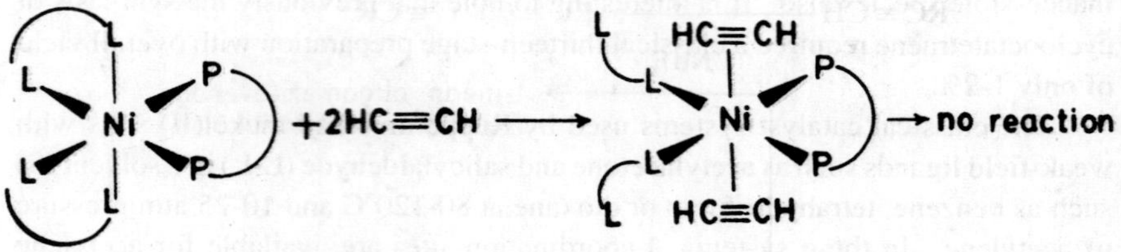

These syntheses are amenable to extensive variation and adaption. Substituted ring systems can be obtained from appropriately substituted acetylenes, while linear oligomers can also be produced under suitable conditions. These features are briefly illustrated in the following pages. Both symmetrical and unsymmetrical alkynes may sometimes lead to the formation of novel cyclic compounds which may be difficult to obtain by other routes. As shown above, the nature of the products can also be influenced by the choice of the catalyst system. For example, as mentioned above also, less labile ligands such as PPh_3 (e.g., in $Ni(CN)_2(PPh_3)_2$ or $Ni(CO)_2(PPh_3)_2$) tend to provide only three vacant sites, leading to mainly cyclotrimerization:

$$3HC \equiv CH \xrightarrow{Ni(CO)_2(PPh_3)_2} C_6H_6$$

Other catalysts such as $NbCl_5$ or WCl_6 can also led to the conversion of acetylene into benzene.

It may be added at this stage that Ziegler-Natta systems also catalyze the oligomerization and polymerization of acetylene:

$$3HC \equiv CH \begin{cases} \xrightarrow{TiCl_4/Al_2R_6} C_6H_6 \text{ (main product)} \\ \xrightarrow{Ti(OBu^n)_4/Al_2Et_6} \text{``polyacetylene'' (linear polymer)} \end{cases}$$

In addition to the formation of polyacetylene using Ziegler-Natta catalysts, linear oligomerization can also be achieved using suitable catalysts:

$$HC \equiv CH \xrightarrow[NH_4Cl]{Cu_2Cl_2} HC \equiv C - CH = CH_2$$

The reaction with monosubstituted acetylenes can also lead to both cyclic and linear oligomers. For cyclotrimerized products in principle two possible isomers: 1, 2, 4- or 1, 3, 5-trisubstituted benzene (which are not easily obtained by other routes) can be produced. Formation of either of these in high selectivity is possible by a suitable choice of the catalyst, reaction conditions and nature of the substituents, as illustrated below:

$$2C_6H_5C \equiv CH \xrightarrow[Cu_2(OAc)_2]{RhCl(PPh_3)_3 \text{ or}} trans - C_6H_5CH = CH - C \equiv CC_6H_5$$

$$RC \equiv CH \xrightarrow{Cu(OAc)_2} RC \equiv C - C \equiv CR$$

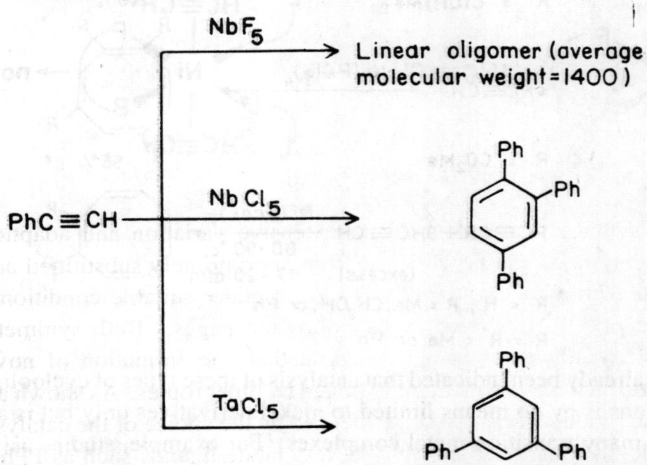

Cyclotrimerization of disubstituted acetylenes is much less favoured and requires some special catalysts, e.g.:

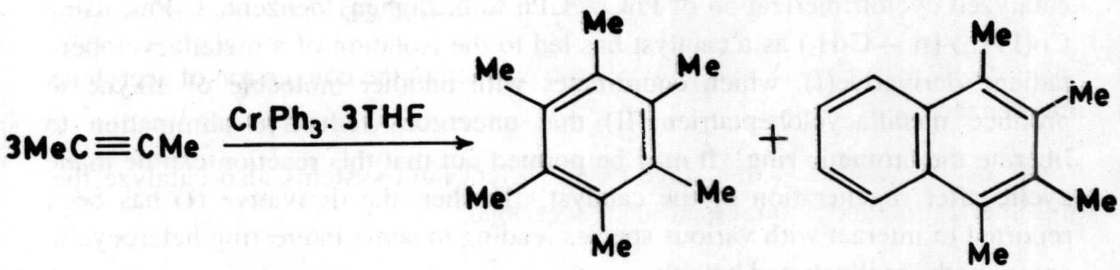

$$3RC \equiv CR \xrightarrow[TiCl_4/AlBu_3^n]{Co_2(CO)_8 \text{ or}} C_6R_6 \text{ (hexasubstituted benzene)}$$

By contrast, cooligomerization of disubstitued acetylenes can often be readily achieved with acetylene in the presence of catalysts, e.g.:

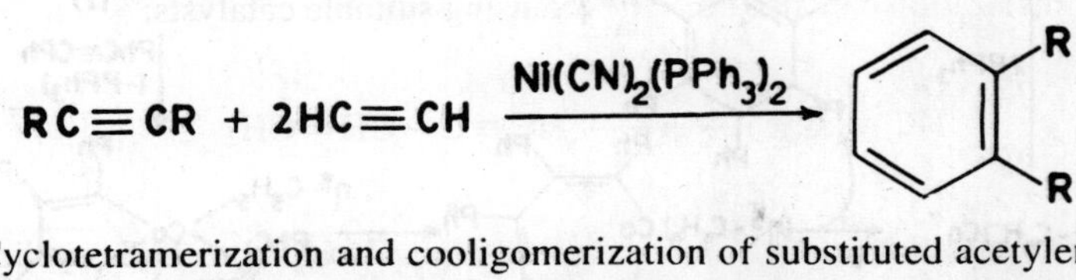

Cyclotetramerization and cooligomerization of substituted acetylenes is frequently observed in the presence of ligand-free nickel catalysts, e.g., $Ni(acac)_2$ and $NiCl_2$—$NaBH_4$ or catalysts containing ligands which are readily displaced by alkyne, e.g., Ni(cod)$_2$ (where cod = cycloocta-1, 5-diene) and Ni(PCl$_3$)$_4$. A few illustrative examples are shown below:

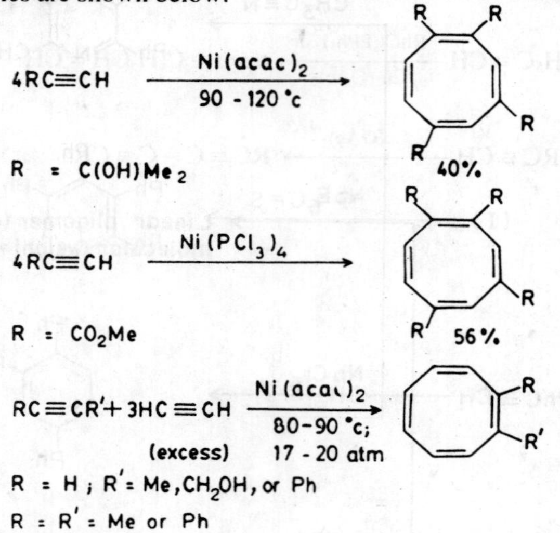

It has already been indicated that catalysis of these types of cyclooligomerization reactions is by no means limited to nickel derivatives only but is a common feature of many transition metal complexes. For example, studies using cobalt, rhodium and palladium derivatives {such as CoL$_2$(η^5—C$_5$H$_5$) (L = CO or PPh$_3$), Rh(CO)$_2$(η^5—C$_5$H$_5$) and PdCl$_2$(PhCN)$_2$} have allowed the isolation of intermediates. It may be interesting to mention that a detailed mechanistic study of the catalyzed cyclotrimerization of PhC ≡ CPh to hexaphenylbenzene, C$_6$Ph$_6$, using Co(PPh$_3$)$_2$(η^5—C$_5$H$_5$) as a catalyst has led to the isolation of a metallacyclopentadiene derivative(I), which coordinates with another molecule of alkyne to produce metallacycloheptatriene(II) that undergoes reductive elimination to liberate the aromatic ring. It may be pointed out that this reaction can be made cyclic after regeneration of the catalyst. Further, the derivative (I) has been reported to interact with various species leading to some interesting heterocyclic compounds, as illustrated below:

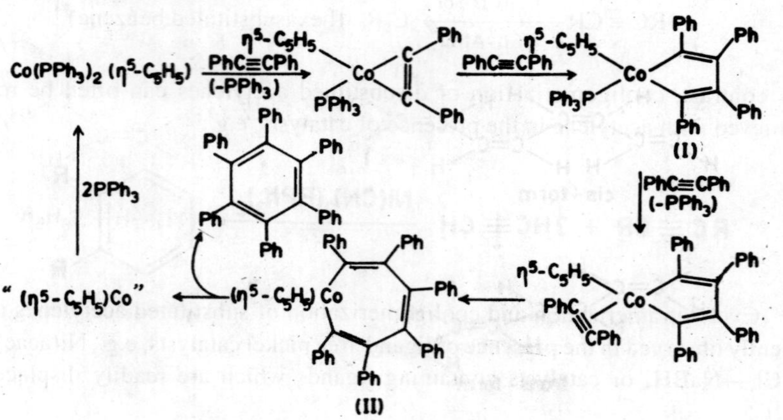

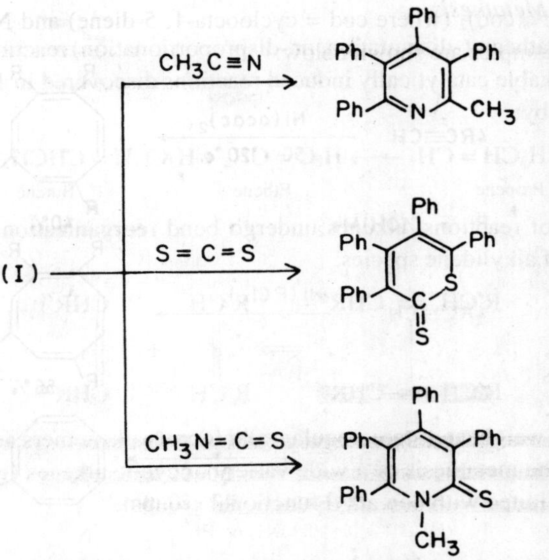

Further, a mechanistic study of the Ziegler-Natta catalytic systems has revealed that the alkyl group of the Al_2R_6 catalyst attaches itself to the end of the final polymer. This shows that the polymerization proceeds *via* the insertion of acetylene into the metal-alkyl bond. At lower temperatures, poly-*cis*-acetylene is formed, but the *trans* form is the main product at higher temperatures. These findings are in conformity with the model that acetylene coordinates initially in the side on manner followed by insertion into the M—C bond.

The formation of some alkylbenzene also in the cyclotrimerization of acetylene with $TiCl_4/Al_2R_6$ indicates this cyclization also to occur *via* an insertion into the M—C bond as shown below:

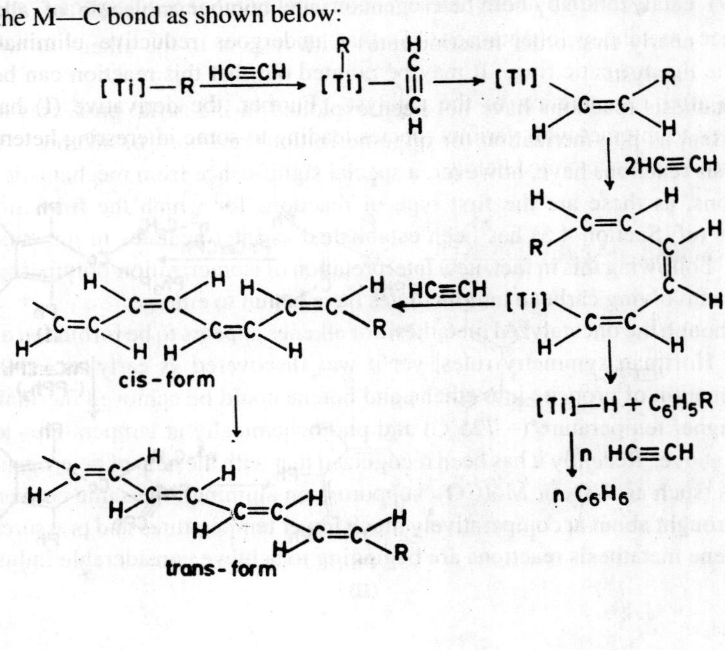

6.4.2.7 Alkene Metathesis

The alkene-metathesis (-dismutation, or -disproportionation) reactions are some of the most remarkable catalytically induced reactions discovered in 1960's and can be exemplified by:

$$\underset{\text{Propene}}{2CH_3CH = CH_2} \longrightarrow \underset{\text{Ethene}}{H_2C = CH_2} + \underset{\text{Butene}}{H_3CCH = CHCH_3}$$

In these types of reactions, alkenes undergo bond reorganization resulting in a redistribution of alkylidene species:

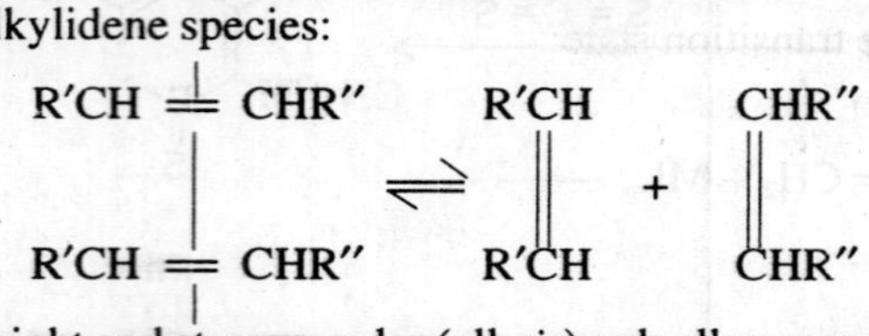

High molecular weight and stereoregular (all *cis*) polyalkenemers are the products obtained from the metathesis of a wide variety of cyclic alkenes (generally other than those substituted with non alkyl functional groups):

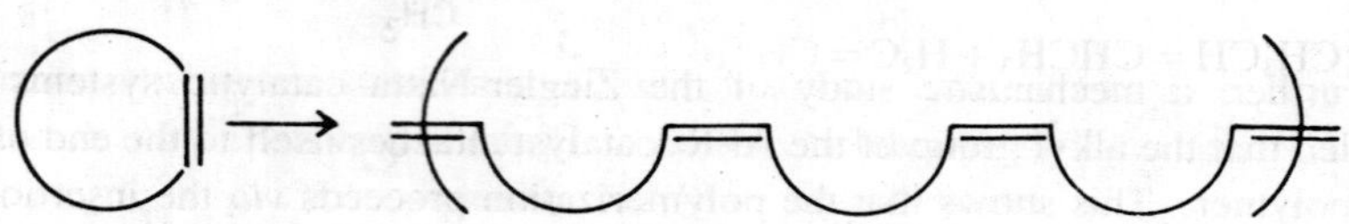

The salient features of transition metal catalyzed metathesis reactions are:

(i) low activation barriers

(ii) fast reaction rates,

(iii) thermal neutrality ($\Delta H = 0, \Delta S = 0$),

(iv) catalyzation by both heterogeneous and homogeneous species, and

(v) nearly first order reaction rates with respect to the alkene as well as catalyst.

Metathesis reactions have not been exploited in industrial processes to the same extent as polymerization (or oligomerization) reactions of alkenes. These metathesis reactions have, however, a special significance from mechanistic considerations, as these are the first type of reactions for which the formation of carbenes (cf. Section 4.5) has been established as intermediates in the catalytic process. Following this in fact, new interpretation of isomerization/polymerization reactions involving carbene intermediates have begun to emerge.

Although the uncatalyzed metathesis of alkenes appears to be forbidden on the basis of Hoffman symmetry rules, yet it was discovered as early as 1930 that transformation of propene into ethene and butene could be achieved thermally at rather higher temperature (~ 725°C) and photochemically at temperatures lower than the above. Recently it has been recognized that with the help of heterogeneous catalysts (such as WO_3 or $Mo(CO)_6$ supported on alumina), the same conversion can be brought about at comparatively much lower temperatures and pressures. In fact, alkene metathesis reactions are beginning to achieve considerable industrial importance in the production of polymerization quality ethene and high purity butene. More recently, with the help of homogeneous catalysts (such as a mixture of WCl_6 and $AlCl_2Et$ in ethanol), alkene metatheses have been carried out remarkably well even at ambient temperatures. The mechanisms of such reactions are still not fully established, although reliable evidence has been accumulating about the same, as can be assessed by the discussion that follows.

Initial suggestions involved a concerted mechanism, which required a ''quasi-cyclobutane'' type transition state:

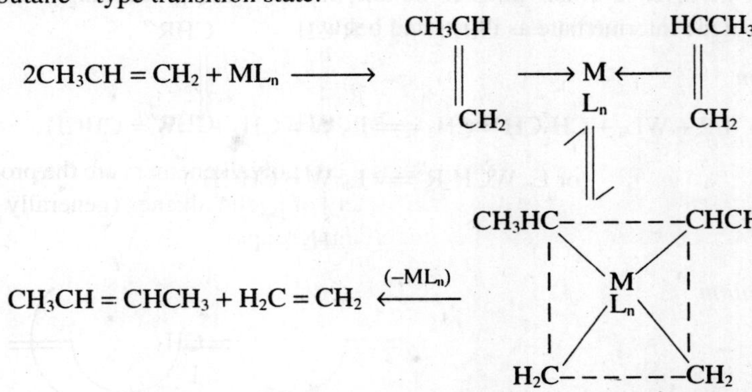

This suggestion has received considerable support on theoretical considerations. Since the interactions between alkenes and d-orbitals of the metals tend to eliminate the Hoffman symmetry restrictions, a low activation energy can be predicted for the process. However, the mechanism lacks experimental evidence of the formation of proposed cyclobutanes. The actual non-reactivity of cyclobutanes under metathesis conditions as well as formation of alkylidene units is inconsistent with the proposed mechanism and throws further doubts on its validity.

A mechanism suggested later involved the rearrangement of an intermediate dialkene complex through a metallo-cyclopentane type intermediate (formed by oxidative addition of alkenes):

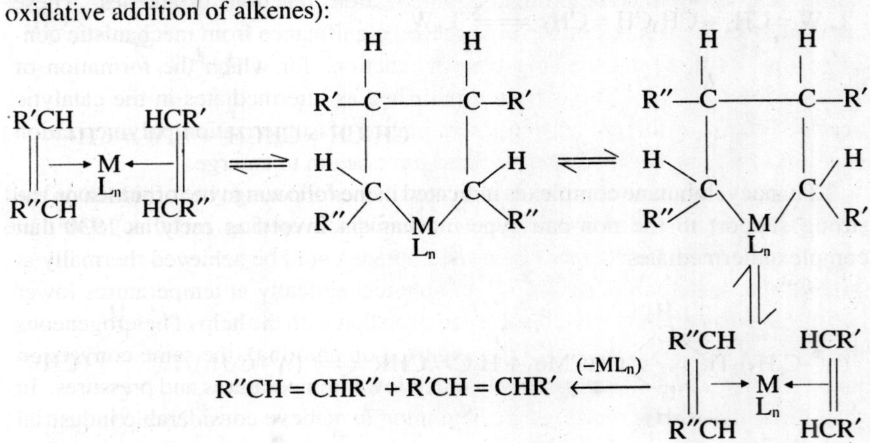

The metal catalyst according to the latter mechanism also appears to provide two adjacent coordination sites for alkenes which may then facilitate further reaction. Although complexes of the above type are known, yet the latter mechanism also appears to be doubtful on the same ground as the former one, i.e., indicating a pairwise exchange of alkylidene moieties.

The currently accepted mechanism for alkene metathesis (indicating non-pairwise exchange) was proposed by the French chemists Herrisson and Chauvin in 1970; it involves a chian transfer of alkylidene units, proceeding via metallacyclobutane intermediate as illustrated below:

Initiation

$$1/2 + WL_n + CH_3CH=CH_2 \rightleftharpoons L_n\,W=CH_2 + L_nW=CHCH_3$$

$$(\text{or } L_n\,WCH_2R \rightleftharpoons L_n\,\underset{\displaystyle |\atop \displaystyle H}{W}=CHCH_3)$$

Propagation

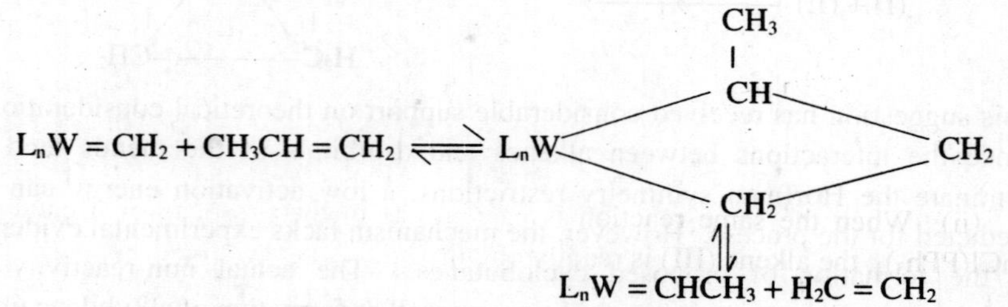

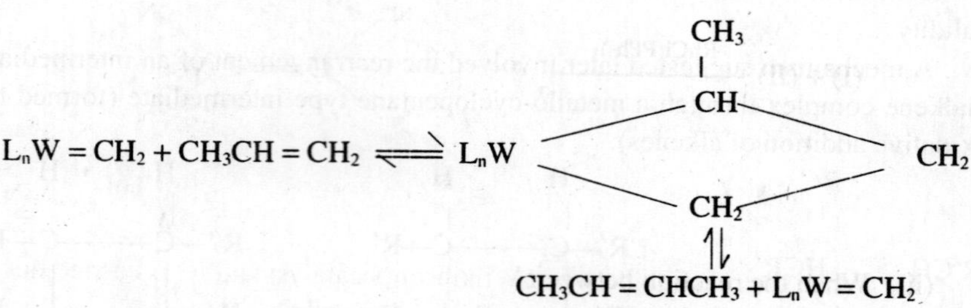

Titanacyclobutane complexes indicated in the following type of reactions lend strong support to the non-pair type mechanism involving metallacyclobutane complex intermediates:

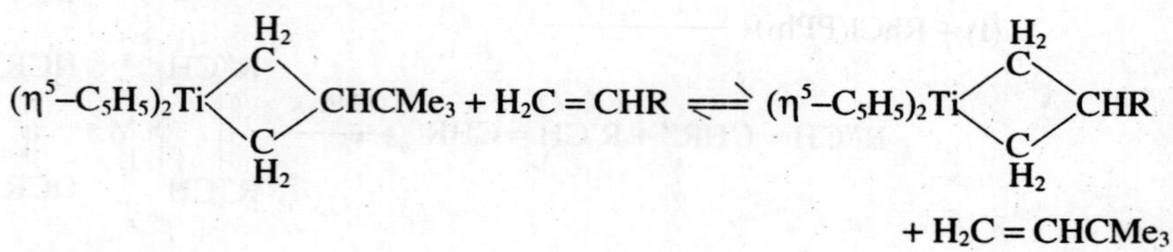

This mechanism can be further supported by the following types of evidence on the observed dismutation patterns of electron-rich alkenes,

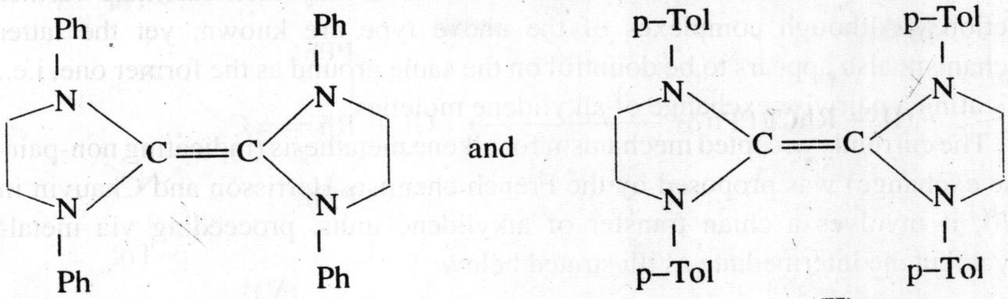

(I) (II)

(i) When a mixture of electron rich alkenes(I) and (II) is heated alone, no tendency of a dismutation reaction resulting in the formation of III appears to take place:

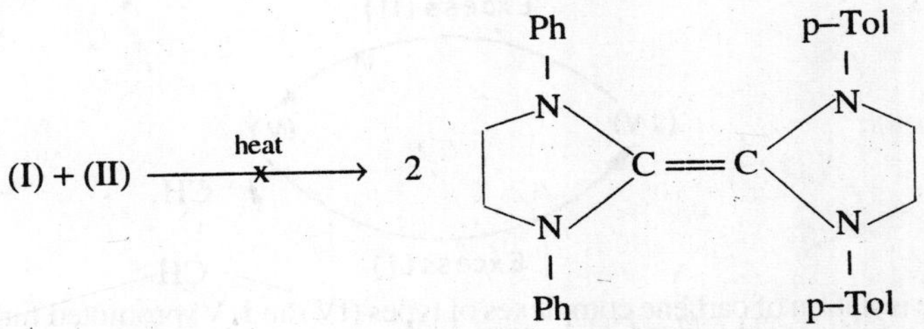

(III)

(ii) When the same reaction is carried out in the presence of a catalyst, $RhCl(PPh_3)_3$, the alkene (III) is readily obtained:

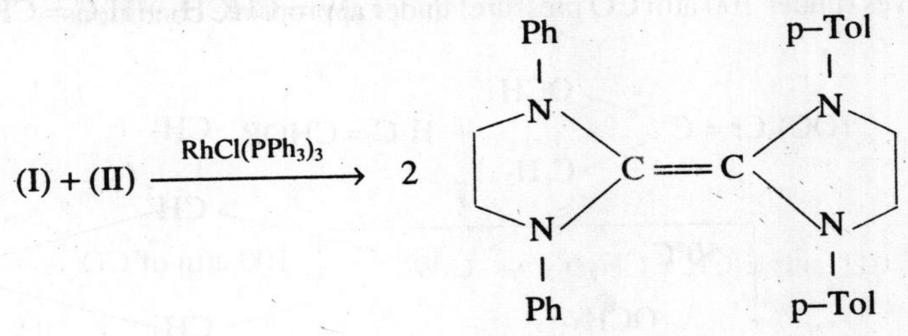

(iii) When the reaction between Wilkinson's catalyst and (I) is carried out in 1 : 1 molar ratio a complex of the type (IV) is obtained:

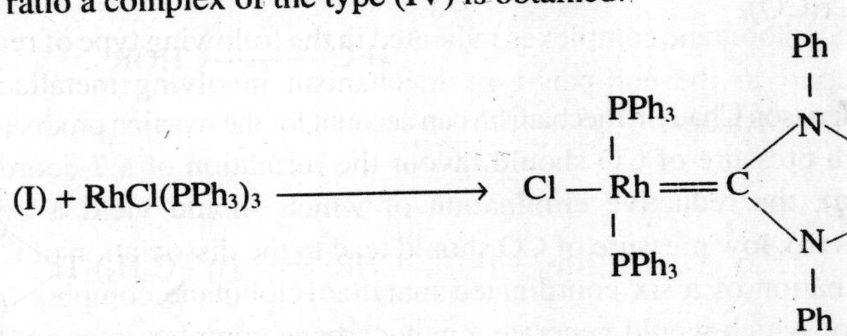

(IV)

Similarly interaction of $RhCl(PPh_3)_3$ and (II) yields the complex (V):

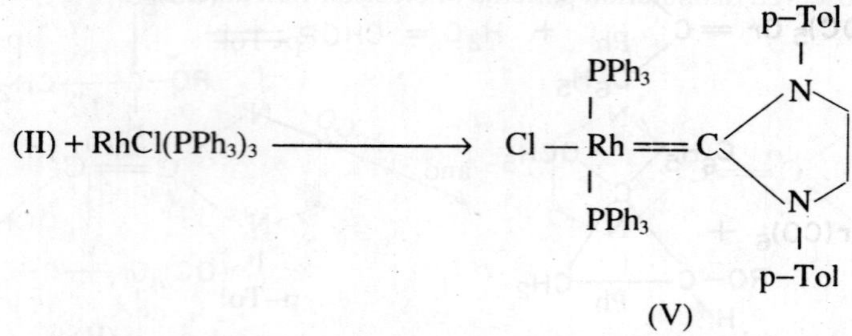

(V)

(iv) Both the complexes (IV) and (V) are themselves catalysts and treatment of (IV) with an excess of (II) gives (V). Similarly when (V) is treated with an excess of (I), the product (IV) can be obtained:

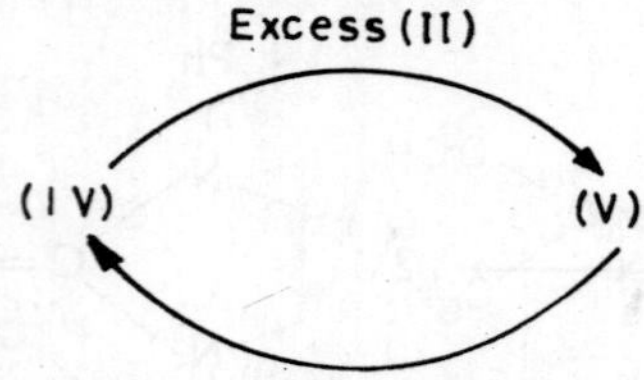

The isolation of carbene complexes of types (IV) and (V) promoted further efforts for the isolation of the reaction intermediates postualted in the Herisson-Chauvin mechanism. For example, interaction of phenyl (methoxy) carbene complexes with vinyl ethers, $CH_2 = CHOR$ yielded (at 50°C) α-methoxystyrene or cyclopropane derivatives (under 100 atm CO pressure) under appropriate conditions:

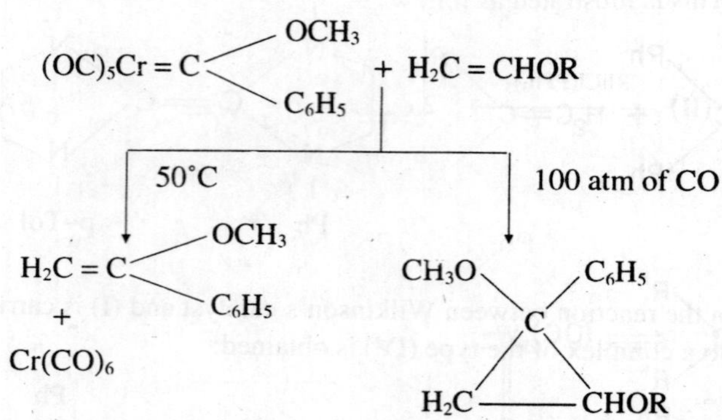

The Herisson Chauvin mechanism can account for the reaction products (Fig. 6.16).

A high pressure of CO should favour the formation of a 7-coordinate intermediate (a), the reductive elimination of which should yield a cyclopropane derivatives (b), low pressure of CO should lead to the dissociation of CO resulting in the formation of a six coordinated matallacyclobutane complex (c), the ring opending of which would generate a new carbene complex intermediate (d) and α-methoxy styrene.

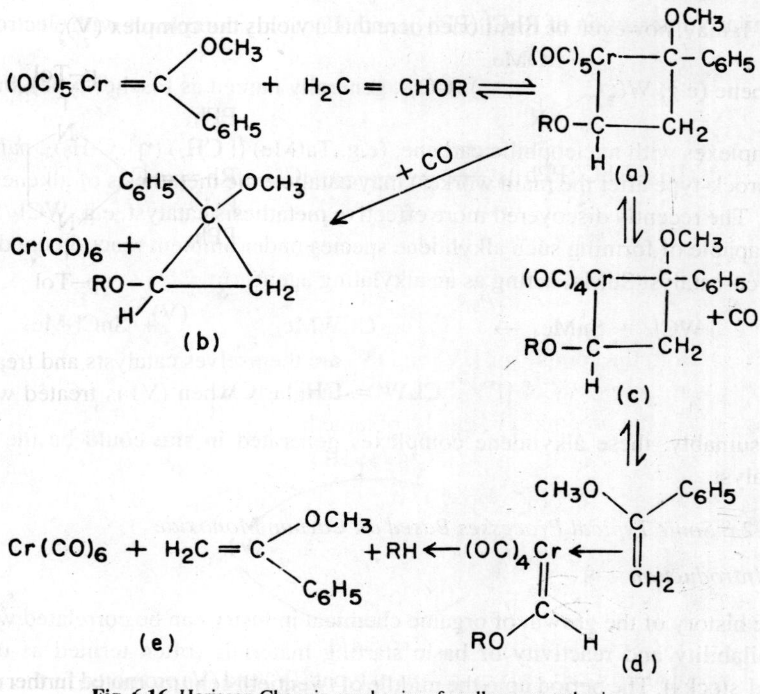

Fig. 6.16 Herisson Chauvin mechanism for alkene metathesis.

The major breakthrough in the direction of carbene mechanism for metathesis of alkenes is the isolation and demonstration of the reactivity of $W(=CPh_2)(CO)_5$ towards a large number of alkenes to produce metathesis products and cyclopropanes. This is illustrated as follows:

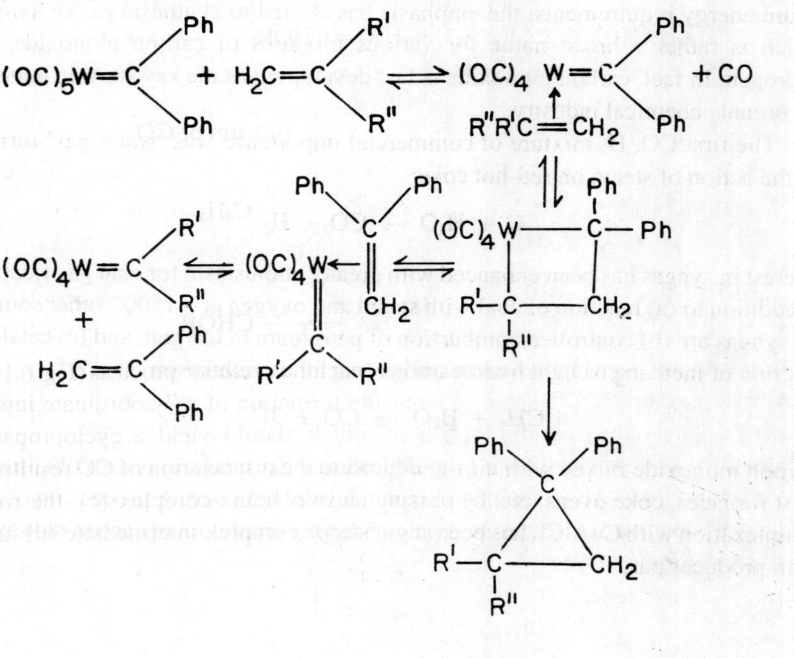

It may, however, be mentioned here that not only complexes with electrophilic carbene (e.g. $W(:C(OMe)(R))(CO)_5$, generally named as Fischer-type), but also complexes with nucleophilic carbene, (e.g. $Ta(Me)(:CH_2)(\eta^5\text{-}C_5H_5)_2$, named as Schrock-type after the main worker) may catalyze the metathesis of alkenes.

The recently discovered more effective metathesis catalyst, e.g. $WCl_6/SnMe_4$ is capable of forming such alkylidene species under ambient reaction conditions, the co-catalyst $SnMe_4$ acting as an alkylating agent:

$$WCl_6 + SnMe_4 \rightarrow Cl_4WMe_2 + SnCl_2Me_2$$
$$Cl_4WMe_2 \rightarrow Cl_4W=CH_2 + CH_4$$

Presumably, these alkylidene complexes generated in situ could be the actual catalyst.

6.4.2.8 *Some Typical Processes Based on Carbon Monoxide*

(i) *Introduction*

The history of the growth of organic chemical industry can be correlated with the availability and reactivity of basic starting materials (often termed as organic feed-stocks). The period upto the middle of twentieth century can be in this regard, characterized as the 'acetylene period', as a number of processes were based on highly reactive but rather expensive acetylene (from coal). The next two to three decades can be appropriately considered as 'petroleum era', as alkenes, a petroleum product, have been rather favourite feed-stocks during this period. With the crisis in petroleium industry and the need to preserve its diminishing stock for future energy requirements, the emphasis has shifted to synthesis gas or syngas, which is rather a broad name for various mixtures of carbon monoxide and hydrogen. In fact, carbon monoxide is fast developing as the key starting material for organic chemical industry.

The first CO, H_2 mixture of commercial importance was 'water gas' formed by the action of steam on red-hot coke:

$$C + H_2O \rightarrow CO + H_2$$

Interest in syngas has been enhanced with greater enthusiasm for coal gasification. In addition to (a) reaction of coal with steam and oxygen at ~ 1500°, other sources for syngas are (b) controlled combustion of petroleum in oxygen, and (c) catalytic reaction of methane or light hydrocarbons (naphtha) with steam:

$$CH_4 + H_2O \rightarrow CO + 3H_2$$

Carbon monoxide mixed with nitrogen (producer gas mixture) is also obtained in blast furnaces, coke ovens, etc. by passing air over hot coke. A process involving complexation with $CuAlCl_4$ has been suggested for separation of carbon monoxide from producer gas.

In the following sections, an attempt will be made to present a brief account of a few common processes based on carbon monoxide.

(ii) *Fischer—Tropsch Process*

Historically, Fischer and Tropsch as early as 1922 studied the conversion of CO and H_2 to hydrocarbons and oxygenated products, with the help of heterogeneous catalysts.

The Fischer-Tropsch reactions offer an attractive potential for the organic chemical industry. Employing hydrogen of the syngas as reducing agent, the reaction may be viewed as the reduction of carbon monoxide (followed by polymerization) over a variety of catalytic surfaces (e.g. Fe, No, Ni, or Ru on alumina or other surfaces as well as cobalt-chromium or other oxides):

$$CO + nH_2 \longrightarrow \begin{cases} HCHO \\ CH_3OH \\ CH_4 + H_2O \\ HOCH_2CH_2OH \\ CH_3(CH_2)_nCH_2OH \\ CH_3(CH_2)_nCH = CH_2 \\ CH_3(CH_2)_nCH_3 \end{cases}$$

All the illustrative reactions indicated above, except the first one, would require molar ratio of H_2 and CO more than 1, and hence a number of processes, e.g., 'water gas shift reactions' (cf. Section 6.4.2.8.3) have been developed for enriching the hydrogen component of the syngas.

The methanation (production of primarily methane) reaction appears to be catalyzed best by nickel and is currently used for removal of small amount of CO from hydrogen.

The synthesis of methanol by the reduction of CO:

$$CO + 2H_2 \xrightarrow{CuO} CH_3OH$$

is now a commercial process and employs Cu-promoted ZnO catalyst at ~250° and 50 atm pressure. The process holds extra-ordinary potential as methanol can now be converted industrially into gasoline. This latter process has been developed recently by Mobil Oil Company and utilizes a framework of silicates (Zeolite ZSM-5) with specially contrived cavity and channel dimensions.

The zeolite acts as an acid catalyst and in effect removes a molecule of water from methanol. The carbene 'CH_2' produced may be stabilized by zeolite to be available for insertion in another molecule of methanol. The pore-size of the zeolite catalyst is so chosen that it is just right for a 'gasoline' molecule allowing the insertion of CH_2 groups to continue upto the formation of a C_6 to C_{10} chain.

Although considerable work is being carried out on zeolite catalysts in many laboratories, including the N.C.L. Poona, yet the mechanism of the process is not quite clear. However, the over-all reaction can be represented as:

$$nCH_3OH \xrightarrow[-\,nH_2O]{ZSM-5} (CH_2)_n$$

In spite of its wide applications, the mechanism of the Fischer-Tropsch reactions in general is still at a speculative stage. In complicated systems involving heterogeneous catalysis, recourse is often made to mimick a catalytic metal surface with a metal cluster compound with groups which appear to be playing some role in actual catalytic operation also. In this connection it is worth mentioning that transition metal cluster compounds (by virtue of the proximity of multiple reaction sites on the faces or edges of the cluster) offer the prospect of a new class of slective homogeneous Fischer-Tropsch catalysts. In Fischer-Tropsch process, reduction of carbon monoxide presumably proceeds via formyl and hydroxymethyl intermediates to form methane. Evidence has been obtained for the above course of reaction by employing triosmium dodecacarbonyl (as a mimick catalyst) when the progress of the reaction can be represented as below:

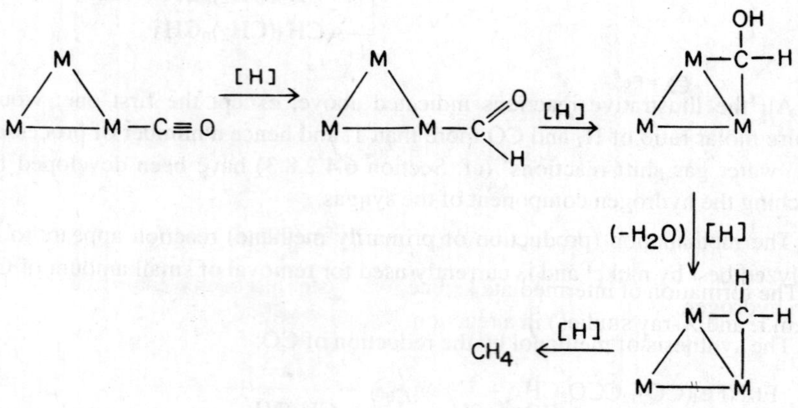

Indirect evidence for the formation of formyl types derivatives has been obtained by the X-Ray diffraction of a metalloformyl complex obtained by the reaction of carbon monoxide with rhodium octaethyl porphyrin hydride, in which the C—H and C—O bond distances are in good agreement with organic formyl compounds.

An alternative ''Carbide'' mechanism has also been proposed recently; this is based on the observation of the remarkable reactivity of the encapsulated carbon atom in $Fe_6C(CO)_{16}^{-2}$ when exposed by loss of two vertices from the orginal cluster, as demonstrated in the synthesis of (carbomethoxy) methylodyne cluster $Fe_4(CO)_{12}CCO_2CH_3$ in an oxidative framentation in methanol:

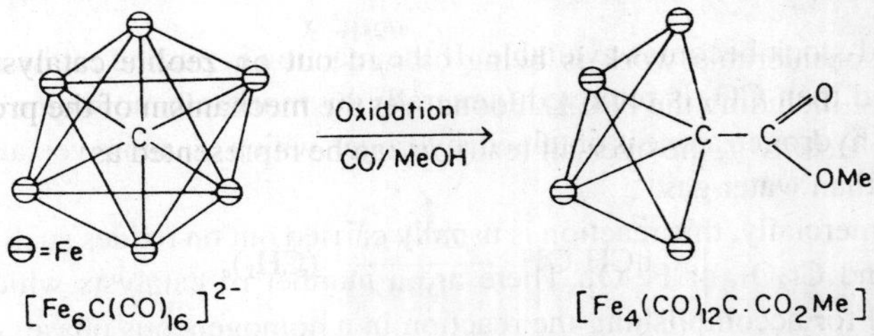

The above facile reaction demonstrated for the first time the unique reactivity of encapsulated carbon atoms with low coordination number. A possible mechanism has, therefore, been suggested (by J.S. Bradley and co-workers in 1981) for the formation of $Fe_4(CO)_2CCO_2Me$ via carbonylation of the carbon atom in a η^4-carbide intermediate to yield a ketenediyl cluster, $Fe_4(CO)_{12}CCO$, which then reacts with solvent methanol to produce the desired product:

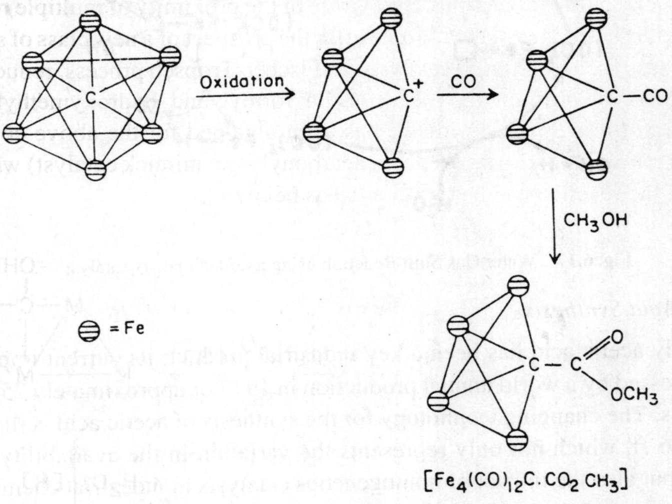

The formation of intermediate ketenediyl cluster has been established (by i.r., [1]H n.m.r. and X-ray studies) in a reaction described below:

$$Et_4NFe_4(CO)_{12}CCO_2CH_3 + CF_3SO_3H \xrightarrow[N_2]{\text{cyclohexane}} (Fe_4C(CO)_{13})^+$$

3. *Water Gas Shift Reaction*

Water gas shift reaction is the name given to any process by which the following reaction can be pushed to the right:

$$CO + H_2O \xrightarrow{\text{metal oxides}} CO_2 + H_2$$

A number of reasons can be cited for the enthusiasm to develop this process (cf. Section 6.4.2.8 (ii)) by which the proportion of hydrogen in the syngas can be increased. Small organic molecules contain about 3 to 4 times as many hydrogen atoms as carbon atoms and hence, if the H_2/CO mole ratio can be increased to 2, a better feed-stock becomes available. If the reaction is pushed to the consumption all CO and then CO_2 is removed (generally by mono-ethanolamine scrubbing in industry), hydrogen alone is obtained, hydrogen which is a more versatile industrial chemical than water gas.

Commercially, this reaction is usually carried out on oxides such as a mixture of ZnO and Cr_2O_3, or Fe_3O_4. There are a number of catalysts which are being developed for accomplishing the reaction in a homogeneous phase. A simplified catalytic cycle may be represented as below:

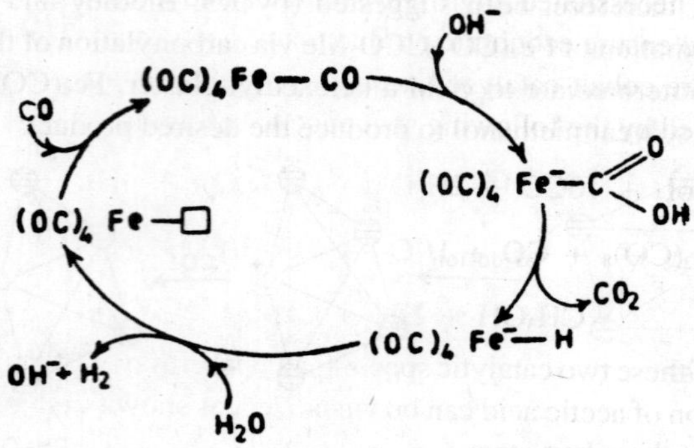

Fig. 6.17 Water Gas Shift Reaction using a homogeneous catalyst.

4. Acetic Acid Synthesis

Historically acetic acid has been a key industrial product; its current importance can be assessed by a world annual production in 1977 of approximately 250000 of metric tons. The changing technology for the synthesis of acetic acid is illustrated in Figure 6.18 which not only represents the variation in the availability of raw materials but also the impact of homogeneous catalysis in industrial chemistry.

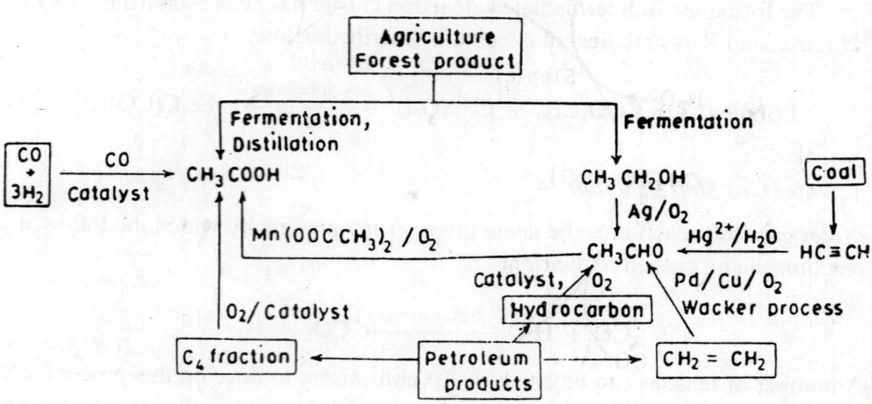

Fig. 6.18 Industrial synthesis of acetic acid.

The technology for the manufacture of acetic acid has undergone a phenomenonal change ever since efforts have been initiated to utilise on a commercial scale Reppe's discovery (1941) that insertion of CO can be brought about in CH_3OH with the help of suitable catalysts:

$$CH_3OH + CO \xrightarrow{\text{catalyst}} CH_3COOH$$

In the early 1960's, BASF built a cobalt based plant, in which the cobalt is introduced in the form of cobalt (II) iodide. The low reactivity and stability of cobalt catalyst necessitated high temperatures and pressures (200–250°C and 500–700 atm) to ensure economic production of acetic acid. The active species in the high pressure cobalt catalyzed acetic acid synthesis are probably $CoH(CO)_4$ and CH_3I, formed by the following sequence of reactions:

$$2CoI_2 + 10CO + 2H_2O \rightarrow Co_2(CO)_8 + 4HI + 2CO_2$$

$$Co_2(CO)_8 + CO + H_2O \rightarrow 2CoH(CO)_4 + CO_2$$

$$CH_3OH + HI \rightarrow CH_3I + H_2O$$

The role of these two catalytic species in the insertion of CO in CH_3OH leading to the production of acetic acid can be visualized, as shown in Figure 6.19 in terms of the following three key-steps:

(a) cobalt-alkyl bond formation; (b) insertion of CO into cobalt-alkyl bond; and (c) cleavage of cobalt-acyl bond:

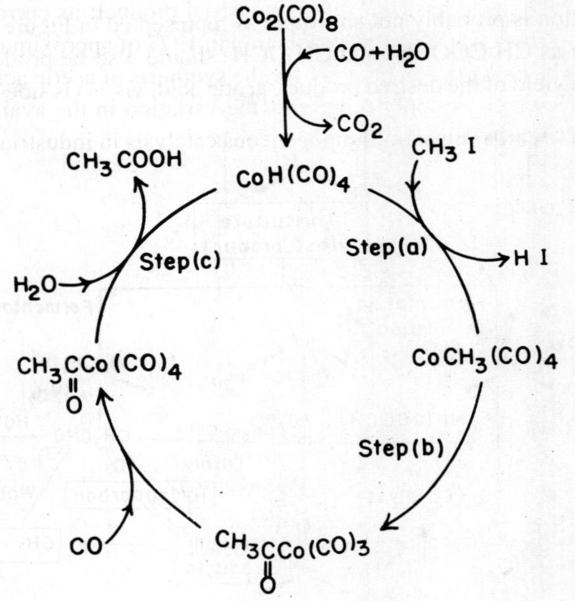

Fig. 6.19 Synthesis of acetic acid in the presence of $Co_2(CO)_8$ (catalyst) and CH_3I (co-catalyst)

In the early 1970's Monsanto Company commercialized the use of rhodium catalysts in the synthesis of acetic acid, and this appears to be technologically the most successful example of homogeneous catalysis:

$$CH_3OH + CO \xrightarrow[180^\circ C,\ 30\text{–}40\ atm.]{RhI_2(CO)_2^-} CH_3COOH$$

This is a highly selective and extremely rapid (rates are in the enzymatic range) reaction producing acetic acid in ~90% yields. The current estimated production of acid by Monsanto's process is 106 tons per annum. The mechanism of the Monsanto acetic acid process has been studied in detail. The active species appear to be methyl iodide (formed from CH_3OH and HI; $CH_3OH + HI \rightarrow CH_3I + H_2O$) and diiododicarbonylrhodium formed by the interaction of a rhodium (III) complex with carbon monoxide:

$$RhI_3 + CO + H_2O \rightarrow RhI_2(CO)_2^- + CO_2 + 2H^+ + I^-$$

The proposed catalytic mechanism involves a cycle of four basic steps shown in Figure 6.20.

(a) oxidative addition, (b) migratory insertion of alkyl to carbonyl, (c) coordination of CO, and (d) reductive elimination of acetyl iodide. Finally, water hydroyzes acetyl iodide ($CH_3COI + H_2O \rightarrow CH_3COOH + HI$) to complete the cycle.

In the above catalytic loop, rhodium changes reversibly between the + 1 and + 3 oxidation states. The catalyst is so efficient that the reaction can proceed under an atmosheric pressure. However, a pressurized system is used in order to enhance the reaction rate and improve the over-all economy of the process.

The reaction is probably not so simple as represented in Figure 6.20 because products such as CH_3COOCH_3 and CH_3OCH_3 should also be produced by side-reactions. The yield of the desired product, acetic acid, vis-a-vis side-products like

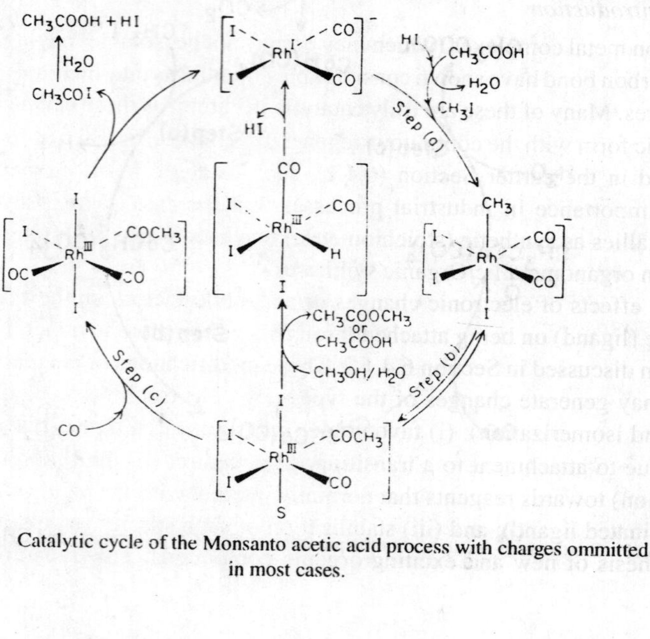

Fig. 6.20 Catalytic cycle of the Monsanto acetic acid process with charges ommitted in most cases.

ether and ester can be considerably improved by carrying out the process under equilibrium conditions in tri-n-octylphosphine oxide as solvent.

The use of platinum group carbonyls as homogeneous catalysts is increasing at a fast rate. Another example of a process developed (by Texaco) employing such catalysts is the homologation (i.e. increasing the chain length) of carboxylic acids and is shown in Figure 6.21 illustrating the conversion of acetic into propionic acid using ruthenium carbonyl iodide as catalyst:

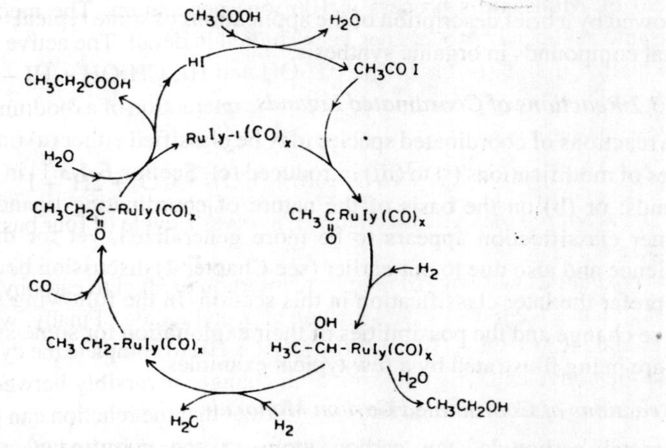

Fig. 6.21 Homologation of acetic acid.

6.4.3 Synthetic (Stoichiometric) Reactions (involving transition metal organometallics)

6.4.3.1 Introduction

Transition metal complexes (which may or may not be isolable) having at least one metal-carbon bond have shown considerable potential as intermediates in synthetic procedures. Many of these are truly catalytic in nature, as these can be represented in a cyclic form with the continuous regeneration of the original active catalyst. As discussed in the earlier Section (6.4.2), such catalytic reactions have achieved greater importance in industrial processes, but the role of transition metal organometallics as synthetic (stoichiometric) reagents is also beginning to make an impact in organometallic/Organic syntheses.

The effects of electronic changes or/and steric factors on the reactivity of a substrate (ligand) on being attached through a carbon atom to a metal centre have also been discussed in Section 6.4.1.2. These modifications in the reactivity of the ligand may generate changes of the types (e.g., electrophilic, nucleophilic, coupling, and isomerization): (i) favourable (activation) to a higher reactivity of the ligand due to attachment to a transition metal centre; (ii) inhibition of reactivity (protection) towards reagents that normally interact with the parent substrate (i.e. uncoordinated ligand); and (iii) stabilization of such species as could be used for the synthesis of new and exciting organic compounds. The transition metal organometallics could thus contribute to the generation of species such as carbanions, carbenes, carbocations, hydride ions, free radicals and other highly reactive species.

The above possibilities have attracted the attention of synthetic chemists and have been extensively exploited in designing novel strategies for organic syntheses. In the following discussion, an attempt is being made to illustrate by only a few well chosen examples the synthetic potentials of transition metal organometallics, focussing the attention mainly on various types of reactions which ligands attached to transition metal centres undergo. This account would be followed by a brief description of the applications of some typical organotransition metal compounds in organic synthesis.

6.4.3.2 Reactions of Coordinated Ligands

The reactions of coordinated species may be classified either (a) on the basis of the types of modifications (i) to (iii) introduced (cf. Section 6.4.3.1) in the coordinated ligands; or (b) on the basis of the nature of coordinating ligand. Although the former classification appears to be more generalized, yet for the sake of convenience and also due to our earlier (see Chapter 4) discussion being ligand-wise, we prefer the later classification in this section. In the following subsections, the above change and the possibilities of their exploitation for some synthetic processes are being illustrated by a few typical examples.

1. *Reactions of Coordinated Carbon Monoxide*

In metal carbonyls, the carbon atom of the coordinated carbonyl group ($M \leftarrow :C \equiv O:$) becomes an electrophilic centre, which facilitates attack by nucleophiles such as H^-, OH^-, H_2O, OR^-, NR_2^- and $OCOR^-$ on the carbon atom. A few typical examples are presented below:

(a) *Hydride*. The attack of hydride, e.g., from $NaBH(OMe)_3$ or $NaBH_4$ in THF on coordinated carbonyl groups can be represented as below:

$$M-C\equiv O \rightarrow \underset{(a)}{M-\overset{\overset{\large O}{\|}}{C}-H} \longrightarrow \underset{(b)}{M-CH_2OH} \rightarrow \underset{(c)}{M-CH_3}$$

(a) formyl (b) hydroxymethyl (c) methyl

These possibilities are illustrated by the following reactions:

(a) $Fe(CO)_5 \longrightarrow (CO)_4Fe-(CHO)^-$

(b) $(\eta^5-C_5H_5)(NO)Re(CO)_2^+ \longrightarrow (\eta^5-C_5H_5)(NO)Re(CO)(CH_2OH)$

(c) $(\eta^5-C_5H_5)(PPh_3)W(CO)_3^+ \longrightarrow (\eta^5-C_5H_5)(PPh_3)W(CO)_2(CH_3)$

(b) *Water, hydroxide and alkoxide anions*. The oxygen atom of the co-ordinated CO undergoes exchange with $H_2^{18}O$ and leads to oxidation of CO to CO_2

$$Rh^{III}-CO + H_2O^* \longrightarrow Rh^{III}-C\begin{cases}{=}O^* \\ {-}OH\end{cases} + H^+$$

$$\downarrow$$

$$Rh^{I} + CO_2 + H^+$$

Similarly, the following sequence of steps involving the reaction of hydroxide with a metal carbonyl (A) to give metallocarboxylic acid (B), followed by loss of CO_2 to generate a metal species (C) appears to be a key feature in many reductions in which carbon monoxide acts as a reducing agent:

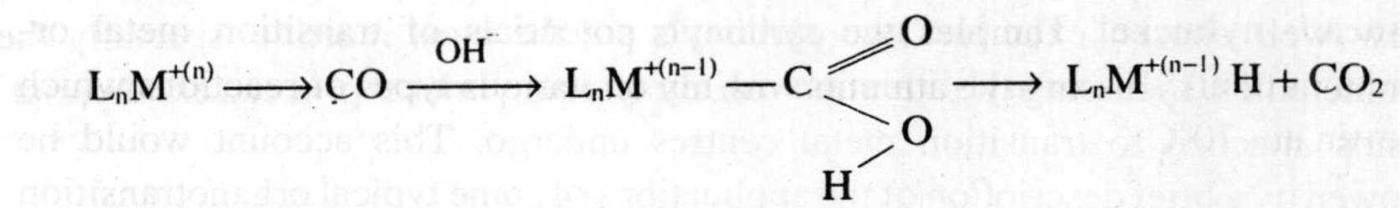

(A) (B) (C)

For example, evidence for hydroxide ion attack at a metal bound carbon monoxide ligand has been obtained by an oxygen-exchange studies in the following reaction:

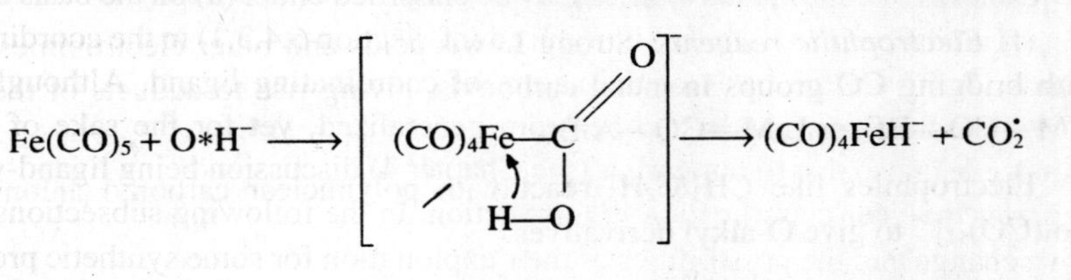

Carbon monoxide coordinated to metals (e.g. Mn, Re, Fe, Ru, Os, Co, Rh, Ir, Pd, Pt and Hg) reacts with alkoxide ions to yield alkoxy-carbonyl complexes:

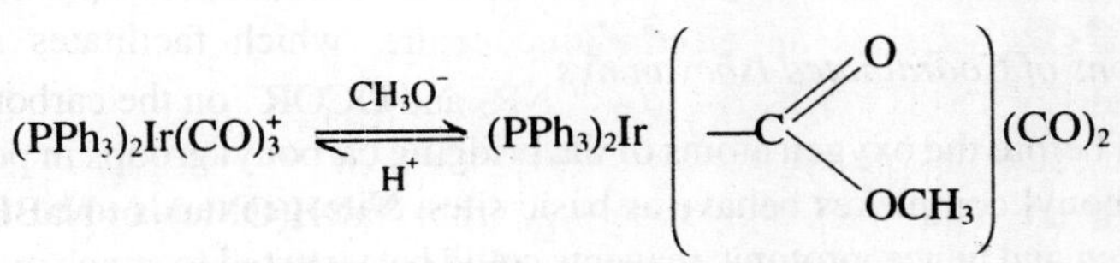

(c) *Ammonia and amines*. The reaction of cationic carbonyls on ammonia and amines results in carboximado or carbamoyl metal complexes, e.g.,

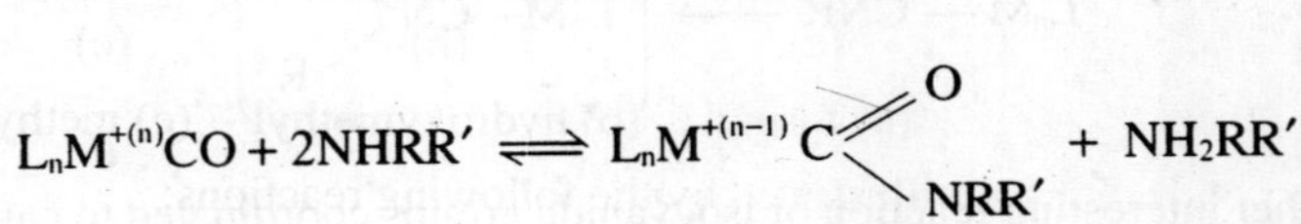

(d) *Tertiary amine oxides*. The reaction of trimethyl amine oxide on metal carbonyls results in the conversion of CO to CO_2 and provides a labile site on the metal carbonyls, as the amine coordinated in the product species can be readily displaced:

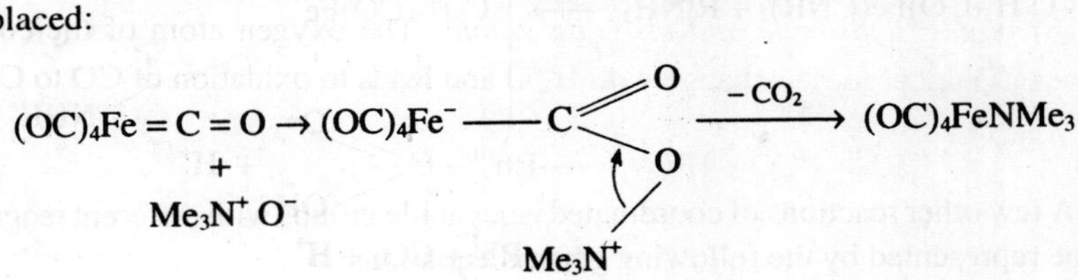

(e) *Metal Alkyls.* Of the metal alkyls, the reactions of which have been studied with carbonyl groups attached to transition metals, lithium alkyls have received maximum attention. These reactions have been used for (i) the synthesis of carbene complexes (cf. Section 4.5.2.3) using hexacarbonyl of chromium, molybdenum and tungsten, and (ii) organic synthesis by means of pentacarbonyliron and tetracarbonylnickel. The last two carbonyls are the most reactive, specially the reactions of alkyl — or aryl-lithiums with tetracarbonylnickel proceed exothermally even at –70°C:

$$Ni(CO)_4 + LiR \longrightarrow Li^+ \left[(OC)_3Ni{-}C{\overset{\displaystyle O}{\underset{\displaystyle R}{\lessgtr}}} \right]^-$$

(f) *Electrophilic reagents.* Strong Lewis acids and other electrophiles react with bridging CO groups in metal carbonyls, giving rise to adducts of the type $L_nM{-}CO{-}BF_3$ or $L_nM{-}CO{-}AlEt_3$.

Electrophiles like $CH_3^+/C_2H_5^+$ react with polynuclear carbonyl anions like $[Fe_3(CO)_{11}]^{2-}$ to give O-alkyl derivatives:

$$[PPN]_2[Fe_3(CO)_{11}] + RX \xrightarrow{CH_2Cl_2} [PPN]_2[Fe_3(CO)_{10}(COR)] + [PPN]X$$

(where PPN = triphenylphosphinenium and R = CH_3 or C_2H_5)

2. *Reactions of Coordinated Isocyanides*

As shown before the oxygen atoms of the bridging carbonyl groups in polynuclear metal carbonyl complexes behave as basic sites. Nitrogen tends to be more basic than oxygen and hence, protonic reagents could be expected to attack more readily isocyanide groups in these are also present:

$$L_nM - CNR \xrightarrow{H+} \left[L_nM{-}CN{\overset{\displaystyle H}{\underset{\displaystyle R}{<}}} \right]^+$$

Another interesting reaction of isocyanide groups coordinated to cationic and neutral iron complexes with primary amines yields diaminocarbene derivatives:

$$\eta^5{-}C_5H_5(CO)Fe(CNR)_2 + R'NH_2 \longrightarrow \left[C_5H_5(CO)Fe{\overset{\displaystyle CNR}{\underset{\displaystyle C{\overset{\displaystyle NR}{\underset{\displaystyle NHR'}{\lessgtr}}}}{<}}} \right]^+$$

A few other reactions of coordinated isocyanide groups with different reagents can be represented by the following general equations:

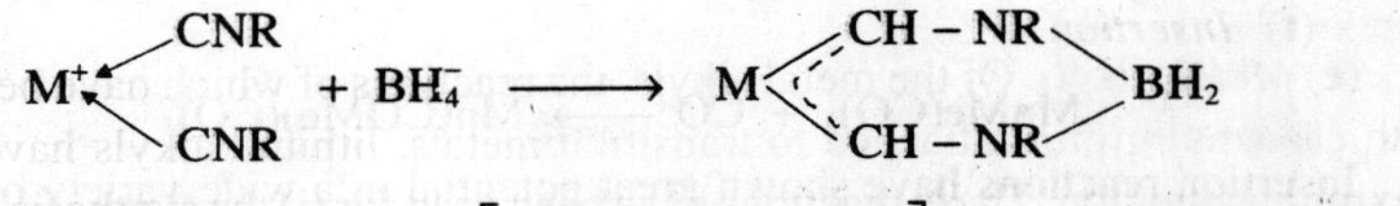

$$M^{+} \leftarrow CNR + OH^{-} \longrightarrow [M^{+} - C(OH) = NR] \longrightarrow M\text{–}CO\text{–}NHR$$

$$M(C \equiv NMe_4^{2+} + 4MeNH_2 \xrightarrow{H_2O} [M\{C(NHMe)_2\}_4]^{2+}$$
(M = Pd, Pt)

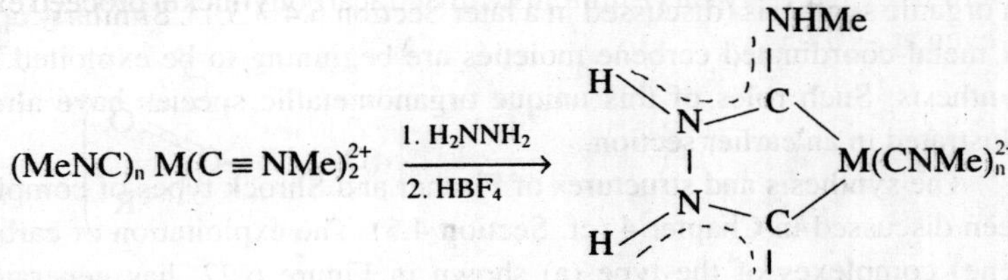

(M = Pd, Pt, n = 2; M = Fe, n = 4)

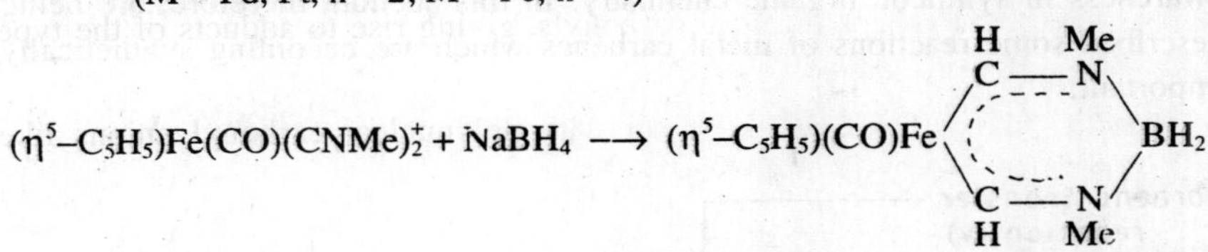

$$(cis\text{–}PtCl_2)(PhNC)(PEt_3) + EtOH \longrightarrow cis\left[PtCl_2\left(=C\begin{matrix}OEt\\NHPh\end{matrix}\right)PEt_3\right]$$

3. *Reactions of Alkyls bonded to Metals*

As these have been discussed already in Section 4.2.3.2, we are just mentioning below for recapituation the different types of possible reactions with one illustrative example of each:

(a) β-*Elimination*:

$$Pt(Bu)_2(PPh_3)_2 \longrightarrow Pt(H)(Bu)(PPh_3)_2 + CH_2 = CHCH_2CH_3$$
$$\longrightarrow n - C_4H_{10} + Pt(PPh_3)_2$$

(b) *Homolytic Fission*: $HgPh_2 \longrightarrow Hg + 2\,Ph$

(c) *Reductive Elimination*: $AuMe_3(PPh_3) \downarrow Au^{I}Me(PPh_3) + C_2H_6$

(d) *Binuclear elimination*

$$2Cu(Bu)(PBu_3) \longrightarrow 2Cu + 2PBu_3 + n - C_4H_{10} + I\text{—}C_4H_8$$

(e) α-*elimination*

$$Ta(CH_2CMe_3)_3Cl_2 + 2LiCH_2CMe_3 \longrightarrow 2LiCl + Ta(CH_2CMe_3)_5$$
$$\downarrow$$
$$Ta(CH_2CMe_3)_3\left[=C\begin{matrix}H\\CMe_3\end{matrix}\right] + CMe_4$$

(f) *Insertion*

$$MnMe(CO)_5 + CO \longrightarrow Mn(COMe)(CO)_5$$

Insertion reactions have shown great potential in a wide variety of catalytic (and also in stoichiometric) reactions.

4. *Reactions of Coordinated Carbenes*

Carbonyl derivatives of transition metals are widely employed as versatile reagents in organic synthesis (discussed in a later Section 6.4.3.3.1). Similarly applications of metal-coordinated carbene moieties are beginning to be exploited in organic synthesis. Such roles of this unique organometallic species have already been illustrated in an earlier section.

The synthesis and structures of Fischer and Shrock types of complexes have been discussed in Chapter 4 (cf. Section 4.5). The exploitation of carbonyl (carbene) complexes of the type (a) shown in Figure 6.22, has generated greater awareness in synthetic organic chemistry. In this section, therefore, are being described some reactions of metal carbenes which are becoming synthetically important.

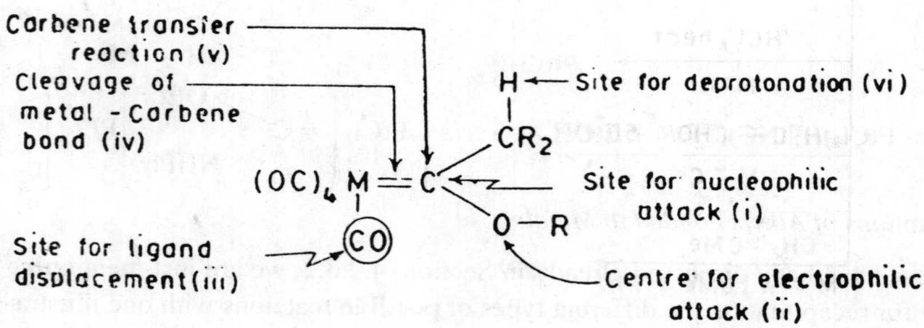

Fig. 6.22 Reactivity pattern of carbonyl (carbene) complexes of transition metals.

In the above, scheme, routes (i), (iii), (iv) and (v) have proved of considerable synthetic utility in the formation of new carbon bonds. Synthetic strategies involving transition metal carbene complexes in organic syntheses generally involve the following types of steps: (a) formation of metal-carbene bond, e.g., subsequent modification of metals by further reaction and (b) finally the cleavage of the metal-carbene bond in a variety of ways, generating new organic products. Some typical examples illustrating the versatility of the transition metal carbene complexes (as reaction intermediates or transition states) in organic synthesis are shown in the Scheme 6.12.

A few more examples of the applications of carbenes in organic synthesis are described in the following pages.

Fischer-type carbene ligands can be decomplexed: (i) thermally, to give alkenes corresponding to the dimerization of the carbene ligand, (ii) thermally, in the presence of an electron deficient alkene or diazomethane, generating

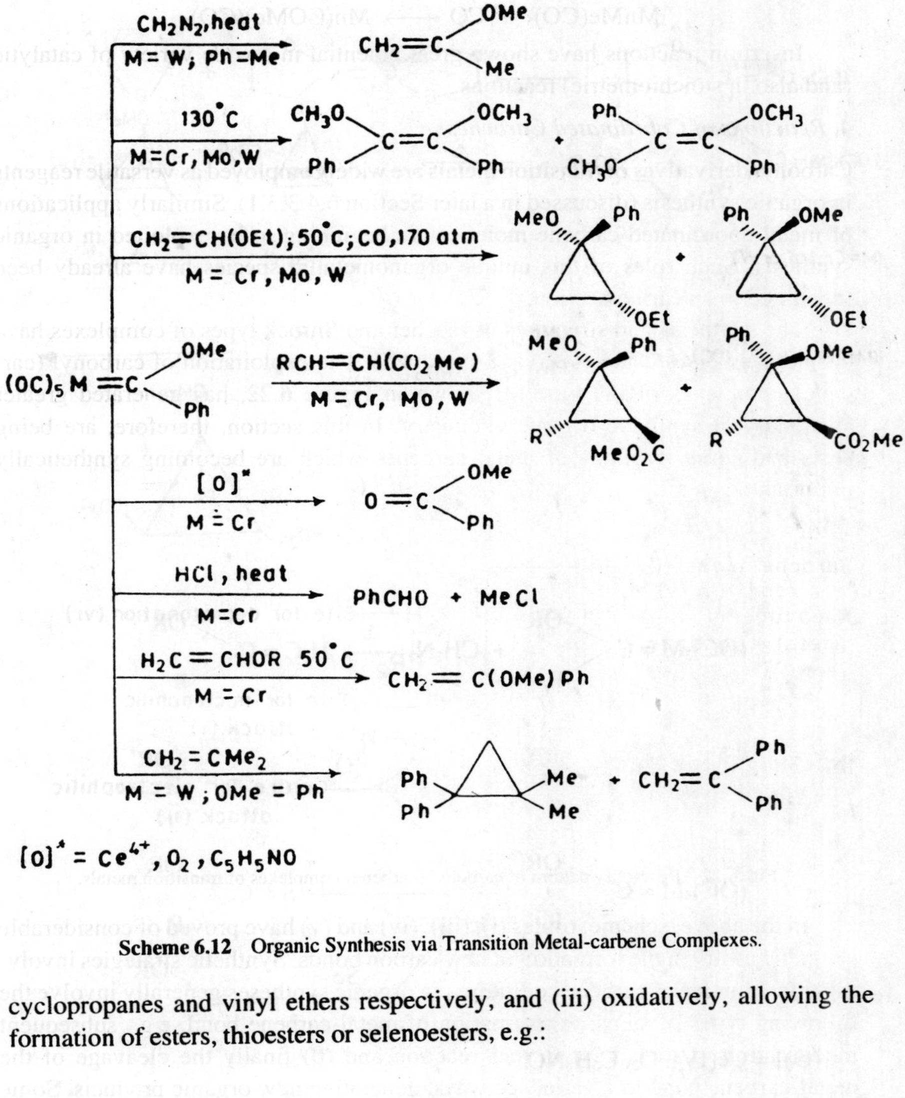

Scheme 6.12 Organic Synthesis via Transition Metal-carbene Complexes.

cyclopropanes and vinyl ethers respectively, and (iii) oxidatively, allowing the formation of esters, thioesters or selenoesters, e.g.:

(M = Cr, Mo, or W)

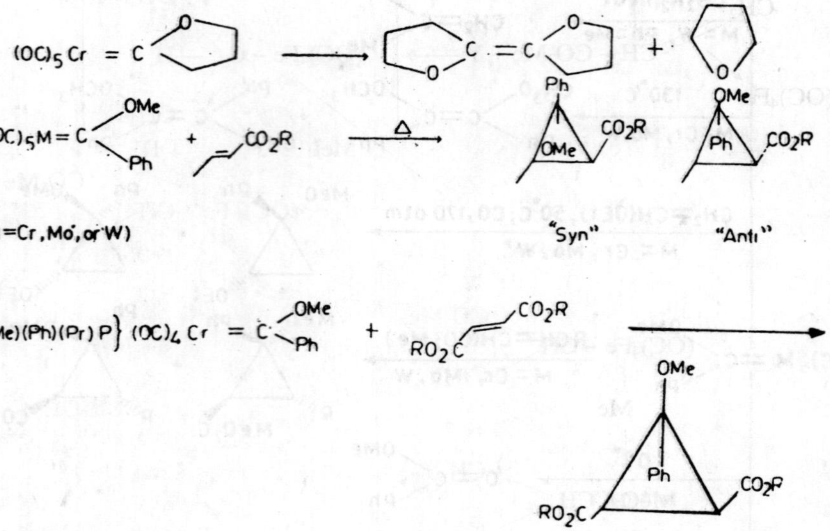

$$(OC)_3M=C(OR')(R) + CH_2N_2 \xrightarrow{\Delta} H_2C=C(OR')(R)$$

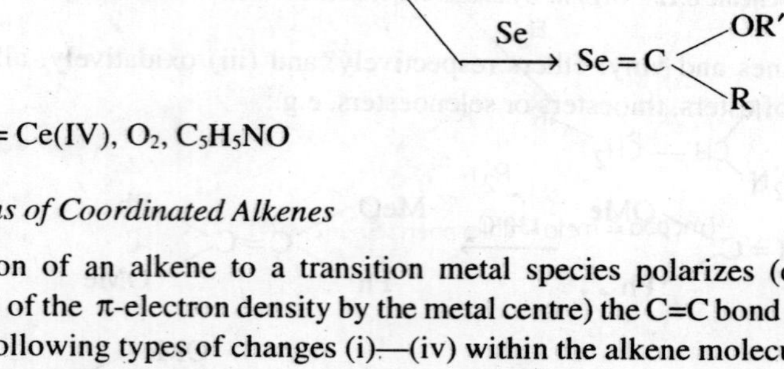

$[(O)] = Ce(IV), O_2, C_5H_5NO$

5. *Reactions of Coordinated Alkenes*

Coordination of an alkene to a transition metal species polarizes (due to the withdrawal of the π-electron density by the metal centre) the C=C bond and brings about the following types of changes (i)—(iv) within the alkene molecule:

(i) Pronounced susceptibility for nucleophilic addition of alkenes coordinated to transition metal moiety, which leads to facile generation of new carbon-carbon, carbon-nitrogen, and carbon-oxygen bonds:

$$Fe(CO)_2(\eta^2\text{–}C_2H_4)(Cp) + H^- \longrightarrow Fe(C_2H_5)(CO)_2(Cp)$$

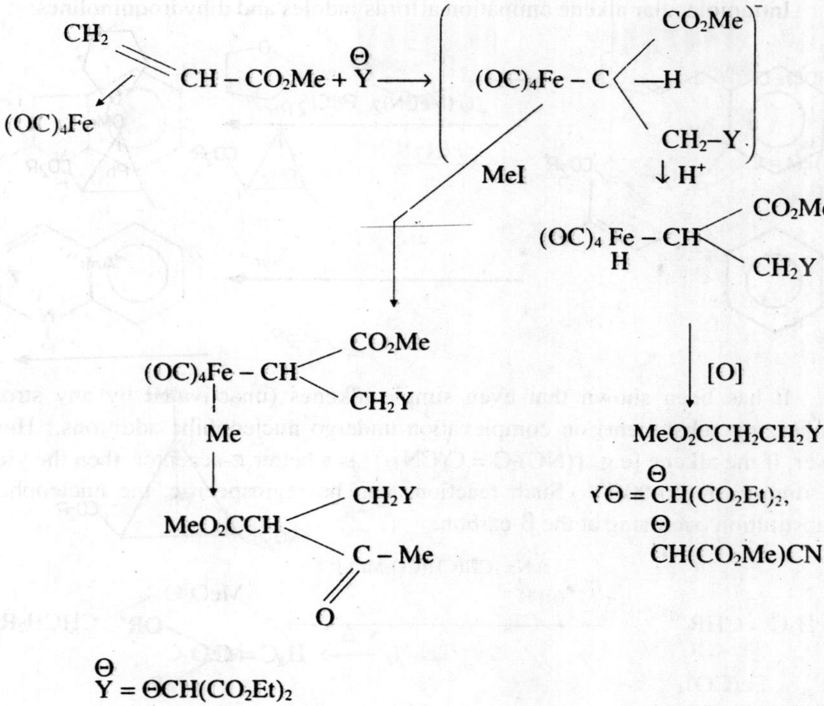

Palladium (II) chloride catalyses the interaction of alkenes with amines leading to the stereospecific formation of amines, aminoacetates, diamines and aziridines:

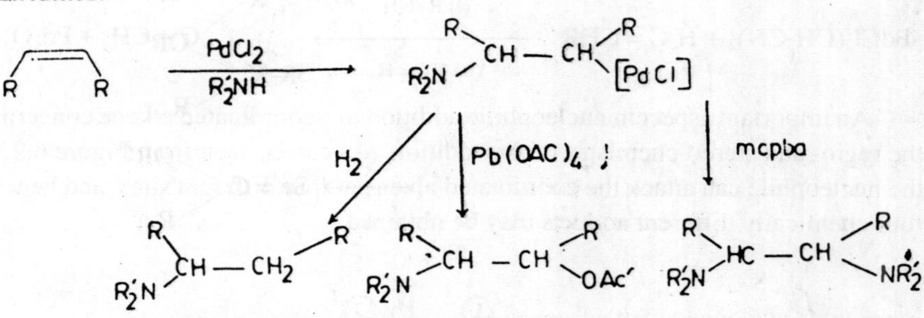

(mcpba = meta- chlorobenzoic acid)

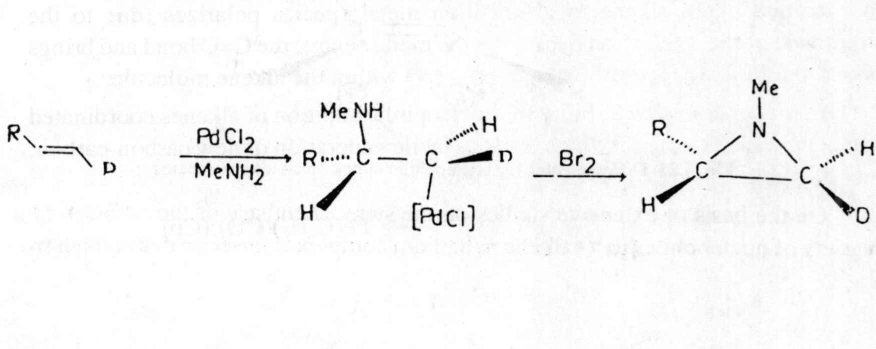

Intramolecular alkene amination affords indoles and dihydroquinolines:

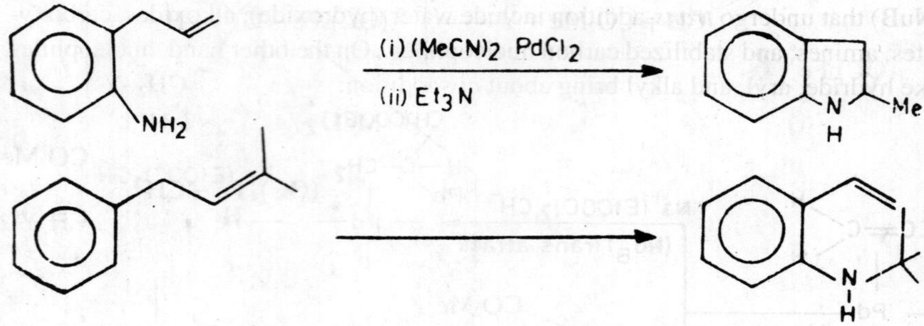

It has been shown that even simple alkenes (unactivated by any strong polarizing substituent) on complexation undergo nucleophilic additions. However, if the alkene [e.g. $\{(NC)_2C = C(CN)_2\}$] is a better π-acceptor, then the yield is improved (70-90%). Such reactions can be regiospecific, the nucleophilic substitution occurring at the β-carbon:

$$\underset{(R = H,\ CO_2Me)}{\underset{Fe(CO)_4}{\overset{}{H_3C = CHR}\downarrow}} \xrightarrow[\text{(ii) } H^+]{\text{(i) } Na^+[CH(CH(CO_2Me)_2]^+} (MeO_2C)_2CHCH_2R$$

In other cases (e.g. in amination of alkenes), regiospecific attack occurs at the more hindered alkene carbon in the presence of palladium:

$$PdCl_2(CH_3CN)_2 + H_2C = CHR \xrightarrow[\text{(ii) } H_2 \text{ or } H^+]{\text{(i) } R_2NH} (R)(R_2N)C - CH_3 + Pd(O)$$

An important aspect in nucleophilic addition to a coordinated alkene concerns the regio- and stereo-chemistry of the addition. As can be seen from Figure 6.23, the nucleophile can attack the coordinated alkene at four different sites, and hence four chemically different adducts may be obtained:

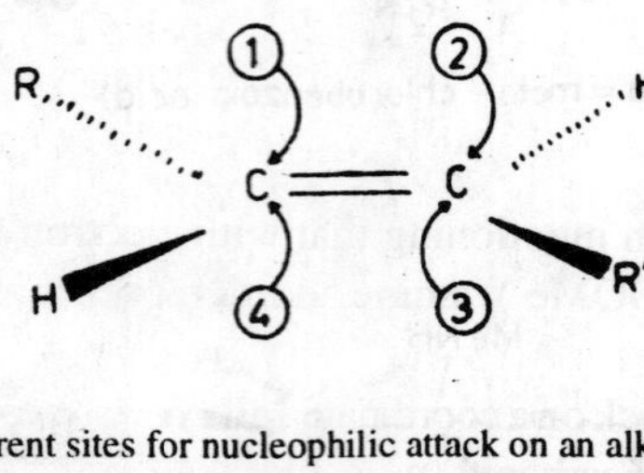

Fig. 6.23 Different sites for nucleophilic attack on an alkene species

On the basis of extensive studies on the stereochemistry of the addition of a variety of nucleophiles to π- alkene palladium complexes, one can distinguish two different classes of nucleophiles: (i) nucleophiles (NuA) that add with *cis* attack via an intramolecular migration from the metal to carbon; (ii) The nucleophiles (NuB) that undergo *trans*-addition include water (hydroxide), alkoxides, carboxylates, amines, and stabilized carbon nucleophiles. On the other hand, nucleophiles like hydride, aryl, and alkyl bring about *cis*-addition:

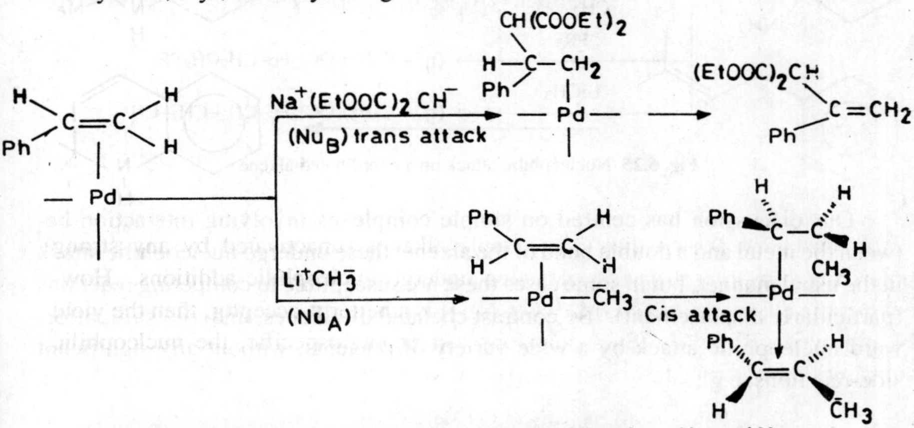

Fig. 6.24 Different behaviour of coordinated nucleophiles of type Nu_A and Nu_B on alkenepalladium complexes.

Another example illustrating regiospecific addition of nucleophiles at the carbon bearing the donor substituent is shown as follows:

$$(\eta^5-C_5H_5)\,(CO)_2\,Fe^+ \leftarrow \begin{matrix} HC\diagup^{R} \\ \| \\ CH_2 \end{matrix} + H_2NCH_2C_6H_5 \rightarrow$$

$$(\eta^5-C_5H_5)\,(OC)_2\,Fe^+CH_2CH(R)NH_2CH_2Ph$$

These two latter regiospecific additions have been explained on the hypothesis that donor substituents should facilitate slippage away from the substituent (as shown in a and b):

(a) (b)

However, it is worth mentioning that with electron withdrawing alkene substituents ($R = CN^-$ or $COOMe^-$), amine addition occurs at the carbon atom remote from the substituent.

The nucleophilic attack on a coordinated alkene has been exploited by introducing cationic species such as dicarbonyl (η^5–cyclopentadienyl) iron complex, which leads to the transformation of cationic η^5-cyclopentadienyl (olefin) iron dicarbonyl complexes into stable, neutral σ-alkyl iron complexes shown in Figure 6.25:

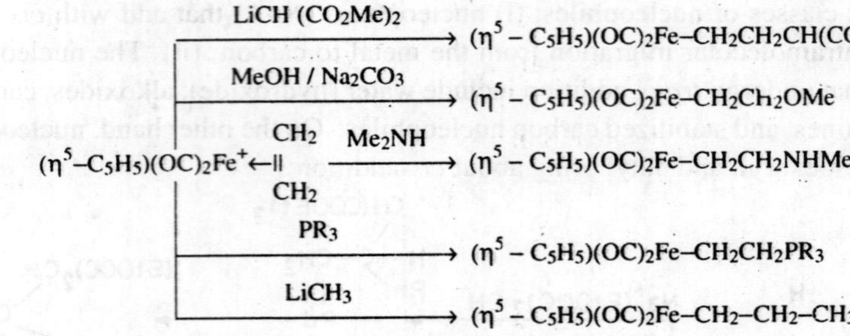

Fig. 6.25 Nucleophilic attack on a coordinated alkene

Our discussion has centred on simple complexes involving interaction between the metal and a double bond of the alkene; these undergo nucleophilic attack in the usual manner, but in some cases these are susceptible to competing reactions (particularly displacement). By contrast chelated di-alkenes, undergo straightforward nucleophilic attack by a wide variety of reagents, without any significant side-reactions, e.g.:

$$(M = Pd, Pt; Nu := \overset{\ominus}{O}R, RNH_2, \overset{\ominus}{N}_3, \overset{\ominus}{C}H(CO_2R)_2, \overset{\ominus}{C}H(COR)_2$$

The generated σ-alkyl-π-alkene complexes resist further attack at the remaining coordinated alkene. The attack is always *trans,* from the face opposite to the metal.

Although nucleophilic attack on chelating dialkene complexes has so far found little use in organic synthesis, the related alkyl amine systems are in fact quite useful, e.g., (see the reactions given in Figure 6.25).

(ii) Protection of the coordinated alkene bond against electrophiles and facilitation of normal electrophilic attack at uncoordinated alkene bond within the

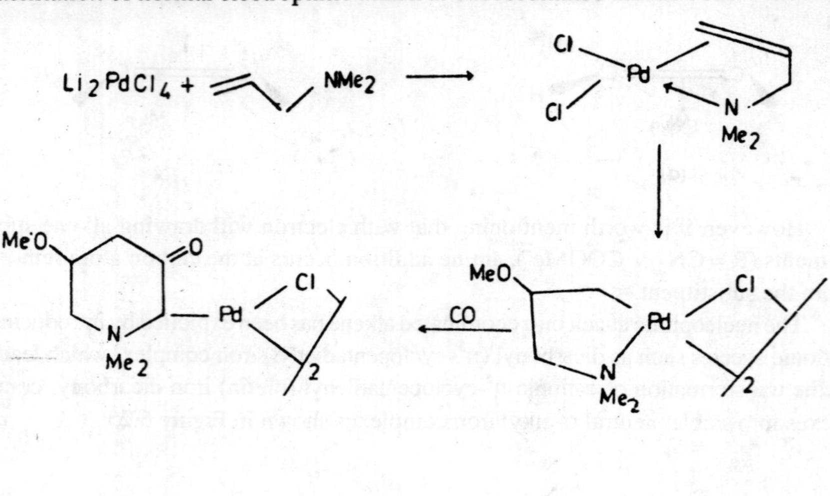

molecule result in organometallic complexes: these on treatment with sodium iodide yield new alkenes:

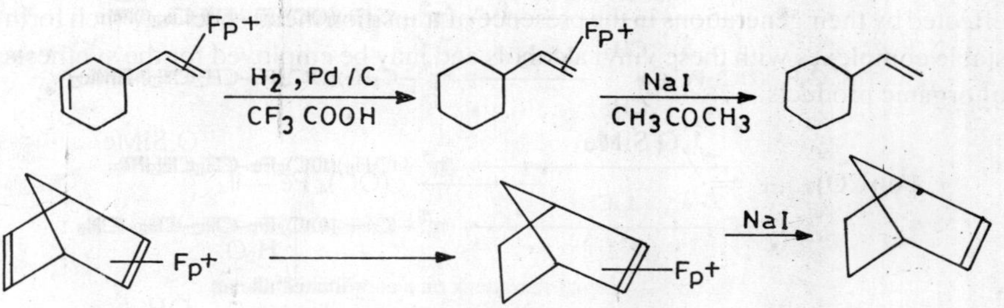

$F_p^+ = Fe(CO)_2(\eta^5-C_5H_5)$

(iii) Stabilization of unstable and reactive species such as 7-norbornadienones, strained acetylenes, and cyclobutadiene can be achieved by their coordination to a transition metal fragment which are then available for the synthesis of the corresponding organic derivatives:

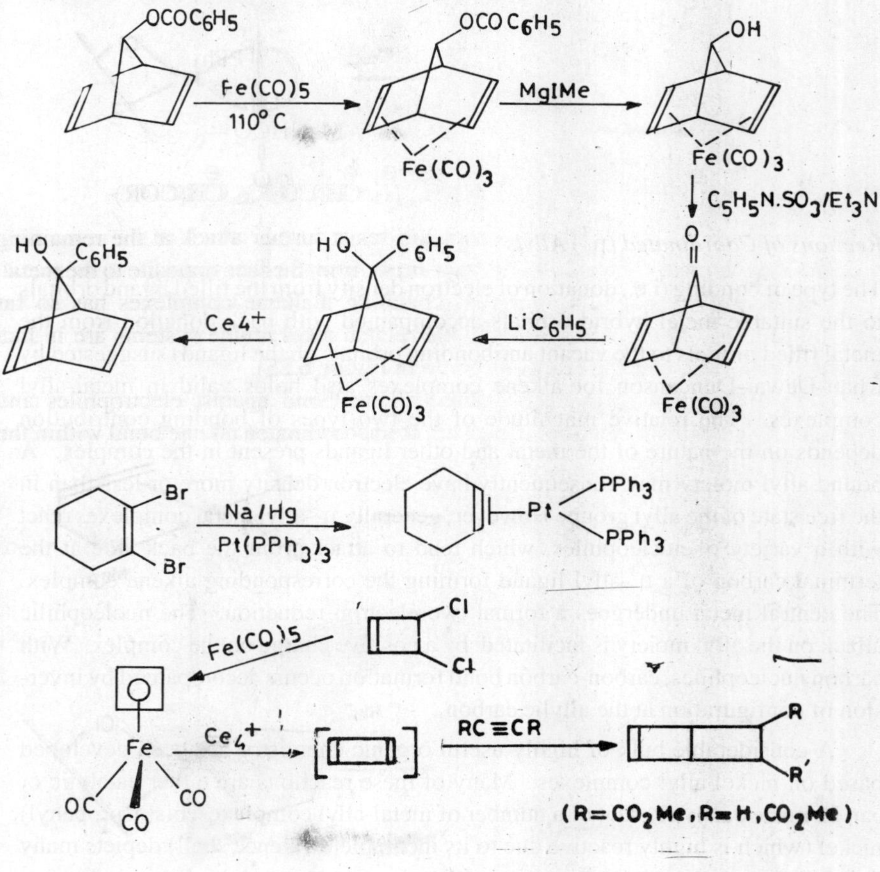

(iv) Stabilization of thermodynamically disfavoured tautomers like those of vinyl alcohols (which are not capable of free existence as they are approximately 12.9–14.6 Kcal/mole less stable than their tautomeric aldehyde forms) can be effected by their generations in the presence of transition metal species, which form stable complexes with these vinyl alcohols and may be employed for the synthesis of organic products:

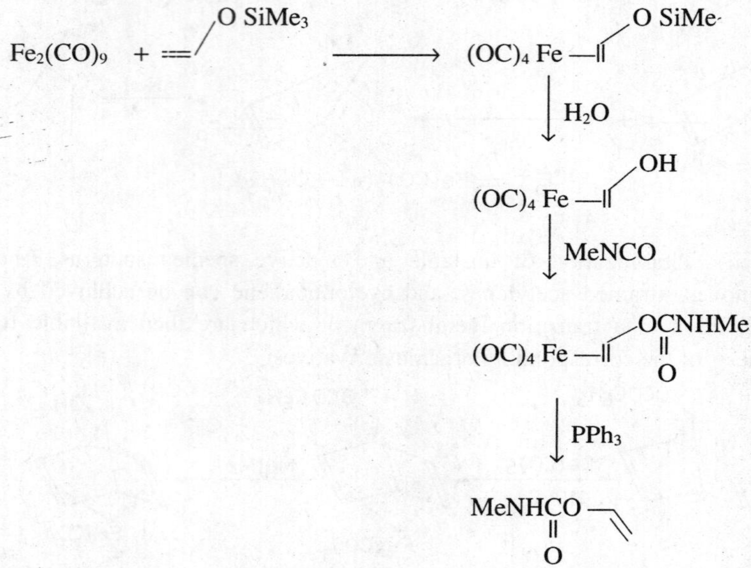

Rections of Coordinated (η^3) Allyls

The type of bonding (i.e., donation of electron density from the filled ligand orbitals to the suitable metal hybrid orbitals accompanied with back donation from the metal filled orbitals to the vacant antibonding orbitals on the ligand) susggested by Chatt-Dewar-Duncanson for alkene complexes also holds valid in metal-allyl complexes. The relative magnitude of the two types of bonding contribution depends on the nature of the metal and other ligands present in the complex. A bound allyl moiety may consequently have electron density more or less than in the free state of the allyl group. However, generally η^3-allyl metal complexes react with a variety of nucleophiles, which tend to attack from the back side at the terminal carbon of a η^3-allyl ligand forming the corresponding alkene complex. The central metal undergoes a formal two-electron reduction. The nucleophilic attack on the allyl moiety is facilitated by a positive charge on the complex. With carbon nucleophiles, carbon-carbon bond formation occurs accompanied by inversion of configuration at the allylic carbon.

A considerable bulk of highly useful organic chemistry has been developed based on nickel allyl complexes. Many of these reactions are either catalytic or can be made catalytic. Out of a number of metal-allyl complexes, bis(π-propenyl) nickel (which is highly reactive due to its incomplete valence shell) depicts many interesting reactions, amongst which the following (cf. Scheme 6.13) may be cited as typical examples.

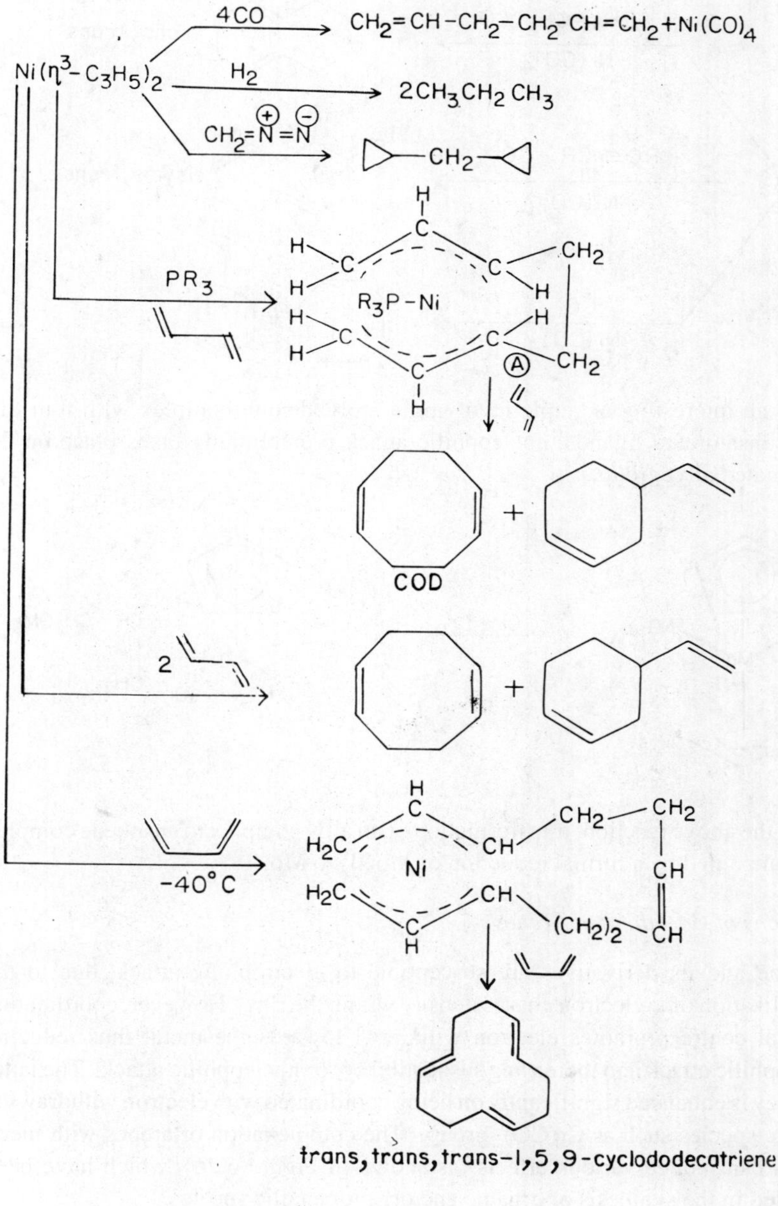

Scheme 6.13 Some synthetically useful reactions of Ni (η^3–C_3H_5)$_2$.

Ethene and alkynes react with the product (A) in the presence of excess carbon monoxide to give ten-membered ring compounds, whereas the reaction of (A) with excess carbon monoxide results in formation of a mixture of six- and eight-membered rings:

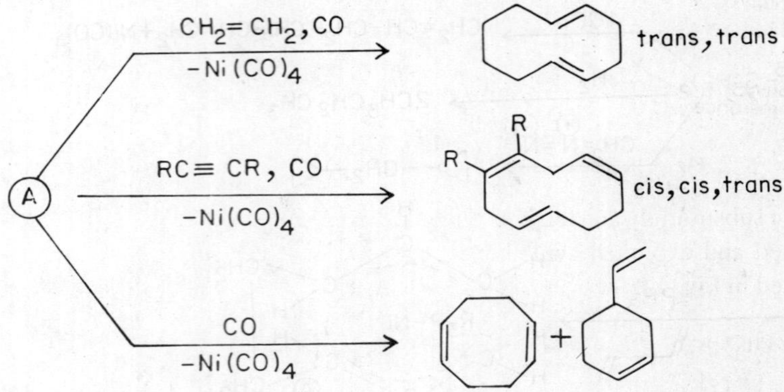

In an interesting example involving a molybdenum complex with four different unsaturated ligands, nucleophilic attack preferentially takes place on the coordinated allyl group:

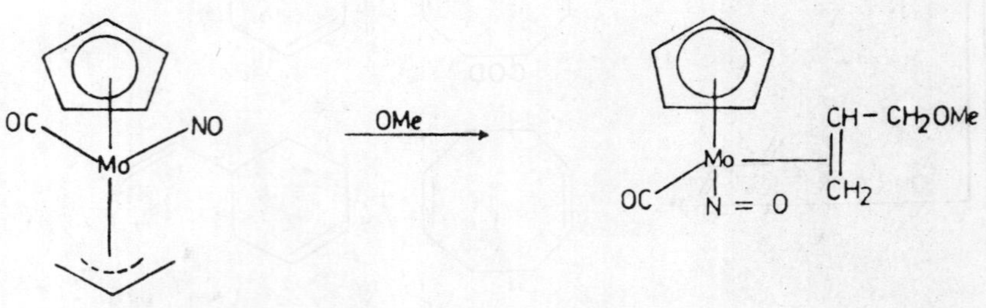

In the above reaction, transformation of an allyl complex to an alkene complex is accompanied by a formal reduction of Mo(II) to Mo(0).

Reactions of Coordinated Arenes

Benzene and its derivatives are susceptible to electrophilic attack, due to the delocalization of π-electron charge density within the ring. However, coordination to metal centre promotes electron withdrawl towards the metal thus reducing electrophilic attack and increasing susceptibility for nucleophilic attack. The latter tendency is enhanced significantly on being coordinated with electron withdrawing cationic species such as $Cr(CO)_3$ group. The complexation of arenes with metal centres can induce various effects (as shown in Figure 6.26), which have been exploited in the synthesis of organic and organometallic species.

Although the arene ring in these complexes is non-reactive towards electrophilic aromatic substitution, it is quite reactive towards nucleophilic

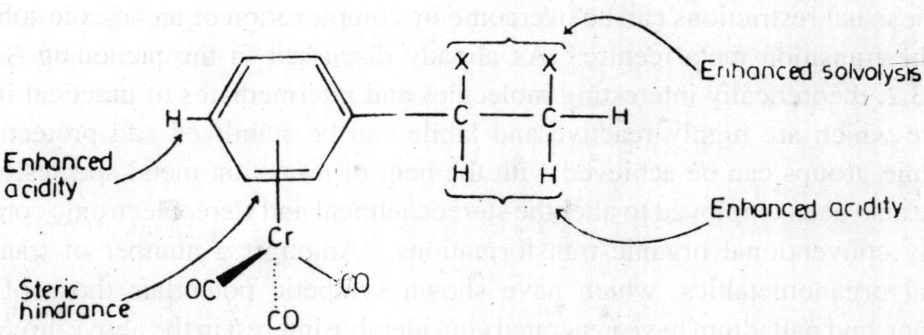

Fig. 6.26 The reactivity pattern of coordinated arenes to $Cr(CO)_3$.

aromatic substitution, and it is this type of reaction which has been most extensively studied and exploited synthetically in a variety of ways, some of which are illustrated in Fig. 6.27:

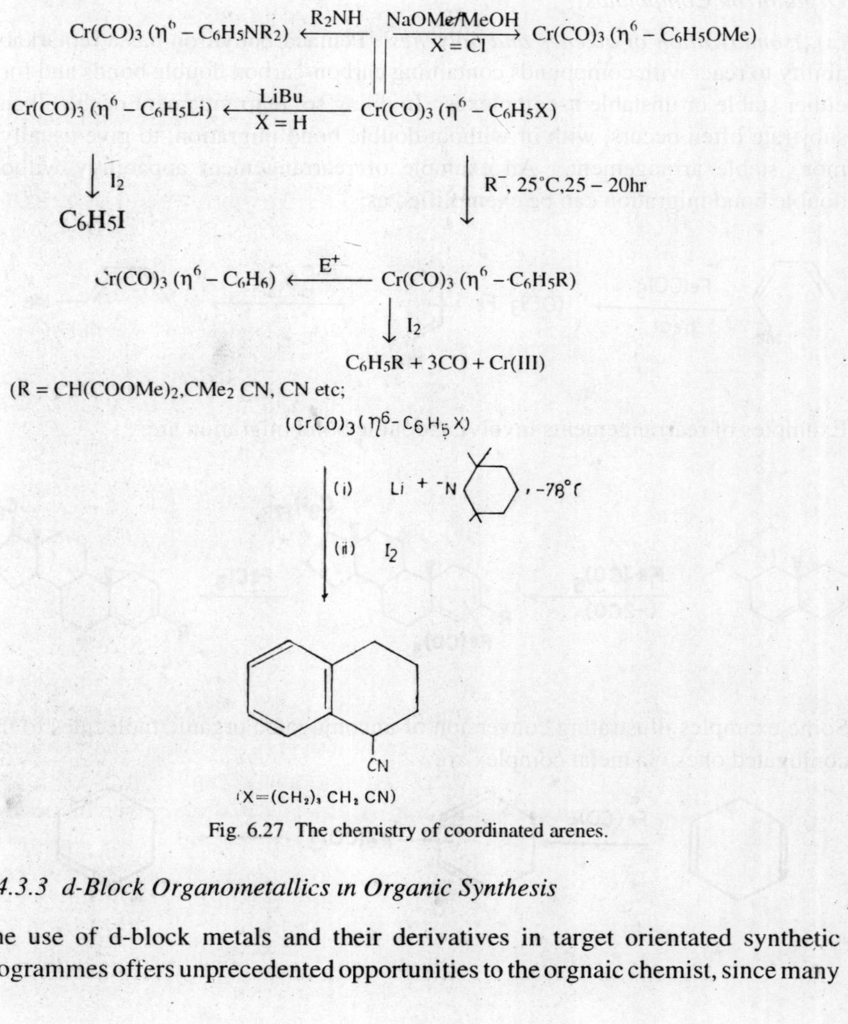

Fig. 6.27 The chemistry of coordinated arenes.

6.4.3.3 *d-Block Organometallics in Organic Synthesis*

The use of d-block metals and their derivatives in target orientated synthetic programmes offers unprecedented opportunities to the orgnaic chemist, since many of the usual restrictions can be overcome by complexation of an organic substrate to the transition metal centre. As already discussed in the preceding Section 6.4.3.2, theoretically interesting molecules and intermediates of practical importance which are highly reactive and labile can be stabilized and protection of alkenic groups can be achieved with the help of transition metal species, which have also been employed to alter the stereochemical and stereoelectronic course of many conventional organic transformations. Amongst a number of transition metal organometallics, which have shown synthetic potentials those of iron, copper, and palladium have generated considerable interest in the above directions.

The applications of transition metal organometallics are growing at an extremely fast pace in ever increasing novel directions. In the following sections is presented a brief glimpse of the synthetic uses of transition metal compounds, beginning with iron.

Organoiron Compounds

(a) *Isomerization of alkenes and polyenes* Pentacarbonyliron has a remarkable ability to react with compounds containing carbon-carbon double bonds and form either stable or unstable π-complexes. In doing so, rearrangement of the alkenic substrate often occurs, with or without double bond migration, to give usually a more stable arrangement. An example of rearrangement apparently without double bond migration can be exemplified as:

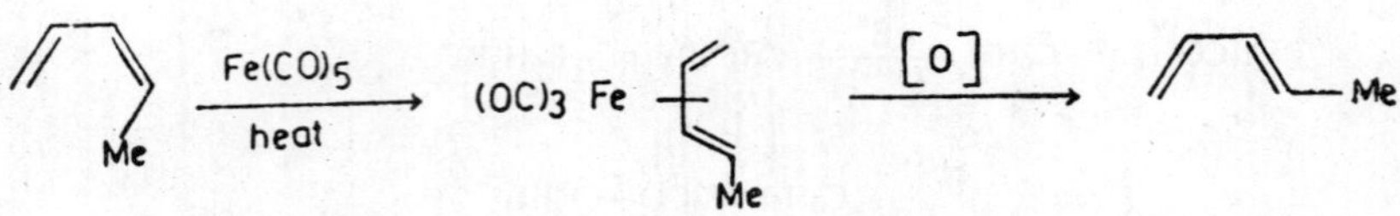

Examples of rearrangements involving double bond migration are:

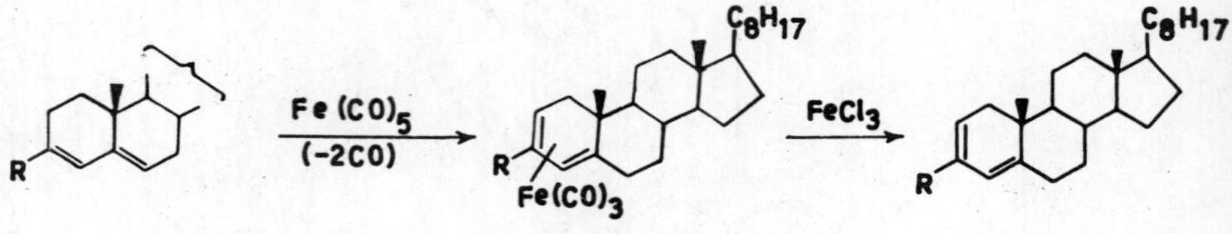

Some examples illustrating conversion of unconjugated organic molecules to the conjugated ones via metal complex are:

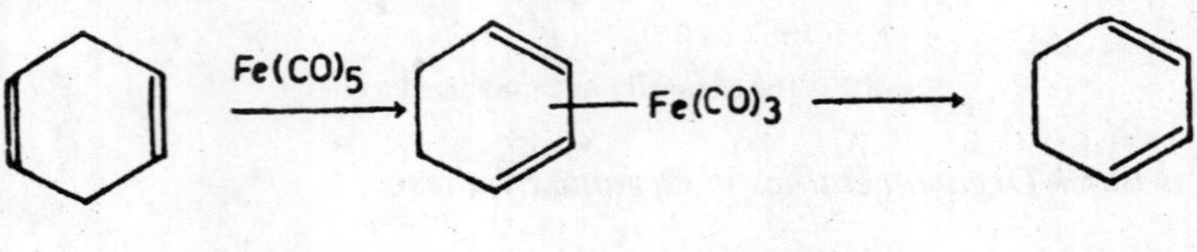

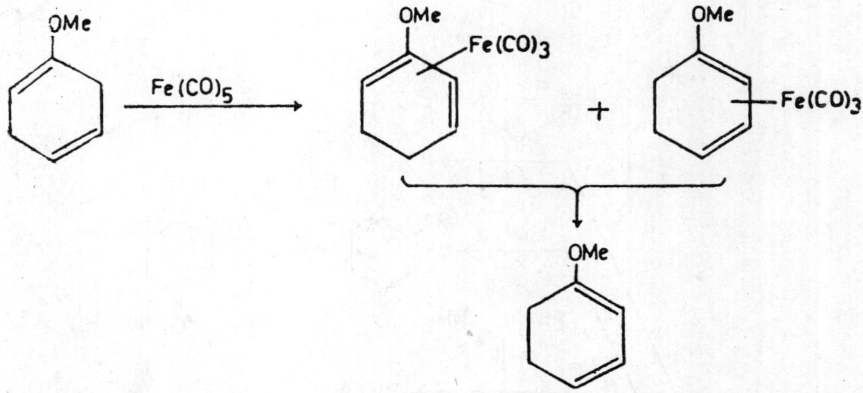

Further examples of interest are the conversion of the vinylcyclohexene to the ethylcyclohexadiene, e.g.,

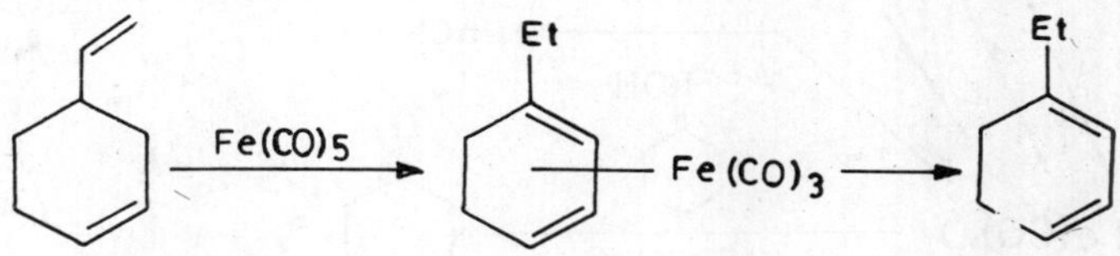

and the conversion of 1-5-cyclooctadiene to 1, 3-cyclooctadiene:

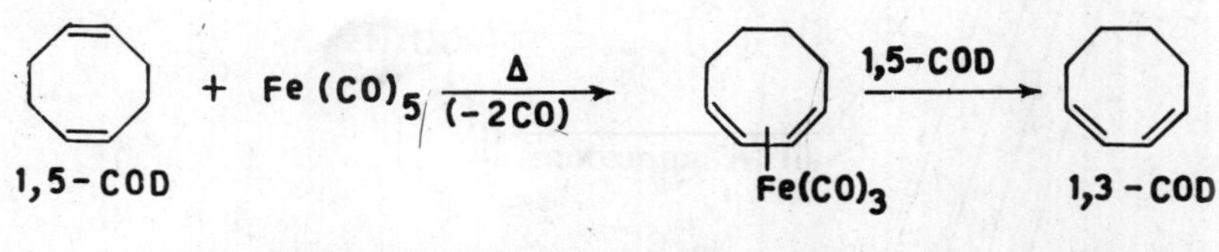

Some other potentially useful conversions via iron carbonyl complexes involving (a) small rings opening, (b) dehydration, (c) dehydrosulphuration, (d) deoximation, and (e) carbonyl insertion reactions are illustrated in Scheme 6.14.

(b) *Disodium Tetracarbonylferrate (II) as Synthon.* Nucleophilicity (which varies with the nature of the metal and the substituents) of the carbonylate anions has been exploited in many important organic transformations. Among a large number of carbonylate anions, those of iron, the tetracarbonyl ferrate(II) which is isolable as a dioxane adduct of its sodium salt $Na_2Fe(CO)_4 \cdot 1.5C_4H_8O_2$ (by the reduction of pentacarbonyliron with sodium benzophenone ketyl in boiling dioxane) has been shown in depict greater synthetic utility (Scheme 6.15):

Organocopper Compounds

The field of organocopper chemistry has grown drammatically during the past two decades. In fact, these reagents have become so important that modern textbooks on introductory organic chemistry expose students to organocopper chemistry. We shall, therefore, present in this section an introductory account of the strategies of

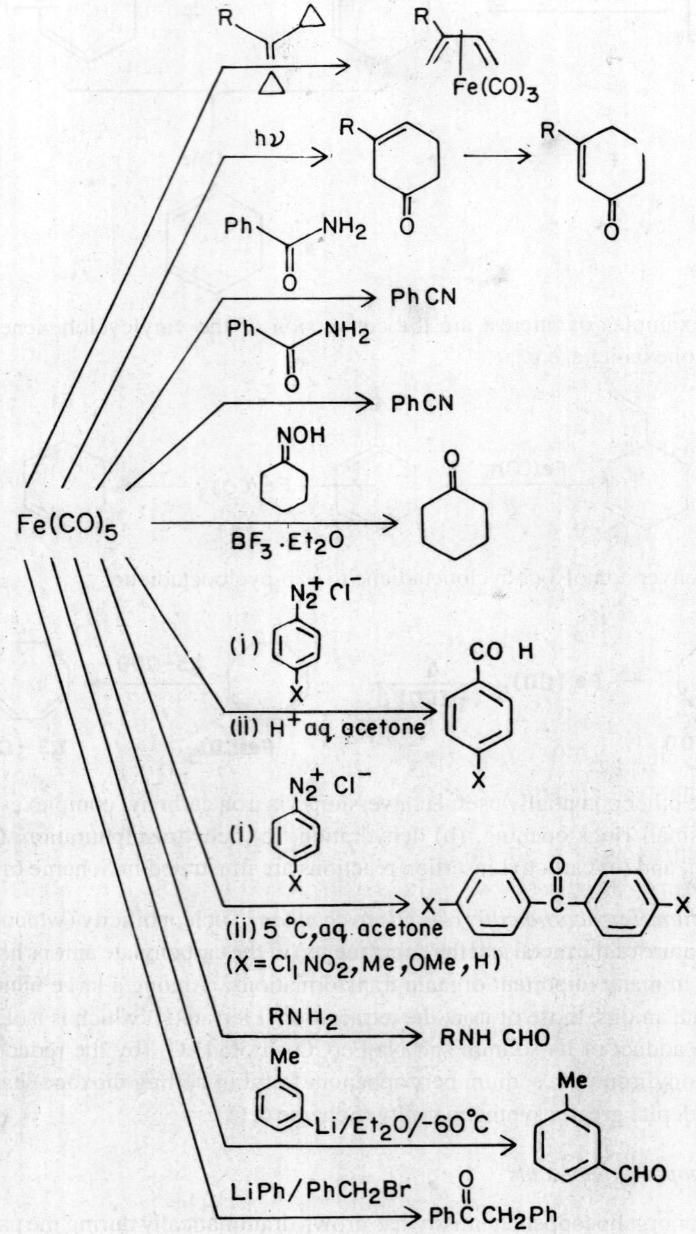

Scheme 6.14 Applications of iron pentacarbonyl in organic synthesis.

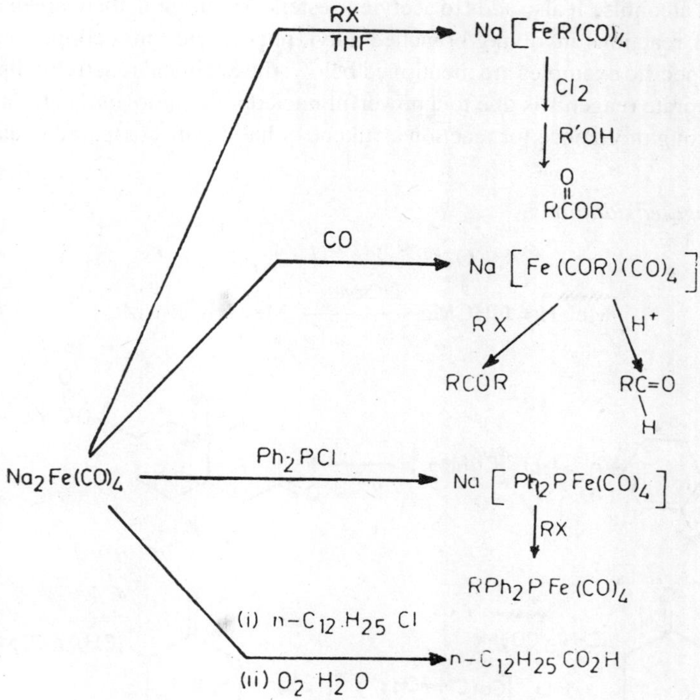

Scheme 6.15 Some reactions of $Na_2Fe(CO)_4$.

organic synthesis using organocopper moieties. Only such reactions would be touched upon which have proved to be valuable in organic synthesis and involve well defined organocopper complexes or where there is substantial grounds for believing that such reactions occur via organocopper intermediates. In this connection, it may be relevant to mention that there are a number of well known synthetic reactions in organic chemistry, which are catalyzed by copper salts, but there is no well established evidence that organocopper species are intermediates and hence these are not included here.

Until 1970's organomagnesium, organolithium, and to a lesser extent, organocadmium derivatives were the main organometallic systems employed in organic synthesis. This situation has changed drastically in the last 20 years. Organocopper compounds have proved to be specially of great importance. One of the first examples of copper catalysts is a reagent prepared from methyllithium and cuprous iodide:

$$2LiCH_3 + CuI_2 \rightarrow Li[Cu(CH_3)_2] + LiI$$

The reagent adds selectively at the β-carbon of an α, β-unsaturated carbonyl system, and also has a unique ability to effect replacement of most types of halogen atoms by methyl groups. Lithium dimethylcuprate reacts with allylic acetates by an S_N^2 mechanism. It has been found to react with epoxides to give open ring alkylated alcohols. It also adds to acetylene esters. In addition, there are examples of similar reactions involving branched alkyl, phenyl and vinyl copper reagents. Several specific examples are mentioned below; the chemical reactivity displayed by the cuprate reagents is due to a powerful nucleophilicity toward carbon, along with a strong preference for reaction at alkene or halide site compared to carbonyl groups:

A. *Conjugated additions*

$$MeCH=CH\overset{\overset{O}{\|}}{C}Me \xrightarrow{LiCuMe_2} Me_2CHCH_2\overset{\overset{O}{\|}}{C}Me$$

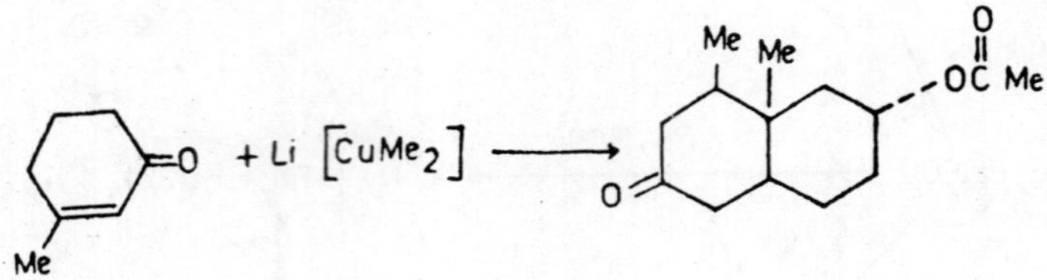

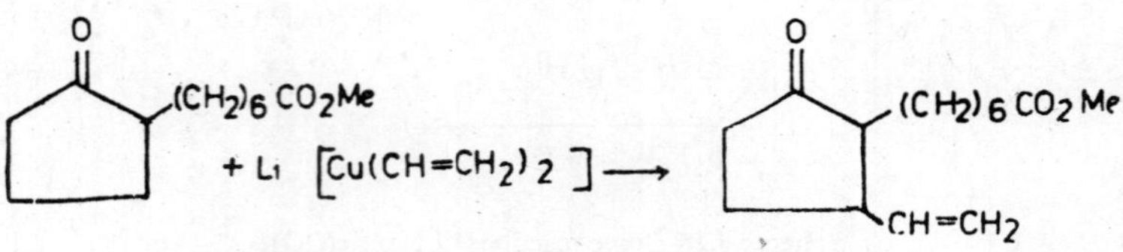

B. *Halogen substitution*

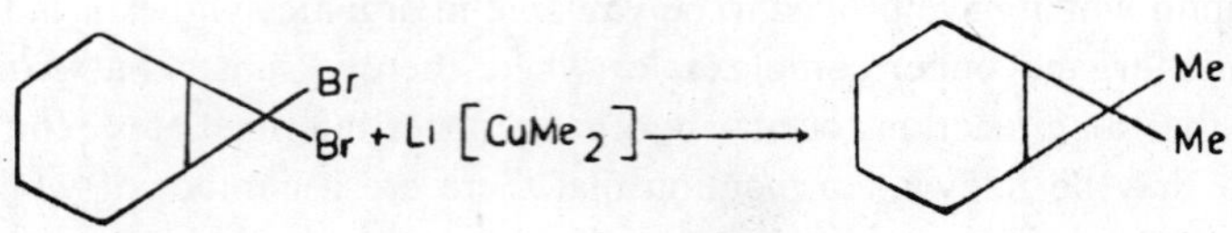

$$C_5H_{11}Br + Li\,[CuBu_2^t] \longrightarrow Bu^t(CH_2)_4CH_3$$

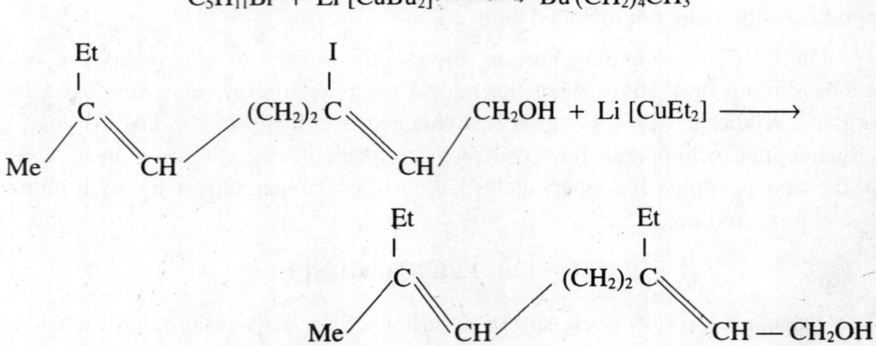

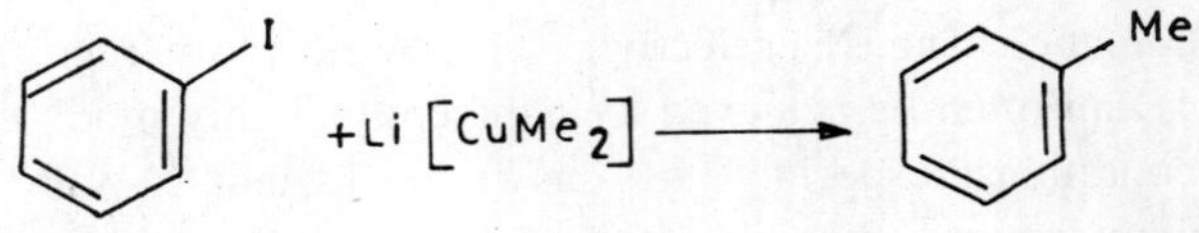

$$trans\text{–}PhCH = CHBr + Li\,CuMe_2 \longrightarrow trans\text{–}PhCH = CHMe$$

This method of replacement of halide by alkyl is generally more satisfactory, compared to displacement by the Grignard or lithium reagent.

C. *Alkylation of epoxides* In these reactions the methyl group is introduced at the less hindered carbon of the epoxide:

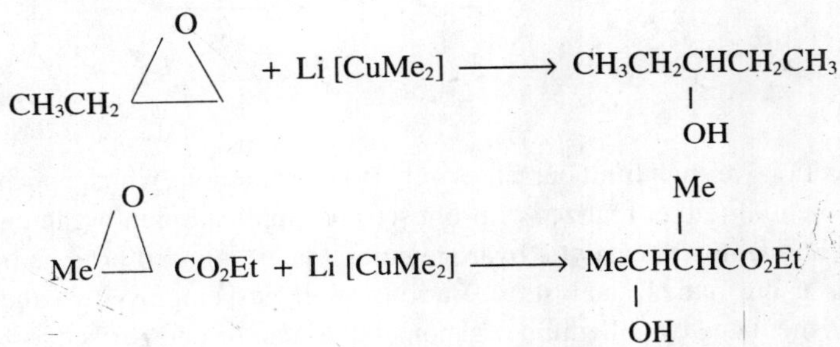

D. *Alkylation of Allylacetates can be illustrated by the following reaction*

$$Me_2C = \underset{\displaystyle O_2CMe}{\underset{|}{C}}HCHCH_2Bu^t + Li[CuMe_2] \rightarrow Bu^tCH = CHCH_2Bu^t$$

E. *Ketones from acid chlorides*. The cuprate reagents can act as an alternative to alkylcadmium reagents in generating ketones from acid chlorides:

$$Bu^t\overset{\displaystyle O}{\overset{\|}{C}}Cl + LiPr^i + CuI \rightarrow Bu^t\overset{\displaystyle O}{\overset{\|}{C}}Pr^i$$

$$N \equiv C(CH_2)_{10}\overset{\displaystyle O}{\overset{\|}{C}}Cl + Li[CuMe_2] \rightarrow N \equiv C(CH_2)_{10}\overset{\displaystyle O}{\overset{\|}{C}}Me$$

A significant feature of cuprate reagents is their tolerance toward groups such as cyano or ester present in the acid chloride molecules.

Organopalladium Compounds

The discovery of the industrial synthesis of acetaldehyde (Wacker process) by the palladium catalyzed oxidation of ethylene was one of the great triumphs in metal mediated organic syntheses and has since stimulated continued efforts for searching other reactions assisted by palladium.

The complexities and subtleties encountered in organopalladium chemistry have impeded the adaptation of palladium mediated reactions in organic synthesis.

However, it is these features that have generated great awareness and interest in this field. Dramatic changes in selectivity (chemoselectivity, regioselectivity and stereocontrol) can often be achieved by apparently slighly modifications in experimental conditions, especially by choice of ligands. With the growing knowledge about factors responsible for such effects, their predictibility and hence, their applications have been increasing. The synthetic chemistry of palladium organometallics rests on two major types of intermediates: π-allyl systems (A) and, σ-palladium species (B).

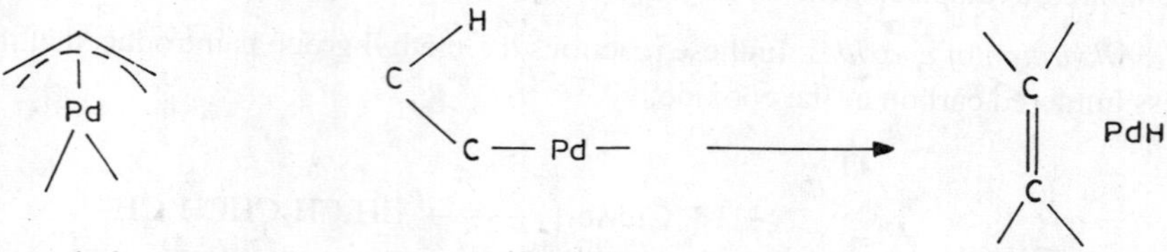

(A) (B)

In this section we shall limit ourselves only to such reactions which appear to have a high potential (either realized or promised) for application in organic synthesis and will not deal with catalytic hydrogenation (an established process in organic synthesis using nickel catalysts). Various strategies employed in the organic synthesis mediated by palladium organometallics can be categorized as follows:

(a) *Isomerization reactions* Nonconjugated dienes may be isomerized by using a variety of palladium complex, e.g:

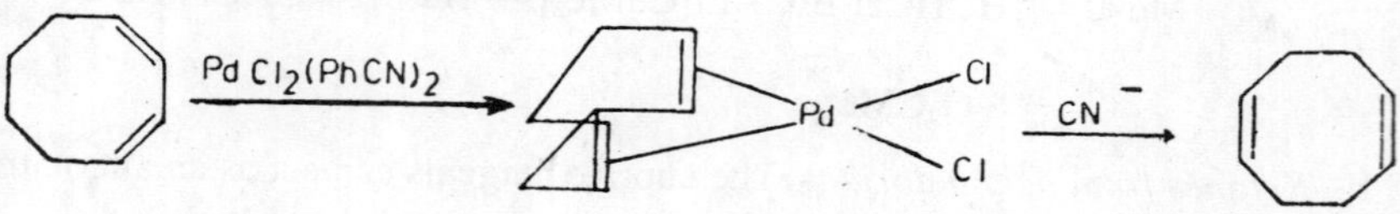

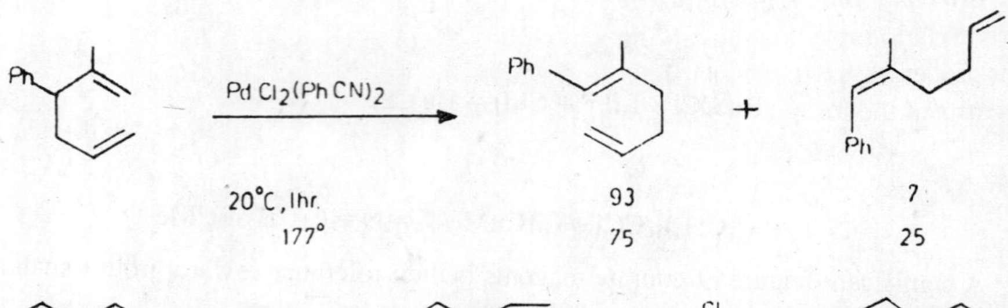

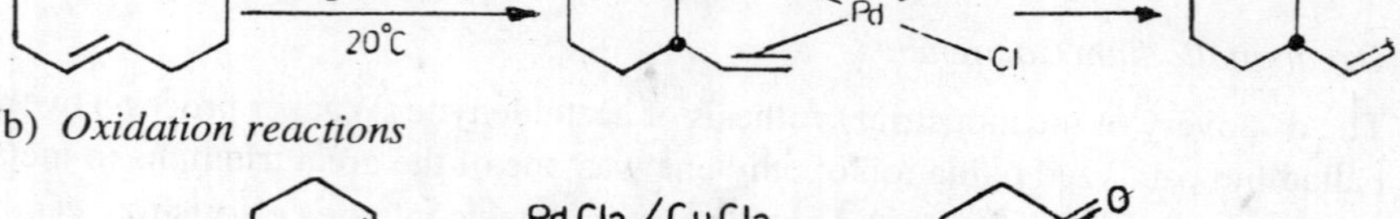

(b) *Oxidation reactions*

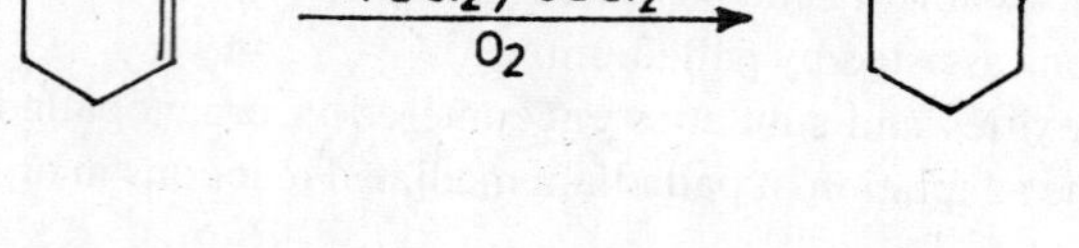

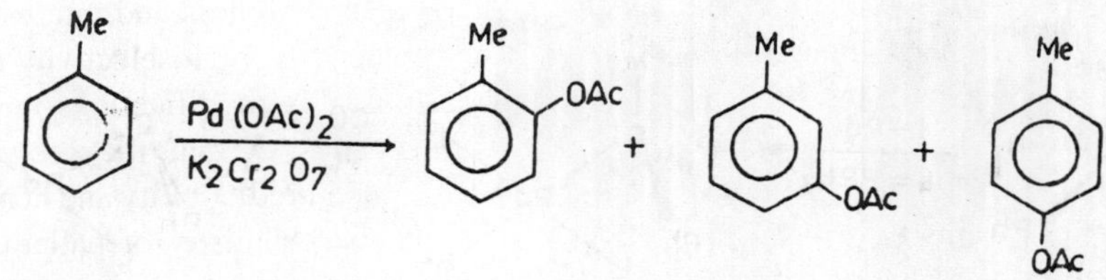

1 : 3 : 1

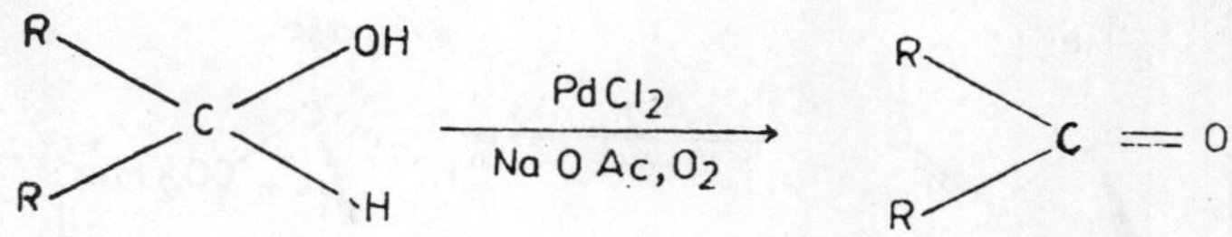

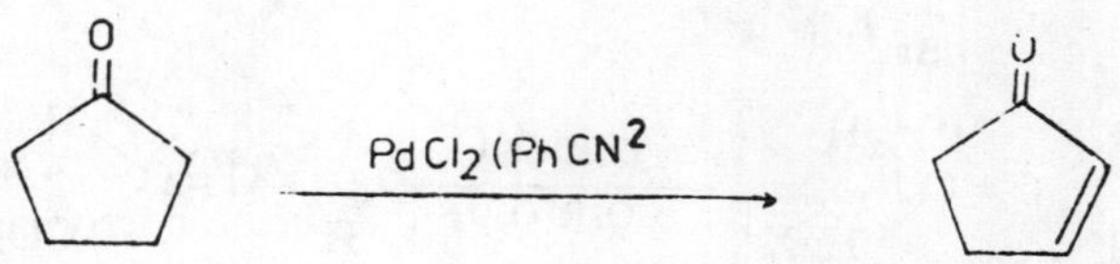

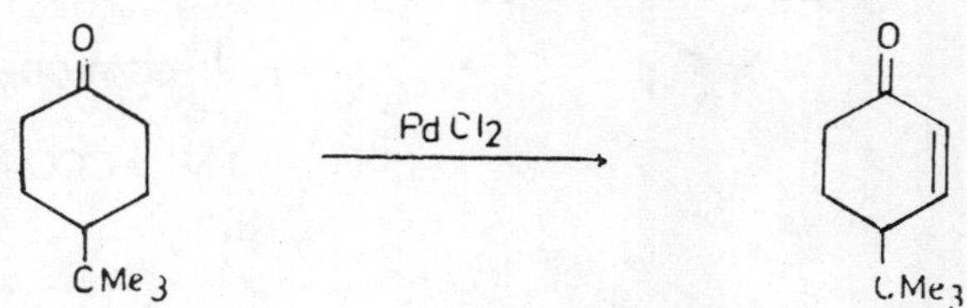

(c) *Carbonylation reactions (a major asset of organopalladium reactions)* These carbonylation reactions involving palladium compounds are emerging as important processes for the synthesis of fine organic chemicals and can be rationalized in terms of the formation of alkyl and acyl intermediates, e.g.,

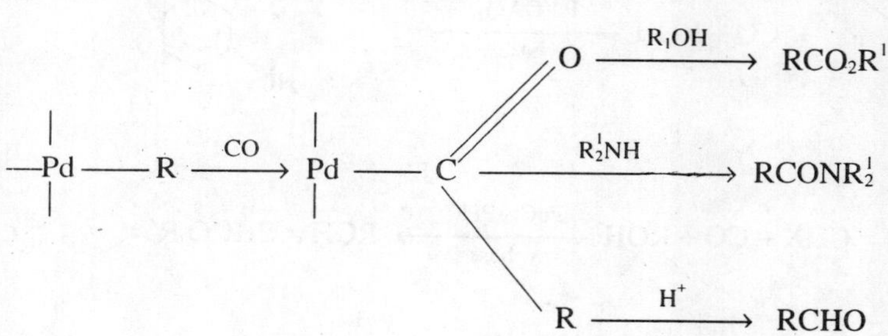

Other typical examples are:

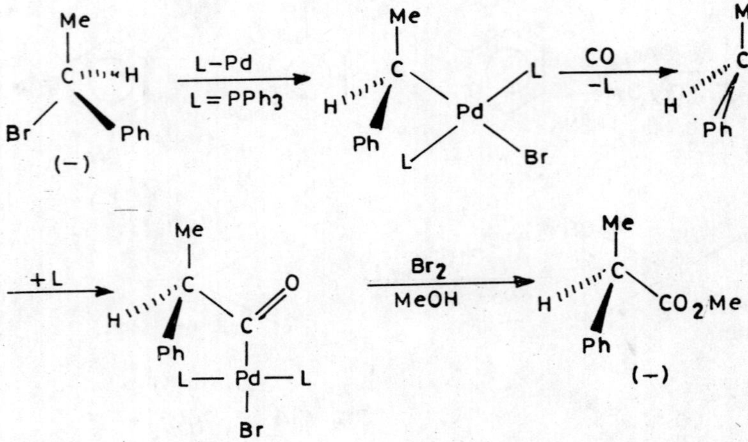

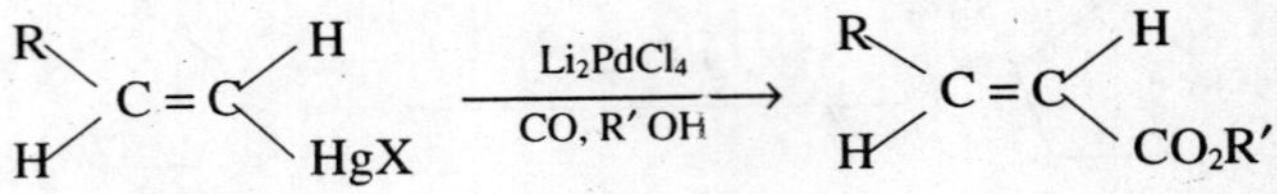

$$\text{PhC} \equiv \text{CHgX} \xrightarrow{\text{PdCl}_2} \text{PhC} \equiv \text{CPdCl} \xrightarrow{\text{CO/MeOH}} \text{PhC} \equiv \text{CCO}_2\text{Me}$$

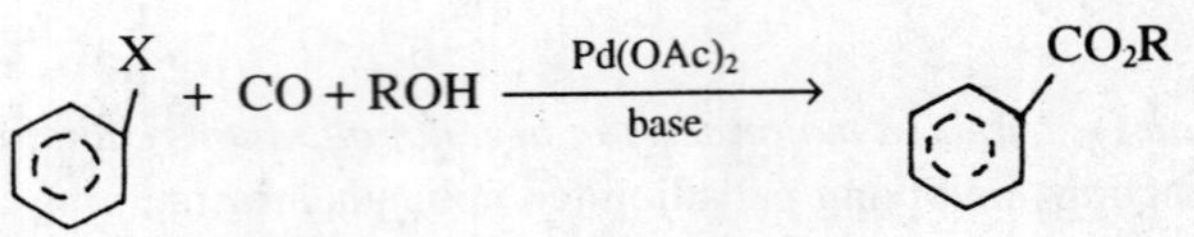

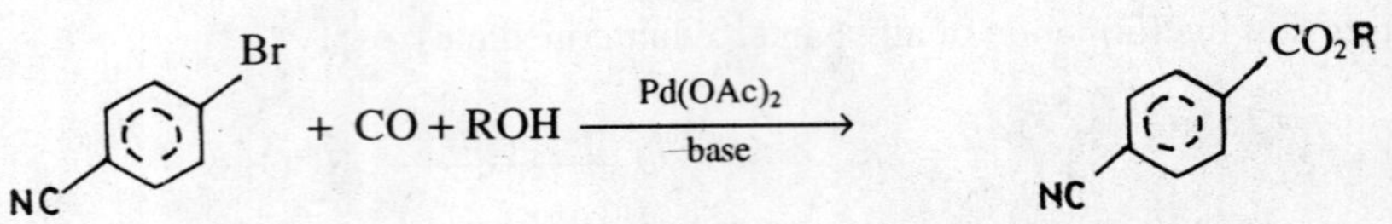

$$\text{RCH} = \text{CHX} + \text{CO} + \text{ROH} \xrightarrow[\text{base}]{\text{PdCl}_2(\text{PPh}_3)_2} \text{RCH} = \text{CHCO}_2\text{R}'$$

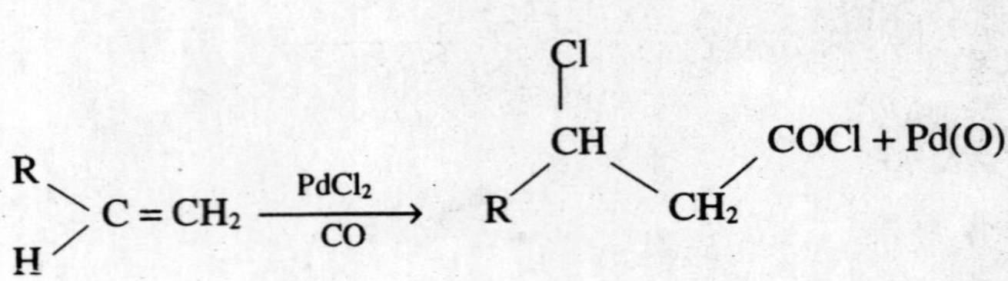

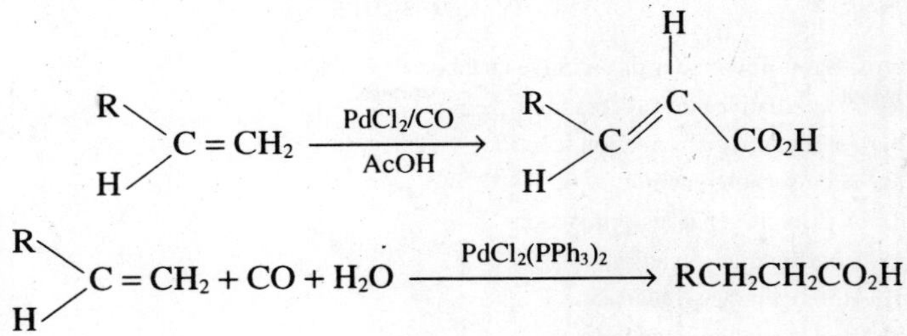

(d) *Nucleophilic addition reactions* Nucleophilic addition occurs readily to (alkene) $PdCl_2$ complexes. The intermediate η^1-alkyl derivatives are generally too unstable to be isolated and must be reacted further *in situ*. Nevertheless an intermediate η^1-alkyl derivatives has been isolated as illustrated in the following cases:

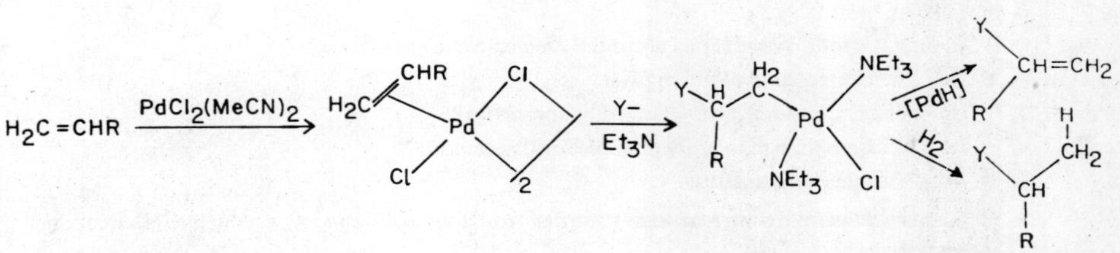

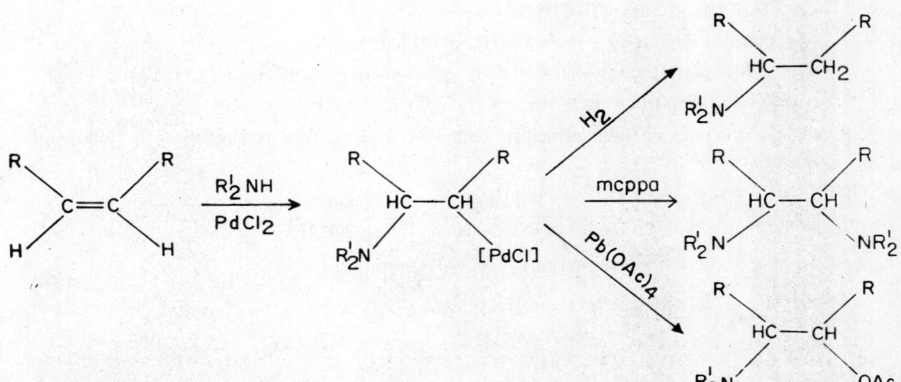

mcpba = *meta*-chlorobenzoic acid

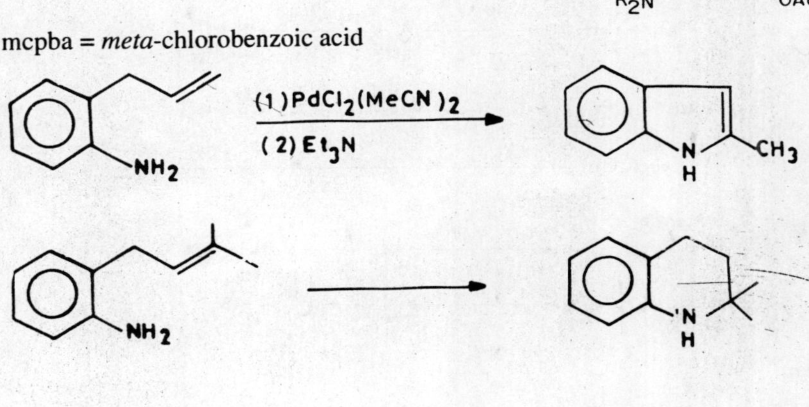

Study Questions

1. Define the following terms and give an example of each:
   (a) An oxidative addition reaction
   (b) An reductive elimination reaction
   (c) An insertion reaction
   (d) A β-hydrogen transfer reaction
   (e) Coordinative unsaturation
   (f) Homogeneous catalysis
   (g) Hydrogenation reaction
   (h) Catalytic cycle
   (i) A catalyst precursor
2. Discuss the reasons for the catalytic efficiency of transition metals.
3. Write a mechanism for the catalytic hydroformylation of $CH_3CH{=}CH_2$.
4. Propose a mechanism for the Wacker process, when the reaction is carried out with ethylene in acetic acid solvent.
5. Propose a mechanism for the polymerization of propene to polypropene by a Ziegler-Natta catalyst.
6. Attempt the following queries about the Ziegler-Natta catalyst:
   (a) The composition of the catalyst
   (b) Which alkenes are polymerized by this catlyst?
   (c) What modification(s) are possible for this catalyst?
   (d) The role of the catalyst.
7. Categorize the following reactions as formal oxidative-addition and/or reductive elimination reactions:
   (a) $Ni(CO)_4 + CH_2{=}CH{-}CH_2Cl \longrightarrow Ni(\eta^1-C_3H_5)Cl(CO)_3 + CO$
   (b) $RhCl(PPh_3)_3 + H_2 \longrightarrow Rh(H)_2Cl(PPh_3)_2 + PPh_3$
   (c) $(Ph_3P)_2(H)(Cl)Rh{-}CH_2CH_2CH_3 \longrightarrow (Ph_3P)_2RhCl + CH_3CH_2CH_3$
   (d) $[PdCl_3(C_2H_4)]^- + H_2O \longrightarrow CH_3CHO + Pd + 2HCl + Cl^-$
8. Sketch structures of the intermediate complexes and the final product(s) of the following reactions:

(a) 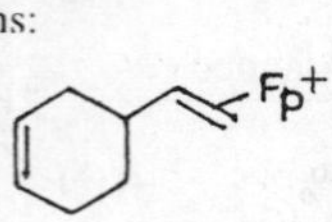 $\xrightarrow[CF_3COOH]{H_2,\ Pd/C} A \xrightarrow[CH_3COCH_3]{NaI} B$

$Fp^+ = Fe(CO)_2(\eta^5-C_5H_5)^+$

(b) $Fe_2(CO)_9 + H_2C{=}CH(OSiMe_3)$

$$\longrightarrow A \xrightarrow{H_2O} B \xrightarrow{MeNCO} C \xrightarrow{PPh_3} D$$

(c) $Ni(\eta^3-C_3H_5)_2 \xrightarrow{PR_3} A \xrightarrow[CO]{H_2C{=}CH_2} B + C$

(d) $Cr(CO)_3(\eta^6-C_6H_5Cl) \xrightarrow{NaOMe/MeOH} A \xrightarrow{I_2} B$

(e) $Na_2Fe(CO)_4 \xrightarrow[THF]{RX} A \xrightarrow{Cl_2} \xrightarrow{R'OH} B$

(f)  $\xrightarrow{PdCl_2(PhCN)_2} A \xrightarrow{CN^-} B$

(g) $C_6H_5C \equiv CHgX \xrightarrow{PdCl_2} A \xrightarrow{Co/MeOH} B$

(f) $$R(H)C=CHR \xrightarrow[PdCl_2]{R_2NH} A \xrightarrow{Pb(CH_3COO)_4} B$$
$$A \xrightarrow{H_2} C$$

9. Discuss briefly the possible mechanisms for the metathesis of propene.
10. Discuss the following:
    (a) Organometallic reactions related to homogeneous catalysis.
    (b) Oxidative-addition reactions of transition metal complexes.
    (c) Synthesis of organic compounds mediated by transition metal complexes.
    (d) Reactions of coordianted ligands

Supplementary Reading

1. '*Organic Synthesis by Means of Transition Metal Complexes*', by J. Tsuji, Springer-Verlag, Berlin (1975).
2. '*Organometallics in Organic Synthesis*', by E. Negishi, Vol. 1, John Wiley, New York (1980).
3. '*Organometallics in Organic Synthesis*' by J.M. Swan and D.St. C. Black, Chapman and Hall, London (1974).
4. '*Transition Metal Intermediates in Organic Synthesis*', by C.W. Bird, Academic Press, New York (1967).
5. '*Organotransition Metal Chemistry*: *Applications to Organic Synthesis*' by S.G. Davies Pergamon Press, Oxford (1982).
6. '*Silicon in Organic Synthesis*' by E. Colvin, Butterworths, London (1981).
7. '*Aspects of Mechanism and Organometallic Chemistry*', by J.H. Brewster, Plenum, New York (1979).
8. '*Homogeneous Catalysis, The Application and Chemistry of Catalysis by Soluble Transition Metal Complexes*', by G.W. Parshall, John Wiley, New York (1980), 2nd ed., by G.W. Parshall and S.D. Ittel, John Wiley, New York (1992).
9. '*Organometallic Mechanisms and Catalysis*, by J.K. Kochi, Academic Press, New York (1978).
10. '*Principles and Applications of Homogeneous Catalysis*', by M. Tsutsui and A. Nakamura, John Wiley, New York (1980).
11. '*Transition Metal Organometallics in Organic Synthesis*, edited by H. Alper, Vols. I (*Transition Metal Olefin Complexes*, by A.J. Birch and I.D. Jenkins, pp. 1-82; *Metal Carbene Complexes in Organic Synthesis, by C.P. Casey*, pp. 189-233) and II (*Alkyne Complexes and Cluster Derivatives and Synthetic Reagents* by D. Seyferth et al., pp. 1-63; *Arene Complexes in Organic Synthesis*, by G. Jaouen, pp. 65-120), Academic Press, New York (1976, 1978)
12. '*Comprehensive Organometallic Chemistry I*', edited by G. Wilkinson F.G.A. Stone and E.W. Abel, Vol. 7 (*Compounds of the Alkali and Alkaline Earth Metals in Organic Synthesis*, by B.J. Wakefield, pp. 1-110; *Organoboron Compounds in Organic Synthesis*, by H.C. Brown, pp. 111-142; *Hydroboration: General Considerations*, pp. 143-160, *Hydroborating Agents*, pp. 161-197, Hydroboration of Hydrocarbons, pp. 199-237, *Hydroboration of Functional Derivatives* of Alkenes, pp. 229-254, by M. Zaidlewicz; *Compounds of Aluminium in Organic Synthesis*, by J.R. Zietz, Jr., G.C. Robinson and K.L. Lindsay, pp. 365-464; *Compounds of Thallium in Organic Synthesis*, by A. McKillop and E.C. Taylor, pp. 464-513; Organosilicon Compounds in Organic Synthesis, by P.D. Magnus, T. Sarkar and S. Djuric, pp. 515-559; *Compoundss of Zinc, Cadmiun, Mercury and of Copper in Organic Synthesis*, by W. Carruthers, pp. 661-722) and 8 (*Organic Syntheses where Carbon Monoxide is the unique source of Carbon*, by R.P.A. Sneeden, pp. 19-100; *Addition of Hydrogen and Hydrogen Cyanide to Carbon-Carbon Double and Triple Bonds*, by B.R. James, pp. 285-369; *Alkene and Alkene*

Oligomerization, Cooligomerization and Tellomerization Reactions, by W. Keim, A. Behr and M. Roper, pp. 371-462; *Asymmetric Synthesis using Organometallic Catalysts,* by H.B. Kagan, pp. 463-498; *Alkene and Alkyne Metathesis Reactions,* by R.H. Grubbs, pp. 499-551, Pergamon, New York (1982).

13. '*Hydrogenation Reactions Catalysed by Transition Metal Complexes* by B.R. James, in *Advances in Organometallic Chemistry,* edited by F.G.A. Stone and R. West, Vol. 17, pp. 319-389, Academic Press, London (1979).

14. '*Ziegler-Natta Catalysis',* by H. Sinn and W. Kaminsky, in *Advances in Organometallic Chemistry,* edited by F.G.A. Stone and R. West, vol. 18, pp. 99-142, Academic Press, London (1980).

15. '*Homogeneous Catalysis of Hydrosilation by Transition Metals'*, by J.L. Speier, in *Advances in Organometallic Chemistry,* edited by F.G.A. Stone and R. West, vol. 17, pp. 407-444, Academic Press, London (1979).

16. '*Hydroformylation'* by R.L. Pruett, in *Advances in Organometallic Chemistry,* edited by F.G.A. Stone and R. West, vol. 17, pp. 1-56, Academic Press, London (1979).

17. '*Olefin Metathesis',* by N. Calderon, J.P. Lawrence and E.A. Ofstead, in Advances in *Organometallic Chemistry,* edited by F.G.A. Stone and R. West, vol. 17, pp. 449-488, Academic Press, London (1979).

18. '*The Fischer-Tropsch Reaction',* by C. Masters, in Advances in *Organometallic Chemistry,* edited by F.G.A. Stone and R. West, vol. 17, pp. 61-99, Academic Press, London (1979).

19. '*The Chemistry of Transition Metal Carbene Complexes and their Role as Reaction Intermediates'* by D.J. Cardin, B. Cetinkaya, M.J. Doyle, and M.F. Lappert, *Chemical Society Reviews,* vol. 2, pp. 99-144 (1973).

20. '*Advanced Inorganic Chemistry',* 5th edn., by F.A. Cotton and G. Wilkinson, pp. 1186-1223 and 1224-1282, John Wiley, New York (1988).

21. '*Inorganic Chemistry'* by K.F. Purcell and J.C. Kotz, pp. 962-979, W.B. Saunders Company London (1977).

22. '*Principles and Applications of Organotransition Metal Chemistry',* by J.P. Collman L.S. Hegedus, J.R. Norton and R.G. Finke University Science Books, Mill Valley, California (1987).

23. 'The Chemistry of the Metal-Carbon Bond', edited by F.R. Hartley and S. Patai, vol. 3, John Wiley Chichester/New York, Vol. I (1982); Vol. II (1985); Vol. III (1985) and Vol. IV (1987).

24. 'Metallo-Organic Chemistry, by A.J. Pearson, John Wiley, Chichester/New York (1985).

25. 'Homogeneous Catalysis with Compounds of Rhodium and Iridium', by R.S. Dickson, D. Reidel Publishing Company, Dordrecht/Boston (1985).

26. 'Homogeneous Transition Metal Catalysis', by C. Masters, Chapman and Hall, New York (1981).

27. 'New Pathways for Organic Synthesis', by H.M. Colquhoun, D.J. Thompson, and M.V. Twigg, Plenum Press, New York (1984).

28. 'Catalytic Chemistry' by B.C. Gates, John Wiley, New York (1992).

29. '*Comprehensive Organometallic Chemistry II*', edited by E.W. Abel, F.G.A. Stone and G. Wilkinson, Vol. 11 (Volume editor A. McKillop, *Main-group Metal Organometallics in Organic Synthesis*, pp. 1-601), Pergamon, New York (1995).

30. '*Comprehensive Organometallic Chemistry II*', edited by E.W. Abel, F.G.A. Stone and G. Wilkinson, Vol. 12 (Volume editor L.S. Hegedus, *Transition Metal Organometallics in Organic Synthesis*, pp. 1-1232), Pergamon, New York (1995).

7

Biological Applications and Environmental Aspects of Organometallic Compounds

7.1 INTRODUCTION

As outlined in the first chapter, pronounced biological effects of organometallic compounds were recognized almost from the beginning of their discovery; salient chronological landmarks in this direction can be indicated as below:

1837 Bunsen's first report on cacodyl. "Arsenic rooms" also began receiving attention of Gmelin and others about this time.

1849 Preparation of $Zn(C_2H_5)_2$ and $Zn(CH_3)_2$ by Frankland and his observations on the poisonous nature of their vapours.

1890 Mond prepared $Ni(CO)_4$, leading to industrial uses of metal carbonyls and concomitant health problems.

1891 Gosio established the role of alkylarsenics in the poisoning caused by "arsenic rooms".

1908 Ehrlich laid the foundation of chemotherapy by the use of organoarsenic compounds.

1914-1918 World War I and the use of Lewisite (β-chlorovinyl-dichloroarsine) and other organoarsenicals as poison gases.

1923 Introduction of tetraethyllead as a gasoline additive and recognition of its toxicity.

1933 Challenger reported the formation of $As(CH_3)_3$ and $Te(CH_3)_2$ by moulds, which could be considered as the starting point of the recognition of the process of "biological methylation" and its potential impact on environment.

1954 "Stalinon" disaster in France traced out to $Sn(C_2H_5)_2I_2$, which was used as a medicine under the trade name of "Stalinon".

1955-1961 The first cases of "Minimata" disease (methylmercury poisoning). Coenzyme form of Vitamin B_{12} containing Co—C bond was reported.

1968	Realization of the possibility of generation of methylmercury compounds from inorganic mercury by biological methylation through microorganisms.
1978-1984	Better understanding about wide occurrence and speciation of organometallic compounds in the environment (cf. references 5-8, 12, 15, 16 and 18 of Supplementary Readings).

The pronounced biological effects of organometallics led to their wide applications in fields like medicine and agriculture. The ever-increasing practical uses of organometallics as biocides in agriculture, as antiknock agents in automobiles and as catalytic agents in a number of industrial (specially biotechnological) processes have led to an increasing concern about their environmental aspects also.

In this chapter a brief account of the applications of organometallics in medicine, agriculture and industry will be followed by a few salient features of their role in environment. While the initial anxiety about the potential dangers caused by most toxic organomercury and lead moieties does not continue to be as intense as it was two decades ago, yet the possibility of alkylation (particularly methylation) of a large number of inorganic species and the extra-ordinary stability of these alkylated species under ambient conditions have added novel dimensions to the interest in the role of the organometallics in the environment.

It might not be out of place to emphasize that the work in these directions is still rather slow due to

(i) the difficult analytical problems involved in detecting and estimating such minute (often less than few parts per trillion) quantities in an extremely complex mixture;

(ii) the inter- and multi-disciplinary cooperation essential for an effective study of such problems; and

(iii) the heavy costs (sometimes of the order of millions of dollars/rupees) involved in providing a sound data base for the toxicity of each organometallic species.

7.2 ORGANOMETALLICS IN MEDICINE

As indicated earlier also, systematic investigations on the chemotherapeutic applications of organometallics began with the classical work of Paul Ehrlich on organoarsenicals although some organomercurials had been off and on used before these in the treatment of syphilis and other infectious diseases. The salient medical uses of a few organometallics are summarized group-wise in the following section.

7.2.1 Organometallics of Group I to III Elements

Most of the organic derivatives of elements of groups I to III used commonly in medicine are not truly organometallic products and in almost all of these (e.g. calcium aspirin), the biological activity is due to the organic moiety present; the presence of the metal sometimes facilitates a slow release of the active organic component. However, some silver and gold containing organic compounds do exhibit medicinal activity which is ascribed to the metal. As an example of a silver compound being currently used in medicine is silver sulphadiazine (2-sulphanilamidopyrimidinesilver (I)). Gold-organic compounds containing Au-S bonds were earlier investigated for treatment of tuberculosis, but are now mainly used as anti-inflammatory drugs.

Of the truly organometallic compounds of the first three groups, only organomercury and -boron compounds deserve special mention.

7.2.1.1 Organomercury Compounds in Medicine

Although the use of organomercurials, which were the first organometallics to be used in chemotherapy of syphilis has been superseded by modern antibiotics, several of these are widely used as antiseptics (Fig. 7.1) and in diuretics (Fig. 7.2):

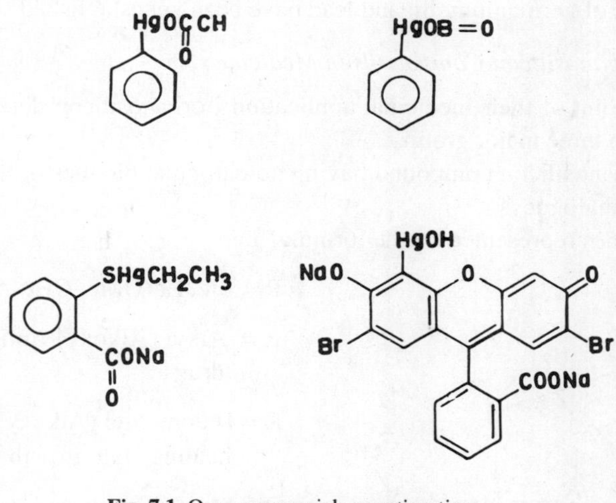

Fig. 7.1 Organomercurials as antiseptics.

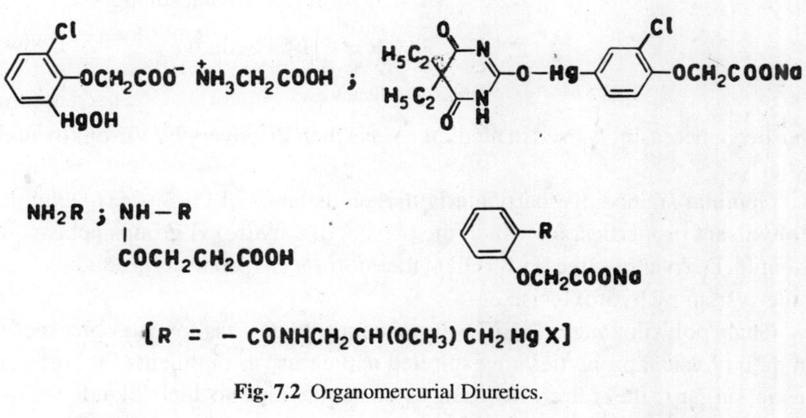

Fig. 7.2 Organomercurial Diuretics.

7.2.1.2 Organoboron Compounds in Medicine

Besides other medicinal uses, numerous organoboron compounds (mainly boronic acid, $RB(OH)_2$, series) have been investigated for cancer therapy. Boron-10 (19.6% natural abundance) has an unusually large cross-sectional capture area for neutrons. Capture of a slow moving neutron (*ca.* 0.025 eV energy) causes a fission, releasing a fair amount of energy, which would be lethal to the cells. With the concentration of boron species in tumour tissue, this fission should provide a way for its destruction. Although aliphatic boronic acids have shown promise, yet these have high toxicity. Efforts are, therefore, being concentrated to aryl-boronic acids. The choice of boronic acids appears to be due to their marked resistance to oxidative degradation in biological systems.

7.2.2 Group IV Organometallics in Medicine

Amongst the group IV organometallics, the applications of organosilicon derivatives have been increasing at a fast rate, whereas very few uses of organometallic compounds of germanium, tin and lead have been yet established.

7.2.2.1 Organosilicon Compounds in Medicine

From the point of their medicinal applications, organosilicon derivatives can be divided into three major groups:

(i) Organosilicon compound having no carbon anologues or having biologically inert analogues.

Silatranes represented by the formula:

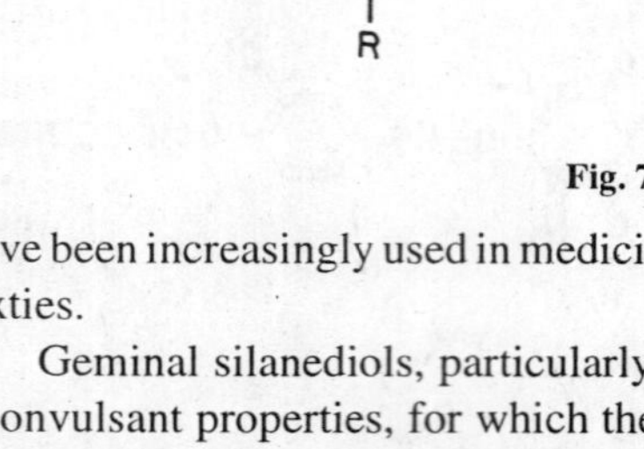

R = Me, Et (OMe, OEt) Anticoagulants

R = Alkyl (Alkoxy) Anti-tumour drugs

R = Halomethyl (Alkoxy) Wound-healing Hair growth factors

R = Me/Analgesic and sedative

R = Ph Local Anaesthetic

R = CH_2Cl Sedative

Fig. 7.3 Silatranes

have been increasingly used in medicine since their discovery by Voronkov in early sixties.

Geminal silanediols, particularly diphenylsilanediol $Ph_2Si(OH)_2$, exhibit anticonvulsant properties, for which the presence of hydroxyl groups appears to be essential. Derivatives like $Ph_2Si(OR)_2$, therefore, act as prodrugs producing the diol in the system by hydroxlysis.

Chain polysiloxanes, $(R_2SiO)_n$ do not appear to possess any biological activity, but their physical properties are exploited in their use as ointments for burns or for plastic surgery; these are also used as cosmetics due to their blandness, water repellency and lubricity. Cyclic siloxanes, e.g., 2, 6-*cis*-diphenyl hexamethylcyclotetrasiloxane (called cisobitan) exhibits oestrogenlike antigonadotropic activity in rats and analgesic effects in humans.

It may not be out of place to mention an interesting use of the compounds,

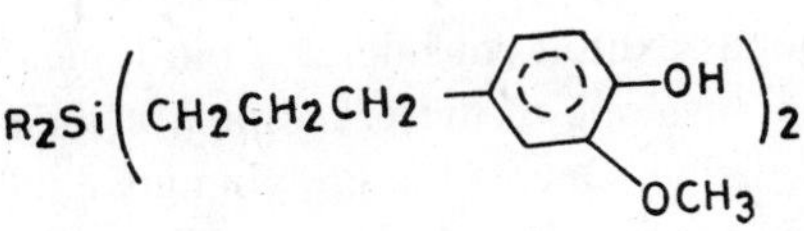

as dentistry cements.

(ii) *Silylated derivatives of bioactive organic compounds* The presence of $SiMe_3$ group, for example in testosterone, renders it soluble in non-polar solvents. This lipophilicity facilitates its penetration across the lipophilic membrances and the easy hydrolysability of the O–Si bond should make it behave as an effective prodrug.

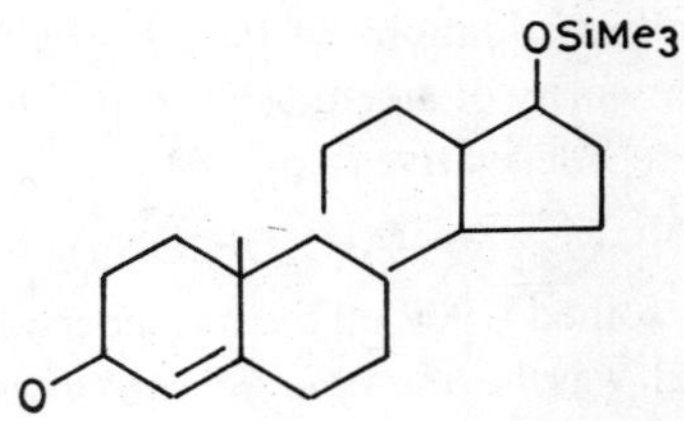

Fig. 7.4 Testosterone (Silylated).

(iii) *Sila-pharmaca* The replacement of carbon atom(s) in organic drugs by silicon has aroused considerable interest. For example, in view of the possibility of (p → d)π interactions of silicon with aromatic rings, *p*-trimethylsilylphenoxy (O—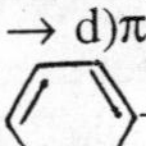—$SiMe_3$) behaves as a better leaving group than its carbon analogue *p-tert*-butylphenoxy O—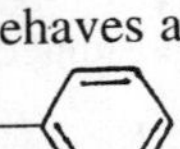—CMe_3 group in phosphate esters of the type, $(EtO)_2P(O)(OC_6H_4CMe_3)$, which makes the silicon product a much better inhibitor of cholinesterases.

The replacement of C—O—C or CH in an organic drug by Si—O—C and Si—H bonds respectively imparts in the silicon products >Si—OC< or >Si—H, a new property of hydrolysability without generally affecting the bioactivity of the original drug. The silanols obtained thus tend to exhibit weaker histaminolytic and stronger anticholingesic properties. For example, the silicon analogues of antihistamines of the benzhydryl ether type,

$$\begin{matrix} XC_6H_4 & & CH_3 \\ & E & \\ C_6H_5 & & OCH_2CH_2NR_2 \end{matrix}$$

(where E=C or Si) renders them effective as short duration drugs in treatment of allergies, as in about 10–15 minutes these are hydrolyzed to inactive silanols. On the other hand although compounds containing >Si—H bonds may be too hydrolyzable to be suitable for drug use in their own right, yet their lipophilicity coupled with this hydrolyzability makes them potentially useful precursors (prodrugs) of the bioactive silanols.

7.2.2.2 Organometallic Derivatives of Germanium, Tin and Lead in Medicine

Although a large number of the title compounds exhibit considerable bio-logical activity (*cf.* use of organotins as pesticides), yet none of these, in spite of their potential, have been yet used actually in medicine. One of the growing areas of interest regarding their potentialities is in antitumour activity (*cf.* Section 7.2). The use of organotin derivatives received a serious set back by more than 100 deaths being caused by the use (1954) of 'stalinon' (a preparation of diethyltin iodide and isolinoleic esters) in France for staphylococcal infections.

7.2.3 Group V Organometallics in Medicine

The science and practice of chemotherapy received a great impetus early in its history by the systematic investigations of Paul Ehrlich on organoarsenicals as antiparasitic agents. The majority of arsenic (V) compounds used in the treatment of trypanosomiasis were derivatives of Atoxyl (i.e., mono-sodium *p*-aminophenylarsonate, $H_2N-C_6H_4-AsO(OH)ONa$).

The name Atoxyl was coined (1907) in the misunderstanding that the compound was not toxic. Ehrlich's goal was the '*therapia sterilans magna*' – a compound whose single massive dose could destroy the infecting parasite. Early in his investigations, he realized that "*Corpora non agunt nise fixata*" (bodies do not act unless fixed) and therefore, he set himself to the synthesis of a variety of new organoarsenic derivatives with different substituents on the aromatic ring as well as on the arsenic atom. Of all these new compounds, the 606th of his preparations proved to be most efficacious and was named 'Salvarsan' by Ehrlich. For a long time, this compound was believed to be an arsenic analogue of azobenzene, but it is now represented by the formula 7.5(a).

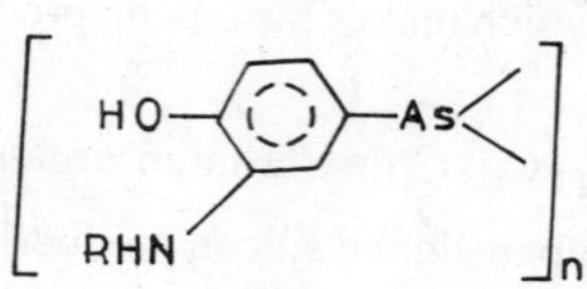

Fig. 7.5 (a) Salvarsan, when R = H, and
(b) Neosalvarsan, when R = CH_2OSO_2Na.

'Salvarsan' was widely used as a pharmaceutical for treatment of infections by spirochetes or trypanosomes, especially syphilis. Sometime later, the water soluble derivative 'Neosalvarsan' (no. 914) was preferred. However, both of these had to be stored under nitrogen in air-tight ampoules as they were oxidised by oxygen.

Research on organoarsenicals continued after Ehrlich's death in 1915. It was soon realised that salvarsan was oxidized in the body to 'oxopharsine' hydrochloride (7.6), which was the form in which this pharmaceutical under the trade name 'Mapharsen' began to be preferred, particularly in view of its stability to air. Mapharsen and allied arsenicals continued to be the main drug for treatment of syphilis until the 1940's, when these were gradually replaced by penicillin.

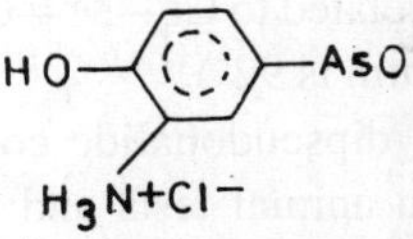

Fig. 7.6 Mapharsen.

Although organoarsenicals are no longer used in the treatment of syphilis, they continue to receive attention for their anti-parasitic activity. For example, the compound,

$$H_2NC(O)C_6H_4As(SCMe_2 \cdot CHNH_2 \cdot COOH)_2$$

has been reported as an efficacious trypanocide and filaricide in 1973.

Two derivatives of phenyl-stibonic acid, i.e., stibamine (7.7a) and stibosan (7.7b) received limited therapeutic uses, but organobismuth compounds have never found acceptability in medicine, although derivatives of bismuthyl BiO^+ and antimonyl SbO^+ ions continue to find applications in the treatment of various parasitic infections:

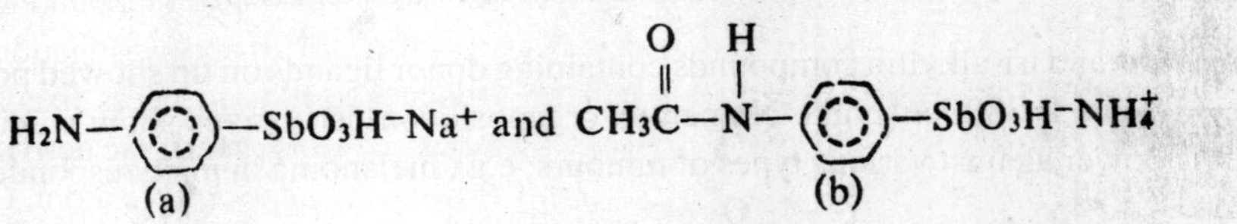

Fig. 7.7 (a) Stibamine and (b) Stibosan.

7.2.4 Anti-tumour Drugs

Although the activity of organolead compounds such as Pr_3PbF, Bu_3PbBr, $(C_6H_{11})_6Pb_2$ and Ph_4Pb towards mouse cancers had been noticed by E. Krause as early as 1929, the discovery of significant anti-cancer activity of cisplatin (*cis*-dichlorodiammineplatinum (II), Fig. 7.8(a) has almost half a century later led to an extensive investigation on the biological properties of compounds with the general formula $(C_5H_5)_2MCl_2$ (Fig. 7.8b), which has some resemblance with the cisplatin

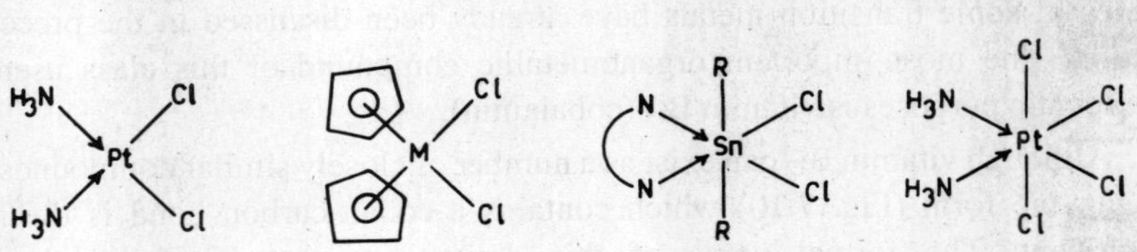

Fig. 7.8(a); (b); (c); (d) Cisplatin and similar compounds in cancer therapy.

molecule. As reviewed by Lippard in 1982, derivatives of the above class (with M == Ti, V, Nb, Mo and W) show a high activity against Ehrlich ascites tumour in mice by a mechanism similar to that of cisplatin (i.e., by binding to strands of DNA leading to its denaturation). It has been suggested (1980) that the anti-tumour activity in such compounds is related to Cl—M—Cl bond angle being less than 95° (*cf.* Cl—Pt—Cl angle in cisplatin is 92°).

Diorganotin dihalide and dipseudohalide complexes of type 7.8c have exhibited anti-tumour activity in animal tests and in this respect, these appear to resemble the active platinum (IV) complexes of the type 7.8d.

Two organogermanium compounds, i.e., a derivative of 1, 1-diethyl-l-germacyclohexane (known as spirogermanium; Fig. 7.9a) and carboxyethyl-germanium sesquioxide (Fig. 7.9b) have recently (1981-82) shown anti-cancer activity in clinical trials;

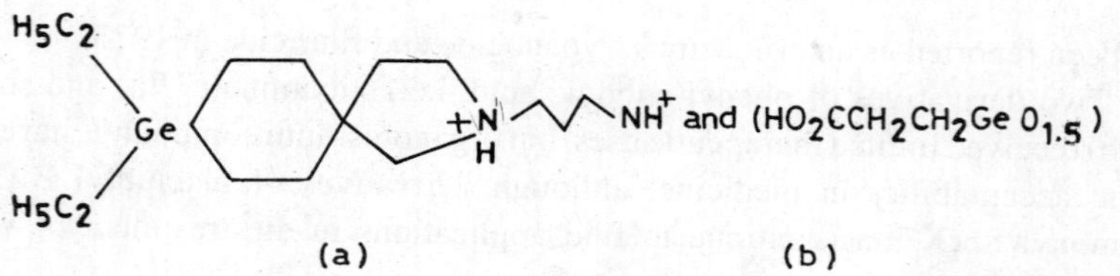

Fig. 7.9 Organogermanium derivatives in cancer therapy.

Di- and tri-alkyltin compounds containing donor ligands on tin showed potency against P-388 leukemia. Numerous organosilicon derivatives show carcinostatic power against various types of tumours; e.g., melanoma in mice responded to treatment with $(CH_3)_3Si(CH_2)_4\overset{\overset{\large O}{\|}}{C}NHC_6H_5$.

As research work on controlling cancer is receiving increasing attention, it is likely that more efficacious organometallic moieties would gradually play a more significant role in cancer therapy.

7.2.5 Transition Metal Organometallics and Vitamin B_{12}

The rapid advances in the chemistry of organotransition metal compounds since the 1950's have not been accompanied by a corresponding activity in their biological properties. The anti-tumour properties of biscyclopentadienyl complexes of some transition metals have already been discussed in the preceding section. The most important organometallic compound of this class used for therapeutic purposes is vitamin B_{12} (cobalamin).

Although vitamin B_{12} can exist as a number of closely similar compounds, the co-enzyme form (Fig. 7.10) which contains a cobalt-carbon bond is the most significant. The formal name of this derivative is α-(15, 6-dimethylbenzimidazolyl) cobalt-5′-deoxyadenoxylcobamide, but it is generally known by a more common name, i.e., 5′-deoxyadenosylcobalamin. According to substituent

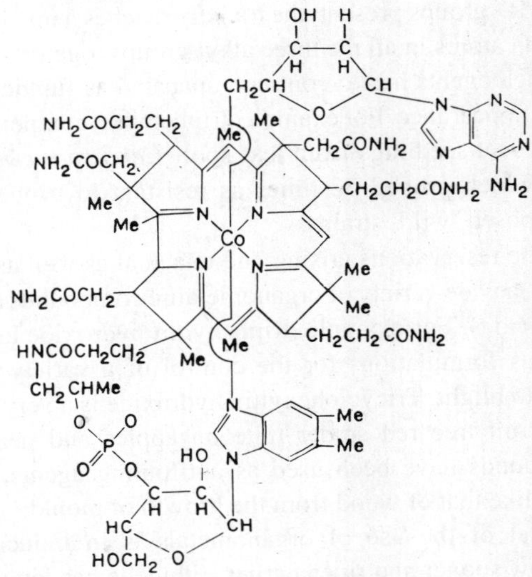

Fig. 7.10 Vitamin B_{12}.

which replaces the deoxyadenosyl group, other forms of vitamin B_{12} are given names like methylcobalamin, aquocobalamin and cyanocobalamin, the last form being the one in which vitamin B_{12} is administered generally.

Vitamin B_{12} was originally termed as the antipernicious anaemia factor and was first isolated form liver. Perhaps methylcobalamin, in which cobalt is bonded to a methyl group, has been the B_{12} derivative which has received the greatest attention. It occurs in human milk, enhances the immune system in rats and has been shown to protect against growth of nervous lesions in rats suffering from diabetes. However, the most important role of methyl-cobalamin is its capacity to act as a methyl transfer agent; it has been shown to be an intermediate in the microbial methylation of arsenic and mercury.

7.3 ORGANOMETALLIC COMPOUNDS IN AGRICULTURE AND HORTICULTURE

A large number of organo-mercury and -tin compounds have been used as fungicides, algacides, herbicides and pesticides in agriculture.

For a long time, organomercurials e.g., methylmercury dicyandiamide increased so much in popularity that many hundreds of tons of these were used in late 1950's only in seed dressing before storage. However, with the occurrence of Minimata tragedy, these are getting out of use and are being replaced by organometallic derivatives of tin and some other metal(loid)s.

In spite of a considerable amount of work on the fungitoxicity of organometallics say organotin compounds, knowledge in this direction continues to be rather empirical. For example, while comparing the toxicity of a large number of mixed trialkyltin acetates, $RR'R''Sn(OOCCH_3)$ it has been shown that irrespective of the actual R, R′ and R″ groups present, the toxicity reaches a maximum when the total number of carbon atoms in all the three alkyl groups together is 9 to 12.

Work on such agents has to continue unabated as fungicide resistant strains often make their appearance. For example, triphenyltin compounds have been used to protect sugar beet seedlings from leaf spot (*Cercospora beticola*), but strains were observed which proved 32 times as resistant to triphenyl compounds as previously unexposed 'wild' strains.

In spite of the reservations arising due to actual as well as potential environmental pollution, a wide variety of organometallic compounds are finding increasing uses as pesticides. For example, triphenyltin hydroxide and acetate are both used in numerous formulations for the control of a variety of fungal growths particularly potato blight. Tricyclohexyltin hydroxide is a very effective acaricide for controlling fruit tree red spider mite on apples and pears. Besides, these organotin compounds have been used as antifouling agents, i.e., in protecting various surfaces like that of wood from the growth of mould.

Another facet of the use of organometals is in inducing the fruits and vegetables to grow sooner and ripen earlier. Ethylene has long been used for this purpose, but the compound *ethephon* (2-chloroethylphosphonic acid) has become a highly preferred organometal (loid) for this purpose.

It has been shown that 2-chloroethylphosphonic acid hydrolyses in situ with the generation of ethylene:

$$ClCH_2CH_2PO_3H_2 + H_2O \rightarrow C_2H_4 + HCl(aq) + H_3PO_4(aq)$$

Organosilicon compounds like *alsol* $[ClCH_2CH_2Si(OCH_2CH_2OCH_3)_3]$ also appear through a similar hydrolytic mechanism producing ethylene. A boron analogue $ClCH_2CH_2B(OR)_2$ has also been proposed as a plant growth regulator. Carbethoxygermanium sesquioxide has also been shown to stimulate the growth of rice seedlings and other plants. Silicone oils have been applied to the surface of plant leaves to control the loss of water through transpiration thus aiding agriculture in arid zones.

In addition to plant growth accelerators, organometallic derivatives are also used as herbicides to control (i.e., retard or destroy) the undesired plants (weeds) which compete with useful food or other plants. In addition to the popular phosphorus derivatives like N-phosphonomethylglycine $(HO_2CCH_2N\ HCH_2PO_3H_2)$, methylarsenic compounds like monosodium salt of methylarsenic acid have been widely used (under the trade name of Daconate). The series of organotins, $(n\text{-}C_4H_9)_3SnCH_2R$ (where R is a quaternized amino group) showed herbicidal activity with no phyto-toxicity. Trin-butyltin fluoride was shown to be selective in controlling weeds without demage to corn or rice.

7.4 ORGANOMETALLICS IN INDUSTRY

The commercial usefulness of the organometallic derivatives of the main group elements, such as the tetramethyl- and tetraethyl-lead additives for petrol and the silicon polymers, is now well-established. The use of transition metal organometallics as catalysts in a number of process (e.g., hydroformylation, hydrogenation, hydrosilylation and oligomerization of alkenes and alkynes) is growing at a phenomenal pace (cf. Section 6.4.2). Nitrogen fixation is another important process of promise in which organometallics are expected to play a vital role.

Nitrogen is essential to life, but it is an inert gas (dinitrogen) which must be converted to a more amenable form (ammonia or nitrate) before it can be used by plants and animals for their growth. This conversion, called nitrogen fixation, forms an essential part of the nitrogen cycle and replenishes the supply of usable nitrogen in soil, which would otherwise be lost to the atmosphere by denitrification. The fixation of nitrogen by organometal species is of great topical interest. Some organometallic compounds have been able to activate molecular nitrogen and form complexes. The following is a well known example:

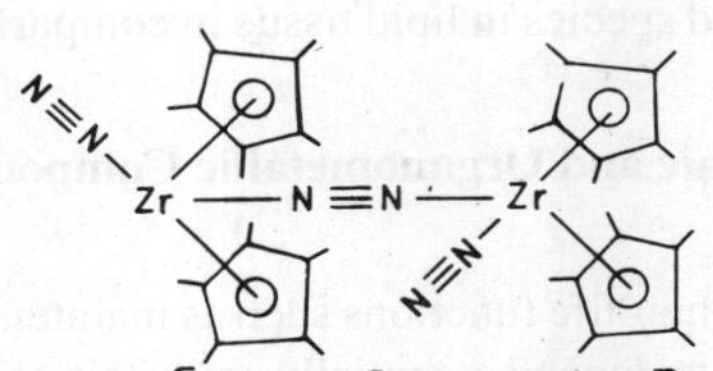

$[\{ZrN_2(\eta^5-C_5Me_5)_2\}_2N_2]$

7.5 ENVIRONMENTAL ASPECTS OF ORGANOMETALLIC COMPOUNDS

Until the mid-1950's any suggestion about the existence of organometallic compounds in the environment would not have received much consideration. The organometallics were generally artifacts created in the laboratories and were in most cases, highly reactive towards water or/and oxygen and hence, their survival under natural conditions except for brief periods was not considered feasible. Challenger's work (1933) indicating the formation of $As(CH_3)_3$ through biological intermediation remained unknown or unheeded. The Minimata tragedy (1955) and the interest in environment generated as its result forced this attitude to change. Investigators soon (1968) confirmed that methyl mercury compounds, could be formed through biological processes and remained in natural waters for long periods of time.

Studies on the environmental aspects of organometallic chemistry deal with the transformation of these compounds, i.e., the formation and destruction of metal-carbon bonds, transfer reactions and redox reactions, etc. These can be given a generic name '*speciation*'. It is now well established that in addition to purely chemical reactions, micro-organisms can transform many inorganic and organic compounds into organometallic species. Further, under natural conditions organometal(loid)s tend to move from one location to another again either by purely physicochemical (*abiotic*) or through the intermediation of organisms (*biotic*).

With the fastly increasing applications of organometallics in different directions, e.g., as biocides, or as gasoline additives, the possibility of their availability in the environment also increases. In addition to their toxic effects during actual use, attnetion has to be focussed on their toxicity and reactions (even at points other than those of their actual application) of their decomposition products persisting for considerable lengths of time under natural ambient conditions. A special feature of this last aspect is the potential of their decomposition products themselves forming new organometallic moieties (e.g. by alkylation) in the environment.

The possibility of the conversion of inorganic species like mercury (II) salts to organometallic derivatives (e.g., half-life of methylmercury species in human body has been reported to be ~70 days compared to 4-5 days for mercuric salts) renders the environment exposed to higher toxicity for longer periods than that from the inorganic species from which they might have been derived. This higher toxicity can usually be explained on the basis of characteristics such as greater solubility of the methylated species in lipid tissue in comparison to that of metallic ions.

7.5.1 Cycling of Inorganic and Organometallic Compounds in the Environment

For being able to perform their life functions such as maintenance, self-growth and reproduction, biota appear to depend essentially on certain elements and preferably tend to avoid some other elements in view of their toxicity (*cf.* Fig. 7.11):

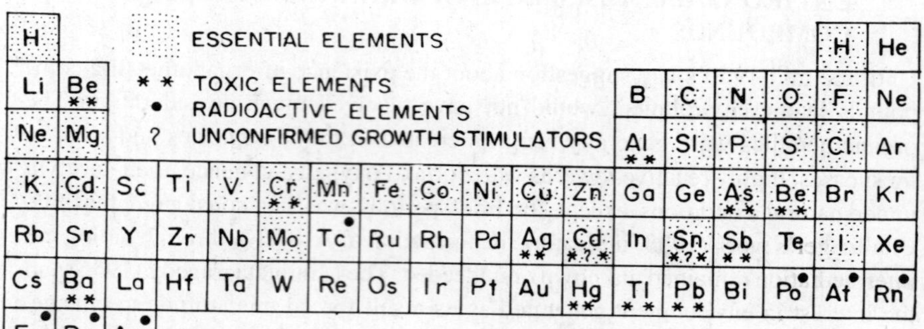

Fig. 7.11 Essential and Toxic Elements in the Environment.

The essential elements shown in figure 7.11 also exhibit toxicity generally at higher concentrations. These twentyseven essential and eleven toxic elements with four common members between the two groups form a large variety of compounds.

Out of the eightysix naturally occurring elements, the majority are metallic in character. These are locked up in nature in the form of insoluble inorganic compounds in rocks and sediments. Before biota can use them, these have to be mobilized by weathering processes involving abiotic chemical (degradation of organic material and redox) reactions and biologically mediated interactions. This mobilization generally converts insoluble substances into soluble ones which are then converted into coordination, organometallic and organometalloidal compounds. These then undergo interactions (inimical or beneficial) with biota either on the outer surface of cells (if exposed) or after their transportation through membranes into the cellular portion. The ensuing biochemical reactions produce new chemical species which may be excreted into the environment and thus become available to other biota. Alternatively, these organometallic moieties associated with cellular material may be deposited upon the death of the organism in sediments and thus rejoin the pool of non-available substances awaiting mobilization. This cycle is depicted in Fig. 7.12:

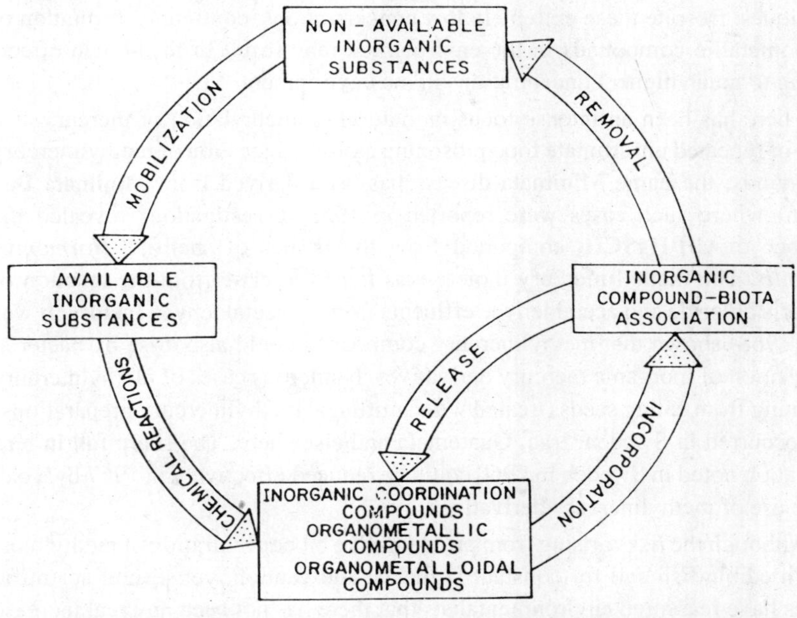

Fig. 7.12 Cycle of Inorganic and Organometallic Materials in the Environment.

Environmental processes involving organometallics are a complex mixture of biological and abiotic processes which are often interdependent and perhaps, with synergetic interactions. Although the transformations and fate of organic chemicals have been extensively studied, the true role of metals and metalloids in the biogeochemical cycles is just beginning to be explored.

In view of the largely aqueous nature of our environment, hydrolyticaly stable organometallic species, e.g., those of Co, Pd, Pt, Au, Hg and of most metalloids assume a higher significance. Methylation being a common reaction in biological systems, may also proceed abiotically and generate methyl-metal (metalloid) bond. Examples of such compounds, which have actually been detected in our environmental samples, are methyl- and dimethyl-mercury, dimethyltin and tetramethyllead species along with dimethyl-selenium, selenoamino acids, methylarsonic and dimethylarsinic acids. Out of these, the first three are of somewhat greater significance and would be discussed in a little greater detail along with the methylation process. It may, however, be clarified that compounds with methyl groups as the organic moieties are not the only organometallic compounds found in nature and derivatives with more complex moieties have also been identified in biota.

The main practical problem in this field has been the extremely low levels at which some of these actually occur in the environment. Methylmercury is found in concentrations less than parts per million level ($\mu g\ g^{-1}$) and the tin derivatives, at parts per trillion ($pg\ g^{-1}$) level. Detection of such minute quantities in natural environment presents extremely difficult analytical problems and doubts on the analytical aspects have also been often raised and lead to more refinement of techniques. Despite these extremely low concentrations, continuous formation of organometallic compounds in the environment may result in feed-chain effects leading to much higher concentrations in the environment.

There has been an intense focus on natural biomethylation of mercury as a result of repeated unfortunate food piosoning cases by ingestion of methylmercury compounds; the name Minimata disease has been derived from Minimata Bay (Japan) where such cases were reported in 1953. Investigations revealed the presence of CH_3HgSCH_3 compound from the tissues of shellfish *Hormomya mutabilis*. The methyl mercury moiety was traced to arise from the reaction of mercuric chloride and acetaldehyde effluents from an acetaldehyde factory. It was later (1968) shown that methylmercury compounds could also arise by bacterial methylation of inorganic mercury derivatives. Numerous cases of methylmercury poisoning from eating seeds (treated with antifungal methylmercury preparations) have occurred in Sweden, Iraq, Guatemala and elsewhere. The sharp fall in bird population noted in Sweden in 1960 could be reduced effectively in 1967 by a ban on the use of methylmercury derivatives.

Although the risks arising from problems like bioconcentration of methylmercury in edible fish call for constant vigilance and caution, yet several scientific studies have reassured environmentalists that there has not been any real increase in mercury concentrations in human tissues and fish during the twentieth century with only a little difference in blood concentrations in polluted compared to non-polluted locations.

The well-established and increasingly controversial use of organolead compounds as gasoline additives for spark igniton engines has been extensively scrutinized in recent years from environmental point of view. Most of organolead in gasoline is converted into inorganic (lead halide) species specially in the presence of organic halide scavengers added to prevent the build up of lead in engines, but a fraction (~ 10%) is emitted in the organic form.

In spite of extensive research, the role of biomethylation in the environmetal lead cycle is not clearly understood as yet. These organolead species (generally $PbMe_4$ and $PbEt_4$) from automobile emissions tend to accumulate in the atmosphere of the cities. Even on decomposition to lead, the fine metallic particles so formed tend to be chemically more reactive than the metal in the bulk form and tend to get either oxidised to lead (II) salts or form methyllead moieties on coming in contact with reagents like iodomethane. The facile disproportionation and redistribution tendencies of alkyllead compounds make them potentially important intermediates, providing pathways for other lead derivatives although they themselves are short-lived and do not generally attain a high concentration. One possible mode of cycling of alkyllead compounds in the environment is shown in Figure 7.13.

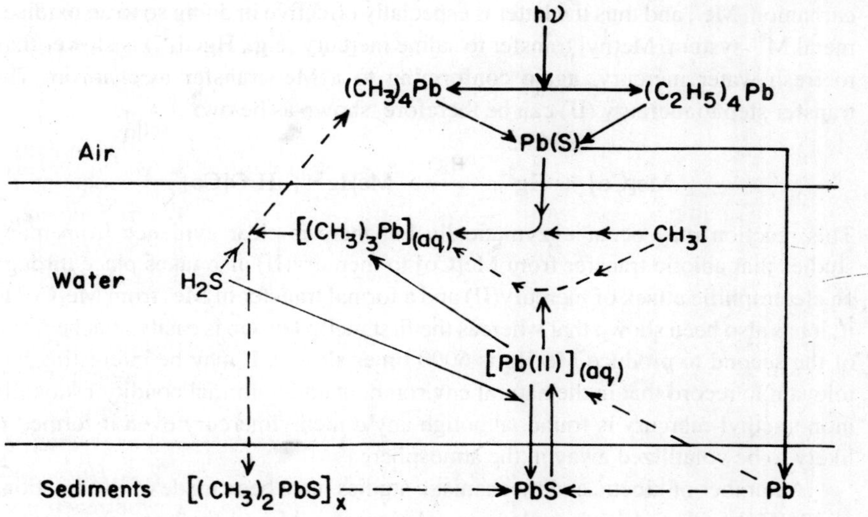

Fig. 7.13 Model for role of organolead in lead geocycle; primary and secondary (hypothesized) routes are represented by solid and dotted lines respectively.

The existence of a tin cycle has been investigated extensively since 1970's in view of the larger quantities of organotins entering the environment through their increasing applications as pesticides, etc. A number of pathways, e.g., oxidative methylation of tin(II) (J.M. Wood, 1977); environmental methylation of tin (P.J. Craig, 1980); and biomethylation of stannic oxide and trimethyltin hydroxide by sediment microorganisms (1981) have been suggested in this direction.

7.5.2 Environmental Methylation

The conversion of an inorgnaic metal species into a methylmetal derivative on exposure to a typical environmental situation (e.g., sea or lake sediment) may be termed 'environmental methylation'. This may occur as (i) an enzymatic 'biological' process, e.g., methylation of mercury by organisms employing methionine synthe to produce methionine from homocysterine; (ii) a non-enzymatic process such as methylations by the algal metabolite, iodomethane or by man-made methylating agent, e.g., Me_3Pb^+ species.

Amongst the naturally occurring methylating agents, methylcobalamin (Me[Co], a Vitamin B_{12} Coenzyme) (Fig. 7.10 with methyl group attached to cobalt in place of 5′-deoxyadenosyl group) has aroused considerable interest. While the other main natural methylating systems, S-adenosylmenthionine and ^{5}N-methyltetrahydrofolate essentially transfer methyl groups as Me^+, methylcobalamin (Me[Co]) is the only species capable of transferring the methyl group as a carbanion, Me^-, and thus the latter is especially effective in doing so to an oxidised metal M^{n+} system. Methyl transfer to saline mercury (e.g., $HgCl_4^{2-}$) is slower than to fresh water mercury, again conforming to a Me^- transfer mechanism. The transfer step to mercury (II) can be therefore, shown as below:

$$\text{Me[Co]} + Hg^{2+} \xrightarrow{H_2O} MeHg^+ + H_2O[Co]^+$$

This reaction may occur enzymatically but there is clear evidence from many studies that abiotic transfer from Me[Co] to mercury(II) also takes place through an electrophilic attack of mercury(II) and a formal transfer of Me^- from Me[Co] to it; it has also been shown that whereas the first methyl group is easily attached, that of the second to produce Me_2Hg is 6000 times slower. It may be interesting and relevant to record that in the natural environment under normal conditions mainly monomethyl-mercury is found, although any dimethylmercury even if formed is likely to be volatilized away in the atmosphere.

A number of kinetic and mechanistic studies have been made of the reactions of Me[Co] with tin, lead, palladium, platinum, gold and thallium compounds. There is also distinct evidence that Me[Co] does play a part in natural methylations, but there is no definite proof for its enzymatic role in methylations. Under different conditions, Me[Co] may transfer a methyl group as Me° or Me^+ also. Whereas the Me^- route is preferred for transferring to metals in higher oxidation states, the free redical (or even Me^+) transfer would be more facile to reduced metals (Pb°, Sn°) or lower oxidation states (Sn^{2+}) or to the complexed metals. Much more detailed investigations are required for elucidating the actual methylation mechanisms to metal systems complexed to organic species in their natural environment.

In addition to the well-known biological methylating agents, S-adenosylmethionine (SAM) and ^{5}N-methyltetrahydrofolate (^{5}N-THF), a number of other methylating agents, which do not appear to involve cobalamin Me[Co], have been suggested, but these require much more intensive investigation. For example, *E. Coli*, which does not produce (although it can utilize) Me[Co], has been shown to methylate mercury. An abiotic photochemical methylation of mercury(II) acetate has been shown to yield methylmercury. Transalkylation between metal species, a well known organometallic phenomenon of synthetic utility, has been cited as being responsible for mercury methylation by methyl-lead and -tin species present as pollutants in some systems, e.g.,

$$Me_3Sn^+ + Hg^{2+} \rightarrow Me_2Sn^{2+} + MeHg^+$$

Mention of a few other possibilities, worthy of further probing may not be out of place here. The natural products, ethylene and acetylene, may convert mercury(II) chloride to alkylmercury and mercury, Iodomethane (CH_3I) has been shown recently to be a natural metabolite of some marine algae, capable of methylating inorganic mercury, lead and tin species.

Study Questions

1. The chemotherapic properties of organometallic compounds could be attributed either to their organic component or to the metal involved. Comment on the therapeutic properties of (i) mercurochrome; (ii) salvarsan; (iii) silatrane; and (iv) cisplatin from the above/other points of view.
2. Discuss the structure of vitamin B_{12}.
3. Citing a few examples, give a brief account of applications of organometallic compounds in agriculture and horticulture.
4. With the possibility of methylation of inorganic derivatives, the concern about the environmental aspects of organometallic derivatives has increased manifold. Explain!

Supplementary Reading

1. '*Industrial Applications of Organometallic Compounds*' by J.H. Harwood, Chapman and Hall, London, 1963.
2. 'The Chemistry of Vitamin B_{12} and Related Inorganic Model Systems', by D.G. Brown, Progress Inorganic Chemistry, *19*, 105-106, 1973.
3. 'Transition Metal complexes in Cancer Chemotherapy' by M.J. Cleare, Coordination Chemistry Review, *12*, pp. 349-405, 1974.
4. 'Concerning the biosynthesis of Vitamin B_{12}', by A.I. Scott, Tetrahedron *31*, 2639-52, 1975.
5. 'Organometals and Organometalloids, Occurrence and Fate in the Environment', eds. F.E. Brinckman and J.M. Bellama, American Chemical Society Symposium Seires No. 82, Washington, D.C. 1978.
6. 'The Biogeochemistry of Lead in the Environment', ed., J.O. Nrigau, Vol. I & II, Elsevier. Amsterdam, 1978.
7. The Biogeochemistry of Mercury in the Environment , ed., J.O. Nrigau, Elsevier, Amsterdam, 1978.
8. 'Metal Pollution in Aquatic Environment', eds. U. Forstner and G.TW. Wittmann, Springer Verlag, Berlin, 1979.
9. 'Synthesis and Properties of Bioactive Organosilicon Compounds' by R. Tacke and U. Wannagat, Topics in Current Chemistry, *84*, pp. 1-75, 1979.
10. 'Biological Activity of Silatranes', by M.G. Voronkov, Topics in Current Chemistry, *84*, pp. 77-135, 1979.
11. 'Vitamin B_{12}: Retrospect and Properties', by A.W. Johnson, Chemical Society Reviews, *9*, pp. 125-40, 1980.
12. 'Lead in the Marine Environment', eds. M. Branicka and Z. Konrad, Pergamon Press, New York, 1980.
13. 'Trends in Organosilicon Biological Research', by R.J. Fessenden and J.S. Fessenden, in Advances in Organometallic Chemistry, eds. F.G.A. Stone and R. West, Vol. XVIII, pp. 275-296, Academic Press, New York, 1980.
14. 'Inorganic Chemistry of Vitamin B_{12}', by J.M. Pratt,Academic Press, New York, 1982.
15. 'Environmental Aspects of Organometallic Chemistry', by P.J. Craig, in Comprehensive Organometallic Chemistry, ed. G. Wilkinson et al, Pergamon, Vol. II, 979-1020, 1982.

16. 'The Biological Methylation of Metals and Metalloids', by J.S. Thayer and F.E. Brinckman, in Advances in Organometallic Chemistry, eds. F.G.A. Stone and R. West, Vol. XX, pp. 313-, Academic Press, New York, 1982.

17. 'Organometallics in Medicine', by A.J. Crowe, Chemistry and industry, pp. 304-310, 1983.

18. 'Organometallic Compounds and Living Organisms', by J.S. Thayer, Academic Press, London, 1984.

19. Organometallic Compounds in the Environment, by P.J. Craig, Longman Group Ltd., England, 1986.

20. 'Organometallic Chemistry (An Overview)' by J.S. Thayer, VCH Publishers, New York/Weinheim, 1988.

Appendix 1.1

Consequences and Development of an Invention*

KARL ZIEGLER**

The awarding of the Nobel Prize for Chemistry for the year 1963 is related to the precipitous expansion of macromolecular chemistry and its industrial applications, which began precisely ten years ago at Max-Plank-Institute for Coal Research, in Mülheim/Ruhr. The suddenness with which this began, and the rapidity with which it was propagated are comparable to an explosion. The energy carriers in this case were the ingenuity, activity, creative imagination and bold concepts of the many unnamed chemists, designers and entrepreneurs in the world who have fashioned great industries from our humble beginnings.

If today I stand with my colleague Natta***, who has been particularly effective in promoting this explosive wave, in the limelight of distinction, and do wish to manifest, with this address, my appreciation for the honor bestowed upon me, I must begin by thanking these many anonymous persons. They, too, deserve this distinction.

The new development had its inception near the end of 1953, when I, together with Holzkamp, Breil and Martin, observed—during only a few days of an almost dramatic course of events—that ethylene gas will polymerize very rapidly with certain catalysts that are extremely easy to prepare, at 100, 20 and 5 atmospheres and, finally, even at normal pressure, to a high molecular weight plastic.

The catalyst is prepared simply by simultaneously pouring, with exclusion of air, two liquid materials into about two liters of a gasoline-like hydrocarbon, after which ethylene is introduced, while stirring. The gas is absorbed quickly; within an hour one can easily introduce 300-400 liters of ethylene into the two liters of

* Extract from 1963 Nobel Lecture.

** Max-Planck—Institute for Kohlenforschung Mülheim/Ruhr.

*** Professor G. Natta shared 1963 Nobel Prize with Professer K. Ziegler and delivered his lecture on the topic entitled ''From the stereospecific polymerization to the asymmetric autocatalytic synthesis of macromolecules''.

liquid. At the same time, a solid substance precipitates, in such a way that after approximately one hour the material becomes doughy and can scarcely be stirred any more. If the brown catalyst is then destroyed, by the addition of some alcohol and by the introduction of air, the precipitate becomes snow-white and can be filtered off. In its final state it will accumulate, in amounts of 300-500 g, as a dry, white powder.

The results of this experiment greatly surprised us, and, later on, many others, since upto that time ethylene had been considered extremely difficult to polymerize. The "polythene" of the Imperial Chemical Industries, a product which had been known for some seventeen years, was being prepared under pressures of 1000-2000 atmospheres, and at a temperature of 200°C. Our experiment thus destroyed a dogma. It led, in addition, to a polyethylene which differed quite markedly from the high-pressure product. Low-pressure polyethylene not only has a better resistance to elevated temperatures and a higher density, but is also more rigid.

The catalyst employed in the experiment described was prepared by mixing aluminium triethyl, or diethyl aluminium chloride, with titanium tetrachloride. However, this is only one example, taken from the countless series of "organometallic mixed catalysts". Most generally they will from, as we found, whenever standard organometallic compounds, preferably those of aluminium, but also many of other metals, are brought into contact with compounds of certain heavy metals. Those of titanium, zirconium, vanadium, chromium, molybdenum, cobalt and the like are especially effective. Since there are many different metal alkyls and many different heavy metal compounds, and since, furthermore, components can be mixed together in varying proportions, and by different methods, and because all this can have an effect, often a truly decisive effect, on the nature of the catalytic activity, it is easy to understand why this field has grown to practically limitless proportions.

In place of the metal alkyls, one can also use metal hydrides, or the metals themselves, whereas metal alkyls probably will still form during the catalyzed processes.

Our catalysts then became known, at the turn of the year 1953/4, to our friends in industry and to their foreign colleagues, in Frankfurt, the Ruhr, Manchester, and, last but not the least, Milan. Shortly thereafter the knowledge junped over to the U.S.A. as well, and ultimately our findings became available to all. The consequences have been characterized, elsewhere, by the statement that revelation of the Mulheim catalysts had the same effect as the starting gun of a race in which the laboratories of the interested industries had been entered. However, representatives of purely scientific chemistry also participated.

Because of the magnitude of the new field, arrival at further stages, or the order of such arrivals, was necessarily dependent upon contingencies. Indeed, many important observations were made within short spaces of time, independently of one another, and at different places. Let me illustrate this with two examples: It was pure chance that in November of 1953 the first of the catalysts in which our invention was clearly recognizable happened to be a relatively weak-acting combination of an aluminium alkyl with a zirconium compound, by which ethylene could be polymerized only under a few atmospheres pressure, and with which propylene, already tested the day after our critical experiment with ethylene, would not polymerize at all. Then, for a number of weeks, we were absorbed in experimenting with normal-pressure polymerizations of ethylene by means of titanium-containing catalysts. Early in 1954 we recognized the possibility of copolymerizing ethylene and propylene, after which we succeeded, at Mulheim, in polymerizing propylene with more effective catalysts, but—and this we did not know at the time—a short while after my colleague Natta of Milan had already observed this. In a first substantiation of his observation, and in an act of fairness, Natta had referred to the catalyst used as a ''Ziegler catalyst'', and that is how this expression found its way into the literature[2]. It is surely understandable that I myself prefer to speak of them as ''Mülheim catalysts''.

The second example: Near the close of 1955 work was being done in many places on the polymerization of butadiene with our catalysts. But no one had observed that in addition to the desired high polymers, a very interesting trimer of butadiene, namely 1, 5, 9-cyclododecatriene, was being produced. Güunther Wilke, of my insitute, became aware of this, and showed how one can guide the reaction entirely in this new direction. While endeavouring to explain the formation mechanism of cyclodocatriene. Wilke discovered a way to redirect this reactions at will, either toward a dimerization to an eight-membered ring, or by a co-reaction with ethylene-toward a cooligomerization to a ten-membered ring[3]. The result was that the Mülheim catalysts also achieved importance for polycondensation plastics such as Nylons 8, 10 and 12, into which the ring compounds can be transformed.

These cyclizations constituted the third surprising development afforded the scientific community by the organometallic mixed catalysts, if I assign number one to the new polyethylene process. I saved the second surprise for later, and I must go into that now. From the middle of 1954 on, it began to be obvious that the Mulheim catalysts were capable of polymerizing in a structurally specific, as well as a stereo-specific manner. This realization is an essential contribution of my colleague Natta[4]. He had often pondered over the mechanism of the polymerization, and very successfully strove to ''train'' the catalysts in such a way that they would possess extremely high specificity.

In order to make the sequence of events which led to such a fruitful invention more understandable, I shall have to go back exactly forty years. Shortly after my graduation, having been a student of Karl von Auwers at the University of Marburg/Lahn, I began my independent scientific work with experiments for testing the theory of so-called free radicals. I incidentally found, in 1923, a new method for the formation of organic compounds of the alkali metals potassium and sodium[5], which brought my attention to the metal alkyls as an interesting, highly diversified field, that has continued to fascinate me, over and over again, up to the present. The new catalysts grew out of this, as a side sprout, in 1953.

A few years later, in 1927, Bähr an dI[6–8] made the discovery important for the further development that alkali alkyls can be added with ease to butadiene or styrene, at room temperature. Repetition of the process leads first to oligomers, in a ''stepwise organometallic synthesis'', and finally to polymers and high polymer reaction products.

$$\text{R—A} + \text{CH}_2\text{—CH=CH=CH}_2 \rightarrow \text{R—CH}_2\text{—CH=CH—CH}_2\text{—A}$$
$$\downarrow$$
$$\text{R—(C}_4\text{H}_6)_n\text{—A} \leftarrow \text{R—CH}_2\text{—CH=CH—CH}_2\text{—CH}_2\text{—CH=CH—CH}_2\text{A}$$

(arbitrarily designated as pure 1, 4 structures)

Oligomers when $n = 2, 3, 4$ etc.

Polymers when n is considerably higher.

This first contact of mine with ''macromolecular chemistry'' later gave impetus to many investigations by third parties, recently butyllithium has also been suggested for industrial polymerizations of isoprene. At first, however, another, indirect result of our work was of more importance. Secondary observations suggested the conclusion that metallic lithium should be amenable to a reaction analogous to the one by which Grignard compounds are formed from magnesium:

$$\text{RCl} + \text{Mg} = \text{RMgCl}$$
$$\text{RCl} + 2\text{Li} = \text{RLi} + \text{LiCl}$$

With Columus[9], I was able to confirm this in 1930, and that is how the organolithium compounds became easily accessible.

In Mulheim/Ruhr, where I have been working since 1943, Gellert and I succeeded in transferring the technique of a ''stepwise organometallic synthesis'' from butadiene to ethylene[10]. In this instance the reaction leads from lithium alkyl directly to the higher straight-chain lithium alkyls, and hence also to alcohols carboxylic acids, and the like.

$$\text{C}_2\text{H}_5\text{—Li} \xrightarrow{\text{C}_2\text{H}_4} \text{C}_2\text{H}_5\text{—CH}_2\text{—CH}_2\text{—Li} \xrightarrow{\text{C}_2\text{H}_4} \text{C}_2\text{H}_5\text{—(CH}_2\text{—CH}_2)_n\text{—Li}$$
$$\xrightarrow{\text{O}_2} \text{C}_2\text{H}_5\text{—(CH}_2\text{—CH}_2\text{—)}_n\text{ OH} \xrightarrow{\text{CO}_2} \text{C}_2\text{H}_5\text{—(CH}_2\text{—CH}_2)_n\text{CO}_2\text{H}$$

Contrary to what holds true for butadiene, however, there is a limit here to the growth which a chain can undergo, since, for ethylene addition, the temperatures required are such that the lithium alkyls will readily decompose to lithium hydride + olefin. This certainly seemed to justify the following conclusion: If, in such decompositions, it is a question of a reversible reaction, as we had reason to believe, then lithium hydride and lithium alkyls should, under proper conditions, function as catalysts for the polymerization or rather, oligomerization, of ethylene to the higher α-olefins.

We did find such a reaction in principle, but it was so complicated by secondary and subsequent reactions that we could do nothing with it. Then when I had already decided to give up these efforts, my coworker, H.G. Gellert, conducted one more experiment—and the last, he was convinced—with just recently discovered lithium aluminium hydride. This led immediately to the desired higher α-olefins. As the decisive turning point, this resulted in the realization that the alkali metal was not the crucial issue at all, and that everything we already knew about the lithium alkyls, and all that we had anticipated besides, with respect to the chemistry of the olefinic hydrocarbons, could be achieved with a great deal more ease through use of organo-aluminium compounds[11]. That is:

(1) There are genuine equilibria: aluminium alkyl $\rightleftarrows$ aluminium hydride + olefin, lying, as a rule, entirely to the left, so that, in reverse of the situation with lithium, it is possible to synthesize the aluminium alkyls from hydride olefin.

(2) In the case of aluminium, too—and this came as real surprise—at moderately high temperatures a stepwise organometallic synthesis, or as we now call it, a ''growth'' or propagation reaction takes place, leading to the higher aluminium alkyls; thus, a synthesis of the higher straight-chain primary monofunctional aliphatic compounds, particularly the fatty alcohols became possible.

(3) Furthermore, we have, from about 1950 on, a catalytic oligomerization of the ethylene to higher α-olefins.

Here the organometallic synthesis appears as the partial reaction of a completely understood, homogeneous intermediate reaction catalysis: After a certain number of addition steps the intermediate product decomposes to a hydride and an olefin whereupon, after the addition of ethylene to the hydride the cycle is repeated.

$$C_2H_5Al + (n-1)\,C_2H_4 = C_2H_5\,(C_2H_4)_{n-1} - Al$$
$$C_2H_5(C_2H_4)_{n-1} - Al = AlH + C_2H_5(C_2H_4)_n - CH = CH_2$$
$$AlH + C_2H_4 = AlC_2H_5$$
$$\overline{\qquad\qquad\qquad\qquad\qquad\qquad\qquad\qquad}$$
$$nC_2H_4 = (C_2H_4)_n$$

Such a reaction was encountered in its most primitive form with propylene, for which the homogeneous catalysis leads, without supplementary chain growth, almost exclusively to a well-defined dimer[12].

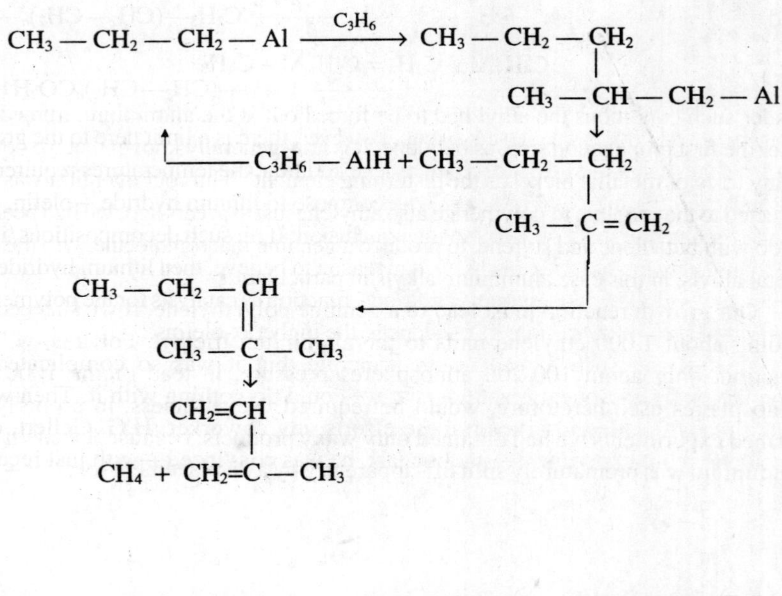

Recently this reaction has achieved significance for high molecular weight chemistry as well, since the cracking of isohexene, following the shifting of the double bond, produces isoprene, in addition to methane[13]. The transition of all these reactions into industrial applications was finally accomplished by the so-called ''direct synthesis'' of aluminium alkyls from aluminium, hydrogen and olefins, discovered by us at approximately the same time as the new polyethylene process. Aluminium hydride, from which the aluminium, trialkyls are quite easy to obtain through the addition of olefins, cannot be prepared directly from the metal and hydrogen. However, in already-prepared aluminium trialkyls, aluminium will dissociate with hydrogen to dialkyl aluminium hydrides which, with ethylene, will give 1.5 times the original amount of aluminium triethyl,

$$\begin{aligned} Al + 2Al(C_2H_5)_4 + 1\tfrac{1}{2}H_2 &= 3HAl(C_2H_5)_2 \\ 3HAl(C_2H_5)_2 + 3C_2H_4 &= 3Al(C_2H_5)_3 \\ \hline Al - 1\tfrac{1}{2}H_2 + 3C_2H_4 &= Al(C_2H_5)_3 \end{aligned}$$

so that any amount of aluminium trialkyl can be prepared without difficulty[14].

Up to the year 1952 we had frequently conducted ''growth'' reactions based on aluminium triethyl, and thought we were thoroughly acquainted with such reactions. But when, together with E. Holzkamp[15], I attempted to apply this type of reaction to aluminium tripropyl the formation of chains, to our great surprise, did not materialize at all. On the contrary, we obtained propylene—from the propyl aluminium in addition to aluminium triethyl and α-butylene. Even starting from aluminium triethyl our reaction now yielded nothing but butylene and the unchanged aluminium compound.

The explanation was obvious: A catalyst in trace amounts must have gotten into this series of experiments, leading to an uncommonly rapid acceleration of the displacement reactions:

$$C_3H_7Al + C_2H_4 = C_2H_5Al + C_3H_6$$

and

$$C_4H_9Al + C_2H_4 = C_2H_5Al + C_4H_8$$

Under such conditions the alkyl had to be forced out at the aluminium, immediately after the first propagation step, as butylene. It is now generally known that we detected a tiny trace of metallic nickel as the disturbing element[1]. Thus our attention was again directed to the problem to polymerize also ethylene just as, years ago, we had been able to do with butadiene and styrene, to produce a genuine macromolecule with the aid of metal alkyls, in this case aluminium alkyls in particular.

Our growth reaction must lead to a genuine polyethylene, if we succeeded in adding about 1,000 ethylene units to the aluminium triethyl. For this, with our reaction, only about 100-200 atmospheres pressure, instead of the 1000-2000 atmospheres used heretofore, would be required. Nevertheless, in properly performed experiments we had obtained only waxy products, because the chain at the aluminium was prematurely split off-apparently by a displacement reaction—as an olefin, with the re-formation of ethyl at the aluminium, an occurrence known to chemists working in the high molecular weight field as a "chain transfer reaction".

$$C_2H_5—(CH_2—CH_2)_n\,Al + C_2H_4 = C_2H_5—(CH_2—CH_2)_{n-1}—CH=CH_2 + C_2H_5Al$$

To the extent that catalyst traces, as we now might well suspect, had been involved here also in effecting the displacement, there existed the prospect that completely "aseptic" procedures could eventually lead to a true polyethylene. In order to provide the essentials for the "asepsis", we began, in the middle of 1953, to systematically investigate substances which have effects somewhat similar to those of nickel. We found, instead, the polymerization-promoting organometallic mixed catalysts, and, in particular, we achieved a low-pressure polymerization of ethylene, and with this I have again arrived at my starting point.

The important consequences of the discovery of organometallic mixed catalysts, evident even to the layman, have led many to regard me, nowadays, as a "macromolecular chemist", and, in fact, even as a plastics expert. I have intentionally set my work in this field within a much broader framework, to show you that I am a "macromolcular chemist" only peripherally and that I am not at all a plastics expert. Rather I have always looked upon myself as a pure chemist. Perhaps that is also why the impact of the invention has been so enduring. The new knowledge has, after all, not come from macromolecular chemistry. It is the metal alkyls that have insinuated themselves into the chemistry of macromolecules to effectively fertilize this field. Typical of the course I have followed from those early beginnings of forty years ago until today, is the fact that I have never started with anything like a formally presented problem. The whole effort developed quite spontaneously, from a beginning which was actually irrational in nature; through an unbroken causal series of observations, interpretations of findings, rechecking of the interpretations by new experiments, new observations, etc. My method resembled a meandering through a new land, during which interesting prospects kept opening up, during which one could frequently view part of the road to be travelled, but such that one never quite knew where this trip was actually leading. For decades I never had the slightest notion that successful technological and industrial applications were also to be encountered during the journey.

Twice this path seemed seriously blocked. The first time was before the transition from lithium hydride to lithium aluminium hydride and from the lithium to aluminium compounds had been accomplished. The second was when our growth reaction suddenly, in a truly mysterious way, refused to go any more. In both cases a capitulation in the face of these difficulties would undoubtedly have broken the red thread of continuity which can now be followed clearly.

But a much more formidable impediment might have presented itself. In order to illustrate this, I must elaborate on the paradox that the critical concluding stages of the investigations I have reported took place in an institute for "coal research".

When I was called to the institute für Kohlenforschung in 1943, I was disturbed by the objectives implied in its name. I was afraid I would have to switch over to the consideration of assigned problems in applied chemistry. Since ethylene was available in the Ruhr from coke manufacture, the search for a new polyethylene process, for example, could certainly have represented such a problem. Today I know for certain, however, and I suspected at the time, that attempt to strive for a set goal at the very beginning, in Mülheim, would have completely dried up the springs of my creative activity. As a matter of fact: Giving up my preoccupation with organometallic compounds in favour of the other, ''bread-and-butter'' problems of coal chemisty—many of my coleagues were of the opinion, at the time, that this would be the natural consequence of my removal to Mülheim—would have cut the leads which I already held invisibly in my hand, and which were to lead me safely to the results that proved of such importance also for the Ruhr industry.

As a condition of my transfer to Mülheim I stipulated that I was to have complete freedom of action in the entire field of the chemistry of carbon compounds, without regard to whether any direct relation to coal research was or was not recognizable. The acquiescence of my stipulation was in accord with the principles of the then Kaiser-Wilhelm, and now Max-Planck Society, of which my institute is a part. As far as the German coal mining industry which supported my institute is concerned, this was an act of great foresight on their part which did in fact provide the conditions for every thing that occurred, particularly the present circumstance that my institute, and I with it, have now received this very great distinction.

The institute, however,—what is it, in its distinctive spiritual and intellectual substance, other than the totality of its active people. I began this address with an expression of gratitude to the many people in the world whom I know only slightly, or not all, and who have developed great industries from our beginnings. I will end the address by expressing my heartfelt thanks to the many, every well known members of my institute who have stood by me faithfully throughout all these years, and who share with me the prize for which I have been singled out.

References

1. K. Ziegler, E. Holzkamp, H. Breil and H. Martin, *Angew. Chem.*, 57 (1955) 541.
2. K. Ziegler, E. Holzkamp, H. Breil and H. Martin, *Angew. Chem.*, 67 (1955) 426.
3. G. Wilke, *Angew. Chem.* 75 (1963) 10; *Angew Chem., Intern, Ed.*, 2 (1963) 105.
4. G. Natta, *Angew. Chem.*, 76 (1964) 553 (Nobel Lecture).
5. K. Ziegler and F. Thielmann, *Ber.*, 56 (1923) 1740.
6. K. Ziegler and K. Bähr, *Ber.*, 61, (1928) 253.
7. K. Ziegler and H. Kleiner, *Ann. Chem.*, 473 (1929) 57.
8. K. Ziegler, F. Dersch and H. Wollthan, *Ann. Chem.*, 511 (1934) 13.
9. K. Ziegler and H. Colonius, *Ann. Chem.*, 479 (1930) 135.
10. K. Ziegler and H.G. Gellert, *Ann. Chem.*, 567 (1950) 195.
11. K. Ziegler, *Angew. Chem.*, 68 (1956) 721, 724.
12. K. Ziegler, *Angew. Chem.*, 64 (1952) 323, 326.
13. V.F. Anhorn *et. al. Chem. Eng. Prog.*, 57 (1961) 43.
14. K. Ziegler *et al Ann Chem.*, 629 (1960) 1.
15. K. Ziegler *et al., Ann Chem.*, 629 (1960) 121, 135.

Appendix 1.2

The Long Search for Stable Transition-Metal Alkyls*

G. WILKINSON**

Chemical compounds in which there is a single bond between a saturated carbon atom and a transition-metal atom are of unusual importance. Quite aside from the significance and role in nature of the cobalt to carbon bonds in the vitamin B12 system and possible metal-to-carbon bonds in other biological systems, we need only consider that during the time taken to deliver this lecture, many thousands, if not tens of thousands of tons of chemical compounds are being transformed or synthesized industrially in processes which at some stage involve a transition metal-to-carbon bond. The non-chemist will probably be most familiar with polyethylene or poly-propylene in the form of domestic utensils, packaging materials, children's toys and so on. These materials are made by Ziegler-Natta or Phillip's catalysis using titanium and chromium, respectively. However, transition-metal compounds are used as catalysts in the synthesis of synthetic rubbers and other polymers, and of a variety of simple compounds used as industrial solvents or intermediates. For example, alcohols are made from olefins, carbon monoxide and hydrogen by use of cobalt or rhodium catalysts, acetic acid is made by carbonylation of methanol using rhodium catalysts and acrylonitrile is dimerized to adiponitrile (for nylon) by nickel catalysts. We should also not forget that the huge quantities of petroleum hydro-carbons processed by the oil and petrochemical industry are re-formed over platinum platinum-rhenium or platinum-germanium supported on alumina.

In all of these processes, every single molecule, at some point in the catalytic cycle, is involved in the formation of a metal-to-carbon single bond. Of course, catalytic processes require that the metal-to-carbon bond be unstable—or, more accurately, that it be labile and able readily to undergo chemical reactions such as

* Nobel lecture, 1973.

** Professor, Department of Chemistry, Imperial College of Science and Technology, London.

$$M-CR_3+CO \longrightarrow M-\overset{\overset{\displaystyle O}{\|}}{C}-CR_3$$

$$M-CR_3+H_2 \longrightarrow MH+HCR_3$$

Today, however, I am concerned not with catalysis but with the synthesis of simple, stable metal compounds that have single bonds to saturated carbon. The synthesis of stable metal alkyls, and indeed, the nature of the transition metal-to-carbon bond, is a problem that has been with us for a long time.

The first attempts to make such compounds were shortly after Frankland's epoch-making discovery of diethylzinc. Thus in 1859 Buckton wrote, 'a rich harvest can scarcely fail to be reaped from submitting to the action of diethylzinc the metallic compounds of the other groups(1). However, he failed with the transition metals, silver, copper and platinum, as indeed did other workers in the 1800s.

We must not forget, however, that a transition-metal organo compound, though of a quite different type had been made much earlier by the Danish pharmacist Zeise. By the interaction of alcohol and chloroplatinic acid he had made the salt $K[PtCl_3(C_2H_4)]$, known now as Zeise's salt. There followed much discussion of the exact constitution of this substance, but it was to be over 125 years before the structure was determined and the way in which ethylene is bound to the metal fully understood.

The first stable alkyls to be isolated were those of platinum (2) and gold (3) by William Pope and his coworkers Peachy and Gibson at the Municipal School of Technology in Manchester. The structure of the trimethylplatinum compound $[PtCl(CH_3)_3]_4$ was not to be determined by X-ray diffraction for another 40 years. At the meeting of the Chemical Society in London on March 21, 1907, at which the discovery was announced, the Chairman, Sir Henry Roscoe, complimented the authors on opening out an entirely new branch of investigation 'which might indeed be said to be a wonderful find'. This promise took a very long time indeed to be fulfilled. Even in 1955 it could be written (4): "It will be apparent from this overall picture of alkyls and aryls of the transition metals that the often-heard generalization that they are much less stable and accessible than those of non-transition metals is quite true".

The main reason for the failure to isolate stable compounds, despite evidence that alkyls, or aryls were present in solution at low temperatures, was that the interaction of transition-metal halides with Grignard reagents of lithium alkyls usually gave either coupled products or products from decomposition of coupled species. Typically, ferric chloride was used to make coupled products, for example, diphenyl:

$$C_6H_5MgBr \xrightarrow{FeCl_3}$$ 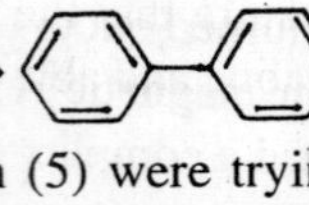

It was this coupling that Kealy and Pauson (5) were trying to utilize, to make dihydrofulvalene, which led to their synthesis of dicyclopentadienyl iron.

$$C_5H_5MgBr \xrightarrow{FeCl_3} Fe(C_5H_5)_2 \text{ not } C_5H_5—C_5H_5$$

It might be interesting to note in passing that my own conclusion about the structure of this iron compound was based on two points. Firstly, my knowledge of the instability of transition-metal alkyls and aryls, secondly my intuition concerning the nature, uncertain at that time, of the binding of ethylene in Zeise's salt and of butadiene in Reihlen's compound, $[Fe(C_4H_6)(CO)_3]$. I was convinced that ethylene was bound 'sideways' (I) and that butadiene in its *cis* form could act as a chelate (II), both double bonds being bound to the metal.

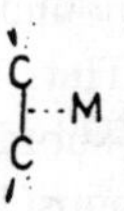

(I)

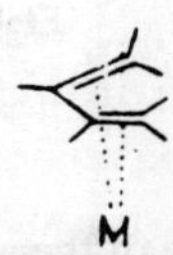

(II)

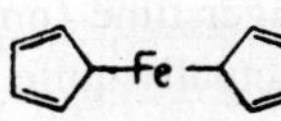

(III)

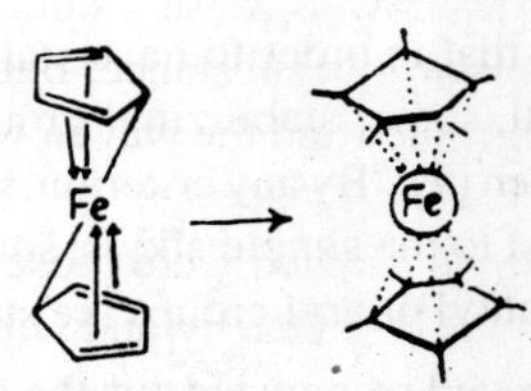

(IV) (V)

These structures were later confirmed by others. Thus I wanted to involve both double bonds of the cyclopentadienyl ring in the structure originally written by Kealy and Pauson, and Miller. Tebboth and Tremaine (III), in a butadiene-like fashion (IV) resulting in a 'sandwich' structure.

Having read Linus Pauling's famous book. *The Nature of the Chemical Bond*, and heard about resonance, this meant that I could write various resonance forms of IV, which directly led to the idea that all the carbon atoms were equivalent as in V, that is, to the well-known structure of the molecule now known as ferrocene.

During the next few years a number of stable compounds in which alkyl or aryl groups were present were synthesized. The first was a phenyltitanium alkoxide $Ti(C_6H_5)(OC_3H_7)_3$ (6) and others soon followed. However, in essentially all of the compounds a special type of ligand was present. These ligands were what are referred to as π-bonding ligands. Examples are π C_5H_5, CO, PEt_3, etc. Representative types of these alkyls are VI-IX.

Compound VII turned out to be especially interesting. We observed only two signals in the proton magnetic resonance spectrum, whereas the σ-C_5H_5 group alone should have had a complex spectrum; the infrared spectrum, however, was

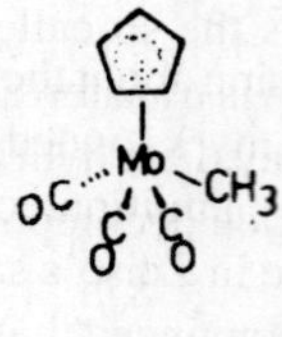

(VI)

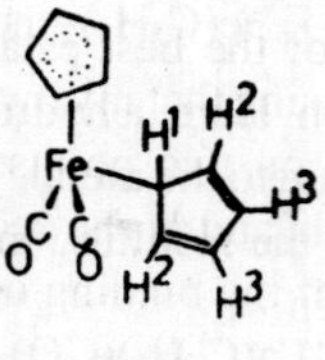

(VII)

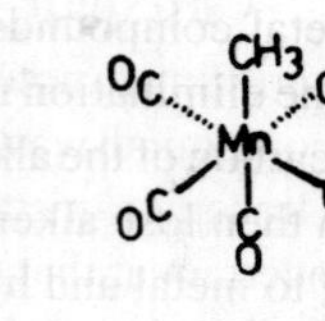

(VIII)

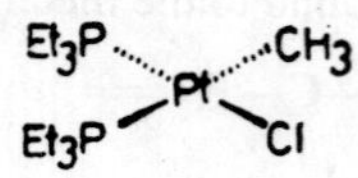

(IX)

consistant with VII. Because of the analogy with photography, where, if one takes a picture of a moving wheel with a short exposure time (infrared), the picture is sharp, whereas if one uses a longer time (nmr) it is blurred, I had to draw the conclusion that the σ-C_5H_5 was actually quite slowly rotating via a 1:2 shift. This was the first recognition of what are now known as fluxional molecules, this particular type being called 'ring whizzers'.

So the view developed that in order to have stable compounds with alkyl or aryl groups bound to a metal, some 'stabilizing' group also had to be present. As recently as 1968, it was written (7): 'By any criterion, simple transition metal alkyls are unstable' and 'In contrast to the simple alkyls, some metal complexes bearing other ligands in addition to alkyl or aryl groups are strikingly stable'.

However, it was not too well recognized that the presence of such 'stabilizing' ligands, is no guarantee of stability, because other factors are involved. Indeed, the activity of many metal complexes with π-bonding ligands present in catalytic reactions depends on the lability of the metal-to-carbon bonds.

Despite all the intense study of metal-carbon bonds and the arguments about their stability, it is remarkable that only a few bond energies are known. The available thermodynamic data show that metal-carbon bonds are not exceptionally weak and the bond energies are quite comparable to those of non-transition metal-to-carbon bonds. It could be argued that since the compounds for which bond energies have been determined have 'stabilizing' π-bonding ligands present, the metal-carbon bond energies are abnormally high, but this view cannot be sustained. There is no reason to assume this to be so nor is there any reason to assume that carbon would differ appreciably in its capability to bond to transition metals compared to other second-period elements such as oxygen and nitrogen or to the halogens). For both oxygen and nitrogen, metal compounds in high oxidation states are well known, for example, the alkoxides, such as $V(OR)_4$, and dialkylamides, such as $W(NR_2)_6$. So, accepting that there is no thermodynamic reason for the instability of simple alkyls, the conclusion was that they are kinetically unstable [labile].

There are several ways by which a metal alkyl can decompose, but for transition-metal compounds one of the best established is the so-called hydride transfer alkene elimination reaction. Here, a hydrogen atom is transferred from the second or β-carbon of the alkyl chain to the metal. The intermediate hydrido-alkene complex can then lose alkene and the resulting metal hydride decompose further, for example, to metal and hydrogen.

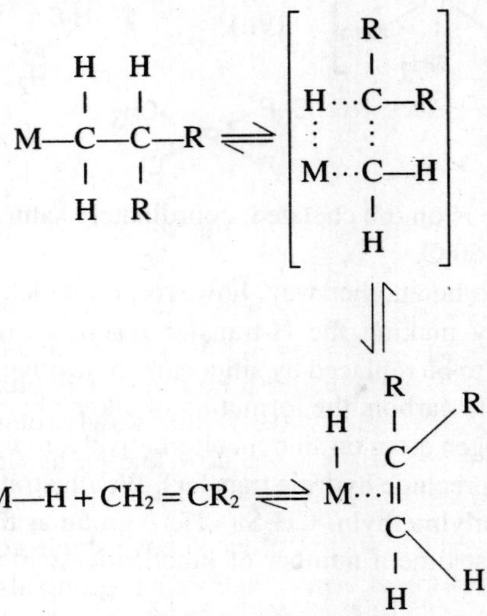

The reverse reaction, namely the generation of a metal alkyl from a metal hydride plus alkene, is the key reaction in many catalytic cycles involving alkenes, hydrogen and metal species. It is involved for example in homogeneous hydrogenation of unsaturated organic compounds and in the hydroformylation reaction, in which aldehydes or alcohols are synthesized from alkenes, carbon monoxide and hydrogen.

Two relevant studies on this decomposition reaction are illustrative. Firstly, a comparison (8) of the relative stability of dialkylmanganese compounds made *in situ* shows that those alkyls, for example $Mn(CH_3)_2$ or $Mn(CH_2C_6H_5)_2$, that cannot readily undergo this reaction are most stable. Secondly, a comparison (9) of the decomposition products of two similar alkyls, one that has a β-hydrogen, $Bu_3PCuCH_2CH_2CH_3$ and one that has not, $Bu_3PC_uCH_2C(CH_3)_2C_6H_5$, shows that the former decomposes via H-transfer-alkene elimination, whereas the second decomposes by homolytic fission and a free radical pathyway.

It will be observed that the H-transfer scheme involves a change in coordination number of the metal. Thus the alkyl group occupies only one coordination site, whereas in the intermediate hydrido alkene, two sites are involved, one for M—H and one for olefin coordination. Thus, one way in which an alkyl could be stabilized against decomposition is by coordinative saturation of the metal. If the sites required for the reaction to proceed are occupied by firmly bound ligands, then there is no possible pathway for decomposition. The stability of the alkyls with π-bonded ligands referred to above are prime examples of this situation. Other examples are the alkyl of the substitution-inert octahedral metal ions, Cr(III), Co(III) and Rh(III) such as the Werner-type complex ion X.

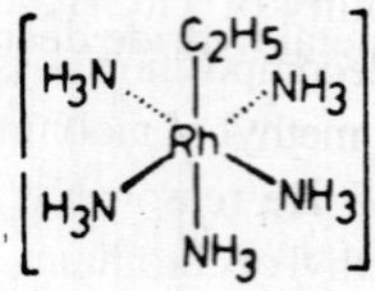

(X)

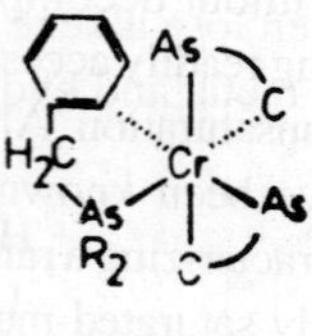

(XI)

A different type is on the chelated, coordinately saturated alkyl XI, which is thermally stable to 350 C.

There seemed to be another way, however, by which stable alkyls could be obtained, namely by making the H-transfer reaction impossible. Thus, if the β-carbon atom were to be replaced by silicon or some other element that could *not* form a double bond to carbon, the formation of alkene becomes impossible, even if there were a hydrogen atom on silicon; alternatively, to have no hydrogen on the β-atom [would also preclude hydride transfer]. We illustrated this concept (10) by use of the trimethylsilylmethyl ($-CH_2Si(CH_3)_3$) group as the alkyl, and were able to isolate for the first time a number of kinetically stable [inert] alkyls such as $Cr\{CH_2Si(CH_3)_3\}_4$.

A number of other groups, including, of course, the carbon analogue, neopentyl, $-CH_2C(CH_3)_3$, fit the required criterion of having no β-hydrogen, and at the present alkyls are known with the following groups.

It may be noted that some of these alkyls notably those of titanium, have been recognized as Ziegler-Natta-type catalysts for the polymerization of alkenes (II).

— $CH_2Si(CH_3)_3$

— CH_2CR_3

— $CH_2C_6H_5$

— $CH_2P(CH_3)_3$

— CH_2 \ $P(CH_3)_2$ / — CH_2

— CH_2SR

— $C(C_6H_5)_3$

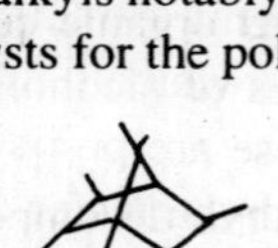

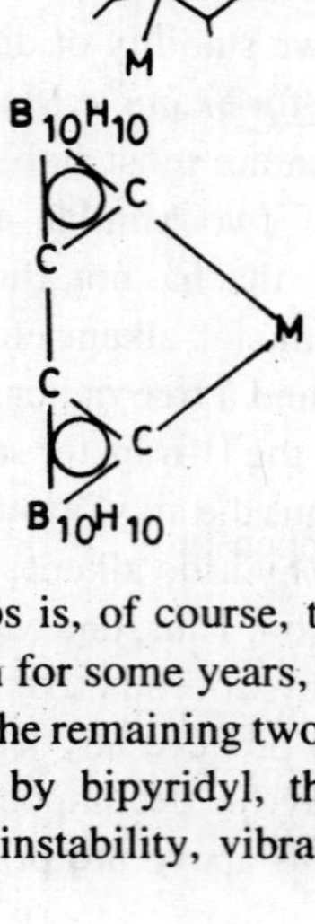

The simplest of all alkyl groups is, of course, the methyl group. Although tetramethyltitanium has been known for some years, it decomposes above –70°C very readily. However, by blocking the remaining two vacant coordination sites of the tetrahedral alkyl, for example by bipyridyl, the thermal stability can be substantially increased. Despite its instability, vibrational spectroscopic studies

(12) of $Ti(CH_3)_4$ suggest that the Ti—C bond strength, as measured by its force constant, is quite comparable to those of the tetramethyls of Si, Ge, Sn and Pb. By contrast; the least stable of those, tetramethyl lead, can be distilled at its boiling point (100°C) without decomposition. The instability of $Ti(CH_3)_4$ is thus clearly kinetic, there being readily accessible pathways for decomposition possible because of its coordinative unsaturation. Although the chloride methyl of niobium and tantalum, $MCl_2(CH_3)_3$, have been known for some time and are reasonably stable, and we succeeded in characterizing a rather unstable tetramethyl of chromium, it appeared that if a coordinatively saturated methyl could be synthesized, there was good reason to expect it to be stable. The obvious candidate was tungsten, for which the hexaalkoxides and dialkylamides as well as the halides are known. We succeeded in synthesizing lexamethyltungsten by the classical reaction of methyl-lithium with tungsten hexachloride(13). The reaction sequence is a very complex one and in order to obtain any $W(CH_3)_6$ it is necessary to use only half of the theoretical quantity of methyl-lithium; probably to avoid the formation of a methyl anion of the type known for other metals, for example, $[Li_4Mo_2(CH_3)_8]^-$. The final step probably also involves the disproportionation of a reduced species:

$$W(V) \longrightarrow W(IV) + W(VI)$$

Once obtained, hexamethyltungsten is reasonably stable and can readily be characterized spectroscopically. It also has a number of interesting chemical reactions. It soon became evident that the compound, though octahedral, is not coordinatively saturated, and that some of the reactions are very facile because coordinative unsaturation allows initial coordination of the reagent. Using tertiary phosphines, we were able to isolate seven-coordinate adducts, $W(CH_3)_6PR_3$. One of the unusual, reactions was that with nitric oxide, which quantitatively gives a compound $W(CH_3)_4\{ON(CH_3)NO\}_2$, which contains two *N*-nitroso-*N*-methyl hydroxylaminato rings. These are probably generated by initial coordination of nitric oxide followed by methyl transfer, as in the sequence:

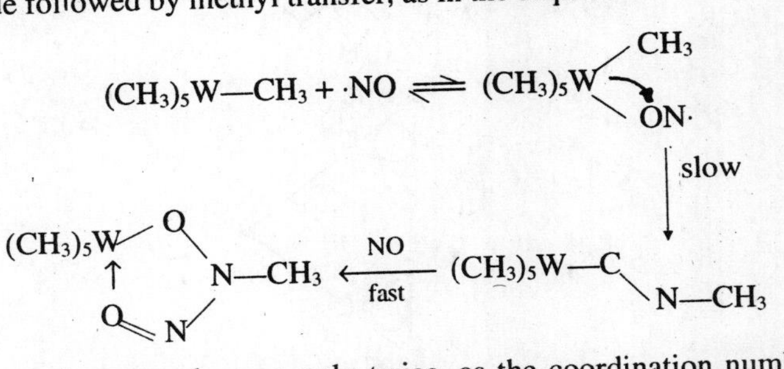

This sequence can happen only twice, as the coordination number reaches eight, which appears to be the maximum for tungsten in the VI oxidation state. Although the synthesis of other methyls is in principle possible, there is more art to it than science. The only other methyl in a very high oxidation state we have yet made is oxotetramethylrhenium(VI), $ReO(CH_3)_4$, but oxo or similar methyls of tunsgten, molybdenum, osmium, etc., may well be stable if suitable synthetic methods are found.

So finally, in conclusion, we can say that the effort of well over 100 years to synthesize stable transition-metal alkyls has finally succeeded. The long established view that the transition metal-to-carbon bond is weak is now untenable and must be discarded. We can expect other types of transition-metal alkyls to be made in due course and can hope that in addition to their own intrinsic interest some of them may find uses in catalytic or other syntheses. The use of titanium and zirconium alkyls in alkene polymerization and the use of alumina treated with hexamethyltungsten for alkene metathesis (14) give good grounds for optimism.

References

1. Buckton G.B. (1859), *Proc. R. Soc.*, **9**, 309.
2. Pope, W.J., and Peachy, S.J. (1907), *Proc. Chem. Soc.*, **23**, 86; *J. Chem. Soc.* (1909), 371.
3. Pope, W.J., and Gibson, C.S. (1907), *Trans. Chem. Soc.*, **91**, 2061.
4. Cotton, F.A. (1955), *Chem. Rev.*, **55**, 551.
5. Kealy, T.J., and Pauson, P.L. (1952), *Nature*, **168**, 1039; for the independent earlier discovery of $FeC_{10}H_{10}$, see Miller, S.A., Tebboth, J.A., and Tremaine, J.F., (1952), *J. chem. Soc.*, 632.
6. Herman, D.F., and Nelson, W.K. (1953), *J. Am. Chem. Soc.*, **75**, 3877.
7. Parshall, G.W., and Mrowca, J.J. (1968), *Adv. Organometal, Chem.*, **7**, 157.
8. Tamura, M., and Kochi, J. (1971), *J. Organometal, Chem.*, **29**, 11.
9. Whitesides, G.M., *et al.* (1972), *J. Am. Chem. Soc.*, **94**, 232.
10. Yagupsky, G., Mowat, W., Shortland, A., and Wilkinson, G. (1970), *Chem. Commun.*, 1369; *J.C.S. Dalton* (1972), 533.
11. For example, Ballard, D.G.H. (1971), *23rd Internat. Congress Pure appl. Chem.*, Butterworths.
12. Eysel, H.H., Siebert, H., Groh, G., and Berthold, H.J. (1970), *Spectrochim, Acta* **26A**, 1595.
13. Shortland, A.J., and Wilkinson, G. (1973), *J.C.S. Dalton*, 872.
14. Mowat, W., Smith. I., and Whan, D.A. (1974), *Chem. Commun.*, 34.

Appendix 1.3

The Borane Adventure—Past, Present and Future*

HERBERT C. BROWN**

Abstract. Forty-nine years ago, diborane was a chemical rarity, available only in two laboratories in the world. The requirements of research during World War II led to the discovery of practical synthetic methods of diborane and to the discovery of sodium borohydride. These turned out to be excellent reducing agents in organic chemistry. Exploration of these reducing characteristics led to the discovery of hydroboration. Hydroboration made organoboranes readily available. These organoboranes have proven to be the most versatile intermediates now available for organic synthesis. A recent development makes possible the synthesis of essentially any compound containing a chiral centre in both optical isomers in essentially 100% optical purity. Thus this research program has taken boranes from a material available only in trace quantities to a reagent widely used in organic synthesis, greatly assisting the chemist in overcoming synthetic difficulties.

It is great pleasure to return to India after an absence of some seventeen years. I wish to express my deep appreciation to the Council of Industrial and Scientific Research for the invitation which made this visit possible.

Each Nobel Laureate is requested to come to Stockholm to receive the Award and to give a lecture based on his work. I used the occasion to trace my research program in boranes from its inception in 1936, as an investigation initiated for my Ph.D. degree, to the year of the Award, 1979. For my present visit to India, I decided to bring this lecture up-to-date by describing a fascinating new development, a promising synthesis of optically pure enantiomers. This is the ''Future'' in my title.

In 1934 the City of Chicago opened three City Junior Colleges. I went to one of these: Wright Junior College. Once I had this opportunity for a college education, my progress was fast. I graduated Wright in 1935, won a scholarship to the University of Chicago, and received my B.S. degree there in 1936.

---

* This is an updated version of his Nobel Lecture (1979), delivered by Professor H.C. Brown at the University of Rajasthan and other centres during his visit to India in 1985.

** Professor, Richard B. Wetherill Laboratory, Purdue University, West Lafayette, Indiana, USA.

At the time, my ambitions did not extend to a higher degree. I wanted to obtain a job and marry my classmate, Sarah Baylen. However, Fate—in the form of Professor Julius Steiglitz-intervened. He was then a leading organic chemist in the U.S., past President of the American Chemical Society, and Professor Emeritus at the University of Chicago. I was a student in his renowned course, Advanced Organic Chemistry. He asked me why I had not applied for a teaching assistantship, which would make it possible for me to go on to the Ph.D. degree. I told him of my plans to marry—I did not think I could handle both a marriage and a Ph.D. He strongly recommended that Sarah and I defer our marriage so that I could continue study.

After discussing the matter with Sarah, we decided to follow his advice. But not all of it. One morning when he was not looking, we went down to City Hall with a fellow student sworn to secrecy as witness and we were married "secretly". But we were so ignorant about such matters that we were not aware that such marriages are routinely published in the daily newspapers. Consequently, our secret marriage was not secret for long. Fortunately, we survived.

I. Introduction

I decided to do my doctorate research with Professor H.I. Schlesinger in the area of boron hydrides.

In 1936 diborane, B_2H_6, was a rare substance, prepared in less than gram quantities in only two laboratories—that of Alfred Stock at Karlsruhe, Germany, and of H.I. Schlesinger at the University of Chicago. The existence of the simplest hydrogen compound of boron, not as BH_3, but as B_2H_6 was considered to constitute a serious problem for the electronic theory of G.N. Lewis.[1] The reactions of diborane were under study at the University of Chicago by Professor H.I. Schlesinger and his research assistant, Anton B, Burg, in the hope that a knowledge of the chemistry would aid in resolving the structural problem.

Why did I decide to undertake my doctorate research in the exotic field of boron hydrides? As it happened on my graduation in 1936, my girlfriend, Sarah Baylen, presented me with a graduation gift, Alfred Stock's book, *The Hydrides of Boron and Silicon.*[1] I read this book and became interested in the subject. How did it happen that she selected this particular book? This was the time of the Depression. None of us had much money. It appears that she selected as her gift the most economical chemistry book ($2.00) available in the University of Chicago bookstore. Such are the developments that can shape a career!

After the usual discussion of possible problems with H.I. Schlesinger and Anton B. Burg, I decided to explore the reaction of diborane with aldehydes, ketones and esters, based upon some preliminary unpublished experiments of Anton B. Burg.[2]

Once I mastered the high-vacuum techniques developed by Stock for work with diborance, it did not take me long to explore the reactions of diborane with aldehydes, ketones, esters and acid chlorides. It was established that simple aldehydes and ketones react rapidly with diborane at 0°C (even at –78°C) to produce dialkoxyboranes (1).

$$2R_2CO + 1/2(BH_3)_2 \rightarrow (R_2CHO)_2BH \quad (1)$$

These dialkoxyboranes are rapidly hydrolyzed by water to give the corresponding alcohols (2).

$$(R_2CHO)_2BH + 3H_2O \rightarrow 2R_2CHOH + H_2 + B(OH)_3 \quad (2)$$

The reactions with methyl formate and ethyl acetate were slower, but quantitative reductions were achieved. No appreciable reaction was observed with chloral, acetyl chloride and carbonyl chloride.[3]

My Ph.D. thesis was completed in 1938 and the contents were published in 1939.[3] At the time the organic chemist had available no really satisfactory method for reducing the carbonyl group of aldehydes and ketones under such mild conditions. Yet little interest in this development was evinced. Why?

In 1939 diborane was a very rare substance, prepared in only minor amounts in two laboratories in the world, handled only by very specialized techniques. How could the synthetic organic chemist consider using such a rare substance as a reagent in his work?

It would be nice to report that one of the three authors had the foresight to recognize that the development of practical methods of preparing and handling diborane would make this reductive procedure of major interest to organic chemists throughout the world. But this was not the case. The problem was later solved, but primarily because of the requirements of research supporting the war effort, and not because of intelligent foresight.

II. The Alkali Metal Hydride Route to Diborane and Borohydrides

In 1939 Anton B. Burg transferred to the University of Southern California and I became research assistant to Professor Schlesinger. In the Fall of 1940 he was requested to undertake for the National Defense Research Committee a search for new volatile compounds of uranium of low molecular weight. As his research assistant, I became his lieutenant in this was research program.

Just prior to this development, aluminium borohydride,[4] $Al(BH_4)_3$, beryllium borohydride,[5] $Be(BH_4)_2$, and lithium borohydride,[6] $LiBH_4$, had been synthesized in our laboratories. The lithium derivative was a typical non-volatile, salt-like compound, but the aluminium and beryllium derivatives were volatile, the most volatile compounds known for these elements. Accordingly, we undertook to synthesize the unknown uranium (IV) borohydride (3).

$$UF_4 + 2Al(BH_4)_3 \rightarrow U(BH_4)_4\uparrow + 2AlF_2(BH_4)\downarrow \quad (3)$$

The synthesis was successful.[7] Moreover, the product, $U(BH_4)_4$, had a low molecular weight (298) and adequate volatility. We were requested to supply relatively large amounts of the material for testing.

The bottleneck was diborane. We had six diborane generators operated by six young men. Each generator could produce 0.5 g of diborane per 8-hour working day, a total product — when all went well — of 3 g per day, or 1 kg per year! Clearly, we had to find a more practical route to diborane.

We soon discovered that the reaction of lithium hydride with boron triflouride in ethyl ether solution provided such a route[8] (4).

$$6LiH + 8BF_3 \cdot OEt_2 \xrightarrow{Et_2O} (BH_3)_2\downarrow + 6LiBF_4\downarrow \qquad (4)$$

We could now prepare diborane in quantity and transform it into uranium (IV) borohydride by simple reactions (5, 6).

$$LiH + 1/2(BH_3)_2 \xrightarrow{Et_2O} LiBH_4 \qquad (5)^9$$

$$AlCl_3 + LiBH_4 \xrightarrow{\Delta} Al(BH_4)_3\uparrow + 3LiCl\downarrow \qquad (6)^{10}$$

$$UF_4 + 2Al(BH_4)_3 \longrightarrow U(BH_4)_4\uparrow + 2AlF_2(BH_4)\downarrow \qquad (3)$$

Unfortunately, we were informed that lithium hydride was in very short supply and could not be spared for this synthesis. We could have to use sodium hydride instead.

Unfortunately, with the solvents then available, the direct use of sodium hydride was not successful. However, a new compound, sodium trimethoxy-borohydride,[11] readily synthesized from sodium hydride and methyl borate, solved the problem (7).

$$NaH + B(OCH_3)_3 \longrightarrow NaBH(OCH_3)_3 \qquad (7)$$

It proved to be very active and provided the desired transformations previously achieved with lithium hydride (4-6).

At this stage we were informed that the problems of handling uranium hexafluride had been overcome and there was no longer any need for uranium borohydride. We were on the point of disbanding our group when the Army Signal Corps informed us that the new chemical, sodium borohydride, appeared promising for the field generation of hydrogen. However, a more economical means of manufacturing the chemical was required. Would we undertake a research program with this objective?

We soon discovered that the addition of methyl borate to sodium hydride maintained at 250°C provided a mixture of sodium borohydride and sodium methoxide[12] (8).

$$4NaH + B(OCH_3)_3 \xrightarrow{250^\circ C} NaBH_4 + 3NaOCH_3 \qquad (8)$$

This provides the basis for the present industrial process for the manufacture of sodium borohydride.

III. Reactions with Complex Hydrides

In the course of search for a solvent to separate the two reaction products, acetone was tested. Rapid reduction of the acetone was observed[9] (9).

$$\begin{array}{l} NaBH_4 + 4R_2C{=}O \rightarrow NaB(OCHR_2)_4 \\ \qquad\qquad\qquad\qquad\quad \downarrow H_2O \\ \qquad\qquad\qquad\quad NaB(OH)_4 + 4R_2CHOH \end{array} \tag{9}$$

In this way it was discovered that sodium borohydride is a valuable reagent for the hydrogenation of organic molecules.

At this stage I departed the University of Chicago for Wayne University (Detroit). With the much smaller opportunities for research at this institution, I concentrated on my program dealing with steric strains.[13,14]

At the University of Chicago the alkali metal hydride route was successfully extended for the synthesis of the corresponding aluminium derivatives. Thus lithium aluminium hydride was synthesized in 1945 by the reaction of lithium hydride and aluminium chloride in ether solution[15] (10).

$$4LiH + AlCl_3 \xrightarrow{Et_2O} LiAlH_4 + 3LiCl \tag{10}$$

The discovery of sodium borohydride[9] in 1942 and of lithium aluminium hydride[15] in 1945 brought about a revolutionary change in procedures for the reduction of functional groups in organic molecules.[16] As first described by W.G. Brown and his coworkers, there is a major difference in the behaviour of these two reducing agents.[16] Lithium aluminium hydride is an exceedingly powerful reducing agent, capable of reducing practically all functional groups. On the other hand, sodium borohydride is a remarkably mild reducing agent, readily reducing only aldehydes, ketones and acid chlorides.[16] Consequently, we had available two reagents which exhibited extremes in their reducing capabilities (Fig. 1).

In 1947 I came to Purdue University with the opportunity for markedly expanding my research program. I decided to explore means of increasing the reducing properties of sodium borohydride and of decreasing the reducing properties of lithium aluminium hydride. In this way the organic chemist would have at his disposal a full spectrum of reducing agents–he could select that reagent which could be most favourable for the particular reduction required in a given situation.

The investigations of several research groups soon established that changes in the metal ion from sodium or potassium to lithium,[17a–c] to magnesium,[17e'd] and to aluminium[17e–g] greatly increase the reducing power of the borohydride moiety[17](11)

$$\underset{\text{increasing reducing power}}{\overrightarrow{NaBH_4 < LiBH_4 < Mg(BH_4)_2 < Al(BH_4)_3}} \tag{11}$$

	$NaBH_4$									$LiAlH_4$
Aldehyde	+									+
Ketone	+									+
Acid chloride	R									+
Lactone	–									+
Epoxide	–									+
Ester	–									+
Acid	–									+
Acid salt	–									+
tert-Amide	–									+
Nitrile	–									+
Nitro	–									+
Olefin	–									–

Fig. 1 Sodium borohydride and lithium aluminum hydride as extremes in a possible spectrum of hydridic reducing agents.

On the other hand, the reducing power of lithium aluminium hydride could be diminished by the introduction of alkoxy substituents[18,19] (12).

$$\underset{\text{decreasing reducing power}}{\xleftarrow{LiAlH(Ot\text{-}Bu)_3 < LiAlH(OMe)_3 < LiAlH_4}} \qquad (12)$$

Indeed, it has proven possible to enhance the reducing power of a borohydride ($LiEt_3BH$)[20] until it exceeds that of sodium borohydride and to diminish the reducing power [$K(i\text{-}PrO)_3BH$)][21] so that it is even less than that of the parent borohydride (Fig. 2).

Finally, we discovered major differences between reduction by electrophilic reagents, such as diborane[22] and aluminium hydride,[23] and by nucleophilic reagents, such as sodium borohydride[16] and lithium aluminium hydride[16] (Fig. 3). It is often possible now to reduce group A in the presence of B, or group B in the presence of A, by a careful choice of reagents. This is nicely illustrated by the synthesis of both [R]- and [S]- mevalonolactone from a common precursor[24,25] (Fig. 4).

It should be pointed out that these studies were greatly facilitated by many exceptional coworkers, among whom I would like to mention especially: Nung Min Yoon and S. Krishnamurthy.

EFFECT OF SUBSTITUENTS

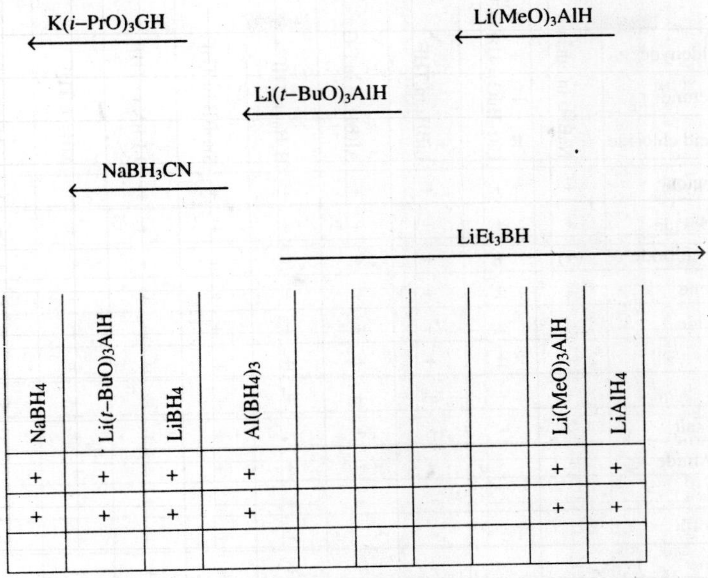

Fig. 2 Alteration of the reducing power of the two extreme reagents, sodium borohydride and lithium aluminium hydride.

IV. Hydroboration

In the course of these studies of selective reductions, a minor anomaly resulted in the discovery of hydroboration. My coworker, Dr. B. Subha Rao, was examining the reducing characteristics of sodium borohydride in diglyme catalyzed by aluminium chloride.[17] He observed that the reduction of ethyl oleate under our standard conditions, 4 moles of hydride per mole of compound, one hour at 25°, took up 2.37 equivalents of hydride per mole of ester. This contrasted with a value of 2.00 for ethyl stearate. Investigation soon established that the reagent was adding an H—B⟨ bond to the carbon-carbon double bond to form the corresponding organoborane.[26]

Exploration of this reaction soon established improved procedures for carrying it out. Of special value was the discovery that the addition of diborane to alkenes was markedly catalyzed by ethers.[27] In the presence of such ethers, the reaction is practically instantaneous and quantitative (13).

$$\rangle C=C\langle + H-B\langle \rightarrow H-\overset{|}{\underset{|}{C}}-\overset{|}{\underset{|}{C}}-B\langle \qquad (13)$$

	$NaBH_4$ in ethanol	$Li(t\text{-}BuO)_3AlH$ in THF	$LiBH_4$ in THF	$Al(BH_4)_3$ in DG	B_2H_6 in THF	Sia_2BH in THF	9-BBN in THF	AlH_3 in THF	$Li(MeO)_3AlH$ in THF	$LiAlH_4$ in THF
Aldehyde	+	+	+	+	+	+	+	+	+	+
Ketone	+	+	+	+	+	+	+	+	+	+
Acid chloride	(+)	+	+	+	–	–	+	+	+	+
Lactone	–	±	+	+	+	+	+	+	+	+
Epoxide	–	±	+	+	+	±	±	+	+	+
Ester	–	±	+	+	±	–	±	+	+	+
Acid	–	–	–	+	+	–	±	+	+	+
Acid salt	–	–	–	–	–	–	–	+	+	+
tert-Amide	–	–	–	–	+	+	+	+	+	+
Nitrile	–	–	–	–	+	–	±	+	+	+
Nitro	–	–	–	–	–	–	–	–	+	+
Olefin	–	–	–	–	+	+	+	–	–	–

Fig. 3 Variation in reduction characteristics with electrophilic and nucleophilic reagents.

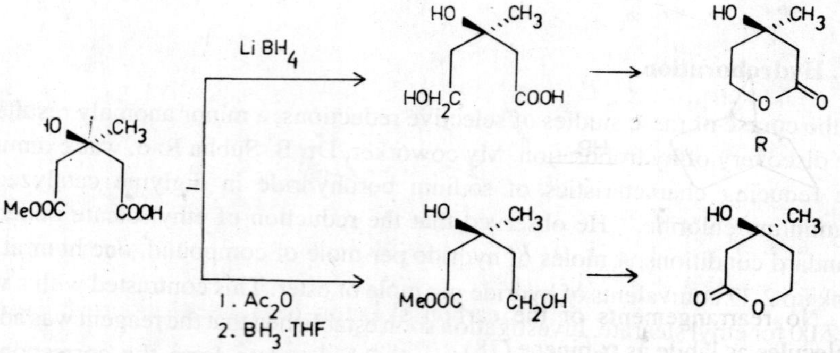

Fig. 4 Synthesis is of both (*R*)- and (*S*)-mevalonolactone by selective reduction of a common precursor.

(My parents were far-seeing in giving me the initials H.C.B.)

Dr. Subha Rao established that oxidation of such organoboranes, *in situ* with alkaline hydrogen peroxide, proceeds quantitatively, producing alcohols with the precise structure of the organoborane[26,27] (14).

$$\mathrm{H{-}\overset{|}{\underset{|}{C}}{-}\overset{|}{\underset{|}{C}}{-}B\!\!<\ \xrightarrow[\sim 40^{\circ}C]{H_2O_2}\ H{-}\overset{|}{\underset{|}{C}}{-}\overset{|}{\underset{|}{C}}{-}OH} \qquad (14)$$

At this stage in the development, Dr. B.C. Subha Rao returned to India, after spending five years with me. Fortunately, an equally competent and productive coworker, Dr. George Zweifel, soon joined my group. Although trained at the E.T.H. in Zurich as a carbohydrate chemist, he expressed a deep interest in the possibilities of the hydroboration reaction and progress was extraodinarily rapid.[28]

It was soon established that the addition proceeds in an anti-Markovnikov manner (15).

$$CH_3CH_2\overset{\overset{CH_3}{|}}{C}=CH_2 \xrightarrow{HB} \underset{99\%}{CH_3CH_2\overset{\overset{CH_3}{|}}{\underset{\underset{H}{|}}{C}}CH_2B\langle} + \underset{1\%}{CH_3CH_2\overset{\overset{CH_3}{|}}{\underset{\underset{B\diagup\diagdown}{|}}{C}}CH_3} \quad (15)$$

The reaction involves a *cis*-addition of the H—B⟨ bond (16).

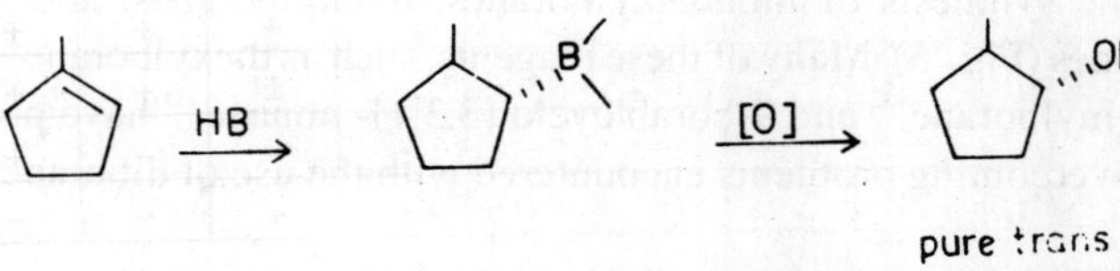

The addition takes place preferentially from the less hindered side of the double bond (17).

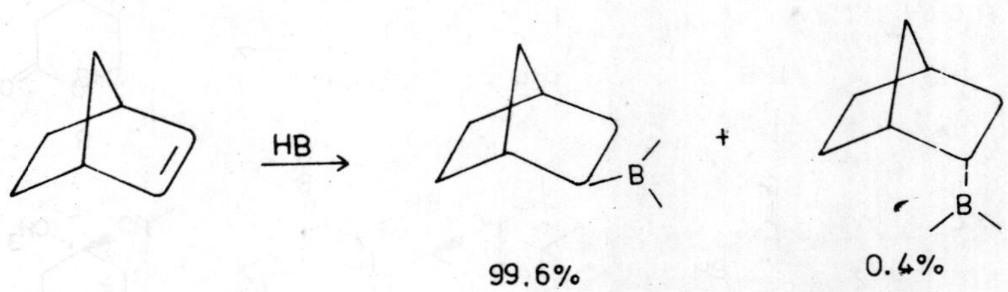

No rearrangements of the carbon skeleton have been observed, even in molecules as labile as α-pinene (18).

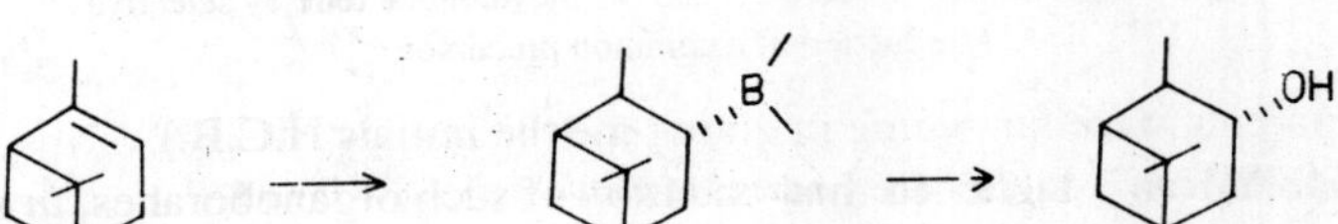

Most functional groups can tolerate hydroboration (19).

$$CH_2=CHCH_2CO_2R \xrightarrow{HB} \rangle B-CH_2CH_2CH_2CO_2R \quad (19)$$

Now the organic chemist can conveniently synthesize reactive intermediates containing functional groups and utilize those intermediates to form products with new carbon-carbon bonds.

Standardized procedures for hydroboration have been developed and are fully described, as well as the utilization of the organoborane products for organic syntheses.[29] (To conserve space, references will be given only to developments which have appeared since the publication of this book.)

V. New Hydroborating Agents

The hydroboration of simple olefins generally proceeds directly to the formation of the trialkylborane[28,29] (20).

$$3CH_3C(CH_3)=CH_2 + BH_3 \xrightarrow{THF} [(CH_3)_2CHCH_2]_3B \qquad (20)$$

However, in number of instances it has been possible to control the hydroboration to achieve the synthesis of monoalkylboranes, dialkylboranes, and cyclic and bicyclic boranes (Fig. 5). Many of these reagents, such as thexylborane[30] disiamylborane,[31] dipinylborane,[32] and 9-borabicyclo [3.3.1]- nonane,[33] have proven to be valuable in overcoming problems encountered with the use of diborane itself.

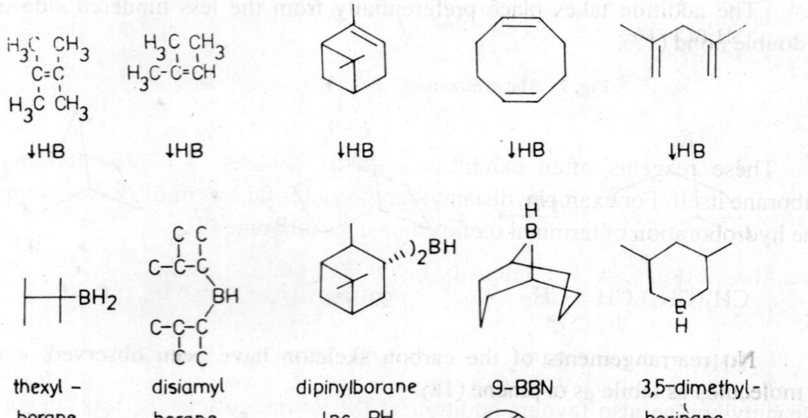

Fig. 5 The synthesis of partially alkylated boranes.

In a number of cases, hydroboration with heterosubstituted boranes has also proven valuable. Research in this area was greatly facilitated by exceptional contributions from S.K. Gupta, N. Ravindran, and S.U. Kulkarni. For example, catecholborane[34] and the chloroborane etherates[35] (Fig. 6), permit the synthesis of boronic and borinic acid esters, as well as the synthesis of the simple mono- and

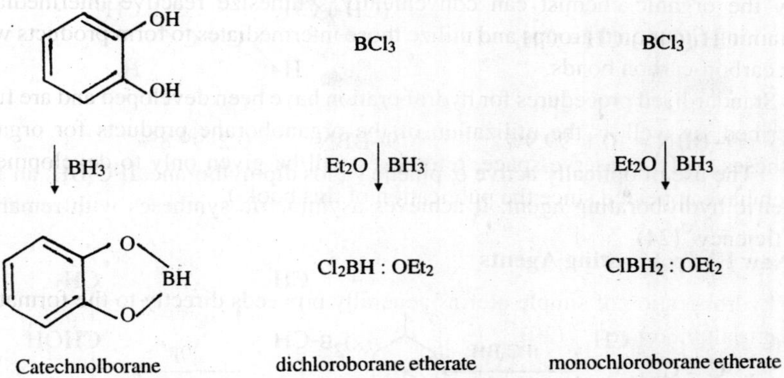

Fig. 6 The synthesis of heterosubstituted boranes.

dialkylchloroboranes, $RBCl_2$ and R_2BCl. The corresponding haloborane-dimethyl sulfides are stabler and easier to work with[36] (Fig. 7).

$Cl_3B{:}SMe_2$	$Br_3B{:}SMe_2$	I_2
+	+	+
$2H_3B{:}SMe_2$	$2H_3B{:}SMe_2$	$2H_3B{:}SMe_2$
↓	↓	↓
$3H_2BCl{:}SMe_2$	$3H_2BBr{:}SMe_2$	$H_2BI{:}SMe_2$

Fig. 7 The synthesis of heterosubstituted boranes.

These reagents often exhibit marked advantages in hyuroboration over diborane itself. For example, disiamylborane yields far less of the minor isomer in the hydroboration of terminal olefins than does diborane (21).

$$\begin{array}{llll} & CH_3(CH_2)_3CH = CH_2 & \quad & CH_3(CH_2)_3CH = CH_2 \\ & \quad\uparrow \qquad\quad \uparrow & & \quad\uparrow \qquad \uparrow \\ BH_3 & \quad 6 \qquad 94\% & Sia_2BH & \quad 1 \qquad 99\% \end{array} \tag{21}$$

Disiamylborane also favours addition of the boron atom to the less substituted position of a 1, 2-dialkylethylene (22).

$$\begin{array}{llll} & (CH_3)_2CH\text{—}CH = CH\text{—}CH_3 & \quad & (CH_3)_2CH\text{—}CH = CH\text{—}CH_3 \\ & \qquad\qquad\uparrow \quad \uparrow & & \qquad\qquad\uparrow \quad \uparrow \\ BH_3 & \qquad\quad 43 \quad 57\% & Sia_2BH & \qquad\quad 3 \quad 97\% \end{array} \tag{22}$$

9-BBN exhibits as even greater selectivity (23).

$$CH_3(CH_2)_3CH=CH_2 \qquad (CH_3)_2CH(H)C=C(H)CH_3 \qquad (23)$$

9–BBN ↑ 0.1 ↑ 99.9% 9–BBN ↑ 0.2 ↑ 99.8%

The use of optically active α-pinene yields dipinylborane, IPC_2BH, an asymmetric hydroborating agent. It achieves asymmetric syntheses with remarkable efficiency[32] (24).

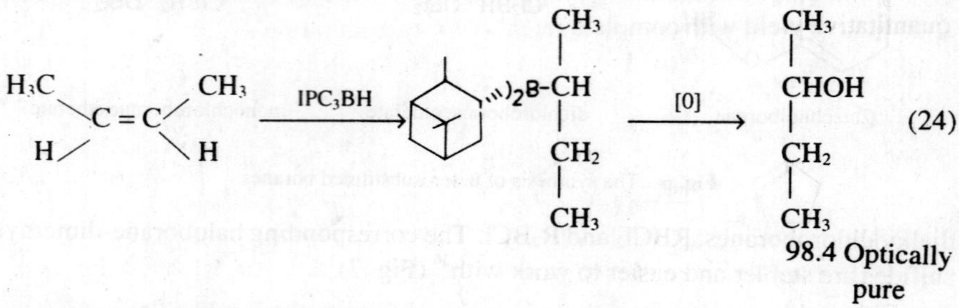

(24)

VI. The Versatile Organoboranes

At the time we were exploring the hydroboration reaction, many individuals expressed skepticism to me as to the wisdom of devoting so much research effort to this reaction. After all, hydroboration produced organoboranes. Relatively little new chemistry of organoboranes had appeared since the original classic publication by Frankland in 1862.[37] They took the position that the lack of published material in this area meant that there was little of value there.

In this case it is now clear that this position is not correct. After our exploration of the hydroboration reaction had proceeded to the place where we felt we understood the reaction and could apply it with confidence to new situations, we began a systematic exploration of the chemistry of organoboranes. This research, facilitated by a host of exceptionally capable coworkers, among whom may be mentioned M.M. Rogic, M.M. Midland, C.F. Lane, A.B. Levy, R.C. Larock, and Y. Yamamoto, made it clear that the organoboranes are among the most versatile chemical intermediates available to the chemist.

It is not possible here to give more than a taste of the rich chemistry. For a more complete treatment, the reader must go elsewhere.[29,38]

Simple treatment of the organoborane with halogen in the presence of a suitable base produces the desired organic halide[39,40] (25, 26).

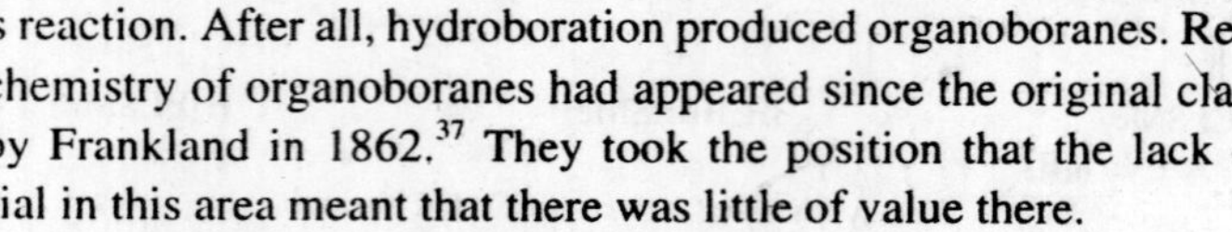

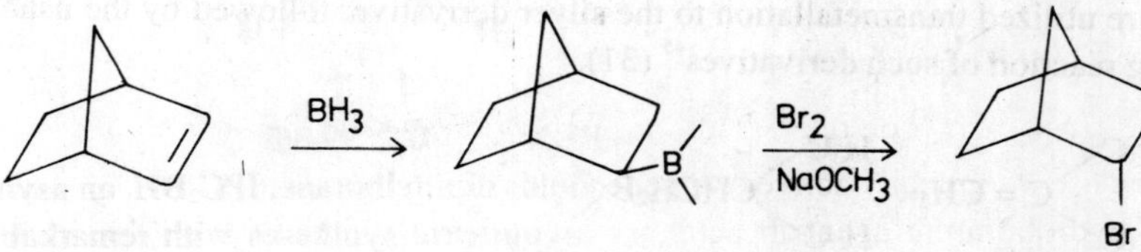

(26)

Oxidation with alkaline hydrogen peroxide produces the alcohol in essentially quantitative yield with complete retention of configuration[41] (27).

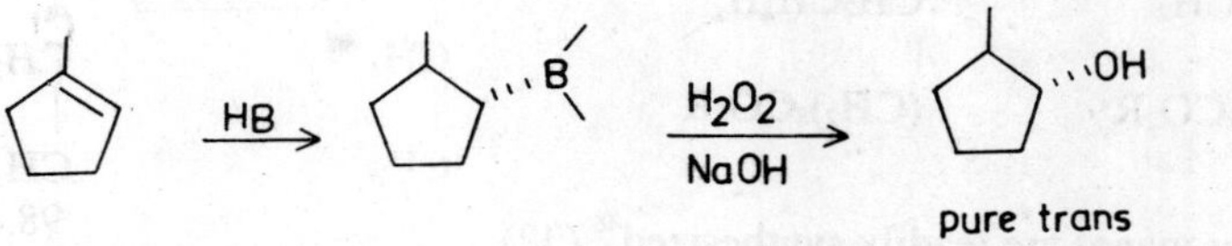

(27)

Either chloroamine or *O*-hydroxylaminesulfonic acid can be used to convert organoborances into the corresponding amines[42] (28).

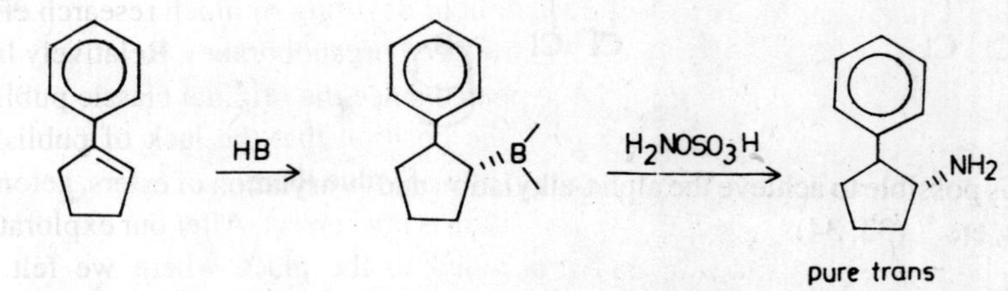

(28)

The reaction of organoboranes with organic azides proceeds sluggishly with the more hindered organoboranes. Fortunately, this difficulty can be circumvented with the new hydroborating agents[43] (29).

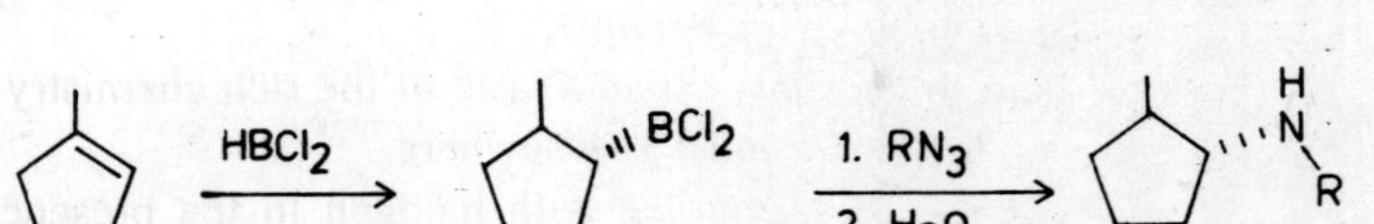

(29)

Other organometallics can be synthesized from the organoboranes[44] (30).

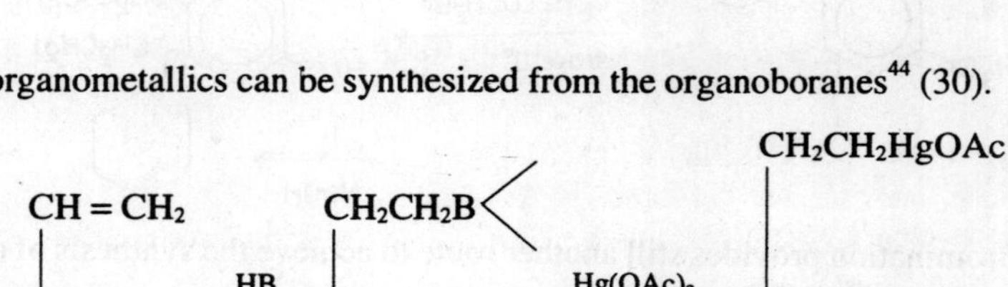

(30)

The organoboranes can also be utilized to form carbon-carbon bonds.[29] One procedure utilized transmetallation to the silver derivative, followed by the usual coupling reaction of such derivatives[45] (31).

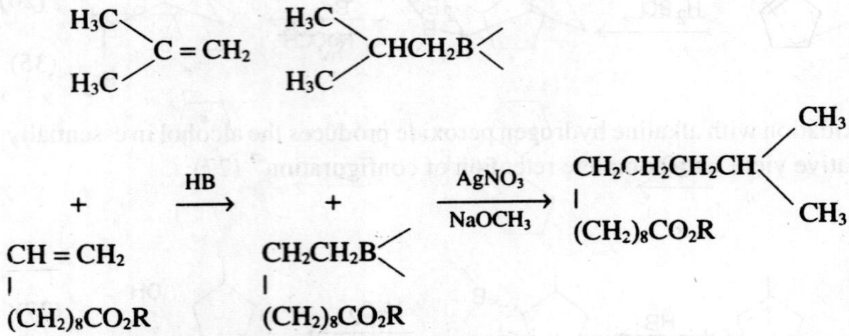

(31)

Cyclopropanes are readily synthesized[46] (32).

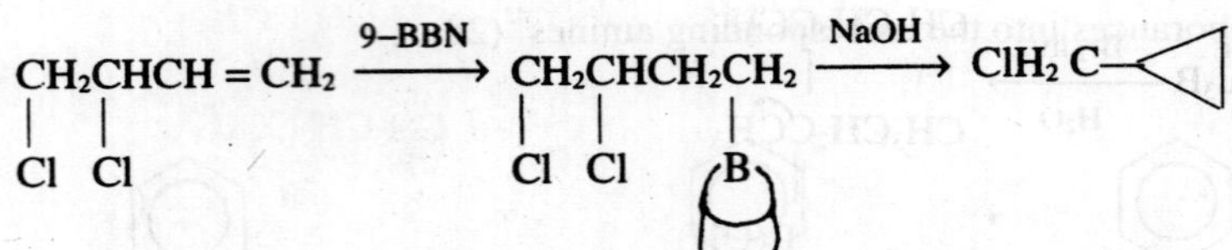

(32)

It is possible to achieve the alpha-alkylation and —arylation of esters, ketones, nitriles, etc.[47] (33, 34).

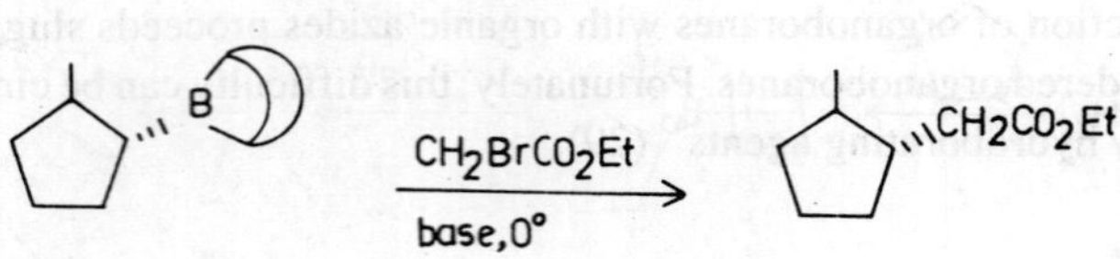

(33)

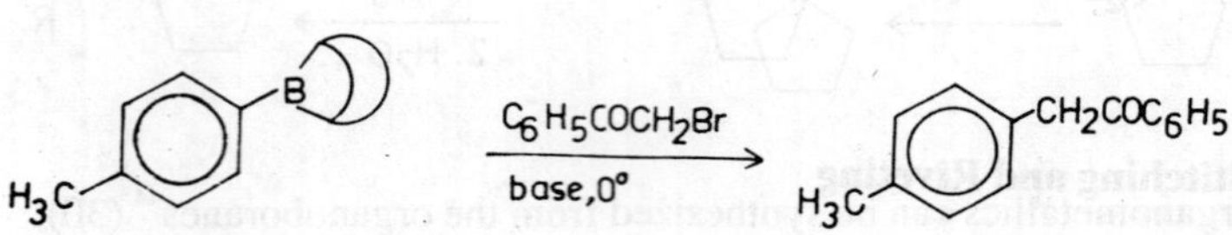

(34)

α-Bromination provides still another route to achieve the synthesis of desired carbon structures[48] (35).

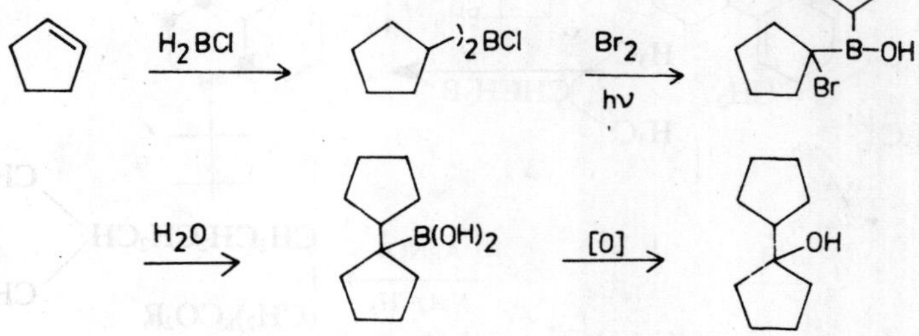

(35)

By means of this reaction, it is possible to combine three *sec*-butyl groups[49] (36).

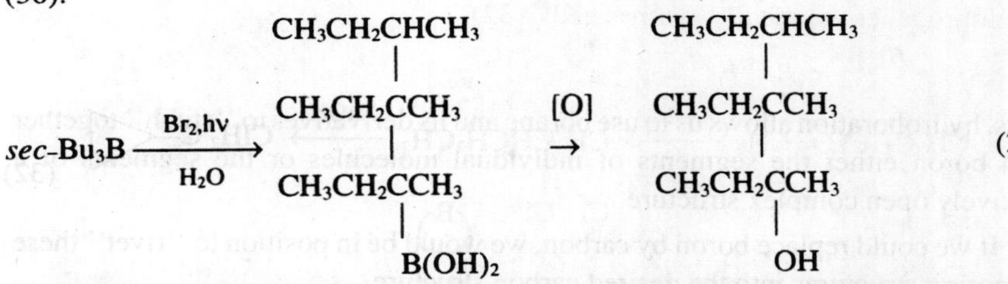

(36)

Finally, it is possible to utilize this reaction to synthesize derivatives not realizable through the Grignard reaction[50] (37).

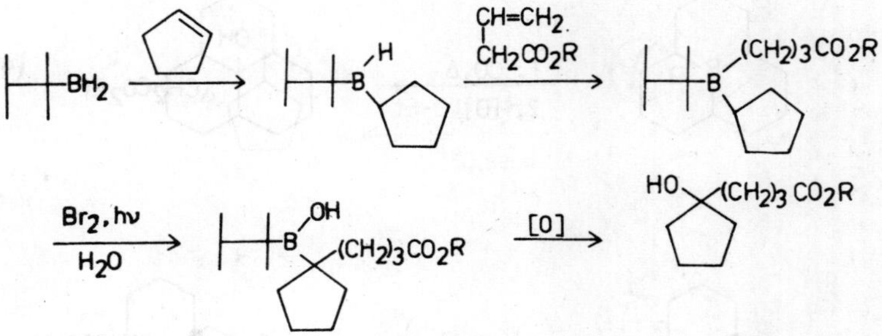

(37)

VII. Stitching and Riveting

The hydroboration reaction allows the chemist to unite to boron under exceptionally mild conditions either three different olefins (38), or to cyclize dienes (39), or trienes (40).

$$3CH_3CH{=}CHCH_3 \xrightarrow{BH_3} (CH_3CH_2\overset{\overset{CH_3}{|}}{C}H—)_3B \qquad (38)$$

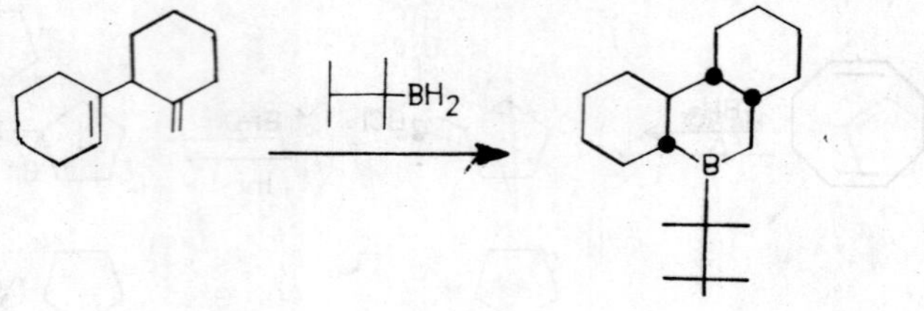

(39)

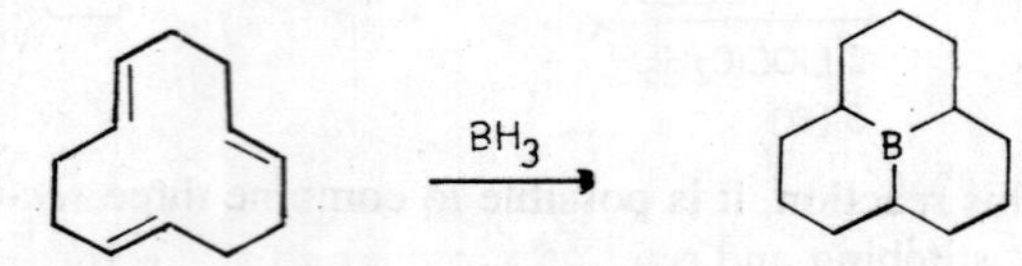

(40)

Thus, hydroboration allows us to use borane and its derivatives to "stitch" together with boron either the segments of individual molecules or the segments of a relatively open complex structure.

If we could replace boron by carbon, we would be in position to "rivet" these temporary strucutres into the desired carbon structure.

In fact, there are now three different procedures which can be used in this way: carbonylation[51] (41), cyanidation[52] (42), and the DCME reaction[53] (43).

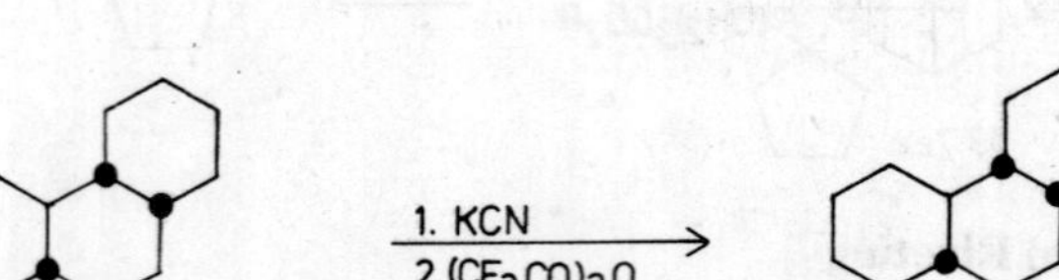

(41)

(42)

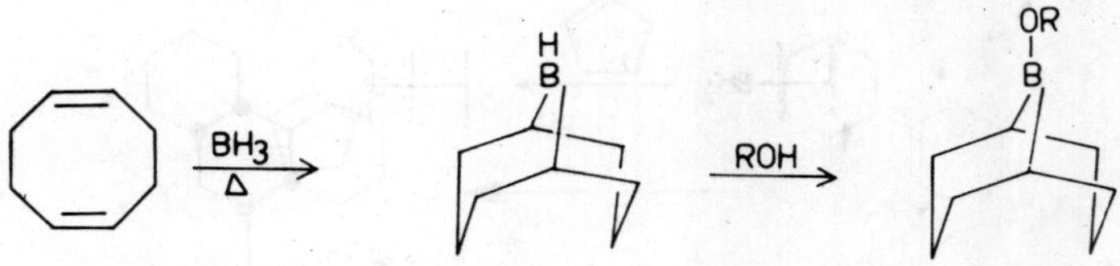

(43)

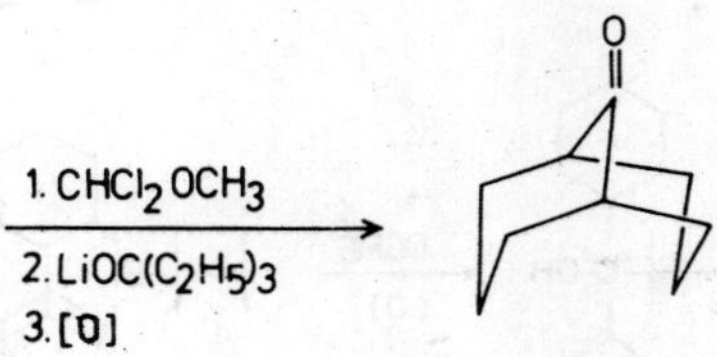

Consequently, stitching and riveting provides an elegant new procedure for the synthesis of complex structures. Its versatility is indicated by the synthesis of an exceptionally hindered tertiary alcohol[54] (44) and by the annelation reaction[55] (Fig. 8).

VIII. Chiral Synthesis via Organoboranes

The remarkable versatility of hydroboration/organoborane chemistry for organic synthesis is indicated by Figure 9. This represents the ''past'' and the ''present'' of my lecture.

Now I would like to discuss a recent development which indicates one important ''future' for this chemistry.

As indicated earlier, the hydroboration of optically active α-pinene (93% ee) yielded diisopinocampheylborane (or dipinylborane).[56] Hydroboration-oxidation of *cis*-2-butene yielded active 2-butanol in 87% ee[56] (45, 46).

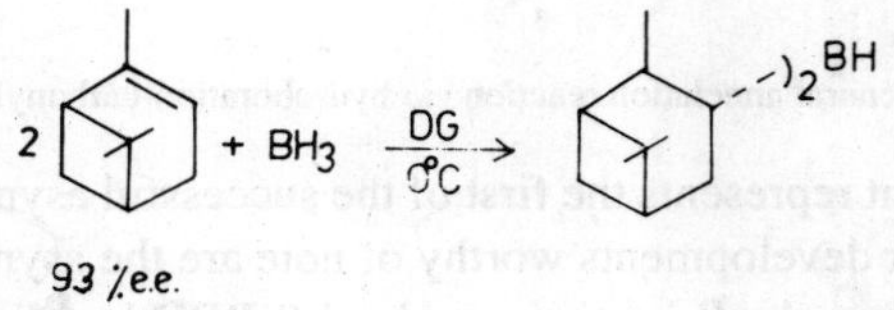

(45)

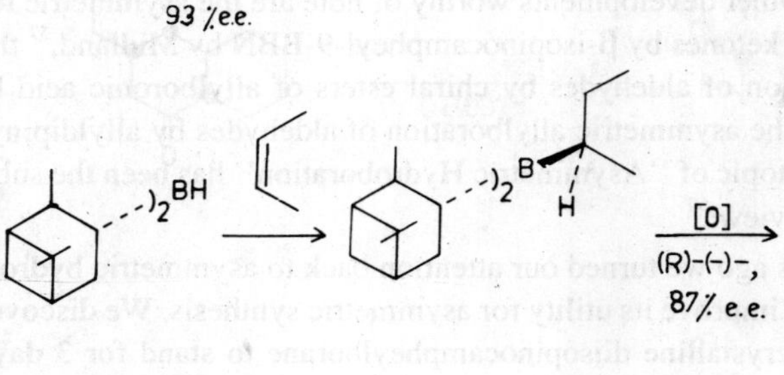

(46)

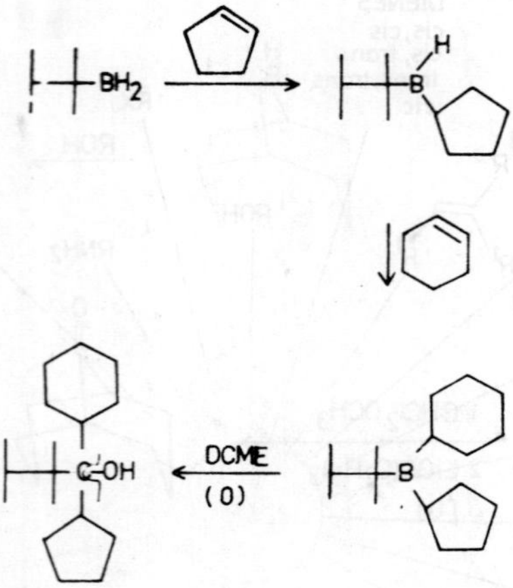

(44)

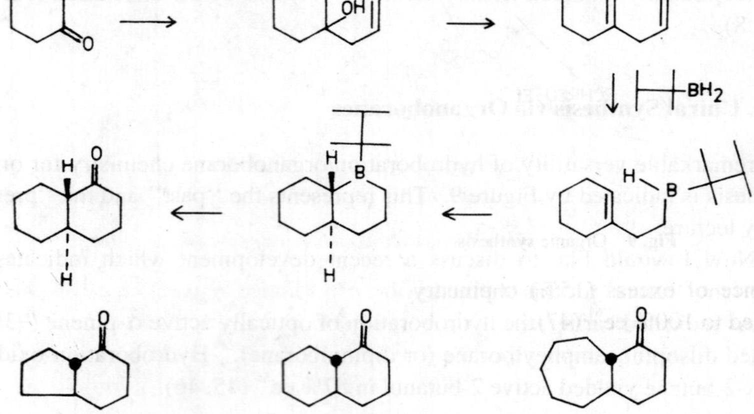

Fig. 8 General annelation reaction *via* hydroboration-carbonylation.

This development represents the first of the successful asymmetric syntheses in recent years. Other developments worthy of note are the asymmetric reduction of aldehydes and ketones by β-isopinocampheyl-9-BBN by Midland,[57] the asymmetric allylboration of aldehydes by chiral esters of allylboronic acid by R.W. Hoffmann[58] and the asymmetric allylboration of aldehydes by allyldipinylborane by Jadhav.[59] The topic of ''Asymmetric Hydroboration'' has been the subject of a recent detailed review.[60]

Several years ago we turned our attention back to asymmetric bydroboration to see if we could improve its utility for asymmetric synthesis. We discovered that by allowing the crystalline diisopinocampheylborane to stand for 3 days in the

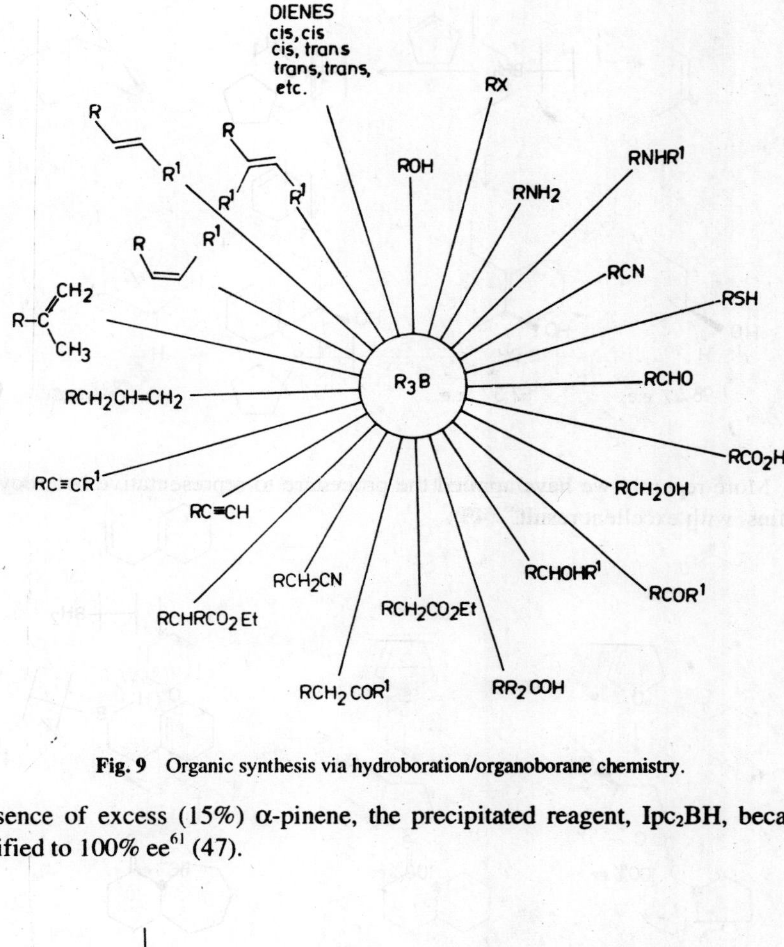

Fig. 9 Organic synthesis via hydroboration/organoborane chemistry.

presence of excess (15%) α-pinene, the precipitated reagent, Ipc_2BH, became purified to 100% ee[61] (47).

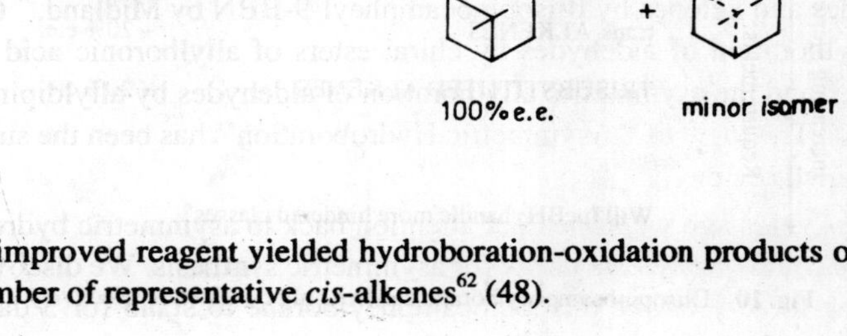

(47)

This improved reagent yielded hydroboration-oxidation products of high ee with a number of representative *cis*-alkenes[62] (48).

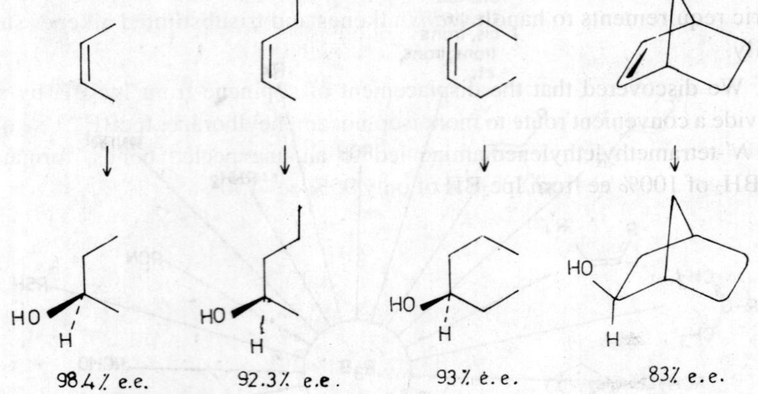

(48)

More recently we have applied the procedure to representative heterocyclic olefins, with excellent result[63] (49).

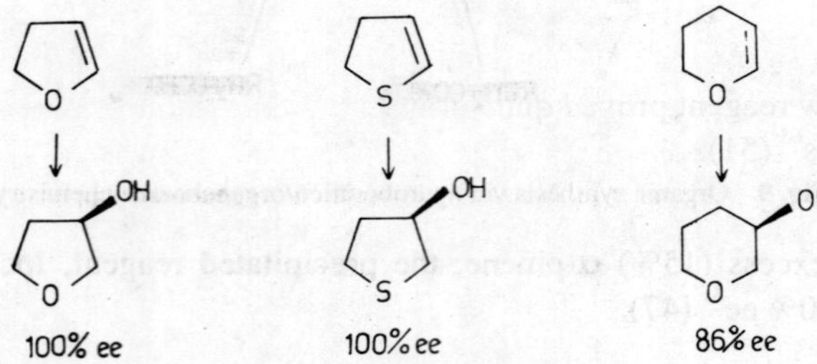

(49)

Unfortunately, the result with 2-methyl-l-alkenes, *trans*-alkenes, and trisubstituted alkenes were less satisfactory[60] (Fig. 10).

Increasing Steric Requirements ↓

2-METHYL-1- ALKENE	~ 20% e.e.
cis-ALKENES	~ 100% e.e.
trans-ALKENES	~ 20% e.e.
TRISUBSTITUTED ALKENES	~ 20% e.e.

Will $IpcBH_2$ handle more hindered classes?

Fig. 10 Diisopinocampheylborane handles only *cis*-alkenes effectively.

We concluded that we required an asymmetric hydroborating agent of lower steric requirements to handle *trans*-alkenes and trisubstituted alkenes satisfactorily.

We discovered that the displacement of α-pinene from Ipc_2BH by amines provide a convenient route to monoisopinocampheylborane, $IpcBH_2$. Use of *N*, *N*, *N*′, *N*′-tetramethylethylenediamine led to an unexpected bonus, formation of $IpcBH_2$ of 100% ee from Ipc_2BH of only 95% ee[64] (50).

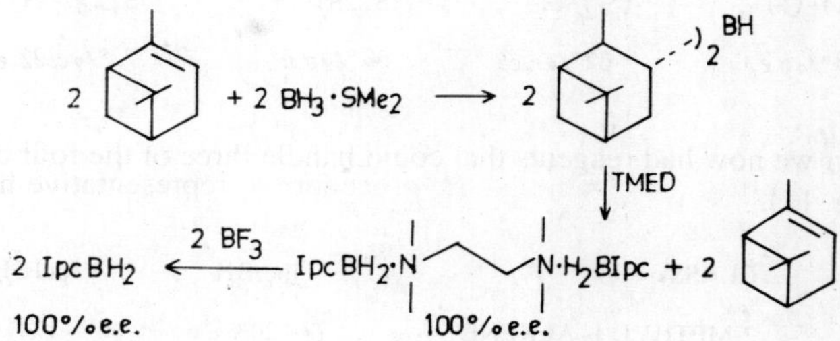

(50)

This new reagent proved quite effective for the asymmetric hydroboration of *trans*-alkenes[65] (51).

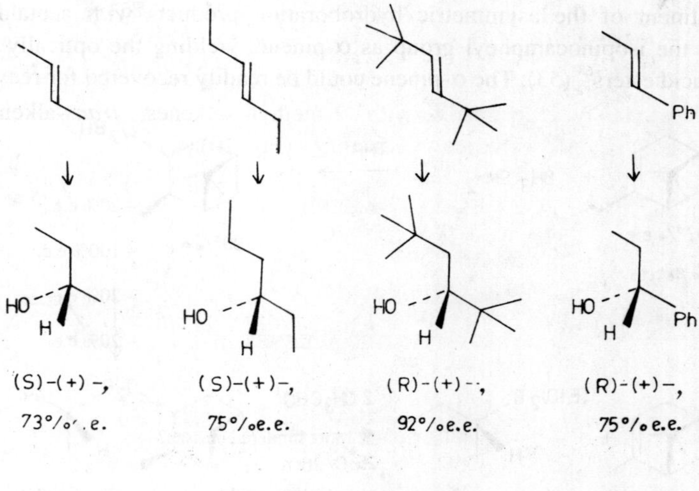

(51)

It also proved applicable to representative trisubstituted alkenes[66,67] (52).

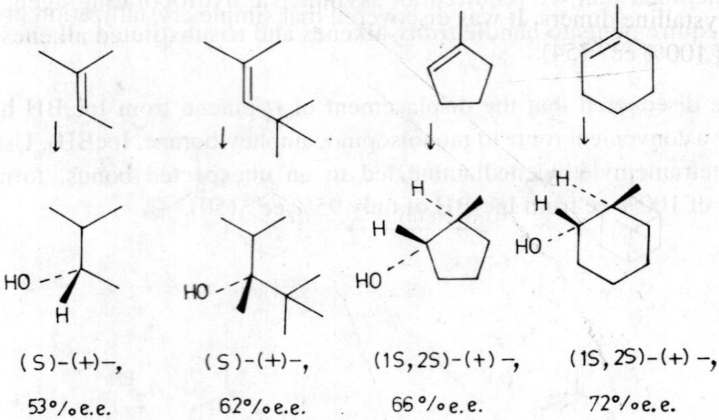

(52)

Consequently, we now had reagents that could handle three of the four classes of alkenes[60] (Fig. 11).

Increasing Steric Requirements ↓

CLASS	Ipc_2BH	$IpcBH_2$
2-METHYL-1-ALKENE	~ 20% e.e.	~ i%
cis-ALKENES	~ 100% e.e.	~ 25%
trans-ALKENES	~ 20% e.e.	70-90% e.e.
TRISUBSTITUTED ALKENES	~ 20% e.e.	60-100% e.e.

Fig. 11 Range of applicability of Ipc_2BH and $IpcBH_2$

Two recent developments proved of major importance. First, we discovered that treatment of the asymmetric hydroboration products with acetaldehyde removed the isopinocampheyl group as α-pinene, yielding the optically active boronic acid esters[68] (53). The α-pinene could be readily recovered for recycle.

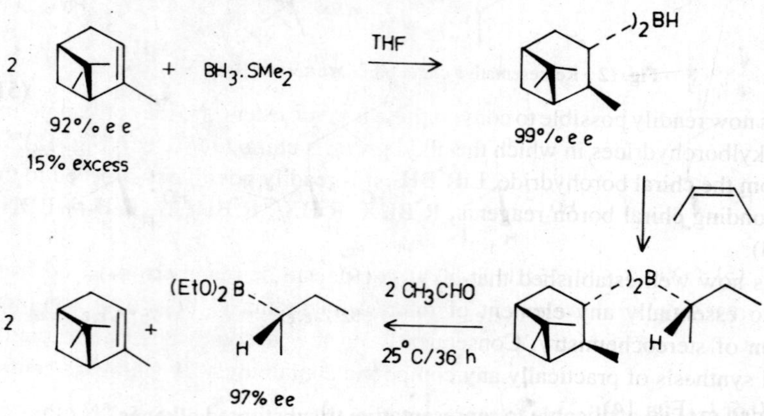

(53)

Secondly, the reaction products from $IpcBH_2$ and hindered alkenes, $IpcR^*BH$, exist as crystalline dimers. It was discovered that simple crystallization provided product of 100% ee[70] (54).

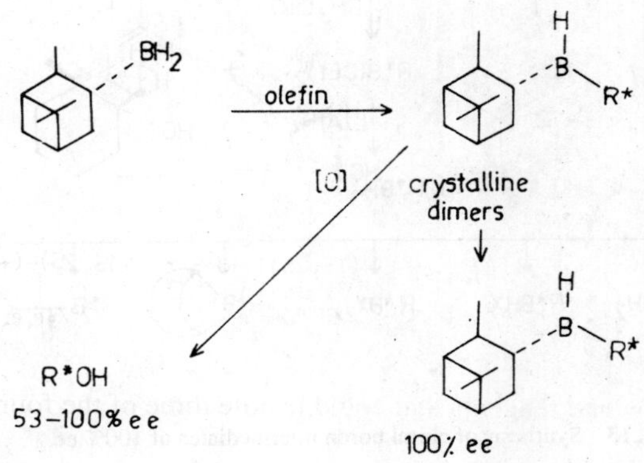

(54)

We could now readily prepare many chiral boronic acid esters of 100% ee (Fig. 12).

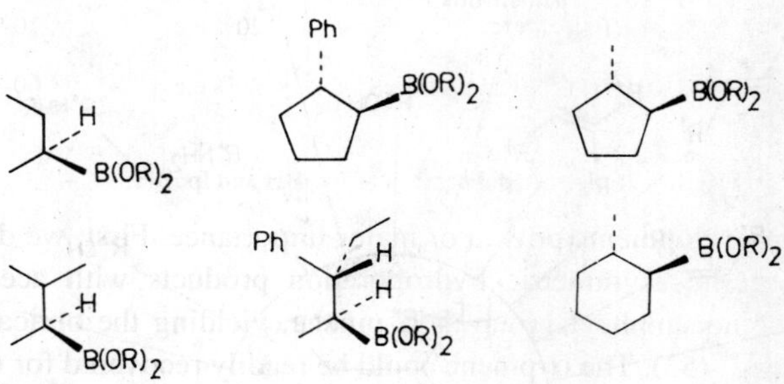

Fig. 12 Representative chiral boronic esters of 100% ee.

It is now readily possible to convert these boronic esters into the corresponding monoalkylborohydrides in which the alkyl group is chiral (100% ee) (Fig. 13).[70]

From the chiral borohydride, LiR^*BH_3, it is readily possible to proceed to the corresponding chiral boron reagents, R^*BHX, R^*BX_2, R^*BHThx, R^*B–9–BBN[71] (Fig. 13).

It is now well established that alkyl groups can be readily transferred from boron to essentially any element of interest, including carbon, with complete retention of stereochemistry. Consequently, it now becomes possible to plan a rational synthesis of practically any compound containing an asymmetric center, either *d*- or *l*- (Fig. 14).

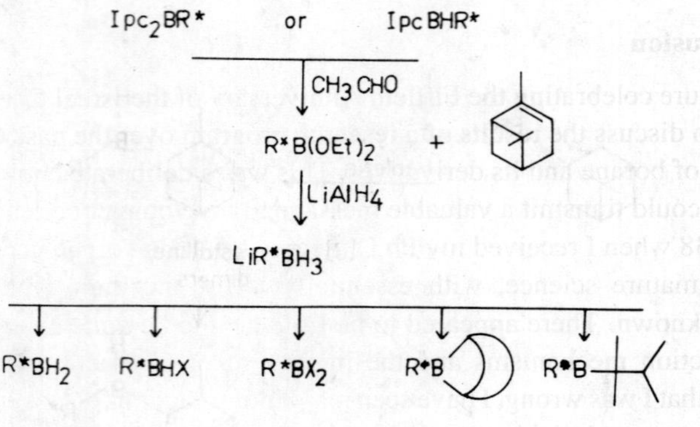

Fig. 13 Synthesis of chiral boron intermediates of 100% ee.

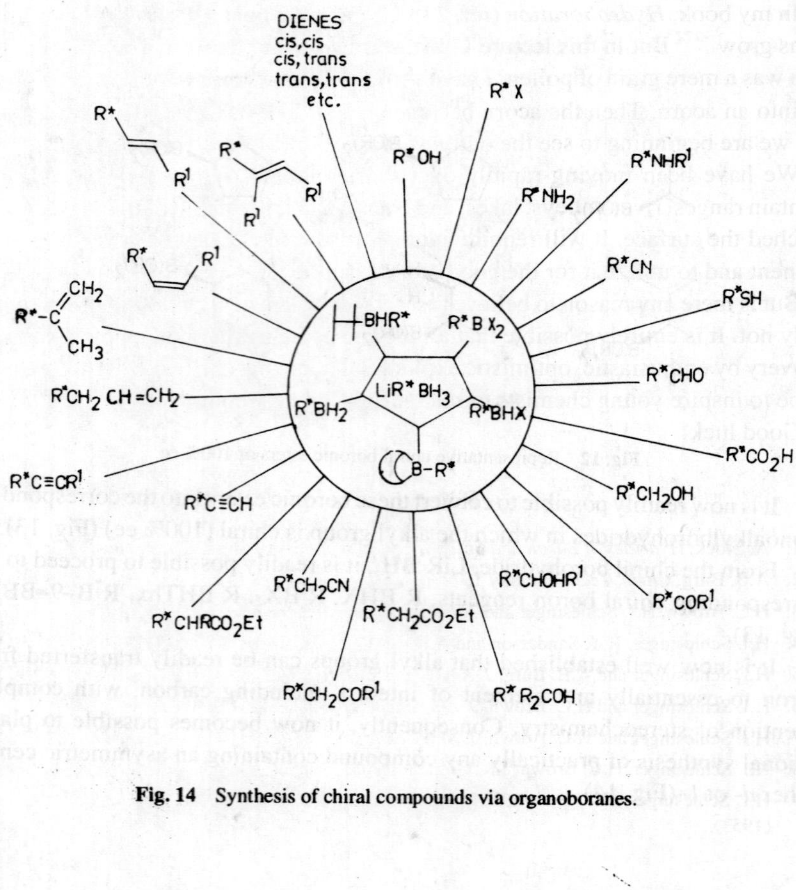

Fig. 14 Synthesis of chiral compounds via organoboranes.

Clearly there is an enormous amount of work facing organic chemists as they explore this new development.

IX. Conclusion

In this lecture celebrating the Fiftieth Anniversary of the Israel Chemical Society, I elected to discuss the results of a research program over the past 48 years on the chemistry of borane and its derivatives. This was a deliberate choice. I felt that in this way I could transmit a valuable message to my younger colleagues.

In 1938 when I received my Ph.D. degree, I felt that organic chemistry was a relatively mature science, with essentially all of the important reactions and structures known. There appeared to be little new to be done except the working out of reaction mechanisms and the improvement of reaction products. I now recognize that I was wrong, I have seen major new reactions discovered. Numerous new reagents are available to us. Many new structures are known to us. We have at hand many valuable new techniques.

I know that many of the students of today feel the same way that I did in 1938. But I see no reason for believing that the next 50 years will not be as fruitful as in the past.

In my book, *Hydroboration* (ref. 28), I quoted the poet: "Tall oaks from little acorns grow."[72] But in this lecture I have started further back, to a time when the acorn was a mere grain of pollen. I have shown how that grain of pollen developed first into an acorn. Then the acorn became an oak. The oak tree became a forest. Now we are beginning to see the outlines of a continent.

We have been moving rapidly over that continent, scouting out the major mountain ranges, river valleys, lakes, and coasts. But it is evident that we have only scratched the surface. It will require another generation of chemists to settle that continent and to utilize it for the good of mankind.

But is there any reason to believe that this is the last new continent of its kind? Surely not. It is entirely possible that all around us lie similar continents awaiting discovery by enthusiastic, optimistic explorers. I hope that one result of this lecture will be to inspire young chemists to search for such new continents.

Good luck!

References

1. A. Stock, *Hydrides of Boron and Silicon*, Cornell University Press, Ithaca, New York, 1933.
2. A.B. Burg, *Chem. Tech.*, January 1977, p. 50.
3. H.C. Brown, H.I. Schlesinger and A.B. Burg, *J. Am. Chem. Soc.*, **61**, 673 (1939).
4. H.I. Schlesinger, R.T. Sanderson and A.B. Burg, *J. Am. Chem. Soc.*, **62**, 3421 (1940).
5. H.I. Schlesinger and A.B. Burg, *J. Am. Chem. Soc.*, **62**, 3425 (1940).
6. H.I. Schlesinger and H.C. Brown, *J. Am. Chem. Soc.*, **62**, 3429 (1940).
7. H.I. Schlesinger and H.C. Brown, *J. Am. Chem. Soc.*, **75**, 219 (1953).
8. H.I. Schlesinger, H.C. Brown, J.R. Gilbreath and J. Katz, *J. Am. Chem. Soc.*, **75**, 195 (1953).
9. H.I. Schlesinger, H.C. Brown, H.R. Hoekstra and L.R. Rapp, *J. Am. Chem. Soc.*, **75**, 199 (1953).

10. H.I. Schlesinger, H.C. Brown and E.K. Hyde, *J. Am. Chem. Soc.*, **75**, 209 (1953).
11. H.C. Brown, H.I. Schlesinger, I. Sheft and D.M. Ritter, *J. Am. Chem. Soc.*, **75**, 192 (1953).
12. H.I. Schlesinger, H.C. Brown and A.E. Finholt, *J. Am. Chem. Soc.*, **75**, 205 (1953).
13. H.C. Brown, *J. Chem. Soc.*, 1248 (1956).
14. H.C. Brown, *Boranes in Organic Chemistry,* Cornell University Press, Ithaca, New York, 1972.
15. A.E. Finbolt, A.C. Bond, Jr. and H.I. Schlesinger, *J. Am. Chem. Soc.*, **69**, 1199 (1947).
16. W.G. Brown, *Org. React.*, **6**, 469 (1951).
17. (a) R.F. Nystrom, S.W. Chaikin and W.G. Brown, *J. Am. Chem. Soc.*, **71**, 3245 (1949). (b) R. Paul and N. Joseph, *Bull. Soc. Chim. France*, 550 (1952). (c) J. Kollonitsch, O. Fuchs and V. Gabor, *Nature*, **173**, 125 (1954). (d) *Idem., Ibid.*, **175**, 346 (1955). (e) H.C. Brown and B.C. Subha Roa, *J. Am. Chem. Soc.*, **77**, 3164 (1955). (f) J. Kollonitsch and O. Fuchs, *Nature*, **176**, 1081 (1955). (g) H.C. Brown and B.C. Subha Rao, *J. Am. Chem. Soc.*, **78**, 2582 (1956).
18. H.C. Brown and P.M. Weissman, *Israel, J. Chem.*, **1**, 430 (1963).
19. H.C. Brown and P.M. Weissman, *J. Am. Chem. Soc.*, **87**, 5614 (1964).
20. H.C. Brown, S.C. Kim and S. Krishnamurthy, *J. Org. Chem.*, **45**, 1 (1980).
21. C.A. Brown, S. Krishnamurthy and S.C. Kim, *J.C.S. Chem. Commun.*, 373 (1973).
22. H.C. Brown, P. Heim and N.M. Yoon, *J. Am. Chem. Soc.*, **92**, 1637 (1970).
23. H.C. Brown, P.M. Weissman and N.M. Yoon, *J. Am. Chem. Soc.*, **88**, 1458, (1966).
24. F.C. Huang, L.F.H. Lee, R.S.D. Mittal, P.R. Ravikumar, J.A. Chan, C.J. Sih, E. Caspi and C.R. Eck, *J. Am. Chem. Soc.*, **97**, 4144 (1975).
25. See H.C. Brown and S. Krishnamurthy, *Tetrahedron*, **25**, 567 (1979) for a detailed review of our program in selective reductions.
26. H.C. Brown and B.C. Subha Rao, *J. Am. Chem. Soc.*, **78**, 5694 (1956).
27. H.C. Brown and B.C. Subha Roa, *J. Org. Chem.*, **22**, 1136 (1957).
28. H.C. Brown, *Hydroboration*, W.A. Benjamin, New York, 1962.
29. H.C. Brown (with techniques by G.W. Kramer, A.B. Levy and M.M. Midland), *Organic Synthesis via Boranes*, Wiley-Interscience, New York, 1975.
30. E. Negishi and H.C. Brown, *Synthesis*, 77 (1974).
31. Chapter 13, ref. 28; Chapter 3, ref. 29.
32. H.C. Brown and N.M. Yoon, *Israel J. Chem.* **15**, 12 (1976/77).
33. H.C. Brown and C.F. Lane, *Heterocycles*, **7**, 454 (1977).
34. H.C. Brown and S.K. Gupta, *J.Am. Chem. Sac.*, **97**, 5249 (1975).
35. H.C. Brown and N. Ravindran, *J. Am. Chem. Soc.*, **98**, 1785, 1798 (1976).
36. H.C. Brown, N. Ravindran and S.U. Kulkarni, *J. Org. Chem.*, ***44***, 2417 (1979).
37. E. Frankland, *J. Chem. Soc.*, **15**, 363 (1862).
38. A. Pelter and Smith in *Comprehensive Organic Chemistry*, Eds. Sir Derek Barton and W. David Ollis, Vol. 3, Pergamon Press, Oxford, 1979.
39. H.C. Brown, M.W. Rathke and M.M. Rogic, *J. Am. Chem. Soc.*, **90**, 5038 (1968).
40. H.C. Brown and C.F. Lane, *J. Am. Chem. Soc.*, **92**, 6660 (1970).
41. G. Zweifel and H.C. Brown, *Org. React.*, **13**, 1 (1963).
42. M.W. Rathke, N. Inoue, K.R., Varma and H.C. Brown, *J. Am. Chem. Soc.*, **88**, 2870 (1966).
43. H.C. Brown, M.M. Midland and A.B. Levy, *J. Am. Chem. Soc.*, **95**, 2394 (1973).
44. R.C. Larock and H.C. Brown, *J. Am. Chem. Soc.*, **92**, 2467 (1970).
45. H.C. Brown, C. Verbrugge and C.H. Synder *J. Am. Chem. Soc.*, **83**, 1002 (1961).
46. H.C. Brown and S.P. Rhodes, *J. Am. Chem. Soc.*, **91**, 2149 (1969).
47. H.C. Brown and M.M. Rogic, *Organomet. Chem. Syn.*, **1** 305 (1972).

48. H.C. Brown, Y. Yamamoto and C.F. Lane, *Synthesis*, 303 (1972).
49. C.F. Lane and H.C. Brown, *J. Am. Chem. Soc.*, **93**, 1025 (1971).
50. H.C. Brown, Y. Yamamoto and C.F. Lane, *Synthesis*, 304 (1972).
51. H.C. Brown and E. Negishi, *J. Am. Chem. Soc.*, **89**, 5478 (1967).
52. A. Pelter, K. Smith, M.G. Hutchings and K. Rowe, *J. Am. Chem. Soc., Perkin I*, 129 (1975).
53. B.A. Carlson and H.C. Brown, *Synthesis*, 776 (1973).
54. H.C. Brown, J.J. Katz and B.A. Carlson, *J. Org. Chem.*, **38**, 3968 (1973).
55. H.C. Brown and E. Negishi, *J.C.S. Chem. Commun.*, 594 (1968).
56. H.C. Brown and G. Zweifel, *J. Am. Chem. Soc.*, **83**, 486 (1961).
57. M.M. Midland in *Asymmetric Synthesis*, Ed., J.D. Morrison, Vol. 2, Academic Press, New York, 1983, pp. 45-69.
58. T. Herold, U. Schrott and R.W. Hoffmann, *Chem. Ber.*, **114**, 359 (1981).
59. H.C. Brown and P.K. Jadhav, *J. Am. Chem. Soc.*, **105**, 2092 (1983).
60. H.C. Brown and P.K. Jadhav in *Asymmertic Synthesis*, Ed., J.D. Morrison, Vol. 2, Academic Press, New York, 1983, pp. 1-43.
61. H.C. Brown and N.M. Yoon, *Israel J. Cham.*, **15**, 12 (1977).
62. H.C. Brown, M.C. Desai and P.K. Jadhav, *J. Org. Chem.*, **47**, 5065 (1982).
63. Research in progress with J.V.N Vara Prasad..
64. H.C. Brown, J.R. Schwier and B. Singaram, *J. Org. Chem.*, **43**, 4395 (1978).
65. H.C. Brown and P.K. Jadhav, *J. Org. Chem.*, **46**, 5047 (1981).
66. H.C. Brown and N.M. Yoon, *J. Am. Chem. Soc.*, **99**, 5514 (1977).
67. H.C. Brown, P.K. Jadhav and A.K. Mandal, *J. Org. Chem.* **47**, 5074 (1982).
68. H.C. Brown, P.K. Jadhav and M.C. Desai, *J. Am. Chem. Soc.*, **104**, 4303 (1982).
69. H.C. Brown and B. Singaram, *J. Am. Chem. Soc.*, **106**, 1797 (1984).
70. Manuscript submitted, T.E. Cole and B. Singaram.
71. Research in progress with T.E. Cole and B. Singaram.
72. See Chapter 20, Epilog. ref. 28.

Appendix 1.4

Building Bridges between Inorganic and Organic Chemistry*

ROALD HOFFMANN**

Robert B. Woodward, a supreme patterner of chaos, was one of my teachers. I dedicate this lecture to him, for it is our collaboration on orbital symmetry conservation, the electronic factors which govern the course of chemical reactions, which is recognized by half of the 1981 Nobel Prize in Chemistry. From Woodward I learned much: the significance of the experimental stimulus to theory, the craft of constructing explanations, and the importance of aesthetics in science. I will try to show you these characteristics of chemical theory may be applied to the construction of conceptual bridges between inorganic and organic chemistry.

Fragments

Chains, rings, substituents—those are the building blocks of the marvelous edifice of modern organic chemistry. Any hydrocarbon may be constructed on paper from methyl groups (CH_3), methylenes (CH_2), methynes (CH), and carbon atoms (C). By substitution and the introduction of heteroatoms all of the skeletons and functional groupings imaginable, from ethane to tetrodotoxin, may be obtained.

The last thirty years have witnessed a remarkable renaissance of inorganic chemistry, and the particular flowering of the field of transition-metal organometallic chemistry. The complexes **1—9** are a selection of some of the simpler creations of the laboratory in this rich and ever-growing field.

Structures **1—3** illustrate at a glance one remarkable feature of transition-metal fragments. Here are three tricarbonyliron complexes of organic moieties cyclobutadiene, trimethylenemethane, and 1-hydroxy-1, 3-butadiene—which on their own would have little kinetic or thermodynamic stability. Yet complexed to

* Nobel Lecture, 1982.

** Professor Department of Chemistry, Cornell University, USA.

$Fe(CO)_3$ these molecules are relatively stable: they exist in a bottle. The inorganic fragment is not merely a weakly attached innocent bystander: it transforms essentially and strongly the bonding relationship in the molecule.

Structures **4—6** contain the ubiquitous cyclopentadienyl ligand (Cp), two of them in the archetypical ferrocene, one in $CpMn(CO)_3$, two bent back in $Cp_2Ti(CO)_2$. Structures **7—9** introduce us to the simplest representatives of the burgeoning class of clusters—assemblages of two or more metal atoms embellished with external ligands.

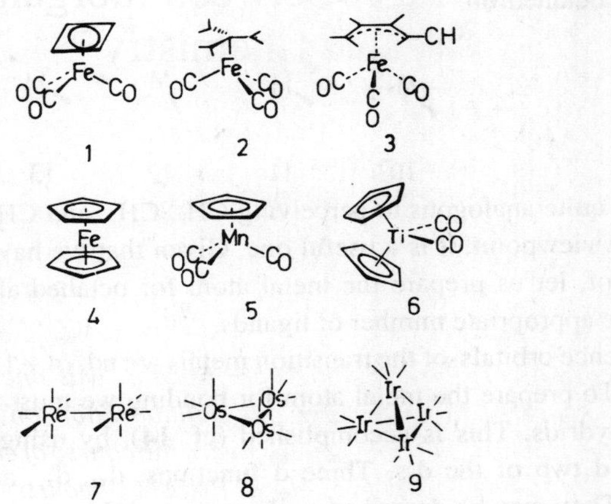

If we seek order, unity, a way of thinking about these complexes, it is not difficult to perceive that the molecules contain as building blocks transition metal-ligand fragments, ML_n, such as $M(CO)_5$, $M(CO)_4$, $M(CO)_3$, or MCp. It must be said immediately that there is nothing special about the carbonyl ligand. It is merely a representative and common component of organometallic complexes. Phosphanes, olefins, alkyls will do as well.

To reconstruct the complexes **1—9**, we need to know the electronic structure of the fragments. For the simple qualitative picture of the bonding in these molecules that we seek, we do not need to know every last detail of the electronic structure of each molecule. It will suffice that we know the frontier orbitals of the fragments—the higher occupied and the lower unoccupied levels—in other words the valence active orbitals of each fragment. It is *Kenichi Fukui* who taught us the importance of the frontier orbitals. We shall soon see that it is the resemblance of the frontier orbitals of inorganic and organic moieties that will provide the bridge that we seek between the subfields of our science.

Over the last eight years my coworkers and I have built up a library of the orbitals of ML_n fragments[1-3]. We have done so using entirely qualitative, approximate molecular orbital calculations of the extended Hückel type (a procedure for its time, developed with another of my teachers, *William N. Lipscomb*) and

and symmetry arguments (the value of which I first learned from still another of my teachers, *Martin P. Gouterman*). Molecular orbital theory, *Robert S. Mulliken's* great contribution to chemistry, is fundamental to our approach, be it in the construction of the very orbitals of the fragments, their changes on molecular deformation, or the interaction of several such fragments to restore the composite molecule. Yet when I seek the simplest of all possible ways to tell you of the orbitals of these fragments, I am led back to the valence bond picture introduced into chemistry by *Linus Pauling*[4].

Let us go back to the building blocks. The general fragments ML_n **10—13**, may be viewed in many ways. One convenient approach is to consic r them as pieces of an octahedron.

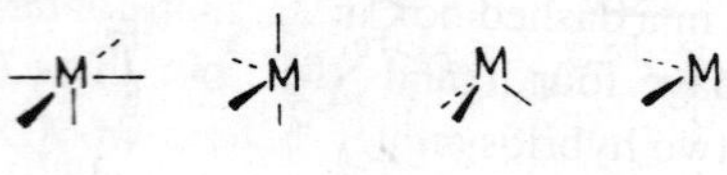

10 11 12 13

This is quite analogous to perceiving CH_3, CH_2, and CH in a tetrahedron. If not a unique viewpoint, it is a useful one. Given that we have an octahedron, or pieces thereof, let us prepare the metal atom for octahedral bonding, and then introduce the appropriate number of ligands.

The valence orbitals of the transition metals are nd, (n + 1)s and (n + 1)p, with n = 3, 4, 5. To prepare the metal atom for bonding we must form six equivalent octahedral hydrids. This is accomplished (cf. **14**), by using al. f the s and p functions and two of the d's. Three d functions, d_{xz}, d_{xy}, and d_{yz}, are left unhybridized. They may be described as the t_{2g} set of the crystal field, ligand field, or molecular orbital theories of an octahedral complex[5]

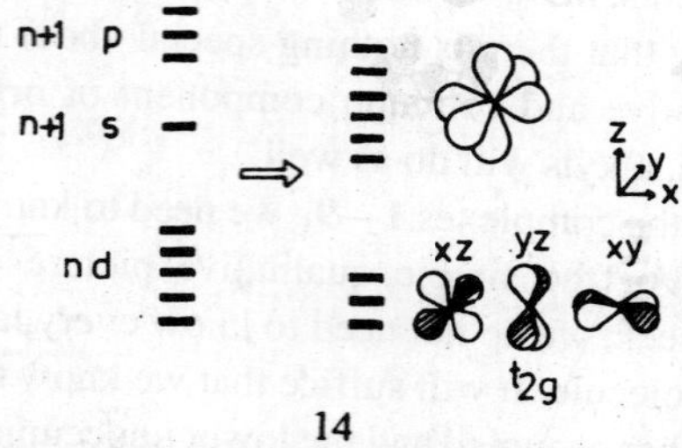

14

To form an octahedral complex we would introduce six ligands to make use of the six octahedral hybrids. Perhaps it is appropriate to digress here and make clear our ligand convention, which is to consider the ligand always as an even-electron Lewis base. While acceptor character or Lewis acidity is a desirable feature in a ligand, Lewis basicity or donation is essential. We see the basicity in the form of the lone pairs of CO, PH_3, and CH_3^- (**15—17**), in bidentate four-electron ligands, be they ethylenediamine (**18**) or butadiene (**19**), or in the electronically tridentate $C_5H_5^-$ (Cp^-, **20**), the equivalent of three two-electron bases[6].

Let six two-electron ligands approach the metal atom prepared for octahedral bonding (cf. **21**). Sizable σ overlaps lead to formation of strongly metal-ligand σ bonding combinations, and their strongly metal-ligand σ* antibonding counterparts. The six electron pairs of the ligands enter the six bonding combinations. Any electrons the metal contributes enter the t_{2g} orbital left behind. Indeed, for $Cr(CO)_6$ with its 6 metal electrons we attain a nice closed shell configuration, a situation we have learned to associate with relative kinetic and thermodynamic stability in organic chemistry.

What if we have not six ligands coming in, but only five? This situation is depicted in **22**. Five hydrids interact strongly, are removed from the frontier orbital region, just as all six were in **21.** One hybrid, the one pointing toward where no ligand is, remains roughly untouched, relatively low-lying in energy. The frontier orbitals, enclosed in a dashed box in **22**, now contain the t_{2g} set plus one hybrid.

What if we have four ligands (ML_4) or three (ML_3)? Much the same thing happens. In ML_4 two hybrids are left behind, in ML_3 three. We have thus reached the simplest of all possible pictures of the electronic structure of the ML_n fragments with n = 5, 4, 3, namely that given in **23—25**. The ML_n fragment's frontier orbitals

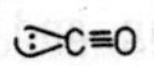

15

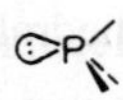

16

17

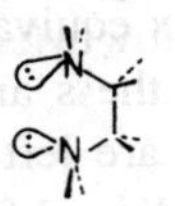

18

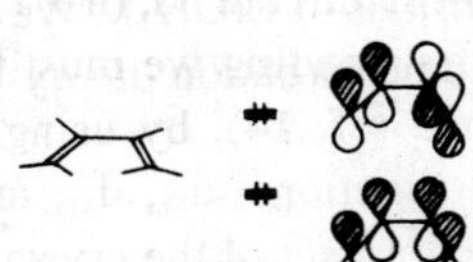

19

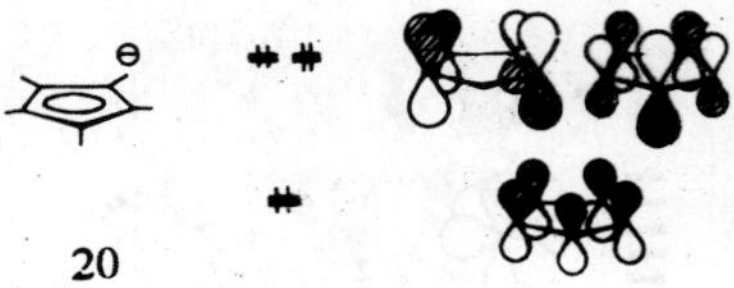

20

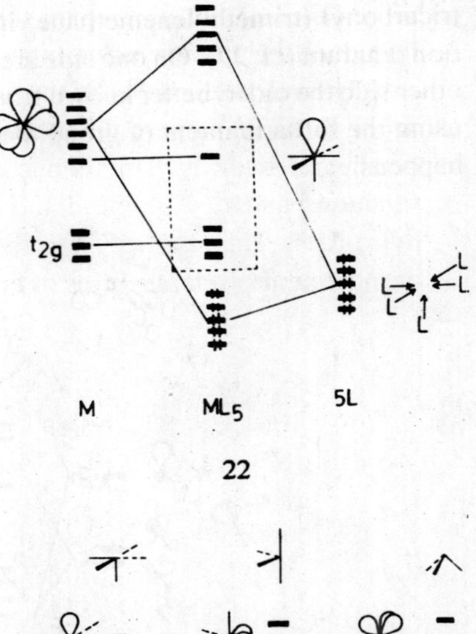

22

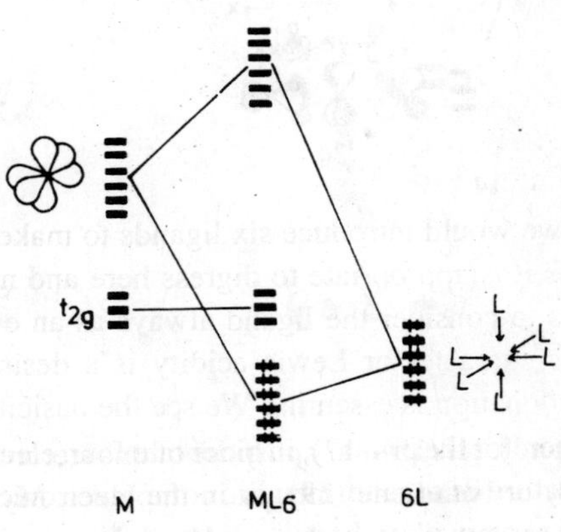

21

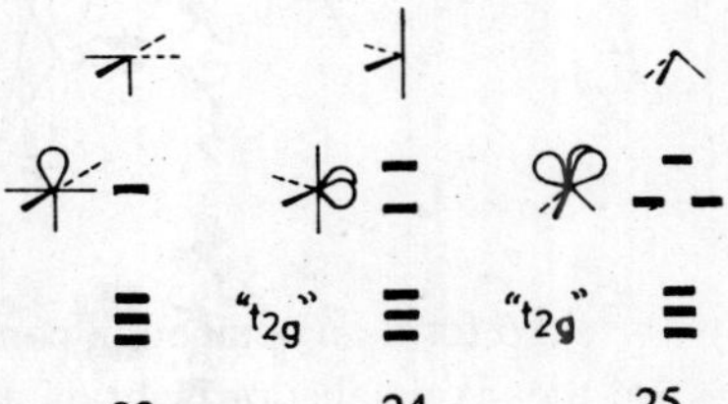

23 24 25

4	5	6	7	8	9	10
Ti	V	Cr	Mn	Fe	Co	Ni
Zr	Nb	Mo	Tc	Ru	Rh	Pd
Hf	Ta	W	Re	Os	Ir	Pt

26

consist of the descendants of an octahedral t_{2g} set at relatively low energy, and above them 6–n (one for n = 5, two for n = 4, three for n = 3) hybrids pointing towards the missing octahedral vertices.

What remains is to decide how many electrons to place into these frontier orbitals, and here the ever-useful Mendeleev Table, modified in **26** for electron counting purposes, tells us that Fe in oxidation state zero will have eight electrons in $Fe(CO)_4$ or $Fe(CO)_3$, and so will Co^{I} or Ni^{II}.

The reader had best beware. The account given here is simplified, as much as I dare simplify it. In that process, perforce, is lost the beautiful detail and complexity that makes $Fe(CO)_3$ different from $FeCl_3^{3-}$. There is a time for detail and there is a time for generality. The reader of my papers will know that my coworkers and I do not stint on detail, whether it is in explication or in perusal of the literature. But the time now, here, is for building conceptual frameworks and so similarity and unity take temporary precedence over difference and diversity.

Recall that the reason for building up the frontier orbitals of inorganic fragments is that we wish to use these orbitals in the construction of organometallic and inorganic complexes. We are now ready for that task. For instance, if we want tricarbonyl (trimethylenemethane) iron we construct a molecular orbital interaction diagram (cf. **27**). On one side are the orbitals of $Fe(CO)_3$ just discussed; on the other side the older, better known frontier orbitals of $C(CH_2)_3$. We interact the two, using the full armament of group theory and perturbation theory[7] to follow what happens.

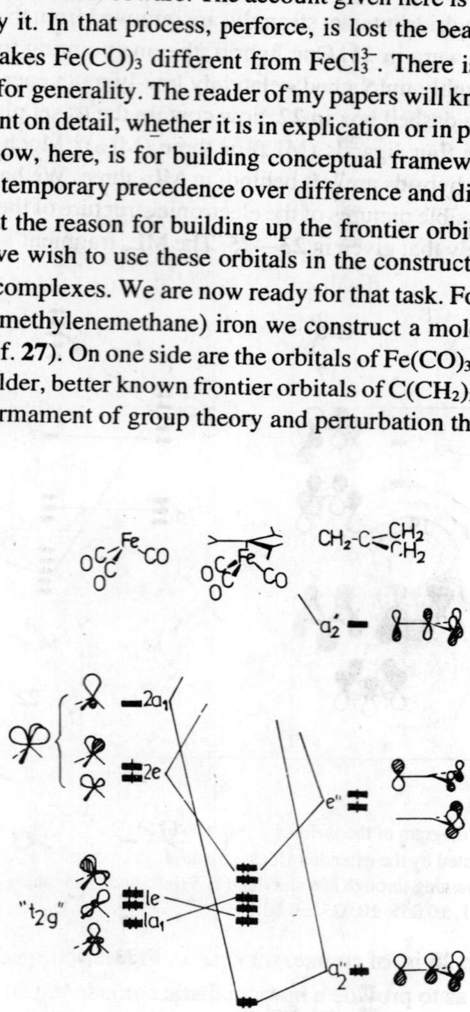

I will not trace this argument any further, for the primary purpose of this lecture is not the description of the electronic structure of organo-metallic complexes. My coworkers and I have done this comprehensively elsewhere[1,8]. Instead, I wish to describe a bridge between organic and inorganic chemistry that becomes possible the moment we gain knowledge of the orbitals of the ML_n fragments.

The Isolobal Analogy

Consider the d^7-fragment, $Mn(CO)_5$ (or $CO(CN)_5^{3-}$). Above the three lone pairs in the t_{2g} set this doublet molecule has a single electron in a hybrid pointing away from the ML_5. The similarity to CH_3, the methyl radical, is obvious (cf. **28**).

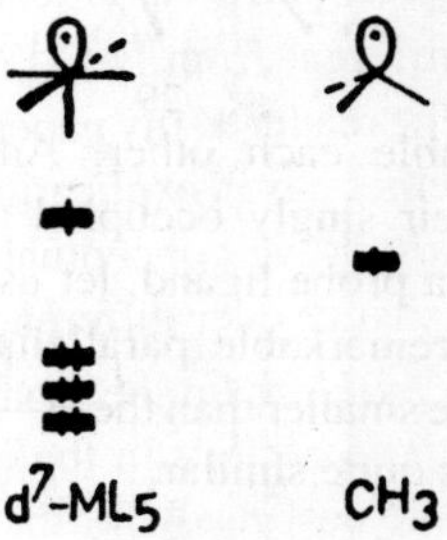

28

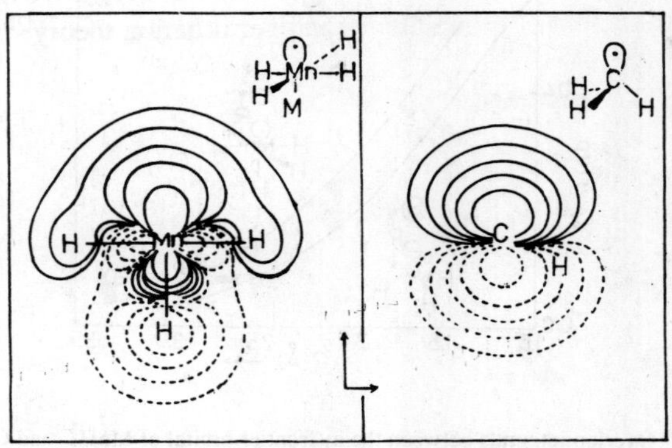

Fig. 1 **Contour diagram of the isolobal a_1 orbitals of MnH_5^{5-} (left) and CH_3 (right), as computed by the extended Hückel method. The contours of ψ, plotted in a plane passing through Mn and three H's (left) and C and one H (right), are ±0.2, ±0.1, ±0.055, ±0.025, ±0.10.**

The representation 28 is, of course, schematic, Figure 1 shows the a_1 orbitals of MnH_5^{5-} and CH_3, so as to provide a more realistic comparison.

If d^7-ML_5, is like CH_3 then they should both behave similarly. Let us think about what a methyl radical does. It dimerizes to ethane and starts radical chains. $Mn(CO)_5$ or $Co(CN)_5^{3-}$ do similar things. They dimerize to $Mn_2(CO)_{10}$ or $Co_2(CN)_{10}^{6-}$ (cf. 29), and each has a rich radical-type chemistry[9]. One can even codimerize the organic and inorganic fragments to give $(CO)_5MnCH_3$. That may

not be the preferred way to make this quite normal organometallic alkyl complex in the laboratory, but the construction on paper is quite permissible.

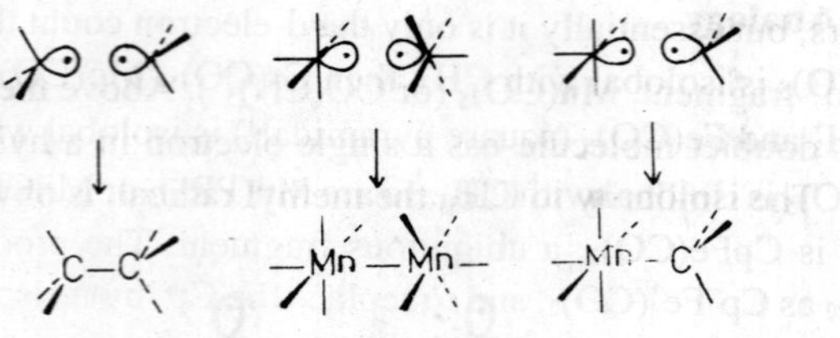

29

CH_3 and d^7-ML_5 resemble each other. Another way we can see that resemblance, traceable to their singly occupied a_1 orbitals, is to compare the overlap of both orbitals with a probe ligand, let us say a hydrogen atom. This is done in Figure 2. Note the remarkable parallelism of the two over-laps. The H—CH_3 overlap is everywhere smaller than the H—MnL_5 overlap, but the dependence of both on the distance is quite similar.

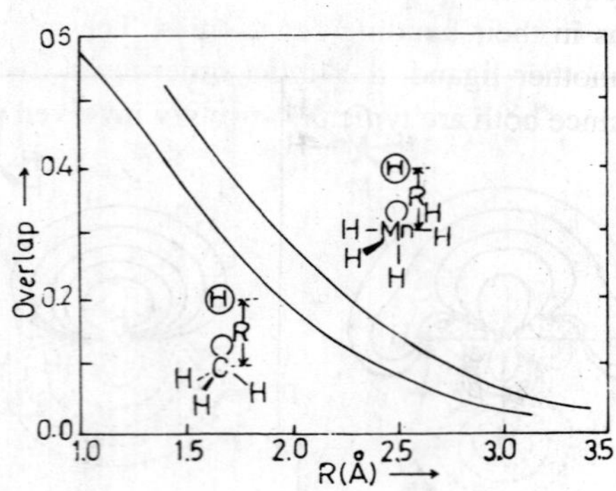

Fig. 2 Overlap integrals between the a_1 frontier orbital of MnH_5^{5-} and CH_3 and a H 1s orbital at a distance *R* from the Mn or c atom.

A word is needed to describe the resemblance of the two fragments, CH_3 and d^7-ML_5. They are certainly not isostructural, nor are they isoelectronic. However, both possess a froniter orbital which looks approximately the same for the two fragments. We will call two fragments *isolobal* if the number, symmetry properties, approximate energy and shape of the frontier orbitals and the number of electrons in them are similar—not identical, but similar[10]. Thus CH_3 is isolobal with $Mn(CO)_5$. We will introduce a symbol for the isolobal relationship: "two-headed" arrow with half an orbital below. Thus,

$$CH_3 \longleftrightarrow Mn(CO)_5$$

Let's extend the definition a little:

(1) If $Mn(CO)_5$ is isolobal with CH_3, so are $Tc(CO)_5$ and $Re(CO)_5$ as well as $Fe(CO)_5^+$. The shape of the a_1 hybrid will vary slightly with different principal quantum numbers, but essentially it is only the d-electron count that matters.

(2) If $Mn(CO)_5$ is isolobal with CH_3, then $Cr(CO)_5$, $Mo(CO)_5$ or $W(CO)_5$ are isolobal with CH_3^+ and $Fe(CO)_5$ (square pyramidal!) is isolobal with CH_3^-.

(3) If $Mn(CO)_5$ is isolobal with CH_3, so are $Mn(PR_3)_5$ or $MnCl_5^{5-}$ or any d^7-ML_5 species. And so is $CpFe(CO)_2$, a ubiquitous fragment. The procedure here is to write $CpFe(CO)_2$ as $Cp^-Fe^+(CO)_2$, and to replace the Cp^- by its isolobal equivalent of three carbonyls, reaching $Fe(CO)_5^+$, which is isoelectronic with $Mn(CO)_5$.

Let us go on to the ML_4 fragment. It is clear that a d^8-ML_4 species, *e.g.*, $Fe(CO)_4$, is isolobal with a methylene or (carbene, CH_2).

$$Fe(CO)_4 \leftarrow\!\!\!\circ\!\!\!\rightarrow CH_2$$

As **30** reveals, both fragments have two electrons in delocalized a_1 and b_2 orbitals which are the equivalent of two localized hybrids. There are explicable differences in the ordering of the two combinations[11]. The different ordering has, however, no grave consequences—recall that we are not so much interested in the fragments themselves as in their bonding capabilities. The moment we interact $Fe(CO)_4$ or CH_2 with another ligand, the initial ordering of a_1 and b_2 becomes relatively unimportant since both are typically strongly involved in the bonding.

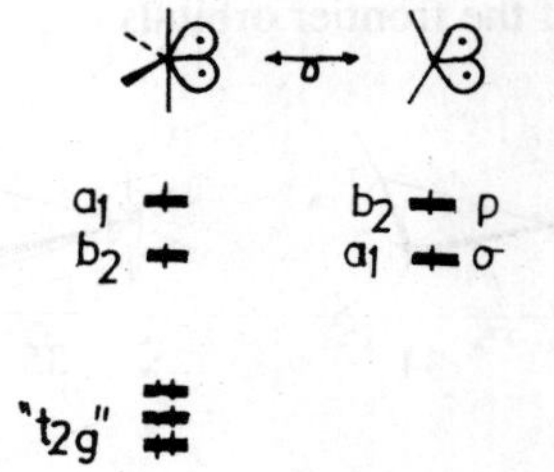

30

Dimerize conceptually, the isolobal fragments $Fe(CO)_4$ and CH_2. One gets the known ethylene, **31**, the tetracarbonyliron carbene complex, **32**, derivatives of which are known[12], and $Fe_2(CO)_8$, **33**. The last molecule is unstable, and so far has been observed only in a matrix[13]. We come here to a cautionary note on the isolobal analogy. The isolobal analogy carries one between organic and inorganic molecules of similar electronic structure. But there is no guarantee that the result of such an isolobal mapping (**31** ← 0 → **33**) leads one to a molecule of great kinetic stability. It might, or it might not.

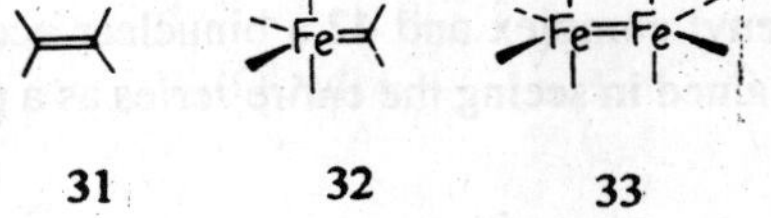

31 **32** **33**

Lest the reader be concerned about this limitation of the analogy, let me remind him or her of what happens as one proceeds from ethylene down Group IV. Si, Ge, Sn, Pb substitution leads to olefin analogues, but they are kinetically and thermodynamically so unstable that it has taken great effort to provide evidence for their fleeting existence.

$Fe_2(CO)_8$, **33**, has π and π^* levels similar to those of ethylene. But the low energy of its π^* makes this molecule coordinatively unsaturated. It can, for instance, add another CO to reach the stable enneacharbonyldiiron. More interesting, as we will soon see, is the strategy of stabilizing the unstable $Fe_2(CO)_8$ by making a complex of it, just as is routinely done for unstable organic molecules (**1**—**3**).

$Fe(CO)_4$, $Ru(CO)_4$ or $Os(CO)_4$ may be trimerized in various combinations with methylene, as in **34**—**37**. These cyclopropanes, ranging from all-organic to all-inorganic, are known. But note that when I show the ''all-metallic'' three-membered ring I have to go to Os. It is well known that the ground state structure of $Fe_3(CO)_{12}$ is **38**, with two bridging carbonyls[14]. Another limitation of the isolobal analogy is exposed here: the unbridged Fe structure analogous to **37** is certainly not much higher in energy than **38**, but nevertheless the lowest energy structure is bridged. Movement of some ligands (*e.g.* carbonyls, but not phosphanes) in and out of bridging sites is an experimental reality, a facile process, for transition metal complexes, especially of the first transition series. Such easy terminal to bridging interconversions are rare in organic chemistry, with the exception of carbonium ions. Bridging in inorganic structures, when it does occur, does not cause a major perturbation in the nature of the frontier orbitals.

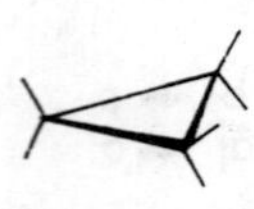

34

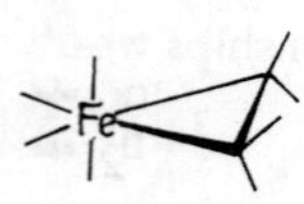

35

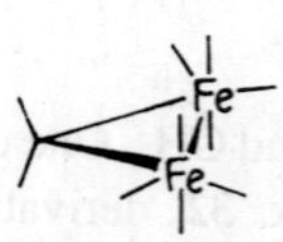

36

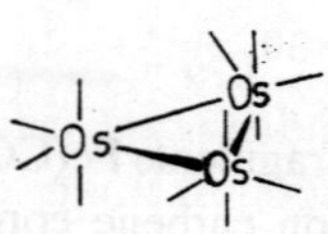

37

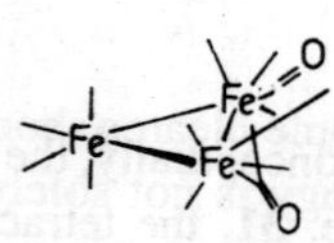

38

Consider next the d^9-ML_3 fragment, *e.g.* $Co(CO)_3$. This is isolobal with a methylidyne (or carbyne) CH, as shown in **39**. Once again there are differences, though of no great significance, in the a_1 versus e energy ordering between the two fragments. Their similarity is revealed most strikingly by the existence of the entire series of mixed organic and inorganic tetrahedranes, **40**—**44**. To be sure, **41** can be called a cyclopropenyl complex and **42** a binuclear acetylene complex, but I believe something is gained in seeing the entire series as a progression of isolobal substitution.

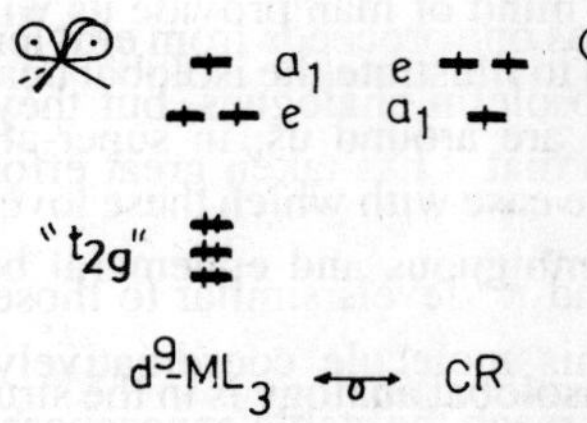

39

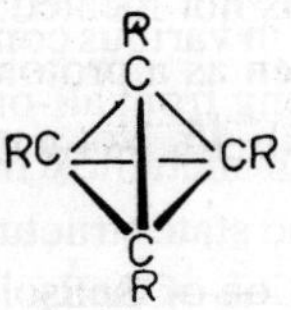

40

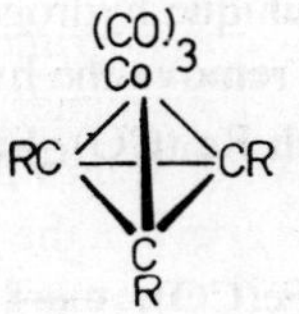

41

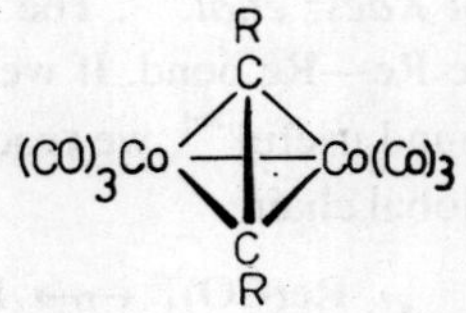

42

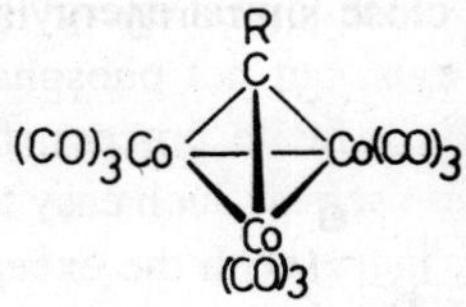

43

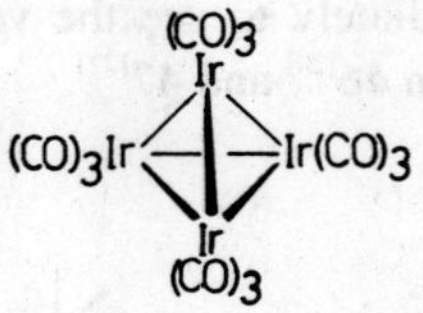

44

The fundamentals of the isolobal analogy have now been exposed. Just how far-reaching the relationships written here

$$d^7\text{–}ML_5 \xleftrightarrow{\circ} CH_3$$

$$d^8\text{–}ML_4 \xleftrightarrow{\circ} CH_2$$

$$d^9\text{–}ML_3 \xleftrightarrow{\circ} CH$$

are, will become clearer shortly. For the moment it is important to note that the isolobal analogy is not solely the creation of my research group. In his fruitful explorations of the reactivity of d^8-square planar complexes *Halpern* often made use of the similarity of such an ML_4 entity to a carbene[9a]. He did the same for d^7-ML_5 and organic free radicals[9a]. *Dhal*, in a beautiful series of structural studies of transition metal clusters, saw clearly the relationship of the orbitals of an ML_n fragment to a chalcogen or pnicogen atom, which of course are easily related to CR[15,16]. And most importantly, *Wade*[17] and *Mingos*[18] independently developed a comprehensive and elegant picture of the electronic structure of transition-metal clusters by relating them to the polyhedral boron hydrides (which *Lipscomb* and I studied—the circle closes!). It is a trivial step from BH to CH^+. All of these workers saw the essence of the isolobal analogy.

Structural Implications of the Isolobal Analogy

How quickly the hands and mind of man provide us with the problem of choice! The molecules I would need to illustrate the isolobal analogy at work did not exist thirty years ago. Now they are around us, in super-abundance. I have made a selection, based in part on the ease with which these lovely molecules illustrate the principles, in part on the ambiguous and ephemeral basis of recent (1981) appearance in the literature.

One obvious use of the isolobal analogy is in the structural sense. The analogy allows us to see the simple essence of seemingly complex structures. I should like to show you some examples centering on the ML_4 fragment.

Last year there appeared a structure of the cluster $HRe_3(CO)_{12}Sn(CH_3)_2$ from the work of *Kaesz et al.*[19]. The unique hydrogen was not located; presumably it bridges one Re—Re bond. If we remove the hydrogen as a proton, a convention we have found useful[8n], we reach $Re_3(CO)_{12}Sn(CH_3)_2^-$ **45.** Not a usual molecule, but the isolobal chain

$$Re(CO)_4^- \longleftrightarrow Fe(CO)_4 \longleftrightarrow CR_2 \longleftrightarrow SnR_2$$

allows us immediately to see the very close similarity of this structure to the previously known **46**[20] and **47**[21].

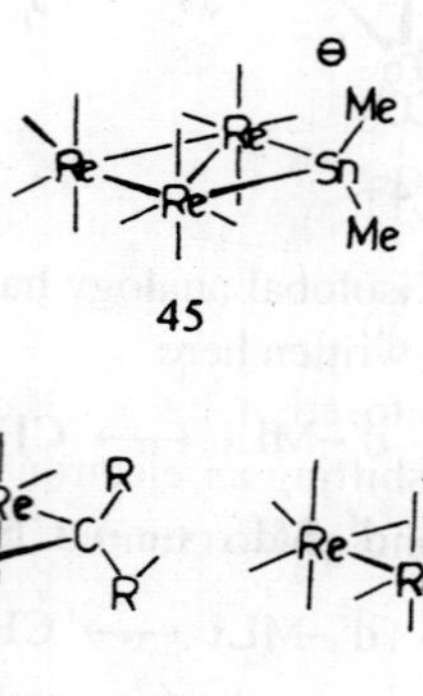

45

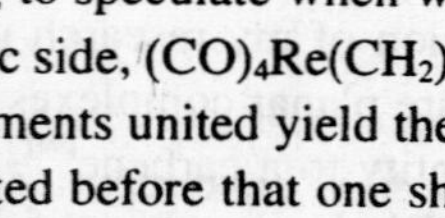

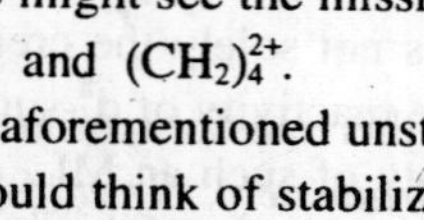

46 47

It is interesting to speculate when we might see the missing members of the series on the organic side, $(CO)_4Re(CH_2)_3^+$ and $(CH_2)_4^{2+}$.

Two ML_4 fragments united yield the aforementioned unstable $Fe_2(CO)_8$ system, **48**. It was hinted before that one should think of stabilizing this species by complexation. A pretty example is at hand, **49**[22]. Two $Fe_2(CO)_8$ units are complexed by a tin atom! Note the pinning back of the equatorial carbonyls, analogous

48

to the bending back of hydrogens in a transition metal complexed olefin. Alternatively, and interestingly, this can be considered as a spiropentane.

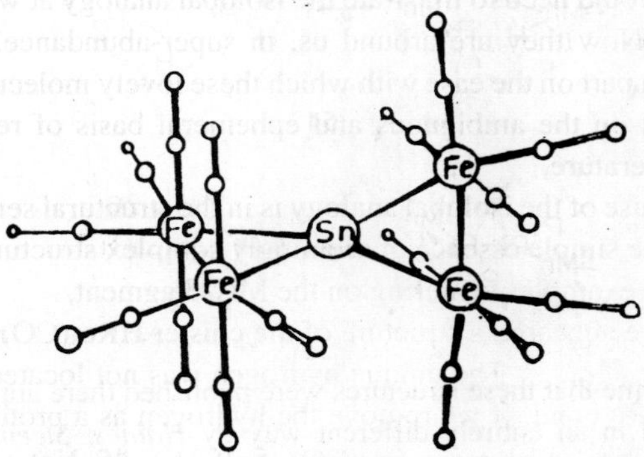

49

Earlier last year, *Lewis, Johnson el al* [23] published a synthesis and structure of $Os_5(CO)_{19}$. The structure, **50**, appears terribly complicated until one realizes it is really **51**, a typical trigonal bipyramidal $Os(CO)_5$ derivative, with two of the equatorial carbonyls substituted by olefines, or rather by the $Os_2(CO)_8$ olefin analogues. The ''olefin'' orientation is just as it should be[1k, 0].

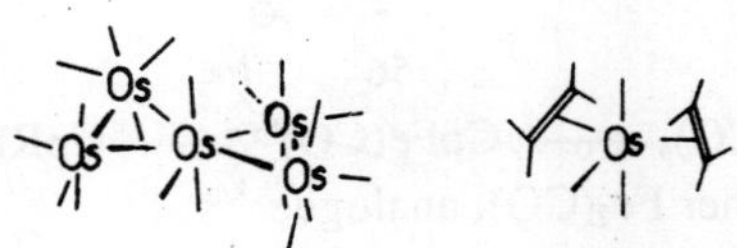

50 51

Another system isolobal to ethylene and $Fe_2(CO)_8$ is the ''mixed dimer'' $(CO)_4FeCH_2$. Upon formally shifting an electron from the metal to the carbon, a bit of alchemy, one gets to a phosphido complex **52**.

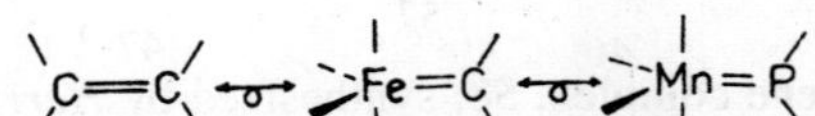

52

The reason for this transformation sequence is that complexes of $Mn(CO)_4PR_2$ have been made. *Braunstein, Grandjean et al.* have reported a remarkable set of structures, among them the three shown with their isolobal analogues in **53—55**[24]. In each structure we can see the obvious ethylene like $(CO)_4MnPR_2$ entity.

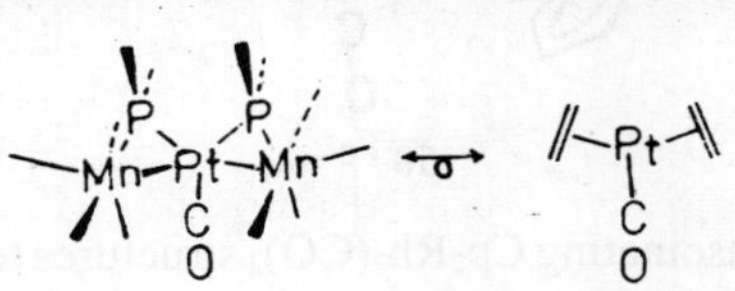

53

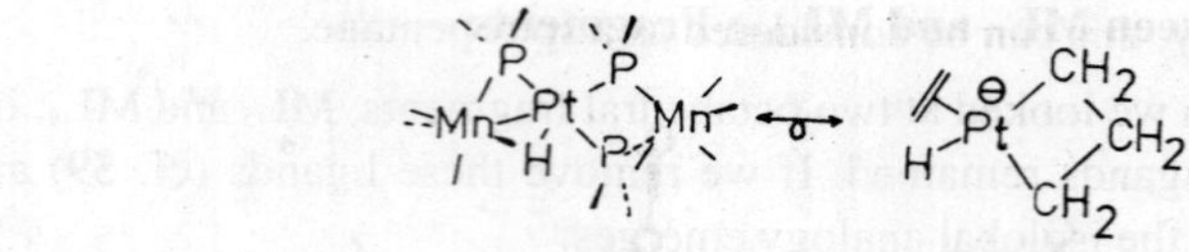

54

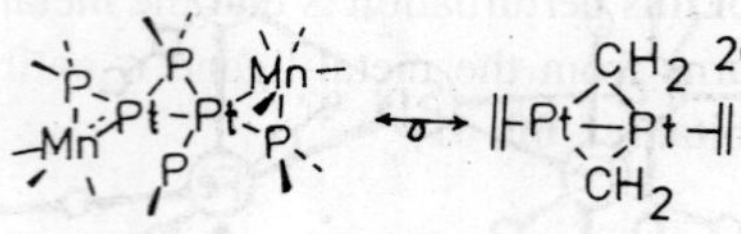

55

At the same time that these structures were published there appeared a structure of **56**, synthesized in an entirely different way by *Haines, Steen,* and *English*[25]. Unbridge the two semibridged carbonyls, do a bit more electronic alchemy relating Mn—Pt to Fe—Rh, and the relationship to **55** becomes crystal clear.

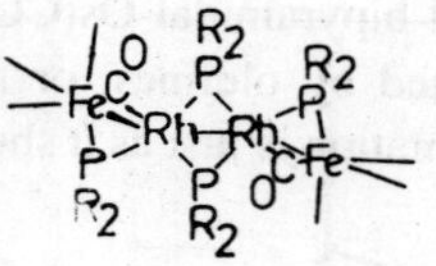

56

Realizing that $Fe(CO)_4$ ←o→ $CpFe(CO)^-$ ←o→ CpRhCO, we see immediately, that **57** is yet another $Fe_2(CO)_8$ analogue.

57

Thus, the methylene complex, **58**, synthesized by *Herrmann* is a two-thirds inorganic cyclopropane[26]

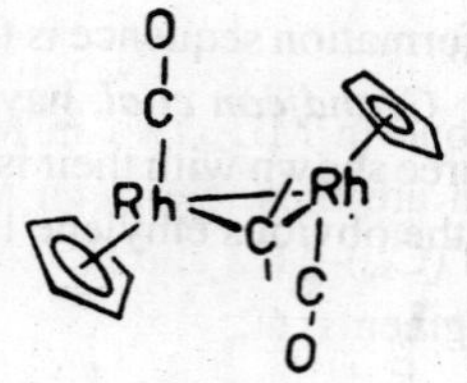

58

There are a few more fascinating $Cp_2Rh_2(CO)_2$ structures to be shown, but first we need to examine one extension of the isolobal concept.

The Relationship between ML_n and ML_{n-2} Fragments

Earlier in the discussion we looked at two octahedral fragments, ML_5 and ML_4, in which a pair of axial ligands remained. If we remove these ligands (cf. **59**) an interesting extension of the isolobal analogy emerges.

If the z-axis is oriented along the direction of the vanishing ligands, then it is clear that the main result of this perturbation is that the metal d_{z^2}, atomic orbital is lowered in energy. It returns from the metal-ligand σ-antibonding manifold to become a non-bonding orbital (cf. **60, 61**)[1o, 3c]

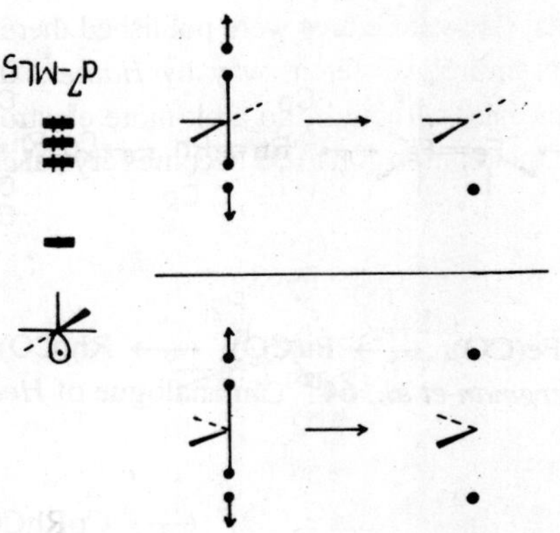

59

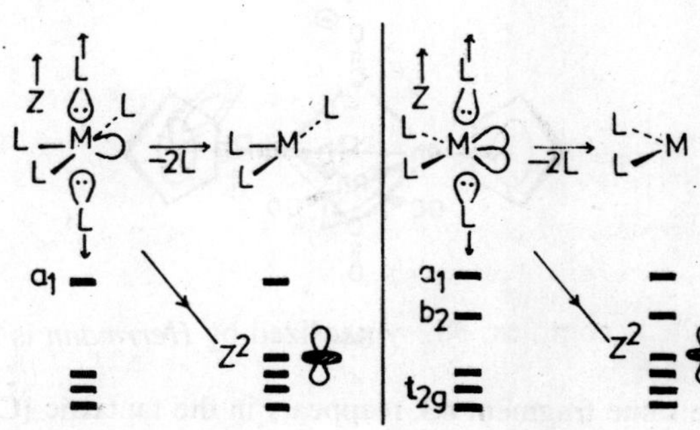

60 61

The high-lying orbitals (one in ML_5, two in ML_4) remain[27]. The obvious relationships that emerge then are those between a d^n–ML_5 and a d^{n+2}(C_{2v})- or T-shaped ML_3; and between d^n(C_{3v})–ML_4 and d^{n+2}–ML_2. Or to put it explicitly in terms of the most common fragments, **62**.

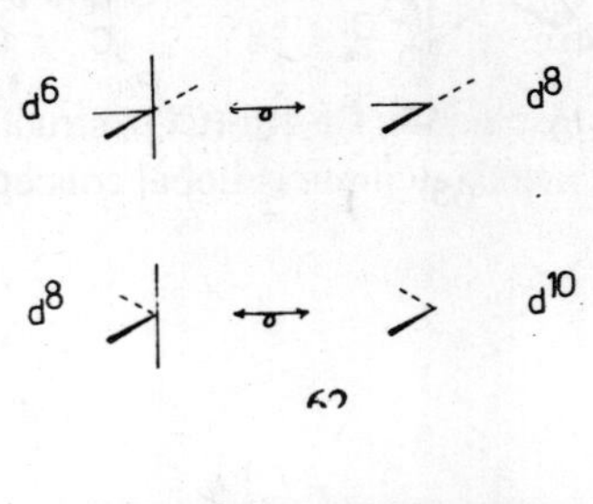

62

So we now add some further diversity to the non-iso-morphic mapping which is the isolobal relationship:

$$CH_3^{\oplus} \leftrightarrow Cr(CO)_5 \leftrightarrow PtCl_3^{\ominus}$$

$$CH_2 \leftrightarrow Fe(CO)_4 \leftrightarrow Ni(PR_3)_2$$

An obvious application is to olefin complexes; the similarity of $(CO)_4Fe(C_2H_4)$ to $(PR_3)_2Ni(C_2H_4)$, and that of $(CO)_5Cr(C_2H_4)$ to Zeise's salt emerges directly[1o, x].

Returning to the $[CpRh(CO)]_2$ analogue of ethylene, we can now think about some other structures. First, it turns out that moving the carbonyl groups into the bridge does little to change the ethylene-like nature of the dirhodium fragment, **63**[1u].

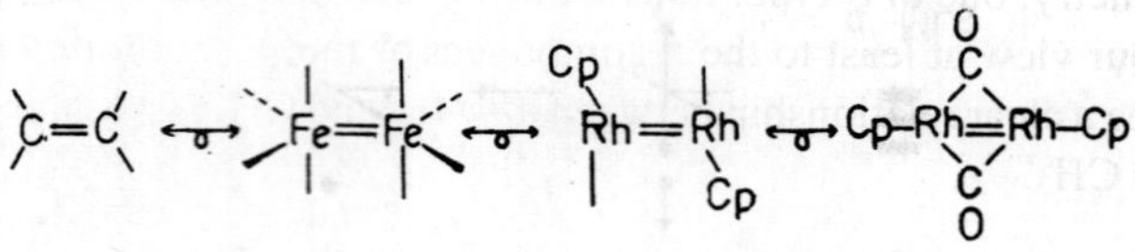

63

Since $CH_2 \leftrightarrow Fe(CO)_4 \leftrightarrow Pt(CO)_2 \leftrightarrow Rh(CO)_2^-$, it is possible to see in the compound of *Bergman et al.*, **64**[28], an analogue of *Herrmann's* compound **58**.

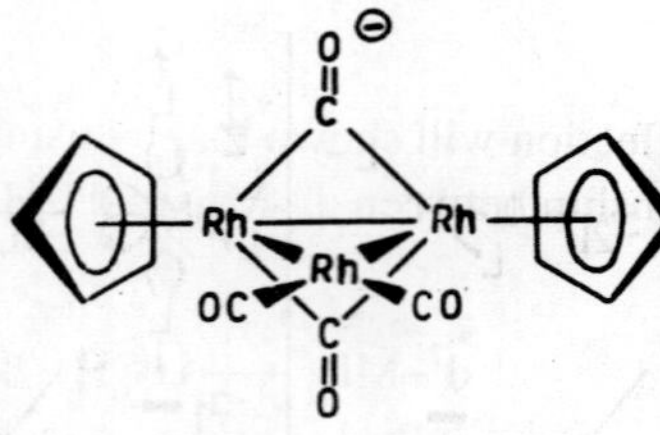

64

Essentially the same fragment **63**, reappears in the fantastic $[CpRh(CO)]_4Pt$ structure, **65**, of *Stone et al.*[29] ($Cp = \eta^5\text{–}C_5Me_5$) and can be related to the $(RC{\equiv}CR)_2Pt$ structure, **66**, earlier synthesized by the same group[30].

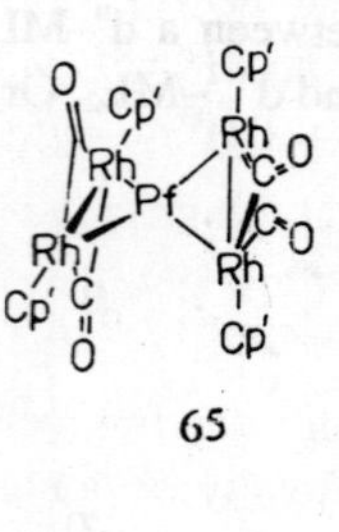

65

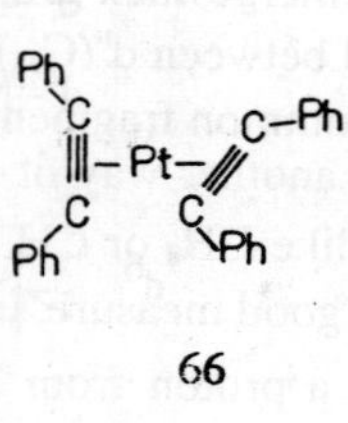

66

Into the t_{2g} Shell

It turns out that not only is d^9–$Co(CO)_3$ isolobal with CH, but so is d^5–$CpW(CO)_2$. To see how this comes about let us first relate the Cp complex to a simple ML_n.

$$CpW(CO)_2 \leftarrow\!\!\circ\!\!\rightarrow CpCr(CO)_2 \leftarrow\!\!\circ\!\!\rightarrow Cr(CO)_5^{\oplus}$$

As was shown earlier, $Cr(CO)_5^+$ is isolobal with CH_3^{2+}. That is not a very productive analogy. So let us examine $Cr(CO)_5^+$ in more detail. The electronic structure of an ML_5 fragment was given earlier. It is repeated in more detail on the left of **67**. The ML_5–CH_3 analogy concentrates on the hybrid of $\sigma(a_1)$ symmetry. But the t_{2g} set, even if it is less ''directional'' than the hybrids, has extent in space and well-defined symmetry properties. In particular, two of the t_{2g} orbitals are of π pseudosymmetry, one of δ-type. If, as we are forced to by the electron deficiency, we extend our view at least to the π component of the t_{2g} set (dotted lines at right in **67**), we see a clear relationship between d^5–ML_5 and CH, just as there is between d^7–ML_5 and CH_3.

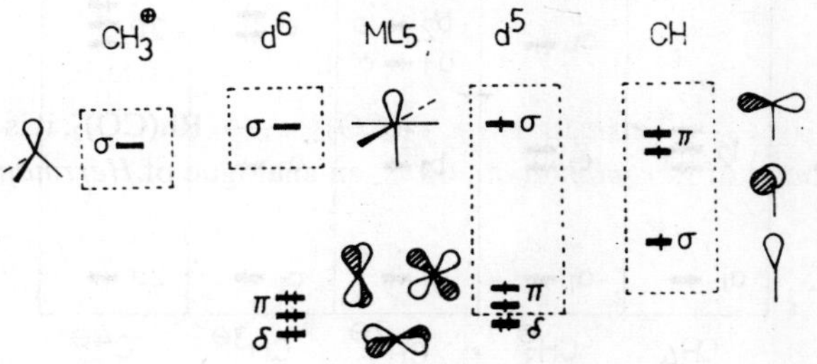

67

A little further reflection will shown that by using one half of the π set of the t_{2g} we can get a relationship between d^6–ML_5 and CH_2.

To summarize:

$$d^7\text{–}ML_5 \leftarrow\!\!\circ\!\!\rightarrow CH_3$$

$$d^6\text{–}ML_5 \leftarrow\!\!\circ\!\!\rightarrow CH_2$$

$$d^5\text{–}ML_5 \leftarrow\!\!\circ\!\!\rightarrow CH$$

or to put it another way

$$d^6\text{–}ML_5 \leftarrow\!\!\circ\!\!\rightarrow CH_3^{\oplus}$$

$$d^6\text{–}ML_5 \leftarrow\!\!\circ\!\!\rightarrow CH_2$$

$$d^6\text{–}ML_5 \leftarrow\!\!\circ\!\!\rightarrow CH^{\ominus}$$

This gives us another way of looking at things, a *deprotonation analogy*. In what way is CH_3^+ like CH_2 or CH^-? Let us draw out their orbitals schematically, including CH_4 for good measure, in **68**.

Taking away a proton from each molecular fragment does not change its ability to function as a donor (though its quality or donor strength will be *very* different). Each fragment, from CH_4 to C^{4-}, is in principle an eight-electron donor.

The recapitulate: the isolobal analogy is not a one-to-one mapping. A d^6-ML_5 fragment is isolobal with CH_3^+, CH_2 and CH^-. This is why the d^5-$CpW(CO)_2$ is isolobal with CH.

The isolobal analogy for low d-electron count metals has been exploited most notably in the work of *Stone's* group at Bristol. Just four compounds from their many beautiful examples are shown in **69—72**[31].

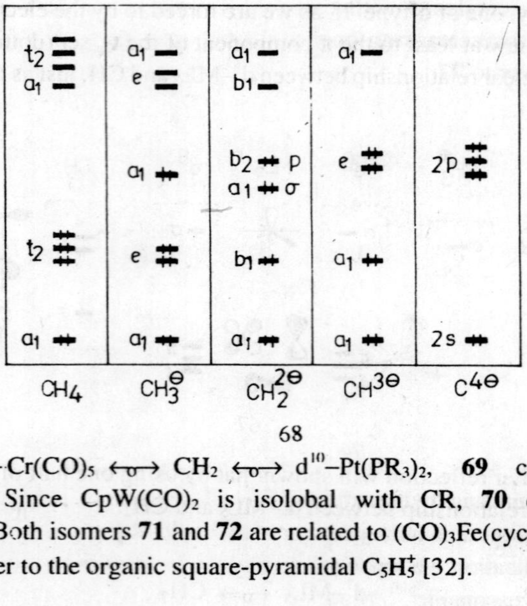

68

Since d^6-$Cr(CO)_5 \leftarrow\!\!o\!\!\rightarrow CH_2 \leftarrow\!\!o\!\!\rightarrow d^{10}$-$Pt(PR_3)_2$, **69** corresponds to cyclopropane. Since $CpW(CO)_2$ is isolobal with CR, **70** corresponds to cyclopropene. Both isomers **71** and **72** are related to $(CO)_3Fe$(cyclobutadiene), **1**, or for that matter to the organic square-pyramidal $C_5H_5^+$ [32].

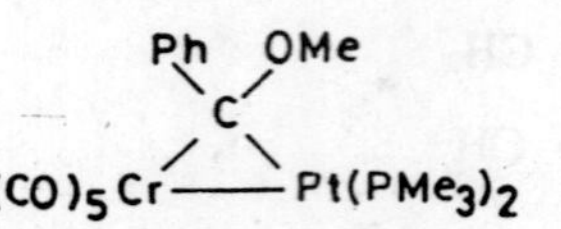

69

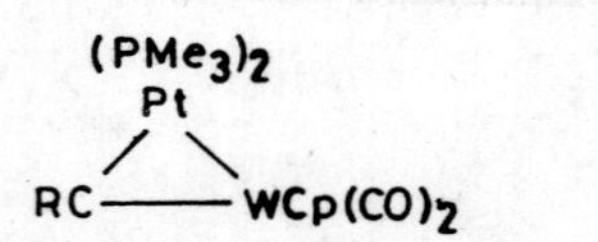

70

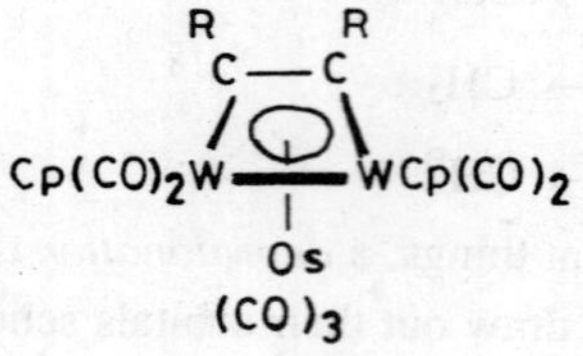

71

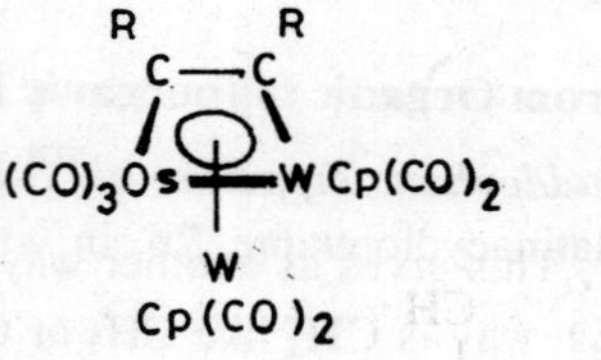

72

From Inorganic to Organic Chemistry

The psychological direction of the isolobal analogy in general has been to make one feel more comfortable about the structures of complex inorganic molecules by relating them to known, presumably simpler, organic molecules. It is interesting to reverse this process and think about as yet unsynthesized organic structures related to known inorganic ones. The mapping from one realm of chemistry to the other must be accompanied by the warning already given: there is no guarantee that the "product" of an isolobal transformation is as stable, kinetically or thermodynamically, as the "reactant"[33].

$Fe(CO)_3$ is isolobal with CH^+. Thus, **1** is related to $C_5H_5^+$, **73**[32], and the ubiquitous ferroles. **74**[34], are related to $C_6R_6^{2+}$, **75**[35]. Another product of the interaction of acetylenes with carbonyliron species is the "flyover bridge" molecule, **76** a binuclear ring-opened fulvene complex[34]. The isolobal replacement carries over to **77**.

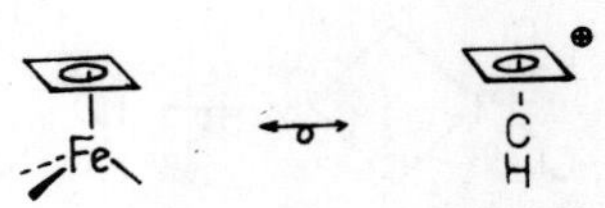

1 73

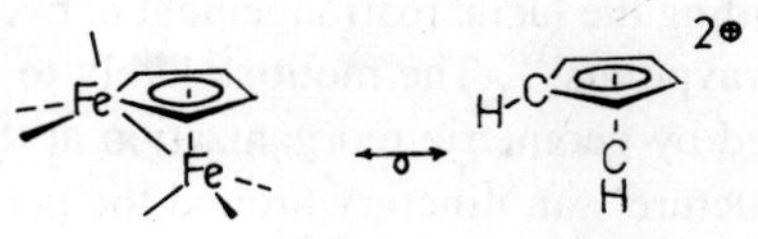

74 75

This is an unusual hydrocarbon $C_8H_8^{2+}$ of C_2 symmetry, a hypothetical doubly homoallylic cation. It is not a geometry one would normally have thought of for a heptafulvene dication, yet once reached by the isolobal mapping it appears to be geometrically reasonable[36]. More mappings of this type await exploitation.

76 77

From Organic to Inorganic Reaction Mechanisms

Puddephatt, Tipper et al.[37] have discovered a remarkable rearrangement of a platinacyclobutane, **78**, in which a carbon atom adjacent to the metal, with its

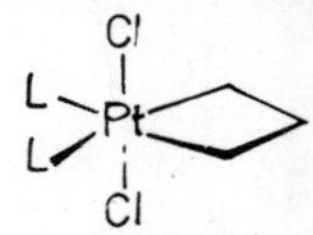

78

substituents, exchanges in a very specific way with the carbon atom opposite the metal[37]. The labeling experiment of *Casey et al.* showing this phenomenon most directly is given in **79**[38]. How does this happen?

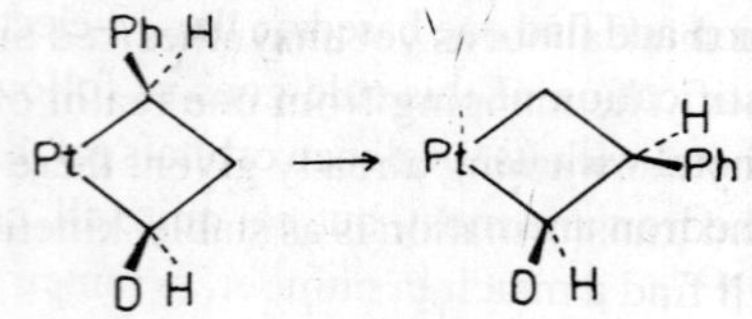

79

Kinetic evidence for a primary dissociative step to $Cl_2(py)Pt(CH_2)_3$ exists[37]. Suppose the ML_3 fragment can distort from its original T-shape to a C_{3v} geometry. Since d^8–$PtCl_2(py)$ is isolobal with CH^+, we can see a relationship to the cyclobutyl cation, **80**.

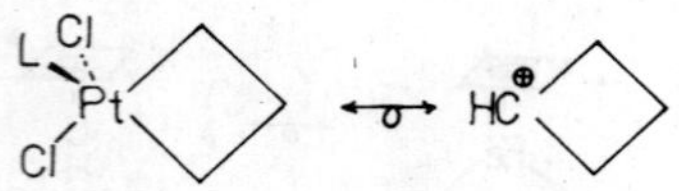

80

This association immediately brings to mind the entire complex of speculations and facts surrounding the facile rearrangement of cyclobutyl cations thorugh ''bicyclobutonium'' waypoints[39]. The motions likely to occur are shown in **81**. Ligand loss is followed by geometric reorganization at the metal, approach to a ''bicyclobutonium structure'' an itinerary around the periphery of a Jahn-Teller wheel through ''cyclopropyl carbinyl'' waypoints, and exit through an isomeric ''bicyclobutonium'' structure. This is but one instance among many where the isolobal analogy is useful in moving between organic and inorganic reaction mechanisms.

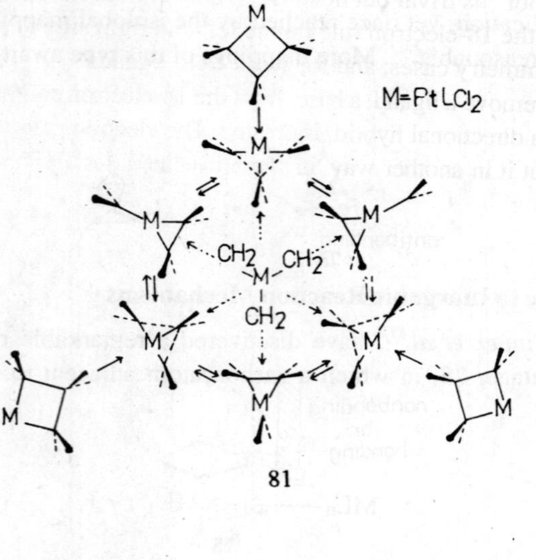

81

Beyond the Octahedron

The octahedron was a most useful starting point for generating fragment frontier orbitals, thereby engendering the isolobal analogy. But the octahedral polytype is not unique for six-coordinate complexes, and higher coordination numbers are feasible. We seek another more far-ranging derivation and find one based on the 18-electron rule.

An (unorginal) justification of this rule goes as follows: Consider n ligands, $n \leq 9$, approaching a metal with its 9 valence orbitals (cf. **82**). A little group theory shows that for the octahedron and most, but not quite all, coordination geometries the n ligand orbitals will find a match in number, symmetry properties, and extent in space among the hybrid sets that can be formed from the nine metal orbitals. The exceptions are very well understood[40]. Given this match, n M–L σ* antibonding combinations will go up, and 9–n metal orbitals will remain relatively untouched, approximately non-bonding. The 18-electron rule then is the statement: "Thou shalt not fill antibonding orbitals". Filling bonding (n)+nonbonding (9–n) orbitals leads to 9 electron pairs or 18 electrons.

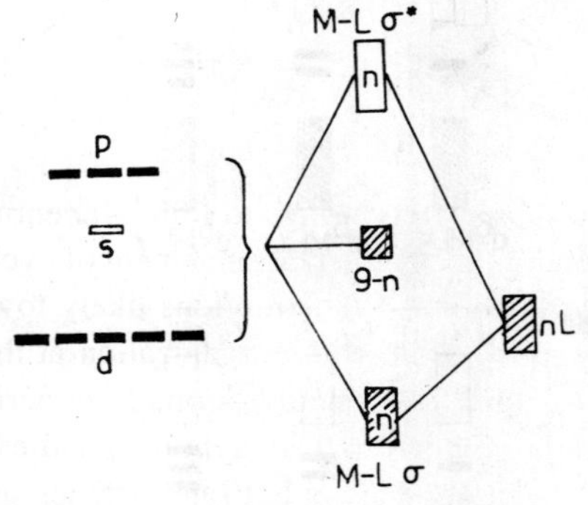

82

This "proof" is trival but not silly. Upon a little reflection it will lay bare the limitations of the 18-electron rule on the left and right side of the transition series, for special symmetry cases, and for weak field ligands.

We next remove a ligand, a base, from the 18-electron complex. A localized hole on the metal, a directional hybrid, is created. The electron pair leaves with the ligand (cf. **83**). To put it in another way, in some localized description of the bonding, one

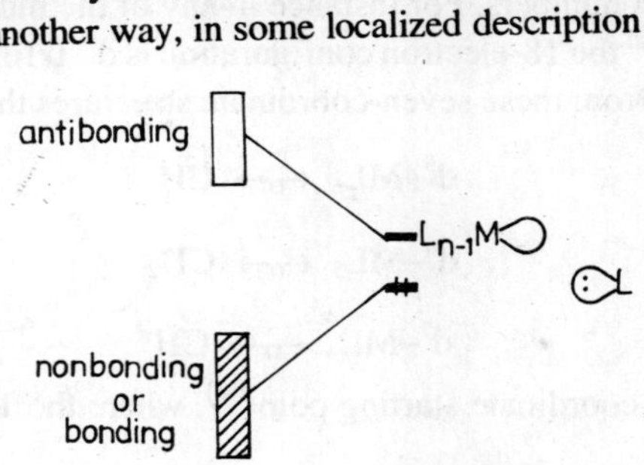

$$ML_n \longrightarrow ML_{n-1} + L$$

83

M–L σ bond was formed by interaction of a ligand pair of electrons with a metal-based hybrid. Reversing the propcess, breaking the bond frees that hybrid.

A parallel analysis for main group elements leads to the octet rule, since only s and p are considered as valence orbitals. Hybrids are freed by removing ligands, so that CH_3^+ has one vacant directional orbital, CH_2^{2+} has two such.

The parallel between ML_n and EL_n fragments (M = transition metal, E = main group element) is derived from the generation of similar hybrid patterns on removal of ligands from 18- or 8-electron configurations. For instance, if the octahedral polytype is used as a starting point, the 18-electron rule is satisfied for a d^6-ML_6. Then d^6-ML_5 will have one hybrid and no electrons in the gap between antibonding and bonding or non-bonding levels, just like CH_3^+, **84**: d^6-ML_4 will have two empty hybrids, so will CH_2^{2+}. The common form of the isolobal analogy follows from this.

84

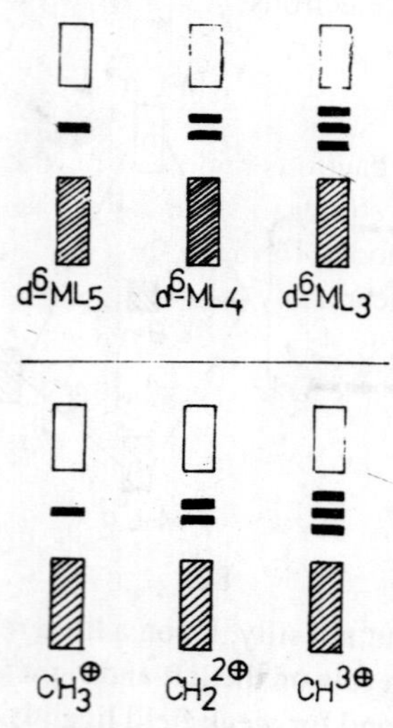

The advantage of this alternative derivation is that it is easily extended to higher coordination numbers. For instance in any of the multitude of seven-coordinate geometries[8g] the 18-electron configuration is d^4. It follows immediately for fragments derived from these seven-coordinate structures that

$$d^5\text{–}ML_6 \leftarrow\!\!\!\circ\!\!\!\rightarrow CH_3$$

$$d^6\text{–}ML_5 \leftarrow\!\!\!\circ\!\!\!\rightarrow CH_2$$

$$d^7\text{–}ML_4 \leftarrow\!\!\!\circ\!\!\!\rightarrow CH$$

From an eight-coordinate starting point[8p], where the 18-electron configuration is d^2.

$$d^3\text{–}ML_7 \leftarrow\!\!\!\circ\!\!\!\rightarrow CH_3$$

$$d^4\text{–}ML_6 \leftarrow\!\!\overset{}{o}\!\!\rightarrow CH_2$$

$$d^5\text{–}ML_5 \leftarrow\!\!\overset{}{o}\!\!\rightarrow CH$$

The conclusions may be summarized in Table 1. Note once again the non-isomorphic, ''many-to-one'' nature of the isolobal analogy. Also, the results of the previous section on ''Into the t_{2g} Shell'' are contained in the present discussion.

Table 1. Isolobal Analogies

Organic Fragment	Transition Metal Coordination Number on which Analogy is Based				
	9	8	7	6	5
CH_3	d^1-ML_5	d^3-ML_7	d^5-ML_6	d^7-ML_5	d^9-ML_4
CH_2	d^2-ML_7	d^4-ML_6	d^6-ML_5	d^8-ML_4	d^{10}-ML_3
CH	d^3-ML_6	d^5-ML_5	d^7-ML_4	d^9-ML_3	

L = neutral two electron ligand.

Miscellanea

The general rules in hand, the cautions understood, one can proceed to apply the isolobal analogy. Indeed my strategy has been to show the applications as I introduced extensions of the model. Here are some additional examples:

In a recent study of binuclear acetylene complexes the discussion focused on four structural types, **85**–**88**[41].

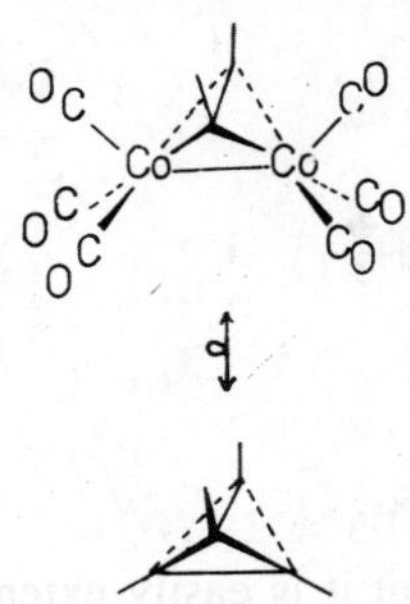

85

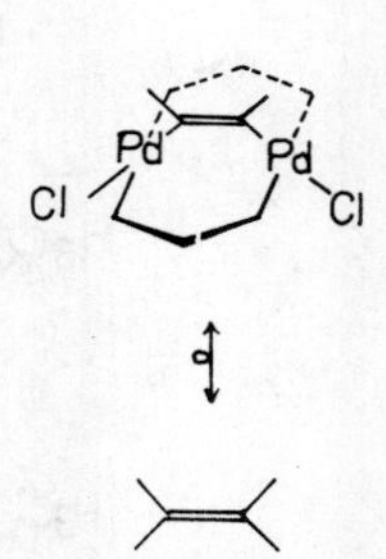

86

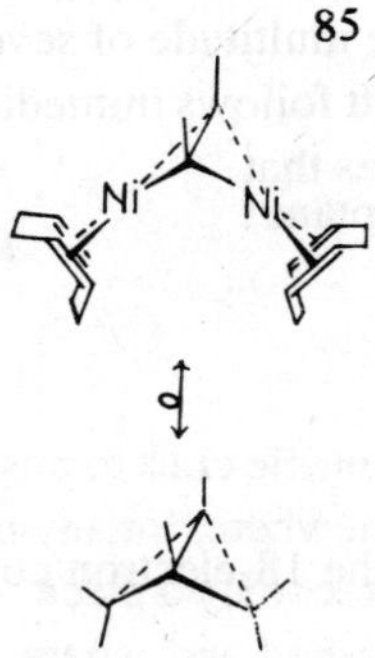

87

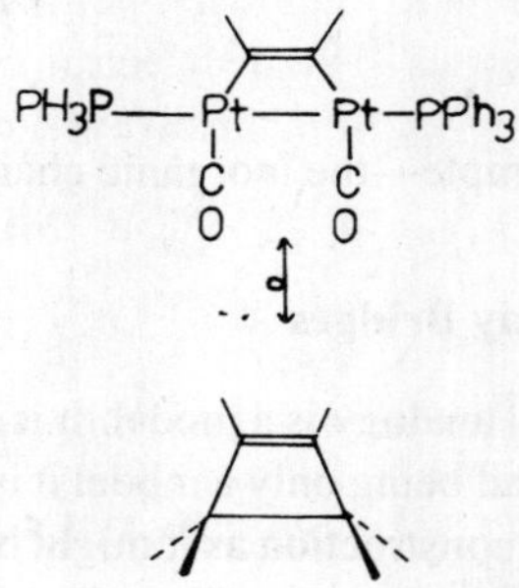

88

That these were isolobal with tetrahedrane, an olefin, bicyclobutane and cyclobutene was not only a curiosity but actually made easier for us the complicated analysis of the interconversions of these molecules. And the isolobal analogy points to the synthesis of the as yet unknown "isomers" in the series, dimetallacyclobutadienes and butadienes, already known in complexed form.

Finally for amusement, consider the chain, **89**, assembled by *Vahrenkamp et al.*[42] (no implication is made in the simplified drawing of the chain stereochemistry). There had better be a relationship to *n*-heptane! And so there is. We proceed using Table 1 as needed. $CpCr(CO)_3$ appears in a seven-coordinate guise here, so does $CpCr(CO)_2$; while $Co(CO)_3$ and $Fe(CO)_4$ are in a five-coordinate environment. It follows that:

$$CpCr(CO)_3 \underset{\circ}{\longleftrightarrow} d^5\text{–}ML_6 \underset{\circ}{\longleftrightarrow} CH_3$$

$$AsMe_2 \underset{\circ}{\longleftrightarrow} CH_2^{\ominus}$$

$$CpCr(CO)_2 \underset{\circ}{\longleftrightarrow} d^5\text{–}ML_5 \underset{\circ}{\longleftrightarrow} CH^{\oplus}$$

$$Co(CO)_3 \underset{\circ}{\longleftrightarrow} d^9\text{–}ML_3 \underset{\circ}{\longleftrightarrow} CH_2^{\oplus}$$

$$Fe(CO)_4 \underset{\circ}{\longleftrightarrow} d^8\text{–}ML_4 \underset{\circ}{\longleftrightarrow} CH_3^{\oplus}$$

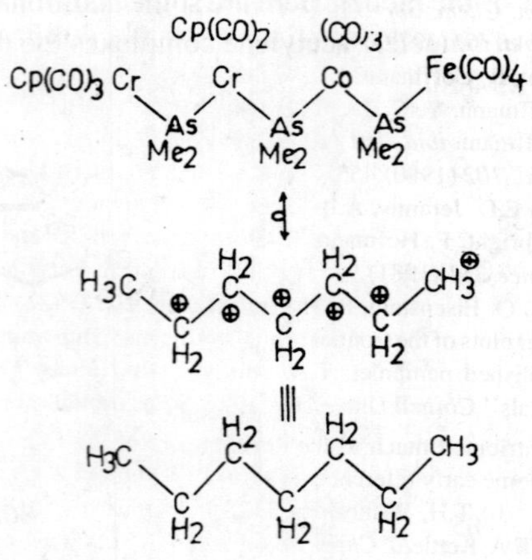

89

Now it is simple—the inorganic chain **89** is really *n*-heptane!

One of Many Bridges

The isolobal analogy is a model. It is the duty of our scientific craft to push it to its extremes, and being only a model it is certain to fail somewhere, for any model, as ingenious a construction as it might be, is bound to abstract only a piece of reality. The reader has seen just how far the model can be pushed and where it breaks down.

The pleasing aspect of this particular model is that it brings together different subfields of our central science. We are separated, split as under—organic, inorganic, physical, biological, analytical chemists—by the very largesse of our creation. The variety of molecules we create, and the methods we use to study them breed jargon and specialization. Yet underneath the seeming complexity there must be a deep unity. I think this approach would have pleased *R.B. Woodward.*

***N.B.** The authors feel gratified that this conclusion also is in conformity with the basic assumption of 'unity in diversity' an objective of 'unified approach' on which this whole book has been based.

References

1 Some selected papers from my group on this subject are the following: (a) M. Elian, R. Hoffmann, *Inorg. Chem. 14* (1975) 1058; (b) R. Hoffmann. P. Hofmann, *J. Am. Chem. Soc. 98* (1976) 598; (c) J.W. Lauther, R. Hoffmann, *ibid* 98 (1976) 1729; (d) J.W. Lauher, M. Elian, R.H. Summerville, R. Hoffmann, *ibid.* 98 (1976) 3219; (e) M. Elian. M.M.-L. Chen, D.M.P. Mingos, R. Hoffmann, *Inorg. Chem.* 15 (1976) 1148; (f) R.H. Sumerville, R. Hoffmann, *J. Am. Chem. Soc. 98* (1976) 7240; *101* (1979) 3821: (g) R. Hoffmann, M.M.-L. Chen, D.L. Thorn, *Inorg. Chem. 16* (1977) 503; (h) T.A. Albright, P. Hofmann, R. Hoffmann, *J. Am. Chem. Soc. 99* (1977) 7546; (i) D.L. Thorn, R. Hoffman, *Inorg. Chem. 17* (1978) 126; (j) Nguyen Trong Anh, M. Ellian, R. Hoffmann, *J. Am. Chem. Soc. 100* (1978) 110; (k) R. Hoffmann, T.A. Albright, D.L. Thorn, *Pure Appl. Chem. 50* (1978) 1; (l) A. Dedieu, R. Hoffmann, *J. Am. Chem. Soc. 100* (1978) 2074; (m) B.E.R. Schilling, R. Hoffmann, D.L. Lichtenberger, *ibid. 101* (1979) 585; (n) B.E.R. Schilling, R. Hoffmann, *ibid. 101* (1979) 3456; (o) T.A. Albright, R. Hoffmann, J.C. Thibeault D.L. Thorn, *ibid. 101* (1979) 3801; (p) T.A. Albright, R. Hoffmann, Y.-C. Tse, T. D'Ottavio, *ibid.* 101 (1979) 3812; (q) A. Dedieu, T.A. Albright, R. Hoffmann *ibid. 101* (1979) 3141; (r) S. Shaik, R. Hoffmann, C.R. Fisel, R.H. Summerville *ibid. 102* (1980) 4555; (s) R.J. Goddard, R. Hoffmann, E.D. Jemmis, *ibid. 102* (1980) 7667; (t) E.D. Jemmis, A.R. Pinhas, R. Hoffmann, *ibid. 102* (1980) 2576: (u) A.R. Pinhas, T.A. Albright, P. Hofmann, R. Hoffmann, *Helv. Chem. Acta 63* (1980) 29; (v) R. Hoffmann, *Science 211* (1981) 995; (w) K. Tatsumi, R. Hoffmann. *J. Am. Chem. Soc. 103* (1981) 3328; (x) O. Eisenstein, R. Hoffmann, *ibid. 103* (1981) 4308; 102 (1980) 6148; (y) Beautiful contour plots of the frontier orbitals of the most important fragments have been made in a rare unpublished pamphlet: T.A. Albright "A Holiday Coloring Book of Fragment Molecular Orbitals" Cornell University, 1977. Some orbitals are reproduced in [3a].

2. Others have contributed much to the development of the fragment formalism in inorganic chemistry. For some early references see: (a) J.K. Burdett, *J. Chem. .Soc. Faraday Trans. 11 70* (1974) 1599; (b) T.H. Whitesides, D.L. Lichtenberger, R.A. Budnik, *Inorg. Chem. 14* (1975) 68; (b) S.F.A. Kettle, *J. Chem. Soc. A 1966* 420; *Inorg. Chem.* 4(1965) 1661; (d) M.L.H. Green: *Organometallic Compounds*, Vol. 2, Methuen, London 1968, p. 115: (e) K. Wade, *Chem Commun 1971*, 792; *Inorg. Nucl. Chem. Lett. 8* (1972) 559, 563; *Electron Deficient Compounds*, Nelson, London 1971; (f) D.M.P. Mingos, *Nature (London) Phys. Sci. 236* (1972) 99; (g) P.S. Braterman, *Struct. Bonding (Berlin) 10* (1972) 57; (h) F.A. Cotton, W.T. Edwards, F.C. Rauch. M.A. Graham, R.N. Perutz, J.J. Turner, *J. Coord. Chem.* 2(1973) 247: (i) D.V. Korolkov, H. Miessner, *Z. Phys. Chem. (Leipzig) 253* (1973) 25.

3. Three recent overviews of the fragment formalism by some of the most active workers in the field: (a) J.K. Burdett: *Molecular Shapes* Wiley Interscience, New York 1980: (b) D.M.P. Mingos in G. Wilkinson, F.G. A. Stone, E.W. Abel: *Comprehensive Organometallic Chemistry* Pergamon, Oxford 1981; (c) T.A. Albright, *Tetrahedron 38* (1982) 1339.

4. (a) L. Pauling: *The Nature of the Chemical Bond* 3rd Ed., Cornell University Press, Ithaca, NY 1960; *Die Natur der chemischen Bindung.* 3. Aufl., Verlag Chemie Weinheim 1976; (b) in recent times Pauling has returned to the problem of the electronic structure of transition

metal complexes. See *e.g.*: *L. Pauling Proc. Nat. Acad. Sci. USA 72* (1975) 3799, 4200; *73* (1976) 274, 1403, 4290; *74* (1977) 2614, 5235; *75* (1978) 12, 569; *Acta Crystallogr. Sect. B 34* (1978) 746.

5. For a leading reference see: C.J. Ballhausen: *Introduction to Ligand Field Theory* McGraw-Hill, New York 1962.

6. An excellent introduction to the conventions and achievements of organometallic chemistry is J.P. Collman, L.S. Hegedus: *Principles and Applications of Organotransition Metal Chemistry* University Science Books, Mill Valley 1980.

7. For an introduction to the methology see E. Heilbronner, H. Bock: *The HMO Model and its Application*: Wiley, London/Verlag Chemie, Weinheim 1976; W.L. Jorgensen, L. Salem: *The Organic Chemist's Book of Orbitals*, Academic Press, New York 1973; R. Hoffmann, *Acc. Chem. Res. 4* (1971) 1.

8. (a) N. Rösch, R. Hoffmann, *Inorg Chem. 13* (1974) 2656; (b) R. Hoffmalnn, M.M.L. Chen, M. Elian, A.R. Rossi, D.M.P. Mingos, *ibid. 13* (1974) 2666; (c) A.R. Rossi, R. Hoffmann, *ibid. 14* (1975) 365: (d) P.J. Hay, J.C: Thibeault, R. Hoffmann, *J. Am. Chem. Soc. 97* (1975) 4884; (e) R. Hoffmann, J.M, Howell, A.R. Rossi, *ibid. 98* (1976) 2484; (f) S. Komiya, T.A. Albright, R. Hoffmann. J. Kochi, *ibid 98* (1976) 7255; *99* (1977) 8440; (g) R. Hoffmann, B.F. Beier, E.L. Muetterties, A.R. Rossi, *Inorg. Chem. 16* (1977) 511; (h) B.L. Barnett, C. Krüger, Y.H. Tsay, R.H. Summerville, R. Hoffmann, *Chem. Ber. 110* (1977) 3900; (i) R. Hoffmann, D.L. Thorn, A.E. Shilov, *Koord. Khim. 3* (1977) 1260; (j) D.L. DuBois, R. Hoffmann, *Nouv. J. Chem. 1* (1977) 479; (k) D.L. Thorn, R. Hoffmann, *J. Am. Chem. Soc. 100* (1978) 2079; *Nouv. J. Chim. 3* (1979) 39; (l) T.A. Albright, R. Hoffmann, *Chem. Ber. 111* (1978) 1578; *J. Am. Chem. Soc. 100* (1978) 7736; (m) T.A. Albright, R. Hoffmann, P. Hofmann, *Chem. Ber. 111* (1978) 1591; (n) R. Hoffmann, B.E.R. Schilling, R. Bau. H.D. Kaesz, D.M.P. Mingos, *J. Am. Chem. Soc. 100* (1978). 6088; (o) P.K. Mehrotra, R. Hoffmann, *Inorg. Chem. 17* (1978) 2187; (p) J.K. Burdett, R. Hoffmann, R.C. Fay, *ibid. 17* (1978) 2553; (q) B.E.R. Schilling, R. Hoffmann. *J. Am. Chem. Soc. 100* (1978) 7224; *Acta. Chem. Scand. B. 33* (1979) 231; (r) H. Berke, R. Hoffmann, *J. Am. Chem. Soc. 100* (1978) 7224; (s) B.E.R. Schilling, R. Hoffmann, J.W. Faller, *ibid. 101* (1979) 592; (t) A.R. Pinhas, R. Hoffmann, *Inorg. Chem. 18* (1979) 654; (u) C. Krüger, J.C. Sekutowski, H. Berke, R. Hoffmann, *Z. Natur forsch B 33* (1978) 1110; (v) E.D. Jemmis, R. Hoffmann. *J. Am. Chem. Soc. 102* (1980) 2570; (w) A. Stockis, R. Hoffmann, *ibid. 102* (1980) 2952; (x) K. Tatsumi, R. Hoffmann, M.H. Whangbo *J. Chem. Soc. Chem. Commun. 1980,* 509; (y) K. Tatsumi, R. Hoffmann, *Inorg. Chem. 20* (1981) 3781; (z) R.J. McKinney, D L Thorn, R. Hoffmann, A. Stockis, *J. Am. Chem. Soc. 103* (1981) 2595; (aa) K. Tatsumi, R. Hoffmann, A. Yamamoto, J.K. Stille, *Bull. Chem. Soc. Jpn. 54* (1981) 1857; (ab) P. Kubácek, R. Hoffmann, *J. Am. Chem. Soc. 103* (1981) 4320; (ac) O. Eisenstein, R. Hoffmann, A.R. Rossi, *ibid. 103* (1981) 5582; (ad) M. Kamata, K. Hirotsu, T. Higuchi, K. Tatsumi, R. Hoffmann, S. Otsuka, *ibid. 103* (1981) 5772; (ae) D.N. Cox, D.M.P. Mingos, R. Hoffmann, *J. Chem. Soc. Dalton Trans. 1981* 1788; (af) D.M. Hoffman, R. Hoffmann, *Inorg. Chem. 20* (1981) 3543.

9. (a) J. Halpern, *Discuss Faraday Soc. 1968*, 467; *Adv. Chem. Ser. 70* (1968) 1; (b) J. Kwiatek, J.K. Seyler, *ibid. 70* (1968) 207.

10. The term "isolobal" was introduced in [le], but the concept, as we will see below, is older.

11. In methylene a_1 is below b_2, because the latter is a pure p orbital, while the former has some s character. In $Fe(CO)_4$, the a_1 and b_2 can be thought of as being derived from the e_g orbital (z^2, $x^2 - y^2$) of the octaherdon. Removal of two *cis*-ligands stabilizes $x^2 - y^2$ (b_2) more than it does z^2(a_1). Where the ligands removed *trans,* to reach the squareplanar ML_4 fragment, the situation would be reversed Cf. also [3].

12. E.O. Fischer, H.J. Beck, C.G. Keilter, J. Lynch, J. Müller, E. Winkler, *Chem. Ber. 105* (1972) 162; R. Pfiz, J. Daub, *J.Organomet. Chem. 152* (1978) C32–C34 H. LeBozec, A. Gorgues, P.H. Dixneuf. *J. Am. Chem. Soc. 100* (1978) 3946; M.F. Lappert, P.L. Pye, *J. Chem Soc. Dalton Trans. 1977*, 2172.

13. M. Poliakoff, J.J. Turner, *J. Chem. Soc. A. 1971*, 2403.

14. C.H. Wei, L.F. Dahl, *J. Am. Chem. Soc. 91* (1969) 1351.

15. A. S. Foust, M.S. Foster, L. F. Dahl, *J. Am. Chem. Soc. 91* (1969) 5631 and references therein.

16. See also J.E. Ellis, *J. Chem. Educ. 53* (1956) 2 for an extensive exposition of this analogy.

17. K. Wade, *Adv. Inorg. Chem. Radiochem. 18* (1976) 1; *Chem. Br. 11* (1975) 177; cf. also [2e].

18. D.M.P. Mingos, *Adv Organomet. Chem. 15* (1977) 1; R. Mason, D.M.P Mingos, *MTP Int. Rev. Sci. Phys. Sci. Ser. II 11* (1975) 121; D.M.P. Mingos, *Trans. Am. Crystallogr. Assoc. 16* (1980) 17; cf. also [2f] and [3b].

19. B.T. Huie, S.W. Kirtley, C.B. Knobler, H.D. Kaesz. *J. Organomet. Chem. 213* (1981) 45.

20. E.O. Fischer, T.L. Lindner, H Fischer, G. Huttner, P. Friedrich, F.R. Kreissl Z. *Naturforsch. B. 32* (1977) 648.

21. R. Bau, B. Fontal, H.D. Kaesz, M.R. Churchill, *J. Am. Chem. Soc. 89* (1967) 6374.

22. J.D. Cotton, J. Duckworth, S.A.R. Knox, P.F. Lindley, I. Paul, F.G.A. Stone, P. Woodward, *Chem. Commun. 1966* 253; J.D. Cotton, S.A.R. Knox, I Paul, F.G.A. Stone, J. *Chem. Soc. A. 1967*, 264.

23. D.H. Farrar, B.F.G. Johnson, J.N. Nicholls, P.R. Raithby, M.J. Rosales, *J. Chem. Commun. 1981*, 273.

24. P. Braunstein, D. Matt, O. Bars, M. Louër, D. Grandjean, J. Fischer, A. Mitscher, *J. Organomet. Chem. 213* (1981) 79.

25. R.J. Haines, N.D.C.T. Steen, R.B. English, *J. Chem. Soc. Chem. Commun. 1981*, 587.

26. W.A. Herrmann, C. Krüger, R. Goddard, I. Bernal, *Angew. Chem. 89* (1977) 342 *Angew. Chem. Int. Ed. Engl. 16* (1977) 334; *P. Hofmann, Angew. Chem. 91* (1979) 591; *Angew. Chem. Int. Ed. Engl. 18* (1979) 554.

27. There are some further details, not discussed here but treated elsewhere [1o, 1x, 3c]. In particular, the ML_2 will have another relatively low-lying orbital when L is a π acceptor.

28. W.D. Jones, M.A. White, R.G. Bergman, *J. Am. Chem. Soc. 100* (1978) 6770.

29. (a) M. Green, J.A.K. Howard, R.N. Mills, G.N. Pain, F.G.A. Stone, P. Woodward, *J. Chem. Soc. Chem. Commun. 1981*, 869; (b) For some other compounds containing a similar unit see L.M. Cirjat, J.S. Huang, Z.H. Zhu, L.F. Dahl. *J. Am. Chem. Soc. 102* (1980) 6623.

30. N.M. Boag, M. Green, D.M. Grove, J.A.K. Howard, J.L. Spencer, F. G.A. Stone, *J. Chem. Soc. Dalton Trans. 1981*, 2170.

31. (a) F.G.A. Stone, *Acc. Chem. Res. 14* (1981) 318 and references therein; L. Busetto M. Green, J.A.K. Howard, B. Hessner, J.C. Jeffery, R.M. Mills, F.G.A. Stone P. Woodward, *J. Chem. Soc. Chem. Commun. 1981*, 1101; T.V. Ashworth, M.J. Chetcuti, L.J. Farrugia, J.A.K. Howard, J.C. Jeffery, R. Mills, G.N. Pain, F.G.A. Stone, P. Woodward *Reactivity of Metal-metal Bonds. ACS Symposium Series 155* in M. H. Chisholm; Washington 1982, p. 299, (b) For further molecules of type 71 and 72 see E. Sappa, A.m. Manotti Lanfredi, A. Tiripicchio, *J. Organomet Chem. 221* (1981) 93; J.R. Shapley, J.T. Park, M.R. Churchill, C. Bueno, H.J. Wasserman, *J. Am. Chem. Soc. 103* (1981) 7385; G. Jaouen, A. Marinetti, B. Mentzen, R. Mutin, JK.-Y. Saillard, B.G. Sayer, M.J. McGlinchey, unpublished results.

32. (a) W.D. Stohrer, R. Hoffmann, *J. Am. Chem. Soc. 90* (1972) 1661; (b) R.E. Williams, *Inorg. Chem. 10* (1971) 210; *Adv Inorg. Chem. Radiochem. 18* (1976) 67; (c) S. Masamune, M. Sakai, H. Ona, *J. Am. Chem. Soc. 94* (1972) 8955; S. Masamune, M. Sakai, H. Ona, A.L. (1972) 8956; (d) H. Hart, M Kuzuya, *ibid. 94* (1972) 8958.

33. See in this context: J. Chandrasekhar. P. V. R. Schleyer, H.B. Schlegel, *Tetrahedron Let. 1978*, 3393.

34. Cf. [1i] for the relevant literature citations.

35. H, Hogeveen, P.W. Kwant, *Acc. Chem. Res. 8* (1975) 413. For some related main group structures see: P. Jutzi, F. Kohl, P. Hofmann, C, Krüger, Y.H. Tsay, *Chem. Ber. 113* (1980) 757.

36. But apparently not a stable structure after all: A. Sevin, A. Devaquet, *Nouv. J. Chim. 1* (1977) 357; T. Clark, P. v. R. Schleyer, *ibid. 2* (1978) 665.

37. R.J. Al-Essa, R.J. Puddephatt, P.J. Thompson, C.F.H. Tipper, *J. Am. Chem. Soc. 102* (1980) 7546 and references therein.

38. C.P. Casey, D.M. Scheck, A. J. Shusterman, *J. Am. Chem. Soc. 101* (1979) 4233.

39. Cf. M. Saunders, H.U. Siehl, *J. Am. Chem. Soc. 102* (1980) 6868 and references therein.

40. The exceptions include $Zr(BH_4)_4$, $W(RCCR)_3(CO)$, Cp_3M and Cp_3MR, Cp_4U, UO_2L_6; among others. For the relevant references see S.Y. Chu, R. Hoffmann, *J. Phys. Chem. 86* (1982) 1289 and [3a].

41. D.M. Hoffman, R. Hoffmann, C.R. Fisel, *J. Am. Chem. Soc. 104* (1982) 3858: D.M. Hoffman, R. Hoffmann, *J. Chem. Soc. Dalton Trans.* in press.

42. H.J. Langenbach, E. Keller, H. Vahrenkamp, *J. Organomet. Chem. 191* (1980) 95.

Appendix 2.1

D-3. NOMENCLATURE OF ORGANOMETALLIC COMPOUNDS*

D-3.0 Introduction

In these rules the term "organometallic compound" includes all compounds in which a carbon atom is joined directly to an atom of any element other than hydrogen, carbon, nitrogen, oxygen, flourine, chlorine, bromine, iodine and astatine, although the above elements may also be present in the molecule.

For expediency, in this Subsection of rules the term "metal atom" includes the atoms of all the elements except those named above.

There are special rules for naming organic compounds containing sulfur, selenium and tellurium (Section C), phosphorus, arsenic and in some cases antimony and bismuth (Subsection D-5), silicon (Subsection D-6) and boron (Subsection D-7) but the following rules may also be applied.

Although many organometallic compounds exist in associated molecular forms and contain structural solvent, their names are generally based on the stoichiometric compositions of the compounds, the solvent, if any, being ignored.

EXAMPLES:

Butyllithium	$(C_4H_9Li)_n$
Triethylaluminium	$[(C_2H_5)_3Al]_2$

Where the structure is known and it is desired to emphasise this, the compounds may be named as coordination compounds (see Subsection D-2).

EXAMPLES:

Di-μ-ethyl-tetraethyldialuminium

or

Di-μ-ethyl-bis(diethylaluminium)

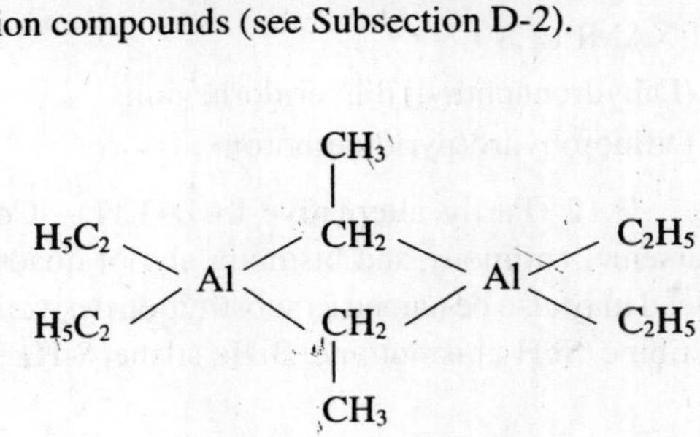

---

* Reproduced from the IUPAC publication "Nomenclature of Organic Chemistry" (Pergamon Press, 1979). The section headings and cross-references to other sections have been left unchanged to facilitate consultation of the original text for the nomenclature of allied compounds.

catena-(Di-μ-methyl-beryllium) 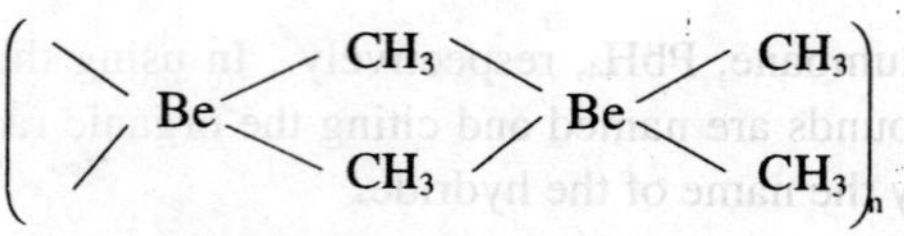

Bis(diethyl ether) iodo-(methyl) magnesium** $[MgI(CH_3)\{(C_2H_5)_3O\}_2]$

D-3.1 *Organometallic compounds in which the metal is bound to organic radicals and hydrogen only*

3.11—Compounds consisting solely of individual metal atoms joined to the carbon atom of one or more organic radicals and/or to one or more hydrogen atoms are named by citing such radicals or hydrogen atoms in alphabetical order, followed by the name of the metal. The presence of hydrogen attached to the metal atom must always be indicated by the prefix "hydrido"†. No break is made between the radical names and that of the metal.

EXAMPLES:

Phenylarsenic	$(C_6H_5As)_n$
Methyllithium	$(CH_3Li)_n$
Vinylsodium	$CH_2{=}CHNa$
Diethyltellurium	$(C_2H_5)_5Te$
Tribenzylarsenic	$(C_6H_5CH_2)_3As$
Tetraphenyltelluriur	$(C_6H_5)_4Te$
Terabutylgermanium	$(C_4H_9)_4Ge$
Penta-2-naphthylantimony	$(C_{10}H_7)_5Sb$
Butylethylmethylpropyltin	$CH_3(C_2H_5)(C_3H_7)(C_4H_9)Sn$
Ethynyltriphenyltin	$(C_6H_5)_3SnC{\equiv}CH$
Ethylhydridoberyllium	C_2H_5BeH
Dihydridophenylarsenic	$C_6H_5A_5H_2$
Hydridomethylpropylantimony	$(CH_3)(C_3H_7)SbH$
Trihydridophenylgermanium	$C_6H_5GeH_3$
Benzylhydridophenylgold	$(C_6H_5)(C_6H_5CH_2)AuH$

EXAMPLES:

(Dihydronaphthyl) dihydridorhenium	$(C_{10}H_9)ReH_2$
Difluorohydro(pyridine)boron	$(C_5H_5N)BHF_2$

3.12 (Partly alternative to D-3.11)—Compounds of tervalent phosphorus, arsenic, antimony and bismuth, and of quadrivalent silicon, germanium, tin and lead may also be named as substitution derivatives of phosphine, PH_3, arsine, AsH_3, stibine, SbH_3, bismuthine, BiH_3, silane, SiH_4, germane, GeH_4, stannane, SnH_4, and

**Although the use of parentheses around simple ligands is not required, it can avoid the possibility of misinterpretation.

† Hydrogen attached to carbon or boron is named only when necessary and then by the prefix "hydro".

plumbane, PbH_4, respectively*. In using this alternative nomenclature the compounds are named and citing the organic radicals in *alphabetical* order followed by the name of the hydride.

EXAMPLES:

Triethylarsine	$(C_2H_5)_3As$
Ethylmethyldiphenylgermane	$(C_6H_5)_2(C_2H_5)(CH_3)Ge$
Tetraphenylplumbane	$(C_6H_5)_4Pb$
Ethylarsine	$C_2H_5AsH_2$
Diphenylstannane	$(C_6H_5)_2SnH_2$

D-3.2—*Organometallic compounds having uncharged ligands*

3.21—Organometallic compounds containing uncharged ligands are named by coordination nomenclature, all ligands (anions, radicals and molecules) being cited in *alphabetical* order before the name of the metal.

EXAMPLES:

(Pyridine)tris(trifluoromethyl) antimony — $Sb(CF_3)_3(C_5H_5N)$

trans-Chloro(methyl)bis(triethylphosphine) platinum

$$\begin{array}{c} P(C_2H_5)_3 \\ \downarrow \\ CH_3 - Pt - Cl \\ \uparrow \\ P(C_2H_5)_3 \end{array}$$

trans-Bis[diethyl(phenyl)phosphine] bis-(pentachlorophenyl) iron

$$\begin{array}{c} C_6Cl_5 \\ | \\ (C_2H_5)_2(C_6H_5)P \rightarrow Fe \leftarrow P(C_2H_5)_8(C_6H_5) \\ | \\ C_6Cl_5 \end{array}$$

Bis(2, 5-dimethyl-2, 5-diphosphahexane) hydridophenylruthenium
or
Bis[ethylenebis[dimethylphosphine]]-hydridophenylruthenium

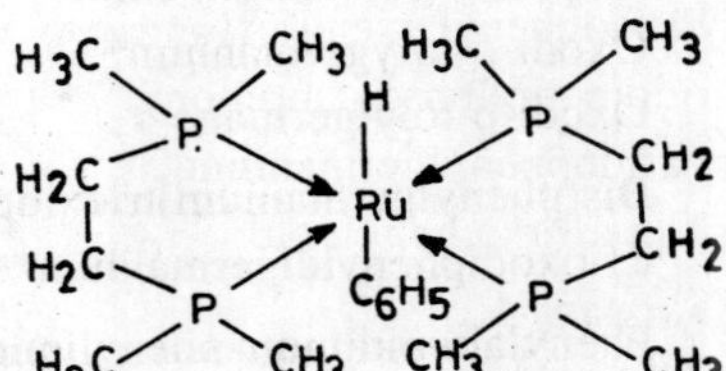

(2, 5-Dimethyl-2, 5-diphosphahexane) (2, 5-dimethyl-2, 5-diphosphahexyl) hydridoruthenium
or
[[[[2-Dimethylphosphino)ethyl)]-methylphosphino]methyl]-p′, p′, C]-[ethylenebis[dimethylphosphine]]-hydridoruthenium

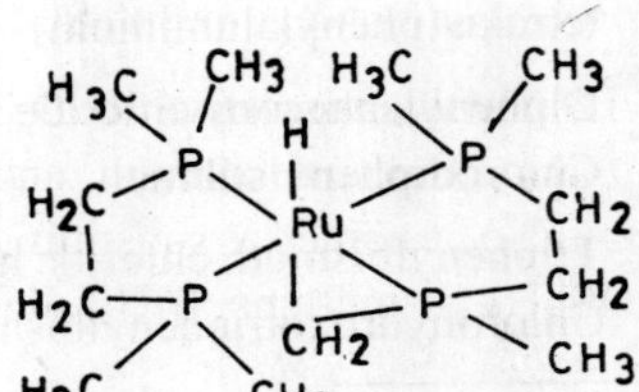

* In naming hydrides the ending "—ane" should only be used for hydrides of B, C, Si, Ge, Sn, Pb and according to Rule D-4.11, for hydrides of unbranched chains of identical atoms of P, As, Sb, Bi, S, Se and Te.

D-3.3—*Organometallic compounds having anionic ligands*

3.31 (Partly alternative to D-3.12)—Organometallic compounds containing anionic ligands are named by methods a, b or c:

(a) by stating the name(s) of the organic radical(s) followed by the name of the metallic element and the anion name(s), note being taken of the stoichiometric composition only. The names of the radicals and anions are each cited in alphabetical order.

(b) by adding, as prefixes, all the radical and anion names to the name of the metal, note being taken of the molecular structure. The radical and anion names are cited together in *alphabetical* order. This method may be used for monomolecular compounds, when the structure is known, and for polymeric compounds by addition of the prefix poly, with further addition of information about bridging (see Rule D-2.71).

(c) where applicable, according to Rule D-3.12.

EXAMPLES:

	Name	Formula
(a)	Methyltin chloride dihydride	CH_3SnH_2Cl
(b)	Chlorodihydridomethyltin	
(c)	Chloro(methyl) stannane*	
(a)	Methylmagnesium iodide	$(CH_3MgI)_n$
(b)	Poly[iodo(methyl) magnesium]*	
(a)	Phenylmercury hydroxide	C_6H_5HgOH
(b)	Hydroxo(phenyl)mercury*	
(a)	Diethylaluminium chloride	$\{(C_2H_5)_2AlCl\}_2$
(b)	Di-μ-chloro-bis(diethylaluminium)	
(a)	Di-p-tolylgermanium oxide	$(p\text{-}CH_3C_6H_4)_2GeO$
(b)	Oxodi-p-tolygermanium**	
(c)	Oxodi-p-tolylgermane	
(a)	Bis(phenylgermanium)trioxide	$(C_6H_5Ge)_2O_3$
(b)	Trioxodiphenyldigermanium****	
(a)	Phenylaluminimum phenylimide	$\{C_6H_5Al(NC_6H_5)\}_4$
(b)	Tetrakis-μ_3-(phenylimido) tetrakis(phenylaluminium)	
(a)	Diphenylantimony chloride	$(C_6H_5)_2SbCl$
(b)	Chlorodiphenylstibine	
(a)	Triphenylbismuth chloride hydroxide	$(C_6H_5)_3BiCl(OH$
(b)	Chlorohydroxotriphenylbismuth	

* Although the use of parenthess around simple groups like methyl and phenyl is not required, it can avoid the possibility of misinterpretation.

** When the polymeric nature of such compounds is indicated as $[(CH_3C_6H_4)_2\ GeO]_n$ the (b) name may be modified as follows: Poly [μ-oxo-di-p-tolygermanium].

*** Since the structure indicated does not show the relationship between the oxygen atoms, a(c) name is not possible. Similarly, a polymer name including the citation of bridging groups cannot be written; the polymeric nature of this example must be shown just by prefixing ''Poly'' to name(b).

3.32—Compounds of the general formula $R_xM(O)_y(OH)_z$, where M is a metal atom, are designated by functional names of the types: hydroxide, oxdie unless other functional names, such as phosphonic acid and seleninic acid are permitted in special nomenclature systems (see Subsections C-6.4, C-7, D-5, D-6 and D-7). However, the use of these systems for organoantimony organotellurium and their extention to organogermanium, tin and lead compounds must be carefully considered since most of these compounds are usually amphoteric polymers.

EXAMPLES:

Phenylantimony dihydroxide oxide	$C_6H_5SbO(OH)_2$
Methyltin hydroxide oxide	$CH_3SnO(OH)$
Diphenylantimony hydroxide oxide	$(C_6H_5)_2SbO(OH)$
Ethyltellurium hydroxide oxide	$C_2H_5TeO(OH)$
Triethyltellurium hydroxide oxide	$(C_2H_5)_3TeO(OH)$

D-3.4—*Organometallic radicals*

3.41—Radicals derived from organometallic compounds and having the free valence (or valences) on the metal atom, are named by prefixing the names of the organic ligands to the modified name of the metal given in Table II (Appendix, p. 462).

EXAMPLES:

Methyltellurio	CH_3Te-
Hydridodimethylstannio	$(CH_3)_2HSn$–
Dimethylstannio	$(CH_3)_2Sn<$
Methylstannio*	$CH_3Sn\equiv$
Methylstannio*	CH_3Sn—
Triphenylstannio	$(C_6H_5)_3Sn$–
Dichlorodimethylantimonio	$(CH_3)_2Cl_2Sb$-
Chloro(methyl)antimonio	$(CH_3)ClSb$-

3.42 (Partly alternative to D-3.11)—Compounds containing univalent organometallic radicals attached to carbon may be named by citing their radical names before the name of the parent compound. The radical name takes its place in alphabetical order with those of the other substituents.

EXAMPLES:

1, 1 2, 2-Tetraphenyl-l, 2-disodioethane

$$\begin{array}{ccc} (C_6H_5)_3C & — & C(C_6H_5)_2 \\ | & & | \\ Na & & Na \end{array}$$

1-Cupriopropyne	$CH_3C\equiv CCu$
Lithiotriphenylmethane	$(C_6H_5)_3CLi$
(Methylmagnesio)benzene	$CH_3MgC_6M_5$

* The use of ''multivalent'' organometallic radicals as substituents in organic nomenclature is not completely free from ambiguity and is under study by both the Inorganic and Organic Nomenclature Commissions.

3.43—Names of univalent radicals may be derived from the hydrides phosphine, arsine, stibine, bismuthine, silane, germane, stannane and plumbane as follows:

Phosphino	$—PH_2$	Silyl	$—SiH_3$
Arsino	$—AsH_2$	Germyl	$—GeH_3$
Stibino	$—SbH_2$	Stannyl	$—SnH_3$
Bismuthino	$—BiH_2$	Plumbyl	$—PbH_3$

EXAMPLES:

Dimethylstannyl	$—SnH(CH_3)_2$
Methylphenylarsino	$—As(CH_3)(C_6H_5)$
Triphenylplumbyl	$—Pb(C_6H_5)_3$

3.44—Multivalent radicals derived from the hydrides in Rule D-3.43 are named systematically by adding diyl, triyl, etc., to the name of the parent hydride.

EXAMPLES:

Methylstibinediyl	$>SbCH_3$
Methylstannanetriyl	$\geq SnCH_3$

D-3.5—*Organic radicals containing characteristic groups*

3.51—An organometallic compound may be named by adding the name of the organic part as a prefix to that of the metal atom regardless of the nature of the organic characteristic group present.

EXAMPLES:

Bis(4-carboxyphenyl)mercury	$(4\text{-}HOOCC_6H_4)_2Hg$
Triethyl(2-formylethyl)tin	$(C_2H_5)_2SnCH_2CH_2CHO$
(Carboxymethyl) triphenylgermanium	$(C_6H_5)_3GeCH_2COOH$
(4-Carboxyphenyl)phenylmercury	$4\text{-}HOOCC_6H_4HgC_6H_5$
(4-Carboxyphenyl)methyltellurium	$4\text{-}CH_3TeC_6H_4COOH$
Diphenyl(4-sulfophenyl)bismuth	$4\text{-}HO_3SC_6H_4Bi(C_6H_5)_2$
(1, 2-Dicarboxyethyl)hydridodimethyltin	$(CH_3)_2HSnCHCOOH$ with CH_2COOH on the CH

3.52 (Partly alternative to D-3 3.51)—An organometallic compound containing an organic characteristic group that can be expressed as a suffix may be named according to the Rules of Section A, B and C, hydrogen atoms of the organic parent compounds being replaced by metal-containing radicals formed according to Rules D-3. 41, D-3.43 and D-3.44

EXAMPLES:

3-(Triethylstannio)propanal 3-(Triethylstannyl)propanal	$(C_2H_5)_3SnCH_2CH_2CHO$
(Triphenylgermanio)acetic acid (Triphenylgermyl) acetic acid	$(C_6H_5)_3GeCH_2COOH$
4-(Phenylmercurio)benzoic acid	$4-(C_6H_5-Hg)C_6H_4-COOH$
4-(Diphenylbismuthino)benzene-sulfonic acid 4-(Diphenylbismuthino) benzene-sulfonic acid	$4\text{-}[(C_6H_5)_2Bi]C_6H_4\text{–}SO_3H$
(Hydridodimethylstannio)-succinic acid (Dimethylstannyl) succinic acid	$(CH_3)_2SnH\text{–}CH\text{–}COOH$ with $CH_2\text{–}COOH$ on the CH
Tris(hydroxomercurio) acetic acid	$(HOHg)_3C\text{–}COOH$
[4-(Dichlorodimethylantimonio) phenyl] trimethylammonium chloride	$[4\text{-}\{(CH_3)_2SbCl_2\}C_6H_4\{N(CH_3)_3\}]Cl$
Bis(triphenylstannio)carbodümide Bis(triphenylstannyl)carbodimide	$(C_6H_5)_3Sn\text{–}N{=}C{=}N\text{–}Sn(C_6H_5)_3$

3.53 (Partly alternative to D-3.51)—If two or more identical organic structures, each containing a characteristic group which can be designated by a suffix, are each linked to one metal atom by a single bond, the resulting organometallic compound may be named according to the principles of Rule C-72.

The metal is designated, in prefix, by a radical name formed by citing other groups and radicals as prefixes to the modified name of the metal as given by Table II (Appendix, p. 461). This radical name is followed by the appropriate numerical affix, then by the name of the organic parent compound.

Multivalent radicals derived according to Rule D-3.44 may also be used.

EXAMPLES:

4-4′-Mercuriodibenzoic acid $(4-HOOCC_6H_4)_2Hg$

2, 2′-(Ethylmethylstannio)-dipropionic acid
2, 2′(Ethylmethylstannanediyl)-dipropionic acid

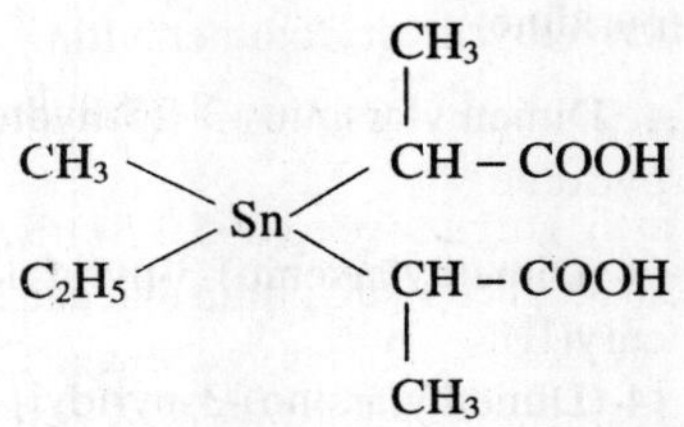

D-3.6—*Organometallic compounds with two or more metal atoms*

3.61—Compounds in which rings are linked by metal atoms may be named:

(a) by using "to" radical names (Rule D-3.41 and Table II, p. 461) as prefixes to the name of a cyclic compound chosen according to the seniority order defined in Rule C-14.I,

(b) by coordination nomenclature, the metal encountered last in Table I (p. 459) being cited as nuclear atom.

If there is a further choice for the parent cyclic compound (method a) or for the nuclear atom (method b), this is made so that the substituents (method a) or the ligands (methods b), are named with the smallest possible number of successive substitutions.

EXAMPLES:

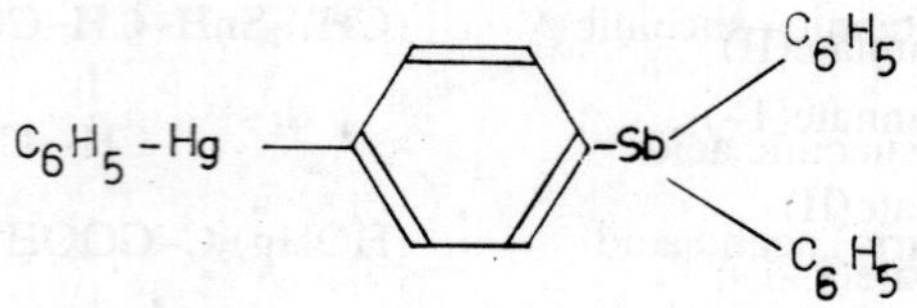

(a) 1-(Diphenylantimonio)-4-(phenylmercurio)benzene

1-(Diphenylstibino)-4-(phenylmercurio)benzene

(b) [4-(Diphenylantimonio)phenyl]phenylmercury(II)

[4-(Diphenylstibino)phenyl]phenylmercury(II)

Note: These names are preferred to alternative names, such as phenyl [[4-(phenylmercurio)phenyl] antimonio]benzene, which express more substitutions in the ligands.

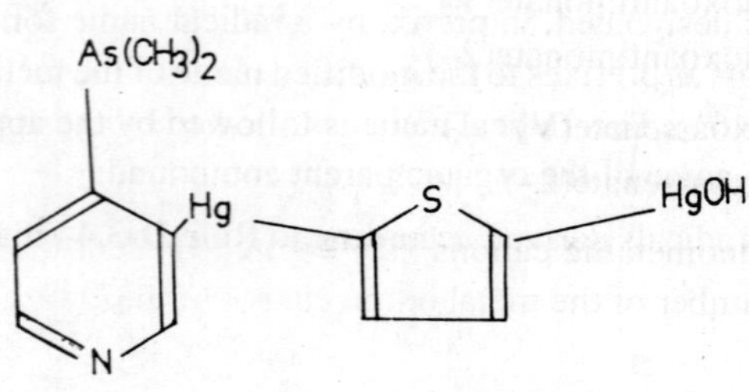

(a) 4-(Dimethylarsenio)-3-[{5-hydroxomercurio)-2-thienyl)-mercurio] pyridine

4-(Dimethylarsinio)-3-[(5-hydroxomercurio)-2-thienyl]-mercurio] pyridine

(b) [4-(Dimethylarsenio)-3-pyridyl] [5-(hydroxomercurio)-2-thienyl] mercury(II)

[4-(Dimethylarsino)-3-pyridyl] [5-(hydroxomercurio)-2-thienyl] mercury(II)

D-3.7—*Organometallic ions*

3.71—Organometallic anions may be named by modifying the ending of the appropriate organometallic radical to end in "—ate". The oxidation number of the metal may be indicated by the Stock system(a), or the charge on the anion by the Ewens-Bassett system(b) (see Subsection D-2).

EXAMPLES:

(a)	Hexaethynylchromate(III)	$[Cr(C{\equiv}CH)_6]^{3-}$
(b)	Hexaethynylchromate(3–)	
(a)	Pentahydroxophenyltellurate(IV)	$[C_6H_5Te(OH)_5]^{2-}$
(b)	Pentahydroxophenyltellurate(2–)	
(a)	Triphenylstannate(II)	$[Sn(C_6H_5)_3]^-$
(b)	Triphenylstannate(1–)	
(a)	Triethylzincate(II)	$[Zn(C_2H_5)_3]^-$
(b)	Triethylzincate(1–)	
(a)	Triethylhydridoaluminate(III)	$[AlH(C_2H_5)_3]^-$
(b)	Triethylhydridoaluminate(1–)	
(a)	Trihydroxotris(trifluoromethyl)–antimonate (V)	$[(CF_3)_3Sb(OH)_3]^-$
(b)	Trihydroxotris(trifluoromethyl)antimonate(1–)	
(a)	Dioxophenylgermanate(IV)	$[C_6H_5GeO_2]^-$
(b)	Dioxophenylgermanate(1–)	
(a)	Methyltrioxoantimonate(V)	$[CH_3SbO_3]^{2-}$
(b)	Methyltrioxoantimonate(2–)	
(a)	Ethyltrioxoarsenate(V)	$[C_2H_5AsO_3]^{2-}$
(b)	Ethyltrioxoarsenate(2–)	

3.72—Organometallic cations may be named according to Rule D-3.31 and the oxidation number of the metal or the charge of the cation is indicated.

EXAMPLES:

(a)	Tricyclohexyltin(IV)	$[(C_6H_{11})_3Sn]^+$
(b)	Tricyclohexyltin(1$^+$)	
(a)	Bis(cyclopentadienyl)iron(III)	$[(C_5H_5)_2Fe]^+$
(b)	Bis(cyclopentadienyl)iron(1$^+$)	
(a)	Hydridotrimethylarsenic(V)	$[AsH(CH_3)_3]^+$
(b)	Hydridotrimethylarsenic(1$^+$)	

3.73 (Partially alternative to D-3.72)—Cations of the types $[MR_4]^+$ (with M = P, As Sb or Bi) or $[TeR_3]^+$, where R is a hydrogen atom or an organic radical.

may be named as substitution derivatives of the following parent cations (see also Rule C-82.1):

Phosphonium	$[PH_4]^+$
Arsonium	$[AsH_4]^+$
Stibonium	$[SbH_4]^+$
Bismuthonium	$[BiH_4]^+$
Telluronium	$[TeH_3]^+$

EXAMPLES:

Tetramethylarsonium	$[As(CH_3)_4]^+$
Dimethylarsonium	$[AsH_2(CH_3)_2]^+$
Tetra-p-tolybismuthonium	$[Bi(4\text{–}CH_3C_6H_4)_4]^-$
Triphenyltelluronium	$[Te(C_6H_5)_3]^+$

3.74—Organometallic salts are designated by citing the name(s) of the cation(s) followed by the name(s) of the anion(s).

EXAMPLES:

Sodium trihydroxotris(trifluoromethyl) antimonate(V)	$Na[(CF_3)_3Sb(OH)_3]$
Potassium dioxo(phenyl)germanate(IV)*	$K[C_6H_5GeO_2]$
Tricyclohexyltin(IV)tetrafluoroborate	$[(C_6H_{11})_3)Sn][BF_4]$

D 3.75—When an ion is not named as a constituent of a salt, it is generally advisable to add "ion", "cation" or "anion" to the names formed according to Rules D-3.71, D-3.72 and D-3.73. This is recommended when Stock numbers are used:

EXAMPLES:

Hexamethylchromate(III) ion (or anion)

Tricyclohexyltin(IV) ion (or cation)

Tetramethylarsonium ion (or cation)

* Although the use of parentheses around such simple ligands is not required, it can avoid the possibility of misinterpretation.

Index